Encyclopedia of Geomorphology

Encyclopedia of Geomorphology

Volume 2

J–Z

Edited by A.S. Goudie

LONDON AND NEW YORK

International Association of Geomorphologists

First published 2004
by Routledge
11 New Fetter Lane, London EC4P 4EE
Simultaneously published in the USA and Canada
by Routledge
29 West 35th Street, New York, NY 10001

Routledge is an imprint of the Taylor & Francis Group

Typeset in Times by Newgen Imaging Systems (P) Ltd, Chennai, India
Printed and bound in Great Britain by
TJ International Ltd, Padstow, Cornwall

British Library Cataloguing in Publication Data
A catalogue record for this book is available from the British Library

Library of Congress Cataloging in Publication Data
Encyclopedia of geomorphology / edited by A.S. Goudie.
p. cm.
Includes bibliographical references and index.
1. Geomorphology–Encyclopedias. I. Goudie, Andrew.

GB400.3.E53 2003
551.41′03–dc21 2003046892

ISBN 0–415–27298–X (set)
ISBN 0–415–32737–7 (volume one)
ISBN 0–415–32738–5 (volume two)

To the Founders of the International Association of Geomorphologists and to the IAG Senior Fellows:

1989: Harley J. Walker (USA)
1993: Hanna Bremer (Germany), Ross Mackay (Canada), Anders Rapp (Sweden)
1997: Denys Brunsden (UK), Richard Chorley (UK), Luna Leopold (USA)
2001: Stanley A. Schumm (USA), Torao Yoshikawa (Japan)

Contents

Illustrations

Figures

Plates

Tables

J

JOINTING

Joints are cracks in rocks formed by stress that results from tectonic events, cooling, or isostatic rebound. They range in length from millimetres to kilometres. In outcrop, joints can be small hairline cracks only millimetres in length or long open fissures a metre or more across. They may be open or filled; common fillings are soil and clays. Joints can also be sealed as a result of hydrothermal activity. They are distinguished from faults (see FAULT AND FAULT SCARP) by the lack of movement between the two sides of the joint. Most joints are tension fractures.

Joints occur in sets, groups of nearly parallel fractures that formed under the same stress regime. At least three joint sets commonly occur in most outcrops: in sedimentary and schistose/foliated metamorphic rocks. There is typically one set parallel to layering and two mutually perpendicular sets normal to layering. In massive rocks like granite, one set is usually horizontal and the others are vertical or steeply dipping. Columnar joints in basalts (e.g. the Giants Causeway, Northern Ireland) are a special example of cooling joints. These joints bound areas where cooling fronts meet so that a hexagonal crack pattern forms where the boundaries converge.

Joints exert significant control on landform shape. Most granite landforms, such as domes (e.g. Stone Mountain, Georgia, USA and Ayers Rock, Australia) and TORs (e.g. Haytor Rocks, Dartmoor, UK), are joint controlled. Other joint-controlled landforms not dependent upon lithology include geos and gulls. A geo is a deep, narrow cleft or ravine (see GORGE AND RAVINE) along a rocky sea coast that is flooded by the sea. Valley orientation in general often parallels major joints. A gull is a joint that opens on escarpments because of tension produced by cambering (see CAMBERING AND VALLEY BULGING). Grikes in LIMESTONE PAVEMENTs (e.g. the Burren, Co. Clare, Eire) are joints enlarged by solution that separate clints, the raised portions of these pavements.

Joints are also very important with respect to both MECHANICAL WEATHERING and CHEMICAL WEATHERING, and slope stability. Joints are the most important zones of weakness in any given rock mass, and provide primary access for moisture to enter rock. A dense pattern of closely spaced joints will thus hasten chemical weathering, leaving upstanding areas where joints are more widely spaced. The WEATHERING FRONT often occurs along a horizontal joint. In addition, most limestone CAVEs originate as joints, and sink holes commonly occur at joint intersections. Corestones are remnants of joint blocks, often occurring in a matrix of weathered debris or as boulder fields after the weathered matrix has been stripped. Furthermore, the stability of a given slope can often depend on the orientation of the joints. Sliding is likely to occur if the joints dip toward the slope face, for instance, and toppling occurs along vertical joints oriented parallel to the slope. Joints also commonly form the slip surface in larger rotational landslides. DRAINAGE PATTERNs are also often controlled by the joint pattern of the underlying rocks, e.g. rectangular drainage patterns in limestone, and drainage density may reflect joint density.

Further reading

Engelder, T. (1987) Joints and shear fractures in rock, in B.K. Atkinson (ed.) *Fracture Mechanics of Rocks*, 27–69, London: Academic Press.

Gerrard, A.J.W. (1974) The geomorphological importance of jointing in the Dartmoor granite, in E.H. Brown and R.S. Waters (eds) *Progress in Geomorphology*, Institute of British Geographers Special Publication 7, 39–50.
——(1988) *Rocks and Landforms*, London: Unwin Hyman.
Hencher, S.R. (1987) The implications of joints and structures for slope stability, in M.G. Anderson and K.S. Richards (eds) *Slope Stability*, 145–186, Chichester: Wiley.
Twidale, C.R. (1982) *Granite Landforms*, Amsterdam: Elsevier.

SEE ALSO: bedrock channel; bornhardt; granite geomorphology; inselberg

JUDY EHLEN

K

KAME

The term kame is derived from the Scottish word *kaim* and was introduced by Jamieson (1874) to mean a short, steep-sided mound or ridge of water laid sands and gravels deposited from melting ice. As a single term it cannot be applied to a specific landform, depositional process or sediment type. A variety of landforms resulting primarily from glacifluvial deposition in association with melting and buried ice are now recognized as kame forms. They include kames, kame terraces, kame complexes, and kame and kettle topography. These landforms occur in subglacial and ice-marginal environments (Holmes 1947; Price 1973; Bennett and Glasser 1996; Benn and Evans 1998).

Kames form subglacially or ice-marginally as small hills or ridges that may occur as isolated features or in a group of similar forms. Some form through deposition of sand and gravel at the base of a glacier by meltwater descending from the ice surface in a moulin, or from an englacial water body. Stream deposition within a channel bounded by ice walls may produce a short kame ridge. Other kames form as small delta-kame surfaces at the ice margin or terminus, and frequently contain beds of sand and gravel, debris flow diamicton and laminated silts and clays. Melting of adjacent and underlying ice results in normal faulting of the sediments on mound margins and sagging and folding of beds. Clay and silt beds often have slump, flow and load structures.

Kame terraces are usually formed either at the margin of the glacier between the debris-mantled decaying ice and the valley slope, at the terminus of the glacier along the ice front or between the decaying ice and an obstructing moraine. A marginal kame terrace surface may slope slightly down-glacier or towards the valley slope. The primary terrace may be fragmented into several sections due to collapse of underlying ice. Several kame terraces can be developed on a slope as the glacier margin recedes. The depression formed between the ice margin and valley slope diverts the meltwater laterally along the ice margin. The depression may be temporarily filled with water to form a narrow marginal lake. Glacifluvial sediment transported along the ice margin and from the ice edge is deposited either onto the buried ice on the floor of the depression or into the lake, or both.

The sediments in the kame terrace vary rapidly in texture and structure laterally and vertically. They consist predominantly of horizontal to gently inclined alternating beds of sand and small gravel due to glacifluvial deposition, but may exhibit laminated silts and deltaic foreset beds due to deposition in relatively deep water. On the ice proximal side of the terrace interbedded masses of debris flow diamicton (flow till) from the ice margin may occur while on the valley side paraglacial deposition of alluvial fans and debris flows can occur due to high discharges and snowmelt in adjacent valleys and on slopes. Melting of buried ice during and subsequent to sediment deposition results in formation of kettle holes that are more frequent near the former glacier margin and sometimes contain small lakes. Removal of support from the sediment by melting of marginal and buried ice results in collapse of the terrace edges to give a crenulated form, due to debris flows and landslides in the ice contact zone. Normal fault, fold, slump and flow structures may be formed within the sediments.

Kame terraces of the terminal zone usually form steep-sided ridges or steeply bounded low plateau surfaces. They may form where ribbon lakes are developed along the ice edge and are filled primarily with glacifluvial sands and gravels. They can also form where a small lake is impounded between the decaying ice and a bounding moraine. A small stream may flow from the ice bringing sand and gravel but much of the sediment consists of laminated silts interbedded with flow diamictons from the ice surface. Melting of underlying and removal of supporting marginal ice usually results in the sediments exhibiting fault, fold, slump, flow and load structures.

Kame complexes occur as a number of steep-sided mounds or ridges usually within a limited area. They result from the deposition of sands and gravels by meltwaters descending from within the ice to the glacier bed where the deposits accumulate in numerous cavities. Since deposition of kame complex sediments mainly takes place subglacially, the glacifluvial cores of the ridges are often discontinuously draped by diamicton resulting from meltout and flowage from the overlying ice during its final decay. The sediments may show faulting, folding, sag, slump and flowage structures.

Kame and kettle topography is recognized in a landscape by the occurrence of many steep-sided hills or short ridges closely juxtaposed and separated by relatively deep circular, oval or elongate depressions. Several depressions may contain ponds. Kame and kettle topography differs in origin from kame complexes by the glacifluvial sedimentation having occurred over abundant buried ice at and beyond the glacier margin. Ten to over twenty metres of glacifluvial sands and gravels are often deposited and the dips and directions of the beds are highly variable being related to deposition into depressions in locally stagnant ice. Melting of the buried ice causes normal faulting and folding of the sediments and a characteristic inversion of the topography with the kames occupying the former sites of deposition and the kettle depressions and lakes the sites of the thickest buried ice masses.

References

Benn, D.I. and Evans, D.J.A. (1998) *Glaciers and Glaciation*, London: Arnold.

Bennett, M.R. and Glasser, N.F. (1996) *Glacial Geology: Ice Sheets and Landforms*, Chichester: Wiley.

Holmes, C.D. (1947) Kames, *American Journal of Science* 245, 240–249.

Jamieson, T.F. (1874) On the last stage of the Glacial Period in North Britain, *Quarterly Journal of the Geological Society of London* 30, 317–338.

Price, R.J. (1973) *Glacial and Fluvioglacial Landforms*, Harlow: Longman.

ERIC A. COLHOUN

KAOLINIZATION

Kaolin ($Si_2Al_2O_5(OH)_4$) is a clay mineral which can be produced by prolonged weathering during which chemical elements are progressively lost. The more complex clay montmorillonite ($Al_{1.7}Mg_{0.3}Si_{3.9}Al_{0.1}O_{10}(OH)_2$) is comprised of not only the basic crystal lattice structure of aluminium and silicon atoms but it also contains magnesium. Kaolin is simply comprised of the more resistant silicon and aluminium, the more soluble elements such as magnesium having been lost through weathering. Thus kaolinization represents a process of the loss of the more soluble chemical elements producing a much simpler crystalline clay over time and can be produced by the prolonged weathering of feldspars, micas and other primary aluminosilicate minerals which may originally have contained magnesium, calcium, sodium, iron and potassium. Kaolinization thus refers to the changes from primary rock aluminosilicate minerals through weathering and their transformation to kaolin as a residual clay mineral.

Kaolin is formed under conditions of slight acidity and free drainage which remove most of these other cations. Free drainage and high rainfall are important factors in the removal of the soluble cations and in areas with annual rainfall of below around 500 mm montmorillonite clays dominate, between 500 and 1,500 mm kaolin clays dominate, with iron oxides also occurring at rainfalls above 1,500 mm (Thomas 1974). The SOLUBILITY of kaolinite is least at pH 6 but increases both as pH rises and falls from this value; so as pH departs from 6 and under strongly oxidizing conditions in the humid tropics, intense weathering tends to remove most of the silica, leaving the oxides of iron or aluminium as residual minerals rather than kaolin.

The distribution of kaolinized mineral material through a vertical weathering profile characteristically shows an increase towards the surface with a concomitant decrease in primary minerals. In a study of weathering profiles in Ghana (Ruddock

1967), some 60 per cent of particles finer than 40 μm were kaolinite at a depth of some 5–6 m and this dropped to around 30 per cent at around 30 m depth while feldspar constituted 25 per cent of the particles over 40 μm in size at 30 m and was absent at 5–6 m. The zone of greatest kaolinization may be around the zone of the water table, with kaolin often accumulating at the base of the vadose zone. The presence of clays at this depth may reduce vertical permeability to water and thus facilitate lateral water movement.

It has been suggested that the presence of kaolin is indicative of past tropical weathering conditions. What is more the case is that kaolin is indicative of prolonged weathering with slightly acid conditions, high rainfall and good drainage to facilitate cation loss. While it is thus true that kaolin is present in currently tropical areas, the presence of kaolin is only indicative of prolonged weathering under appropriate and stable conditions. It should be remembered that kaolin is also produced by metamorphosis in proximity to molten magmas.

References

Ruddock, E.C. (1967) Residual soils of the Kumasi district in Ghana, *Geotechnique* 17, 359–377.

Thomas, M.F. (1974) *Tropical Geomorphology*, London: Macmillan.

STEVE TRUDGILL

KARREN

Karren are surface solutional weathering features of varied size and shape, found on karstifiable (see KARST) rocks, usually limestones, dolomites and dolomitic limestones, but including gypsum, rock salt and silicates. The French word is *lapié* but the German *karren* generates terms now widely accepted. Wherever karstifiable rocks exist there will generally be karren features. Early research, in the nineteenth century, focused on observations in Alpine Europe. Later knowledge of karren extended to many climatic zones. Considerable experimental and field research to examine form and conditions of formation has been carried out.

Karren extend from the nanoscale of individual features to landform complexes measured in metres or tens of metres, the latter often having smaller karren on their surfaces. This rock sculpturing by DISSOLUTION, involves complex, and, sometimes, complete surface patterning. Removal (negative) forms are fundamental, e.g. pits, rills, or runnels, created directly by process (Ford and Lundberg 1987). Other karren are remnant positive forms, e.g. flachkarren (clints) and spitzkarren. Microscale forms, e.g. pits and rillenkarren, are on a centimetre scale or less. Large solution runnels, carved pinnacles, karrenfelder, giant grikelands, LIMESTONE PAVEMENTs and other complex polygenetic assemblages lie at the other extreme. Features vary in basic shape, e.g. some are linear, others circular. The largest linear solution forms can be tens of metres long although the commonest are tens of centimetres long. Width of features is also usually in tens of centimetres. Restrictions on dimensional development are often due to downslope available rock distance. Depths likewise vary considerably due partly to topographic opportunity. Small-scale surface roughness varies, with some karren characteristically smooth, others spikey or rough (Plate 66). Karren-like forms are also found underground.

Karren development is principally affected by: the type and intensity of chemical processes attacking the rock; rock characteristics including intrinsic lithology, disposition (structure) and relation to surrounding topography; the nature of any cover material; and the time available, and the changes in conditions during that time, for processes to act.

Karren classification has been attempted. Some schemes examine processes: Bögli's (1960, trans. in Sparks 1971) involves the effects of the main carbonate solution processes, depending on whether the limestone is exposed to the air, partly

Plate 66 Spitzkarren in the Triglav area of the Julian Alps, Slovenia

covered, or entirely covered by soil. This cover factor fundamentally influences solutional processes on limestone, affecting the amount of CO_2 available for solution and the time period and speed of solution. Many karren are examples of BIOKARST as they are fundamentally the result of biological corrosion.

However, many karren are not simple genetically, for they reflect past as well as present processes, and many forms are polygenetic. Thus Ford and Williams's 1989 classification is subdivided to allow for genetic factors, rather than a purely genetic system. They retain descriptive karren terminology, finding form intuitively useful. They distinguish circular karren forms, those which are linear and controlled by structure, linear forms controlled hydrodynamically, and polygenetic forms. However, these distinctions may be blurred; for example, after karren initiate hydrodynamically, field evidence suggests that small-scale lithological structures affect development. Weaknesses affect fluid movement, and may interrupt karren, causing capture or change. The importance of lithology at the microscopic scale has been stressed by researchers such as Goudie *et al.* (1989). Slope significantly influences karren, especially the pattern and complexity of branching networks.

It is useful to describe the main forms. *Rillenkarren* are tiny gravitomorphic packed channels, starting from a crest and extinguishing downslope. Width is about 1 to 3 cm, and length a few tens of cm. They are separated by sharp ridges.

Rinnenkarren and *rundkarren* are Hortonian channels with an unrunnelled catchment surface above their commencement points. They enlarge downslope depending on available rock. Width varies but rundkarren often stabilize at 20 to 30 cm and rinnenkarren are narrower. High in the long section, rundkarren develop parabolic cross sections, deepening downstream before stabilizing at the bottom end. Occasionally overdeepening develops where the runnel flows into a grike. Rundkarren are smooth features with well-rounded crests, whereas rinnenkarren are distinguished by sharper crests. The former result from covered conditions and soil removal makes them visible. Rinnenkarren generate much debate, but are considered to be formed in free or, possibly, half-free conditions. Flow forms may meander (*meanderkarren*) depending partly on slope. More complex branching of rundkarren is found on gentler slopes.

Trittkarren are step or heel-print-like features, about 10 to 30 cm in scale, essentially features of very gentle slopes. Their headwall is arc-shaped, they are flat floored and they open downslope (Vincent 1983). Trittkarren appear related to the ripples forming from sheet flow across a gentle sloping rock surface (White 1988).

Spitzkarren are peak-shaped features remaining from surface solution widespread over horizontal or gently sloping surfaces. Their sides are carved by rillenkarren. Their diameter is typically 50 cm and their height about 10 cm. They tend to merge into other positive features, pinnacles in particular, and are essentially polygenetic (Plate 66).

Kluftkarren are clefts, fissures or grikes. These are the major splits into limestone surfaces formed by widening, deepening and eventually merging of small solution features developing along linear weakness in the rock. Mature examples may run considerable lengths, e.g. several metres. Depth varies with bed thickness. By definition grikes should have broken through at least the top bed of an outcrop. This distinguishes them from runnels, which may develop into grikes (Plate 67, and see LIMESTONE PAVEMENT, Plates 73 and 74).

Kamenitzas are solution basins or pans (see WEATHERING PIT). Size varies enormously, but small ones have flat floors and scalloped edges about 3 cm deep, with diameters of several cm or more. Many are tens of cm in diameter with the largest examples measurable in metres. Development is, however, limited, as when the pan becomes drained either at the surface over

Plate 67 Sloping limestone pavement (12° to 15°) on Farleton Fell, Cumbria, UK: showing strong karren formation on a wide range of clints across several limestone beds

a rim, or from beneath via solutionally opened cracks in the floor. The feature may become more complex; perhaps part of a staircased runnel, or of a complex hole into which several runnels drain. Kamenitzas occur in both free and covered conditions. True kamenitzas enlarge outwards at their rims where solution conditions remain ideal. Kamenitzas may merge into kluftkarren, or coalesce, leaving peaky sharp Spitzkarren.

Kluftkarren also merge into larger weakness-oriented features, e.g. bogaz, lapiés wells, etc. Several joints crossing focuses processes, resulting in mill-like features possibly several metres across and complex in form. This is demonstrated in limestone pavement areas in the UK, and in high mountain areas such as north Norway and the Alps, where large amounts of glacial meltwater may have enhanced them.

Terminology cannot encompass the full variety of natural sculptured forms. Features grade into each other, cut across, develop idiosyncratically according to local conditions and deteriorate when ideal formation conditions change. Destruction of rundkarren, formed under soil, illustrates this: on soil removal their smooth surface undergoes etching in subaerial conditions into sharper rougher features, e.g. kamenitzas or rillenkarren. Destruction may include mechanical weathering effects involving freeze and thaw, for example, and ends with broken rubble fields over underlying intact rock layers. However, another change in conditions could find water flow, or even a return to soil cover: both would alter the karren again. In karst areas such changes of local conditions can happen especially easily and quickly because of capture of active-process locations.

Rock type is fundamental to karren development. Strong, pure limestones or dolomites produce the best development. On rock salt, karren form easily due to high solubility but only persist in relative aridity. Gypsum karren are intermediate in persistence as gypsum's solubility is such that karren develop in humid conditions, but these conditions also favour their destruction. Silicate rocks are only slightly soluble; only in prolonged warm and wet conditions will karren, like other karst features, form. Sandstones in temperate areas may display weak karren but development is only significant in the wet tropics. Research has placed a general timescale for development of limestone karren in thousands of years, gypsum features in hundreds of years and rock salt karren in tens of years (Mottershead and Lucas 2001).

Other lithological factors influence particular karren forms. Research has considered karren initiation at tiny weaknesses or variations in the rock (Moses and Viles, in Fornos and Gines 1996). Chance unevenness allows rainwater to pond and start surface solution, simple plants may then develop. If soil develops so do higher plants and accelerated biological corrosion becomes very important, especially in warm, wet locations. The forms produced can be very striking. On bare rock, flow processes down any slope, however smooth, result eventually in flow-concentration into channels. A slight slope gives flow rather than ponding, a slight dip can give ponding before flow.

Sense can be made of karren by considering their form, mode of origin and development conditions. However, field situations demonstrate that, although 'perfect' examples of types are found, there will always be a wide spectrum, and merging of features is both possible and common even without changes in external factors such as climate.

References

Ford, D.C. and Lundberg, J. (1987) A review of dissolution rills in limestone and other soluble rocks, *Catena Supplement* 8, 119–140.

Ford, D.C. and Williams, P.W. (1989) *Karst Geomorphology and Hydrology*, London: Unwin Hyman.

Fornos, J.J. and Gines, A. (eds) (1996) *Karren Landforms*, Palma: Universitat de les Illes Balears.

Goudie, A.S., Bull, P.A. and Magee, A.W. (1989) Lithological control of rillenkarren development in the Napier Range, Western Australia, *Zeitschrift für Geomorphologie N.F. Supplementband* 75, 95–114.

Mottershead, D. and Lucas, G. (2001) Field testing of Glew and Ford's model of solution flute evolution, *Earth Surface Processes and Landforms* 26, 839–846.

Sparks, B.W. (1971) *Rocks and Relief*, London: Longman.

Vincent, P.J. (1983) The morphology and morphometry of some arctic Trittkarren. *Zeitschrift für Geomorphologie N.F. Supplementband* 27(2) 205–222.

White, W.B. (1988) *Geomorphology and Hydrology of Karst Terrains*, New York: Oxford University Press.

SEE ALSO: biokarst; dissolution; karst; limestone pavement.

HELEN S. GOLDIE

KARST

Karst is terrain with distinctive hydrology and landforms arising from a combination of high rock solubility and well-developed secondary porosity (Ford and Williams 1989). It is commonly associated with carbonate rocks such as limestone, marble and dolomite and is well known for features such as caves, enclosed depressions, fluted rock outcrops, underground rivers and large springs. It takes its name from a limestone region in the northern Adriatic inland of Trieste on the Slovene–Italian border where such features are particularly well developed. The region is called *kras* (Slovenian) or *carso* (Italian), but this was Germanicized to *Karst* in the period of the Austro-Hungarian Empire, when the first scientific studies were made of the region's geomorphology and hydrology. By extension, other areas with similar features are also referred to as karst, and this includes places where it is developed on other soluble rocks such as gypsum and rock salt (see GYPSUM KARST; SALT (EVAPORITE) KARST). This discussion focuses on karst in carbonate rocks. These outcrop over about 12 per cent of the ice-free continental areas (Figure 95), with well-developed karst covering about 7–10 per cent of the continental area. Karst also develops beneath the surface when karst rocks are interbedded with other lithologies; this is known as *interstratal karst*. Limestone (as opposed to karst) geomorphology refers to landscapes developed on carbonate rocks, and includes landforms that are not necessarily produced by karst processes (e.g. coral atolls, glacial troughs).

Cvijić (1960) defined different morphological types of karst, including *holokarst* where karst reaches its fullest development in thick carbonate rocks that extend below sea level, for example in the extensive limestones of the Dinaric region; *merokarst* where karst development is evident but rather poorly expressed, by virtue of not very suitable lithologies or rather thin limestones, such as in the chalk of northern France or in some areas of Jurassic limestones such as the Swabian

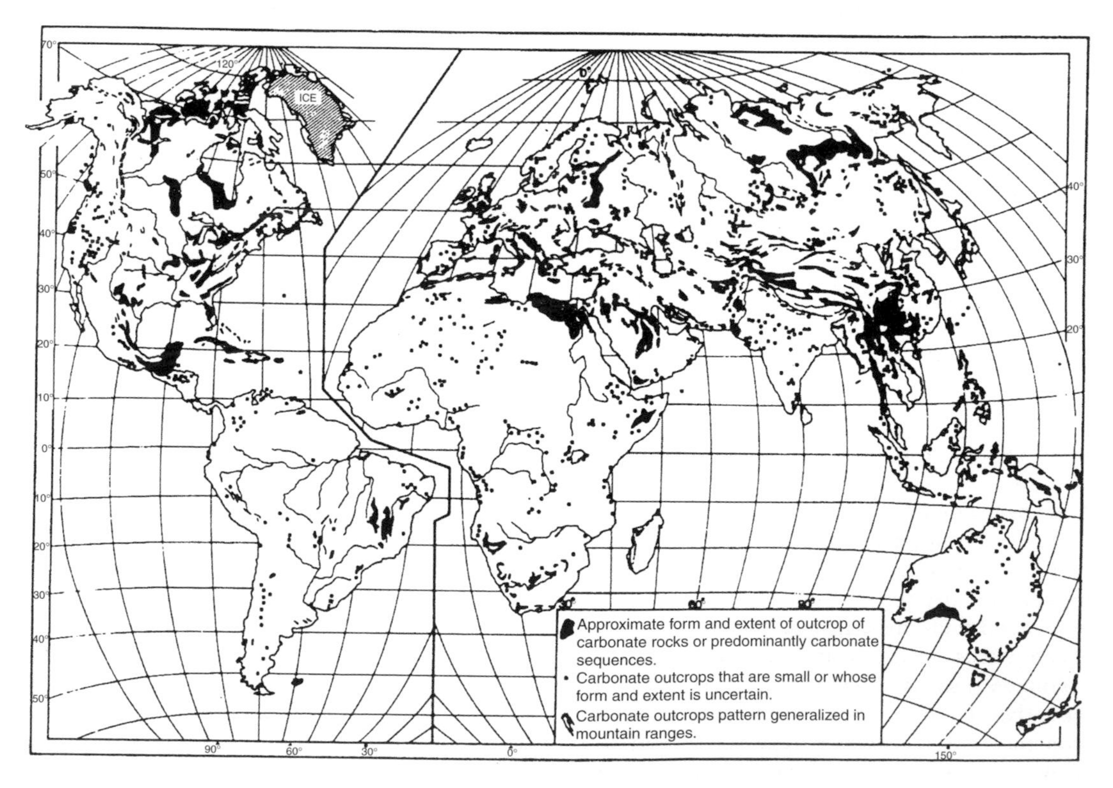

Figure 95 Global distribution of carbonate rock outcrops. Karst occurs over most of this area and also in subcrop beneath various coverbed lithologies (from Ford and Williams 1989.)

Alb of Bavaria or the Cotswold Hills of Britain; and transitional types where the carbonate rocks are quite thick and well karstified, but underlain by or interbedded with non-carbonate formations, as in the Causses of France. However, these terms are seldom used. When the imprint of now dry river valleys is evident in the landscape, it is sometimes referred to as *fluviokarst*.

The occurrence of pure carbonate rocks with high solubility is insufficient to produce karst, because the structure, density and thickness of the rock are also important. The carbonate rocks in which karst reaches its best expression are thick, dense, pure (> 90 per cent calcium carbonate) and massive with low primary porosity, but with well-developed secondary porosity along fissures such as joints, faults and bedding planes. Even pure soluble rocks such as coral and chalk have relatively poorly developed karst, because their very high primary porosity (which can be 30–50 per cent) leads to diffuse groundwater flow, which is not conducive to extensive cave and closed depression development, though they have some. Impure carbonate rocks such as argillaceous limestones hardly support karst, partly because they tend to have high primary porosity but especially because insoluble residues inhibit the growth of secondary porosity by clogging groundwater pathways as they form. Nevertheless, there is a spectrum of rock types and degrees of purity, with a corresponding spectrum of karst development.

Chemical processes

The dominant process that produces karst features is the solution of the rock by rainwater. The chemical process is DISSOLUTION (or corrosion); in a carbonate karst context the process can be summarized as:

$$CaCO_3 + H_2O + CO_2 = Ca(HCO_3)_2$$

Calcium carbonate, $CaCO_3$, dissolves in the presence of water and carbon dioxide ($H_2O + CO_2 = H_2CO_3$ or carbonic acid) to yield the more soluble calcium bicarbonate, $Ca(HCO_3)_2$, which is readily transported away in solution.

The process of carbonate rock solution can be conceptualized as operating in a situation in which two hydrological and geochemical subsystems interact. The hydrological cycle provides the main source of natural energy that powers the system and drives the evolution of karst, because water is the solvent that dissolves karst rock and then carries it away in solution. Geochemical processes control the rate of dissolution (the speed with which solid rock is converted into ions in solution), which in a carbonate karst depends very strongly on strength of acidification by dissolved carbon dioxide during its passage through the atmosphere and soil layer before making contact with the limestone. The concentration of CO_2 in the open atmosphere is about 0.03 per cent by volume, whereas it is commonly 2 per cent in the soil and can even reach 10 per cent. A factor of 100 in the concentration of CO_2 results in ~5 times increase in the solutional denudation rate (White 1984). Although this is important, the amount of rainfall is even more significant, the wettest places in the world having the fastest rate of limestone solution. For example, limestone denudation by solution processes has been estimated as high as $760\,m^3\,a^{-1}\,km^{-2}$ (cubic metres per year per square kilometre of limestone outcrop) in very wet places such as parts of Papua New Guinea where rainfall can reach 12,000 mm per year, but as low as $5\,m^3\,a^{-1}\,km^{-2}$ in some arid zones like the Nullarbor Plain in southern Australia with rainfall of less than 350 mm per year. The amount of solutional attack on the limestone rock therefore depends on the concentration of the solute (determined by biogeochemical processes) and the volume of solvent (determined by the rainfall). The solute load of a karst spring is the product of its discharge and the concentration of limestone salts in solution.

Biochemical and physical processes associated with various organisms also assist in weathering limestones, especially in the intertidal zone, and produce a suite of landforms known as BIOKARST.

Landscape development

A conceptual model of the karst system is presented in Figure 96. Karst evolution is explained by White (1988), Ford and Williams (1989) and Gabrovšek (2002), and is summarized in Williams (2003). In order for major karst landforms such as enclosed depressions and caves to develop, the rock removed in solution must be carried right through the body of karst rock and be discharged at springs. Thus the development of an underground plumbing system is a necessary precursor to surface landform evolution. When a continuous conduit of 5–15 mm extends right through the rock, the drainage can become turbulent and

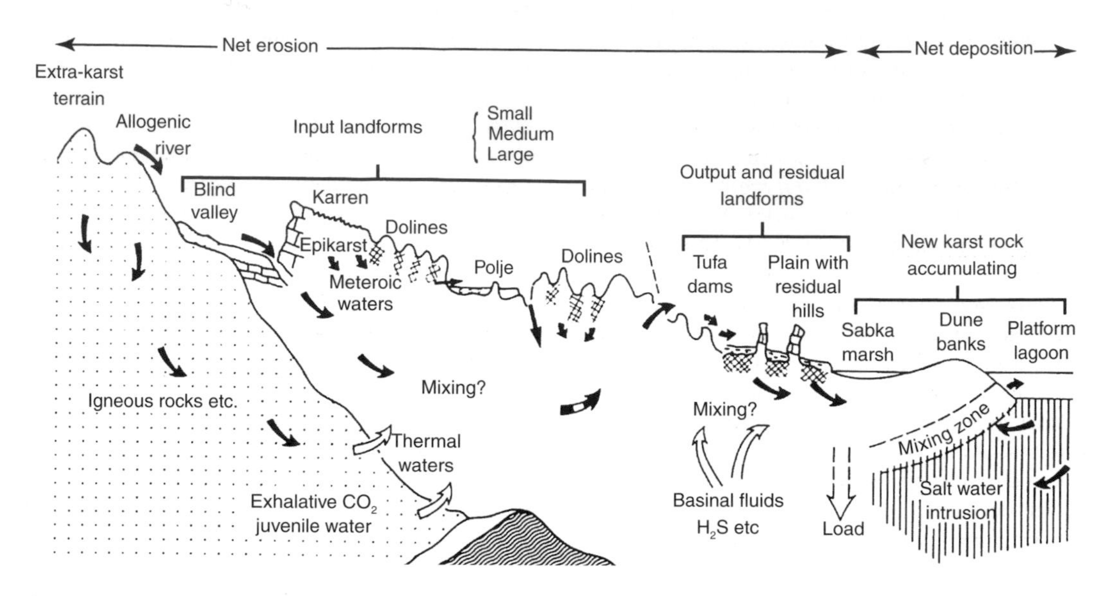

Figure 96 A conceptual model of the comprehensive karst system (from Ford and Williams 1989.)

then can begin to discharge fine insoluble particles. It takes the order of 5,000 years for a conduit of 1 km length to develop up to this size (White 1988). But once the hydraulic threshold of laminar to turbulent flow has been passed, cave enlargement proceeds more rapidly. Cave passages to 3 m diameter can develop in about 10^4–10^5 years.

The development of subterranean karst hydrology also depends on the manner in which water enters the karst. Rainwater that falls directly onto the limestone outcrop is known as *autogenic* recharge. It infiltrates diffusely into the rock via countless fissures. By contrast, rain that falls onto impervious non-karstic rocks but later flows onto the karst is known as *allogenic* recharge. It runs off as an organized stream, which sinks underground at the end of a BLIND VALLEY soon after encountering the limestone. Places where streams disappear underground are called *swallow holes*, *swallets*, *stream-sinks* or *ponors*. Sites where water cascades steeply into open pits are sometimes called pot-holes or abîmes (French).

Autogenic waters are mainly acidified by carbonic acid, but allogenic waters may have flowed from peat bogs (and hence contain organic acids) or have encountered sulphide minerals when draining from shales and hence contain sulphurous acid. They also tend to have a greater mechanical load, which can abrade limestone and help to incise cave floors. Streams sinking along the allogenic input boundary converge underground and emerge at springs at the output boundary of the system, thereby establishing dendritic subterranean drainage networks. The flow in these conduits is turbulent, as opposed to the much slower flow in tiny interconnecting pores and fissures which is laminar. Processes of CAVE network development (*speleogenesis*) are discussed by White (1988), Ford and Williams (1989), Gillieson (1996) and Klimchouk *et al.* (2000).

Some karsts have very extensive cave systems. The longest in the world is the Mammoth–Flint Ridge–Roppel–Procter System in Kentucky, USA. It comprises an interconnected system of essentially horizontal dendritic passages at different levels, totalling over 530 km in length, but developed in only about 100 m vertical stratigraphic thickness of limestone. The world's deepest known caves are Voronja Cave, Arabika massif, West Caucasus, at over 1,750 m deep, Reseau Jean Bernard (1,602 m) in France and Lamprechtsofen–Vogelschacht (1,535 m) in Austria.

Caves can be some of the world's oldest landforms, because they are located deep underground and so are protected for a long time from surface denudation. Thus sediments in the Mammoth Cave system have been dated by radioactive decay of cosmogenic isotopes to 3.5 Ma (Granger *et al.* 2001), and fossil hominid and animal remains in cave sediments in South Africa have been dated by palaeomagnetism to a similar age.

Once the input–output connections are established diffuse autogenic recharge infiltrates the bedrock. Most of the dissolutional attack takes place on bedrock just beneath the soil, close to where CO_2 is generated and consequently where the percolating water attains its greatest aggressivity. Thus up to about 90 per cent of the corrosion is accomplished in the top 10 m or so of the limestone outcrop. Since water penetrates underground mainly by means of joints and faults, these fissures become more widened by corrosion near the surface than they are at greater depth. The surface of karst is therefore very permeable, but permeability (the capacity to transmit water) decreases with depth. This highly corroded superfical zone is termed the EPIKARST (or alternatively the *subcutaneous zone*).

Before rainwater drains underground it flows across rocky outcrops on the surface. The resulting corrosion of these outcrops yields a small-scale solution sculpture of vertically fluted rock and widely opened joints, collectively known as *lapiés* (French) or *Karren* (German). Rocky spires produced in this way can sometimes be tens of metres high, although individual KARREN are normally smaller. These forms are particularly common above the tree-line where soil and vegetation are thin or absent, but karren also develop beneath a soil and vegetation cover. When the bedrock has been scoured by glaciation and stripped of loose debris, postglacial weathering produces bedding-plane surfaces with joints opened by dissolution. In northern England the whole surface is known as a LIMESTONE PAVEMENT, with the widened joints called *grikes* and the intervening blocks called *clints*. Large solutionally widened joint corridors are known as *bogaz* and complex networks of such features are sometimes known as labyrinth karst. Areally extensive expanses of any or all of these features, especially above the tree-line, constitute a *karrenfeld*.

Sometimes part of the carbonate dissolved near the surface is precipitated further down in pores and fissures in the bedrock. This is particularly common in porous limestones such as coral, and results in several metres of the bedrock near the surface being hardened by being made less permeable, a process known as CASE HARDENING. This is encouraged in hot climates in particular by evaporation near the surface. Carbonate crusts produced by secondary precipitation of carbonate are known as CALCRETE or *caliche*. Sometimes induration of the rock by case hardening proceeds in calcareous dune limestones (AEOLIANITE) at the same time as karst development is occurring. This produces a style of landscape known as SYNGENETIC KARST. After further percolation the seepage waters may emerge into cave passages. Since cave air usually has CO_2 concentrations similar to the open atmosphere, the emergence of supersaturated percolation waters results in CO_2 degassing and in the precipitation of calcite in the form of stalactites, stalagmites and flowstones (collectively termed SPELEOTHEMs).

Whereas the most characteristic subterranean features of karst are caves, the most typical surface landforms are closed depressions, especially DOLINEs, which are enclosed bowl or saucer-shaped hollows, usually of a few hundred metres in diameter and some tens of metres deep. When dolines occupy all the available space, the surface has a relief like an egg-tray and is known as *polygonal karst*, but this does not always develop and often dolines are dispersed or in clusters across an undulating surface. Polygonal karsts can have doline densities ranging from 4–55 per square kilometre. The particularly large and correspondingly deep solution dolines of some tropical and subtropical karsts are also called *cockpits*, a Jamaican term.

Solution dolines (Plate 68) develop in the subcutaneous zone and drain water centripetally to enlarged fissures that discharge it vertically to the deep groundwater system (Figure 97). Small

Plate 68 The hallmark landform of karst: solution dolines near Waitomo, New Zealand

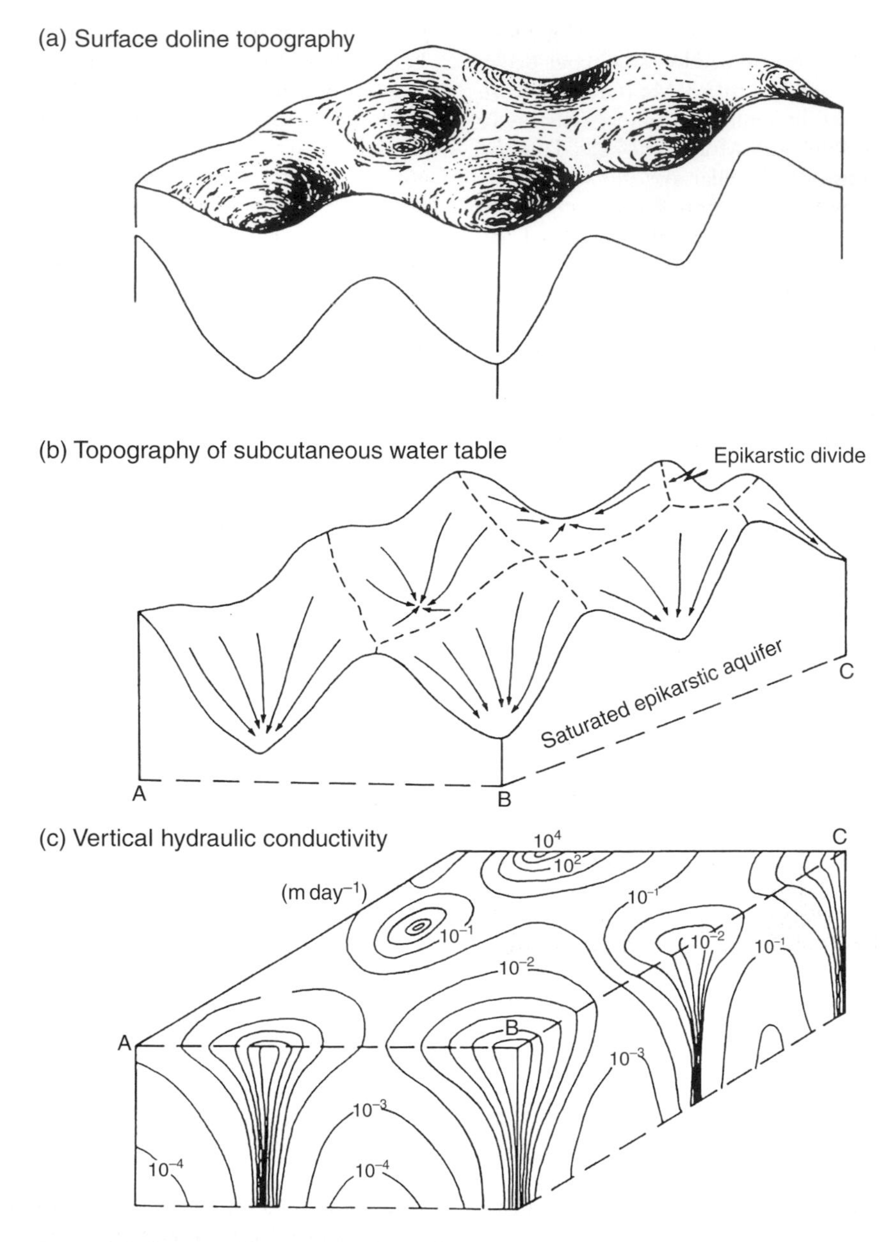

Figure 97 A model showing the relationship of solution doline development to flow paths in the epikarst (subcutaneous zone) and vertical hydraulic conductivity (from Williams 1985.)

allogenic streams also form enclosed basins, where they disappear underground. Large allogenic streams penetrate further into the karst in well-defined valleys before they sink, and produce landforms known as BLIND VALLEYs, because their valleys usually terminate abruptly in a cliff or steep slope. The sinking streams give rise to caves, and if a cave roof is close to the surface it sometimes collapses, producing a cylindrical or crater-like depression termed a *collapse doline*. A collapse that exposes an underground river is sometimes called a *karst window*. Some caves can be completely unroofed by progressive collapse, producing a gorge of cavern collapse (though not all gorges in karst are produced in this way, many being produced by antecedent drainage). Where doline collapse intersects the water table, the steep-sided enclosed depression holds a lake, such features being called CENOTEs in the Yucatan Peninsula.

As dolines evolve, they often enlarge laterally and coalesce, producing compound closed depressions known as *uvalas*. Where the rate of vertical incision of dolines is significantly greater than the rate of solutional denudation of the intervening land, the inter-doline areas develop into hills. This is particularly common in humid tropical and subtropical karsts, where residual hills can be so well developed that they visually dominate the landscape, giving rise to a style of landscape called *cone karst*. In China, *fengcong* is the term used to describe such karsts (Plate 69).

Many karst areas have developed on rocks that have been folded and faulted. These tectonic influences considerably complicate karst evolution and are of major significance in guiding groundwater flow and denudation of the surface. Faulted terrains often provide the conditions in which the largest enclosed karst depressions – known as POLJEs – are developed, some exceeding 100 km^2 in area. The term 'polje' is of Slav origin and means 'field', probably because it was the largest area of flat tillable land in the karst. Many examples of these features are found in the Dinaric karst, where poljes are often located in faulted basins. Ford and Williams (1989) define three types: border polje, structural polje and baselevel polje – according to the dominant influences in their evolution (Figure 98).

Sometimes a particularly large blind valley encloses a basin of a square kilometre or more with a well-developed flat, floodplain floor, and it may receive more than one allogenic stream sinking at different points. Such large enclosed depressions at the edge of karsts are known as *border poljes*. Because of their relationship to sinking streams the floors of poljes often flood, particularly when the discharge of the inflowing river is greater than can be absorbed by the stream-sink(s). There is no clear demarcation between blind valleys and border poljes. They are transitional forms, the larger ones with particularly flood-prone flattish floors being called poljes.

Poljes may also be found in the interior of karsts, where structural dislocations have produced tectonic depressions with inliers of relatively impervious rocks. In these cases, the inlier acts as a dam on regional groundwater movement, forcing it to emerge as springs on the upstream side of the barrier. Water then flows across the impermeable inlier, to sink in ponors on the downstream side, the intervening region being developed into an alluviated plain. These features are known as *structural poljes*.

Genetically distinct from the above is the *base-level polje*, which is a very large enclosed depression entirely in karst rock that has been incised by solution down to the level of the *epiphreatic* zone (the zone of fluctuation of the water table). Such poljes are typically located close to the outflow boundary of a karst. They have swampy floors and can be envisaged as windows on the water table. Hence they inundate when the regional water table (or piezometric surface) rises in the wet season.

When vertical denudation eventually reduces the bottom of closed depressions to the level of the regional water table, they can incise no further, so instead they widen their floors. As a result residual hills between dolines become isolated, such

(a)

(b)

Plate 69 Two examples of humid subtropical karst in Guangxi province, China. The skylines are dominated by the conical forms of the hills, but enclosed depressions occur between them. When depression floors reach the water table they widen at their base, isolate the hills and extend the floodplain surface

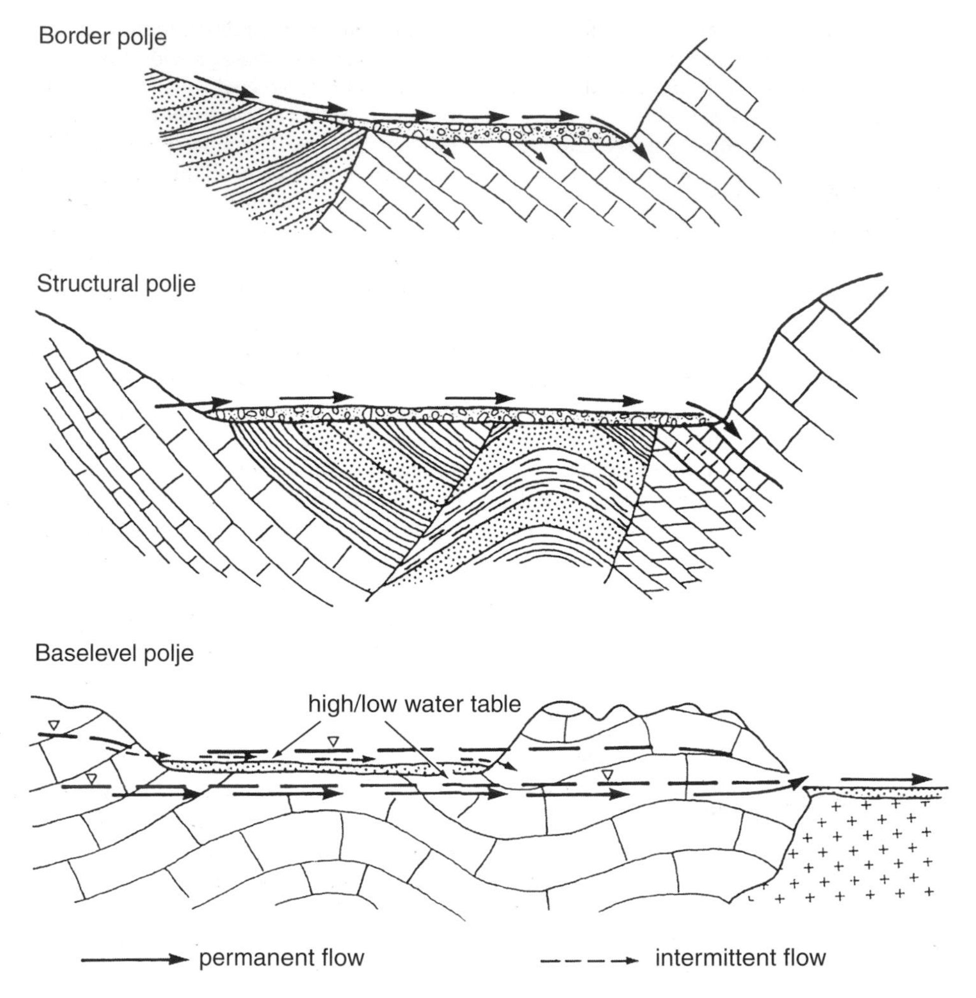

Figure 98 Three main types of polje. These are the largest enclosed depressions found in karst landscapes. Their flat floors are prone to flood and may cover many square kilometres in area (from Ford and Williams 1989.)

landforms being called *hums* in Europe. The lower slopes of such hills, which are usually of a rounded conical shape, can be over-steepened by undercutting and may collapse at their base, a process brought about by the corrosional attack of swamp waters – made particularly vigorous if allogenic rivers periodically flood the intervening plains. This is especially common in tropical humid karsts, where landscapes of steep isolated hills are referred to as *tower karst* (*Kegelkarst*, German), superb examples being known in southern China (where it is called *fenglin*). In the Caribbean isolated karst hills produced in this way are called *mogotes*.

During the end stages of karst denudation caves are drained and dismembered and their remnant passages are left at various elevations within residual karst hills, until eventually even the residual hills are removed by solution and only a corrosion plain is left. A superb example of a corrosion plain is the Gort lowland of counties Clare and Galway in western Ireland, where Pleistocene glaciations have stripped away the mantle of residual soil, alluvium and loose rock to reveal the karstified bedrock beneath.

Uplift can rejuvenate karst systems. But whereas in the first cycle of karstification the rock was unweathered and had only primary porosity, in the second cycle there is an inheritance of landforms on the surface and secondary porosity underground. Thus a new phase of karst evolution would exploit the inherited features and develop them further.

Modelling karst evolution

Various attempts have been made to model karst landscape development. Early conceptual models of karst evolution were presented by Grund (1914) and Cvijić (1918) (see translations into English in Sweeting 1981), but it was not until the late twentieth century that models became quantitative.

Ford and Ewers (1978) used a physical laboratory model to elucidate the development of protocaves and successive flow paths. White (1984) developed a theoretical expression showing the relationship between chemical and environmental factors in the solutional denudation of limestones and he also developed a model of the development of cave passages (White 1988).

Ahnert and Williams (1997) developed a 3-dimensional model of surface karst landform development that started with a terrain in which proto-conduit connections were already established and then showed how the relief might develop given different assumptions about starting conditions, such as randomly disposed sites of greater permeability or random variations in initial relief (Figure 99). This model illustrates sequential steps in the development of doline and polygonal karst and reveals the importance of divergent and convergent flow paths in explaining the development of residual cones between incising depressions.

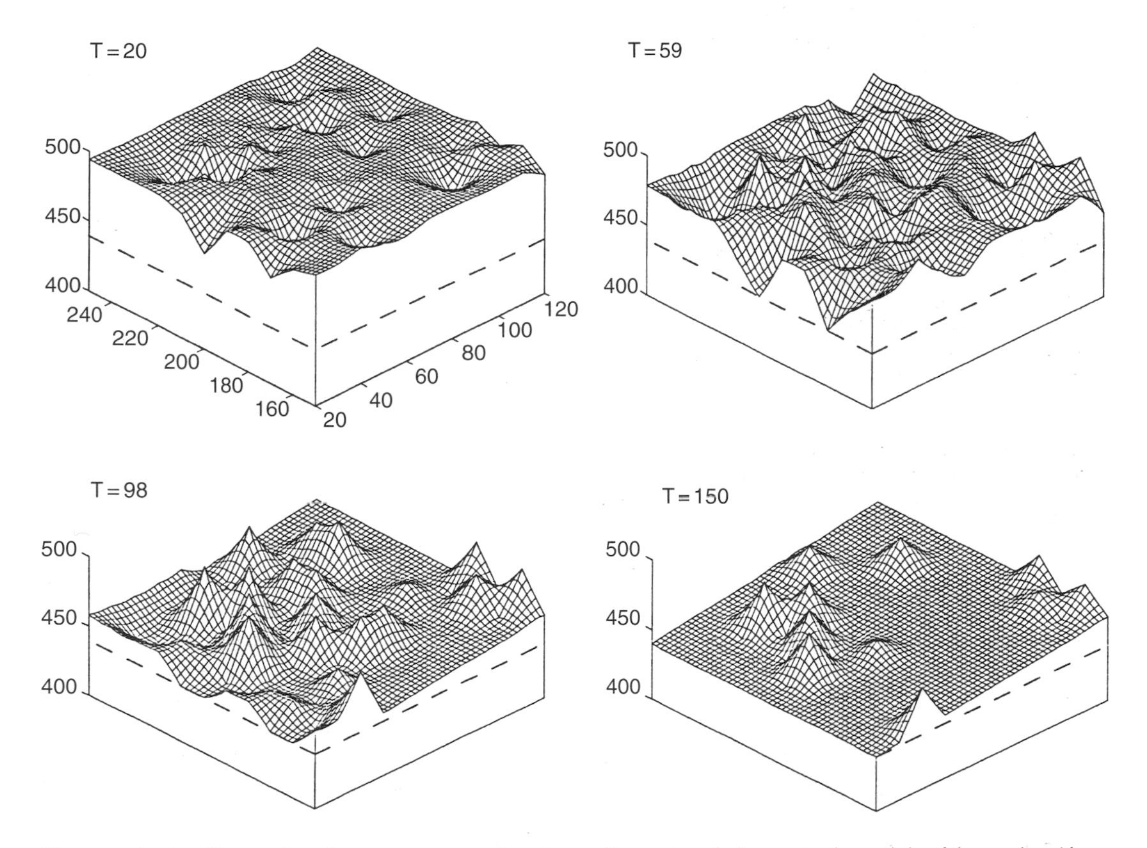

Figure 99 An illustration from one run of a three-dimensional theoretical model of karst landform development. The model shows an undulating surface with dolines at time 20 (T = 20), the development of polygonal karst by T = 59 (when some doline bottoms attain base level (shown by dashed line), the commencement of isolation of residual hills by T = 98, and the development of a corrosion plain with isolated hills by T = 150. The corrosion plain has a gradient towards the left because of the slope (hydraulic gradient) of the water-table (from Ahnert and Williams, 1997.)

Other more recent models associated with dissolution and the processes governing the evolution of karst are presented in Klimchouk *et al.* (2000) and Gabrovšek (2002).

References

Ahnert, F. and Williams, P.W. (1997) Karst landform development in a three-dimensional theoretical model, *Zeitschrift für Geomorphologie, Supplementband* 108, 63–80.

Cvijić, J. (1960) *La Géographie des Terrains Calcaire*, Monographies tome 341, No. 26, Belgrade: Academie Serbe des Sciences et des Arts.

Ford, D.C. and Ewers, R.O. (1978) The development of limestone cave systems in the dimensions of length and depth, *Canadian Journal of Earth Science* 15, 1,783–1,798.

Ford, D.C. and Williams, P.W. (1989) *Karst Geomorphology and Hydrology*, London: Chapman and Hall.

Gabrovšek, F. (ed.) (2002) *Evolution of Karst: from Prekarst to Cessation*, Postojna–Ljubljana, Inštitut za raziskovanje krasa, ZRC SAZU, Založba ZRC (Zbirka Carsologica).

Gillieson, D. (1996) *Caves: Processes, Development, Management*, Oxford: Blackwell.

Granger, D.E., Fabel, D. and Palmer, A.N. (2001) Pliocene-Pleistocene incision of the Green River, Kentucky, determined from radioactive decay of cosmogenic ^{26}Al and ^{10}Be in Mammoth Cave sediments, *Geological Society of America Bulletin* 113(7), 825–836.

Klimchouk, A.B., Ford, D.C., Palmer, A.N. and Dreybrodt, W. (eds) (2000) *Speleogenesis: Evolution of Karst Aquifers*, Huntsville, AL: National Speleological Society Inc.

Sweeting, M.M. (ed.) (1981) *Karst Geomorphology*, Benchmark Papers in Geology 59, Stroudsburg, PA: Hutchinson Ross.

White, W.B. (1984) Rate processes: chemical kinetics and karst landform development, in R.G. LaFleur (ed.) *Groundwater as a Geomorphic Agent*, 227–248, Boston: Allen and Unwin.

——(1988) *Geomorphology and Hydrology of Carbonate Terrains*, Oxford: Oxford University Press.

Williams, P.W. (1985) Subcutaneous hydrology and the development of doline and cockpit karst, *Zeitschrift für Geomorphologie*, 29, 463–482.

——(2003) Karst evolution, in J. Gunn (ed.) *Encyclopedia of Caves and Karst*, New York: Routledge.

PAUL W. WILLIAMS

KETTLE AND KETTLE HOLE

Kettle holes are depressions formed by the melt of discrete blocks of glacier ice that have been partially or completely buried by GLACIFLUVIAL sediments. Kettle holes have been reported from many present-day proglacial environments. Ice blocks originate in three ways: (1) detachment from the glacier snout due to ablation (Rich 1943), (2) transport on to the outwash plain or 'sandur' by glacial streams or rivers (Maizels 1977) and (3) release and transport on to the sandur surface during jökulhlaups (glacier OUTBURST FLOODS) (Fay 2002). Kettled or pitted sandur is glacial outwash in which numerous kettle holes have formed.

Kettle holes may be inverse-conical or steep-walled in shape. Inverse-conical kettles form due to the melt of partially or totally buried ice blocks and develop by sediment slumping and avalanching down the kettle walls. Inverse-conical kettles formed by the melt of partially buried blocks may possess raised diamict (see DIAMICTITE) rims (rimmed kettles) and/or diamict mounds in the base of the depression. Steep-walled kettles form by collapse of overlying sediment into voids created by the *in situ* melt of completely buried ice blocks. Ice blocks transported on to the sandur by water are often progressively buried by glacifluvial sediments. However, during a jökulhlaup, sediment deposition can be so rapid that small ice blocks are incorporated within the flow and deposited simultaneously with flood sediments.

The physical properties of the sediment in which kettle holes form control their collapse sequence (Fay 2002). Shallow kettles with vertical or inward-dipping walls, whose base is a coherent block of sediment, form in coarse, clast-supported sediments. In coarse sediments dominated by matrix-support or in entirely fine-grained sediments, deeper kettles with steeply dipping or overhanging walls form, often through sudden roof collapse. Steep-walled kettle holes may form over small, buried blocks, or over larger, buried ice blocks which melt irregularly. Steep-walled kettles can develop into inverse-conical kettle holes by slide or avalanche of sediment into the kettle hollow.

Since the development of kettle holes is similar over both stagnant glacier ice and flood-related ice blocks, it may be difficult to differentiate between flood-related kettles and kettles produced by non-fluvially driven processes. However, on a palaeoflood surface, a flood origin for kettle holes is indicated by a distinct radial pattern of kettle holes on prominent outwash fans reflecting flow expansion, and/or a decrease in kettle size down sandur relating to a progressive decrease in stream power.

References

Fay, H. (2002) Formation of kettle holes following a glacial outburst flood (jökulhlaup), Skeiðarársandur, southern Iceland, in A. Snorrason, H.P. Finnsdóttir and M. Moss (eds) *The Extremes of the Extremes: Extraordinary Floods*, IAHS Publication 271, 205–210.

Maizels, J.K. (1977) Experiments on the origin of kettle-holes, *Journal of Glaciology* 18, 291–303.

Rich, J.L. (1943) Buried stagnant ice as a normal product of a progressively retreating glacier in a hilly region, *American Journal of Science* 241, 95–99.

Further reading

Maizels, J.K. (1992) Boulder ring structures produced during jökulhlaup flows - origin and hydraulic significance, *Geografiska Annaler* 74A, 21–33.

HELEN FAY

KNICKPOINT

Knickpoint (nickpoint) refers to a substantially steepened section of a stream long profile. In cyclic interpretations of landscape (Davis 1899) knickpoints carried a new 'cycle' of erosion inland replacing an older, more elevated low relief surface and producing a gorge (see GORGE AND RAVINE). For knickpoint erosion in alluvial material, soil and rill systems, and in the laboratory, the term 'headcut' is used (Bennett 1999), or for extended steep reaches, knickzone (Downs and Simon 2001). The term WATERFALL is synonymous with knickpoint. The steepened valley walls below the knickpoint subsequently collapse by mechanical failure following stress release (Philbrick 1970), enhanced groundwater drainage, and by stream erosion undercutting the valley wall. Despite morphological similarities between knickpoints in cohesive clay materials and those in bedrock, they may not retreat by similar processes. Both cases depend upon the headwall being continually re-steepened with the bulk of any detrital material present at the base of the knickpoint being removed. In weak material incapable of maintaining a cap to the headwall the knickpoint face may rotate as it retreats: longitudinal slope diminishing until the knickpoint is removed.

Niagara Falls has a mean flow rate (before abstraction for power generation) of 5,730 $m^3 s^{-1}$. It retreated at a rate 1.5 $m yr^{-1}$ in the last decades of the nineteenth century (Philbrick 1970); a figure consistent with the estimated rate for the entire postglacial period (14,000 calendar years). When flow was only 10 per cent of the present value (Tinkler *et al.* 1994) for 5,000 years, the recession rate was between 0.10 and 0.15 $m yr^{-1}$, and with a reduced elevation for the headwall. Although 1.5 $m yr^{-1}$ is high, the recession rate of the 12 m St Anthony Falls on the Missouri may have been comparable in the nineteenth century (Winchell 1878; Sardeson 1908). The inference for both waterfalls is that the headwall retreated in an essentially parallel fashion during periods of thousands of years.

Recession rates for other major knickpoints are known only for a few systems at present. In south-east eastern Australia (van der Beek *et al.* 2001) rates of the order of 1,000 m per million years have been estimated, and half the rate that has been reported in adjacent areas of Australia (Nott *et al.* 1996; Seidl *et al.* 1997). In southern Africa Derricourt (1976) has suggested rates for the recession of the Batoka Gorge, below the Victoria Falls, have varied between 0.09 and 0.15 $m yr^{-1}$ over the last million years or so. Stranded and stationary surface knickpoints have been reported in karst terrains (Fabel *et al.* 1996; Youping and Fusheng 1997).

The process of knickpoint recession is far from clear. There is no published evidence that for large waterfalls headwall undercutting takes place in the plunge pool. In non-cohesive sediments with low slope the entire headcut may be submerged but this has only rarely been described for large bedrock forms (Rashleigh 1935). More probably headwall recession above the plunge pool level proceeds by subaerial weathering in a very damp environment; by sapping, from water fed to the vertical face from the upper river bed (Krajewski and Liberty 1981); and by stress release close to rock face weakening joints and bedding plains prior to block release. In the upper caprock zone (if present) accelerating water approaching the waterfall edge eventually exerts enough force to tear blocks out of the undermined capping beds (Philbrick 1970). At full flow, on waterfall faces less than vertical, erosional wear by water and entrained sediment may effect rock removal (Bishop and Goldrick 1992; and examples in Rashleigh 1935). Knickpoints in alluvial sediments can progress much faster (Simon and Thomas 2002) cite rates of 0.7 to 12 $m yr^{-1}$.

References

Bennett, J.S. (1999) Effect of slope on the growth and migration of headcuts in rills, *Geomorphology* 30, 273–290.
Bishop, P. and Goldrick, G. (1992) Morphology, processes and evolution of two waterfalls near Cowra, New South Wales, *Australian Geographer* 23, 116–121.
Davis, W.M. (1899) The Geographical Cycle, *Geographical Journal* 14, 481–504.
Derricourt, R.M. (1976) Retrogression rate of the Victoria Falls and the Batoka Gorge, *Nature* 264, 23–25.
Downs, P.W. and Simon, A (2001) Fluvial geomorphological analysis of the recruitment of large woody debris in the Yalobusha River network, Central Mississippi, USA, *Geomorphology* 23(1–2), 65–91.
Fabel, D., Henricksen, D. and Finlayson, B.L. (1996) Nickpoint recession in karst terrains: an example from the Buchan karst, southeastern Australia, *Earth Surface Processes and Landforms* 21, 453–466.
Gregory, J.W. (1911) Constructive waterfalls, *Scottish Geographical Magazine* 27, 537–546.
Krajewski, J.L. and Liberty, B.A. (1981) Present dynamics of Niagara, in I. Tesmer and J.C. Bastedo (eds) *Colossal Cataract: Geological History of Niagara Falls*, 63–93, Albany: University of Toronto, State University of New York.
Nott, J., Young, R. and McDougall, I. (1996) Wearing down, wearing back, and gorge extension in the long-term denudation of a highland mass: quantitative evidence from the Shoalhaven catchment, southeast Australia, *Journal of Geology* 104, 224–232.
Philbrick, S.S. (1970) Horizontal configuration and the rate of erosion of Niagara Falls, *Geological Society of America Bulletin* 81, 3,723–3,732.
Rashleigh, E.C. (1935) *Among the Waterfalls of the World*, London: Jarrolds.
Sardeson, F.W. (1908) Beginning and recession of Saint Anthony Falls, *Geological Society of America Bulletin* 19, 29–52.
Seidl, M.A., Finkel, R.C., Caffee, M.W., Hudson, G.B. and Dietrich, W.E. (1997) Cosmogenic isotope analyses applied to river longitudinal profile evolution: problems and interpretations, *Earth Surface Processes and Landforms* 22, 195–209.
Simon, A. and Thomas, A. (2002) Processes and forms of an unstable alluvial system with resistant, cohesive streambeds, *Earth Surface Processes and Landforms* 27(7), 699–718.
Tinkler, K.J., Pengelly, J.W., Parkins, W.G. and Asselin, G. (1994) Postglacial recession of Niagara Falls in relation to the Great Lakes, *Quaternary Research* 42, 20–29.
van der Beek, P., Pulford, A. and Braun, J. (2001) Cenozoic landscape development in the Blue Mountains (SE Australia): lithological and tectonic controls on rifted margin morphology, *Journal of Geology* 109, 35–56.
Winchell, N.H. (1878) The recession of the Falls of St Anthony, *Journal of the Geological Society* 34, 886–901.
Youping, T. and Fusheng, H. (1997) Research on waterfall calcareous tufa mats from Xiangzhigou, Guizhou, *Carsologica Sinica* 16(2), 145–154.

Further reading

Wohl, E. (2000) *Mountain Rivers*, Washington, DC: American Geophysical Union.

KEITH J. TINKLER

L

LAGOON, COASTAL

There are three main meanings to the term lagoon. The most usual is to describe a stretch of salt water separated from the sea by a low sandbank or coral reef. Another meaning is that for a small freshwater lake near a larger lake or river. The third usage is for an artificial pool used for the treatment of effluent or to accommodate an overspill from surface drains. This entry is soley concerned with the first of these meanings.

Coastal lagoons are mostly to some degree estuarine, are usually shallow, and have generally been partly or wholly sealed off from the sea by the deposition of spits or barriers, by localized tectonic subsidence, or by the growth of coral reefs. They range in size from over 10,000 km^2 to less than 1 ha. They occur on about 12 per cent of the length of the world's coastline (Bird 2000). They are best formed on transgressive coasts, particularly where the continental margin has a low gradient and sea-level rise is slow. They are ephemeral features and their depths and areas become reduced by sedimentation from inflowing rivers, as well as by accumulation of sediment washed in from the sea, wind-blown material, and chemical and organic deposits. Indeed, they can be classified on the basis of whether they are infilling or increasing in size (Nichols 1989). The infill of some lagoons, particularly those that are parallel to the shore, may involve the development of cuspate divisions that divide the lagoon into a series of segments. These divisions have been attributed *inter alia* to winds blowing along the length of the lagoon producing waves which build spits that isolate the lagoon into separate basins.

The entrances between lagoons and the sea vary in origin and form. Some are residual gaps that persisted between spits or barrier islands where the lagoon was never completely sealed off from the sea. Others are caused by breaching either by storm waves or by floods from on land. Their configuration is the outcome of a contest between the inflow and outflow of currents, which keeps entrances open, and the effects of onshore and longshore drift of sediment, which tends to seal them off. Lagoon entrances tend to be larger, more numerous and more persistent on barrier coastlines where relatively large tidal ranges generate strong currents.

A good review of the diversity of coastal lagoon morphologies and evolution is provided by Cooper (1994).

References

Bird, E.C.F. (2000) *Coastal Geomorphology: An Introduction*, Chichester: Wiley.

Cooper, J.A.G. (1994) Lagoons and microtidal coasts, in. R.W.G. Carter and C.D. Woodroffe (eds) *Coastal Evolution*, 219–265, Cambridge: Cambridge University Press.

Nichols, M.M. (1989) Sediment accumulation rates and sea-level rise in lagoons, *Marine Geology* 88, 201–219.

A.S. GOUDIE

LAHAR

An Indonesian word originally used by Escher (1922) for a hot volcanic mudflow that had been generated by an eruption through a crater lake. The term specifically applied to the 1919 volcanic mudflows originating from the crater lake of Mt Kelut, on Java, which inundated over 130 km^2 of the surrounding lowlands, with the loss of over 5,000 lives. The word quickly gained acceptance as a general term for a mudflow on the flanks of a volcano, irrespective of its triggering mechanism.

With increased understanding of water-saturated flow processes, the term has evolved to include both mudflows and debris flows, and later hyperconcentrated flows, on the flanks of a volcano. This has led to the more generally accepted definition: a rapidly flowing mixture of rock debris and water (other than normal stream flow) from a volcano (Smith and Fritz 1989). This definition describes the flow but not the resultant deposit.

Lahars are denser than normal stream flow because of their high sediment loads, causing them to move faster due to greater gravity forces and damped internal energy losses. The highly concentrated slurry usually exhibits a yield strength, behaving like wet cement but also exhibiting features of cohesionless grain flow. At lowering velocities, high concentration lahars may quickly halt, 'freezing' the coarser fraction within its finer matrix as a single massive sedimentary unit. The matrix strength, buoyancy and grain-dispersive pressures help support the coarser particles (Iverson 1997). Thus, resultant deposits are poorly sorted and may show reverse or no grading because little or no time was available for settling to occur. Thin, fine-grained sole-layers are attributed to shear and cataclasis within the basal flow. Lahars may travel as successive flow surges, which in high sediment concentration flows tend to continually shunt and overtop previous deposits. Clay-poor lahars may exhibit a frontal cliff or snout and show coarse bouldery levees at their margins.

A feature of lahars is 'bulking', when on the steep slopes of a volcano the flow typically erodes loose sediment over which it is flowing to increase its volume several times. On lower gradient slopes the reverse may happen leading to progressive dilution of the flow and formation of a HYPERCONCENTRATED FLOW or a normal flood (Pierson 1998). A water wave ahead of the lahar has been observed in some instances, which may represent a solitary wave or soliton (Cronin *et al.* 2000) or may be caused by under-ramping of the dense flow behind (Manville *et al.* 2000).

Lahars, armoured with their coarse bouldery loads, are a devastating volcanic hazard, capable of killing large numbers of people and removing all structures in their path. In recorded history at least 64,000 persons have been killed by lahars (Neall 1996). Lahars may vary in size from small volume events ($<$0.1 million m^3) to large-scale collapses of volcanic edifices ($>$3,500 million m^3). Measured peak discharges of historical lahars range from 100 to $>$100,000 m^3 s^{-1}. Lahars are capable of travelling up to 150 km/h and may travel for hundreds of kilometres down valleys. This mobility is attributed to positive pore-fluid pressures, which greatly decrease internal friction within the flow (Hampton 1979). Hence the most hazardous lahar zones are adjoining river courses draining volcanoes, particularly those draining crater lakes.

Lahars may be a more common hazard at stratovolcanoes than the geological record attests. Of twenty lahars observed in the Whangaehu catchment at Mt Ruapehu, New Zealand, during 1995, only one resultant deposit is preserved. This highlights the preservation potential of only larger events.

Lahars may be generated by both eruptive and non-eruptive mechanisms. Lahar triggering eruptive mechanisms include phreatic or phreatomagmatic explosions (often from hydrothermally altered and structurally weakened edifices), displacement of waters by eruptions through a crater lake, pyroclastic flows admixing with water in rivers or lakes, and volcanic melting of snow and ice. Non-eruptive triggering mechanisms include collapse of the walls of crater lakes, and heavy rains falling on recently erupted materials.

In the most recent large-scale volcanic disaster of modern times more than 23,000 persons were killed by lahars from Nevado del Ruiz, Colombia, in 1985. Relatively small eruptions at the summit produced pyroclastic surges, which quickly gouged and melted 0.06 km^3 of snow and ice to form lahars peaking at 48,000 m^3 s^{-1} and totalling 40–60 million m^3.

Large and extensive prehistoric lahar deposits are reported from many stratovolcanoes. One of the first to be recognized was the Osceola Mudflow from Mt Rainier, Washington State, USA, which filled proximal valleys 85–200 m deep before spreading over 350 km^2 of the Puget Sound lowlands about 5,600 years ago.

Mitigative measures to reduce damage from lahars include adequate real-time warning systems (such as acoustic flow monitors), reducing the level of water in crater lakes by engineering methods (such as tunnels at Mt Kelut, Indonesia), dams to reduce gradient and encourage sediment deposition, reduction of lake levels in hydro dams to accommodate sudden inflows, construction of embankments to divert flow away from assets at most risk, or as at Mt Pinatubo, Philippines, continually elevating houses above each successive lahar deposit.

References

Cronin, S., Neall, V., Lecointre, J. and Palmer, A. (2000) Dynamic interactions between lahars and stream flow; a case study from Ruapehu volcano, New Zealand: reply, *Geological Society of America Bulletin* 112, 1,151–1,152.
Escher, B.G. (1922) On the hot 'lahar' (mud flow) of the Valley of Ten Thousand Smokes (Alaska), *Proceedings Koninklijke Akademie van Wetenschappen, Amsterdam* 24, 282–293.
Hampton, M.A. (1979) Buoyancy in debris flows, *Journal of Sedimentary Petrology* 45, 753–758.
Iverson, R.M. (1997) The physics of debris flows, *Reviews of Geophysics* 35, 245–296.
Manville, V., White, J.D.L. and Hodgson, K.A. (2000) Dynamic interactions between lahars and stream flow; a case study from Ruapehu volcano, New Zealand; discussion, *Geological Society of America Bulletin* 112, 1,149–1,151.
Neall, V.E. (1996) Hydrological disasters associated with volcanoes, in V.P. Singh (ed.) *Hydrology of Disasters*, 395–425, Dordrecht: Kluwer.
Pierson, T.C. (1998) An empirical method for estimating travel times for wet volcanic mass flows, *Bulletin of Volcanology* 60, 98–109.
Smith, G.A. and Fritz, W.J. (1989) Volcanic influences on terrestrial sedimentation, *Geology* 17, 375–376.

VINCENT E. NEALL

LAKE

Lakes are defined as bodies of slow moving water surrounded by land. They represent approximately 2 per cent of the Earth's surface but contain only about 0.01 per cent of the world's water (Wetzel 2001). While they are temporary features of the landscape on a geological timescale, they can exist for very long periods and therefore strongly influence the human development of a region. They can also provide a record of the region's environmental history in their sedimentary record. The study of lakes is called limnology and limnologists characterize lakes in a number of ways, including their geologic origin, mixing behaviour and nutrient status. While these classifications appear to be distinctly disciplinary, the geology, physics and chemistry of lakes interrelate significantly and all act to regulate the biological dynamics in lakes.

The depressions, or basins in the Earth's surface, that collect water to become lakes can be formed by several geologic and geomorphic processes. Catastrophic geologic origins include tectonic activity and volcanism. The deepest of the Earth's lakes are caused by tectonic faulting, whereas the clearest of lakes are found in the craters of old volcanoes. The majority of the Earth's lakes (72 per cent) have been created by glaciation (Kalff 2002). High alpine cirque (see CIRQUE, GLACIAL) lakes, lowland KETTLES AND KETTLE HOLES and glacial ice-scour lakes are abundant in the regions once covered by ice. Other lakes are generated by the modification of drainage systems, or the impoundment of flowing waters by natural disasters such as landslides or human activity such as damming. Riverine or fluvial lakes, such as those developed on floodplains, deltas and blocked valleys, comprise 10 per cent of the world's lakes and are the dominant lake type in low latitudes (Kalff 2002). Chemical dissolution of rocks also generates basins that collect water. Wind and ocean shoreline processes create barriers which act to block fresh water while animals and meteorites cause terrestrial depressions creating specific lake types. Some lake basins are excavated by wind (see PAN). The fact that lakes are most often created by processes that have dominated that landscape results in lakes of similar origin being regionally grouped. This is why there seems to be a preponderance of Lake Districts around the world, where bodies of water that appear and behave similarly due to their shared origin are identified as a lake grouping (e.g. Cumbria Lakes, England; Finger Lakes, New York; Great Lakes, North America). Such groupings provide opportunities for regional-scale research and have facilitated the study of lake types and processes.

The shape of the lake basin, or its morphometry, is a function of the lake origin but over time it does change as sedimentation and shoreline infilling occurs as part of the natural succession process. Zonation in lakes is also a function of morphometry. The nearshore region of lakes is called the littoral zone while the open water deeper portion is called the pelagic zone. Littoral zones support rooted plants, some of which are visible above the water surface and are called emergent plants. The littoral extends to the depth of water able to support rooted plants and it represents that region to which light can penetrate, allowing photosynthesis of these primary producers. The littoral zone may be extensive in a shallow dish-shaped lake or it may represent only a small nearshore area if the slope is very steep (see Figure 100). Primary productivity (plant growth) in the littoral is usually greater than in the pelagic zone as it is more closely linked to the catchment from which it receives nutrients for plant growth.

Lakes are classified physically by their mixing behaviour which is a function of their thermal structure. Solar energy enters the lake at the water surface and is attenuated as it moves down through the water column. Heat is therefore transferred predominantly to the surface waters. Warm water lies above the colder, denser water creating a separation or stratification of water layers. The warmer surface waters are known as the epilimnion while the cooler bottom waters are called the hypolimnion. The point where the temperature transition is most extreme between these two layers is called the thermocline and that portion of the lake is termed the metalimnion (see Figure 100).

Water is an unusual liquid in that it is most dense at 4 °C, meaning that water both warmer and cooler than 4 °C will be more buoyant and rise to a layer above. Water in its solid form, ice, is less dense than water and thus floats at the lake surface. The unique density characteristics of water ensure that lakes do not freeze from the bottom up and that most temperate lakes have bottom waters with moderate under-ice temperatures (4 °C) to support life.

Depending on the temperature (and therefore density) differences between the epilimnion and hypolimnion the two parcels of water may be very resistant to mixing. During periods of strong stratification the two compartments of water do not interact, or mix, and therefore their exchange of materials is restricted. The movement of both soluble and low density particulate material entering a lake can be confined to surface waters if the density differences at the metalimnion are extreme. When denser particulate matter settles through the metalimnion and enters the hypolimnion it will be stored and/or decomposed, but until the stratification is reduced all these materials (e.g. nutrients, pollutants, particles) will remain in the bottom waters. Chemical changes associated with the decomposition of the particles, such as reduced oxygen levels, are now confined to the bottom waters as there is little exchange of water and chemicals across the thermocline. In well-stratified, productive lakes oxygen depletion or anoxia often occurs in the bottom waters due to the oxygen demand of the organic-rich sediments (see Figure 100). Anoxic conditions at the sediment–water interface also generate chemical alterations in the sediment, resulting in increased exchanges between sediments and overlying water.

When stratification breaks down, due for example to the cooling of surface waters with changing seasons, and the water column becomes isothermal (one temperature) there is little resistance to mixing. Wind energy at the water surface can mix the full water column which is called 'turnover'.

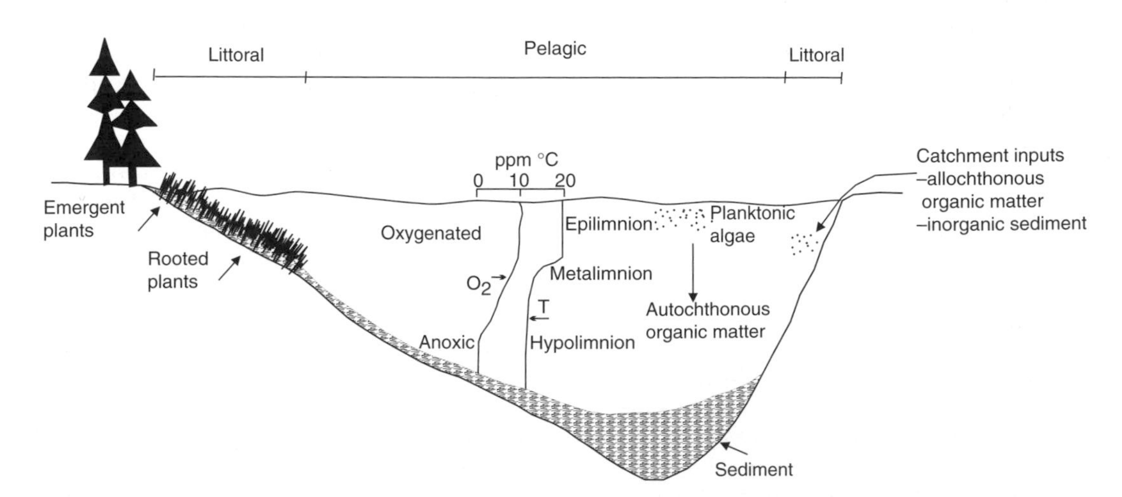

Figure 100 Limnological terms for the zonation, thermal structure and particulate sources to lakes. The general distribution of lake primary production and resultant oxygen profiles are also presented

This full lake mixing brings nutrients and chemicals from the hypolimnion to mix with the surface waters. Lakes in different climatic regions exhibit different annual patterns of stratification and so can be classified in this way. Lakes which mix twice a year are common in the temperate regions where cold winters give way to springtime heating generating isothermal conditions as the lake warms to 4 °C. Again in the autumn the lakes cool, causing destratification and an isothermal situation which allows autumnal winds to mix the bottom with the surface waters. This mixing, twice annually, is termed dimictic. Monomictic lakes mix only once per year and are found in high elevation and high latitude areas, while polymictic lakes, commonly located in equatorial regions, mix many times a year. Note that mixing is important in the transfer of both nutrients and pollutants in lakes. A less common but environmentally significant situation is a meromictic lake which has a bottom water layer that is chemically different from the rest of the water body. This generates a large enough density difference that even when the system is isothermal a chemical density gradient restricts this bottommost layer from mixing. This is often caused by salinity differences or groundwater inputs of differing chemistry. In this case the exchanges between the bottom sediments underlying the meromictic layer are restricted from delivering to the full water column.

Nutrient status is one variable used to chemically classify lakes. The terms oligotrophic, mesotrophic, eutrophic and hypereutrophic represent the spectrum of conditions from nutrient-poor through to nutrient-rich systems. In general, primary productivity increases across the spectrum with hypereutrophic lakes representing the stage where organic nutrient inputs (carbon, nitrogen, phosphorus) are high. Organic inputs to lakes can be derived from within the lake, termed autochthonous, or they can be delivered from the catchment or atmosphere, when they are called allochthonous. In-lake primary production of free-floating photosynthetic plants, or planktonic algae, and littoral rooted plants are the main sources of autochthonous organic matter. Catchment inputs from river inflows represent the main allochthonous contribution of organic matter. If the organic matter in lakes is not either eaten by an organism or decomposed during settling it will accumulate in the bottom sediments. Organic sedimentation rates increase over the trophic range as productivity increases.

The bottom sediments are also comprised of inorganics which have been eroded from the landscape and delivered via river inflows or shoreline erosion. As well, some autochthonous inorganics can be generated by chemical precipitation and biogenic processes. Sedimentation rates vary as a function of the location, size and activities in the catchment as well as the in-lake productivity, but in most natural temperate lakes ranges between 0.1 and 2 mm per year (Kalff 2002).

Globally much research has focused on the process and remediation of cultural eutrophication (Cooke *et al.* 1993). This accelerated change in tropic status occurs when unnaturally high phosphorus loads are received by the lake over relatively short time periods (decades). It is termed cultural eutrophication because the sources of the growth nutrient, phosphorus, are associated with human activities in the catchment, such as agriculture and sewage treatment. Increased primary productivity in both the littoral and pelagic zone, increased sedimentation rates, hypolimnetic oxygen depletion and the coarsening of fish species are often associated with anthropogenically induced alterations of tropic status. Remediation and management approaches for these problems are presented in Cooke *et al.* (1993).

As lakes are bodies of water that receive and store material from the surrounding catchments and the atmosphere they are of interest to geomorphologists as the accumulated sediment can reflect regional changes over time. Catchment erosion rates associated with changing land use, sediment source tracing, climatic variations, historic pollutant loads, flood records and vegetation patterns can be detected by evaluating different characteristics of the accumulated lake sediments. Sediment, collected by coring down through the accumulated material, can be horizontally sliced to differentiate sediments from specific time periods. Paleolimnology, or the use of lake sediments for reconstructing past events, requires some means of dating the accumulated material (see DATING METHODS) and a variety of methods exist (e.g. ^{210}Pb, ^{14}C, ^{137}Cs, thermoluminescence) but the precision and accuracy of each is restricted to specific time intervals. Given this, and the fact that lake sediments are temporally and spatially variable, it is very important to design the collection of cores and the analytical methods to suit the questions being addressed. Dearing and Foster (1993) provide a useful discussion of the problems, errors and implications of using sediment cores in geomorphic research. An earlier text by Hakanson and Jansson (1983) introduces the topic of lake sedimentology and provides information on physical, chemical and biological aspects of sediment.

Since 1970 the focus of limnological research has moved away from viewing the lake as a closed system upon which to do ecological research and more effort has been placed on linking catchment processes to lake conditions (Kalff 2002). Lakes are intimately connected to their catchments and therefore the role of lakes in geomorphological research and the role of geomorphologists in the interdisciplinary study of lakes are significant.

References

Cooke, G.D., Welch, E.B., Peterson, S.A. and Newroth, P.R. (1993) *Restoration and Management of Lakes and Reservoirs*, Boca Raton, FL: Lewis.

Dearing, J.A. and Foster, I.D.L. (1993) Lake sediments and geomorphological processes: some thoughts, in J. McManus and R.W. Duck (eds) *Geomorphology and Sedimentology of Lakes and Reservoirs*, 5–14, Chichester: Wiley.

Hakanson, L. and Jansson, M. (1983) *Lake Sedimentology*, Berlin: Springer-Verlag.

Kalff, J. (2002) *Limnology: Inland Water Systems*, Upper Saddle River, NJ: Prentice Hall.

Wetzel, R.G. (2001) *Limnology: Lake and River Ecosystems*, 3rd edition, Fort Worth: Academic Press.

Further reading

Hutchinson, G.E. (1957) *A Treatise on Limnology, Vol. 1. Geography, Physics and Chemistry*, New York: Wiley.

Vallentyne, J.R. (1974) *The Algal Bowl: Lakes and Man*, Special Publication 22, Department of the Environment Fisheries and Marine Service, Ottawa.

ELLEN L. PETTICREW

LAND SYSTEM

Land systems are an integral part of the land evaluation process. The first land systems studies were devised and carried out by Australia's Commonwealth Scientific and Industrial Research Organisation (CSIRO) as part of their *Land Research Series* (Christian and Stewart 1968; Stewart 1968). A hierarchical classification of land was used. The smallest areal unit is the *land facet*, which is a relatively homogeneous area of land in terms of slope and soil; such as the midslope unit of a CATENA. Repetitive sequences of land facets form *land units*, which are the fundamental mapping unit of the land systems approach. Land units comprise a distinctive pattern of land facets (slopes, soils and vegetation), are essentially catenary in nature and classically were illustrated by block diagrams with explanatory annotations. *Land systems*, then, are larger areas comprising groups of land units representing recurring patterns of soils, topography and vegetation.

Work commenced in Australia in the late 1940s and continued to the 1970s, later refined by government agencies in some States. Similar surveys were carried out by the British Ministry of Overseas Development, particularly in Africa (Land Resources Development Centre 1966 onwards), and by the French scientific organization ORSTOM and the Netherlands International Institute for Aerial Survey and Earth Science (ITC) in many countries (Nossin 1977).

The objective was to provide rapid, cost-effective and objective assessments of land resources, initially for agriculture and forestry but more recently as frameworks for regional or national planning, in remote areas for which there was little if any base information. There were no maps depicting contours, soils or vegetation and little geological or climatic information for these areas. The primary data source was small-scale (1:80,000), monochrome aerial photography, flown primarily for military purposes from the 1940s. Thus the method of land systems surveys was essentially an air photo interpretation exercise, supported by ground truthing. Patterns of tone and texture were delimited on the aerial photographs. Offroad ground traverses were then designed to visit examples of each pattern on the ground in order to record the vegetation, soils, topography and geology. Since the tonal and textural patterns primarily reflect vegetation, plus areas of bare ground, rock outcrops and water bodies, vegetation associations were particularly important in characterizing land units and land systems.

However, several members of the early land systems teams were British-trained geomorphologists (including Mabbutt, Ollier, Twidale and Young). Not surprisingly, therefore, geomorphology played a large part in the written interpretations of the land systems. That geomorphology was heavily influenced by the then prevailing Davisian cycle of erosion which, as pointed out by Chorley (1965: 35–36) in a different context, has little if any bearing on the prediction of land suitability for agricultural or other land uses. This is also largely true of genetic interpretations of soils.

The first land system study in Australia was of the Katherine–Darwin region (Christian and Stewart 1953). As a result of the survey a relatively small area was identified as being possibly suitable for agriculture, and a research station was established to conduct detailed field investigations and experiments. A similar outcome was obtained from the East Kimberley survey. Both research stations are still operational, and in the East Kimberley the eventual outcome was the development of the Ord irrigation scheme. While the entire area of Papua New Guinea was eventually covered by land systems surveys, the spatial coverage of over twenty reports in the Land Research Series in Australia was more scattered. Debatably, the developments outlined above were the only non-educational, positive outcomes of the series of Land Research reports.

Weaknesses of the technique included: its reconnaissance-type of approach (which ironically was also its strength); its static coverage (processes were not included); its geomorphic base with a strong, denudation chronology flavour; its exclusive focus on the biophysical environment; and the crudity of the data. Nevertheless, land systems may provide a useful basis for later and more dynamically oriented work, especially when there is little background information on the environment or on ecological conditions. In Victoria, for example, geomorphic elements have been refined as part of a land systems review to provide a suitable framework for describing the spatial attributes of land (Rowan 1990).

Only in the Hunter Valley of New South Wales was a CSIRO land systems survey conducted in an area already well-developed for agriculture. The Hunter Valley survey covered a smaller area than previous surveys and that, plus the larger amount of information which was available for the region compared with the more remote regions, made it possible to provide greater detail (Story *et al.* 1963). But that survey also exposed the limitations of focusing on the biophysical environment. In most areas, planning for future land uses does not take place in unoccupied country but in places which are already occupied and the land already subject to land-use practices. Existing uses, therefore, must be incorporated in any survey designed to assist planning for future land uses.

The South Coast study

Realization of this self-evident truth led to the eventual abandonment by CSIRO of its limited land systems surveys (although the basic method is still used widely in various contexts), and the development of two separate but related lines of further work. One was a focus on databases, which eventually merged with the currently thriving area of research into and applications of Geographical Information Systems in land use planning; and the other was the *South Coast study* (also in New South Wales) (Basinski 1978). This was a very 'geographical' survey, in the sense that its approach would be familiar to any regional geographer. Not surprisingly, geographers were prominent amongst the team members – as physical geographers had been in the earlier land systems survey teams.

The South Coast study, a large, multidisciplinary project, conducted the familiar, integrated, land systems survey of the biophysical environment of the 6,000 km^2 region. Importantly, however, present land use, population demographics and socio-economic aspects such as settlements, transport networks, and the economy and social structure of the region were also included. The study explicitly set out to provide a 'rational basis for planning decisions on a wide variety of land uses', as well as investigating methods of providing and analysing biophysical and socio-economic data for the region (Basinski 1978). This survey was a precursor of 'land suitability' surveys and regional planning and, more particularly, of CSIRO's Siroplan approach to land suitability evaluation for integrated, regional land use planning (Cocks *et al.* 1983).

Recent approaches to land evaluation

The term 'land systems' continues to be used and modified in some jurisdictions, particularly by agricultural agencies (at least in Australia). But it is the mapping unit of the Land Research Series, the *land unit*, which provides the basis of several approaches; notably FAO's land suitability evaluation (FAO 1976) and land use planning frameworks (FAO 1993), as well as regional planning, at least in Western Australia, in the form of the basic planning and land management unit (example: WAPC 1996).

Improved technology and data availability have made possible considerable improvements in land evaluation methods and the quality of the output, and Davidson (2002) has provided a useful review of recent developments in the assessment of land resources. The availability of high resolution, satellite-borne remote sensing imagery, the rapid development of geographic information systems (or science) (GIS), and geostatistics (spatial data analysis, modelling and fuzzy set algebra) have meant that current methods of land evaluation would be unrecognizable to the early practitioners. Nevertheless, there is still a need to map areas of land ('land units') and to obtain field-derived data. The quality of existing, mapped soil data (and other land attributes) in many developing countries necessitates field-based surveys. Unfortunately there is a widespread tendency amongst GIS practitioners to consider that computer manipulation and interpolation of existing data is all that is required. There is a renewed need to educate natural land resource assessors to obtain data which are relevant to the problem in hand and not to 'make do' with – often inappropriate or inaccurate but readily available – existing data.

References

Basinski, J.J. (ed.) (1978) *Land Use on the South Coast of New South Wales: a study in methods of acquiring and using information to analyse regional open space options, Volume 1, General Report*, Melbourne: CSIRO.

Chorley, R.J. (1965) A re-evaluation of the geomorphic system of W.M. Davis, in R.J. Chorley and P. Haggett (eds) *Frontiers in Geographical Teaching*, 21–38, London: Methuen.

Christian, C.S. and Stewart, G.A. (1953) *General Report on Survey of Katherine–Darwin Region, 1946*, CSIRO Australia Land Research Series No. 1, Melbourne.

Christian, C.S. and Stewart, G.A. (1968) Methodology of integrated surveys, in *Aerial Surveys and Integrated Studies*, 233–280, Paris: UNESCO.

Cocks, K.D., Ive, J.R., Davis, J.R. and Baird, I.A. (1983) SIRO-PLAN and LUPLAN: an Australian approach to land-use planning. I. The SIRO-PLAN land-use planning method, *Environment and Planning B* 10, 331–345.

Davidson, D.A. (2002) The assessment of land resources: achievements and new challenges, *Australian Geographical Studies* 40, 109–128.

FAO (1976) *A Framework for Land Evaluation*, Soils Bulletin 32, Rome: Food and Agriculture Organization of the United Nations.

FAO (1993) *Guidelines for Land-use Planning*, FAO Development Series 1, Rome: Food and Agriculture Organization of the United Nations.

Land Resources Development Centre (1966) *Land Resources Studies, No. 1 England*, Surbiton, Surrey: Land Resources Development Centre, Ministry of Overseas Development.

Nossin, J.J. (ed.) (1977) *Proceedings ITC Symposium. Surveys for Development*, Amsterdam: Elsevier.

Rowan, J.N. (1990) *Land Systems of Victoria*, Department of Conservation and Environment, Land Conservation Council, Victoria, Australia.

Stewart, G.A. (ed.) (1968) *Land Evaluation*, Melbourne: Macmillan.

Story, R., Galloway, R.W., van de Graaff, R.H.M. and Tweedie, A.D. (1963) *General Report on the Lands of the Hunter Valley*, CSIRO Australia Land Research Series No. 8, Melbourne.

WAPC (1996) *Central Coast Regional Strategy*, Perth: Western Australian Planning Commission.

ARTHUR CONACHER

LANDSCAPE SENSITIVITY

The sensitivity of a landscape to change is the likelihood that a given change in the controls of a system will produce a sensible, recognizable and sustained response. The sensitivity is a function of the propensity for change which is measured by the size of the impulse required to initiate change.

The necessary impulse is determined by the number, type and magnitude of the barriers to change. The barriers to change are complex and include the resistance of the rocks, their structure and their resilience (strength resistance); the slope, relative relief and elevation that determine the potential and mobilized energy (morphological resistance); the distance to BASE LEVEL, sources or energy or other barriers (location resistance); the ability of the system to transmit energy, waste and water, the density of pathways, stream density and joint frequency (transmission resistance); and the strength of the linkages between components of the system and the degree of hillslope-channel coupling (structural resistance).

A system also has a varying capacity to absorb the impulses of change. There are shock absorbers, filters, void spaces and energy drains. For example a BEACH disposes of wave energy by moving particles and doing work (displacement, friction, heat) and by drainage.

Sensitivity is also determined by the inheritance or degree of influence of previous system states, process systems and landforms. Of particular interest are systems that have experienced a very efficient sediment flux regime but which after environmental change or re-specification of the controls finds itself too flat or exhausted to permit further vigorous change. Large events may have a dramatic effect on a hillslope and remove all available (weathered) material. The barrier to further change is then the time (efficiency of preparation processes) needed to 'ripen' the system.

There are two limiting types of system. Mobile, fast-responding landforms have a high sensitivity, react quickly and relax to new states with facility. If they have high 'permeability' they may act as an energy filter so that change is 'skin deep' and fluctuates about an average form (e.g. a beach profile). They may also store small impulses and accumulate them into a large impulse that may cross an internal threshold value. Such systems are often morphologically complex and the correlative deposits are fragmentarily preserved. They are usually capable of rapid restoration.

Slow-responding systems may be insensitive because they are too flat, too far from boundary (energy) changes, propagate events slowly, have large storage or low concentrations of sediment transport or progressive weakening axes. Change does not take place easily so that, once their form is established, they may require large or effective events to achieve adjustment. However, if change does take place the results may be dramatic. For example, gully incision into a plain, or landsliding following progressive softening. Generally there is a persistence of relief and pattern, stagnancy of development, a palimpsest of forms and 'traditional' development in which the inherited landforms continue to develop as before even though the controlling environment may have changed.

Further reading

Begin, Z.B. and Schumm, S.A. (1984) Gradational thresholds and landform singularity: significance for Quaternary studies, *Quaternary Research* 21, 267–274.

Brunsden, D. (1990) Tablets of Stone: toward the ten commandments of geomorphology, *Zeitschrift für Geomorphologie N.F. Supplementband* 79, 1–37.

——(2001) A critical assessment of the sensitivity concept in geomorphology, *Catena*, 42, 99–123.

Brunsden, D. and Thornes, J.B. (1979) Landscape sensitivity and change, *Transactions Institute of British Geographers* NS4, 463–484.

Chorley, R.J., Schumm, S.A. and Sugden, D.E. (1984) *Geomorphology*, London: Methuen.

Schumm, S.A. (1985) Explanation and extrapolation in geomorphology: seven reasons for geologic uncertainty, *Transactions Japanese Geomorphological Union* 6-1, 1–18.

Thomas, D.S.G. and Allison, R.J. (1993) *Landscape Sensitivity*, Chichester: Wiley.

Thomas, M.F. (1994) *Geomorphology in the Tropics: A Study of Weathering and Denudation in Low Latitudes*, Chichester: Wiley.

——(2001) Landscape sensitivity in time and space, *Catena*, 42, 83–99. (Also see entire Volume 42 of *Catena* 2001.)

DENYS BRUNSDEN

LANDSLIDE

Landslides belong to a group of geomorphological processes referred to as MASS MOVEMENT. Mass movement involves the outward or downward movement of a mass of slope forming material, under the influence of gravity. Although water and ice may influence this process, these substances do not act as primary transportational agents. Landslides are discrete mass movement features and are distinguishable from other forms of mass movement by the presence of distinct boundaries and rates of movement perceptibly higher than any movement experienced on the adjoining slopes. Thus this group of processes includes falls, topples, slides, lateral spreads, flow and complex movements as classified by Cruden and Varnes (1996). Widespread diffuse forms of mass movement such as creep, subsidence, rebound and sagging are generally not treated as landslides.

The criteria used to distinguish different types of landslide generally include: movement mechanism (e.g. slide, flow), nature of the slope material involved (rock, debris, earth), form of the surface of rupture (curved or planar), degree of disruption of the displaced mass, and rate of movement (see MASS MOVEMENT).

Form and behaviour

There are several morphological features that can be recognized to a greater or lesser extent in most landslides (Figure 101). The uppermost part is the depletion or concave zone (erosional, generating or failure zone) where slope material has failed and become displaced downslope. In some cases the displacement may be only a few metres while in others the failure zone will be completely evacuated to expose the surface of rupture and to leave a distinctive scar on the hillslope (Plate 70). The displaced mass may remain close to the failure zone or it may continue to travel downslope leaving a transport track ending in a colluvial accumulation zone or acting as a supply to some other geomorphic agent (e.g. river, sea or glacier). The distance that landslide material travels (runout) is a characteristic of the type of landslide. For example, controlled by the height of fall and volume, rock avalanches can travel at high velocity for several kilometres. Runout distance and velocity for other types of landslide are controlled by factors such as volume, slope angle and morphology, clay content, water content, and surface frictional characteristics of the runout pathway.

Landslide movement may be instantaneous, with failure, transport and deposition taking place in a matter of seconds or minutes. Other landslides are known to have been intermittently active over tens of thousands of years, undergoing successive periods of reactivation. Eight states of

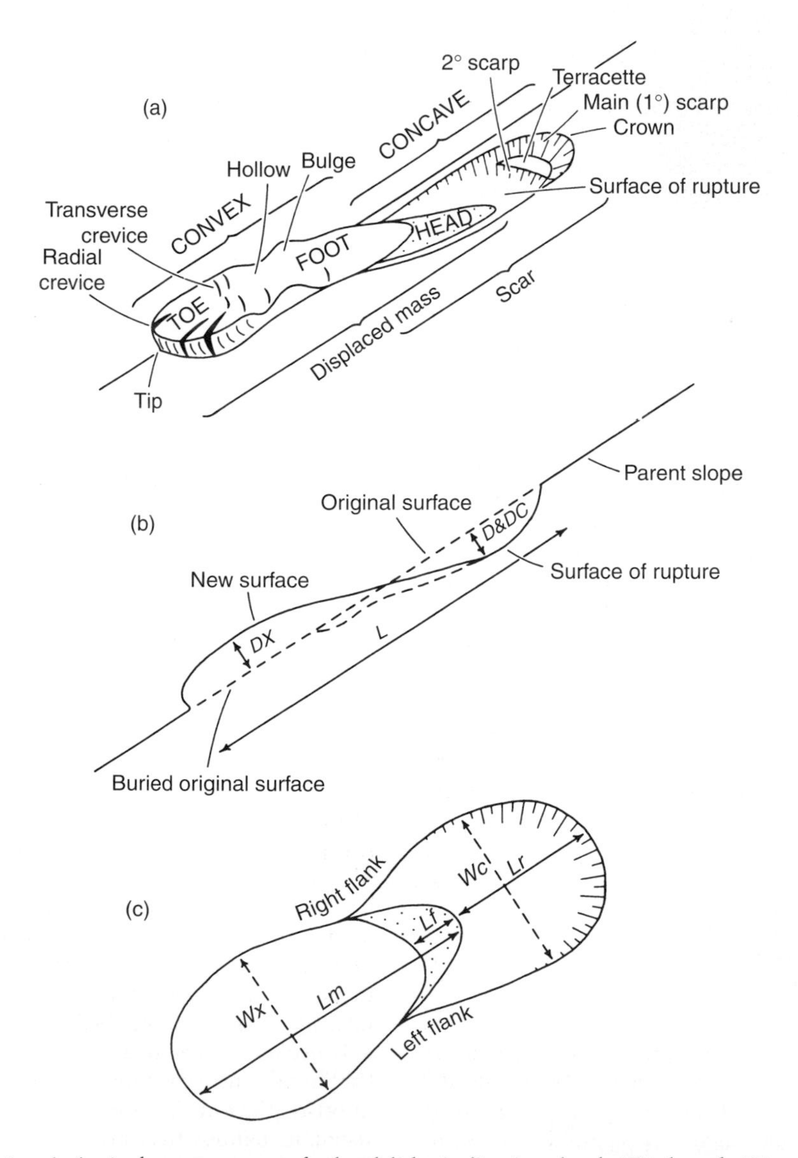

Figure 101 Morphological components of a landslide, indicating depth (*D*), length (*L*), and width (*W*) measurements

Plate 70 Hope rock slide (British Columbia), triggered by an earthquake in 1965 showing the depletion zone almost entirely evacuated of sediment

activity are recognized (Cruden and Varnes 1996) including such categories as 'active' (currently moving) 'dormant' (no movement in the last twelve months but capable of being reactivated) and 'relict' (unlikely to be reactivated under present climatic or geomorphologial conditions).

Rates of movement for different types of landslide are highly variable. Some landslides record only a few centimetres of movement a year, sustaining this rate for decades. Certain debris flows have recorded velocities of 100 kmh^{-1} while large rock avalanches are capable of reaching velocities of 350 kmh^{-1}. Landslide deposits range in texture from dislodged blocks of intact source material to highly comminuted sediments forming a poorly sorted, unstratified diamictite.

Significance

The potential impact of landslides depends not only on velocity but also on volume. One of the largest landslides described is Green Lake Landslide, Fiordland, New Zealand which is estimated to be 13 km^3 in volume. Submarine landslides of a similar magnitude are evident on edges of the continental shelf while recorded events such as the 1989 Ok Tedi mine landslide, Papua New Guinea and the 1970 Huascaran rock avalanche, were both estimated to exceed 50 million m^3 in volume. Large volume, high velocity movements can also create substantial LANDSLIDE DAMS impacting fluvial systems and often posing a significant dam burst hazard.

Landslides are a manifestation of slope instability (see SLOPE STABILITY) and occur when changing conditions on a part of the slope allow shear stress to exceed shear strength. This can be brought about by a decrease in the effectiveness of factors that promote strength (e.g. a reduction in friction caused by increased pore-water pressures) or by an increase in shear stress (e.g. slope steepening). However, when a landslide takes place it changes the relative stability conditions from an unstable to a stable state, by reducing slope angle, height or weight, or by removing susceptible material. If boundary conditions are stable over a long period of geomorphic time, continued landsliding in a particular region may so alter slope conditions that other slower acting processes such as soil creep become dominant.

As most slopes are stable most of the time, landslides when they occur can be seen as a rapid and effective geomorphic response to the appearance of destabilizing changes in boundary conditions, enabling rapid adjustment and an eventual return (or tendency) toward more persistent landscape forms.

Destabilizing conditions (preparatory factors) in natural systems may be instigated by disturbances such as tectonic uplift, oversteepening of slopes by erosional activity, climatic change, deforestation or slope disturbance by human activity. The effectiveness of preparatory factors in reducing stability depends on preconditions such as material properties and slope geometry. The degree of stability afforded by preconditions and preparatory factors defines 'landslide susceptibility'. The occurrence of landslides in space can be related to susceptibility thresholds (e.g. minimum critical slope angle) while occurrence in time can be related to exceeding magnitude thresholds for triggering agents (e.g. rainfall intensity, or seismic shaking).

Landslides represent a significant hazard to life, livelihood, property, infrastructure and resources in many parts of the world. Some individual catastrophic failures have been associated with high death tolls; the 1970 Huascaran rock avalanche in Peru killed 18,000 and deaths in the 1920

Kansu landslide in China are estimated at between 100,000 and 200,000. However, much landslide damage is less obvious, seriously depleting soil resources, reducing primary productivity and destroying property by slow chronic movement.

Reference

Cruden, D.M. and Varnes, D.J. (1996) Landslide types and processes, in A.K. Turner and R.L. Schuster (eds) *Landslides: Investigation and Mitigation*, Special Report 247, Transportation Research Board, National Research Council, 36–75, Washington, DC: National Academy Press.

Further reading

Brabb, E.E. and Harrod, B.L. (eds) (1989) *Landslides: Extent and Economic Significance*, Rotterdam: Balkema.

Bromhead, E., Dixon, N. and Ibsen, M.-L. (eds) (2000) *Landslides in Research, Theory and Practice*, three volumes, London: Thomas Telford.

Crozier, M.J. (1989) *Landslides: Causes, Consequences, and Environment*, London: Routledge.

——(1999) Landslides, in M. Pacione (ed.) *Applied Geography: Principles and Practice*, 83–94, London: Routledge.

Dikau, R., Brunsden, D., Schrott, L. and Ibsen, M.-L. (1996) *Landslide Recognition: Identification, Movement and Causes*, Chichester: Wiley.

Selby, M.J. (1993) *Hillslope Materials and Processes*, Oxford: Oxford University Press.

MICHAEL J. CROZIER

LANDSLIDE DAM

Landslide dams are naturally occurring stream blockages caused by hillslope-derived MASS MOVEMENT. They represent end members in the spectrum of landforms of geomorphic HILLSLOPE-CHANNEL COUPLING, and arise from a temporary or permanent transport-limitation of the fluvial system due to excess lateral sediment delivery. The impounded natural reservoir is referred to as a *landslide-dammed lake*, whereas the term *landslide pond* relates to small water bodies perched on top of LANDSLIDE deposits, which need not necessarily be associated with stream blockage.

Landslide dams commonly form in steep terrain of many upland areas throughout the world, and constitute some of the highest natural dams on Earth. Yet the majority of occurrences are short-lived: 85 per cent of 185 worldwide examples have failed within less than one year, and nearly half within ten days (Costa and Schuster 1988). Typical failure mechanisms comprise overtopping by either lake-level rise or landslide-induced displacement waves, breaching, piping, gravity collapse of the dam face, or artificial spillway control. Rainfall, snowmelt and earthquakes are amongst the main triggers of landslides causing temporary or permanent blockage of rivers. Other mechanisms include volcanic eruptions, fluvial undercutting or, in some instances, anthropogenic activity. Stream blockages formed by LAHARs or pyroclastic flows (see PYROCLASTIC FLOW DEPOSIT) may be regarded as phenomena on a continuum between landslide and volcanic dams.

Landslide dams constitute a variety of GEOMORPHOLOGICAL HAZARDs, which may extend for considerable distances up- and downstream of the initial point of blockage. Long-term impoundment of river channels may cause extensive backwater flooding and associated SEDIMENTATION. Substantial physical impact may be created by sudden dam failures leading to catastrophic OUTBURST FLOODs of landslide-dammed lakes, which can turn into DEBRIS FLOWs, given sufficient amount of valley floor deposits to allow for alluvial bulking. Downstream reaches usually experience massive AGGRADATION, lateral channel instability and AVULSIONs, in the wake of landslide-dam failures. Rapid drawdown of the draining lake reservoir may cause further secondary loss of SLOPE STABILITY.

Landslide dams often create profound geomorphic legacies in the context of long-term landscape evolution of VALLEY floors. These include partly BURIED VALLEYs, drainage reversal, large intramontane alluvial flats resulting from the infill of landslide-dammed lakes, LAKE terraces, spillway gorges (see GORGE AND RAVINE) and RAPIDS, as well as conspicuous step-wise disruption of river long profiles (see LONG PROFILE, RIVER).

The 60-km long Lake Sarez, Tajikistan, is recognized to have been impounded by the world's highest existing landslide dam (~700 m), which has been formed by an earthquake-triggered rock avalanche of $\sim 2 \times 10^9 \mathrm{m}^3$ in volume near the former village of Usoi, in 1911 (Alford *et al.* 2000).

References

Alford, D., Cunha, S.F. and Ives, J.D. (2000) Lake Sarez, Pamir Mountains, Tajikistan: mountain

hazards and development assistance, *Mountain Research and Development* 20, 20–23.
Costa, J. and Schuster, R.L. (1988) The formation and failure of natural dams, *Geological Society of America Bulletin* 100, 1,054–1,068.

Further reading

Korup, O. (2002) Recent research on landslide dams – a literature review with special attention to New Zealand, *Progress in Physical Geography* 26, 206–235.

SEE ALSO: dam

OLIVER KORUP

LARGE WOODY DEBRIS

Large Woody Debris (LWD) refers to fallen trees and detached wood within streams or along stream banks. The term 'large' applies to wood that is big enough to affect stream hydraulics, sediment transport, bank erosion and the resultant stream morphology, or is big enough to provide cover and habitat for aquatic organisms. In practice, LWD often refers to wood that is at least 1–2 m in length and a minimum of 10–30 cm in diameter, although the definition varies among geomorphologists and with stream size. Living trees along stream banks and vegetation outside stream channels are not referred to as LWD, although these types of wood also affect geomorphic processes (see EROSION; FOREST GEOMORPHOLOGY; RIPARIAN GEOMORPHOLOGY).

LWD is widely recognized as a critical component of aquatic ecosystems (Maser *et al.* 1988), although as recently as the 1970s management agencies removed wood from streams to reduce flooding and promote fish passage. LWD provides organic material, traps sediments, creates pools and, in general, increases the habitat diversity of streams. Resource managers now regularly emplace LWD in rivers to promote natural stream processes and recovery.

LWD enters the stream channel as logs, branches and root balls derived from bank failures, windthrow, landslides, debris flows, natural death and breakage, and human disturbances such as logging. LWD in small headwater streams often bridges the channel, only affecting morphology where breakage and decay create local log steps that promote sediment accumulation and channel widening (Nakamura and Swanson 1993). As stream size increases, the LWD falls into the stream rather than spanning it. These intermediate size streams cannot easily transport the wood, which is trapped and accumulates as single pieces or in piles known as 'jams'. The trapped wood forms log steps and pools and increases sediment storage, although channel widening can occur where logs deflect flow against a bank. In larger order streams, channel dimensions significantly exceed the size of LWD, and the wood can be transported by the flow. Larger streams sweep up the LWD as single pieces or jams along banks, on bars, and in backwater eddies (Marcus *et al.* 2002). When deposited along banks, wood in these larger systems can reduce bank erosion and promote pool development within the channel. When deposited in mid-channel, however, jams in larger streams can deflect the current and promote development of secondary channels that widen the overall stream width and increase bank erosion.

References

Marcus, W.A., Marston, R.A., Covard, C.R. Jr, and Gray, R.D. (2002) Mapping the spatial and temporal distribution of woody debris in streams of the Greater Yellowstone Ecosystem, U.S.A., *Geomorphology* 44, 323–335.
Maser, C., Tarrant, R.F., Trappe, J.M. and Franklin, J.F. (eds) (1988) *From the Forest to the Sea: A Story of Fallen Trees*, Pacific Northwest Research Station PNW-GTR-229, Portland, OR: USDA Forest Service.
Nakamura, F. and Swanson, F.J. (1993) Effects of coarse woody debris on morphology and sediment storage of a mountain stream system in Western Oregon, *Earth Surface Processes and Landforms* 18, 43–61.

SEE ALSO: debris flow; hillslope-channel coupling; landslide

W. ANDREW MARCUS

LAVA LANDFORM

Volcanic eruptions produce two main kinds of material – lava and pyroclasts (see VOLCANO). While most eruptive episodes will yield quantities of each, they can, in general, be described as predominantly effusive (producing lavas) or explosive (pyroclastic). The focus here is on the effusive variety and on extrusive rather than intrusive landforms. Lava is always partially molten on eruption but can contain a considerable volume fraction of both crystalline phases (minerals) and gas bubbles (vesicles). First-order controls on the eventual geomorphology of lava flows are provided by eruption rate and total amount of lava emitted, viscosity, the topography over which the lava flows, and the external environment (i.e. atmosphere, water or ice). Viscosity, the ratio of shear stress to strain rate, is a measure of the internal resistance to flow when a force is applied to a fluid. Magma viscosities are highly variable because there are many controlling factors, including temperature and composition of the melt. The presence of crystals increases viscosity. Bubbles, on the other hand, may increase or decrease viscosity depending on their size, properties and flow rate of the magma. Dissolved water also plays a role because it interrupts silicon bonding in the melt, hence reducing viscosity. In short, magma viscosities can vary by many orders of magnitude due to cooling, crystallization, vesiculation or loss of gas, and thereby strongly influence the nature of lava landforms.

Plate 71 Erta 'Ale lava lake in the Danakil Depression (Ethiopia), is the longest-lived active lava lake on Earth. Its dimensions are approximately 80×90 m, and its power output is around 100–200 MW (Oppenheimer and Yirgu 2002)

Lava lakes and flows

Low silica, mafic lavas such as those erupted at Kīlauea (Hawai'i) have typical viscosities on eruption of only 10^2–10^3 Pa s, not much more than that of mayonnaise. When low viscosity lava erupts within a crater, it will tend to be confined within the vent region and crater, forming a lava pond or lake. Active lava lakes are connected to a deeper reservoir of magma, while passive lava lakes are not. Long-lived lava lakes are a comparatively rare phenomenon on Earth. They have been reported at several volcanoes including Kīlauea, Nyiragongo (Congo), Erebus (Antarctica) and Erta 'Ale (Ethiopia) but very few individual examples have reportedly persisted for more than a few decades. These include Halema'uma'u (Hawai'i, 1823–1924) and Erta 'Ale (probably active for at least the past century; Plate 71). Lava lakes have also been observed at oceanic ridges (Fouquet *et al.* 1995), and on the Jovian moon, Io (McEwen *et al.* 2000).

While all the manifestations discussed here are flows of lava and hence lava flows, the term 'lava flow' usually refers to erupting lava that has the opportunity to run down the flanks of a volcano or cross open ground. The expression is used both for active flows during emplacement, and for the resulting landform. In the course of a long-lived eruption, a lava flow field may develop by the superposition of many individual flow units. The current eruption of Kīlauea has yielded over 2 km^3 of lava and built up a flow field around 100 km^2 in area since it began in 1983 (a mean eruption intensity of around 3 m^3 s^{-1}, Plate 72). Areas of higher relief within a flow field that are above the 'high-tide' mark of the erupting lavas are called kīpuka. These 'islands' may preserve mature vegetation destroyed elsewhere by the active flows. Where lava enters the sea, as at Hawai'i, a lava delta or bench may build seawards, though these are often unstable features. As active lava interacts with seawater on the shore in front of the lava bench, steam explosions can construct ephemeral littoral cones.

Gas-rich, low-viscosity magma can erupt in quite spectacular fashion, with lava fountains ('fire fountains') playing to heights of up to several hundred metres. These can occur at individual vents or along fissures, which may become delineated by spatter cones or ramparts. Intense lava fountains can feed clastogenic lava flows when the spatter expelled loses little heat during its transit through the air.

The largest lava eruption of the past millennium was that of Laki (Iceland) in 1783/4.

Plate 72 The current eruption of Kīlauea (Hawai'i) has been in progress since 1983 and has formed a lava flow field some 100 km^2 in area. Most of the lava seen here is of the pāhoehoe variety, though the dark patch on the slope in the background consists of 'a'ā

It emplaced an estimated 14.7 km^3 of mafic lava, 40 per cent of which was erupted in the first twelve days at peak rates exceeding 5,000 m^3 s^{-1} (Thordarson and Self 1993). Bigger still, however, are the flood basalt eruptions or 'large igneous provinces' that punctuate the geological record. These have occurred both on land and in the oceans, and can contain 10^5–10^7 km^3 or more of lava, covering areas of 10^6 km^2, and erupted over timescales of perhaps 1 Myr, e.g. the 65-Myr-old Deccan Traps of India. The resulting landforms are known as lava plateaux. Where they are dissected by erosion, a staircase topography results from differential weathering of the rubbly boundaries and massive inner core of each superposed flow unit. This is the origin of the term 'trap', after an old Swedish word for staircase. Cooling and contraction of the cores of lava bodies can result in columnar jointing, seen spectacularly at the Giant's Causeway (Northern Ireland) and Devil's Postpile (USA).

Active lava flows radiate prodigious quantities of heat near the vent such that they rapidly form a surface crust. This can thicken sufficiently to insulate the core of the flow from thermal losses, lowering the rate of viscosity increase, and thereby promoting longer travel distance. Mafic flows quite often crust over completely, with lava continuing to flow in tunnels, which can grow in cross section by thermal erosion of the walls. When the supply of lava at the vent ceases, the last slug of lava may drain downslope leaving an empty conduit or lava tube. On Kīlauea, much of the lava flow between the Pu'u 'O'o vent, and the coastline where lava pours into the sea (a distance of over 10 km), takes place in a tunnel network, with only sporadic breakouts at the surface.

As silica content and crystallinity of lava increases, and eruption temperature drops, viscosities climb many orders of magnitude. As a result, much thicker accumulations of lava are required to overcome the resistance to flow.

Lava domes and coulées

The most silicic lavas attain viscosities of up to 10^8–10^{10} Pa s. On eruption, viscous flow is strongly resisted and slip may become concentrated along shear planes. Mounds of rock that accumulate around the vent, are called lava domes, and if they show some flow away from the vent, they are termed coulées. Sometimes, lava domes are intruded just beneath the surface causing bulges known as cryptodomes. In the weeks and days before its major 1980 eruption, a cryptodome of around 100 m in height developed on the upper flanks of Mount St Helens (USA). The eruption of Soufrière Hills Volcano (Montserrat), which began in 1995, produced an intermediate composition dome exceeding 10^8 m^3 in volume (Druitt and Kokelaar 2002). Such domes are really composite features, the construct of many individual lobes extruded at the surface of the dome (exogenous growth) and also intruded within it (endogenous growth). The frequent gravitational collapses and explosions that are typical of many dome-building eruptions can represent a severe hazard at volcanoes in populated areas. Frequent block-and-ash flows generated as the result of failure of hot sections of the dome of Merapi volcano (Indonesia) have claimed many lives.

Lava flow surface textures

The spectrum of surface textures of lava flows is impressive, and identical lava compositions can display very different textures due to subtle variations in effusion rate or cooling history. 'A'ā lava is characterized by loose clinker-like rubble, often overlying a more massive core. 'A'ā flows often build longitudinal levees by accretion along their margins. These are sometimes breached by new slugs of lava moving down the channel, and some civil protection efforts have attempted to divert flows by purposefully excavating flow levées. A curious feature of ʻaʻā flows are accretionary lava balls. They form in snowball fashion as a solid core rolls along the surface of a lava flow, picking up sticky molten rock. They can reach several metres in diameter, and gain sufficient momentum to roll some distance ahead of the lava flow front.

In contrast to ʻaʻā, pāhoehoe lava has a comparatively smooth surface composed of interlocking lobes, often adorned with a millimetre–centimetre-scale texture of interwoven threads. It can appear like coils of cord in ropy lava, like intestines in entrail lava, platy in slab lava, or blistered in shelly lava. Toothpaste lava forms by squeezing through cracks in the solid crust of a flow. Pāhoehoe can transform into ʻaʻā lava as it moves downslope due to changes in viscosity or strain rate but the reverse transformation is never observed. Investigations of the shapes of lava flow margins have revealed further insights into pāhoehoe and ʻaʻā flows. The former can be described by a scale-independent, fractal dimension, suggesting that pāhoehoe spreading is controlled by the ratio of the finely balanced forces driving advance and of those imposed by formation of the crust (Bruno *et al.* 1992). In contrast, the spread of ʻaʻā flows is dominated by the driving forces, and, as a result, the shape of their margins is not fractal. These observations have potential application in interpretation of volcanic terrains on other planetary bodies.

More viscous flows often develop a blocky surface texture, consisting of fractured chunks of lava, up to several metres across, with angular facets. Giant pressure ridges called ogives sometimes wrinkle the surface of very viscous flows. Lava domes often develop spines when they are active but these are usually rather ephemeral. After the devastating 1902 eruption of Mont Pelée (Martinique), an exceptional 300-m high spine was extruded.

Smaller features surrounding openings on mafic lava flows include hornitos (literally 'little ovens'), which are chimneys or pinnacles of lava spatter and dribbles squeezed up through openings in the roof of lava tubes. Pāhoehoe flows on low slopes often display elliptical, domed structures known as tumuli. These form when the magma pressure within an active lava tunnel ruptures the overlying surface of the flow field. Fractures usually extend along the length of a tumulus, and often lava squeezes out through these and other cracks on the sides to build a larger feature.

Lavas erupted subaqueously resemble toothpaste squeezed out of a tube. The bulbous lobes that form are known as pillow lavas. These are commonplace along oceanic ridges but can also form in shallow water found in coastal, river and lake environments.

References

Bruno, B.C., Taylor, G.J., Rowland, S.K., Lucey, P.G. and Self, S. (1992) Lava flows are fractals, *Geophysical Research Letters* 19, 305–308.

Druitt, T.H. and Kokelaar, P. (eds) (2002) *The Eruption of Soufrière Hills Volcano, Montserrat, from 1995 to 1999*, London: Geological Society.

Fouquet, Y., Ondreas, H., Charlou, J.L., Donval, J.P., Radfordknoery, J., Costa, I., Lourenco, N. and Tivey, M.K. (1995) Atlantic lava lakes and hot vents, *Nature* 377, 201.

McEwen, A.S., Belton, M.J.S., Breneman, H.H., Fagents, S.A., Geissler, P., Greeley, R., *et al.* (2000) Galileo at Io: results from high-resolution imaging, *Science* 288, 1,193–1,198.

Oppenheimer, C. and Yirgu, G. (2002) Thermal imaging of an active lava lake: Erta ʻAle volcano, Ethiopia, *International Journal of Remote Sensing* 23, 4,777–4,782.

Thordarson, T. and Self, S. (1993) The Laki (Skaftár Fires) and Grímsvötn eruptions in 1783–1785 *Bulletin of Volcanology* 55, 233–263.

Further reading

Francis, P. and Oppenheimer, C. (2004) *Volcanoes*, Oxford: Oxford University Press.

Hawaiian Volcano Observatory, http://wwwhvo.wr.usgs.gov/

Kilburn, C.R.J. and Luongo, G. (eds) (1993) *Active Lavas: Monitoring and Modelling*, London: University College London Press.

Rhodes, J.M. and Lockwood, J.P. (1995) *Mauna Loa Revealed: Structure, Composition, History, and Hazards*, Washington, DC: American Geophysical Union.

Sigurdsson, H., Houghton, B.F., McNutt, S.R., Rymer, H. and Stix, J. (eds) (2000) *Encyclopedia of Volcanoes*, San Diego: Academic Press.

SEE ALSO: volcano

CLIVE OPPENHEIMER

LAWS, GEOMORPHOLOGICAL

The whole concept of a scientific 'law' is far from straightforward and may be seen to become more and more problematic the further we get from the basic physical principles which can be translated into rational equations. Harvey (1969) concluded that all laws are statements of some universality, embedded within a theoretical structure. Haines-Young and Petch (1986: 13) consider that laws 'describe some characteristic or behaviour of all the members of a class of things' but that that something 'would not usually be employed in recognizing it'. They go on (pp. 14–17) to give two, contrasting examples of scientific laws: that described by Coriolis force, governing the deflection of moving bodies to right or left in the northern and southern hemispheres; and Playfair's 'Law of accordant junctions', which states that river valleys meet 'on neither too high or too low a level' (1802: 102). Whereas the first statement is a purely physical construct capable of mathematical formulation, the second is more problematic in that it is not evident how far it was intended to apply to the confluences of river channels or to valley floors (Haines-Young and Petch 1986: 16–17). Nor did Playfair formulate his observation as a law.

Since geomorphology is best seen as a historical science, it has provided many examples of both the 'qualitative' laws such as Playfair's and the 'physical' laws, such as that of Coriolis force. Although A.N. Strahler in his very influential 1952 paper called for geomorphologists to strive for 'the deduction of general mathematical MODELS to serve as quantitative natural laws' (p. 937), there has been rather limited success in that direction, since not all the phenomena of interest to geomorphology have proved equally amenable to mathematical formulation. This is recognized by Rhoads and Thorn (1996: 123) who conclude that:

> [T]he study of complex phenomena poses a problem both for inductively establishing the existence of underlying causal laws by combining mathematical formulations of these laws in predictive models. This situation may account for the fact that geomorphology has not been very successful at developing its own laws, or in using simple models that combine a few basic physical laws to predict the form and dynamics of specific landforms.

They are clearly following Strahler's exhortation in their evaluation of success or failure. Yet this is to ignore some of the most fundamental principles (or laws) which have governed the historical science of the study of landforms since the eighteenth century. We can make a crude distinction between the qualitative expressions of universality made by – most notably – James Hutton, John Playfair, G.K. Gilbert, W.M. Davis and W. Penck; and the quantitative formulations of R.E. Horton, S.A. Schumm, M.A. Melton, J.T. Hack and R.L. Shreve; and J.W. Glen.

By far the most basic proposition accepted by geomorphologists is that first recognized by Hutton in 1785, namely, that there is no need for recourse to extraordinary processes to account for the fashioning of the Earth's surface, providing we accept an almost unbelievably long lapse of time. Hutton's 'Principle of Uniformitarianism' has been much debated and travestied and – as Shea (1982) has forcefully reminded us – it remains a principle to be followed by the geomorphologist, not a 'truth' about the workings of the planet and its processes. Nevertheless, most of us would date the origin of geomorphology to Hutton and, possibly more convincingly, to John Playfair's (1802) extension and elaboration of Huttonian views. It is in the course of that development of an effectively modern geomorphological argument that Playfair described the system of valleys and 'the nice adjustment of their declivities' that came to be known as 'Playfair's Law'. In so far as fluvial valleys do, indeed, tend to show that mutual adjustment, whereas glacially eroded troughs do not, Playfair's Law is, indeed, both universal and causal.

The next significant and widespread development of principles and laws came from the United States in the latter nineteenth century: J.W. Powell's principle of BASE LEVEL; W.M. Davis's CYCLE OF EROSION and the 'laws' of the explanatory significance of 'structure, process and stage'. But there is also the whole set of propositions, explicitly termed laws, which G.K. Gilbert propounded from his great study of the badlands on the rim of the Henry Mountains, Utah, in 1877. Chorley *et al.* (1964: 550–567) discuss these contributions in some detail, which include: the law of uniform slopes; the law of declivities; the law of structures; and the law of divides. All these are seen as universal ideals, towards which a fluvial landscape will tend, but which it will never attain. One may consider the models of SLOPE EVOLUTION of W. Penck to belong in a similar tradition.

By the middle of the twentieth century, the qualitative laws discussed above were felt to be insufficiently precise. R.E. Horton (1945) set out what became known as HORTON'S LAWS of drainage basin composition (which were expanded by S.A. Schumm and M.A. Melton) covering the regularities in stream number, length, area, frequency and slope as basin order changed. To these was later added Hack's Law (Hack 1957), which relates the length of the longest stream in a drainage basin to the area of the basin (see Rodriguez-Iturbe and Rinaldo 1997). Still later, R.L. Shreve (1966) demonstrated that drainage networks could be considered to follow the statistical law of topological randomness. The most fundamental of these laws are, in some sense, related to the universal properties of networks and it is not altogether evident how far they may be related to basin geomorphology. The law which describes the deformation of ice over time – discovered by J.W. Glen in 1955 – may be more closely linked to geomorphological processes as it successfully predicts the extreme sensitivity of secondary creep in ice to changes in shear stress (Paterson 1981).

There would seem some prospect that the development of approaches such as those examined by Rodriguez-Iturbe and Rinaldo (1997) that are based upon advanced computer modelling and the concept of FRACTALs may, in time, be able to provide quantitative expressions of some of the most basic qualitative geomorphological laws (cf. Rodriguez-Iturbe and Rinaldo 1997, Chapter 6). Until that time, it must be accepted that the key laws and principles of the subject remain verbal rather than mathematical. Geomorphology is still a 'consumer' rather than a creator of quantitative physical laws.

References

Chorley, R.J., Dunn, A.J. and Beckinsale, R.P. (1964) *The History of the Study of Landforms*, vol. I, London: Methuen.

Hack, J.T. (1957) Studies of longitudinal profiles in Virginia and Maryland, *US Geological Survey Professional Paper* 294-B.

Haines-Young, R. and Petch, J. (1986) *Physical Geography: Its Nature and Methods*, London: Harper and Row.

Harvey, D. (1969) *Explanation in Geography*, London: Edward Arnold.

Horton, R.E. (1945) Erosional development of streams and their drainage basins: hydro-physical approach to quantitative morphology, *Geological Society of America Bulletin* 56, 275–370.

Paterson, W.S.B. (1981) *The Physics of Glaciers*, Oxford: Pergamon.

Playfair, J. (1802) *Illustrations of the Huttonian Theory of the Earth*, London: Cadell and Davies. Reprinted in facsimile, G.W. White (ed.) (1964), New York: Dover.

Rhoads, B.L. and Thorn, C.E. (1996) Towards a philosophy of geomorphology, in B.L. Rhoads and C.E. Thorn (eds) *The Scientific Nature of Geomorphology*, 115–143, Chichester: Wiley.

Rodriguez-Iturbe, I. and Rinaldo, A. (1997) *Fractal River Basins: Chance and Self-organization*, Cambridge: Cambridge University Press.

Shea, J.H. (1982) Twelve fallacies of uniformitarianism, *Geology* 10, 455–460.

Shreve, R.L. (1966) Statistical law of stream numbers, *Journal of Geology* 74, 17–37.

Strahler, A.N. (1952) Dynamic basis of geomorphology, *Geological Society of America Bulletin* 63, 923–938.

SEE ALSO: base level; complexity in geomorphology; confluence, channel and river junction; Cycle of Erosion; fractal; glacier; Horton's Laws; slope, evolution; unequal slopes, law of

BARBARA A. KENNEDY

LEACHING

Leaching is the removal in solution of constituents by water or other percolating solutions. It is usually applied to chemical DISSOLUTION and removal as a liquid moves through a porous solid such as soil or a rock mass. Leaching is also used in a broader context to apply to the removal of soluble compounds from solid waste, or the extraction of metals or salts from ores. The most important factor influencing ion mobility is the amount of available water, and that in turn depends on a number of other factors. Leaching directly affects the Eh–pH conditions surrounding minerals and thus determines which elements will stay in solution. Other factors affecting the degree of leaching or weathering include texture, permeability, initial carbonate content, subaerial climate, depth of the wetting front, precipitation of secondary carbonates, soil moisture and temperature, plant transpiration, root mat extraction, porosity of the material, and depth to the saturated zone of ground water.

Leaching affects the alteration of rocks and sediments. As rocks and sediments are weathered they are altered in 3-dimensions beneath the land surface. Layers or zones of altered material may form subparallel to the land surface. Leaching is the mechanism by which dissolved ions and clay are transferred from zone to zone. These layers or WEATHERING zones may be defined in various ways. They may differ physically, chemically and mineralogically from adjacent layers and have lateral extent. The vertical section through a stack of these zones is often referred to as a weathering profile. A soil profile is one type of weathering profile in which layers or horizons are designated. A descriptive terminology for geologic weathering zones has long been in use (e.g. Kay 1916; Kay and Apfel 1929; Leighton and MacClintock 1930; Frye *et al.* 1968) and provides a shorthand for field description of material below the solum (the A and B horizons of a soil). Hallberg *et al.* (1978) discuss the importance of standardizing weathering zone descriptive terminology and subdivide the terminology by material. The terminology describes Quaternary sediments in terms of their colour and the presence or absence of soluble carbonate minerals. In the mid-continent USA, a typical weathering profile in unconsolidated Quaternary sediments such as till consists of an oxidized and leached zone, an oxidized and unleached zone, and a reduced (deoxidized in loess) and unleached zone (Ruhe 1975). In weathered rock, zones may be determined based on the ratio of core stones to weathered matrix (Ruxton and Berry 1957). Examples of these types of weathering profiles are illustrated in Figure 102.

A few caveats are necessary when considering the use of weathering zone terminology. Colour is somewhat problematic as a descriptor because of the concepts associated with weathering zone terms such as deoxidized, reduced or unoxidized. They are not synonymous with reduced chemical states. Second, although abrupt contacts between weathering zones may coincide with stratigraphic breaks, the contacts do not indicate that stratigraphic boundaries are present. For example, a single till unit may have a prominent yellowish-brown oxidized weathering zone overlying a light-grey deoxidized zone. The zones give the unit an appearance of two tills but in actuality there is only one. Weathering zones have been related to various hydrologic conditions. In the absence of continued leaching, weathering zones function as closed systems in which chemical weathering ceases.

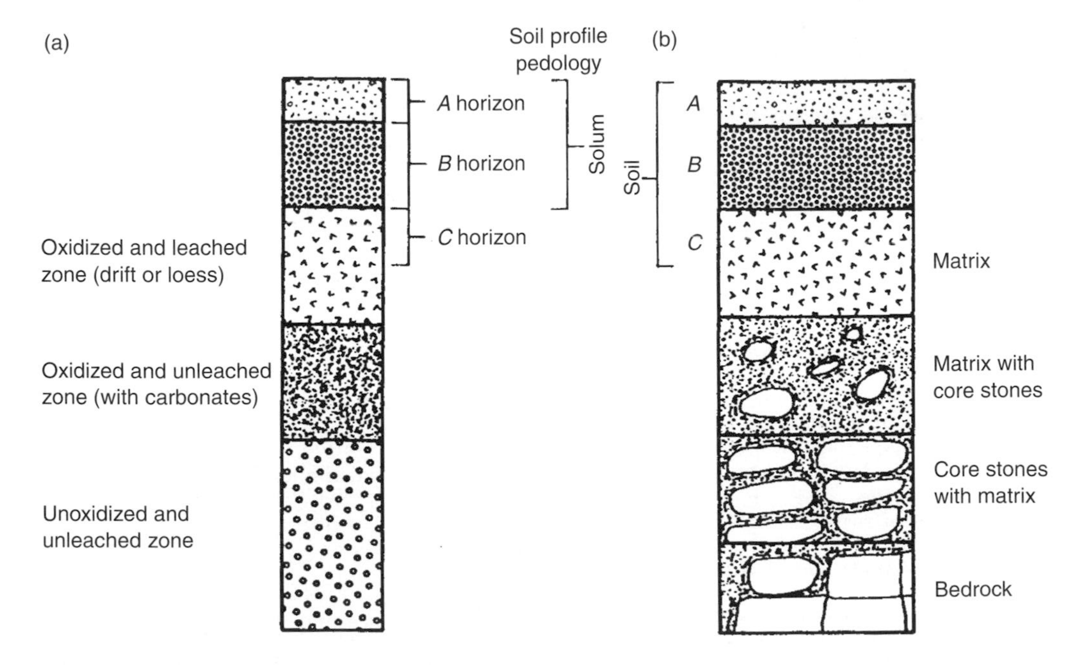

Figure 102 Weathering profiles in sediment (a) and in rock (b) with soil profiles developed in the surface materials for comparison. Modified from Ruhe (1975)

Leaching rates are important in assessing rates of groundwater contamination and plant growth. Organic chemicals are prone to be lost to leaching as their solubility in water increases. The greatest risks for hazardous leaching, rapid and unaltered dispersal away from sources, occurs in highly permeable materials such as sands that have little or no organic material. The leaching capability of introduced chemicals varies but herbicides are usually more mobile than fungicides and insecticides and should be applied with care in areas with very permeable soils.

Cover crops and no-till farming can be effective in reducing some nutrient loss through leaching. For example, the process of crop growth slows percolation and removes nitrate from solution by incorporating it into plant tissue as nitrogen. Unchecked, nitrate leaching through soils and sediments into water supplies is a serious human health and environmental contaminant.

As mentioned earlier, a common determinant for leaching is precipitation, its rate, intensity and duration. A lack of precipitation can lead to an accumulation of mobile constituents such as carbonate and other salts in soils and sediments of arid climates. Conversely, excess precipitation can lead to complete removal of these constituents from sediments in more humid environments. In artificially irrigated crop lands, a leaching requirement (LR) is usually recommended. The LR is the fraction of irrigation water that must be leached through the root zone to control soil salinity at a specific level (Foth 1984). The goal is to sustain a productive soil and produce no change in salinity during irrigation.

References

Foth, H. (1984) *Fundamentals of Soil Science*, New York: Wiley.

Frye, J.C., Glass, H.D. and Willman, H.B. (1968) Mineral zonation of Woodfordian loesses in Illinois, *Illinois State Geology Survey Circular* 427.

Hallberg, G.R., Fenton, T.E. and Miller, G.A. (1978) Standard weathering zone terminology for the description of Quaternary sediments in Iowa, in G.R. Hallberg (ed.) *Standard Procedures for Evaluation of Quaternary Materials in Iowa*, Technical Information Series 8, 75–109.

Kay, G.F. (1916) Some features of the Kansas drift in southern Iowa, *Geological Society of America Bulletin* 27, 115–119.

Kay, G.F. and Apfel, E.T. (1929) The pre-Illinoian Pleistocene geology of Iowa, *Iowa Geological Survey Annual Report* 34, 1–304.

Leighton, M.M. and MacClintock, P. (1930) Weathered zones of the drift-sheets of Illinois, *Journal of Geology* 38, 28–53.

Ruhe, R.V. (1975) *Geomorphology, Geomorphic Processes and Surficial Geology*, Boston, MA: Houghton Mifflin.

Ruxton, B.P. and Berry, L. (1957) Weathering of granite and associated erosional features in Hong Kong, *Geological Society of America Bulletin* 68, 1,263–1,292.

CAROLYN G. OLSON

LEAST ACTION PRINCIPLE

Forms and patterns in natural systems are often the products of optimized circumstances related inherently to operational efficiency. In systems of motion, optimum operational efficiency requires, by the minimization of a system's 'action', the least expenditure of energy for completing a particular task. In other words, out of many possible alternatives nature follows the most 'economical' path. This least action principle (LAP) was formulated originally by numerous mathematical physicists, notably Pierre-Louis Moreau de Maupertuis (1698–1759), Leonhard Euler (1707–1783), Joseph Louis Lagrange (1736–1813), Sir Rowan Hamilton (1805–1865) and Carl Gustav Jacob Jacobi (1804–1851). It contains a curious and subtle twist on Newton's laws, for its variational formulation of motion does not use force and momentum but instead the physical quantities of energy and work. As a result, it often shows structural analogies between various areas of physics and has been found useful in unifying subjects and consolidating theories in various branches of science (Lanczos 1986).

By the end of the nineteenth century, LAP had become a very successful scheme applicable not only in classic mechanics, but also in electrodynamics and thermodynamics, typically through the work of Hermann von Helmholtz (1821–1894). During the twentieth century even more widespread applications were developed and in the 1910s, Albert Einstein (1879–1955) deduced the equations of *general relativity* from LAP. In the 1940s, Richard Feynman (1919–1988) identified the applicability of LAP in quantum physics and since then physicists have found that LAP also underlies the fundamental gauge theories of particle physics, leading to the establishment of what is termed *fundamental* physics (Brown and Rigden 1993). LAP has also been applied widely outside of physics. A notable example is the study of George Zipf (1902–1950), who, within the context of LAP, tried to derive the power-law form of his law for understanding the behaviour of humans on the basis of a principle of least effort (Zipf 1949). With the wide adoption of FRACTAL theory in the 1970s, Zipf's 'law' became ever popular and has been regarded as one of the essential phenomena of nature. It occurs not only in the distribution of words but also in the occurrence of cities, populations, wars, species, coastlines, floods, earthquakes and many other processes and behaviours (Schroeder 1992).

Huang and Nanson (2000, 2002) and Huang *et al.* (2002) have examined the applicability of LAP in geomorphology. Their theoretical inferences and evaluation of a wide range of case studies have shown that LAP governs fluvial systems in the form of MAXIMUM FLOW EFFICIENCY (MFE), providing a soundly based explanation as to why regular bankfull HYDRAULIC GEOMETRY relations occur in very different geographical regions. They also showed that MFE subsumes the previously proposed extremal models in geomorphology of maximum sediment transport capacity and minimum stream power. Further work is examining the application of MFE under various physical constraints (particularly available energy in the form of gradient) to explain the physical conditions for the formation of different river patterns and drainage networks.

SEE ALSO: fractal; hydraulic geometry; maximum flow efficiency; models

References

Brown, L.M. and Rigden, J.S. (1993) *'Most of the Good Stuff' – Memories of Richard Feynman*, New York: American Institute of Physics.

Huang, H.Q. and Nanson, G.C. (2000) Hydraulic geometry and maximum flow efficiency as products of the principle of least action, *Earth Surface Processes and Landforms* 25, 1–16.

Huang, H.Q. and Nanson, G.C. (2002) A stability criterion inherent in laws governing alluvial channel flow, *Earth Surface Processes and Landforms* 27, 929–944.

Huang, H.Q., Nanson, G.C. and Fagan, S.D. (2002) Hydraulic geometry of straight alluvial channels and the principle of least action, *Journal of Hydraulic Research* 40, 153–160.

Lanczos, C. (1986) *The Variational Principles of Mechanics*, New York: Dover.

Schroeder, M. (1992) *Fractals, Chaos, Power Laws: Minutes from an Infinite Paradise*, New York: Freeman.

Zipf, G.K. (1949) *Human Behaviour and the Principle of Least Effort*, Boston, MA: Addison-Wesley.

HE QING HUANG AND GERALD C. NANSON

LEVEE

Among the most prominent products of overbank deposition on alluvial and deltaic FLOODPLAINS and submarine abyssal plains are natural levees bordering channels. A natural levee is a wedge-shaped ridge of water-laid, channel-derived sediment which elevation tapers gently into the flanking floodbasin. Moderately to well-developed levees are present along most channel reaches, the exceptions being new, rapidly migrating, or coarse-grained braided channels.

The morphology of a levee depends upon its age, channel size and, in the case of alluvial and deltaic channels, the maximum height to which waters are ponded during floods or storm surges. Vegetation type, grain size and rate of channel alluviation exert a secondary control. Levee heights above the adjacent floodplain range from a few centimetres on young creeks to 5 m along mature sections of large rivers. Heights of submarine levees typically are greater by a factor of at least 10. Levee widths are also highly variable, ranging from fractions of a channel width to over 10 for alluvial and deltaic cases and 10 to 20 widths for submarine levees. Although data are sparse, levee cross-sectional area appears to scale linearly with channel cross-sectional area. Levee slopes transverse to the channel range widely, from virtually horizontal to near-channel values of 6° for fluvial deltaic levees, and to at most 3.5° for levees on the Amazon submarine fan. In alluvial and deltaic levees one levee of a pair is often significantly higher, wider or steeper than the other, with no systematic variation along a channel; in submarine levees of the northern hemisphere, the right-hand levee (when looking downstream) is often higher, sometimes by as much as three-halves, due to Coriolis forces.

The sedimentological characteristics of levees are highly variable. Generally it can be said that levee deposits fine laterally from coarser sands and silts closer to the channel to distal silts and clays with no vertical textural trends. Sedimentary structures and stratification consist of climbing ripple cross-laminated sands alternating with lenticular and wavy-laminated muds and rhythmically bedded, laminated, thin silts. Levees bounding submarine channels generally consist of silty to clayey spill-over turbidites with occasional thicker fine-grained sandstone beds and hemipelagic and pelagic intervals.

Levees arise by the transfer of suspended sediment from the channel to the floodplain via two mechanisms: diffusion and advection. Diffusion occurs when turbid turbulent eddies along the channel-floodplain boundary spin off onto the floodplain and decelerate, allowing grains to settle at distances determined by channel geometry, floodplain roughness, particle size distribution and flow character. Advection occurs when turbid flows leave the channel as channel water overtops the banks during the rising limbs of floods and on the outsides of meander bends.

References

Cazanacli, D. and Smith, N.D. (1998) A study of the morphology and texture of natural levees; Cumberland Marshes, Saskatchewan, Canada, *Geomorphology* 25, 43–55.

Klaucke, I., Hesse, R. and Ryan, W.B.F. (1998) Seismic stratigraphy of the Northwest Atlantic Mid-Ocean Channel: growth pattern of a mid-ocean channel-levee complex, *Marine and Petroleum Geology* 15, 575–585.

RUDY SLINGERLAND

LICHENOMETRY

Lichenometry is a dating technique using lichen measurements to supply relative or absolute dates for rock surface exposure. Geomorphologists have chiefly applied it to MORAINES and periglacial (see PERIGLACIAL GEOMORPHOLOGY) landforms in arctic and alpine environments. Other dating applications have included MASS MOVEMENTS, SCREE and blockstreams (see BLOCKFIELD AND BLOCKSTREAM), NIVATION, avalanches (see AVALANCHE, SNOW) and snow cover, LIMESTONE PAVEMENTS, seismic movements, river channel deposits, lake levels, shorelines, storm beaches and archaeological features.

The technique's dating range depends on lichen form, competition and the local environment. In temperate regions some foliose forms might survive 150 years; minutely crustose forms can provide dating over 400 to 600 years and dating may exceed 1,000 years at high latitudes. All dates can only be empirically justified unless validated by an independent source. Claims have been made for a dating range of 8,000 to 9,000 years based on radiocarbon dates for organics beneath moraines, but such a range is unlikely due to rock WEATHERING, successive glacier advances and climate change. Lichens only provide minimum dating for the last period of surface exposure while radiocarbon determinations provide maximum dates.

Three main approaches have been developed: the original approach is based on size/age correlations of largest lichens; the other two approaches are based on population size–frequency distributions, one using measurements of populations of largest lichens on uniform-aged surfaces and the other, measurements of whole lichen populations, with a population defined by its individual rock surface.

The original approach

In the original approach, pioneered in 1950 by Roland Beschel, lichen species cumulative growth rates are derived indirectly from size/age correlations of near-circular lichen bodies (thalli) growing on known-age surfaces. The longest axes of largest lichens on each surface are plotted on a graph, thallus size against date. The species cumulative growth rate is assumed to be that described by the curve traced through the largest size/age plots in the distribution. There may be a delay before species start colonizing a fresh surface; establishing this period relies on projection of the plotted growth curve on to its time axis. Absolute dates for a rock surface are obtained by fitting the largest lichen measurement to the local growth curve and relative dating may be achieved by drawing lines of equal maximum growth around areas defined by largest lichens of equal size.

Lichenometry is frequently perceived as a quick, easy, cheap method for use in areas where other dating tools are lacking. However, it has often been misapplied and results (only justifiable empirically) have been at best questionable. A review of the technique (Innes 1985) describes the difficulties surrounding species identification, particularly those in the yellow-green *Rhizocarpon* section *Rhizocarpon* group containing the species most frequently used (and confused) in lichenometric studies and the various methodologies employed, many of which were developed in attempts to resolve uncertainties arising from Beschel's five initial assumptions:

1. Thallus size correlates with age.
2. Colonization occurs soon after rock exposure.

3 Largest thalli are founder members of their populations.
4 Variations in growth rate due to habitat differences are minimized by selection of largest thalli growing in optimal conditions.
5 Accurate species identification in the field.

These are approximate assumptions because species growth rates vary over time and space, sensitively depending on habitat and climate, and identification is notoriously difficult for the non-expert. In addition, although lichens can theoretically colonize almost any bare rock surface, this may not occur until surfaces are WEATHERED. Consequently, both growth rates and delay before colonization should be confirmed at each dating site, a difficult procedure in places where there are no dated surfaces for growth-rate calibration.

Additional problems for this approach introduced during its development, arose as a result of misconceptions of the nature of lichen growth and confusions in terminology. The two terms causing the greatest confusion are 'great period' and 'lichen factor'.

Beschel (1961: 1,046, 1,057) described lichen growth as sigmoidal: beginning very slowly then gathering speed with many thalli passing through a 'great period' (limited to a few decades) before dropping to a long-term constant value. However, geomorphologists, due to a scarcity of younger thalli on surfaces of known age in their field areas, have often missed the initial part of the growth curve and assumed that the 'great period' is represented by linear growth over the whole historical period, possibly lasting some 300 years, with this followed by apparently declining growth, the decline being suggested by a curve drawn through very few lichen-size/radiocarbon-age correlations. However, as noted above, radiocarbon cannot reliably be used to date the latest period of moraine exposure and protracted growth curves constructed on this basis are therefore questionable.

These misconceptions contributed to the confusion over interpretation of the 'lichen factor' a term employed in many studies to describe the average value of maximum growth over 100 years, despite Beschel's (1961: 1,055) assertion that when calculating a 'lichen factor' the 'great period' should not be taken into consideration and that averaging cannot describe growth rates. This is because a straight line drawn from its origin and projected over 100 years will only cut a sigmoid curve in one place. Beschel's intention was that the 'lichen factor' should be used as a measure of hygrocontinentality and he defined the term as a growth velocity gradient produced by low precipitation at high altitudes that could be a helpful indicator when planning mountain reservoirs. And he warned that no standard growth velocity could be expected across a large region.

Population size-frequency approaches

In attempts to resolve some of the uncertainties in the initial assumptions and supply a measure of dating independence, two population size-frequency approaches have been developed. In the first, a composite curve is computed for the frequency distribution of large samples of longest axial measurements of largest lichens growing on uniform-age surfaces. The curve can be decomposed into subpopulations describing discrete events in their relative-age order. Where absolute dates are required, means of the largest lichens on each event surface are either plotted or regressed against dates of historical events to obtain a regional growth curve. Such a curve is, however, questionable (as Beschel warned) and requires careful testing before acceptance.

The advantages of this approach are that it can be used to investigate the history of seismic and large diachronous surfaces; statistical methods, with error limits and modelling can be applied, and anomalous lichens (either coalesced thalli or survivors from an earlier population) are less likely to corrupt the data.

In the second size-frequency approach, longest axial measurements are taken of whole lichen populations, with populations growing on surfaces of similar aspect and lithology and containing not less than thirty individuals, but ideally over 100 (statistical tests can be used to show where the smaller populations may safely be grouped). Changes in the shape of unimodal population size-frequency distributions show the relative age of populations. Bi- or multi-modal distributions indicate surface changes that have partially removed lichen thalli creating space for new colonizers; isolated large thalli may either be survivors or anomalous growths.

The advantages of this approach are similar to those of the largest lichen size-frequency one, but in this case a single boulder's whole colonization history is reflected by its population distribution, and comparison of its history with the histories of other surrounding surfaces provides insights into micro-environmental effects; showing for example, colonization rates differing on surfaces of differing aspects. Hence, colonization delays should be investigated before dating the proximal and distal sides of moraines.

The limitations of these two approaches are that a very large number of lichen measurements are required and the quality of absolute dating depends on the reliability of the independent dating source. Consequently, neither size-frequency approach achieves the desired status of full independence in absolute dating. However, because of large data sets the uncertainties implicit in the initial assumptions are less important.

Lichenometry is a reliable relative dating tool and, used with proper care, can provide revealing insights into land-forming processes especially in locations where the technique can be used within some independent dating framework supplied, for example, by historical records or DENDROCHRONOLOGY. In these circumstances absolute dating, frequencies and rates of change can be established with a reasonable degree of confidence.

References

Beschel, R.E. (1961) Dating of rock surfaces by lichen growth and its application to glaciology and physiography (lichenometry), in G.O. Raasch (ed.), *Geology of the Arctic II. 1st International Proceedings, Arctic Geology*, 1,044–1,062, Toronto: University of Toronto Press.

Innes, J.L. (1985) Lichenometry, *Progress in Physical Geography* 9, 187–254.

Further reading

Beschel, R.E. (1973) Lichens as a measure of the age of recent moraines, *Arctic and Alpine Research* 5, 303–309.

Bull, W.B., King, J., Kong, F., Moutoux, T. and Phillips, W.M. (1994) Lichen dating of coseismic landslide hazards in alpine mountains, *Geomorphology* 10, 253–264.

McCarroll, D. (1993) Modelling late-Holocene snow-avalanche activity: incorporating a new approach to lichenometry, *Earth Surface Processes and Landforms* 18, 527–539.

Winchester, V. and Harrison, S. (1994) A development of the lichenometric method applied to the dating of glacially influenced debris flows in Southern Chile, *Earth Surface Processes and Landforms* 19, 137–151.

VANESSA WINCHESTER

LIDAR

LIDAR light detection and ranging, is a form of airborne scanning altimetry which is of great importance for mapping landforms and landform change, especially at relatively local scales. It can be used to produce DIGITAL ELEVATION MODELS and provides high-resolution data on topography. It has been used in a number of geomorphological applications including cliff and landslide monitoring, the study of tidal channels, assessing subsidence risk, predicting areas subject to storm surges and tsunamis and changes in beach height (see, for example, Adams and Chandler 2002; Brock *et al.* 2002; Stockdon *et al.* 2002).

References

Adams, J.C. and Chandler, J.H. (2002) Evaluation of lidar and medium scale photogrammetry for detecting soft-cliff coastal change, *Photogrammetric Record* 17, 405–418.

Brock, J.C., Wright, C.W., Sallenger, A.H., Krabill, W.B. and Swift, R.N. (2002) Basis and methods of NASA airborne topographic mapper lidar surveys for coastal studies, *Journal of Coastal Research* 18(1), 1–13.

Stockdon, H.F., Sallenger, A.H., List, J.H. and Holman, R.A. (2002) Estimation of shoreline position and change using airborne topographic Lidar data, *Journal of Coastal Research* 18(3), 502–513.

A.S. GOUDIE

LIESEGANG RING

Liesegang rings are alternating concentric iron-rich and iron-depleted shells which are formed by differential chemical leaching and precipitation in rocks exposed to WEATHERING. The rings (shells) are developed in rock blocks, typically 2–20 cm in diameter and 10 cm thick (Liesegang blocks), bound by the tectonically induced joints at the periphery, and joint or bedding planes at the bottom. The Liesegang rings follow the configuration of the outer shape of the blocks. Structurally, the Liesegang blocks can be divided into four types: (a) primitive (single shell), (b) ordinary (multi-shell), (c) compound (multi-pattern), and honeycomb (cylindrical) Liesegang blocks. The most important factors involved in transformation of a bed of rock into Liesegang blocks are: a condensed grid of joint polygons, surface water, the appropriate topographic site of exposure, and composition, texture and thickness of the beds. Field, hand specimen, microscopic and geochemical studies suggest two opposing diffusion direction trends in the course of Liesegang ring formation: one mainly Si-trend (outward), and one mainly Fe-trend (inward).

The effect of recent tectonic movements on the Liesegang blocks manifests itself in various forms. Of interest is the partial replacement of the earlier Liesegang patterns by the later patterns along joints, and formation of compound Liesegang blocks (Liesegang blocks with more than one set of Liesegang patterns). From the study of Liesegang patterns within the compound Liesegang blocks, the configuration of the joint polygon related to each Liesegang pattern and thereby the sense of stress field migration in the area, can be deduced.

Further reading

Shahabpour, J. (1998) Liesegang blocks from sandstone beds of the Hojedk Formation, Kerman, I.R. Iran, *Geomorphology* 22, 93–106.

SEE ALSO: WEATHERING

JAMSHID SHAHABPOUR

Plate 73 Stepped limestone pavement with well-developed karren features, Triglav area, Julian Alps, Slovenia

LIMESTONE PAVEMENT

Limestone pavements are stripped bedding-plane rock surfaces constituting complex polygenetic assemblages of KARREN landforms, or Karrenfeld. The limestone's bedding planes are scoured of mature weathering forms and debris producing a surface on which develop solutionally widened fissures (grikes, Kluftkarren) and residual limestone blocks (clints, Flachkarren).

Adjectives such as 'regular' or 'rectangular' popularly describe clint and grike patterns, reflecting DISSOLUTION along the two main joint directions at right angles usually found in limestone massifs. However, field situations show highly complex and variable dissection patterns. In addition, not all limestone pavement areas are on single bedding planes. They can be benched or stepped across several beds depending on the limestone outcrop (Schichttreppenkarst, sheet-stepped karst) (Plate 73, and see KARREN, Plate 67). Whilst the term 'pavement' implies a walkable surface, there is a spectrum of surface limestone bedding-plane landforms, from highly dissected, almost shattered strewn rock fields, to virtually undissected rock sheets.

Morphometric work (see GEOMORPHOMETRY) comparing numerous pavements identifies typical ranges of clint and grike dimensions (Goldie and Cox 2000). Clints up to 1 m long are commonest, but many are longer, typically up to several metres. Whilst most widths are up to 500 cm, many are over 1 metre. Most grikes are up to 50 cm wide, whilst grike depths have a wider range, between 20 and 150 cm. These ranges, however, do not encompass all limestone pavement surfaces. For example, although 95 per cent of measured grikes were less than 30 cm wide, the absolute range is from 1 cm to over 1 m; although a 'typical' grike width is 13 to 16 cm. Grike depths showed a 70-fold variation from 4 to 274 cm but focus on 100 cm. A rather weak relationship between grike widths and depths suggests that their horizontal and vertical development are not strongly linked.

Clint shape as demonstrated by a clint width to length ratio (theoretically ranging from 0 to 1) is interesting in relation to conventional ideas of pavements, as a ratio of 0.4 is four times more common than a ratio of nearly 1 (indicating square). Very elongated clints are also rare (ratios close to 0). This implies that one joint direction dominates the planform of limestone pavements. A complication, however, is that the two main joint directions in limestone massifs are frequently bisected, resulting in many triangular and diamond-shaped features, causing problems for measurement and interpretation of rectangularity.

Beyond measurability are features of extreme or minimal dissection. Where clints are too narrow or small they move easily and fall over, failing to satisfy the walkability concept. These should be excluded from the limestone pavement definition. Harder to exclude, however, is the undissected extreme where the surface lacks grikes but is certainly walkable. Although lacking the karren forms expected on limestone pavement, such bare rock expanses can really only be called pavement. The more dissected extreme shows a geomorphological transition to a rock field.

Plan morphometric data says nothing about cross-sectional form of clints and grikes, important for understanding pavement weathering sequences. Depth and intensity of surface weathering may not affect clint area but changes their vertical relief (see Figure 103). For example, clints become pinnacled or very well rounded where solutional weathering is advanced (Plate 74).

The main factors producing a good limestone pavement are strong limestone, with some fractures, but not so fractured as to weaken the beds too much. Scouring is necessary to clear weathered debris off bedding planes, and pavements then need to be in a favourable solutional environment, i.e. reasonably wet and possibly covered with soil and vegetation, to develop their characteristic karren. Then that cover needs removal.

Glaciation is the scouring mechanism creating most of the world's limestone pavements. Other scouring agents include marine stripping, scarp retreat in semi-arid conditions and even human removal. In the main, though, distribution of limestone pavements is related to glaciation and thickly bedded, strong, pure limestones. Thus there is much in the younger mountain ranges such as the Alps, Rockies and Himalayas and in shield areas, such as northern Canada. Vast expanses of pavement on dolomites are found in northern Canada.

Rock type influences pavements at several scales. The best, most persistent, pavements develop on pure, mechanically strong and massively bedded limestones and dolomites. Similar landforms can occur on sandstones but weathering and soil development here generally mean that the features do not last except in situations such as waterfall or river beds where constant fluvial action keeps jointed outcrop clear. Pure limestones stay clear as soil formation is exceptionally slow due to the lack of insoluble residue, and the

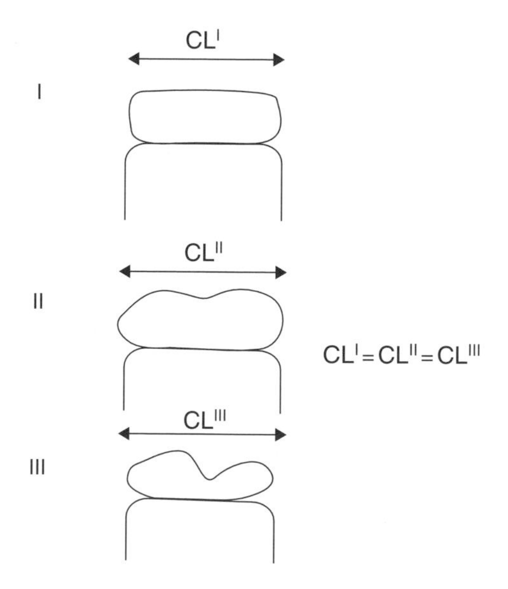

Figure 103 Simple cross-sectional diagrams of three clints demonstrating how plan measures of clint size (CL^I, etc.) can be identical on clints with very different degrees of down-wearing

Plate 74 Limestone pavement with a slight slope (5° to 8° down away from viewer) at Gaitbarrows, Cumbria, UK. There are several stages of grike development from relatively immature short slits to grikes running for tens of metres. Other karren features are less well developed

swallowing of material by developing solutional fissures. Thin or weak limestones (internal weaknesses, either horizontal weaknesses or frequent vertical joints within a given bed) favour weathering both by mechanical and solutional means, permitting easy rock breakup unable to sustain a pavement.

Slope significantly affects both how glacial scour operates and the subsequent pavement surface development by solution (see Plate 67). Sloping layered outcrops provide complex situations for glacial scour. The postglacial surface may be interrupted by areas where shelter from the ice allowed preglacial karstified features to survive. This is found in north-west England and in the Alps, for example. Some pavements may also have characteristics surviving from earlier phases of karstification (see PALAEOKARST AND RELICT KARST).

Another factor affecting pavement form and distribution has been human activity. Some limestone pavements are anthropogeomorphic (see ANTHROPOGEOMORPHOLOGY), indirectly because their exposure results from soil erosion, and directly because of clint removal by humans for resource reasons. Because their botanic as well as landscape value is threatened by this, limestone pavements in the UK are now legally protected (Goldie 1993).

References

Goldie, H.S. (1993) The legal protection of limestone pavements in Great Britain, *Environmental Geology* 21, 160–166.

Goldie, H.S. and Cox, N.J. (2000) Comparative morphometry of limestone pavements in Switzerland, Britain and Ireland, *Zeitschrift für Geomorphologie N.F. Supplementband* 122, 85–112.

Further reading

Ford, D.C. and Williams, P.W. (1989) *Karst Geomorphology and Hydrology*, London: Unwin Hyman.

Fornos, J.-J. and Gines, A. (1996) *Karren Landforms*, Palma: Universitat de Illes Balears.

Sparks, B.W. (1971) *Rocks and Relief*, London: Longman.

Vincent, P.J. (1995) Limestone pavements in the British Isles: a review, *Geographical Journal* 161(3), 265–274.

SEE ALSO: karren; karst; limestone pavement

HELEN S. GOLDIE

LINEATION

Lineations are linear patterns observed on imagery that represent fractures. The fractures may be either joints or faults (see FAULT AND FAULT SCARP). Each lineation does not necessarily represent an individual fracture, and typically represents a fracture zone. Lineations include, but are not limited to, actual joints in bedrock, straight stream segments, linear alignments of natural vegetation, aligned topographic features and linear changes in image tone or texture. Any feature thought to be a lineation should be one of a group of parallel features, and each lineation and group (set) may consist of different types of patterns. For instance, a line of vegetation may continue as a linear pattern of dark-toned soil across a light-toned field, and then as a stream valley.

Lineations are readily apparent on all scales, resolutions and types of imagery – aerial photographs, radar and various types of satellite imagery – on hard copy, as well as in digital form. The delineation of lineations is, however, both art and science (Wise 1982). For best results, and particularly if quantitative results are desired, the imagery should be vertical or near vertical. Certain features, such as those noted above, define lineations, but the interpretation of these features, particularly tonal and textural variations, can be difficult. The eye must be trained to see them, and the more experience one has, the greater the number of lineations that can be identified with confidence on any given image. All hard-copy images should be evaluated meticulously from different angles and with different kinds and angles of illumination. Some lineations will be obvious in sunshine, for example, but cannot be seen in natural light on a grey, overcast day. Topographic lineations are often more easily identified using stereo imagery, but those defined by vegetation or tonal variations can be readily – and more easily – identified monoscopically. Although attempts have been made to automatically delineate lineations on digital imagery, they have not been generally successful (e.g. Ehlen *et al.* 1995). Finally, the individual delineating the lineations should also be careful not to confuse linear man-made features, ancient or modern, with natural lineations.

Stereonet projections can be used to display lineation orientation data so that sets can be identified (dips are assumed vertical). Different patterns are indicative of fracture formation under different tectonic conditions, and the chronology of the different events can be identified using cross-cutting relations between individual lineations and lineation sets. When the lineations are in digital form, quantitative information on length and spacing (frequency) can also be obtained. These types of data are useful with respect to landform evolution and engineering geomorphology, and can provide information on fracturing in areas that are difficult to access in the field. Furthermore, lineations can be directly related to joint or fault data identified on the ground. Several studies have identified lineations on imagery, then located the individual lineations on the ground (e.g. Lattman and Parizek 1964). Other studies have shown statistical links between joints and lineations (e.g. Ehlen 1998, 2001).

Another term often used for lineation is 'lineament'. Lineaments are in fact subsets of lineations: they are topographic features and, for example, do not include linear patterns produced by tonal changes, which are among the most important indicators of fractures on imagery. Examples of lineaments include aligned saddles or ridge lines and aligned stream valleys.

References

Ehlen, J. (1998) A proposed method for characterizing fracture patterns in denied areas, in J.R. Underwood, Jr and P.L. Guth (eds) *Military Geology in War and Peace*, Boulder, CO: Geological Society of America Reviews in Engineering Geology 13, 151–163.

——(2001) Predicting fracture properties in weathered granite in denied areas, in J. Ehlen and R.S. Harmon (eds) *The Environmental Legacy of Military Operations*, Boulder, CO: Geological Society of America Reviews in Engineering Geology 14, 61–73.

Ehlen, J., Hevenor, R.A., Kemeny, J.M. and Girdner, K. (1995) Fracture recognition in digital imagery, in J.J.K. Daemen and R.A. Schultz (eds) *Rock Mechanics, Proceedings of the 35th U.S. Symposium on Rock Mechanics*, 141–146, Brookfield, VT: A.A. Balkema.

Lattman, L.H. and Parizek, R.R. (1964) Relationship between fracture traces and the occurrence of ground water in carbonate rock, *Journal of Hydrology* 2, 73–91.

Wise, D.U. (1982) Linesmanship and the practice of linear geo-art, *Geological Society of America Bulletin* 93, 889–897.

SEE ALSO: jointing; remote sensing in geomorphology

JUDY EHLEN

LIQUEFACTION

Liquefaction is the transformation of granular material from a solid to a liquid by an increase in pore pressure. Liquefaction is associated with earthquakes, with saturated ground, with granular sandy/silty soils, with some spectacular building collapses, and with submarine landslides and waste tip failures. During the 1964 earthquake at Niigata in Japan large apartment buildings were gently tilted by as much as 80° on to their sides, with no structural damage, as a result of liquefaction of a shallow sand bed underlying the structures.

In the Kobe earthquake of January 1995 widespread damage occurred. Massive soil liquefaction was observed in many reclaimed lands in Osaka Bay including two man-made islands. Both islands had been constructed from fill material derived from decomposed granite. The grain size of the fill varied from gravel and cobble-sized particles to fine sand, with a mean grain size of approximately 2 mm. In the Port Island ground, liquefaction of this fill material (the thickness of the submerged fill in this island was about 12 m) resulted in settlement of between 0.5 and 0.7 m. Soil liquefaction also occurred around the port of Kobe and caused extensive damage to many industrial and port facilities such as tanks, wharves, quay walls, cranes, and the collapse of the Nishinomiya Harbour Bridge.

A catastrophic liquefaction failure occurred in a coal waste tip at Aberfan in Wales in 1966. The tips from the Merthyr Vale Colliery were sited on the upper slopes of the valley sides, directly above Aberfan, and some 100 m up on a 13° slope. These slopes consisted of the permeable Brithdir sandstone, with many springs and seeps in evidence. Tip no. 7 was built over these springs and in October 1966 a flowslide developed in the saturated material – a classic liquefaction event. The liquefied tip material ran into the village and killed 144 people, most of whom were children in the village school.

In the Netherlands liquefaction events often concern flowslides in underwater slopes of loose sand. Flowslides are observed after sudden liquefaction of the sand under static loading. The liquefaction is caused by the contraction (packing collapse) occurring simultaneously in a large volume of sand and the impossibility of rapid drainage of the completely saturated pore fluid. The upper 10–30 m of soil in the western part of the Netherlands consists of Holocene layers of clay, peat and sand. Tidal estuaries were formed in Zeeland, the south-west part of the country, from the beginning of the Holocene. The position of the tidal estuaries, however, has shifted in an alternate process of rapid erosion and sedimentation. Rapid sedimentation has resulted in thick layers of loose sand at many locations. Median grain diameter is around 200 μm; with a silt content of perhaps 3 per cent. Liquefaction is accompanied by sudden pore pressure increase as the open sand packing collapses from a metastable state to a more stable state.

Further reading

Ishihara, K. (1993) Liquefaction and flow failure during earthquakes: 33rd Rankine lecture, *Géotechnique* 43, 351–415.

Nieuwenhuis, J.D. and de Groot, M.B. (1995) Simulation and modelling of collapsible soils, in E. Derbyshire, T. Dijkstra, and I.J. Smalley (eds). *Genesis and Properties of Collapsible Soils*, 345–360, Dordrecht: Kluwer.

Smalley, I.J. (1972) Boundary conditions for flowslides in fine-particle mine waste tips, *Transactions of the Institution of Mining and Metallurgy*, section A 81, 31–37.

Ter-Stepanian, G. (2002) Suspension force and mechanism of debris flows, *Bulletin of the International Association of Engineering Geology and the Environment* 61, 197–205.

Waltham, T. (1978) *Catastrophe – The Violent Earth*, London: Macmillan.

SEE ALSO: quickclay

IAN SMALLEY

LITHALSA

The term lithalsa was used for the first time by S. Harris (1993) to describe mineral mounds in Yukon (Canada). They are mounds similar to PALSAS but without any cover of peat. The upheaval of the soil is the result of the formation of segregation ice in the ground. These mounds which, like the palsas, are found in the domain of discontinuous PERMAFROST, were successively called purely minerogenic palsas with no peat cover, mineral palsas, cryogenic mounds, mineral permafrost mounds and palsa-like mounds. Lithalsas are presently known only in Subarctic Northern Québec and Lapland in places where the temperature of the warmest month is between +9 and +11.5 °C and the mean annual temperature between −4 to −6 °C.

Remnants of lithalsas formed during the Younger Dryas exist in Ireland, Wales and on the Hautes Fagnes plateau in Belgium (Pissart 2000). They are closed depressions surrounded by a rampart. These features were first explained as remnants of pingos. They are now regarded as periglacial phenomena which provide more precise palaeoclimatic indications.

References

Harris, S.H. (1993) Palsa-like mounds developed in a mineral substrate, Fox Lake, Yukon Territory, in *Proceedings of the Sixth International Conference on Permafrost*, 5–9 July, Beijing, China, South China University of Technology, vol. 1, 238–243.

Pissart, A. (2000) Remnants of lithalsas of the Hautes Fagnes, Belgium: a summary of present-day knowledge. *Permafrost and periglacial processes* 11, 327–355.

ALBERT PISSART

LITHIFICATION

The process by which unconsolidated sediments become indurated sedimentary rock, originating from Greek and Latin for 'to make rock'. The source of sediments is typically loose material that accumulates upon the surface as a result of subaerial weathering. These are then transported and deposited before the process of lithification begins. Lithification precedes DIAGENESIS and must not be confused with the term (diagenesis refers to the processes and their products affecting rocks during their burial). The process of lithification involves three main components. First, desiccation (drying of sediment) reduces the amount of pore space within the sediment by eliminating the water within the material. Desiccation is particularly significant in fine-grained sediments (clays and silts), where water cannot percolate easily through the pores. Compaction further reduces pore space, aided by the increasing weight from the materials above as surface deposition continues. Surface tension between the individual grains then allows them to act as a consolidated mass.

Second, cementation binds the loose sediment together by filling in the remaining pore spaces. It is typically done through precipitation of cementing agents in solution as water flows through the pore spaces. The most common cementing agents are calcite ($CaCo_3$), dolomite ($CaMg[Co_3]_2$), quartz (SiO_2), and iron oxide (Fe_2O_3) (haematite).

The pH of a solution plays a significant role in determining the cementing agent type. An acidic solution will tend to produce a quartz cementing agent, whereas an alkaline solution will invariably generate a calcitic or dolomitic cement.

Third, crystallization is a specific form of cementation. It is particularly effective on carbonate sediments, whereby crystals form in pore spaces from minerals in solution and bond onto existing crystals within the sediment. This often has the side effect of making some rocks harder.

Further reading

Maltman, A.J. (1994) *The Geological Deformation of Sediments*, London: Chapman and Hall.

STEVE WARD

LOESS

Loess is a sedimentary deposit, largely composed of coarse and very coarse silt, which is draped over the landscape. The silt is largely quartz and, at the soil structure level, the loess particle system has an open packing, which is a result of aeolian deposition during Quaternary times. This open structure gives rise to the main practical problem; when loess ground is loaded and wetted the structure may collapse and subsidence occurs. Construction on loess has to be carefully designed in order to avoid these subsidence problems. The silt-sized particles and the open metastable structure are the main characterizing features of loess deposits.

The literature on loess is vast and complex; the major loess languages are Russian, Chinese, English, German and French. The largest literature is in Russian (see Kriger 1965, 1986; Trofimov 2001) but there are major works in many languages. Scheidig (1934) was for many years the standard work, possibly until supplanted by Pye (1987) and Rozycki (1991). Early literature has been collected (Smalley 1975) and an outline bibliography attempted (Smalley 1980). Loess investigation has had a major part in the activity of the International Union for Quaternary Research (INQUA), which publishes the one specialized journal – *Loess Letter*. The one institute in the world totally devoted to the study of loess is the Xian Laboratory for Loess and Quaternary Geology of the Chinese Academy of Sciences; there is a focus of geotechnical activity at Moscow State University and at the G.A. Mavlyanov Institute of Seismology in Tashkent.

Loess really is an aeolian material; it is the aeolian factor which gives it its defining characteristics, and separates it from other deposits. Loess is a wind-blown silt; the aeolian deposition accounts for the open metastable structure, which accounts for the propensity to collapse when loaded and wetted. The wind-blown deposition mode accounts for its characteristic geomorphological position – the 'draping' over the landscape – of the loess on the ground, thinning away from source. Loess in the air is a dust cloud, travelling in low suspension over a relatively short distance. In parts of the world where the geomorphology is favourable this dust has been depositing for millions of years (but see also Pecsi 1990).

Loess formation

An area rich in controversy and argument; how do loess deposits form? Even in the twenty-first century the tag-end of this debate still carries on – but only in the Russian literature. By the end of the nineteenth century it was fairly widely accepted that the key event in the formation of a loess deposit was deposition by aeolian action. This might be called the mainline loess view; loess is an airfall sediment.

But starting early in the twentieth century and continuing fitfully throughout the entire century there has been an alternative view of loess formation, which has generated a large literature, from proponents and opponents. This can be called the 'soil' theory of loess formation, or the '*in situ*' theory. At its heart is the idea that loess is formed where it is found, by a process of 'loessification', the conversion of non-loess ground into loess ground. This is largely a Russian theory and its chief, and very forceful protagonist, was L.S. Berg. The minor role was played by R.J. Russell in the USA who put forward a very similar theory to explain the Mississippi valley loess in 1944.

Another question, which arose somewhat later in the history of loess investigation, was that of the possible mechanism for the formation of the vast amounts of quartz silt that were required to make large loess deposits. What sources of geo-energy were available for large-scale silt formation? And did this affect the nature and distribution of loess deposits? This question connected in particular to the problem of 'desert' loess.

V.A. Obruchev, early in the twentieth century, placed loess at the fringes of hot, sandy deserts. This desert loess has been something of a problem ever since. B. Butler, a noted Australian soil scientist, suggested that it did not exist, because he could not find it in Australia; and Albrecht Penck, a famous geomorphologist, suggested that peri-Saharan deposits were essentially lacking. So the question arose, could large hot sandy deserts generate enough loess-sized material to form significant loess deposits? In geographical terms it appears that the Chinese and Central Asian deserts do have major loess deposits close by, but the Sahara and Australian deserts do not. There is a loess-like material, called 'PARNA' by Butler, in south-east Australia, but it has modest extent. It has been proposed that the Central Asian and Chinese loess deposits are also near very high mountain regions, and that these are the real source of the loess particles, with the deserts simply acting as holding areas or reservoirs.

Loess distribution

There are some very large deposits, and many widespread smaller ones. The classic deposit is in China, in the north and north-west of the country; this is the 'Yellow Earth', which gives colour to the Yellow River, and played a major part in the development of the Chinese civilization – the only one of the ancient civilizations which has lasted into modern times. Deposits in Lanzhou are believed to be over 300 m, perhaps over 400 m, thick. Although loess is thought of as a Quaternary material there are suggestions that loess deposition in China has been going on since the Miocene, i.e. for over 20 million years.

The other large deposits are in central North America, relating to the Great Plains, and the Dust Bowl; in South America, the Pampas; and in Europe. The European deposits are complex but might be divided into the northern glacially related deposits, and the Danubian deposits which derive their materials from mountain sources. Of the smaller deposits those in North Africa and in New Zealand are of interest. There are deposits near the coast in Tunisia and Libya which are certainly very loess-like, but the particle size is rather large. The Sahara is a great source of wind-blown dust but most of it is very small, which travels in high suspension for great distances. In New Zealand there are significant loess deposits in the North and South Islands, associated with the mountains. New Zealand is the only country which has a monograph devoted to the national loess (Smalley and Davin 1980) and it is to be hoped that more countries will follow this lead.

With the disappearance of the Soviet Union and the revision of geography in the fringe regions new loess-rich countries have emerged, in particular Ukraine and Uzbekistan. Ukraine has large loess cover and in many parts chernozem soils have developed in the loess providing top-grade agricultural land. Also, in Kyiv is the most amazing loess building, the Pecherskaya Lavra, the Caves Monastery, where the monks have excavated into the loess and developed a subterranean complex in the Dnepr loess. In Uzbekistan, in the eastern part, near the Tien Shan mountains, are major loess deposits. The capital city of Tashkent is built on loess and was largely destroyed by a large earthquake in 1966. Loess ground is very vulnerable to earthquake shocks.

Loess stratigraphy

The basic idea of loess stratigraphy was invented by John Hardcastle in Timaru, New Zealand in 1890; this was that a loess deposit can give a good indication of past climates. Loess acted as a 'climate register'. At the beginning of the twentieth century problems of climate change were exciting quite a lot of interest and this drove a large research effort devoted to loess stratigraphy.

Hardcastle's insights were ignored; as in the case of Mendel and genetics, the scientific culture was not prepared for them. They were addressing a problem that had no structure, no framework, no reference points. It appears that stratigraphic ideas re-emerge with Soergel in Germany in 1919 and developed slowly in Europe. The 'eureka' moment came at the INQUA Congress in 1961 in Poland when Liu Tung-sheng presented the work of the Chinese investigators which showed multiple palaeosols in the Luochuan loess. This thick loess section showed alternating layers of loess and palaeosols. The palaeosols indicated warm periods and the loess layers cold periods – a climatic indicator as Hardcastle had suggested. The data from Luochuan suggested many climatic oscillations in the Quaternary period and laid the foundations for continuing investigations. It was a giant step away from the simple four event Quaternary derived from Alpine observations.

Applied geomorphology

Loess drapes the landscape; it is the accessible ground in which engineers operate. The most significant and expensive of the loess ground engineering problems is hydroconsolidation and soil structure collapse, caused by loading and wetting, and leading to subsidence and structural failure. This was classically a problem in the Soviet Union and now occurs in post-Soviet states like Ukraine and Uzbekistan. The building of irrigation canals in loess in Uzbekistan during one of the early five-year plans alerted the Soviet authorities to the vast problems of subsidence. It was, more than anywhere, a Soviet problem. Therefore most of the subsidence literature is in Russian (see Trofimov 2001 for a good review) and tends to conform to the requirements of the *in situ* approach to loess formation.

A major topic in the Russian literature is 'how did collapsibility develop?' This has never been discussed in 'western' literature because the reason for collapse is implicit in the aeolian deposition mechanism which forms loess deposits. In the Russian literature it appears to be the applied geomorphologists and ground engineers who continue to cling to the *in situ* theories of loess formation.

The problem of soil erosion falls within the purview of applied geomorphology. Because of its silty nature loess soil is very prone to erosion, by wind and water. The classic wind erosion events (e.g. the Dust Bowl) tend to be the blowing away of loess soil material; and the major water erosion problems are often loess connected – in particular the loss of soil material in north China. It is soil erosion that makes the Yellow River yellow. In north-west China where the loess is spread over mountainous terrain, there is a considerable landslide problem (Derbyshire *et al.* 2000). In 1920 a large earthquake caused much loss of life because of widespread collapse of loess tunnels which housed a large proportion of the population.

References

Derbyshire, E., Meng, X.M. and Dijkstra T.A. (eds) (2000) *Landslides in the Thick Loess Terrain of North-West China*, Chichester: Wiley.
Kriger, N.I. (1965) *Loess, Its Characteristics and Relation to the Geographical Environment*, Moscow: Nauka (in Russian).
——(1986) *Loess: Formation of Collapsible Properties*, Moscow: Nauka (in Russian).
Pecsi, M. (1990) Loess is not just the accumulation of dust, *Quaternary International* 7/8, 1–21.
Pye, K. (1987) *Aeolian Dust and Dust Deposits*, London: Academic Press.
Rozycki, S.Z. (1991) *Loess and Loess-like Deposits*, Wroclaw: Ossolineum.
Scheidig, A. (1934) *Der Loss und seine geotechnischen Eigenschaften*, Dresden and Leipzig: Steinkopf.
Smalley, I.J. (ed.) (1975) *Loess: Lithology and Genesis, Stroudsburg, PA*: Dowden, Hutchinson and Ross.
——(1980) *Loess – A Partial Bibliography*, Norwich: Geobooks (Elsevier).
Smalley, I.J. and Davin, J.E. (1980) The First Hundred Years: A Historical Bibliography of New Zealand loess 1878–1978, New Zealand Soil Bureau Bibliographical Report 28.
Trofimov, V.T. (ed.) (2001) *Loess Mantle of the Earth and its Properties*, Moscow: Moscow University Press (in Russian).

Further reading

Loess Letter ISSN 0110-7658; published twice a year by the International Union for Quaternary Research.
Collapsing Soils Communique ISSN 1473-0936; published by the Collapsing Soils Commission of the International Association of Engineering Geology and the Environment.
Loess Inform ISSN 0238-065X; published by the Hungarian Academy of Sciences; see in particular issue 2 for 1993 for general discussion.
Pecsi, M. (1968) Loess, in *Encyclopedia of Geomorphology*, 674–678, New York: Reinhold Book Corporation.
Pecsi, M. and Richter, G. (1996) Loss: Herkunft-Gliederung-Landschaften, *Zeitschrift für Geomorphologie, Supplementband* 98.

SEE ALSO: aeolian geomorphology; palaeoclimate; wind erosion of soil

IAN SMALLEY

LOG SPIRAL BEACH

There are several terms for the plan shape of asymmetrical beaches and bays, including log spiral, half-heart, crenulate, hook-shaped and zetaform. BEACHes between headlands consist of: a curved, nearly circular, portion in the lee of the headland, which may be absent in some areas; a logarithmic spiral section; and a nearly linear to curvilinear reach tangential to the downcoast headland. Yasso (1982) proposed that plan curvature is described by a logarithmic spiral law, with the distance from the centre of the spiral (r) to the beach increasing with the angle θ according to:

$$r = e^{\theta \cot \alpha}$$

where: θ is the angle of rotation, or spiral angle, which determines the tightness of the spiral; and

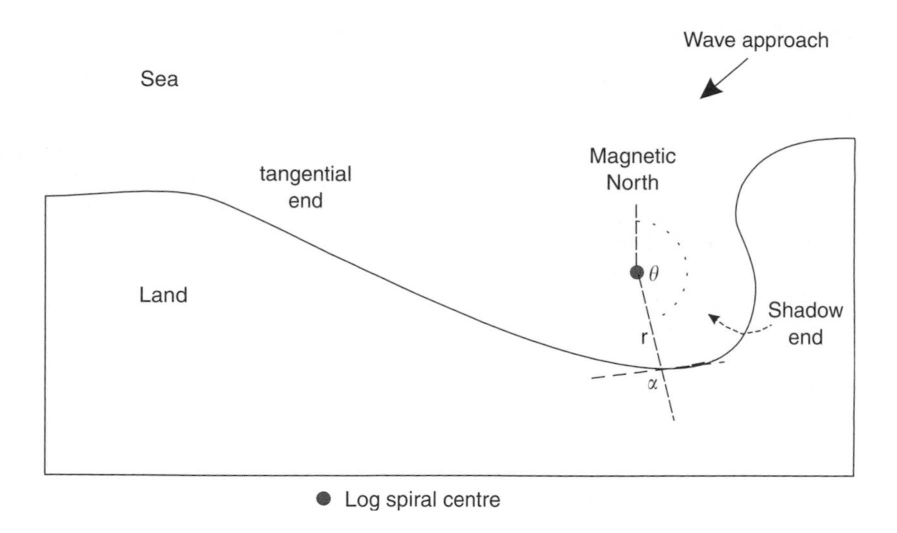

Figure 104 Log spiral beach and Yasso's (1982) logarithmic spiral law

α is the logarithmic spiral constant, the angle between a radius vector and the tangent to the curve at that point – this is constant for a given log spiral (Figure 104).

There may be systematic variations in beach sediment and morphology along log spiral beaches corresponding to longshore changes in wave height and energy. Beaches in California are finer grained and more gently sloping in the sheltered portions of log spiral bays, but they tend to be steep or reflective in southeastern Australia, while sections exposed to the dominant swell and storm waves are gentle or dissipative. The sheltered portion of mixed sand and coarse clastic bays in Alaska generally have low wave energy, small grain sizes, fairly well-sorted sediment, gentle beachface slopes and eroding shorefaces. The central portions have high wave energy, large grain sizes, poor sorting, moderate beachface slopes and shorelines transitional between erosion and deposition. The tangential ends of the bays are similar to the shadow ends, except that the beachfaces are steep and the shorelines depositional (Finkelstein 1982).

Progressive decline in beach curvature downdrift of headlands is usually thought to reflect increasing exposure to wave action, although wave energy is only one of the factors that must be considered. Shorelines try to attain equilibrium conditions determined by offshore wave refraction and diffraction, the distribution of wave energy, rates of longshore sediment transport and the relationships between beach slope, wave energy and grain size. Log spiral beaches formed by oblique waves are generally thought to be the most stable in nature. They are in static equilibrium when the tangential downcoast section is parallel to the wave crests, and as waves diffract into the bay they break simultaneously along the whole periphery. There is no longshore component of breaking wave energy and no littoral drift, and the plan shape, local beach slope and sediment size distribution are constant through time.

References

Finkelstein, K. (1982) Morphological variations and sediment transport in crenulate-bay beaches, Kodiak Island, Alaska, *Marine Geology* 47, 261–281.

Yasso, W.E. (1982) Headland bay beach, in M.L. Schwartz (ed.) *The Encyclopedia of Beaches and Coastal Environments*, 460–461, Stroudsburg, PA: Hutchinson Ross.

Further reading

Trenhaile, A.S. (1997) *Coastal Dynamics and Landforms*, Oxford: Oxford University Press.

ALAN TRENHAILE

LONG PROFILE, RIVER

A graph representing the relationship between altitude (H) and distance (L) along the course of a river, expressed by

$$H = f(L)$$

Any of three functions – exponential, logarithmic or power – can provide a reasonable fit to stream profiles and give rise to smooth, concave-upward curves (see GRADE, CONCEPT OF). Most long profiles tend to be concave, but they are not invariably smooth. Local steepening of channel gradient (see WATERFALL and KNICKPOINT) can result from such causes as more resistant bedrock strata, the introduction of a coarser or larger load, tectonic activity (see, for example, Riquelme *et al.* 2003) and BASE LEVEL changes (Knighton 1998) associated with REJUVENATION. Long profiles of glaciated valleys are often characterized by steps and overdeepenings and by hanging tributaries (MacGregor *et al.* 2000).

Some rivers have convex long profiles, and this may be a characteristic of updoming passive margins to continents or of discharge reductions downstream as is found in arid zone rivers (e.g. in the Namib).

Excessively or overconcave rivers occur in rivers with lower reaches that have become recently infilled by, for example, estuarine sedimentation or which have been affected by glacial diversion and subsequent lengthening of their courses (Wheeler 1979). At a local scale river long profiles may be punctuated by pool and riffle topography, and by the development of dune bedforms (see STEP-POOL SYSTEM).

The inverse relationship between channel gradient and discharge recognized by Gilbert (1877) helps to explain concavity, since tributary inflows cause a downstream increase of discharge which enables the stream's sediment load to be transported on progressively lower slopes. When discharge increases rapidly downstream with increasing contributing area, profile concavity is greater (Wheeler 1979). In addition, the calibre of sediment load is also related to stream gradient (see Richards 1982, for a review). However, whereas discharge is clearly an independent control of stream gradient, causation is less obvious in the gradient–sediment size relationship, for there are complex feedbacks between sediment size and gradient.

References

Gilbert, G.K. (1877) *Report on the Geology of the Henry Mountains*, Washington, DC: United States Geological and Geographical Survey.
Knighton, D. (1998) *Fluvial Forms and Processes, A New Perspective*, 2nd edition, London: Arnold.
MacGregor, K.R., Anderson, R.S., Anderson, S.P. and Waddington, E.D. (2000) Numerical simulations of glacial-valley longitudinal profile evolution, *Geology* 28, 1,031–1,034.
Richards, K.S. (1982) *Rivers. Form and Process in Alluvial Channels*, London: Methuen.
Riquelme, R., Martinod, J., Hérail, G., Darrozes, J. and Charrier, R. (2003) A geomorphological approach to determining the Neogene to recent tectonic deformation in the coastal cordillera of Northern Chile (Atacama), *Tectonophysics* 361, 55–275.
Wheeler, D.A. (1979) The overall shape of longitudinal profiles of streams, in A.F. Pitty (ed.) *Geographical Approaches to Fluvial Processes*, 241–260, Norwich: GeoAbstracts.

A.S. GOUDIE

LONGSHORE (LITTORAL) DRIFT

Longshore sediment transport is the net displacement of sediment parallel to a coastline. Such transport is maximized where WAVES (especially *breaking waves*) and wave-induced, shore-parallel, quasi-steady *longshore currents* (see CURRENT) prevail, i.e. in the *surf zone*. A secondary peak also occurs in the *swash zone*. Here wave energy is finally dissipated in the reversing currents of the wave *uprush* (flow at an angle to the shoreline under oblique wave approach) and *backwash* (flow normal to the shoreline under the force of gravity). This gives rise to a 'zig-zag' motion of water and sediment along the beach face. In both cases the obliqueness of the angle of wave approach is critical to the transport rates. The volume or mass of sediment transported by *longshore currents* and *swash* is termed *littoral drift*. Sediment may move as BEDLOAD, in continuous contact with the bed, or as SUSPENDED LOAD. In theory, particles in suspension are supported by turbulent (Reynolds) stresses within the fluid, while bedload particles are supported by inter-granular impact forces (Bagnold 1963). Debate exists concerning their relative importance. For example, Komar (1998) suggests that bedload comprises upwards of 75 per cent of the total transport, while Sternberg *et al.* (1989) suggest that in the surf zone, where waves are the pre-eminent entraining mechanism, suspended load can account for virtually all the longshore transport. The distinction between the two is, however, purely theoretical; the time-averaged mass concentration of particles decreases rapidly away from the immobile bed without a significant break, as the two support mechanisms merge. Further, since wave-induced, reversing oscillatory motions are the main entraining force, both suspended and bedload reveal strongly episodic behaviour. In the swash zone, the thin flows make the relative role of bedload and suspended load even more difficult to define. The variable directions of wave approach result in reversals of the direction of transport by both the longshore currents and the swash; thus, the *gross* sediment transport alongshore may be very large, while the *net* transport may be significantly smaller. However, transport rates up to several million $m^3 a^{-1}$ are common and this transport has been studied extensively by geomorphologists, coastal oceanographers and engineers concerned with coastal forms and dynamics, harbour siltation and dredging, the effectiveness of groynes as shore protection, etc.

Longshore sediment transport rates

Longshore transport is driven fundamentally by the alongshore component of the *wave momentum flux* or *radiation stress* (Longuet-Higgins and Stewart 1964), which itself creates a quasi-steady longshore current, although the complex interaction between the waves, the currents and the sediment is far from understood. Transport takes place within a combined *wave-current boundary layer*. The wave boundary layer is relatively thin compared to that of the current; consequently the stresses generated by the waves are significantly larger. However, to a first order, waves are purely oscillatory and thus cannot induce significant transport. It was this assumption, that waves entrain (initiate motion) and currents advect sediment, that led to the most common formulation for sediment transport alongshore (Bagnold 1963).

Longshore transport models

The Energetics Model: Inman and Bagnold (1963) defined longshore transport rates as a simple function of the alongshore component of the incident wave energy flux:

$$I_l = (\rho_s - \rho_f)\, g a' Q_l = K\, P_L = K\, (E\, C\, n)_b \sin \alpha_b \cos \alpha_b$$

where I_l is immersed weight transport rate, a′ is a constant of 0.6, Q_l is longshore transport rate, ρ_s and ρ_f are solid and fluid mass densities, K is a dimensionless proportionality coefficient, E is the specific wave energy density, C is the wave celerity or speed, n is the ratio of wave group velocity to wave phase velocity, α is the angle of wave breaking and the subscript b refers to conditions at the point of wave breaking. The constant K, which was originally proposed as an efficiency factor (≈ 0.77 when the root-mean-square wave height is used in the energy calculation; Komar 1971), has subsequently been the source of considerable debate. This proportionality factor is also thought to be dependent upon the grain size, wave steepness, the surf similarity parameter or Irribarren Number, bed slope, etc. The 'constant' K was originally conceived as applying to fully developed transport in an instantaneous sense. A number of authors have used a time-averaged relationship to predict the potential for a time-averaged rate of transport and thus a model for the long-term potential for erosion and deposition (see Greenwood and McGillivray 1980). Komar (1998) gives an extensive review of the energetics model.

Bailard (1981), following Bagnold (1966) and Bowen (1980), expanded the basic model and showed that the total longshore transport rate depends upon the steady current and a number of higher order moments of the velocity field:

$$\langle i_y \rangle = \rho\, c_f u_m^3 \left\{ \left[\frac{\varepsilon_b}{\tan \phi} \right] \left(\delta_v^3 + \frac{\delta_v}{2} + \frac{\tan \beta}{\tan \phi} (u3)^* \tan \alpha \right) + \frac{u_m}{w} \varepsilon_s \delta_v (u3)^* + \frac{u_m^2}{w^2} \varepsilon_s^2 \tan \beta\, (u5)^* \tan \alpha \right\}$$

where c_f is a drag coefficient, u_m is oscillatory velocity magnitude, ϕ is angle of internal friction, α is angle of wave approach, ε_b is bedload efficiency, ε_s is suspended load efficiency, w is sediment fall velocity and the relatively steady currents δ, δ_u and δ_v are defined as:

$$\delta = \frac{\bar{u}}{u_m}\ ;\ \delta_u = \frac{\bar{u}}{u_m} \cos \theta\ ;\ \delta_v = \frac{\bar{u}}{u_m} \sin \theta$$

and the velocity moments ψ_1 and ψ_2 are defined as:

$$\psi_1 = \frac{\langle \tilde{u}^3 \rangle}{u_m^3}\ ;\ \psi_2 = \frac{\langle |\bar{u}|^3 \tilde{u} \rangle}{u_m^4}$$

and the integrals $(u3)^*$ and $(u5)^*$ contain the time-averaged magnitudes of the combined velocities associated with each of the higher moments and are defined as:

$$(u3)^* = \frac{1}{T}\int_0^T (\delta^2 + 2\delta \cos(\theta-\alpha)\cos \sigma t + cos^2 \sigma t)^{\frac{3}{2}}\, dt$$

$$(u5)^* = \frac{1}{T}\int_0^T (\delta^2 + 2\delta\, cos(\theta-\alpha)\cos \sigma t + \cos^2 \sigma t)^{\frac{5}{2}}\, dt$$

where T is the wave period, θ is steady current angle and σ is wave frequency.

The Applied Stress Model uses the concept of 'excess stress' for alongshore transport, i.e. the stress in excess of that used to initiate motion of the sediment. The models use precise physical relationships coupled with semi-empirical expressions to describe the transport process. A typical model of the longshore component of transport, q_l, is that of Grant and Madsen (1979):

$$q_l(t) = 40wD\left(\frac{\frac{1}{2}\rho c_f\left(u^2(t)+v^2(t)\right)}{(s-1)\rho g D}\right)^3 \frac{v(t)}{\sqrt{u^2(t)+v^2(t)}}$$

where t is time, w is mean sediment fall velocity, D is sediment diameter, ρ is fluid density, c_f is coefficient of bed friction, u and v are horizontal cross-shore and longshore fluid velocities, s is specific gravity and g is gravitational constant.

Most of the models have been calibrated using sand-sized material, but clearly longshore transport also includes coarser grained materials. Van Wellen *et al.* (2000) reviews the longshore transport equations for coarse-grained beaches.

Geomorphological significance

Littoral drift is an important part of the *sediment budget*, which is based on mass conservation principles applied to the coastal zone. A major concept in coastal geomorphology is that of the littoral cell, consisting of a zone of sediment supply (may be rivers or shore erosion), a zone of erosion (net loss of sediment), a zone of longshore transport and a zone of accretion (net gain of sediment). An excellent review of the concept is available in Carter (1988). *Gradients* in the rates of longshore sediment transport can be used to model major aspects of shoreline erosion and/or accretion (see Komar 1998).

Several specific geomorphological forms owe their origin to longshore sediment transport. For example, a TOMBOLO is a strip of sediment accumulating between an offshore island and the main shoreline formed as a result of wave refraction around the island. Refraction produces waves, which approach from opposing directions in the lee of the island and induce currents and longshore transport, which converge from two directions. Spectacular *sand spits* may develop as a result of littoral drift being deposited at a re-entrant in the shoreline or across embayments and may stretch for several kilometres. Refraction of waves around the down drift terminus often produce *recurved* or *hooked* spits.

References

Bagnold, R.A. (1963) Mechanics of Marine Sedimentation, in M.L. Hill (ed.) *The Sea: Ideas and Observations*, Vol. 3, 507–528, New York: Wiley Interscience.

——(1966) An approach to the sediment transport problem from general physics, *United States Geological Survey Professional Paper* 422-I.

Bailard, J.A. (1981) An energetics model for a plane sloping beach, *Journal of Geophysical Research* 86, 10,938–10,954.

Bowen, A.J. (1969) The generation of longshore currents on a plane beach, *Journal of Marine Research* 27, 206–214.

——(1980) Simple models of nearshore sedimentation: beach profiles and longshore bars, in S.B. McCann (ed.) *The Coastline of Canada*, 1–11, Ottawa: Geological Survey of Canada.

Carter, R.W.G. (1988) *Coastal Environments. An Introduction to the Physical, Ecological and Cultural Systems of Coastlines*, London: Academic Press.

Grant, W.D. and Madsen, O.S. (1979) Combined wave current interaction with a rough bottom, *Journal of Geophysical Research* 84, 1,797–1,808.

Greenwood, B. and McGillivray, D.G. (1980) Modelling the impact of large structures upon littoral transport in the Central Toronto waterfront, Lake Ontario, Canada, *Zeitschrift für Geomorphologie Supplementband* 34, 97–110.

Inman, D.L. and Bagnold, R.A. (1963) Littoral processes, in M.L. Hill (ed.) *The Sea: Ideas and Observations*, Vol. 3, 529–543, New York: Wiley Interscience.

Komar, P.D. (1971) The mechanics of sand transport on beaches, *Journal of Geophysical Research* 76, 713–721.

——(1998) *Beach Processes and Sedimentation*, Upper Saddle River, NJ: Prentice Hall.

Longuet-Higgins, M.S. and Stewart, R.W. (1964) Radiation stresses in water waves: a physical discussion with applications, *Deep Sea Research* 11, 529–562.

Sternberg, R.W., Shi, N.C. and Downing, J.P. (1989) Suspended sediment measurements, in R.J. Seymour (ed.) *Nearshore Sediment Transport*, 231–257, New York: Plenum.

Van Wellen, E., Chadwick, A.J. and Mason, T. (2000) A review and assessment of longshore transport equations for coarse-grained beaches, *Coastal Engineering* 40, 243–275.

Further reading

Bayram, A., Larson, M., Miller, H.C. and Kraus, N.C. (2001) Cross-shore distribution of longshore sediment transport: comparison between predictive formulas and field measurements, *Coastal Engineering* 44, 79–99.
Komar, P.D. and Inman, D.L. (1970) Longshore sand transport on beaches, *Journal of Geophysical Research* 75, 5,914–5,927.
Longuet-Higgins, M.S. (1970a) Longshore currents generated by obliquely incident waves, 1, *Journal of Geophysical Research* 75, 6,778–6,789.
——(1970b) Longshore currents generated by obliquely incident waves, 2, *Journal of Geophysical Research* 75, 6,790–6,801.
——(1972) Recent progress in the study of longshore currents, in R.E. Meyer (ed.) *Waves on Beaches and Resulting Sediment Transport*, New York: Academic Press.
Van Rijn, L.C. (1998) *Principles of Coastal Morphology*, Amsterdam: Aqua Publications.

BRIAN GREENWOOD

LUNETTE

Transverse and roughly concentric aeolian accumulations (the *bourrelets* of some French workers) that occur on the downwind margins of PANs. Although they had been described before, they were named as such in Australia by Hills (1940), though the basis of his etymology is unclear. They tend to occur in areas where present-day precipitation levels are between about 100 and 700 mm, but their stratigraphy can give a good indication of past changes in climate and hydrological conditions. Some basins may have two or more lunettes on their lee sides and these may have different grain size and mineralogical characteristics. Lunettes may be some kilometres long and in exceptional circumstances may attain heights in excess of 60 m.

Good regional descriptions of lunettes are provided for the High Plains of the USA by Holliday (1997), for Tunisia by Perthuisot and Jauzein (1975), for the Kalahari by Lancaster (1978) and for the Pampas of Argentina by Dangavs (1979).

The materials that make up lunettes can vary from clay-sized material (which in the case of clay dunes (Bowler 1973) can make up 30 to 70 per cent of the total) through to sand-sized material. Equally some lunettes are carbonate-rich (Goudie and Thomas 1986) whereas others are almost pure quartz. Lunettes may also contain appreciable quantities of evaporite minerals derived from the basins to their windward.

Various hypotheses have been put forward to explain lunette composition. Hills (1939) believed that the lunettes were built up when the pans contained water and that they were composed of atmospheric dust captured by spray droplets derived from the water body. Stephens and Cocker (1946) pointed out that this could not account for those lunettes that were not predominantly silty. They also suggested that many of the lunettes were built up of aggregates transported from the floors of pans. Campbell (1968) believed that this deflation hypothesis could indeed account for many lunette features. As she remarked (p. 104) 'the close similarity between the composition of the lunette and its associated lake bed suggested that the two are causally related, i.e. that the material in the lunette was derived from the lake bed'. However, she also recognized that some of the material could be derived from wave-generated beaches and so could be analogous to primary coastal foredunes.

This was a view that was developed by Bowler (1973), who saw sandy facies as being associated with a beach provenance (at times of relatively high water levels) whereas clay-rich facies, which also may have a high content of evaporite grains, formed during drier phases when deflation of the desiccated lake floor was possible. Lunettes can therefore provide evidence for understanding past hydrological changes (Page *et al.* 1994).

References

Bowler, R.M. (1973) Clay dunes: their occurrence, formulation and environmental significance, *Earth-Science Reviews* 9, 315–338.
Campbell, E.M. (1968) Lunettes in South Australia, *Transactions of the Royal Society of South Australia* 92, 83–109.
Dangavs, N.V. (1979) Presence de dunes de argilla fosiles en La Pampa Deprimada, *Revista Association Geologica Argentina* 34, 31–35.
Goudie, A.S. and Thomas, D.S.G. (1986) Lunette dunes in Southern Africa, *Journal of Arid Environments* 10, 1–12.
Hills, E.S. (1939) The physiography of north-western Victoria, *Proceedings of the Royal Society of Victoria* 51, 297–323.

Hills, E.S. (1940) The lunette: a new landform of aeolian origin, *Australian Geographer* 3, 1–7.
Holliday, V.T. (1997) Origin and evolution of lunettes on the High Plains of Texas and New Mexico, *Quaternary Research* 47, 54–69.
Lancaster, I.N. (1978) The pans of the Southern Kalahari, Botswana, *Geographical Journal* 144, 80–98.
Page, K., Dare-Edwards, A., Nanson, G. and Price, D. (1994) Late Quaternary evolution of Lake Urana, New South Wales, Australia, *Journal of Quaternary Science* 9, 47–57.
Perthuisot, J.P. and Jauzein, A. (1975) Sebkhas et dune d'argile: l'enclave endoréique de Pont du Fahs, Tunisie, *Revue de Géographie Physique et de Géologie Dynamique* 17, 295–306.
Stephens, C.G. and Crocker, R.L. (1946) Composition and genesis of lunettes, *Transactions of the Royal Society of South Australia* 70, 302–312.

A.S. GOUDIE

M

MAGNITUDE–FREQUENCY CONCEPT

When the activity of different geomorphic processes is compared on a given timescale some processes appear to operate continuously while others operate only when specific conditions occur (referred to as events). The term eposidicity refers to the tendency of processes to exhibit discontinuous behaviour and occur sporadically as a series of individual events. Episodicity occurs when discontinuity is inherent in the forcing process (e.g. discontinuous rainfall produces episodic gully erosion). It may also occur if the relationship between the forcing process and the geomorphic response is not constant (e.g. the strain resulting from continuous crustal plate convergence may be manifest as either continuous deformation or episodic earthquake activity and coseismic uplift). The point at which the resistance to stress imposed by a forcing process is overcome is marked by a discernible geomorphic response (event) and is referred to as a threshold (see THRESHOLD, GEOMORPHIC). Because of hysteresis effects, the initiating threshold for a geomorphic event may be of a different magnitude to the terminating threshold.

Like many concepts in geomorphology, interpretations of eposidicity are scale dependent. Even processes that are sometimes considered continuous can be interpreted as episodic on different timescales. For example, SOIL CREEP in a given area is commonly portrayed as continuous and ubiquitous in terms of landform evolution. However, on a diurnal or seasonal scale, some forms of soil creep are evidently episodic, operating only when certain temperature or moisture conditions are met. For a given process, large events involving high amounts of concentrated energy (high magnitude) are rare (low frequency) and small events are common and tend to be less episodic.

Historically geomorphology has gone through a number of major paradigm shifts involving the eposidicity and the effectiveness of processes. The earliest interpretations of how landforms were created were based on catastrophic formative supernatural or natural events unrelated to the work of contemporary processes. As the science of geomorphology developed, increasing recognition of the age of the Earth allowed for the possibility that slow acting, contemporary processes, with sufficient time, could account for the development of much of the form evident in the landscape. Paradoxically the ability to date accurately key deposits revealed that certain landforms had been formed under specific and unusual circumstances. This gave rise to the prospect of NEOCATASTROPHISM. The consequential question that heightened interest in the study of magnitude–frequency relations was: were the effects of a rare catastrophic event 'permanent' or, given time, were they overwhelmed by 'normal', small but constantly operating processes responsible for producing a characteristic form? The prospect of identifying a 'characteristic' form in a geomorphic system however requires stability in boundary conditions such as climate, tectonic activity and vegetation cover and thus the existence of some form of DYNAMIC EQUILIBRIUM. Clearly the shorter the period of observation the more likely it is that these conditions pertain.

From the point of view of landform development, the RELAXATION TIME associated with a geomorphic event may be a more appropriate parameter than magnitude for identifying the formative or geomorphically effective event. The relaxation time is the length of time over which the effects of an event can be discerned in the landscape. Although this is clearly a function of event magnitude it also incorporates the influence of terrain resilience, LANDSCAPE SENSITIVITY and ambient conditions (Crozier 1999). The degree of equilibrium and the prospect of identifying characteristic form can then be determined by the 'transient form ratio' (Brundsen and Thornes 1979) which is the ratio of relaxation time of a geomorphic event to its recurrence interval (frequency). Values equal or greater than one indicate a constantly changing system while values less than one represent some form of dynamic equilibrium.

Many studies on this question, including the seminal work by Wolman and Miller (1960) have drawn conclusions based only on the period of instrumental record and the extrapolation of such results to landform evolution must be done with care (Wolman and Gerson 1978). In Wolman and Miller's original study it was shown that most of the work of sediment transportation in rivers was carried out by moderate flow events of a magnitude that recurs on average at least once every five years. The analytical approach leading to this conclusion is illustrated in Figure 105. Curve (a) indicates that the rate of sediment movement is a power function of applied stress, or in this case a surrogate such as discharge. Curve (b) shows a log normal frequency–magnitude distribution of measured flow events. Curve (a × b) is the product of frequency and rate of movement which attains a maximum work value which can be related back to the discharge axis at (x) to identify the magnitude of the event performing the most work. The frequency of this event (y) can be determined with reference to the frequency–magnitude distribution curve. Wolman and Miller's work has been extended to a number of different processes and environments, in some cases producing conclusions at variance with the original findings. Selby (1974), for example, argues that in high-energy hillslope environments and the headwaters of drainage basins, high magnitude landslide events occurring at frequencies less than once every five years dominate geomorphic activity on slopes and low-order stream channels. Similar conclusions have been made by a comprehensive comparison of fluvial transport and mass movement in different parts of the drainage basins (Trustrum *et al.* 1999).

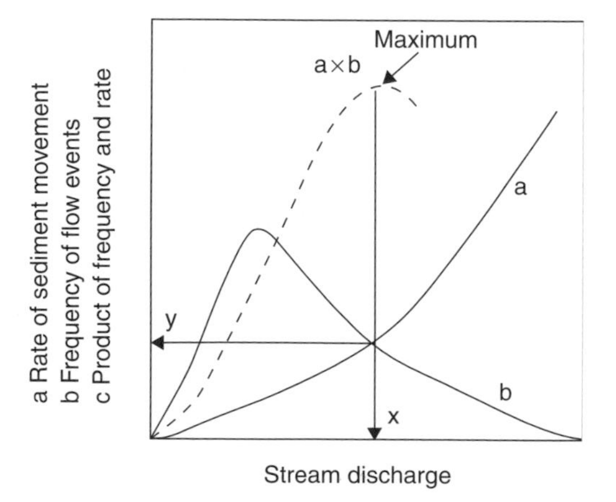

Figure 105 Relations between rate of sediment transport, discharge and the frequency–magnitude distribution of discharge events, c= a × b (after Wolman and Miller 1960)

In strictly geomorphic terms, the magnitude of an event generally refers to the amount of work carried out (e.g. mass of sediment transported) or the degree of landform change experienced (geomorphic response). However, because of difficulty in directly measuring geomorphic events many magnitude–frequency studies are approached indirectly. This involves characterizing events from the behaviour of the forcing agent, rather than from the geomorphic event itself. However, the indirect approach implies a known relationship between geomorphic work and the behaviour of the forcing agent. For example, if this relationship and the initiating thresholds are known, the frequency and magnitude of soil erosion and landsliding can be determined from the rainfall record, aeolian transport from wind speed, coastal changes from the wave regime and fluvial transport from the streamflow regime.

A major problem with the indirect approach is that relationships between the forcing process and geomorphic work are not always close or temporally or spatially stable. For example, stream discharge may not relate closely to sediment transport in a supply-constrained system. Indeed in some streams, sediment load may be related to stream power at certain times and to rates of hillslope sediment supply at others. The nature of sediment control can also change throughout the catchment to the extent that sediment load in steep upper catchments may relate to the magnitude and frequency of landslide events. Furthermore, the parameter used to represent magnitude also needs to be carefully chosen if it is to reflect accurately and consistently the energy of the forcing process. In the case of landsliding for example, the groundwater content of a hillslope is directly responsible for initiating movement, however lack of adequate databases dictates that magnitude–frequency analysis is often carried out using rainfall values; a more appropriate analysis would involve the magnitude and frequency of groundwater levels. The use of arbitrarily defined events such as daily rainfall as opposed to event rainfall introduces another level of inaccuracy into the analysis, as important parameters such as actual intensity and duration are obscured. Besides magnitude and frequency, duration is another parameter of the forcing processes that influences geomorphic response. Clearly, in power-constrained systems, high effective streamflow, maintained for long periods of time, will accomplish more work than the same levels of flow occurring as short duration events. Similarly, studies of the relationship between the magnitude and intensity of earthquakes and the degree of land deformation have shown that the duration of shaking is important in evoking a geomorphic response.

For a given process, average event frequencies can be established from the number of times an event initiating threshold is exceeded or the number of times the threshold is exceeded by a specified magnitude, in a unit of time. More commonly, following the practice of flood frequency analysis, frequencies are expressed as recurrence intervals (return periods) determined by the relationship of the length of record to the ranking or relative magnitude of the event in a series of events. A range of statistical models for extreme value probability distributions can be used for depicting the declining exponential function between magnitude and frequency. Little confidence can be placed on derived event magnitudes with recurrence intervals greater than the length of record.

An important property of recurrence intervals (sometimes overlooked when regional comparisons are made) is that they are generally a function of the size of the area from which they are derived. In other words, the larger the area, all other things being equal, the more likely it will be to experience a specified event and therefore the shorter the derived recurrence interval. Another point of caution is that the statistical derivation of frequency may obscure the clustering of events, which is a behavioural property that can be significant in generating a geomorphic response. Because variation of frequency through time may signal significant environmental change the identification of clustering is an important aspect of magnitude–frequency analysis.

For some processes, magnitude and frequency of occurrence in space (Innes 1985; Hovius *et al.* 1997) mirrors temporal magnitude–frequency relationships. In the case of landslides for example, episodes producing numerous landslides on the one occasion show that large landslides are greatly outnumbered by smaller landslides. These spatial magnitude–frequency relationships are used in hazard assessments as analogues for temporal magnitude–frequency relationships. Establishing magnitude–frequency relationships from landscape evidence needs to take account of the longevity of landscape evidence; clearly the signature of large landslides persists longer than that of smaller landslides.

The magnitude–frequency concept and the analysis it has provoked have been important in informing the discipline about the variability and behaviour of geomorphic processes. It reminds us that within one lifetime we are unlikely to experience the whole range in magnitudes that a particular process is capable of generating. The concept provides a rationale for extrapolating short-term measurements over longer periods, as a way of assessing the long-term rates of geomorphic processes. However changes in boundary conditions limit the extent to which the relationships can be extrapolated. Magnitude–frequency analysis also provides a method for statistically identifying the key events in terms of work and landform response, thereby providing a key variable for characterizing geomorphic systems and predicting other system characteristics. Finally, from a pragmatic point of view, the concept enables the identification of a design or planning event for use in engineering decisions and hazard management.

References

Brunsden, D. and Thornes, J.B. (1979) Landscape sensitivity and change, *Transactions Institute British Geographers* 4, 463–484.

Crozier, M.J. (1999) The frequency and magnitude of geomorphic processes and landform behaviour, *Zeitschrift für Geomorphologie N.F. Supplementband* 115, 35–50.

Hovius, N., Stark, C.P. and Allen, P.A. (1997) Sediment flux from a mountain belt derived from landslide mapping, *Geology* 25, 231–234.
Innes, J.L. (1985) Magnitude–frequency relations of debris flows in northwest Europe, *Geografiska Annaler* 67A, 23–32.
Selby, M.J. (1974) Dominant geomorphic events and landform evolution, *Bulletin International Association Engineering Geologists* 9, 85–89.
Trustrum, N.A., Gomez, B., Page, M.J., Reid, L.M. and Hicks, D.M. (1999) Sediment production and output: the relative role of large magnitude events in steepland catchments, *Zeitschrift für Geomorphologie N.F. Supplementband* 115, 71–86.
Wolman, M.G. and Gerson, R. (1978) Relative scales of time and the effectiveness of climate in watershed geomorphology, *Earth Surface Processes and Landforms* 3, 189–208.
Wolman, M.G. and Miller, J.P. (1960) Magnitude and frequency of forces in geomorphic processes, *Journal of Geology* 68, 54–74.

Further reading

Crozier, M.J. and Mausbacher, R. (eds) (1999) Magnitude and frequency in geomorphology, *Zeitschrift für Geomorphologie N.F. Supplementband* 115.
Selby, M.J. (1993) *Hillslope Materials and Processes*, 2nd edition, Oxford: Oxford University Press.

MICHAEL J. CROZIER

MANAGED RETREAT

Usually defined as the backward realignment of coastal defences so as to allow the reformation of SALTMARSH or mud flat (see MUD FLAT AND MUDDY COAST) on the seaward side. It may also, rarely, be used where other coastal features such as dunes are expected to form in front of a retreating line of defence. The term has now largely been superseded by the preferred phrase '*managed realignment*' which is effectively synonymous but avoids the negative implications to many people of the word 'retreat'.

In low-lying tidally dominated coastal areas, especially in northern Europe and North America, large areas of saltmarsh have been 'reclaimed' by the building of sea walls for agricultural use or for urban and industrial development. Such land claim alters the dynamics of the local tidal flow and sediment budget and may lead to localized increase in tidal energy and to EROSION. Reclaimed areas are often in estuaries; the natural funnel shape of an ESTUARY is transformed to a narrower more parallel one providing less space for the dissipation of tidal energy. The tidal flow tends to move further upriver, the river channel deepens and widens and the mudflat/saltmarsh complex migrates also, maintaining its relation to the local energy gradient (Pethick 2000). Although seawalls were often originally positioned so as to leave in front some marsh and mudflat, these often become eroded away. High tides come up to the sea defences and the threat of flooding at spring tides and during storms increases; the walls themselves may be destroyed. If also SEA LEVELS are rising, the effect is magnified.

Appropriate responses to the threat depend on the type of infrastructure in the hinterland. Raising and strengthening seawalls, while expensive, may be thought the best option if valuable built-up or agricultural land is protected. Where the hinterland consists of low value agricultural land, managed retreat is seen as the best solution to the erosion problem. Reinstatement of the coastal marshes should provide a dispersal area for the tidal flow, a supply of sediment for local coastal processes, and a buffer zone for dispersion of wave and tide energy. Given time and space, it is hoped that the coastline will develop its own contours and maintain itself, altering to suit changing conditions without the need for expensive engineering works.

However, breaching or removal of seawalls does not always lead to the reestablishment of saltmarsh. Some low-level sites inundated either through storm damage or by the deliberate realignment of walls have remained as largely unvegetated mudflats (French *et al.* 2000) over a period of as much as fifty years. While popular with environmentalists and relatively inexpensive compared with hard engineering solutions, managed realignment is less attractive to the owners of the land to be sacrificed.

The success of realignment programmes has been shown to depend on several factors. It appears only to be successful on sites where saltmarsh once existed and where marshes survive nearby as sources of suitable plant propagules. There should be traces of the original saltmarsh creek system (though a channel system may initially be artificially created). The site should be sufficiently high that it is only inundated 400 to 450 times annually – in practice this means above 2.1 m OD or only mud flat is likely to form. It should preferably be neither completely flat nor steeply sloping but with a gentle land to sea slope (Burd 1995).

At least in the EU, managed realignment schemes can attract funding under the saltmarsh option of the Countryside Stewardship scheme (which replaced the Habitat Scheme in January 2000). Detailed Coastal Habitat Management Plans need to be drawn up in all cases before action is taken.

References

Burd, F. (1995) *Managed Retreat: A Practical Guide*, Peterborough: English Nature.

French, C.F., French, J.R., Clifford, N.J. and Watson, C.J. (2000) Sedimentation-erosion dynamics of abandoned reclamation; the role of waves and tide, *Continental Shelf Research* 20 (12–13), 1,711–1,733.

Pethick, J. (2000) Coastal management and sea level rise, *Catena* 42, 307–322.

Further reading

French, P.W. (2001) *Coastal Defences: Processes, Problems and Solutions*, London: Routledge.

PIA WINDLAND

MANGROVE SWAMP

Mangroves are trees or shrubs which grow in sheltered, or low-energy, upper intertidal zones in the tropics and subtropics, replacing SALTMARSH which is found along muddy temperate shorelines. There are more than fifty species of mangroves occurring in two distinct provinces, the Indo-west Pacific which contains the greatest diversity of species, and the West Indian region with far fewer species. In the West Indies, three species of mangrove occur, each tending to occupy a discrete location, *Rhizophora mangle* to seaward, *Avicennia germinans* in more landward locations, and *Laguncularia racemosa* mixed with the other species, or in areas that have been disturbed. In the Indo-west Pacific province there are more species of mangroves, with up to thirty species in the most diverse locations. There is generally a seaward zone, composed of species of *Avicennia* or *Sonneratia*, an intermediate zone dominated by species of *Rhizophora* and *Bruguiera*, and more landward zones in which *Ceriops* occurs with genera such as *Lumnitzera*, *Heritiera* and stunted forms of *Avicennia*. In areas of heavy rainfall mangrove forests merge into tropical forests, whereas in arid regions they often contain a hypersaline mudflat and are flanked by bare or samphire-covered supratidal flats.

Mangroves can grow in freshwater, but they appear to have a competitive advantage over other vegetation when the substrate is saline. Mangroves perform best at salinities that are less than that of seawater, but can survive in salinities of up to 90 parts per thousand. They show a series of adaptations to saline sediments. The root systems of mangroves contain breathing cells called lenticels, and are often distinctive, enabling the plants to survive in upper intertidal habitats which are subject to frequent inundation and are composed of muds or sandy muds which are anaerobic. The enormous prop root systems that characterize *Rhizophora* (Plate 75) are almost impenetrable. *Avicennia* and *Sonneratia* have pencil-like roots termed pneumatophores, whereas other species have knee or buttress roots. In addition many species of mangroves produce viviparous seedlings, the fruit already producing a well-developed root before falling from the tree.

These adaptations to the inhospitable environment in which mangroves live, and the zonation of species which is apparent on many shorelines, led to an early interpretation that mangroves undergo a succession of species culminating in a terrestrial climax vegetation. The root systems slow the movement of water and may trap sediment promoting accretion along the shore. There have been various attempts to measure sedimentation rates beneath mangrove forests. Direct measurements have been made by using systems

Plate 75 The interior of a mangrove swamp in the West Indies. The mangrove *Rhizophora mangle* has a network of prop roots that extend two or more metres up the trunks of the trees. These root systems can substantially reduce the flow of water across the substrate encouraging deposition of mud where suspended in the water column. In other cases, such as the forest shown here, the fibrous roots themselves contribute to a mangrove peat substrate

of stakes throughout the mangroves, or placing a marker layer. These approaches have generally met with less success than in saltmarshes because of the active bioturbation of the muds by a rich and diverse fauna, including crabs and mud skippers. Bioturbation also limits the efficacy of determining sedimentation rates using ^{210}Pb or ^{137}Cs isotopes, but rates of up to 3 mm yr^{-1} are indicated (Lynch *et al.* 1989). Longer term sedimentation rates have been determined using radiocarbon dating and indicate that rates of up to 6–8 mm yr^{-1} can occur. However, the short-term and long-term estimates may not be directly comparable because it is often difficult to discriminate whether root material was formed close to, or at some depth below, the surface. It is also difficult to take compaction into account, or to allow for the variation in sedimentation rate that is likely across the intertidal zone.

Zonation of mangrove species, even if it does occur, need not indicate a temporal succession that involves replacement of successive zones through time. Patterning of species may be a static equilibrium in relation to environmental gradients in habitat factors, such as salinity or water-logging (Lugo 1980). It has proved useful to distinguish various geomorphologically defined habitats within which mangroves grow (Thom *et al.* 1975; Semeniuk, 1985).

There is considerable variation both in the mangrove species that are found, and the elevation at which they grow, up estuaries. Particularly extensive mangrove forests occur in the abandoned parts of deltas where tidal processes are important, such as the Sundarbans west of the mouth of the Ganges–Brahmaputra rivers. Where mangrove swamps are associated with the mouths of large rivers, the substrate on which they are established is generally composed of terrigenous sediments washed from the catchments. The patterning of mangrove species in such deltaic environments reflects an ever-changing series of geomorphological habitats in which mangroves respond opportunistically to habitat change induced by geomorphological processes, such as channel migration and avulsion (Thom 1967).

Mangroves also occur in reef environments where their extent is often a function of stage of reef evolution (Stoddart 1980). In these settings, mangroves either establish over calcareous sediments formed from reef organisms, or they develop over mangrove peat which is derived from the roots of the mangroves themselves. Mangrove islands within the Belize barrier reef, termed mangrove ranges (Plate 76), are complex. Although these islands are dominated by tidal overwash, the patterning of species is not straightforward, with *Rhizophora* adopting a dwarfed scrub form in some locations, and elsewhere reaching a woodland of several metres tall, intergrading into areas of more open *Avicennia* woodland. In places the mangrove forests are dissected by sinuous creeks, and elsewhere there are bare areas which may result from storm damage, but where soil chemistry presently appears to prevent recolonization by mangroves.

It is clear that mangrove swamps have altered in extent considerably. The stratigraphy of mangrove-dominated coasts records the Holocene evolution of coastal environments and there have been significant adjustments as a result of changes in sea level. Throughout much of the Indo-west Pacific region, where sea level achieved a level close to present around 6,000 years ago and has been relatively stable or fallen slightly since, former mangrove sediments often underlie extensive Holocene plains upon which freshwater wetlands or peat swamp forest have established (Woodroffe *et al.* 1985). Former mangrove sediments often represent potential acid-sulphate soils in which pyrite can become oxidized, resulting in extremely acidic waters if drained or excavated. In the West Indies and the Everglades of Florida, where the history of relative SEA LEVEL change through the Holocene appears to have

Plate 76 A mangrove island, called a mangrove range, on the Belize barrier reef. Although dominated by two mangrove species, *Rhizophora mangle* and *Avicennia germinans*, there is a complex vegetation pattern with areas that are bare of vegetation and a network of creeks that dissect the island

been characterized by rise to present, intertidal mangrove peat overlies previously terrestrial environments, such as freshwater sedge peat.

These productive and protective forests are also likely to undergo changes in distribution as a result of sea-level change in the future (Woodroffe 1990). Although this has attracted considerable attention, it is the case that human impact, clearing forests for other land use, or for timber, and most recently for shrimp farming, have generally already had far-reaching impacts. Mangroves are subject to various natural disturbances, for instance storm impact. However, they play an important role as storm protection, especially where surges are experienced as in the Bay of Bengal.

References

Lugo, A.E. (1980) Mangrove ecosystems: successional or steady state? *Biotropica (Supplement)* 12, 65–95.

Lynch, J.C., Meriwether, J.R., McKee, B.A., Vera-Herrera, F. and Twilley, R.R. (1989) Recent accretion in mangrove ecosystems based on ^{137}Cs and ^{210}Pb, *Estuaries* 12, 284–299.

Semeniuk, V. (1985) Development of mangrove habitats along ria shorelines in north and northwestern tropical Australia, *Vegetatio* 60, 3–23.

Stoddart, D.R. (1980) Mangroves as successional stages, inner reefs of the northern Great Barrier Reef, *Journal of Biogeography* 7, 269–284.

Thom, B.G. (1967) Mangrove ecology and deltaic geomorphology: Tabasco, Mexico, *Journal of Ecology* 55, 301–344.

Thom, B.G., Wright, L.D. and Coleman, J.M. (1975) Mangrove ecology and deltaic-estuarine geomorphology, Cambridge Gulf-Ord River, Western Australia, *Journal of Ecology* 63, 203–222.

Woodroffe, C.D. (1990) The impact of sea-level rise on mangrove shorelines, *Progress in Physical Geography* 14, 483–520.

Woodroffe, C.D., Thom, B.G. and Chappell, J. (1985) Development of widespread mangrove swamps in mid-Holocene times in northern Australia, *Nature* 317, 711–713.

Further reading

Alongi, D. and Robertson, A. (eds) (1992) *Tropical Mangrove Ecosystems*, Washington, DC: American Geophysical Union, Coastal and Estuarine Studies.

Hogarth, P.J. (1999) *The Biology of Mangroves*, Oxford: Oxford University Press.

COLIN WOODROFFE

MANTLE PLUME

'A blob of relatively hot, low-density mantle that rises because of its buoyancy' (Condie 2001: 1). Their existence was suggested by J.T. Wilson (1963) who sought to explain the progressive change in age and form of islands along oceanic chains, such as the Hawaiian–Emperor Chain of the Pacific. He proposed that as a lithospheric plate moves across a fixed hotspot (the mantle plume), volcanic activity is recorded as a linear array of volcanic seamounts and islands parallel to the direction in which the plate moves (Plate 77).

As rising mantle plumes reach the base of the lithosphere, they spread laterally to produce plume heads which may have diameters that reach 500 to 3,000 km. The tails of the plumes are only 100–200 km in diameter. The surface manifestations of plumes are large hotspots, which are zones with active volcanism. Those plumes that are 1,500–3,000 km in diameter are termed superplumes (Condie 2001: 2).

Mantle plumes are of fundamental geomorphological importance. First, their associated hotspots help to explain the distribution of volcanic activity in intraplate situations. They also explain the development of volcanic chains on the Pacific plate and elsewhere, and also the development of some seamounts. Second, broad zones of uplift (swells) are sometimes associated with mantle plumes. These can create large domes that then dominate drainage patterns at a regional and subcontinental scale. Cox (1989) argued that the drainage directions and patterns of India, eastern South America and southern Africa were good examples of this effect (Figure 106). Third, many large igneous provinces (LIPs) have a mantle

Plate 77 The Spitzkoppje in central Namibia is a granite mass that was emplaced in the early Cretaceous as a result of magmatic activity associated with the presence of a mantle plume that contributed to the rifting of Gondwanaland. The hotspot is now located in the Southern Ocean in the vicinity of Gough and Tristan da Cunha

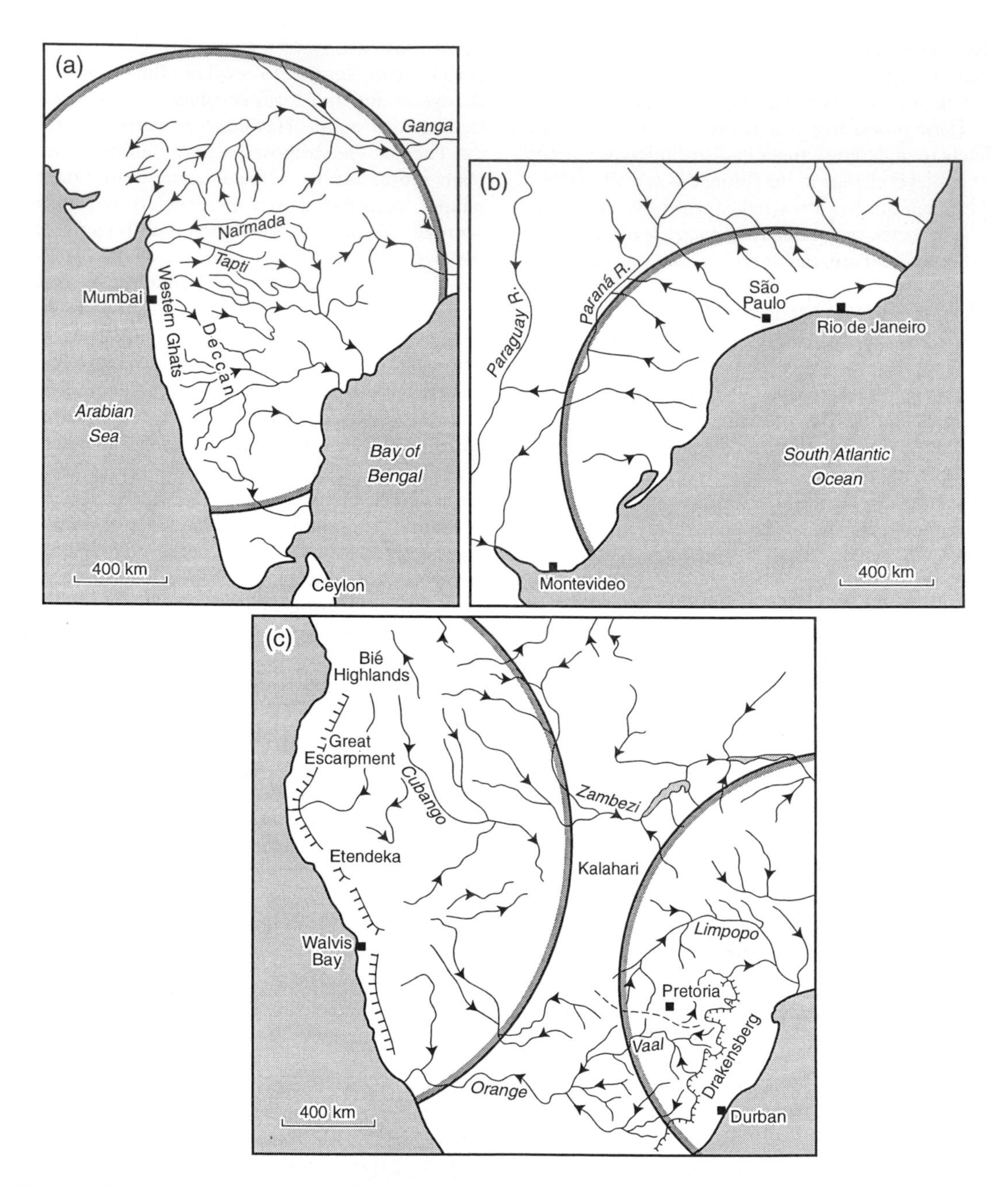

Figure 106 Postulated locations of major domes associated with mantle plumes and showing their relationship to drainage patterns (modified after Cox 1989)

plume origin, and phenomena associated with them include such features as continental flood basalts (e.g. in the Deccan of India), giant dyke and sill swarms (Ernst *et al.* 1995), and large layered intrusions (e.g. the Bushveld Igneous Complex in South Africa). Fourth, mantle plumes may play a major role in the breakup of supercontinents (e.g. Gondwanaland), the formation of passive margins like those in Namibia (Goudie and Eckardt 1999), and the development of basins (e.g. Red Sea, Gulf of Aden, etc.).

References

Condie, K.C. (2001) *Mantle Plumes and their Record in Earth History*, Cambridge: Cambridge University Press.

Cox, K.G. (1989) The role of mantle plumes in the development of continental drainage patterns, *Nature* 342, 873–877.

Ernst, R.E., Head, J.W., Parfitt, E., Grosfils, E. and Wilson, L. (1995) Giant radiating dyke swarms on Earth and Venus, *Earth-Science Reviews* 39, 1–58.

Goudie, A.S. and Eckardt, F. (1999) The evolution of the morphological framework of the Central Namib Desert, Namibia, since the early Cretaceous, *Geografiska Annaler* 81A, 443–458.

Wilson, J.T. (1963) A possible origin of the Hawaiian islands, *Canadian Journal of Physics* 41, 863–868.

A.S. GOUDIE

MASS BALANCE OF GLACIERS

The mass balance of a glacier is the sum of all processes which add mass to a glacier and remove mass from it. Accumulation or addition of mass most commonly takes place in the form of snowfall, often modified by wind and avalanches (see AVALANCHE, SNOW). Melting of snow and ice is the predominant form of ablation or removal of mass, but calving of tidal glaciers, ice avalanching at steep hanging glaciers or evacuation of wind-blown snow in dry areas can locally be of high relative importance. The resulting loss or gain of mass is the direct, undelayed response of a glacier to climatic conditions and must be considered to be a key indicator of climate change (IPCC 2001; Haeberli *et al.* 2002).

Techniques applied to determine glacier mass balance include (a) the direct glaciological method (snowpits and ablation stakes), the geodetic/photogrammetric method (repeated precision mapping), the hydrological method (difference between measured precipitation minus evaporation and runoff) and index methods (snowline mapping, etc.). Long-term mass changes can also be inferred from cumulative glacier length changes using continuity approaches or flow models (Haeberli and Hoelzle 1995; Oerlemans *et al.* 1998).

In humid maritime regions large amounts of ablation are required to compensate for heavy snowfall: the equilibrium line which separates accumulation from ablation areas on glaciers remains at altitudes with relatively warm air temperatures, enabling intense sensible heat flux and strong ice melt during extended ablation seasons. Temperate glaciers at melting temperature, exhibiting high mass turnover and rapid flow, dominate these landscapes. The lower parts of such temperate glaciers commonly extend into grassland and forested valleys where summer warmth and winter snow accumulation prevent the development of PERMAFROST. Ice caps and valley glaciers of Patagonia and Iceland, the western Cordillera of North America and the coastal mountain chains of New Zealand and Norway are features of this type. In contrast, dry continental conditions such as exist in northern Alaska, arctic Canada, subarctic Russia, parts of the Andes near the Atacama desert or in many central Asian mountain chains, force the EQUILIBRIUM LINE OF GLACIERS to elevations with cold air temperatures, short ablation seasons, reduced sensible heat flux and limited amounts of ice melting. In such regions, polythermal or cold glaciers, lying far beyond the treeline and often even beyond tundra vegetation, have a low mass turnover, less rapid flow and are associated with severe periglacial conditions and permafrost (Shumskii 1964).

The goals of long-term mass balance observations are (1) to determine annual ice loss/gain as a regional signal and (2) to better understand processes of energy and mass exchange. A change (δ) in equilibrium line altitude (ELA) induces an immediate change in specific mass balance (b = total mass change divided by glacier area). The resulting change in specific mass balance (δb) is the product of the shift in equilibrium line altitude (δELA) and the gradient of mass balance with altitude (db/dH) as weighed by the distribution of glacier surface area with altitude (hypsometry). The hypsometry represents the local/individual or topographic part of the glacier sensitivity whereas the mass balance gradient mainly reflects the regional or climatic part (Kuhn 1990). As the mass balance gradient tends to increase with increasing humidity and mass turnover (Kuhn 1981), the sensitivity of glacier mass balance with respect to changes in equilibrium line altitude is generally much higher in areas with humid/maritime than with dry/continental climatic conditions (Oerlemans 1993). Rising snowlines and cumulative mass losses lead to changes in average albedo and continued surface lowering. Such effects cause pronounced positive feedbacks with respect to radiative and sensible heat fluxes. In areas of cold firn, atmospheric warming first induces firn warming; mass loss only sets in when the firn has reached melting temperatures and water can leave the system instead of refreezing.

A Global Terrestrial Network for Glaciers as part of the climate-related Global Terrestrial Observing System (GTOS/GCOS) is operated by the World Glacier Monitoring Service (WGMS) which co-ordinates worldwide compilation and dissemination of standardized data on glacier fluctuations. Mass balance measurements are reported in a biennial bulletin (IAHS(ICSI)/UNEP/UNESCO/WMO 2001; http://www.geo.unizh.ch/wgms/). Overall ice loss appears to be strong and probably even accelerating. Anthropogenic greenhouse forcing could have started to exert a predominant influence on this development and may lead to complete deglaciation of many mountain regions of the world within decades (Dyurgerov and Meier 1997a,b; Haeberli *et al.* 2002).

References

Dyurgerov, M.B. and Meier, M.F. (1997a) Mass balance of mountain and subpolar glaciers: a new global assessment for 1961–1990, *Arctic and Alpine Research* 29/4, 379–391.

Dyurgerov, M.B. and Meier, M.F. (1997b) Year-to-year fluctuations of global mass balance of small glaciers and their contribution to sea level, *Arctic and Alpine Research* 29/4, 392–402.

Haeberli, W. and Hoelzle, M. (1995) Application of inventory data for estimating characteristics of and regional climate-change effects on mountain glaciers: a pilot study with the European Alps, *Annals of Glaciology* 21, 206–212. Russian translation: *Data of Glaciological Studies* 82, 116–124.

Haeberli, W., Maisch, M. and Paul, F. (2002) Mountain glaciers in global climate-related observation networks, *WMO Bulletin* 51/1, 18–25.

IAHS(ICSI)/UNEP/UNESCO/WMO (2001) *Glacier Mass Balance Bulletin* No. 6 (Haeberli, W., Hoelzle, M. and Frauenfelder, R. (eds)), World Glacier Monitoring Service, University and ETH Zurich.

IPCC (2001) *Climate Change 2001 – The Scientific Basis*, Contribution of Working Group I to the Third Assessment Report of the Intergovernmental Panel on Climate Change, Cambridge: Cambridge University Press.

Kuhn, M. (1981) Climate and glaciers, *IAHS Publication* 131, 3–20.

——(1990) Energieaustausch Atmosphäre – Schnee und Eis, in *Schnee, Eis und Wasser der Alpen in einer wärmeren Atmosphäre*, Mitteilungen der Versuchsanstalt für Wasserbau, Hydrologie und Glaziologie 108, 21–32, ETH Zürich.

Oerlemans, J. (1993) A model for the surface balance of ice masses: part I, alpine glaciers, *Zeitschrift für Gletscherkunde und Glazialgeologie* 27/28, 63–83.

Oerlemans, J., Anderson, B., Hubbard, A., Huybrechts, P., Johannesson, T., Knap, W.H., Schmeits, M., Stroeven, A.P., van de Wal, R.S.W., Wallinga, J. and Zuo, Z. (1998) Modelling the response of glaciers to climate warming, *Climate Dynamics* 14, 267–274.

Shumskii, P.A. (1964) *Principles of Structural Glaciology*, trans. D. Kraus, New York: Dorer.

WILFRIED HAEBERLI

MASS MOVEMENT

A mass movement is the downward and outward movement of slope-forming material under the influence of gravity. The process does not require a transporting medium such as water, air or ice. The term LANDSLIDE often is used as synonymous for mass movement phenomena. However, in a pure sense the term landslide is used as a generic term describing those downward movements of slope-forming material as a result of shear failure occurring along a well-defined shear plane.

There are numerous classifications of mass movements. Most are based on morphology, mechanism, type of material and rate of movement (e.g. Varnes 1978; Hutchinson 1988; Cruden and Varnes 1996). The classification and description used in this entry was developed by a European research group (Dikau *et al.* 1996) (Table 27). The terminology is based upon the classifications of the International Geotechnical Societies' UNESCO Working Party on World Landslide Inventory (WP/WLI) (UNESCO 1993).

Fall

Alternative terms: rockfall, stone fall, pebble fall, boulder fall, debris fall, soil fall.

A fall is a free movement of material from steep slopes. Different types of falls are described by material and failure processes. The term rockfall is often used as the general term without further reference to the material involved. Whalley (1984) and Flageollet and Weber (1996) provide summaries.

Falls occur in various sites such as coastal cliffs, steep riverbanks, edges of plateau, mountain faces or escarpments. They may also occur on artificial embankments (road outcrops). Joints and faults produce planar sheets to form wedge-shaped hollows and boundary vertical joints. Falls can form a fan-shaped cone at the base of the slope. These

Table 27 Classification of slope movements

Type	Rock	Debris	Soil
Fall	Rockfall	Debris fall	Soil fall
Topple	Rock topple	Debris topple	Soil topple
Slide (rotational)	Single (slump) Multiple Successive	Single Multiple Successive	Single Multiple Successive
Slide (translational)	Block slide Rock slide	Block slide Debris slide	Slab slide Mudslide
Lateral spreading	Rock spreading	Debris spread	Soil (debris) spreading
Flow	Rock flow (sackung)	Debris flow	Soil flow
Complex (with runout or change of behaviour downslope, note that nearly all forms develop complex behaviour)	e.g. rock avalanche	e.g. flow slide	e.g. slump-earthflow

Source: Dikau *et al.* (1996)
Note: A compound landslide consists of more than one type e.g. rotational-translational slide. This should be distinguished from a complex landslide where one form of failure develops into a second form of movement i.e. a change of behaviour downslope by the same material

TALUS slopes should be distinguished from accumulations arising from a large fall, a complex rock avalanche (STURZSTROM), which result in an accumulation of debris of all sizes that can block a river, leading to flooding hazards.

Falls are influenced by slope aspect and angle, size and shape of jointed rocks, strike angle, status and deformation of rocks and vegetation cover. Debris and soil falls originate in material which has already been detached from the bedrock. In solid rock the separation process may take time and it arises from internal and external factors which are often combined.

The main implication of fall processes for a planner is to ensure that there are suitable surveys of those areas which are likely to produce rockfalls (Table 27). The main product will be a planning hazard zoning map and suitable monitoring systems. Warning systems are also in operation on various other sites with this type of hazard.

Topple

Alternative terms: rock topple, debris topple, soil topple, tilting blocks.

A topple consists of a forward rotation of a mass of rock, debris or soil about a pivot or hinge on a hillslope. The toppling may culminate in an abrupt falling or sliding, but the form of movement is tilting without collapse. Goodman and Bray (1976) and Dikau *et al.* (1996) provide summaries.

There are several processes responsible for toppling failure such as progressive weathering or erosion resulting in a weakening or loss of elastic underlying material, swelling and shrinking of clay-rich material because of soil moisture changes, and deepening or undercutting of slopes by erosion providing sufficient stress to cause unloading decompression.

The primary driving force for topple failure is the detachment of a column so that the load is transferred to a narrow base of weaker rock. Slope height is an important controlling parameter as is the width of the supporting base. Topples in rocks usually require high cliffs, whereas topples in debris and soil fail on lower cliffs. The formation of tension cracks is caused by severe undercutting by fluvial agents, sea wave action or man-made scarps. Joint and bedding plane water pressures are a vital contribution to failure at the base of the column.

Slide

The term 'slide' is used for a movement of material along a recognizable shear surface. The shear surface type and the number of shear surfaces are used to divide the slide group.

SLIDE (ROTATIONAL)

Alternative terms: slump, rotational slip, rotational slide.

Rotational slides occur as a rotational movement on a circular or spoon-shaped shear surface. They differ in the degree of disintegration in the slide masses and the depositional features in the toe areas. Varnes (1978) and Buma and van Asch (1996) provide summaries.

Varnes (1978) defines a single rotational slide or slump as a 'more or less rotational movement, about an axis that is parallel to the slope contours, involving shear displacement (sliding) along a concavely upward-curving failure surface, which is visible or may reasonably be inferred'. A rotational slide has a small degree of internal deformation although sometimes soil slump material liquefies and transforms into a flow at its toe. In the terminology the American usage 'earthflow' is replaced in European literature by 'mudslide'. Rotational slides can vary from an area of a few square metres to large complexes of several hectares.

Given the relatively small degree of internal deformation, the slump matrix will essentially be the same as the surrounding undisturbed slope. Soil slides generally consist of fine-textured, cohesive materials, like consolidated clays, weathered marls and mudstones. Rotational rock slides often develop in formations of interbedded strong and weaker materials, e.g. marls and limestones or sandstones. In general, rotational sliding produces a disrupted, anomalous drainage pattern. Ponds and peaty areas may develop in depressions between slump units.

Movement of rotational slides starts with initial failure followed by rotation. It may disintegrate into several discrete blocks. In the head area, these blocks tilt backwards while sliding downhill and often flattening or even slope reversal occurs. Sliding along the flanks causes longitudinal and diagonal shear stresses. The lowest part of the slump mass moves over the toe of the failure surface, thereby bulging, cambering, overriding and producing transverse tension cracks.

Movement rates of slumps can vary by several orders of magnitude, between a few centimetres a year to several metres per month, while soil slumps can attain velocities up to 3 metres per second. Tilted trees (generally backwards in the head area, forwards in the foot and toe areas) can reveal the presence of rotational slides.

Typical causal situations are undercutting by waves or streams, excavation and other construction activities. Common triggering mechanisms are earthquakes, explosions and sudden increases of overburden or high water tables following periods of rainfall or snowmelt.

SLIDE (TRANSLATIONAL)

A translational slide is a non-circular failure which involves translational motion on a near planar slip surface. The movement is largely controlled by surfaces of weakness within the structure of the slope-forming material. Translational slides may occur in three types of material: rock, debris and soil. Depending on the slope angle and the velocity, slides will either stay as a discrete block on the failure surface or break into debris.

Block slide

Alternative terms: planar rock slide (rock block slide), slab slide (common usage for soil/earth block slide).

Large block slides are often part of extensive compound landslides involving rotational slides either at the toe or the head and occasionally mudslides at the edges of the landslide. Due to the geometry of the slip surface the mass may only move through the development of internal shears and displacements. Hutchinson *et al.* (1991) and Ibsen *et al.* (1996b) provide summaries.

Block slides are found in stiff, fissured or overconsolidated clays often in combination with stronger rock formations. The basal surface needs only a very low-angle for displacement, because the driving forces are usually very large. Large block slides are particularly sensitive at the toe and sometimes require only slight toe erosion to trigger failure. Movements involve deep settlement at the head, whilst the margins of the block slide are generally totally destroyed.

Block slides can be initiated or reactivated where construction loads the slope or where excavation undercuts or unloads the toe of an area of strongly developed joints or bedding planes that dip outward toward the natural slope. Block slides may move continuously in frequent pulses. In large slides movement is primarily controlled by wet year sequences and extreme or intense rainfall events. Though velocities are frequently low, the masses involved can be very great and extremely difficult or impossible to stabilize.

The primary cause of block slides is the presence of an abrupt change of rock type or a bedded rock sequence which provides a weak stratum dipping gently towards the slope. Strong discontinuities parallel to and into the face which clearly define a potential movement area are also helpful. The second condition concerning block failure is that the slope may be unloaded by erosion or excavation to a point where the potential failure surface crops out above or close to the base level.

Slab slide

Alternative terms: debris block slide, soil block slide, earth block slide, sheet slide (shallow translational failure in dry, cohesionless soils).

Slab slides are translational failures in slopes composed of coherent, fine soils or coarser debris with a fine matrix. Weathered soils, especially those derived from clays, mudrocks and silty-clays, are commonly involved. The weathered material normally moves on a shear zone close to a surface of unweathered or lightly weathered bedrock, a pedogenic horizon or a structural surface. A slab slide is dominated by the pedological or geological structure and frequently fails along discontinuities. Hutchinson (1988) and Ibsen *et al.* (1996a) provide summaries.

Slab slides are extremely susceptible to seasonal changes in groundwater levels or to loading at the head or unloading at the toe. Movement increases in wet months and may cease in dry periods. If the ground freezes and seasonally thaws the saturated conditions can initiate movement. Slab slides also occur at permeable/impermeable soil junctions. Movement takes place on low angle shear surfaces and is normally parallel to the ground surface.

Slab slides are caused by geometrical changes to the slope, water regime changes or human activity. They are dominated by the relationship between regolith depth and slope angles which determines the critical depth for failure. Melting of PERMAFROST is a special cause of saturated ground overlying an impermeable horizon.

Rock slide

A rock slide is a translational movement of rock which occurs along a more or less planar or gently undulating surface (Varnes 1978). It is typical for mountain slopes or rock exposures where the slope angle is close to, or parallel to, the dip of the rock. The movement is controlled by planar structural discontinuities, such as faults, joints and layering and the presence of weaker formations within the rock mass. Sorriso-Valvo and Gulla (1996) and Erismann and Abele (2001) provide summaries.

Rock slides are characterized by well-defined head scarps and flanks, a pronounced scar generally left with little or no debris, and a mass of debris that accumulates in the track or at the base. In case of a sliding rock mass a planar slope surface is developed. If the rock slides well away from the depletion zone, the scar and flanks may remain visible.

There are different mechanisms of movement of rock slides. If the movement is slow (mm to m/day) the whole mass may disaggregate because of differences in velocity along the yield surface. The frequency of events and magnitude of each single slide may vary. In velocity slides the mass disaggregates during movement, transforming it into a rock avalanche or a debris flow.

The essential prerequisites are steep slopes, intense jointing, bedding or fault planes dipping towards the open face and the preparation of the slope by unloading and weathering and the development of joint water pressures. The triggers are the undercutting of the toe support and earthquakes. The fundamental cause of a rock slide is the presence of a rock mass which produces such a stress that the resistance of the intact rock or the friction mobilized on existing discontinuities is exceeded.

Rock slides have a wide range in volume and velocities and pose considerable hazards to human settlements and lives. The destructive power of rock slides can be enormous in the case of rapid rock slides on steep slopes.

Debris slide

Alternative terms: shallow translational slides, sheet slides, soil slips.

Debris slides are failures of unconsolidated material which breaks up into smaller parts as the slide advances downslope. The material involved is mostly COLLUVIUM and weathered material of fractured rocks masses (i.e. flysch formations, shales and slates). The failure surface usually develops at the contact between the REGOLITH cover and the bedrock, and is roughly parallel to the ground surface. Clark (1987) and Corominas *et al.* (1996) provide summaries.

The speed of sliding and degree of runout tend to increase with slope angle and decrease with clay content (Hutchinson 1988). Velocities of up to 16 m/sec have been recorded. Many translational debris slides turn into debris flows. This occurs where water is available, and where the topography favours the convergence of both debris and water into concavities and channels. On very steep slopes, debris slides can reach high velocities. This is common in valleys shaped by glaciers in which morainic sediment is located high above the present river.

Debris slides are often triggered by intense rainfall or by earthquakes. The probability of a debris slide occurring is greatly increased by the destruction of vegetation cover by fires or logging. Sites most likely to provide failures are first-order basins with hollows where regolith can reach the maximum thickness and high slope angles. Failures are often caused by an increase in PORE-WATER PRESSURES following heavy rains which reduce the shear strength of the material. After failure, the breakage of the sliding mass allows the water to escape and the debris to stop. Debris slides can burst explosively out of a slope. Hazard assessments should be based on regolith depth and slope angle.

Mudslide

Alternative terms: earthflow (US usage), mudflow (redundant usage also climatic and temperate mudflow), slump-earthflow (complex, lobate mudslide form with minimal track).

Mudslides are a form of mass movement in which masses of softened silty or very fine sandy debris slide on discrete boundary shear surfaces in relatively slow-moving lobate or elongate forms. Brunsden (1984) and Ibsen and Brunsden (1996) provide summaries.

A mudslide is divided into source, track and lobe and accumulation zone units. The source has a bowl-shaped head. The material in this section is usually soft, weathered debris often with depressions containing water. The track evolves as an elongate or lobate channel through which the material moves. The accumulation zone develops at the base of the slope and normally consists of lobes of debris. Mudslides usually occur in saturated clays which have been described as fissured, mudstones, siltstones and overconsolidated clays with a generally medium plasticity.

Mudslide movement rates range from about 1–25 m/yr and are generally classified as slow mass-movement types. Extreme events range from hundreds of metres per day. Mudslide movement is normally seasonal, as the wetter weather increases the water content to the point where pore water becomes sufficient to generate movement. Mudslides in temperate areas display a pronounced winter–summer cycle. Movement usually commences in the late autumn, peaks in mid-winter and comes to a slow halt by late spring and summer. Heavy rainfall will frequently result in a mudslide surge. Movement can also be generated by undrained loading when there is rapid supply of debris at the head. Other causes are associated with the supply of water such as snowmelt or permafrost melt. Finally unloading of the toe area is important since it presents the development of passive resistance at the toe.

The essential plannning and engineering implication is that this form of slide requires water. Planning must include surveys for drainage installations. The removal of the source of water influx is critical, as is the reduction of pore-water pressure from the mudslide mass. Roads and other linear features are most vulnerable to movement at the lateral shears.

Lateral spreading

Lateral spreading describes the lateral extension of a cohesive rock or soil mass over a deforming mass of softer underlying material.

ROCK SPREADING

Alternative terms: gravitational spreading, gravity faulting, block-type slope movement, cambering and valley bulging.

Rock spreading is the result of deep-seated, plastic deformation in the rock mass, leading to extension at the surface. It may take place in a relatively homogeneous rock, or there might be a fractured capping stratum leading to gravitational stresses. Where the rock is relatively homogeneous, the moving mass breaks up into successive units arranged like horsts and grabens. Rock spreading in homogeneous rock masses is characterized by double ridges, trenches and uphill facing scarps which occurs often in high mountains. Uphill facing scarps evolve from the combination of ridge spreading and erosion of the uphill sides of the tension cracks. Zaruba and Mencl (1982) and Pasuto and Soldati (1996) provide summaries.

Rock spreading is very often associated with toppling, rock falls, slumps and mudslides. The rock masses involved are huge and generally more than one million cubic metres. The velocity is particularly low in comparison with other types of mass movement and seasonality is rarely observed. Lateral spreading phenomena are highly controlled by geological structures and often connected with deep-seated gravitational slope deformations.

Rock spreading is caused by an outward and downward movement of both valley sides along low-angle shear planes causing rock spreading at the ridge top. There may be sliding of blocks with respect to the whole ridge with minor crushing and redistribution of material at the point of the wedges. Alternative wedging along the centre of the ridge in combination with rotation of the peripheral blocks together with crushing at the areas of greatest stress may take place. Rock spreading rates range between 10^{-4} and 10^{-1} metres per year. Large-scale rock spreading is generally not seasonal.

The process depends on the local geological and topographical conditions. The horizontal or subhorizontal overposition of a thick slab of competent rocks, affected by a dense network of tectonic joints, on clayey materials which may behave as a visco-plastic medium, is a prerequisite for the process.

SOIL/DEBRIS SPREADING

Alternative terms: sudden spreading failure, lateral soil spreading, quick clay sliding, quick clay flow, soil liquefaction sliding.

Soil spreading is defined as the collapse of a sensitive soil layer followed by either settlement of the overlying more resistant soil layers, or progressive failure throughout the whole sliding mass. The duration is generally only a few minutes. Considerable lateral movement along the basal mobile zone occurs. The material deformations accompanying the spreading often cause loss of life and severe damage to roads, buildings and embankments. The areas involved are low angled and are ideal for both agricultural use and urban development. Areas of particularly intense hazards are often situated close to the shoreline or the banks of rivers, as toe erosion is a significant factor in destabilizing these slopes. Bjerrum (1955) and Buma and van Asch (1996) provide summaries.

QUICKCLAYS are found in areas which have been depositional marine environments adjacent to Pleistocene ice margins. Morphological features of quickclay slides which have their whole slide mass liquefied are the pear-shaped scar, and the presence of an extensive flow lobe on the lower side of the scar. This lobe can attain lengths of up to 1,000 m. Quickclay slides start in lower slope regions and extend upslope by retrogression. Due to loss of horizontal support, retrogression proceeds rapidly, both upslope and to the flanks. Due to remoulding the clay already subject to movement loses its internal strength and descends the slope as a more or less viscous slurry, and can travel up to several kilometres. In (quick) clay soils, initial failures are mostly caused by extremely high water contents (30–40 per cent) within the clay, oversteepening of the local slope and/or stress caused by loading.

Quickclay slides are initiated within a body of sensitive ('quick') clay. Sensitive clays are defined as clays which need only very little remoulding to transform into a viscous slurry. However, the whole sliding mass does not necessarily liquefy. It may be restricted to a thin layer of sensitive clay at some depth. The sensitivity of the clay is influenced by the dominant type of clay minerals and the depositional environment. Marine sedimentation results in flocculation and an open 'card-house' porosity structure. Strength reduction through interparticle repulsion is caused by leaching of salt in pore water over time.

Sand liquefaction slides are situated on coasts and adjacent to rivers or lakes. Especially sensitive are slopes already subject to mass movement, with disturbed drainage and corresponding high water contents in the soil layers. Varved clay formations are also sensitive to sand liquefaction failures along the boundary between impermeable clay layers and water-bearing fine sand or silt layers under high pore-water pressures. Tension fractures will develop at the head of the slide, dipping towards the sliding mass. The overhanging wall of the sliding block collapses to form a graben. The sliding blocks become subject to tilting, internal fracturing, subsidence, heaving and overthrusting, producing a very hummocky topography. The liquefied layer material escapes through tension cracks and may cause sand boils.

A triggering mechanism of sand liquefaction is a combination of prolonged periods of heavy rainfall or snowmelt, causing high initial pore pressures, and earthquakes. During an earthquake a soil undergoes cyclic stress loading and unloading, and pore-water pressures increase with each cycle. After a certain number of cycles the pore-water pressures become equal to the existent confining pressures and the soil suddenly loses its strength, suffering from considerable deformations without exhibiting resistance. This process is restricted to cohesionless soils such as sand and silt.

Flow

A flow is a landslide in which the individual particles travel separately within a moving mass. They involve highly fractured rock, clastic debris in a fine matrix or small grain sizes. Flow in its physical sense is defined as the continuous, irreversible deformation of a material that occurs in response to applied stress. They are, therefore, characterized by internal differential movements that are distributed within the mass.

ROCK FLOW

Alternative terms are: SACKUNG, sagging, rock creep, deep-seated gravitational creep.

Rock flows are creeping flow-type, deep-seated gravitational deformations affecting homogeneous rock masses. Rock flows are characterized by the high volume of the rock mass (several thousand to millions of cubic metres) and small total displacement rates. Structural elements and landforms associated with rock flows are high angle extensional shear planes in the upper part of the deforming slopes, producing graben-like depressions (trenches), double ridges, ridge depressions and troughs. The slope foot often shows compressional features such as bulging and, sometimes, low-angle shear planes. Mencl (1968) and Bisci *et al.* (1996) provide summaries.

The mechanisms of rock flows are not well known. Mencl (1968) postulates that in the central part of the slope, where the confining pressures are high and the deviator stress is too small to produce shearing, the rock mass deforms through viscous flow. At the uppermost and lowermost parts of the slope, where the confining pressures are low, the rock mass moves along the shearing surfaces. At depth, the high pressures induce plastic deformation without necessarily creating a proper slip surface.

Some rock flows display evidence of a constant rate of creep deformation. Others show a rapid reactivation phase in connection with extreme rainfall or earthquakes. Some rock flows may be characterized by a step-like evolutionary behaviour, including short active phases related to critical events and long-lasting dormant phases involving creep deformation at extremely slow rates. Under particular circumstances, the slow slope deformation can be turned into a catastrophic event leading to the collapse of a large-scale rock avalanche or debris flow. Rock flows can be considered as extremely slow preparatory stages of huge landslides which only in a few cases reach their final evolutionary step.

Rock flows are produced only where slopes are high enough to induce strong gravitational stress in the bedrock. Such conditions are typical of valley slopes in mountain areas and high coastal cliffs. Rock flows do not normally represent a major problem to planning and engineering. In recent years, however, there is a concern that large structures such as dams or hydrological sites may be placed at the toe of these phenomena.

DEBRIS FLOW

Alternative terms: mudflow (old usage), lahar (volcanic mudflow).

Debris flows consist of a mixture of fine material (sand, silt and clay), coarse material (gravel and boulders), with a variable quantity of water, that forms a muddy slurry which moves downslope. The flow moves in surges and includes the erosion of the channel bed and the collapse of bank material. Debris flows usually take place on slopes covered by unconsolidated rock and soil debris. Debris flows are characterized by the source area, the main track and the depositional toe. The flows commonly follow existing drainage ways. Some of the coarse debris will be deposited at the side of the track to form lateral ridges (levees). Deposits are accumulated where the channel gradient decreases or at the toe of mountain fronts. Successive surges will build up a debris fan. Some debris flows are exceptionally large, fluid, and can reach long distances beyond the source area. Debris flow is a gravity induced mass movement between landsliding and water flooding, although with mechanical characteristics very different from either of these processes. Pierson and Costa (1987) and Corominas *et al.* (1996) provide summaries.

Debris flows are a very destructive type of mass movement caused by heavy rain or snowmelt. In alpine environments debris flows are composed of coarser material from mechanical weathering and glacial deposits. Debris material in melting permafrost (near the lower limit of the discontinuous permafrost) has also been considered a debris flow source area. Debris flows which originate on the slope of a volcano contain vulcaniclastic materials and are called LAHARS.

Debris flows consist of large coarse material embedded in a fine-grained matrix. The coarse material is randomly distributed and individual beds are generally poorly sorted. Buoyant forces and dispersive pressures may concentrate boulders at the top of the deposit, forming reverse grading. The frequency of debris flow events is controlled by the rate of accumulation in hollows or channels, and by the recurrence of climatic triggering events.

Debris flows have well-graded deposits with a small clay content, generally less than about 5 per cent. They have a range in volume concentration of solids from approximately 25 to 86 per cent. Sediment concentration is the primary criteria for classification of flows given by Pierson and Costa (1987). There is a continuum from sediment movement in rivers to debris flows. Fluids with large sediment concentrations do not deform until a threshold strength is exceeded and they behave like a non-Newtonian fluid.

Debris flows are commonly triggered by an unusual presence of water created by intense rainfall, rapid snowmelt, and glacier or lake overflows. Debris flows are frequent in topographic concavities or hollows at first-order watersheds. This geometry supports the accumulation of colluvium and the convergence of groundwater flow necessary to cause the failure. Many debris flows start as a translational or rotational slide and then turn into a debris flow.

Series of debris flow waves are frequent and can be produced by breaching of the temporary dams or obstructions in the channel. The front lobe is composed of coarse blocks occasionally mixed with trees. During the progression of the debris flow through a channel, lateral ridges (levees) can develop. Overflowing of channel banks is a significant natural hazard of debris flows. The velocity of the flow depends on the size and sediment concentration, and on the geometry of the path. Observed velocities range from 0.5 to about 20 m/s. Large lahars can travel over a distance of more than 100 km and the rate of movement may reach more than 50 km/h. The erosion that occurs on both the channel floor and banks, cause some debris flows to significantly increase their volume.

The socio-economic impact and the loss of life, property and agriculture can be catastrophic. Even smaller mudflows and debris flows may cause serious damage, especially in mountainous regions. The deposits are also responsible for severe indirect damage and hazards such as damming of rivers or sudden debris supply to river systems. It is essential that potential source areas and runout zones are assessed.

SOIL FLOW (MUDFLOW)

Alternative terms: mudflow, alpine mudflow, earthflow, sandflow.

Soil flows may occur in three forms, wet mudflow, wet sand flow and dry sand flow. The wet forms are special categories of debris flow where the material is of a single and fine grain size and coarse clasts are rare. They are very mobile and can flow downslope quickly. They tend to follow gullies or shallow depressions and then to spread out into a flat fan or even a thin sheet when they reach low gradients. Soil flow conditions are abundant water, unconsolidated material and insufficient protection of the ground (i.e. lack of vegetation). Pierson and Costa (1987) and Schrott *et al.* (1996) provide summaries.

A wet soil flow contains relatively cohesive earth material that has at least 50 per cent sand, silt and clay. Thus, the term soil flow should be used for a flow with a significant lack of coarse-grained material. Typical source areas and starting zones are steep (25°–40°) slopes (e.g. moraines, proglacial zones), volcanic environments and mountain torrents.

A characteristic of soil flows is their ability to travel long distances (some kilometres) over even low slopes, usually following pre-existing drainage patterns. They are often termed viscous slurry flows, but they can be either viscous or inertial flows, depending on the driving forces. The flow behaviour is normally that of a visco-plastic type. The change from a slow creep to visco-plastic flow in clay-rich soils supersedes the destruction of strong bonding and the subsequent decrease in viscosity. Continuous undrained loading causes rapid readjustment of this mass and velocities of up to 10 m per second have been recorded.

Very rarely dry sand flows may occur. These form when a large mass of dry material falls from or over a steep slope and fluidizes on impact. Flow is then of 'rock avalanche' type with a track of uniform depth. Dry sand flows may also occur from dunes or similar sandy deposits. They require the sliding of material at the head and then either the transfer of momentum by cohesionless grain flow (i.e. grain upon grain momentum transfer) or descent over a cliff and fluidization. The first type can be observed in fluviglacial deposits or riverbanks.

Complex

It is common for mass movement processes to combine together and complex landslides occur where the initial failure type changes into another as it moves downslope. Compound landslides include two types of movement which occur concurrently within the same failure.

ROCK AVALANCHE

Alternative terms: STURZSTROM, rockfall avalanche, rock-slide avalanche.

FLOW SLIDE

A flow slide is a structural collapse of slope-forming material with momentary fluidization and is usually referred to as a high magnitude event both in terms of velocity and destruction. The high energy is capable of causing incredible devastation not merely through its impact on humans, by cutting communication and power lines or diverting a river, but also environmentally by littering the surrounding valley with debris. Flow slides are often associated with man-made tips and spoil heaps, although this type of failure may also occur in rock debris of geological origin. The internal structure has very little cohesion and the matrix ranges from clay-size particles to large blocks. Flow slides are a subclass of debris flows. The various causes of flow slides are an initial rotational failure at the head, vibrations or shocks, heavy rainfall, loosely deposited spoil heaps, removal of lateral support and rapid loading. Bishop *et al.* (1969) and Ibsen *et al.* (1996c) provide summaries.

The key characteristics of a flow can lie in their origin in artificial spoil materials but their behaviour is also found in many natural debris flow events. A flow slide is generally composed of loose material, which losses its cohesion with a reduction in strength, becoming a fluidized mass. The fluid may be air or water and, therefore, the dominant mechanism of a flow slide is fluidization or liquefaction. Sliding may occur at the head of the flow slide perhaps in the form of a rotational slide, but there is usually little indication of shearing at the subsequent stages of movement. Flow slides not only fluidize very quickly, but also rapidly consolidate and become solid when they cease moving, creating an additional hazard in the depositional area.

References

Bisci, C., Dramis, F. and Sorriso-Valvo, M. (1996) Rock flow (sackung), in R. Dikau *et al.* (eds) *Landslide Recognition*, 150–160, Chichester: Wiley.

Bishop, A.W., Hutchinson, J.N., Penman, A.D. and Evans, H.E. (1969) *Geotechnical investigations into the causes and circumstances of the disaster of 21st October 1966. A selection of technical reports submitted to the Aberfan tribunal*, Welsh Office, London: HMSO.

Bjerrum, L. (1955) Stability of natural slopes in quick clay, *Géotechnique* 5(1), 101–119.

Brunsden, D. (1984) Mudslides, in D. Brunsden and D.B. Prior (eds) *Slope Instability*, 363–418, Chichester: Wiley.

Buma, J. and van Asch, T. (1996) Slide (rotational), in R. Dikau, *et al.* (eds) *Landslide Recognition*, 43–61, Chichester: Wiley.

Clark, G.M. (1987) Debris slide and debris flow historical events in the Appalachians south of the glacial border, *Review of Engineering Geology* 7, 125–138.

Corominas, J., Remondo, J., Farias, P., Estevao, M., Zezere, J., Diaz de Teran, J., Dikau, R., Schrott, L., Moya, J. and Gonzalez, A. (1996) Debris flow, in R. Dikau *et al.* (eds) *Landslide Recognition*, 161–180, Chichester: Wiley.

Cruden, D.M. and Varnes, D.J. (1996) Landslide types and processes, in A.K. Turner and R.L. Schuster (eds) *Landslides. Investigation and Mitigation*, 36–75, Washington: National Academy Press.

Dikau, R., Brunsden, D., Schrott, L. and Ibsen, M.-L. (1996) *Landslide Recognition*, Chichester: Wiley.

Dikau, R., Schrott, L. and Dehn, M. (1996) Topple, in R. Dikau *et al.* (eds) *Landslide Recognition*, 29–41, Chichester: Wiley.

Erismann, T.H. and Abele, G. (2001) *Dynamics of Rockslides and Rockfalls*, Heidelberg: Springer.

Flageollet, J.C. and Weber, D. (1996) Fall, in R. Dikau *et al.* (eds) *Landslide Recognition*, 13–28, Chichester: Wiley.

Goodman, R.E. and Bray, J.W. (1976) *Toppling of Rock Slopes*, Proceedings of Special Conference on Rock Engineering for Foundation 2, 201–234.

Hutchinson, J.N. (1988) Morphological and geotechnical parameters of landslides in relation to geology and hydrology, General Report, in C. Bonnard (ed.) *Landslides*, Proceedings of the 5th International Symposium on Landslides **1**, 3–35.

Hutchinson, J.N., Bromhead, E.N. and Chandler, M.P. (1991) Investigations of landslides at St. Catherine's Point, Isle of Wight, in R.J. Chandler (ed.) *Slope Stability Engineering Developments and Applications*, Proceedings of the International Conference of Civil Engineers, Isle of Wight, 169–179.

Ibsen, M.-L. and Brunsden, D. (1996) Mudslide, in R. Dikau *et al.* (eds) *Landslide Recognition*, 103–119, Chichester: Wiley.

Ibsen, M.-L., Brunsden, D., Bromhead, E. and Collision, A. (1996a) Slab slide, in R. Dikau *et al.* (eds) *Landslide Recognition*, 78–84, Chichester: Wiley.

Ibsen, M.-L., Brunsden, D., Bromhead, E. and Collision, A. (1996b) Block slide, in R. Dikau *et al.* (eds) *Landslide Recognition*, 64–77, Chichester: Wiley.

Ibsen, M.-L., Brunsden, D., Bromhead, E. and Collision, A. (1996c) Flow slide, in R. Dikau *et al.* (eds) *Landslide Recognition*, 202–211, Chichester: Wiley.
Mencl, V. (1968) Plastizitätslehre und das wirkliche Verhalten von Gebirgsmassen, in *Felsmech. und Ing. Geologie*, Suppl. 4, 1–8.
Pasuto, A. and Soldati, M. (1996) Rock spreading, in R. Dikau *et al.* (eds) *Landslide Recognition*, 122–136, Chichester: Wiley.
Pierson, T.C. and Costa, J.E. (1987) A rheologic classification of subaerial sediment-water flows, Geological Society of America, *Reviews in Engineering Geology* 7, 1–12.
Schrott, L., Dikau, R. and Brunsden, D. (1996) Soil flow (mudflow), in R. Dikau *et al.* (eds) *Landslide Recognition*, 181–187, Chichester: Wiley.
Sorriso-Valvo, M. and Gulla, G. (1996) Rock slide, in R. Dikau *et al.* (eds) *Landslide Recognition*, 85–96, Chichester: Wiley.
UNESCO (1993) *Multilingual Landslide Glossary*, The International Geotechnical Societies' UNESCO Working Party for World Landslide Inventory, BiTech Pubs, Canada.
Varnes, D.J. (1978) Slope movement types and processes, in R.L. Schuster and R.J. Krizek (eds) *Landslides analysis and control*, Transportation Research Board, National Academy of Sciences, Special Report 176, 12–33.
Whalley, W.B. (1984) Rockfalls, in D. Brunsden and D.B. Prior (eds) *Slope Instability*, 217–256, Chichester: Wiley.
Zaruba, Q. and Mencl, V. (1982) Slope movements caused by squeezing out of soft rocks, in Q. Zaruba and V. Mencl (eds) *Landslides and their Control*, 110–120, Amsterdam: Elsevier.

Further reading

Bromhead, E.N. (1992) *The Stability of Slopes*, 2nd edition, New York: Surrey University Press/Chapman and Hall.
Brunsden, D. (1993) Mass movement; the research frontier and beyond: a geomorphological approach, *Geomorphology* 7, 85–128.
Brunsden, D. and Prior, D.B. (eds) (1984) *Slope Instability*, Chichester: Wiley.

SEE ALSO: factor of safety; fluidization; lahar; landslide; landslide dam; method of slices; quickclay; riedel shear; sackung; sensitive clay; sturzstrom

RICHARD DIKAU

MATHEMATICS

Mathematics provides an essential set of tools for the study of geomorphological phenomena. Aspects of mathematics particularly relevant to geomorphology include analysis of numerical relationships among measured quantities, the use of differential equations to describe processes, and descriptive mathematics related to form, such as geometry and topology. The former two aspects are critical when deriving equations to describe geomorphological processes or statistical relations between system variables. The latter is relevant when considering morphology of landscape features.

Prior to the twentieth century, the study of Earth sciences in the western world consisted mainly of qualitative consideration of early Earth history and landform origins. Until the mid-twentieth century, mathematics in geomorphology was limited to a few pioneering studies or engineering problems. Davis (1899), following earlier traditions, influenced geomorphological research for many decades by adopting a qualitative framework to describe regional landscape evolution in his 'CYCLE OF EROSION'. At around this time, Gilbert (1877) developed an innovative, quantitative research agenda based on the application of Newtonian mechanics to the study of landscape processes. This latter approach was not generally adopted at the time, perhaps because of difficulties in measuring processes at the large scales of study that dominated the discipline.

Around the mid-twentieth century, the 'quantitative revolution' in geomorphology emerged. Landmark papers such as Horton (1945) and Strahler (1952) signalled the shift to a quantitative approach. Morphometrics and numerical analysis of processes became dominant research themes. Scales of study decreased, as it is generally easier to measure system parameters at smaller scales. In the ensuing decades, many studies focused on establishing empirical relations between system variables. Physically meaningful expressions, in contrast to empirical relations, must be dimensionally balanced. In fact, what may appear to be fundamental relations in geomorphology do not actually fit this definition. Such equations, therefore, represent scale relations and are not true representations of the underlying physics. Physically based attempts to describe processes make use of mathematical tools, such as differential calculus, to portray fundamental conservation principles (mass, momentum, energy) in temporally evolving geomorphological systems. The Buckingham Pi Theorem, which is based on formal dimensional analysis, can be used in an attempt to derive rational equations that encompass essential system attributes.

The use of mathematical techniques is now firmly entrenched in geomorphological research. Continuing advances in technologies, such as remote sensing and radiometric dating, and the availability of improved data sets, such as DIGITAL ELEVATION MODELS (DEMs), have made it possible to measure and describe geomorphic phenomena at large spatial and temporal scales. Although complexities in the natural environment prohibit exact quantification of geomorphic processes, simplified numerical models of landscape processes have emerged due to advances in computing technologies. Sensitivity analysis, which examines the significance of changes in controlling variables on process operation, can be undertaken within a modelling framework.

References

Davis, W.M. (1899) The Geographical Cycle, *Geographical Journal* 14, 481–504.

Gilbert, G.K. (1877) *Report on the Geology of the Henry Mountains*, US Geographical and Geological Survey of the Rocky Mountain Region, Washington, DC.

Horton, R.E. (1945) Erosional development of streams and their drainage basins; hydrophysical approach to quantitative morphology, *Geological Society of America Bulletin* 56, 275–370.

Strahler, A.N. (1952) Dynamic basis of geomorphology, *Geological Society of America Bulletin* 63, 923–938.

YVONNE MARTIN

MAXIMUM FLOW EFFICIENCY

The operation of fluvial systems can be described by the physical relationships of flow continuity, resistance and sediment transport. However, these three relationships are applicable to any alluvial-channel section and their solution involves more than three variables (width, depth, velocity and slope). As a result, the number of sections satisfying the three relationships is generally very large. In an attempt to understand the physics behind this problem of non-closure solution, Huang and Nanson (2000, 2002) propose a mathematical analytical approach that identifies a mechanism of self-adjustment. Among the numerous channel sections that satisfy flow continuity, resistance and sediment transport, a unique solution of HYDRAULIC GEOMETRY occurs when flow reaches the following inherent, optimal state defined as maximum flow efficiency (MFE):

$$\varepsilon = \frac{Q_s}{\Omega^{\lambda}} = a\ maximum$$

where ε is flow efficiency factor, Q_s is sediment discharge, Ω is stream power or $\Omega = \rho g Q S$, and exponent λ has a value of 0.65–0.85. For two reasons this state can be regarded as a fundamental law governing the adjustment of fluvial systems. First, MFE can be directly derived from the widely applied LEAST ACTION PRINCIPLE (LAP) (Huang *et al.* 2002). Second, mathematically derived MFE channel geometry relations are highly consistent with empirical relations developed from a wide range of observations for stable canals and relatively stable river channels (Huang and Nanson 2000, 2002).

While recognition of the applicability of LAP to river systems in the form of MFE provides a soundly based and computationally economical method for determining stable alluvial channel geometry, its physical implications are much more profound. First, it is an advance over the thermodynamic analogies and the empirical formulas previously used to justify numerous extremal MODELS in geomorphology, for it clarifies confusion regarding which of these hypotheses should be regarded as rationally based. Indeed, MFE subsumes the earlier hypotheses of maximum sediment transport capacity and minimum stream power. Second, it identifies a fundamental cause for the formation of different river channel patterns. It is the balance between available stream power and imposed sediment load that ultimately determines equilibrium channel form. This balance has been shown to exist in the ideal case of a single-thread, straight and fully adjustable channel system (Huang and Nanson 2000, 2002). Ongoing research suggests that when the balance in the ideal system cannot be maintained due to the effect of physical restrictions, such as the imposed valley gradient, then planform or cross-sectional changes will occur to either consume excess energy or to increase transport efficiency over low gradients. As a consequence, meandering, braiding, anabranching and wandering channel patterns will be formed (Huang and Nanson 2002).

References

Huang, H.Q. and Nanson, G.C. (2000) Hydraulic geometry and maximum flow efficiency as products of the principle of least action, *Earth Surface Processes and Landforms* 25, 1–16.

Huang, H.Q. and Nanson, G.C. (2002) A stability criterion inherent in laws governing alluvial channel flow, *Earth Surface Processes and Landforms* 27, 929–944.

Huang, H.Q., Nanson, G.C. and Fagan, S.D. (2002) Hydraulic geometry of straight alluvial channels and the principle of least action, *Journal of Hydraulic Research* 40, 153–160.

SEE ALSO: hydraulic geometry; least action principle; models

HE QING HUANG AND GERALD C. NANSON

MEANDERING

Phenomenology of meandering

Meandering refers to the spontaneous evolution of a single channel to high values of SINUOSITY, or to a channel that shows this pattern (Plate 78). Meandering is one of three basic types of river planform, of which the other two are straight and braided (see BRAIDED RIVER) (or, more generally, anabranching). Individual channels of anastomosed rivers can show meandering (see ANABRANCHING AND ANASTOMOSING RIVER). Meandering with planform shape comparable to that in rivers also occurs in tidal channels, in marine channels produced by density CURRENTs, and in geostrophic ocean-surface currents like the Gulf Stream.

Plate 78 The meandering Pembina River, Alberta, Canada. Flow towards top of image

In classical meandering the width of channel remains constant as the sinuosity increases, so the channel planform is well described by a single sinuous line. The simplest function that gives an adequate description of this curving line is a relative of the common sine wave known as a sine-generated curve. A sine-generated curve is one in which the local direction of flow varies as the sine of distance along the channel. As meandering reaches high sinuosity values, the symmetric sine-generated curve develops an asymmetry that gives the channel path a more saw-toothed shape (Parker and Andrews 1986). The sharp 'corner' of the tooth is on the upstream side of the bend.

Origin of meandering

Observationally, meandering is favoured in fine-grained (sand and finer) rivers with high suspended load relative to bedload and low slopes, but it is known from steep and/or coarse-bedded rivers as well, so the former conditions are apparently not fundamental. Meandering river channels are also often associated with fine-grained and/or vegetated banks and usually have well-developed floodplains.

The first step towards understanding the mechanics of meandering was to understand the effects of streamline curvature on flow in channel bends. Rozovskii was the first to show mechanistically how streamline curvature leads to a secondary flow, that is, a closed circulation of the water across the direction of the vertically averaged flow. The secondary circulation moves fast-moving surface water toward the outer bank and slowly moving bottom water toward the inner bank. The net effect of this circulation pattern is erosion on the outer bank and deposition on the inner bank, leading to development of a bank-attached bar (a point bar) on the inner bank. Thus channel curvature tends to be self-amplifying – an example of positive feedback. As the bend grows, deposition on the inner bank often maintains a rough balance with erosion of the outer bank, keeping the width constant. The growth of the point bar is often recorded in the surface morphology as a set of scroll bars that trace previous positions of the inner bank (Plate 78).

Later work has elaborated on this simple model considerably. First, the dynamics of bend flow is influenced at least as much by mean-flow inertia and by bottom topography as by secondary circulation. Interaction of the secondary circulation and the mean flow also tends to displace the thread of maximum velocity downstream relative to the bend apices, leading to a tendency of the bends to migrate downstream.

A major next step was formal mathematical analysis of the stability of a straight channel, with the aim of providing a mechanistic basis for the occurrence of straight, meandering and braided channel patterns (Fredsoe 1978; Parker 1976) (see BRAIDED RIVER). Meandering is thought to correspond to cases where the main mode of instability is alternate bars, which are bank-attached bars on alternating sides of the channel. Alternate bars are predicted to develop in channels with widths of roughly 15 to 150 times the flow depth. The bars alternately deflect the flow from one bank to the other, producing a sinuous thalweg (planform trace of the deepest part of the channel). This initial sinuosity leads to fully developed meandering via the positive-feedback mechanisms discussed above. A more elaborate stability analysis (Blondeaux and Seminara 1985) shows that the meandering instability is actually a kind of 'resonance' between the original alternate-bar instability and a planform instability of the channel curvature. Given that they apply strictly only to the initial growth of the bend, it is remarkable that these stability analyses correctly predict the wavelength of fully developed meanders: about seven times the channel width.

Stability analysis suggests that the main control on channel pattern is the aspect ratio of the channel. Although the dynamics of channel width are still not entirely understood, it is clear that one of the main controls is the total effective sediment flux that must pass through the channel. Apart from directly controlling the width, the effective sediment flux also influences the aspect ratio indirectly: the slope is proportional to the ratio of sediment discharge to water discharge; as slope decreases, depth tends to increase. Hence reducing the effective sediment flux relative to water discharge reduces the width and increases the depth. The central role of the sediment/water discharge ratio in controlling the plan pattern is consistent with the empirical observation that meandering tends to occur in low-gradient rivers with relatively fine, suspension-dominated sediment flux. However, neither the gradient nor the sediment grain size *per se* is the real controlling factor.

Stability analysis helps explain the origin of meandering. However, since it treats only the initial instability, it does not constrain the final amplitude of the fully developed meanders. The ratio of amplitude to meander wavelength sets the sinuosity of the channel. There is still no complete analysis of the controls on the amplitude of fully developed river meanders. The main process that limits meander growth in rivers is cutoff of the bend by formation of a new, shorter channel across the bend. Thus one could view meander geometry and average sinuosity as the outcome of a competition between sinuosity production by meander growth and sinuosity destruction by cutoff (Howard 1992). The ease with which cutoff occurs appears to be controlled by the erodibility and resistance to flow of the point bar surface. Vegetation on the point bar helps prevent cutoff in several ways. Stems and leaves block flow and provide baffling that aids in the deposition of fine, cohesive sediment, while roots bind deposited sediment. It would be helpful if techniques could be developed to reproduce steady-state, fully developed meandering experimentally, but so far this has proved extremely difficult to do (Smith 1998).

Submarine meandering

Submarine channel systems formed by density currents produce meander patterns that are very similar to river meanders in planform geometry. However, the scale of submarine channels is much larger than that of river channels: depths are typically 100–200 m and widths typically several kilometres. The mechanics of submarine meanders are similar to those of subaerial meanders; the larger scale results mainly from the fact that the density difference between turbid and clear water is much less than that between water and air. Hence much deeper flows are required to provide sufficient force to move sediment particles.

References

Blondeaux, P. and Seminara, G. (1985) A unified bar-bend theory of river meanders, *Journal of Fluid Mechanics* 157, 449–470.

Fredsoe, J. (1978) Meandering and braiding of rivers, *Journal of Fluid Mechanics* 84, 609–624.

Howard, A.D. (1992) Modeling channel migration and floodplain sedimentation in meandering streams, in P.A. Carling and G.E. Petts (eds), *Lowland Floodplain Rivers: Geomorphological Perspectives*, 1–41, New York: Wiley.

Parker, G. (1976) On the cause and characteristic scales of meandering and braiding in rivers, *Journal of Fluid of Mechanics* 76, 457–480.
Parker, G. and Andrews, E.D. (1986) On the time development of meander bends, *Journal of Fluid Mechanics* 162, 139–156.
Smith, C.E. (1998) Modeling high-sinuousity meanders in a small flume, *Geomorphology* 25, 19–30.

CHRIS PAOLA

MECHANICAL WEATHERING

Mechanical weathering causes disintegration of rock, without substantial chemical and mineralogical alteration or decomposition. The culmination of mechanical weathering is the collapse of parent material and diminution of clast size. Rock disintegration is caused by stresses exerted along zones of weakness in the material, which may include pre-existing fractures, bedding planes or intergranular boundaries.

Mechanical weathering processes promote rock breakdown by inducing stresses within the rock; these stresses may be produced by volumetric change in the rock itself or by deposition of, and/or volumetric change in, material introduced into voids in the rock. Volumetric expansion of rock may be induced by temperature changes, wetting and drying or pressure release. Volumetric expansion of material within rock pores and cracks predominantly involves salt and ice crystals. Although mechanical weathering processes may cause rock disintegration by themselves, they often operate in association with CHEMICAL WEATHERING and biological weathering processes.

Expansion of rocks and minerals

A rock unit beneath the land surface is subjected to high compressive stress from the rock and sediment layers above. Once surface erosion removes the overlying rock layers, the rock unit will tend to expand in the direction of stress (or pressure) release. This may result in sheet jointing parallel to the unloading surface. As the rock surface expands in response to the PRESSURE RELEASE, the tensional strength of the rock may be exceeded by the tensile stress due to expansion, causing cracks to develop perpendicular to the rock surface. In this way, the rock unit may be broken into smaller slabs, which increases the surface area available for attack by other weathering processes. Pressure release cracking is most common in granites and metamorphic materials, and may also be seen on massive sandstones. As the surface sheets are lost, curved surfaces are formed; on a large scale (outcrop size) this process is termed exfoliation, on a smaller scale (boulder size) it is termed SPHEROIDAL WEATHERING. Exfoliation plays an important role in the creation of some landforms such as INSELBERGS, tors, arches and natural bridges.

Expansion of a rock surface creates tensile stresses and contraction of the surface creates compressive stresses. This may be induced by changes in the temperature of a rock surface. In a similar fashion to the stresses induced by pressure release, temperature-induced stresses decline in magnitude with increasing distance from the exposed rock surface. Stresses are restricted to the outer few centimetres due to the low conductivity of rock, which prevents inward transfer of heat (Hall and Hall 1991). Physical disruption of rock due to thermal shock occurs during forest fires, although the effectiveness of this process depends on rock composition (Ollier and Ash 1983). Whether or not receipt of insolation and diurnal temperature cycles can drive this process has long been debated. Daily temperatures in hot desert environments may reach 50 °C, and rock surface temperatures can reach 80 °C, with rapid cooling at night. Early anecdotes about rocks cracking in the desert were dismissed as hearsay when early experimental studies suggested the process was not viable. More recent research has suggested that thermal expansion, or insolation weathering, may indeed cause rock disintegration, though its effectiveness may largely be dependent on sufficient moisture within the rock. Igneous rocks, which contain many types of minerals with differing coefficients of thermal expansion, may experience stresses as a result of differential thermal response of minerals to heating and cooling cycles.

Rock minerals may expand when water is introduced into their structure; certain clay minerals typically behave in this way. Clays such as smectites and montmorillonites have the capacity to absorb water into the mineral during periods of wetting, which causes the mineral to swell. Bentonite (Na-montmorillonite), for example, may increase in volume by up to 1,500 per cent due to hydration and swelling. The species of clay mineral determines the degree to which it will expand and contract on wetting and drying (Yatsu 1988).

Expansion of material in voids

Exposed rock surfaces experience cycles of WETTING AND DRYING WEATHERING related to rainfall events and periods of evaporation. Simple wetting and drying of some rocks may cause their breakdown. When water enters a crack, or void, on a rock surface, it will become adsorbed by minerals lining the crack which may show unsatisfied electrostatic bonds. Further ingress of water may induce a swelling pressure within the void. Evaporation will remove all water molecules except those strongly bonded to the mineral surfaces; the sides of the crack, or void, may be pulled together by the attractive forces between water molecules on opposing faces. In this way, cycles of wetting and drying may induce expansion and contraction, which can split susceptible rocks such as shale, schist and even sedimentary rocks. Rocks that may be affected by wetting and drying usually have high clay mineral content, structural weaknesses, high permeability and low tensile strength. Wetting and drying may increase the size and/or number of pores and microcracks in rock, which has important implications for both frost (see FROST AND FROST WEATHERING) and SALT WEATHERING; an increase in water absorption capacity and a reduction in rock strength will accelerate the action of these processes.

Frost weathering involves the breakdown of rock as a result of the stresses induced by the freezing of water. Water experiences a 9 per cent increase in volume upon freezing and, in a closed system, may create pressures (theoretically 250 MPa) that exceed the tensile strength of rock (typically 25 MPa). A closed system may be produced if water in a rock freezes rapidly form the surface downwards, allowing ice to seal water in surface cracks and pores in the rock. Experimental work has shown that numerous mechanisms are involved in frost weathering, not only volumetric expansion of water; the most significant include adsorptive suction, as pore water moves toward the freezing front, and expansion (0.6 per cent from +4 °C to −10 °C) of absorbed water. Many experimental studies have shown the importance of rapid freezing rates (at least 0.1 °C per minute), low minimum temperatures (<-5 °C) and high rock moisture content in determining the efficiency of frost weathering in causing shattering of rock samples (McGreevy and Whalley 1985). Moisture content is particularly important as the presence of air in pore spaces in unsaturated rock allows ice to expand into empty pores and voids and prevents crack growth. Rock properties exert an important control on the efficiency of frost weathering, as texture and structure determine both water absorption capacity and strength. Igneous and metamorphic rocks tend to be most resistant to frost shattering, while shales, sandstones and porous chalk tend to be least resistant. Frost weathering is likely to be most effective in alpine and cold temperate environments where freeze–thaw cycles occur frequently and abundant moisture is available, rather than cold deserts and polar areas. Angular rock fragments produced by frost shattering are termed felsenmeer.

Salts, the chemical compounds formed from reactions between acids and bases, are very important in causing rock breakdown. The effects of salt attack can best be seen in arid, coastal and urban environments, where salts are available and rocks routinely experience desiccating conditions that allow the salts to crystallize. The most common salts found in rocks are halite (NaCl), gypsum ($Ca_2SO_4 \cdot 2H_2O$), sodium sulphate (Na_2SO_4), magnesium sulphate ($MgSO_4$), sodium carbonate (Na_2CO_3) and sodium nitrate ($NaNO_3$) and their hydrated forms. The stresses causing rock breakdown are produced by three mechanisms: crystallization of salts in rock cracks and voids, expansion of salt crystals on hydration and thermal expansion of crystals (Cooke and Smalley 1968).

The most potent cause of salt weathering is salt crystal growth (Goudie and Viles 1997). Crystal growth occurs as a result of a saline solution becoming saturated as evaporation occurs and/or temperature changes, or by mixing of salts in solution, termed the 'common ion effect' (Goudie 1989). The role of salt crystallization in causing breakdown depends on the pore-size distribution and the pore connectivity of the material, as well as its overall strength. A second cause of stress arises from the capacity of certain common salts to take significant quantities of water into their structure. This hydration causes volumetric expansion of the salt and may exert pressure on crack and pore walls. Sodium sulphate, for example, will expand by 313 per cent on hydration. Hydration expansion may occur in response to changes in relative humidity, which, as this is closely related to temperature, may be diurnal. The extent to which salts expand when they are heated depends on their thermal characteristics and the temperature ranges to which they are subjected. Most commonly occurring salts have coefficients of expansion higher than those of rocks, yet simulation studies have not demonstrated differential thermal expansion to be a very effective weathering mechanism in isolation from other effects of salts. One reason may be that rocks only experience thermal cycling in a shallow surface layer, so subsurface salts probably do not experience significant diurnal temperature changes.

Evidence exists for the severe damage caused by salts to many rock types in a range of environments. Weathering forms produced will vary with lithology, though landforms produced by weathering are usually small or minor forms. Salt weathering processes may produce CAVERNOUS WEATHERING, flaking, scaling and GRANULAR DISINTEGRATION of the surfaces of most rock types; porous sedimentary rocks are particularly susceptible. Decay to stone used in buildings and monuments is commonly caused by salt attack, induced by salt-rich environmental conditions (Cooke *et al.* 1993) and by polluted conditions in urban atmospheres (Cooke and Doornkamp 1990).

Other material which can expand in rock voids, causing internal stresses and eventual breakup, include plant material. Growth of plant roots or lichen thallus, for example, in cracks in rock may have a biophysical effect in creating growth stresses. The likely tensile stresses created are, however, smaller (3 MPa) than the tensile strength of most rocks. The impact of ORGANIC WEATHERING is the result of a complex suite of biochemical and biophysical processes, and cannot be explained by mechanical weathering processes alone.

References

Cooke, R.U., Brunsden, D., Doornkamp, J.C. and Jones, D.K.C. (1993) *Urban Geomorphology in Drylands*, Oxford: Oxford University Press.

Cooke, R.U. and Doornkamp, J.C. (1990) *Geomorphology in Environmental Management*, 2nd edition, Oxford: Oxford University Press.

Cooke, R.U. and Smalley, I.J. (1968) Salt weathering in deserts, *Nature* 220, 1,226–1,227.

Goudie, A.S. (1989) Weathering processes, in D.S.G. Thomas (ed.) *Arid Zone Geomorphology*, 11–24, London: Belhaven Press.

Goudie, A.S. and Viles, H. (1997) *Salt Weathering Hazards*, Chichester: Wiley.

Hall, K. and Hall, A. (1991) Thermal gradients and rock weathering at low temperatures: some simulation data, *Permafrost and Periglacial Processes* 2, 103–112.

McGreevy, J.P. and Whalley, W.B. (1985) Rock moisture content and frost weathering under natural and experimental conditions; a comparative discussion, *Arctic and Alpine Research* 17, 337–346.

Ollier, C.D. and Ash, J.E. (1983) Fire and rock breakdown, *Zeitschrift für Geomorphologie* 27, 363–374.

Yatsu, E. (1988) *The Nature of Weathering: An Introduction*, Tokyo: Sozosha.

Further reading

Amoroso, G.G. and Fassina, V. (1983) *Stone Decay and Conservation*, Materials Science Monographs 11, New York: Elsevier.

Bland, W. and Rolls, D. (1998) *Weathering: An Introduction to the Scientific Principles*, London: Arnold.

Ollier, C.D. (1984) *Weathering*, 2nd edition, London: Longman.

Whalley, W.B. and McGreevy, J.P. (1985) Weathering, *Progress in Physical Geography* 9, 559–581.

Whalley, W.B. and McGreevy, J.P. (1987) Weathering, *Progress in Physical Geography* 11, 357–369.

Whalley, W.B. and McGreevy, J.P. (1988) Weathering, *Progress in Physical Geography* 12, 130–143.

SEE ALSO: deep weathering; frost and frost weathering; honeycomb weathering; hydration; organic weathering; pressure release; salt weathering; wetting and drying weathering

ALICE TURKINGTON

MECHANICS OF GEOLOGICAL MATERIALS

Newton's laws of motion provide the basis for the science of classical mechanics, and the mechanics of geological materials involves mostly a special branch of classical mechanics called continuum mechanics. Continuum mechanics utilizes the same physical laws that govern motion of discrete bodies such as planets and billiard balls, but continuum mechanics addresses the internal deformation of bodies that cannot be idealized as discrete points.

Familiar geomorphological examples of deformable bodies that can be analyzed using continuum mechanics include water flowing in streams and soil moving down slopes. In treating such bodies as continua, a basic assumption is that the elemental constituents of the bodies (e.g. molecules of water or grains of soil) are minuscule relative to the scale of observable phenomena of interest. Observable motion of streams or slope debris typically involves momentum exchange amongst billions of water molecules or soil grains, and it is consequently impractical to analyse the movements of individual molecules or grains to predict the behaviour of the aggregate. A logical alternative is to analyse the aggregated behaviour of a mass of grains or molecules by treating them as a single deformable body, or 'continuum'.

Numerous textbooks provide excellent introductions to continuum mechanics. Examples range from mathematically rigorous treatises (e.g. Malvern 1997) to introductory books aimed at the interests of Earth scientists (e.g. Middleton and Wilcock 1994). An oft-overlooked but very informative book is the small classic by Jaeger (1971).

Continuum conservation laws

The central principles in geological continuum mechanics are conservation of mass, momentum and energy. Although both momentum and energy conservation apply in every geomorphological setting, momentum conservation is generally the more useful principle, because momentum is a vector quantity that includes information about the direction of motion, whereas energy is a scalar. However, if thermal effects or phase changes are important (as in melting or formation of ice, for example) energy conservation must be considered explicitly, in addition to conservation of momentum and mass. The discussion below assumes that thermal effects are negligible, and emphasizes purely mechanical behaviour that arises solely from conservation of momentum and mass.

The basic equations of mass and momentum conservation describe behaviour in four dimensions (space + time), and they apply to any continuous body, regardless of its composition or state (solid, liquid or gas). The equations can be written in mathematical vector notation (with a brief English translation beneath) as:

$$\text{mass conservation: } \frac{\partial \rho}{\partial t} + \nabla \cdot \rho \vec{v} = 0$$

local rate of mass increase + rate of mass efflux due to deformation = 0

$$\text{momentum conservation: } \rho \frac{\partial \vec{v}}{\partial t} + \rho \vec{v} \nabla \cdot \vec{v} = \rho \vec{g} + \nabla \cdot T$$

local rate of momentum increase + rate of momentum efflux due to deformation = force imposed by gravity + internal reaction force (stress).

In these equations the dependent variables are ρ, the local mass density within the continuous substance, and $\vec{v}$, the local vector velocity, which can vary as a function of position and time, t. In many geomorphological phenomena, the only imposed driving force is the so-called 'body force' $\rho\vec{g}$ due to gravitational acceleration, $\vec{g}$. Additional imposed forces can be included if necessary.

The final quantity in the momentum-conservation equation is the stress, T, which represents the reaction forces (per unit area) that develop within a deformable body as a consequence of the interaction between the driving force and local accelerations. Unlike a rigid body, in which the action-reaction involving gravitational driving force and acceleration can be represented by a familiar form of Newton's second law, $\rho(d\vec{v}/dt) = \rho\vec{g}$, a deformable, continuous body can react to external forcing by generating internal stress. Thus, the concept of stress is a key one in continuum mechanics, and it should be mastered by all students of physical geomorphology. In general stress is a second-order tensor quantity (commonly represented mathematically by a 3×3 matrix), and the stress tensor is symmetric (which eliminates the need for a separate angular momentum equation in addition to the linear momentum equation above). Textbooks that explain stress and tensors in the context of geological sciences include an excellent one by Means (1976).

An important general feature of stress is known as 'static indeterminacy'. This condition dictates that even if a continuous substance is motionless and the momentum equation above reduces to $\rho\vec{g} + \nabla \cdot T = 0$, the stresses cannot be calculated without specifying a 'constitutive equation' that summarizes the mechanism of stress generation. The only mechanical analyses in which static indeterminacy and the need for constitutive equations can be circumvented are analyses in which stress is assumed to vary in one direction only. These 'one-dimensional' analyses can be useful for building insight but seldom provide accurate models of multidimensional geomorphic phenomena.

Constitutive equations and rheology

Two main branches of continuum mechanics, solid mechanics and fluid mechanics, have developed from observations of the fundamentally different mechanisms by which solids and fluids (liquids and gases) generate stress. In solid mechanics a pivotal observation is that, for sufficiently small deformations, stress is simply proportional to the magnitude of deformation (or, more precisely, proportional to the magnitude of strain, which may differ subtly from deformation). First quantified by Robert Hooke (1635–1703), this observation led to the theory of linear (or 'Hookean') elasticity. A similarly pivotal postulate, first made by Isaac Newton (1642–1727), was that fluids deform such that stress is simply proportional to the *rate* of deformation. Subsequently confirmed by experiments, this postulate led to the linear (or 'Newtonian') theory of viscous fluid flow.

Equations that relate stress to deformation are known as constitutive equations, because they express the influence of a body's constitution (or rheology) on the internal reaction forces that generate stress. In effect, constitutive equations (i.e. rheological models) serve as surrogates for momentum exchange that occurs at scales too small to be resolvable in a particular setting or observation (for example, momentum exchange by colliding water molecules in a stream observed by eye). The constitutive equations for linearly elastic solids and linearly viscous fluids may be written in simple forms as

$$\mathrm{T} = \epsilon D \text{ (linearly elastic behaviour)}$$

$$\mathrm{T} = \eta \dot{D} \text{ (linearly viscous behaviour)}$$

where ϵ represents elastic moduli, η represents dynamic viscosity, D represents deformation, and $\dot{D}$ represents deformation rate. Whole treatises expound the detailed meaning of these equations and the detailed definition of D and $\dot{D}$ (which are tensor quantities, like stress). Here it suffices to emphasize that these well-known constitutive equations express straightforward connections between stress and a measurable macroscopic quantity such as D or $\dot{D}$.

Stresses in Earth-surface fluids such as water and air can be represented with fair accuracy by a simple constitutive equation describing linearly viscous behaviour, and stresses in a solid rock can be represented with similarly good accuracy by linearly elastic behaviour (provided the rock does not fracture). However, many materials encountered in geomorphology are not so simple. Rocks, soils and sediments that undergo large, irreversible deformation (as might occur in a landslide, for example) exhibit neither linearly viscous nor linearly elastic behaviour. Diverse constitutive equations have been proposed to represent such behaviour, but the discussion here introduces only the most significant equation, the Coulomb model.

Experiments demonstrate that stresses in soils, sediments and fragmented rocks undergoing large deformations adjust plastically to maintain limiting values, independent of the deformation magnitude or rate. These limiting 'plastic yield' stresses depend principally on friction due to rubbing and interlocking of adjacent grains, and to a lesser degree on cohesive bonding between grains. On planes of shearing the limiting shear stress τ is described by

$$\tau = \sigma \tan \phi + c \text{ (Coulomb plastic behaviour)}$$

where σ is the normal stress on the shear plane, φ is the angle of internal friction, and c is the cohesion. This equation, first posited by Charles Augustin Coulomb (1736–1806), has withstood repeated testing but does not provide the same straightforward interpretation as the linearly elastic and viscous equations noted above. Nonetheless, the Coulomb model has proved very useful for analysing phenomena such as LANDSLIDES, DEBRIS FLOWs and incipient motion of BEDLOAD particles in streams.

Initial and boundary conditions

In addition to conservation laws and constitutive equations, continuum mechanics requires specification of initial conditions that isolate a phenomenon in time as well as boundary conditions that isolate it in space. (For example, to mechanically analyse the behaviour of a flood, one must specify the initial channel geometry and the distributions of water-surface elevations and velocities prior to the flood onset.) Appropriate specification of these 'auxiliary conditions' can be the crux of successful mechanical modelling, because geomorphological phenomena seldom occur in isolation from the surrounding environment. Nonetheless, mechanical analyses of such 'open' geomorphological systems can provide key insights if auxiliary conditions are specified with care and precision (Iverson 2003).

References

Iverson, R.M. (2003) How should mathematical models of geomorphic processes be judged? in P.R. Wilcock and R.M. Iverson (eds) *Prediction in Geomorphology*, Washington, DC: American Geophysical Union.

Jaeger, J.C. (1971) *Elasticity, Fracture and Flow with Engineering and Geological Applications*, London: Chapman and Hall.
Malvern, L.E. (1997) *Introduction to the Mechanics of a Continuous Medium*, 2nd edition, Englewood Cliffs: Prentice-Hall.
Means, W.D. (1976) *Stress and Strain (Basic Concepts of Continuum Mechanics for Geologists)*, New York: Springer-Verlag.
Middleton, G.V. and Wilcock, P.R. (1994) *Mechanics in the Earth and Environmental Sciences*, Cambridge: Cambridge University Press.

RICHARD M. IVERSON

MEGAFAN

Large fluvial depositional features that have been defined thus (Horton and DeCelles 2001: 44):

> Fluvial megafans constitute volumetrically significant depositional elements of sedimentary basins adjacent to mountain belts. A fluvial megafan is a large (10^3-10^5 km^2), fan-shaped (in plan view) mass of clastic sediment deposited by a laterally mobile river system that emanates from the outlet point of a large mountainous drainage network. Modern fluvial megafans have been recognized in nonmarine foreland basin systems at the outlets of major rivers that drain fold-thrust belts, particularly in the Himalayas and northern Andes. Although fluvial megafans are similar to sediment gravity flow-dominated and stream-dominated alluvial fans in terms of their piedmont setting, planform geometry and sedimentation related to expansion of flow downslope of a drainage outlet, fluvial megafans are distinguished by their greater size (alluvial fans rarely exceed 250 km^2), lower slope, presence of floodplain areas and absence of sediment gravity flows. The term 'terminal fan' is commonly used for a large, distributary fluvial system in which surface water infiltrates and evaporates before it can flow out of the system. A fluvial megafan therefore may be considered 'terminal' in cases where fluvial channels run dry before reaching bodies of water downstream. Although fluvial megafans are clearly related to the emergence of a large mountain river onto a low-relief alluvial plain, their stratigraphic evolution in nonmarine foreland basins may also be critically dependent on variables such as sediment flux, water discharge, drainage catchment size, catchment lithology and subsidence rate, factors that are ultimately controlled by tectonic, climatic and geomorphic processes.

Alternative names include megacone, inland delta, wet alluvial fan and braided stream fan. Some especially impressive megafans occur on the north side of the Ganga plain in India. They are fed from the Himalayas and may have formed in the Late Pleistocene when coarser grained sediment and high sediment and water discharges were available (Shukla *et al.* 2001).

Not all megafans occur in such dramatic settings as those of the Andean and Himalayan forelands. For example, the Okavango Fan of northern Botswana has accumulated in a graben and has been deposited by a low sinuosity/meandering river rather than a braided stream system (Stannistreet and McCarthy 1993).

References

Horton, B.K. and DeCelles, P.G. (2001) Modern and ancient fluvial megafans I. The foreland basin system of the Central Andes, southern Bolivia: implications for drainage network evolution in fold-thrust belts, *Basin Research* 13, 43–63.
Shukla, U.K., Singh, I.B., Sharma, M. and Sharma, S. (2001) A model of alluvial megafan sedimentation: Ganga Megafan, *Sedimentary Geology* 144, 243–262.
Stannistreet, I.G. and McCarthy, T.S. (1993) The Okavango Fan and the classification of subaerial fan systems, *Sedimentary Geology* 85, 115–133.

A.S. GOUDIE

MEGAGEOMORPHOLOGY

The term 'mega-geomorphology' was introduced in 1981 to designate a London conference of the British Geomorphological Research Group (Gardner and Scoging 1983). The original intent was for the term to apply to geomorphology on the scale of plate tectonics, biological evolution and macro-climatic change. Thus, it was to be concerned with entire landscapes, through histories of millions of years, in the context of continental or macro-regional evolution.

In 1985 another conference, 'Global Megageomorphology', was organized in Oracle, Arizona, and this meeting further refined geomorphological study at the largest spatial and temporal scales (Baker and Head 1985). Particularly relevant issues for the Oracle conference were the use of orbital remote sensing procedures to produce global mapping and analyses, studies of continental-scale denudation, the relationship of geomorphology to regional tectonics, global environmental change, and the geomorphology of other rocky planets besides Earth. The last issue is particularly important in both philosophical and historical perspectives. The history of geomorphological study of Earth began with scientists observing their immediate surroundings, and then generalizing to explanations that placed those observations in a larger context. In contrast, the study of other rocky planetary surfaces always begins at the largest spatial scales, through the remote sensing instrumentation of flyby planetary missions. The geomorphological understanding of Mars and Venus, the most Earthlike of the known planets, began at the megascale, while understanding of Earth's surface began at small scales and only evolved to the megascale after a long history of observations at the scales that were most accessible to human observers. A result of this history is the set of theories that currently constrains geomorphological inquiry to somewhat limiting viewpoints (Baker and Twidale 1991; Baker 1993). Megageomorphology affords an opportunity to break the constraints and develop new sets of theories.

The combined issues of scale and viewpoint are highlighted by a survey in the United Kingdom that found for the early 1980s that 75 per cent of geomorphological research concerned small-scale, modern process studies and another 15 per cent concerned Quaternary studies (Gardner and Scoging 1983). This emphasis on small scales of time and space is indicative of a reductionistic epistemological perspective in which one presumes that small-scale studies will integrate over time to generate a theory of the whole. Small-scale, modern process studies also tend to minimize the geomorphological role of rare processes of extreme magnitude because these are both remote from possible direct observation and destructive of most attempts to measure (Baker 1988). In contrast, as shown in Figure 107, the processes responsible for landforms and landscapes operate over a broad range of spatial and temporal scales. Moreover, the responses to these processes add further extensions of time and space to the zone of relevant natural operation for geomorphological change. One must conclude, 'Contemporary process studies are of little worth in evaluating landscape evolution' (Church 1980).

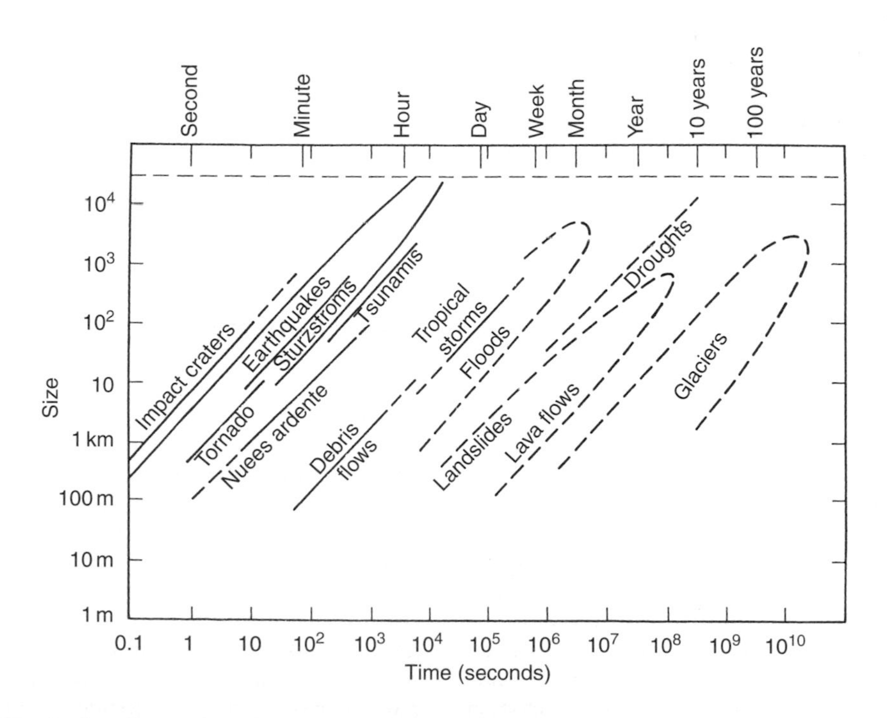

Figure 107 Scale relationships for various geomorphological processes, based on Carey (1962)

Many geomorphological systems are nonlinear, with thresholds that involve negligible responses at the process scales most easily measured in the field, while the relevant processes are both rare and of extreme magnitude (Baker 1988). The debates over the origin of the Channeled Scabland in eastern Washington in the 1920s illustrate this issue because a very large-scale process of immense magnitude was involved (Baker 1978). It is only because the process of megaflooding was recognized and understood for the Channeled Scabland that subsequent work has been able to show that cataclysmic megafloods played the dominant role in landscape development for other parts of Earth, and, somewhat surprisingly, for parts of Mars as well (Baker 2002).

Modern megageomorphology makes extensive use of global observations from spacecraft that employ a variety of imaging and remote-sensing instrumentation, including multispectral imaging, radiometers and radars. Image processing of digitally formated data has revolutionized the ability to study landscapes at the largest spatial scales. The theme of global-scale remote sensing is developed in the book *Geomorphology from Space* (Short and Blair 1986). A full scheme of geomorphology at large spatial scales is readily achieved in regard to the new technologies (Baker 1986).

Other technological advances that are stimulating megageomorphology include the quantitative geochronology of geomorphic surfaces and deposits; the widespread availability of digital topography; and the mathematical modelling of landscape evolution. These elements are creatively employed in the rapidly developing subfield of tectonic geomorphology (Burbank and Anderson 2001). In essence, the tectonic geomorphologist performs thought experiments with the computer, then tests those notions against the topographic response and the ages of elements in the landscape, such as glacial moraines, stream terraces and denudation surfaces. This can be done for entire regions that comprise the major tectonic elements of the planet.

The two major factors for geomorphological evolution at very large scales are tectonic factors arising from forces inside Earth (with the occasional imposition of extraterrestrial impacts) and the denudational factors arising from Earth's atmosphere, as largely summarized in its climate. Both these concerns are critical to any modern sense of global geomorphology (Summerfield 1991). The understanding of Earth tectonics was revolutionized in the 1960s and 1970s by the plate-tectonic theory. During the 1980s and 1990s climate became the central issue for international initiatives to understand global environmental change and the operation of Earth as a system. With the theoretical underpinnings of these two major elements, megageomorphology is now poised for major development as a science.

It is particularly interesting that a form of megageomorphology developed after 1945 in the former Soviet Union. Because of its practical application to mineral and petroleum exploration, the details of this science were subject to state secrecy, and its community of practitioners became somewhat isolated from the greater world scientific community. The focus of this science was the study of morphostructures, which are linear and circular elements of continental landscapes controlled by tectonics and denudation. The morphostructures exhibit hierarchical relationships with one another, and they evolve over geological timescales (Baker *et al.* 1993). Morphostructural analyis deciphers the complex interaction of long-term endogenetic processes with surface relief. It proved especially useful in identifying various concentric, circular or oval, and linear hidden dislocations of basement rocks that on Earth are commonly obscured by sedimentary or volcanic cover, deformations and intrusions. Though not tied to the modern plate-tectonic ideas about Earth's large-scale evolution, morphostructures are surprisingly similar to various quasi-circular shaped upland regions discovered on Venus by the planetary missions of the 1980s and 1990s. These also exist in hierarchical arrangement and may be related to mantle plume tectonics (Finn *et al.* 1994). Merely thinking at the megascale led to some surprising results.

Sharp (1980: 231) observes, 'One of the lessons from space is to "think big"'. This theme, combined with new analytical tools for geomorphological study at very large scales affords an opportunity for discovery and scientific excitement. Unifying models of global tectonics and climate dynamics afford the scientific framework for large-scale studies. Orbital remotes sensing, digital topographic data and geochemical tools for dating Earth history all permit the quantification of geomorphological parameters in greatly expanded temporal and spatial domains. Moreover, the science is immensely stimulated by the discovery of entire new landscapes on the surfaces of other rocky planetary bodies.

References

Baker, V.R. (1978) The Spokane Flood controversy and the Martian outflow channels, *Science* 202, 1,249–1,256.
——(1986) Introduction: regional landforms analysis, in N.M. Short and R.W. Blair, Jr (eds) *Geomorphology from Space: A Global Overview of Regional Landforms*, 1–26, Washington, NASA Special Publication 486.
——(1988) Cataclysmic processes in geomorphological systems, *Zeitschrift für Geomorphologie Supplementband* 67, 25–32.
——(1993) Extraterrestrial geomorphology: science and philosophy of Earthlike planetary landscapes, *Geomorphology* 7, 9–35.
——(2002) High-energy megafloods: planetary occurrences and sedimentary dynamics, in I.P. Martini, V.R. Baker and G. Garzon (eds) *Flood and Megaflood Processes and Deposits: Recent and Ancient Examples*, 3–15, International Association of Sedimentologists Special Publication 32.
Baker, V.R. and Head, J.W., III (1985) Global megageomorphology, in R.S. Hayden (ed.) *Global Geomorphology*, 113–120, Washington, NASA Conference Publication 2,312.
Baker, V.R. and Twidale, C.R. (1991) The reenchantment of geomorphology, *Geomorphology* 4, 73–100.
Baker, V.R., Finn, V.I. and Komatsu, G. (1993) Morphostructural megageomorphology, *Israel Journal of Earth Sciences* 41, 65–73.
Burbank, D.W. and Anderson, R.S. (2001) *Tectonic Geomorphology*, Oxford: Blackwell.
Carey, S.W. (1962) Scale of geotectonic phenomena, *Journal of the Geological Society of India* 3, 97–105.
Church, M.A. (1980) Records of recent geomorphological events, in D. Cullingford, D.A. Davidson and J. Lewin (eds) *Timescales in Geomorphology*, 13–29, New York: Wiley.
Finn, V.S., Baker, V.R., Dolginov, A.Z., Gabitov, I. and Dyachenko, A. (1994) Large-scale spatial patterns in topography at Alpha Regio, Venus, *Geophysical Research Letters* 22, 1,901–1,904.
Gardner, R. and Scoging, H. (eds) (1983) *Mega-Geomorphology*, Oxford: Clarendon Press.
Sharp, R.P. (1980) Geomorphological processes on terrestrial planetary surfaces, *Annual Reviews of Earth and Planetary Sciences* 8, 231–261.
Short, N.M. and Blair, R.W., Jr (eds) (1986) *Geomorphology from Space: A Global Overview of Regional Landforms*, Washington, NASA Special Publication 486.
Summerfield, M.A. (1991) *Global Geomorphology*, Harlow, Essex: Longman.

SEE ALSO: extraterrestrial geomorphology; global geomorphology; tectonic geomorphology

VICTOR R. BAKER

MEKGACHA

A Setswana term for the DRY VALLEY systems which traverse the flat, sandy terrain of the Kalahari region of southern Africa. These broad, shallow, drainage features contain CALCRETE and SILCRETE in their floors and flanks and are also referred to as *laagte, omuramba* or *dum* in other regional languages. The origins of *mekgacha* (singular *mokgacha*) have been ascribed to episodes of permanent or ephemeral fluvial activity during wetter periods of the Quaternary, and, at longer timescales, to RIVER CAPTURE, groundwater sapping or DEEP WEATHERING focused along geological lineaments (Nash *et al.* 1994). Two groups of valleys can be identified (Shaw *et al.* 1992). The first are the exoreic (externally directed) Auob, Nossop, Kuruman and Molopo systems which drain the southern Kalahari and connect to the Orange River. The second are the endoreic (internally directed) systems which focus upon the Okavango Delta and Makgadikgadi Depression. Surface runoff is comparatively rare within the Kalahari and, as a result, flow within *mekgacha* is unusual. However, most exoreic systems are spring-fed and contain water in their headwater sections, with more extensive flooding following prolonged rainfall. In contrast, the endoreic systems are effectively fossil networks and only contain water in seasonal pools. Floods may occur in such networks after exceptional rainfall, but only two have been documented in the historical period (1851 and 1969), both in the Letlhakane valley (Nash and Endfield 2002).

References

Nash, D.J. and Endfield, G.H. (2002) Historical flows in the dry valleys of the Kalahari identified from missionary correspondence, *South African Journal of Science* 98, 244–248.
Nash, D.J., Thomas, D.S.G. and Shaw, P.A. (1994) Timescales, environmental change and valley development, in A.C. Millington and K. Pye (eds) *Environmental Change in Drylands*, 25–41, Chichester: Wiley.
Shaw, P.A., Thomas, D.S.G. and Nash, D.J. (1992) Late Quaternary fluvial activity in the dry valleys (*mekgacha*) of the Middle and Southern Kalahari, southern Africa, *Journal of Quaternary Science* 7, 273–281.

DAVID J. NASH

MELTWATER AND MELTWATER CHANNEL

Meltwater can be produced from the melting of snow (nival meltwater) or GLACIERs (glacial meltwater). The nival and glacial meltwater regimes represent two ends of a spectrum, as in many instances there may be contributions of both within the same catchment. The amount of meltwater produced is determined by the energy balance, a key component of which is solar radiation, thus most melt will occur at the ice or snow surface. Melting can also occur at the base of a glacier due to either geothermal heat or the effects of high pressure. However, melting will predominantly be related to air temperature, thus the temporal pattern of meltwater generation is not constant and varies over daily, annual and longer timescales. In nival regimes highest flows tend to occur in spring (the start of the melt season) as temperatures begin to rise above freezing. Meltwater discharge will usually exhibit a diurnal pattern, with a peak in the afternoon and a low around dawn, related to the cycle of daily temperature change. Over time, as less snow remains to be melted, diurnal fluctuations become more muted and flow will eventually cease altogether. Glacial meltwater production also exhibits a similar type of pattern, except that the melt season is longer and peak meltwater discharge is later. Glacial meltwater discharge displays a gradual rise through the melt season as the melting of snow on the ice surface is then followed by glacier melt. Peak flows in mid/late summer when the entire glacier, devoid of the insulating surface snow cover, contributes to melt. Outside the summer melt season flows can become very low or cease altogether. The specific meltwater regime is very closely tied with the environment, for example, glaciers in more polar environments will tend to have a shorter melt season and more muted diurnal discharge signature than those of mountain glaciers in temperate locations (Tranter *et al.* 1996). In addition to these more regular and predictable discharge regimes, meltwater can also be released catastrophically in OUTBURST FLOODs.

Once snow or ice has melted it can take several pathways with respect to its exit from an ice mass: as surface (supraglacial) channels, pipes or conduits within (englacial) the ice or as flow at the base (subglacial) of the ice (Shreve 1972). Supraglacial channels can be up to a few metres deep and because the ice is smooth water velocities are high (3–7 ms^{-1}). Supraglacial water will either flow off the ice surface or descend vertically into the ice via holes called MOULINs where the water connects with the pipes or conduits of the englacial system. Englacial pipes range in diameter from a few millimetres to a few metres, water can also flow through thin veins between ice crystals, but at a greatly reduced rate. Englacial water will often connect to the subglacial flow system at the base of the glacier. Where a cold-ice basal thermal regime predominates the supraglacial drainage system is often dominant, in a warm-ice situation a complex system involving all three types of drainage is often present. However, the drainage system is not fixed within the ice mass and will change and develop through the melt season as some channels or pipes open up and others close. In addition to this pattern of flow routes, meltwater can also be stored in supraglacial, englacial or subglacial lakes and ponds.

Subglacial meltwater has the biggest significance in terms of glacier dynamics and also in terms of providing evidence of meltwater flow in the landscape once the ice has receded. Subglacial meltwater significantly increases the rate of basal sliding and hence how fast the ice will move. However, if there is a highly permeable substrate at the base of the ice then very little meltwater flow may occur at the ice–bed interface, in which case basal sliding will not be greatly enhanced. A key factor about subglacial meltwater (and also englacial) is that due to the pressure of the weight of the overlying ice, the water can be flowing at much greater pressures than would be experienced under normal atmospheric pressure conditions at the surface. The direction of flow is determined by the ice surface slope and to a lesser extent the bed topography, such that uphill flow is possible (Shreve 1972). As the ice surface steepens, subglacial flow becomes less influenced by bed topography. Flow at the base will thus trend in the general direction of ice flow, often following valley floors and crossing divides at the lowest point.

Subglacial meltwater flow can occur in a distributed or discrete system. Distributed systems include sheet flow or linked cavity networks, discrete systems encompass the full range of channelized flows. Sheet flow is where meltwater exists as a thin continuous layer between the ice and the bed. Research has shown that this type of flow is relatively unstable and that flow is more likely to occur in channels (Shreve 1972). The linked cavity system is where basal hollows are linked by narrow short connections. Channelized systems can be cut down into the bed (Nye channels or N-type), cut up into the overlying ice (Röthlisberger channels or R-type) or display characteristics of both. The discrete (channelized) drainage systems transport meltwater more efficiently than the more circuitous pathways of the distributed system. The type of basal meltwater system thus has an important influence on water pressure and hence glacier motion. Meltwater can switch between these different flow systems and it has been suggested that this can trigger GLACIER surging.

Nye channels cut into bedrock or consolidated sediments leave the most distinctive imprint on the land surface once the ice has retreated. Due to the pressure conditions experienced by subglacial meltwater and the nature of meltwater supply to the subglacial system Nye channels often display very different characteristics to a normal subaerial fluvial system. For example, they can possess an undulating long profile, very steep gradients, lack any significant drainage basin and have an abrupt inception or termination (Glasser and Sambrook Smith 1999). The nature of the substrate beneath the ice will have an important influence on the overall channel morphology. For example, it is not uncommon for a channel to change its morphology over very short distances as it goes from being deep, narrow and incised in bedrock and then wide and shallow when passing over a deformable bed (Plate 79). Channels can range greatly in size, from tens of metres up to 100 kilometres long or from tens of metres to several kilometres wide. Once the channels reach the kilometre scale they are often referred to as TUNNEL VALLEYs rather than Nye channels. Channels can occur as either isolated features or part of a much larger channel network. The most spectacular form of isolated meltwater channel is what is referred to as a chute channel; these often occur on the flanks of a bedrock slope and are very steep and incised. They are thought to form as rapidly descending water within an ice mass reaches the glacier bed and cuts down into it. In contrast to these isolated channels, networks of large tunnel valleys can extend over very large areas and have been reported from many parts of North America and Europe that were covered by ICE SHEETs. The origin of such large features is thought to be due to either the catastrophic release of water stored in large subglacial lakes or that they were cut by normal meltwater flows over extended time periods. By mapping networks of exposed meltwater channels and coupling this with the theory of meltwater flow under pressure it has been demonstrated that reconstructions of the likely extent and dynamics of an ice mass is possible (Sugden *et al.* 1991).

Plate 79 A meltwater channel (flow away from camera) incised ~7 m into bedrock, Cheshire, England. Downstream the channel becomes wider and shallower as it passes over unconsolidated sediment before disappearing completely 500 m from where this photograph was taken. Photograph taken by N.F. Glasser

As well as being erosive, meltwater can also be a significant agent of deposition. For example, an ESKER is a long narrow ridge of sediment often deposited in subglacial channels as a result of high rates of meltwater flow. When subglacial channels completely fill with meltwater they can behave like pipes, allowing water under pressure to move, and subsequently deposit, large volumes of sediment in a single event.

References

Glasser, N.F. and Sambrook Smith, G.H. (1999) Glacial meltwater erosion of the Mid-Cheshire Ridge: implications for ice dynamics during the Late Devensian glaciation of northwest England, *Journal of Quaternary Science* 14, 703–710.

Shreve, R.L. (1972) Movement of water in glaciers, *Journal of Glaciology* 11, 205–214.

Sugden, D.E., Denton, D.H. and Marchant, D.R. (1991) Subglacial meltwater channel system and ice sheet overriding of the Asgard range, Antarctica, *Geografiska Annaler* 73A, 109–121.

Tranter, M., Brown, G.H., Hodson, A.J. and Gurnell, A.M. (1996) Hydrochemistry as an indicator of subglacial drainage system structure: a comparison of alpine and sub-polar environments, *Hydrological Processes* 10, 541–556.

Further reading

Menzies, J. (1995) Hydrology of glaciers, in J. Menzies (ed.) *Modern Glacial Environments*, 197–239, Oxford: Butterworth-Heinemann.
Menzies, J. and Shilts, W.W. (1996) Subglacial environments, in J. Menzies (ed.) *Past Glacial Environments*, 15–136, Oxford: Butterworth-Heinemann.

SEE ALSO: glacifluvial; subglacial geomorphology

GREGORY H. SAMBROOK SMITH

MESA

Mesa (the word is of Spanish origin) is a steep-sided and flat-topped hill or ridge rising above a flat plain, usually built of flat-lying soft sedimentary rocks capped by a more resistant layer, e.g. of shales overlain by sandstones. In volcanic terrains a former lava flow may act as a caprock; likewise, this role may be assumed by a blanket of DURICRUST. Mesas originate due to unequal scarp retreat, in the course of which parts of the plateau become isolated and remain standing in front of the retreating scarp. Lava and duricrust-capped mesas may form due to dissection of a plateau or through the mechanism of relief inversion (see INVERTED RELIEF), then the distribution of mesas indicates the position of a former valley floor or lava flow.

Mesas steadily decrease in size through slope retreat accomplished mainly by various kinds of mass wasting and gully erosion; however, due to the presence of a resistant cap they may be very durable landforms surviving long after the initial scarp retreated. With time, they are reduced to BUTTEs, but there are no agreed criteria as to when a mesa becomes a butte. Mesas are climate-independent landforms although those in desert areas have the most distinctive appearance.

SEE ALSO: caprock; sandstone geomorphology; structural landform

PIOTR MIGOŃ

METHOD OF SLICES

Slope stability is at present routinely analysed by the Limit Equilibrium approach, derived somewhat loosely from the Theory of Plasticity. It is based on the assumption that the pattern of stress in a failing slope can be determined from static equilibrium, without the need to consider stress redistribution due to elastic and inelastic straining. The stresses acting on the boundary of the sliding body ('rupture surface') can then be compared with available strengths, to evaluate equilibrium. The 'FACTOR OF SAFETY' is usually defined as a ratio by which available soil or rock strength may be reduced, without causing failure.

The Limit Equilibrium approach was originally applied to the sliding of rigid portions of a slope, along assumed rupture surfaces. Coulomb was the first to calculate the Factor of Safety of a block above a planar surface. Later, circular ('Swedish Circle') surfaces were analysed, as it was observed that slides in clay are often rotational.

In the 'Method of Slices' (Fellenius 1927), the sliding body, viewed in cross section, is divided into vertical slices. An approximate assumption is made that the vertical stress at the base of each slice is constant and equal to the weight of the material column above it. In the 'Ordinary' ('Fellenius') method, all stresses acting at the vertical boundaries between adjacent columns are neglected. The equilibrium problem then becomes statically determinate and could be solved by simple vector analysis. The Factor of Safety is equal to the ratio of the sum of available strengths on all column bases, to the sum of applied shear stresses. The results are often excessively conservative.

Bishop (1955) realized that, with a circular rupture surface, it is unlikely for the inter-slice *shear* stresses to be very high. He therefore assumed these to be zero and derived the normal and shear stresses at the base of each column from vertical force equilibrium. The Factor of Safety can then be evaluated from the moment equilibrium of the slice assembly, without the need to neglect the normal inter-slice forces. This method neglects horizontal force equilibrium, but nevertheless produces very accurate results when compared with more sophisticated approaches, for circular and some non-circular surfaces.

More detailed methods have been derived, taking into consideration inter-slice shear and all three equilibrium conditions. Recently, a much more sophisticated approach, based on numerical stress–strain analysis of the slope, has been developed, so called 'Stress Reduction Method' (Dawson *et al.* 1999). One of the advantages of this method is that it removes the need to predetermine the shape of the rupture surface. However, it is much more difficult to implement and lacks the long track record of practical experience, inherent in the Method of Slices. Thus, for the foreseeable future, the Method of Slices will continue to be an important tool for the analysis of slope stability. An extension to a three-dimensional 'Method of Columns' has now also been developed (e.g. Hungr *et al.* 1989).

References

Bishop, A.W. (1955) The use of the slip circle in the stability analysis of slopes, *Géotechnique* 5(1), 7–17.

Dawson, E.M., Roth, W.H. and Drescher, A. (1999) Slope stability analysis by strength reduction, *Géotechnique* 49, 835–840.

Fellenius, W. (1927) *Erdstatische Berechnungen mit Reibung und Kohasion*, Berlin: Ernst.

Hungr, O., Salgado, F.M. and Byrne, P.M. (1989) Evaluation of a three-dimensional method of slope stability analysis, *Canadian Geotechnical Journal* 27, 679–686.

OLDRICH HUNGR

MICRO-EROSION METER

Many theories of landform evolution over time can only be tested through a knowledge of erosion rates. Earlier ideas of erosion sequences were constructed through logic and deduction from available morphological and sedimentary evidence. However, the answer to the question: 'could the landform have actually evolved in the timescale envisaged?' must involve a knowledge of rates. Two sets of endeavours at different scales devolve from this question, first large-scale measurements, for example on drainage basins, and smaller scale measurements. The latter may be performed more precisely and over a short space of time but the issue then remains of how far the results may be influenced by conditions at the time of measurement and of how they may be scaled up over a longer time span and a larger scale. Thus many of the data derived have been useful in short-term experimentation but extrapolation over longer time spans, over which conditions may be different to those obtaining at the present, remains an issue (Trudgill *et al.* 1989, 2001).

It was realized by High and Hanna (1970) that a micrometer dial gauge, used in engineering, could provide measurements of up to 0.0001 mm or even greater precision. A measurement is made of the height of a rock surface relative to some fixed datum. The instrument consists of a micrometer dial gauge and attached micrometer probe which is mounted onto a tripod framework. The tripod gives the instrument stability and it rests on three reference studs drilled securely into the rock. The measurement of the surface height of the rock relative to the studs can be made at repeated intervals, yielding results of surface lowering in mm a^{-1}. Initial meters had probes which were fixed and the tripod could be rotated to three positions, yielding three measurement points. Later meters (Trudgill *et al.* 1981; Trudgill 1983) used a traversing mechanism where not only was there a tripod base plate which could be rotated, additionally the dial gauge could itself be placed in several reference positions enabling a much larger number of points to be measured.

In terms of methodological limitations, Spate *et al.* (1985) suggest that erosion of the rock by the tip of the probe could lead to a fictitious annual loss of around 0.019 mm a^{-1}. They reported that when repeated measurements were made successively then surface lowering was recorded, the more so in softer limestones. For replicate readings at one site, on a harder (Buchan) limestone their data showed differences of 0.0001–0.0052 mm and for a softer (Gambier) limestone 0.0090–0.0284 mm. This may be initial surface compaction rather than actual erosion. Their data suggest that up to 0.0126–0.0284 for softer limestones and 0.0016–0.0052 for harder limestones will be an artefact of probe impact for any one measurement rather than actual erosion.

References

High, C.J. and Hanna, K. (1970) A method for the direct measurement of erosion on rock surfaces, *British Geomorphological Research Group, Technical Bulletin* 5.

Spate, A.P., Jennings, J.N., Smith, D.I. and Greenaway, M.A. (1985) The micro-erosion meter: use and limitations, *Earth Surface Processes and Landforms* 10, 427–440.

Trudgill, S.T. (1983) *Weathering and Erosion*, 104–110, London: Butterworth.

Trudgill, S.T., High, C.J. and Hanna, K.K. (1981) Improvements to the micro-erosion meter, *British Geomorphological Research Group, Technical Bulletin* 29, 3–17.

Trudgill, S.T., Viles, H.A., Inkpen, R.J. and Cooke, R.U. (1989) Remeasurements of weathering

rates, St. Paul's Cathedral, London, *Earth Surface Processes and Landforms* 14, 175–196.
Trudgill, S.T., Viles, H.A., Inkpen, R., Moses, C., Goshing, W., Yates, T. *et al.* (2001) Twenty-year weathering re-measurements at St. Paul's Cathedral, London, *Earth Surface Processes and Landforms* 26, 1,129–1,142.

STEVE TRUDGILL

MICROATOLL

Microatolls are discoid colonies of massive corals that grow in the lower intertidal zone on shallow reef flats. Corals are organisms that secrete a calcareous exoskeleton and they are major contributors to a CORAL REEF that may be preserved in the geological record as limestone. Microatolls are corals that grow at the reef-atmosphere interface, adopting a predominantly lateral growth form, constrained in their upward growth by exposure at lowest tides. This upper limit to coral growth reflects physiological factors that inhibit coral polyps when exposed.

Microatolls can be formed by several species of corals, but massive corals, such as *Porites*, are particularly prominent and more likely to be preserved (Plate 80). These growth forms have received particular attention because their flat upper surface is limited in terms of upward growth by exposure and is therefore related to regional sea level. There was initially debate as to whether the distinct form of microatolls might be due to sediment accumulation on top of the coral or nutrient limitation, but it has been demonstrated that it is water level that constrains vertical coral growth (Stoddart and Scoffin 1979). It is possible to examine the pattern of growth because corals form growth bands which are generally annual and which can be detected using X-radiography. The banding within microatolls confirms that growth has been primarily lateral and can indicate periods during which the limit to coral growth has been temporarily raised or lowered, preserved as undulations on the upper surface of the colony (Plate 81).

Plate 80 Microatolls on the reef flat in the Cocos (Keeling) Islands, Indian Ocean. The two large massive corals in the foreground comprise colonies of *Porites* that are no longer living on top, but which have live polyps confined to the margin. Their upper surface is constrained by exposure during lowest tides, and the concentric rim around the margin records a time at which this water level was slightly higher. The microatolls are approximately a metre in diameter and have been growing for several decades

Microatolls have been particularly important in the reconstruction of mid and late Holocene SEA-LEVEL change. The significance of microatolls was recognized during the 1973 Royal Society expedition to the Great Barrier Reef, northeastern Australia. Microatolls were surveyed on the surface of several different reefs and their growth form was used to indicate that sea level during the mid-Holocene had been higher than present (Scoffin and Stoddart 1978). Surveying of sequences of microatolls across transects along the mainland shore of Queensland, and cross-correlation using radiocarbon dating, provided convincing evidence that relative sea level had been above present level by more than a metre around 6,000 years ago, and demonstrated that it had undergone a smooth fall since that time (Chappell 1983). Much of the Indo-west Pacific reef province has experienced relative sea levels in mid and late Holocene that have been slightly above present. Microatolls have sometimes been preserved at a height presently above that of their living counterparts on reef flats in the eastern Indian Ocean, Southeast Asia, northern Australia, and across much of the equatorial Pacific Ocean, and provide evidence of sea-level change particularly on atolls (Smithers and Woodroffe 2000).

However, it is also clear that the banding structure of microatolls can preserve details of other events in the life history of the coral. In areas where storms are experienced, overturning of the colony can occur during individual storms, or microatolls may have responded to the moating of water that can occur behind boulder ramparts formed as a result of storms. In these cases their upper surface may record elevation of water level within impounded moats above that of regional

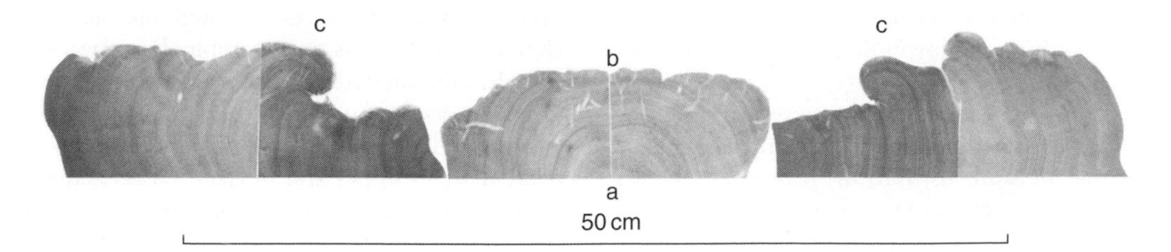

Plate 81 An X-radiograph of a vertical slice through a *Porites* microatoll. The banding indicates the growth form of the coral as the colony has aged. Initially the coral was hemispherical (a), but when it reached a level at which it was exposed too frequently during the lowest tides (b) it ceased upward growth and has extended laterally. Undulations on the surface, which occur symmetrically about the centre of the coral (c), record periods during which this upper limit to coral growth has been slightly higher

sea level. Although in open-water situations microatolls can enable centimetre-scale reconstructions of former sea level, they are subject to misinterpretation if moating has occurred, and it is therefore important to assess the geomorphological setting within which these corals grew before using fossil specimens to draw conclusions about past sea levels.

Fossil microatolls are often more accessible across the Indo-west Pacific reef province than massive hemispherical corals which are likely to have become buried by subsequent reef growth. Microatolls can be sampled along their horizontal growth axis in order to reconstruct a palaeoclimatological proxy record of the chemistry of surface waters during mid and late Holocene. Within living microatolls from the central Pacific Ocean interannual variations in water level indicate patterns of sea-level variation associated with EL NIÑO EFFECTS. Modern microatolls also enable oxygen isotope analyses of their skeleton preserving an important proxy of sea surface temperature which varies in association with the El Niño-related oscillations of sea level (Woodroffe and Gagan 2000). The application of these techniques to fossil microatolls offers the prospect of an insight into the palaeoclimatology of surface waters of more extensive areas within the tropical Indian and Pacific Oceans.

References

Chappell, J. (1983) Evidence for smoothly falling sea levels relative to north Queensland, Australia, during the past 6000 years, *Nature* 302, 406–408.

Scoffin, T.P. and Stoddart, D.R. (1978) The nature and significance of micro atolls, *Philosophical Transactions of the Royal Society of London* B284, 99–122.

Smithers, S.G. and Woodroffe, C.D. (2000) Microatolls as sea-level indicators on a mid-ocean atoll, *Marine Geology* 168, 61–78.

Stoddart, D.R. and Scoffin, T.P. (1979) Microatolls: review of form, origin and terminology, *Atoll Research Bulletin* 224, 1–17.

Woodroffe, C.D. and Gagan, M.K. (2000) Coral microatolls from the central Pacific record late Holocene El Niño, *Geophysical Research Letters* 27, 1,511–1,514.

SEE ALSO: atoll; reef

COLIN WOODROFFE

MICROMORPHOLOGY

The investigation of sediment and soil sequences has relied heavily on characterization of field morphological properties supplemented by data derived from laboratory analysis. Micromorphology, also called micropedology, extends this approach to the microscopic scale through the collection and analysis of undisturbed samples. The manufacture of soil thin sections involves the drying of samples (by air and oven drying or acetone replacement or freeze-drying), impregnation of resin under reduced pressure and curing of blocks, and sawing, lapping and mounting on glass slides. The final results are thin sections *c.*30 μm thick, which can be examined using a petrological microscope. It is possible to examine in thin sections the nature and spatial arrangement of such features as rock fragments, individual mineral grains, weathering products, void space, coatings of clay, organic material, precipitation of carbonate and excrement from soil animals.

W.L. Kubiena is widely regarded as the father figure to soil micromorphology and is best known for his book *The Soils of Europe* (Kubiena 1953). He very much established micromorphology's contribution to soil classification and our understanding of soil formation processes. Other benchmark contributions have been by Brewer and Sleeman (1988) in terms of the first attempt to systematize terminology, FitzPatrick (1993) for pedological interpretation and Bullock *et al.* (1985) for devising an international description system. The provision of full descriptions is extremely instructive since the investigator has to systematically examine all thin section attributes. Basic concepts from the international descriptive system are summarized in Tables 28 and 29. The application of micromorphology can be illustrated by summarizing some examples from soil science, archaeology and Quaternary science.

The application of micromorphology and its associated techniques provide distinctive insights into soil processes as induced by physical, biological and land use mechanisms (Miedema 1997). It has been used to determine the effects of different cultivation techniques on soils (Drees *et al.* 1994) and the interactions of soil and the soil biota (Kooistra 1991). As one example, Davidson *et al.* (2002) investigated the impacts of fauna on an upland grassland soil using micromorphological analysis. The incidence of eight different types of excrement in upper soil horizons was quantified using point counting. In both the organic horizon (H) and the underlying organo-mineral horizon (Ah), the bulk of the soil volume consisted of excrement derived primarily from enchytraeids and earthworms. Micromorphological analysis was thus able to demonstrate that the organic matter in these horizons was primarily derived from a limited range of soil fauna.

The early work on micromorphology as applied to archaeology very much focused on the palaeoenvironmental interpretation of buried soils as a means of providing environmental contexts for archaeological sites. This remains a key concern, but increasing attention is being given to site taphonomy and the wider impact of anthropogenic activity on soil landscapes. Courty *et al.* (1994) highlight the ways by which micromorphology can assist with understanding the relationships between environment and human

Table 28 Terminology in micromorphology

Soil fabric
Total organization of a soil as expressed by the spatial arrangements of the soil constituents (solids and voids)

Soil structure
Size, shape and arrangement of primary particles and voids in both aggregated and nonaggregated material and size, shape and arrangement of any aggregates present

Soil microstructures
Structures evident at magnifications $>\times 5$

Coarse and fine material
A division between coarse and fine material – division e.g. at 10 μm or 2 μm

Basic components
These are the simplest mineral and organic particles seen in thin section. They form the building blocks to the soil organization

Groundmass
General term to describe the coarse and fine material which forms the base material of soil

Micromass
General term for finer material

Pedofeature
A distinct fabric unit; stands out in contrast to adjacent soil material, e.g. clay coating in void, nodules of iron oxides

Table 29 Descriptive framework for micromorphology

1 *Structure*

2 *Groundmass*
Subdivided into coarse material and micromass

3 *Organic components*
(a) plant residues (>5 cells connected in original tissues)
(b) organic fine material (<5 cells and includes amorphous components)
(c) organic pigment staining – whole or part of micromass
Excrement pedofeatures can also be described under this heading

4 *Textural concentration features*
Features associated with an increase in concentration of material of a particular size, e.g. coatings, infillings, cappings

5 *Amorphous concentration features*
Appear amorphous in plane polarized light (PPL), isotropic in cross polarized light (XPL). Three types identified under oblique incident light are:
(a) white or dark brown colours – organic components
(b) black to yellowish brown – oxides and hydroxides of manganese
(c) yellowish brown to reddish brown – oxides and hydroxides of iron
These occur as e.g. nodules, segregations (e.g. mottles) which are impregnative

6 *Crystalline concentration features*
E.g. features consisting of Fe oxides, gypsum, gibbsite

behaviour. They summarize intra-site analysis (anthropogenically structural transformations, animal effects, anthropogenic deposits and spatial and temporal variability) and off-site analysis (land use practices and human-induced soil alteration and effects on landscape dynamics). As an example, they list micromorphological features and fabrics associated with such land use practices as slash and burn, up-rooting, ploughing, manuring, irrigation, horticultural practices and pasturing-herding. Thus ploughing, for example, results in fragmented slaking crusts, dusty silty clay intercalations and coatings, mixing of horizons, loss of the fine fraction, decrease in biological activity and changes in biological fabrics.

Micromorphology has contributed to Quaternary science, primarily through the investigation of buried soils. As an example, Zárate *et al.* (2000) demonstrate the particular contribution that micromorphological analysis can make to the investigation of a 5 m Holocene alluvial record in Argentina. Changes in the ratio of shells:diatoms:non-bioclastic coarse materials (e.g. quartz, feldspars) are used to propose different modes of deposition over time, for example the upper aeolian and alluvial units are dominated by non-bioclastic coarse particles. Two palaeosols are distinguished in thin sections by the presence of black isotropic, partially degraded root fragments, derived from the original vegetation cover. Microstructural changes down the section indicate the extent of pedogenic modification of the sedimentary fabric. The presence of partially welded excrements in the palaeosols indicates the former effect of soil fauna. Overall, the field micromorphological data provide the basis to a pedosedimentary reconstruction for the Holocene in this site in the Argentinian pampas. An alternative approach is to investigate soil formation on surfaces of different age. Srivastava and Parkash (2002) demonstrate the polygenetic nature of soils on the Gangetic Plains through micromorphological analysis of samples collected from surfaces ranging in age from 135,000 to < 500 BP. Of particular interest was the degradation of early clay pedofeatures by bleaching, loss of preferred orientation and the development of a coarse speckled appearance and fragmentation; in contrast, clay pedofeatures from more recent soils were thick, smooth, strongly birefringent and microlaminated.

Kemp (1998) overviews the contribution of micromorphology to palaeopedological research and he stresses the importance of relying on a combination of such features as channels, faunal excrements, calcitic root pseudomorphs, and illuvial clay coatings. He highlights fundamental challenges posed by the polygenetic nature of soils and the fact that in contemporary soils, attributes may not be in equilibrium with current environmental conditions. This leads him to discuss the equifinality problems – the same end result can come from varying combinations of processes. He illustrates this with reference to argillic horizons; in Quaternary studies the traditional view is to regard the accumulation of illuvial clay as occurring under temperate (interglacial), seasonally dry climates under stable forest cover. Kemp (1998) argues that such illuviation can occur under a range of environmental conditions and thus it would be erroneous to propose a particular palaeoenvironmental condition because of evidence of translocated clay.

In summary, micromorphological analysis can yield distinctive information on past processes of soil and sediment formation. The continuing development of image analytical techniques is encouraging more quantitative approaches. However, research using micromorphological analysis is usually combined with other approaches, for example, soil physical and chemical analysis, pollen analysis, magnetic susceptibility, organic geochemistry or microprobe analysis.

References

Brewer, R. and Sleeman, J.R. (1988) *Soil Structure and Fabric*, East Melbourne: CSIRO Publications.

Bullock, P., Fedoroff, N., Jongerius, A., Stoops, G., Tursina, T. and Babel, U. (1985) *Handbook for Soil Thin Section Description*, Wolverhampton: Waine Research Publications.

Courty, M.A., Goldberg, P. and Macphail, R.I., (1994) Ancient people – lifestyles and cultural patterns, *Proceedings 15th World Congress of Soil Science*, 250–269.

Davidson, D.A., Bruneau, P.M.C., Grieve, I.C. and Young, I.M. (2002) Impacts of fauna on an upland grassland soil as determined by micromorphological analysis, *Applied Soil Ecology* 20, 133–143.

Drees, L.R., Karathanasis, A.D., Wilding, L.P. and Blevins, R.L. (1994) Micromorphological characteristics of long-term no-till and conventionally tilled soils, *Soil Science Society of America Journal* 58, 508–517.

FitzPatrick, E.A. (1993) *Soil Microscopy and Micromorphology*, Chichester: Wiley.

Kemp, R.A. (1998) Role of micromorphology in paleopedological research, *Quaternary International* 51/52, 133–141.

Kooistra, M.J. (1991) A micromorphological approach to the interactions between soil structure and soil biota, *Agriculture, Ecosystems and Environment* 34(1–4), 315–328.

Kubiena, W.L. (1953) *The Soils of Europe*, London: Murphy.

Miedema, R. (1997) Applications of micromorphology of relevance to agronomy, *Advances in Agronomy* 59, 119–169.

Srivastava, P. and Parkash, B. (2002) Polygenetic soils of the north-central part of the Gangetic Plains: a micromorphological approach, *Catena* 46, 243–259.

Zárate, M., Kemp, R.A., Espinosa, M. and Ferrero, L. (2000) Pedosedimentary and palaeo-environmental significance of a Holocene alluvial sequence in the southern Pampas, Argentina, *Holocene* 10, 481–488.

DONALD A. DAVIDSON

MILITARY GEOMORPHOLOGY

Military geomorphology refers to the application of geomorphic concepts, principles and technologies to military operations. This subfield links geomorphology and military science. Traditionally, military geomorphology is viewed from a perspective of the powerful influences terrain morphology has on military operations (see Winters 1998). Comparatively little attention is given to the profound effects armed conflict has on the physical landscape.

Warfare causes rapid and widespread terrain alteration. Physical landscape modification by munitions, intense vehicular movement, construction of obstacles and fortifications, and deliberate destruction are common consequences of war. Some of these activities leave erosional or depositional landforms similar to natural processes (see EQUIFINALITY) (Table 30).

Munitions and vehicle manoeuvers can alter the upper soil profile, destroy vegetation and change natural drainage patterns. These effects may persist for decades. The clay-rich landscape near Verdun, France for example, remains a pockmarked anthropogenic surface resembling GILGAI, CRATERS, DRUMLINS and HUMMOCKS nearly a century after artillery pounded the terrain during the First World War. Military defensive structures including castle moats, tank ditches, trenches and bunkers remain on the physical landscape long after their military usefulness is past. CAVES built or modified for military purposes such as in the Tora Bora region of Afghanistan or Gibraltar, remain an intricate part of contemporary landscape morphology.

Deliberate destruction of terrain is not uncommon in war. Governments in conflict, including the Russians in 1812 and 1941–1942, and the

Table 30 Some possible geomorphic effects of military operations

Military activity	Possible geomorphic effects	Example
Vehicular movement	Causes COMPACTION OF SOILS, decreases soil infiltration rates, destroys vegetation and changes erosion and deposition patterns. Forms RILLS, gullies, WADIS, ARROYOS, and track scars.	Tank track scars are preserved in desert pavement of the southwestern United States over 60 years.
Use of artillery, bombs, minefields	Creates blast craters, alteration of the upper soil profile, and destruction of vegetation with subsequent changes to erosion and deposition patterns. Mined areas may remain relatively unchanged by further anthropogenic alteration.	Over 20 million blast craters were produced during the Vietnam conflict. Over 100 million mines remain from conflict in more than 90 countries worldwide.
Construction of bunkers, trenches, defensive fortifications	May form mounds, gullies, wadis, arroyos, moats, CAVEs and canals.	Mounds marking defensive fortifications remain along the Normandy coast of France from the Second World War.

United States Union Army during the American Civil War, employed 'scorched earth' tactics. Crops, vegetation and structures were purposefully destroyed to deny the enemy their use, generating changes to erosion and deposition patterns across large areas. In the Second World War, Allied forces destroyed dams on the River Ruhr in Germany, drastically changing downstream fluvial morphology. Millions of gallons of defoliants devastated tropical rainforest and cropland soils during the Vietnam conflict. In 1991, retreating Iraqi forces set fire to over 730 oil wells in Kuwait, creating oil lakes and a durable 'tarcrete' surface of petroleum sludge, an artificial DURICRUST.

Reference

Winters, H.A. (1998) *Battling the Elements: Weather and Terrain in the Conduct of War*, Baltimore: Johns Hopkins University Press.

SEE ALSO: anthropogeomorphology

DANIEL A. GILEWITCH

MIMA MOUND

Also called prairie mounds and pimple mounds, Mima mounds take their name from Mima Prairie, Thurston County, Washington, USA. Such mounds are characteristically up to around 2 m in height, 25–50 m in diameter, and occur at a density of 50–100 or more to the hectare. There are many hypotheses for their origin (Cox and Gakahu 1986), including that they are erosional residuals, result from depositional processes around vegetational clumps, are the product of frost sorting, have been formed by communal rodents, are degraded termitaria, or have been created by seismic activity or groundwater vortices (Reider *et al.* 1996).

They are found from the Gulf Coast to Alberta, while in southern Africa, where they are termed *Heuweltjies*, they are widely distributed in the drier, western parts (Lovegrove and Siegfried 1989). Similar forms are also known from Argentina and Kenya. These mounds probably have many different origins, but the role of such beasts as mole rats, prairie dogs and gophers should not be underestimated.

References

Cox, G.W. and Gakahu, C.G. (1986) A latitudinal test of the fossorial rodent hypothesis of Mima mound origin, *Zeitschrift für Geomorphologie NF* 30, 485–501.

Lovegrove, B.G. and Siegfried, W.R. (1989) Spacing and origin(s) of Mima-like earth mounds in the Cape Province of South Africa, *South African Journal of Science* 85(2), 108–122.

Reider, R.G., Huss, J.M. and Miller, T.W. (1996) A groundwater vortex hypothesis for Mima-like mounds, Laramie Basin, Wyoming, *Geomorphology* 16, 295–317.

A.S. GOUDIE

MINERAL MAGNETICS IN GEOMORPHOLOGY

Mineral magnetic (or environmental magnetic) analysis provides a means of characterizing soil, sediment or rock samples on the basis of their magnetic properties. As demonstrated by rock magnetic research (Dunlop and Özdemir 1997), soils, unconsolidated sediments and solid rocks all display magnetic properties. These properties can be quantified to reveal information about the types of magnetic minerals present, their concentration and, in some circumstances, their magnetic grain size. The method shares many of the underlying principles of other methods of material characterization such as size, shape, colour, mineralogical or chemical composition and, for the geomorphologist, therefore offers a similar range of potential applications.

The magnetic behaviour of materials can be broadly classified into three types: diamagnetic (i.e. quartz, feldspar, water), paramagnetic (i.e. olivine, pyroxene, biotite) and ferromagnetic (i.e. magnetite, haematite). The first two types are relatively weak phenomena. While such materials produce a measurable magnetic response in the presence of an artificial external magnetic field (due to changes induced in electron motions within the constituent atoms), they are not capable of holding a remanence of that field. Thus, when the field is removed, the electron motions return to their previous behaviour and their magnetic properties cancel, resulting in no net spontaneous magnetization in the material.

In contrast, ferromagnetic materials display much stronger magnetic response in the presence of a magnetic field and can, in some circumstances, retain a memory (remanence) of that field after it is removed (spontaneous magnetization). Ferromagnetic behaviour can, in turn, be classified into sub-types. Of most interest in a geomorphological context are ferrimagnetic (i.e. magnetite, maghemite, greigite) and imperfect (canted) antiferromagnetic (i.e. haematite, goethite) types. These two groups show contrasting magnetic behaviour when subjected to specific laboratory measurements, although their magnetic response also varies with their exact composition and grain size. In addition, many natural samples will contain assemblages of mixed mineral types, concentrations and grain sizes, resulting in potentially complex 'bulk' magnetic behaviour.

Routine measurements (Table 31) are made at room temperature and provide (a) an insight into the magnetic mineralogy, concentration and grain size within a sample and (b) a characterization (or fingerprint) of the sample. In more advanced studies, temperature dependent magnetic properties may also be measured as these may provide more conclusive evidence in terms of (a) above. The advantages of mineral magnetic techniques include the ease of measurement, the ability to process large numbers of samples relatively quickly, their non-destructive nature (for room and low temperature measurements at least) and relatively low cost. However, the most significant advantage is the sensitivity of the instrumentation. In the majority of cases, differences in iron oxide concentrations can be detected that are well below the resolution of other methods such as X-ray diffraction, differential thermal analysis or differential chemical extractions.

Mineral magnetic analysis has a range of applications in studies of the environment (Table 32). In a geomorphological context, two such applications have received particular attention: (1) studies of WEATHERING/pedogenesis and (2) sediment tracing. The alteration and redistribution of iron that takes place within the soil environment is often reflected in corresponding changes in magnetic properties and certain soil types may show diagnostic variations in their magnetic signature with depth (Maher 1986). A common phenomena is the 'enhancement' of topsoil magnetic properties (e.g. higher values than the subsoil), where fine-grained magnetite and maghemite are formed as a result of biogeochemical transformations of iron weathered from other minerals (Dearing 2000). Like other changes induced by weathering and/or pedogenesis, all other things being equal, the degree of alteration of the REGOLITH may increase with time (although not necessarily in a linear fashion). In some circumstances therefore, the level of magnetic alteration may be used as a relative dating method (e.g. White and Walden 1994; Walden and Ballantyne 2002).

Differentiation of topsoil and subsoil materials on the basis of their magnetic properties has been widely used in sediment tracing studies

Table 31 Common room-temperature mineral magnetic parameters and their basic interpretation

Parameter	Interpretation
χ (10^{-6} m^3 kg^{-1})	Initial low field mass specific magnetic susceptibility. This is measured within a small magnetic field and is reversible (no remanence is induced). Its value is roughly proportional to the concentration of ferrimagnetic minerals within the sample.
χ_{fd} (10^{-6} m^3 kg^{-1})	Frequency dependent susceptibility. This parameter measures the variation of magnetic susceptibility with the frequency of the applied alternating magnetic field. Its value is proportional to the amount of magnetic grains whose size means they lie at the stable single domain/ superparamagnetic (<0.1 μm) boundary.
χ_{ARM} (10^{-6} m^3 kg^{-1})	Anhysteretic Remanent Magnetization (ARM) is proportional to the concentration of ferrimagnetic grains in the 0.02 to 0.4 μm (stable single domain) size range. The final result can be expressed as mass specific ARM per unit of the steady field applied (χ_{ARM}).
SIRM (10^{-6} Am^2 kg^{-1})	Saturation Isothermal Remanent Magnetization (SIRM) is the highest amount of magnetic remanence that can be produced in a sample by applying a large magnetic field. The value of SIRM is related to concentrations of all remanence-carrying minerals in the sample but is dependent upon the assemblage of mineral types and their magnetic grain size.
Soft IRM (10^{-6} Am^2 kg^{-1})	The amount of remanence acquired by a sample after experiencing an applied field of 40 mT. At such low fields, the high coercivity, canted antiferromagnetic minerals such as haematite or goethite are unlikely to contribute to the IRM, even at fine grain sizes. The value is therefore approximately proportional to the concentration of the low coercivity, ferrimagnetic minerals (e.g. magnetite) within the sample, although also grain-size dependent.
Hard IRM (10^{-6} Am^2 kg^{-1})	The amount of remanence acquired in a sample beyond an applied field of 300 mT. At fields of 300 mT, the majority of ferrimagnetic minerals will already have saturated and the value is therefore approximately proportional to the concentration of canted antiferromagnetic minerals within the sample.
IRM backfield ratios	Various magnetization parameters can be obtained by applying one or more magnetic 'reverse' or 'backfields' to an already saturated sample. The magnetization at each backfield can be expressed as a ratio of IRM_{field}/ SIRM and can discriminate between ferrimagnetic and canted antiferro-magnetic mineral types.

Sources: After Thompson and Oldfield (1986); Maher and Thompson (1999); Walden *et al.* (1999)

within lake and fluvial sediment systems (e.g. Dearing *et al.* 1985; Dearing 2000) and at a catchment scale, soils based upon different parent lithologies can also be distinguished. Considerable potential also exists for magnetic properties to be used in studies of SOIL EROSION and redistribution on hillslope systems, where they may complement other methods such as artificial radionuclides. Despite the advantages of the method, the user must also be aware of the underlying assumptions and potential problems. Two key issues are (1) the ability to identify and fully characterize the variability of the potential sediment source types and (2) the validity of assuming that the magnetic properties remain unaltered during sediment transport and deposition (Dearing 2000).

References

Dearing, J.A. (2000) Natural magnetic tracers in fluvial geomorphology, in I.D.L. Foster (ed.) *Tracers in Geomorphology*, 57–82, Chichester: Wiley.

Table 32 Environmental applications of mineral magnetic analysis

Application	Sedimentary environment	Example
Sediment correlation	Lacustrine, glacial, loess, fluvial, marine, soil erosion, etc.	Correlation of Heinrich layers in N. Atlantic sediments.
Tracing sediment provenance	Lacustrine, glacial, loess, fluvial, marine, soil erosion, etc.	Source areas of glacial sediment sequences. Soil erosion into river/lake sediment systems. Catchment fire histories.
Weathering/soil forming processes	Contemporary soils, colluvial deposits, PALAEOSOL identification, ALLUVIAL FAN surfaces.	Relative dating of weathered/pedogenic surfaces. Quaternary climate change record in Chinese LOESS sediments.
Artificial tagging of sediment	Fluvial, estuarine.	Tracing movement of fluvial sediments.
Pollution monitoring	Recent organic sediments, urban drainage, atmospheric pollution.	Industrial emissions from coal-fired power stations.

Sources: After Thompson and Oldfield (1986); Maher and Thompson (1999); Walden *et al.* (1999); Dearing (2000)

Dearing, J.A., Maher, B.A. and Oldfield, F. (1985) Geomorphological linkages between soils and sediments: the role of magnetic measurements, in K. Richards, R.R. Arnett and S. Ellis (eds) *Geomorphology and Soils*, 245–266, London: George Allen and Unwin.

Dunlop, D.J. and Özdemir, Ö. (1997) *Rock Magnetism: Fundamentals and Frontiers*, Cambridge: Cambridge University Press.

Maher, B.A. (1986) Characterization of soils by mineral magnetic measurements, *Physics of Earth and Planetary Interiors* 42, 76–92.

Maher, B.A. and Thompson, R. (eds) (1999) *Quaternary Climates, Environments and Magnetism*, 81–125, Cambridge: Cambridge University Press.

Thompson, R. and Oldfield, F. (1986) *Environmental Magnetism*, London: Allen and Unwin.

Walden, J. and Ballantyne, C.K. (2002) Use of environmental magnetic measurements to validate the vertical extent of ice masses at the last glacial maximum, *Journal of Quaternary Science* 17(3), 193–200.

Walden, J., Oldfield, F. and Smith, J.P. (eds) (1999) *Environmental Magnetism: A Practical Guide*, London: Quaternary Research Association.

White, K. and Walden, J. (1994) Mineral magnetic analysis of iron oxides in arid zone soils, Tunisian Southern Atlas, in K. Pye and A.C. Millington (eds) *The Effects of Environmental Change on Geomorphic Processes and Biota in Arid and Semi-arid Regions*, 43–65, Chichester: Wiley.

SEE ALSO: soil geomorphology; tracer

JOHN WALDEN

MINING IMPACTS ON RIVERS

Mining for heavy metals such as lead, zinc, copper, gold and silver has affected river environments since the advent of metallurgy. Pollution issues associated with heavy metal extraction from alluvial sediment or directly from their original host rock are known to have had their greatest impact on river systems since the start of the industrial revolution, *c.*1800. The fate of metal pollutants is similar to the natural sediment load; it may be stored within the channel, on the floodplains or it can be transferred out of the system via the estuaries and the ocean. It is the connectivity of river, estuarine and coastal transport systems coupled to the storage capacity of fluvial environments that determines the distribution of such sediment borne metals in a catchment. A basin-wide assessment of the fate and storage of mining related metal contaminants in the north-east of England (Macklin *et al.* 2000) indicated that only a small proportion has been flushed out into the Humber estuary with the remainder being stored in and along associated river systems.

Because of the protracted residence times of heavy metals within rivers and their floodplains (from 10^1 up to 10^4 years), metal-contaminated sediment may act as major sources of future contamination. Normal channel erosion and sedimentation processes are important in redistributing such contaminants across floodplains and for moving any inchannel material downstream. Fluvial sediments are spatially and temporarily complex, reflecting changes in the frequency and magnitude of river behaviour (erosion and sedimentation) to either extrinsic (e.g. climate or land use changes) or intrinsic (e.g. threshold adjustments) impacts. The distribution of metals in a river system is similarly complex. In more homogenous fluvial environments, it is often possible to determine pre-mining, mining era and post-mining era sediments through the examination of vertical overbank sequences. Here the deepest, oldest units may contain a geochemical imprint of the catchment prior to human disturbance. Moving up through the sediment profile one might encounter changes in sediment metal values that reflect the impact of human activity in a catchment. However, several studies (e.g. Macklin *et al.* 1994; Taylor 1996) have revealed that the imprint of mining activity may not be so simply distributed in floodplain sediments. Pre- and post-industrial anthropogenic deforestation may disrupt the predicted geochemical profile of alluvium either through dilution of the contaminant signal or through the erosion and transfer of sediments from a mineralized catchment that is naturally enriched in heavy metals. Floodplain geomorphology, such as terraces, levees, back channel environments, lakes and cut-offs may also assert a major control on the distribution of metals within floodplain environments such that overbank sediment profiles may show considerable variation in heavy metal concentrations both laterally and vertically.

The transfer of metals in rivers is dominated by four major mechanisms (Lewin and Macklin 1987): (1) hydraulic sorting according to individual particle size and mineral density; (2) chemical dispersal – solution, adsorption, the formation of Fe and Mn complexes and organic uptake (bioaccumulation in plants and animals); (3) dilution with clean uncontaminated sediments; (4) loss and exchange with floodplain sediment. Chemical remobilization and dispersal remains a significant problem particularly with respect to acid mine drainage where low pH values and changes in redox (i.e. reducing) conditions can liberate co-precipitated or adsorbed metals from relatively stable mine spoil into adjacent water and sediment bodies. Changes in water and sediment chemistry can increase the solubility, mobility and bioavailability of metal-contaminated sediment and thus make them more deleterious to the surrounding environment.

The transport of mining waste in river systems may occur in one of two modes: 'passive dispersal' or 'active transformation' (Lewin and Macklin 1987). In the 'passive dispersal' of river-borne metals, the river system remains in equilibrium and the waste is transported alongside the natural sediment load such that there is no significant alteration to the channel and floodplain's morphology.

'Active transformation' is associated with a greatly increased sediment load such that it results in a major transformation of the types, rates and/or magnitudes of geomorphic processes that control the prevailing channel morphology (Miller 1997). Gilbert's (1917) seminal paper described the effects of hydraulic gold mining in the Sierra Nevada (USA) on the tributaries of the Sacramento River between 1855 and 1884. Gilbert (1917) explained how mining waste resulted in rapid rises in the elevation of tributary channel beds during and following the cessation of mining. Aggradation was subsequently replaced by incision when the supply of mining debris declined and sediment was transferred downstream in wave-like form over time. However, the rate of incision (recovery) was markedly slower than that of aggradation (James 1993).

Other impacts of mining debris include the phytotoxic damage of riparian vegetation leading to bank destabilization, changes in sediment supply to the channel zone and ultimately planform metamorphosis from a meandering to a braided channel planform. In the Tasmanian Ringarooma River basin, tin mining during the nineteenth and early twentieth centuries (Knighton 1989, 1991) caused major changes to the width and planform of the Ringarooma River. The channel bed aggraded up to 10 m and channel width increased by up to 300 per cent. The response in the Ringarooma basin was very different to that on the Sacramento River because the source of the mining debris was much more diffuse, with mining sediments being distributed along the length of the river system as opposed to having a more geographically limited and discrete point source. Although the aggradation–incision cycle was evident in the Ringarooma and progressed downstream along with the mining debris sediment wave, the spatial pattern was highly variable due to the input of debris from tributaries all along the system.

The human imprint of land use change may manifest itself in many forms within a river basin. Those related to the direct impact of mining on rivers can result in the active transformation of a channel. This may cause major and highly visible disruption to the physical structure of fluvial environments through bed level adjustments, and cross-sectional and planform changes following the release of substantial volumes of toxic materials directly into the system. Less visible impacts such as the storage of heavy metal contaminants within a riverbed and on floodplains can result in the long-term storage of contaminants long after the primary pollution activity has dissipated. These may provide a latent but potentially insidious secondary source of pollution for future adjacent agricultural and urban land uses.

References

Gilbert, G.K. (1917) Hydraulic mining debris in the Sierra Nevada, *United States Geological Survey Professional Paper* 105.

James, L.A. (1993) Sustained reworking of hydraulic mining sediment in California: GK Gilbert's sediment model reconsidered, *Zeitschrift für Geomorphologie Supplementband* 88, 49–66.

Knighton, A.D. (1989) River adjustment to changes in sediment load: the effects of tin mining on the Ringarooma River, Tasmania, 1875–1984, *Earth Surface Processes and Landforms* 14, 333–359.

——(1991) Channel bed adjustment along mine-affected rivers of northeast Tasmania, *Geomorphology* 4, 205–219.

Lewin, J. and Macklin, M.G. (1987) Metal mining and floodplain sedimentation in Britain, in V. Gardiner (ed.) *International Geomorphology 1986 Part 1*, 1009–1027, Chichester: Wiley.

Macklin, M.G., Ridgway, J., Passmore, D.G. and Rumsby, B.T. (1994) The use of overbank sediment for geochemical mapping and contamination assessment: results from selected English and Welsh floodplains, *Applied Geochemistry* 9, 689–700.

Macklin, M.G., Taylor, M.P., Hudson-Edwards, K.A. and Howard, A.J. (2000) Holocene environmental change in the Yorkshire Ouse basin and its influence on river dynamics and sediment fluxes in the coastal zone, in I. Shennan, and J. Andrews (eds) *Holocene Land–Ocean Interaction and Environmental Change around the North Sea*, 87–96, Special Publications 166, London: Geological Society.

Miller, J.R. (1997) The role of fluvial geomorphic processes in the dispersal of heavy metals from mine sites, *Journal of Geochemical Exploration* 58, 101–118.

Taylor, M.P. (1996) The variability of heavy metals in floodplain sediments: a case study from mid Wales, *Catena* 28, 71–87.

Further reading

Macklin, M.G. (1996) Fluxes and storage of sediment-associated heavy metals in floodplain systems: assessment and river basin management issues at a time of rapid environmental change, in M.G. Anderson, D.E. Walling and P.D. Bates (eds) *Floodplain Processes*, 441–460, Chichester: Wiley.

SEE ALSO: alluvium; floodplain; fluvial erosion quantification; fluvial geomorphology; sedimentation; threshold, geomorphic

MARK PATRICK TAYLOR

MIRE

Definition of the term mire is not straightforward because mires may exist along a continuum from deep water aquatic systems through to terrestrial systems, and their boundaries are not easily defined and identified. Consequently, mires reflect different origins and patterns of development; are found in widely different geographical locations and under different climatic regimes; encompass different sets of controlling forces, and reflect different stages of successional development (Hofstetter 2000).

Mires are essentially peat-accumulating landscape features which, under natural conditions, are found where the water level is continuously near the soil surface resulting in a very narrow aerobic layer. These abiotic conditions favour specific plant and animal species, mosses or micro-organisms, which are adapted to wet and often nutrient poor conditions. The distribution of, and differences in, mire type, vegetation composition and soil type are caused primarily by geology, topography and climate but the formation, persistence, size and function of mires are controlled by *hydrological processes*. The source of the water, its quantity and quality and the mechanism by which it is delivered to the mire combine to influence mire development and character, giving rise to the wide spectrum of different mire types that occur in the landscape (e.g. Gore 1983; Heathwaite and Gottlich 1993; Moore 1984). Indeed, Mitsch and Gosselink (1993) went so far as to state that: 'Hydrology is probably the single most important determinant of the establishment and maintenance of specific types of mires and mire processes.'

There is no uniquely correct way of classifying mires because: (1) they are characterized by continuous gradation of properties with variable discontinuities, (2) their formation is affected by changing climatic, geomorphological and hydrological conditions, and (3) variations in mire types occur on a variety of scales. Thus it is not surprising that terms such as *bog, fen, mire* and *moor* are used widely but often imprecisely and occasionally interchangeably! The source of this imprecision lies with the range of criteria that have been used in attempts to define and classify mires. These criteria include floristic composition, site hydrology, site topography, water chemistry and nutritional status, and peat structure. Such criteria are sometimes used separately and sometimes in combination. The International Mire Conservation Group (www.imcg.net) suggest the following priority characteristics in differentiating mire types:

1 Source of water,
2 Prevailing hydrology-geomorphology,
3 Base content (saturation) or pH,
4 Nutrient availability and C:N, and
5 Prevailing plant communities.

Some of these classification characteristics are examined in more detail below.

Mire hydrology

The one common ingredient of all natural mires is an 'excess' of water, or at least a hydrological balance adequate to create conditions in which the surface is usually waterlogged for at least part of the year. The wet conditions characteristic of mires may result from impeded drainage, high rates of water supply or both. Water supply may consist of telluric water (i.e. water that has had some contact with mineral ground such as river water, surface runoff or groundwater discharge) or meteoric water (i.e. precipitation). Hydrological relationships play a key role in mire ecosystem processes, and in determining structure and growth. Thus different mires have a characteristic hydroperiod, or seasonal pattern of water levels, that defines the rise and fall of surface and subsurface water. An important geoindicator is the water budget of a mire, which links inputs from ground water, runoff, precipitation and physical forces (wind, tides) with outputs from drainage, recharge, evaporation and transpiration. Annual or seasonal changes in the range of water levels affect visible surface biota, decay processes, accumulation rates and gas emissions. Such changes can occur in response to a range of external factors, such as fluctuations in water source (river diversions, groundwater pumping), climate or land use (forest clearing). Waters flowing out of mires are chemically distinct from inflow waters, because a range of physical and chemical reactions take place as water passes through organic materials, such as peat, causing some elements (e.g. heavy metals) to be sequestered and others (e.g. dissolved organic carbon, humic acids) to be mobilized.

Because mire vegetation is largely responsive to primary environmental factors, such as hydrological and hydrochemical factors, hydrological classifications such as the one in Table 33 for British mires (see Heathwaite 1995) often offer the most direct explanation of mire types, because form and biotic characteristics are determined by these features.

Mire morphology

It has long been recognized that differences in topographic situation and water supply

Table 33 Hydrological classification for British mires

Source of water	Extent		
	Small (<50 ha)	Medium (50–1,000 ha)	Large (>1,000 ha)
Rainfall	Parts of basin mires	Raised mires	Blanket mires
Springs	Flushes, acid valley and basin mires	Fen basins, acid valley and basin mires	Fen massif, The Fens, Somerset Levels
Floods	Narrow floodplains	Valley floodplains	Floodplain massif

mechanism profoundly influence mire type. Such considerations formed the basis of the early but long-standing systems of mire classification. Von Post and Granlund (1926) subdivided mires into three types: ombrogenous mires, developed under the exclusive influence of precipitation; topogenous mires, irrigated by telluric water which naturally collects on flat ground and topographic hollows; and soligenous mires developed on slopes and kept wet by a supply of telluric water. Moore and Bellamy (1974) used this principle to subdivide British mires into broad groups based on their physiography which they justified on the basis that mire development is mainly determined by climatic, hydrological and geochemical conditions that control vegetation communities, leading to development of a particular mire type (Table 34).

Mire development

Viewed statically, mire types reflect a water table zonation from more or less open water through to conditions where the water table is rarely if ever above the substrate surface, though rarely far below base. Viewed dynamically, mires reflect successional or hydroseral changes. In the UK, it is difficult to demonstrate different successional stages because most hydroseres began developing at about the same time in the postglacial (Flandrian) period and are often, therefore, at similar successional stages. However, palaeoreconstruction based on the stratigraphy of accumulated peat deposits can be used to demonstrate the dynamic features of mire development over time.

The natural progression of the autogenic mire succession is in the direction of increasing acidity as the growing peat surface becomes progressively isolated from the nutritional effect of ground and soil water and more dependent on rainwater nutrition. Geomorphological criteria, principally climate, topography and substrate geology, are fundamental in directing changes to this autogenic succession. Thus the zonation of topogenous mires characteristically follows the water table gradient round enclosed basins or hollows which concentrate flow and allow the accumulation of a peat where lateral water movement is impeded. Soligenous mires develop where slow lateral gravitational seepage maintains waterlogged conditions at the ground surface. Topography is still important but it is the slow percolation of water through this mire type that distinguishes their type. Water flowing through soligenous mires is typically more oxygenated relative to the stagnant conditions of topogenous mires, and consequently the rate of organic matter decomposition is higher and the depth of peat accumulation lower. Ombrogenous mires develop where precipitation is high relative to evapotranspiration. Topography, whilst important, largely acts to retard runoff from the mire, rather than to concentrate runoff to it from other areas. The dependence on atmospheric inputs alone produces mire habitats that are characteristically of low base status, and where organic matter decomposition is low and the accumulation of unhumified acid peat high relative to other mire types. In the UK, ombrogenous mires are subdivided into raised mires and blanket mires or bogs. Blanket mires develop where the ground surface is

Table 34 Physiographical classification for British mires

Soligenous mires	Moving water/flushes/springs, slow peat formation (due to O_2)
Basin mires	Found in deep hollows, e.g. kettleholes; deep peat; vegetation surface may float; limited groundwater movement
Valley mires	Characterized by water flow along valley axis; broad range in pH, nutrients and vegetation communities
Floodplain mires	Develop on flood-prone alluvium; broad range in pH, nutrients and vegetation communities
Raised mires	Characterized by a peat surface that is isolated from the regional groundwater table; ombrotrophic; domed shape
Blanket mires	Usually develop on impermeable materials in regions with high precipitation and low temperature

permanently wet, initiating peat formation on flat and gently sloping ground. Raised mires occur over a wide range of climatic conditions and may represent a late stage in the autogenic succession of topogenous mires. In lowland Britain, raised mires are recorded in basins, floodplains and at the heads of estuaries, for example Thorne Moors National Nature Reserve which forms part of the Humberhead Peatlands. They are characterized by a raised central mire area where the peat has accumulated to the extent that it becomes isolated from water feeding the mire margins to become solely dependent on rainfall inputs. Here acidification ensues, the rate of decomposition falls, peat accumulation increases, and the mire type shifts from topogenous or soligenous mires to ombrogenous types. Raised mires form characteristic shallow domes of peat where the topography is typically convex, with a gently sloping rand away from its centre towards the surrounding moat-like drainage channel or lagg surrounding the bog.

Mire hydrochemistry

In addition to hydrological controls on mire development, the base and nutrient status of the mire water supply influence the mire type. The water chemistry of mires is primarily a result of geologic setting, water balance (relative proportions of inflow, outflow and storage), quality of inflowing water, type of soils and vegetation, and human activity within or near the mire. Mires dominated by surface-water inflow and outflow reflect the chemistry of the associated rivers or lakes. Mires that receive water primarily from precipitation and lose water by way of surface-water outflows and (or) seepage to ground water tend to have lower concentrations of chemicals. Thus ombrotrophic mires are rainwater-dominated and consequently base-deficient whereas minerotrophic mires are supplied with minerals and nutrients via the mire substrate which is in turn dependent on catchment geology and drainage water quality. Minerotrophic mires may range from oligotrophic through to mesotrophic or eutrophic types depending on the quality of their source water. Thus hydrochemical mire classifications focus largely on the source and quality of water to a mire giving a range of mire types from ombrotrophic raised mires, through transitional mires to minerotrophic mires or fens (Table 35).

On the basis of floristic variation in Swedish mires, Du Rietz (1949, 1954) suggested that mires could be divided into areas fed almost exclusively by precipitation and those in which water supply was supplemented by telluric water. These early concepts are important because they broadly correspond with major habitat differences still recognized today. The term fen is largely used as a synonym for minerotrophic mires and bog to refer to ombrotrophic examples (see Wheeler and Proctor 2000). These colloquial terms are still confused however, particularly as the vegetation of bogs and fens can be very similar. The distinction between these two habitats is based on their respective water sources. Bogs obtain their water from rainfall alone and this water is essentially stagnant, at least in the lower bog layers or catotelm. The water in a fen flows, although this may happen very slowly. Joosten (1998) used base conditions and trophic status to differentiate mire types, see Table 36.

The nutrient supply to bogs is characteristically low although nitrogen may be supplemented by

Table 35 Mean values of the concentration of major ions in waters from European mires

Mire hydrochemistry		Major ions									
		pH	HCO_3	Cl	SO_4	Ca	Mg	Na	K	H	Total
Eutrophic ↓	1	7.5	3.9	0.4	0.8	4.0	0.6	0.5	0.05	0	10.25
	2	6.9	2.7	0.5	1.0	3.2	0.4	0.4	0.08	0	8.28
	3	6.2	1.0	0.5	0.7	1.2	0.4	0.5	0.02	0	4.32
	4	5.6	0.4	0.5	0.5	0.7	0.2	0.5	0.04	0.01	2.85
	5	4.8	0.1	0.3	0.5	0.3	0.1	0.3	0.07	0.03	1.70
	6	4.1	0	0.4	0.4	0.2	0.1	0.3	0.04	0.14	1.58
	7	3.8	0	0.3	0.3	0.1	0.1	0.2	0.04	0.16	1.20
Oligotrophic											

Source: After Moore and Bellamy (1974)

Table 36 Base conditions, trophic conditions and mire types

C/N ratio	> 33				< 20
pH		< 4.8		> 6.4	
	Oligotrophic acid	Mesotrophic acid	Mesotrophic subneutral	Mesotrophic calcareous	Eutrophic
Lowland bog	■				
Mountain bog	■				
Kettlehole mire	■	■			■
Percolation mire		■	■	■	
Surface flow mire		■	■		■
Terrestrialization mire		■	■	■	■
Spring mire		■	■	■	■
Coastal floodplain mire					■
Fluvial floodplain mire					■

Source: After Joosten (1998)

atmospheric enrichment from industrial, urban and agricultural sources. Bog biodiversity is low. Typically the pH is <4.5 compared to fens where the pH range is 4.5–7.5.

Significance of mires in the landscape

In western Europe the majority of natural mires have been degraded through anthropogenic changes in hydrology, both at the regional and local scale, primarily for agricultural purposes. These measures have affected the biotic composition, the soil physical and chemical properties, the carbon and nutrient dynamic, as well as the landscape ecological functions of mires.

The regulatory function of hydrologically undisturbed mires compared to degraded mires has been neglected until recently. Natural mires act as ecotones between terrestrial and aquatic environments and are important owing to their transformation, buffer and sink qualities. For example, minerotrophic mires or fens are connected with their surrounding terrestrial areas via several hydrological pathways such as groundwater inflow, surface runoff, interflow or river water surplus. Nutrients transported with the inflowing water into such mires are transformed or accumulated by several biogeochemical processes. As a result, the nutrient concentration in the outflow can be reduced and water quality improved. Thus lowland mires are often areas of high biological productivity and diversity and mediate large and small-scale environmental processes by altering downstream catchments. For example, lowland mires can affect local hydrology by acting as a filter, sequestering and storing heavy metals and other pollutants, and serving as flood buffers and, in coastal zones, as storm defences and erosion controls. Upland mires can act as a carbon sink, storing organic carbon in waterlogged sediments. Even slowly growing peat may sequester carbon at between 0.5 and 0.7 tonnes ha^{-1} a^{-1}. Mires can also be a carbon source, when it is released via degassing during decay processes, or after drainage and cutting, as a result of oxidation or burning. Globally, upland mires have shifted over the past two centuries from sinks to sources of carbon, largely because of human exploitation.

References

Du Rietz, G.E. (1949) Huvudenheter och huvudgränser i svensk myrvegetation, *Srensk Botanisk Tidskrift* 43, 274–309.

——(1954) Die mineralbodenwasser-zeigegrenze als Grunlage einer natürlichen Zweigliederung der nord- und Mitteleuropäischen Moore, *Vegetatio* 5/6, 571–585.

Gore, A.J.P. (1983) *Ecosystems of the World, 4B: Mires, Swamp, Bog, Fen and Moor, Regional Studies*, Amsterdam: Elsevier.

Heathwaite, A.L. (1995) The hydrology of British mires, in J. Hughes and A.L. Heathwaite (eds) *Hydrology and Hydrochemistry of British Mires*, 11–20, Chichester: Wiley.

Heathwaite, A.L. and Gottlich, Kh. (1993) *Mires – Process, Exploitation and Conservation*, Chichester: Wiley.

Hofstetter, R.F. (2000) *Universal Mire Lexicon*, Greifswald: International Mire Conservation Group.

Joosten, J. (1998) *Mire Classification for Nature Conservation*, IMCG Working paper, Greifswald: International Mire Conservation Group.
Mitsch, W.J. and Gosselink, J.G. (1993) *Mires*, New York: Van Nostrand Reinhold.
Moore, P.D. (ed.) (1984) *European Mires*, London: Academic Press.
Moore, P.D. and Bellamy, D.J. (1974) *Peatlands*, London: Elek. Science.
Von Post, L. and Granlund, E. (1926) Sodra Sveriges tortillangar I, *Sver. Geol. Unders.* 19 C 335, Stockholm.
Wheeler, B.D. and Proctor, M.C.F. (2000) Ecological gradients, subdivisions and terminology of north-west European mires, *Journal of Ecology* 88, 187–203.

Key websites

RAMSAR: http://www.ramsar.org
Irish Peatland Conservation Council: http://www.ipcc.ie
Society of Mire Scientists: http://www.sws.org
International Peat Society: http://www.peatsociety.fi
British Ecological Society Mires Research Group: http://www.britishecologicalsociety.org/groups/mires/index.php

LOUISE HEATHWAITE

MOBILE BED

A fluid, such as air or water, flowing over cohesionless sediment has the ability to entrain solid particles. The bed surface becomes mobile when the shear stress applied on the particles by the flow exceeds the critical shear stress of the sediment mixture. The initiation of particle motion is a stochastic phenomenon that depends on the average fluid motions and, because the dimensions of sediment particles usually are relatively small compared to the dimensions of the flow, on the magnitude of turbulent deviations from the average (Nelson *et al.* 2001). It also depends on the position of a particle on the bed, which determines its exposure to the flow (Kirchner *et al.* 1990; Buffington *et al.* 1992), and the relative proportion of each size fraction in a mixture (Wilcock 1993). The shear stress at which particle motion is initiated in heterogeneous sediment may be approximated by Shields's relation, if the median grain size is used to characterize the entire sediment mixture (Kuhnle 1993; Buffington and Montgomery 1997).

As the threshold condition for the initiation of particle motion is approached there is an abrupt increase in the rate of sediment movement. Particle movement is neither uniform nor continuous over the bed, because turbulent sweeps, the structures responsible for particle motion, move groups of particles intermittently at random locations on the bed (Drake *et al.* 1988; Williams *et al.* 1990). As the shear stress (and rate of sediment transport) increases the sweeps become more laterally stable and longitudinal streaks form on the bed, and in heterogeneous sediment a pattern of alternating coarse- and fine-grained stripes emerges (McLelland *et al.* 1999). At higher shear stresses the coarser sediment becomes more mobile and the stripes are replaced by flow-transverse BEDFORMS (Gyr and Müller 1996).

Sediment initially was thought to move in sliding layers, with the most rapidly moving layer positioned adjacent to the flow, but it was soon recognized that only the surficial grains move. In the absence of significant scour or bedform development, the depth of the active layer is of the order of 0.4 to $2D_{90}$ (where D_{90} is the size for which 90 per cent of the surficial bed material size distribution is finer). The sediment in transport is termed the bed material load. Bed material either can be swept up into the main part of the flow by turbulence, and transported in suspension, or it may move, by rolling/sliding or SALTATION, as BEDLOAD in a layer immediately above the bed. This layer is of the order of two to four grain diameters thick in water, and a few tens of centimetres thick in air. As the flow intensity increases above the critical value, particles first move by rolling. Saltation rapidly becomes the dominant type of motion as the flow intensity increases further, and at still higher flow intensities suspension begins to dominate. There is a clear physical difference between the two basic modes of transport (Abbott and Francis 1977). The weight of a saltating particle is supported by the bed, whereas the flow supports the weight of a suspended particle. However, the two modes of transport cannot easily be differentiated on the basis of particle size, and there is a continual exchange of particles between the bedload and SUSPENDED LOAD. There is also an important difference between the movement of particles by saltation, in air and in water. In air, once saltation commences subsequent movement is induced by the impact of particles hitting the bed, rather than by the hydrodynamic forces that act on static particles, as is the case in water. The difference arises because the submerged density of sediment particles is substantially greater than the density of air at atmospheric pressure, whereas it is less than twice the density of water.

Particles transported in suspension move at the velocity of the flow. Particles comprising the bedload continually move in and out of storage on the bed, and their pattern of motion can be characterized as a series of relatively short steps of random length, each of which is followed by a rest period of random duration (Habersack 2001). The sensitivity of travel distance to particle size decreases as size decreases below the median diameter of the substrate (Church and Hassan 1992), but the virtual velocity of particles in water is only of the order of metres per hour, compared to the flow velocity which may be of the order of metres per second (Haschenburger and Church 1998). This is because each particle spends a negligible time in motion compared to the time spent at rest. In the case of sediment that is deposited on the lee side of a bedform, the velocity at which the particles move is much slower and is determined by the rate of movement of the bedform (Grigg 1970; Tsoar 1974).

References

Abbott, J.E. and Francis, J.R.D. (1977) Saltation and suspension trajectories of solid grains in a water stream, *Philosophical Transactions of the Royal Society of London* A284, 225–254.

Buffington, J.M. and Montgomery, D.R. (1997) A systematic analysis of eight decades of incipient motion studies, with special reference to gravel-bedded rivers, *Water Resources Research* 33, 1,993–2,029.

Buffington, J.M., Dietrich, W.E. and Kirchner, J.W. (1992) Friction angle measurements on a naturally formed gravel streambed: implications for critical boundary shear stress, *Water Resources Research* 28, 411–425.

Church, M.A. and Hassan, M.A. (1992) Size and distance of travel of unconstrained clasts on a streambed, *Water Resources Research* 28, 299–303.

Drake, T.S., Shreve, R.L., Dietrich, R.L., Whiting, P.J. and Leopold, L.B. (1988) Bedload transport of fine gravel observed by motion-picture photography, *Journal of Fluid Mechanics* 192, 193–217.

Grigg, N.S. (1970) Motion of single particles in alluvial channels, *Journal of the Hydraulics Division*, American Society of Civil Engineers 96, 2,501–2,518.

Gyr, A. and Müller, A. (1996) The role of coherent structures in developing bedforms during sediment transport, in P.J. Ashworth, S.J. Bennett, J.L. Best and S.J. McLelland (eds) *Coherent Flow Structures in Open Channels*, 227–235, Chichester: Wiley.

Habersack, H.M. (2001) Radio-tracking gravel particles in a large braided river in New Zealand: a field test of the stochastic theory of bed load transport proposed by Einstein, *Hydrological Processes* 15, 377–391.

Haschenburger, J.K. and Church, M.A. (1998) Bed material transport estimated from the virtual velocity of sediment, *Earth Surface Processes and Landforms* 23, 791–808.

Kirchner, J.W., Dietrich, W.E., Iseya, F. and Ikeda, H. (1990) The variability of critical shear stress, friction angle, and grain protrusion in water-worked sediments, *Sedimentology* 37, 647–672.

Kuhnle, R.A. (1993) Fluvial transport of sand and gravel mixtures with bimodal size distributions, *Sedimentary Geology* 85, 17–24.

McLelland, S.J., Ashworth, P.J., Best, J.L. and Livesey, J.R. (1999) Turbulence and secondary flow over sediment stripes in weakly bimodal bed material, *Journal of Hydraulic Engineering* 125, 463–473.

Nelson, J.M., Schmeeckle, M.W. and Shreve, R.L. (2001) Turbulence and particle entrainment, in M.P. Mosley (ed.) *Gravel-Bed Rivers* V, 221–248, New Zealand Hydrological Society, Wellington.

Tsoar, H. (1974) Desert dunes, morphology and dynamics, El Arish (northern Sinai), *Zeitscrift für Geomorphology Supplementband* 20, 41–61.

Wilcock, P.R. (1993) Critical shear stress of natural sediments, *Journal of Hydraulic Engineering* 119, 491–505.

Williams, J.J., Butterfield, G.R. and Clark, D.G. (1990) Rates of aerodynamic entrainment in a developing boundary layer, *Sedimentology* 37, 1,039–1,048.

BASIL GOMEZ

MODELS

'The sciences', wrote mathematician John von Neumann (1963), 'mainly make models.' Model building is as much a part of twenty-first-century geomorphology as it is in any science. A model, in its most general sense, is a simplified or idealized representation of an existing or potential reality. Examples of models range from architects' miniatures to quantum theory. In geomorphology, models serve as representations of Earth surface processes and landforms, and as such they embody the theory that underpins the science. All geomorphologists rely on models of one sort or another. However, the word 'model' is used in a variety of contexts in geomorphology, and it is useful to distinguish between three general forms: conceptual models, hardware (or experimental) models and mathematical models.

Conceptual models of landform origins must surely predate the word 'geomorphology'. As the formal scientific field of geomorphology began to take shape in the late nineteenth century, influential figures such as William Morris Davis, Walther Penck and G.K. Gilbert developed conceptual models of landscape systems that provided a guiding impetus for research. Davis's 'geographical cycle' is a classic example of a conceptual model in geomorphology. It provided an explanation for many observed landforms, made predictions about their course of evolution, provided guiding assumptions for the interpretation of particular landform elements (such as low-relief surfaces) and influenced the type of questions posed by researchers. Although many of Davis's ideas have not stood the test of time, the creation and progressive refinement of conceptual models are still fundamental parts of geomorphology. Unavoidably, our ideas about how geomorphic systems operate will always guide the type of questions we choose to ask and the kind of interpretations we make (Brown 1996).

Hardware models represent geomorphology's experimental side. A hardware model is a physical representation, often (but not always) scaled down, of a particular geomorphic system. G.K. Gilbert's (1914) flume experiments at Berkeley, which led to his classic paper 'The transportation of debris by running water', represent one of the first experimental studies in geomorphology. Gilbert's data are, in fact, still used today, and have been complemented by many other flume studies of sediment transport. The literature is replete with examples of experimental models of geomorphic systems. Phenomena that have been studied experimentally include drainage basin evolution, bedrock landsliding, soil creep, rock weathering, alluvial fans, and subaerial and subaqueous debris flows, to name a few. In some cases, laboratory experiments have coupled geomorphic processes with tectonic, eustatic, and/or depositional processes. In some instances, hardware models operate on the same spatial scale as the geomorphic system in question; the US Geological Survey experimental debris flow flume near Bellingham, Washington, USA is one such example (Major and Iverson 1999). More commonly, the physical system is scaled down, which can introduce problems in preserving basic scaling relationships between physical properties such as fluid viscosity and gravity. Nonetheless, hardware modelling continues to be an important source of information and insight into a wide range of geomorphic systems.

A mathematical model, like a conceptual model, acts as a simplified or idealized representation of reality that provides a framework for guiding and interpreting observations. Seen in this light, a mathematical model can be understood as a quantitative hypothesis or set of linked hypotheses. A mathematical model has an obvious advantage over purely conceptual models in its precision, lack of ambiguity and ability to satisfy basic constraints such as continuity of mass, momentum and energy. At the same time, mathematical models, like hardware models, allow for a degree of experimentation – in the sense of testing the behaviour of a system (one comprising a set of mathematical-logical postulates and assumptions, rather than a physical construct) that has been built as an analogy for a natural system. The use of ocean and atmospheric general circulation models to test palaeoclimate hypotheses (e.g. Cane and Molnar 2001) is a good example of this type of experimentation.

Although there is no generally agreed classification of types mathematical model, the categories suggested by Kirkby *et al.* (1992) provide a useful framework. They distinguish between black-box (statistical or empirical) models, process models, mass-balance models and stochastic models. There is significant overlap among these categories, and indeed many mathematical models combine elements of several of these.

Mathematical models in geomorphology began to emerge in the post-Second World War era. Many of these early models were descriptive or empirical (i.e. 'black box') in nature. R.E. Horton's drainage network laws (now known as HORTON'S LAWS), for example, provided a quantitative description of river network topology, while the hydraulic geometry equations of Leopold and Maddock (1953) provided a similar description of river channel changes through space and time. These and many other morphometric models are essentially statistical in nature.

Beginning in the 1960s, such statistical models were complemented by process models. Where a black-box model represents relationships in a purely empirical form, a process model attempts to describe the mechanisms involved in a system. For example, a black-box model of soil erosion would be based on regression equations obtained directly from data, whereas a process model would attempt to represent the mechanics of overland flow and particle detachment. Process models often overlap with mass-balance (or energy-balance) models, in the sense that equations for processes are used to model the transfer of mass or energy among different *stores*, where a store could represent anything from the water in a lake, the population of a species in an ecosystem, the energy stored as latent heat in an atmospheric column, the carbon mass in a tree, or the depth of soil at a point on a hillslope.

Process models have been widely used to study landform evolution. Usually phrased in terms of continuum mechanics, these landform evolution models provide a link between the physics and chemistry of geomorphic processes, and the shape of the resulting topography. Among the pioneers in landform 'process-response' modelling in the late 1960s and early 1970s were F. Ahnert and M.J. Kirkby. The latter showed, for example, that the convexo-concave form of hillslopes can be predicted from simple laws for sediment transport (Kirkby 1971).

Many process models are deterministic, meaning that for a given set of inputs they will predict a unique set of outputs. Often, however, the inputs to a particular geomorphic system are highly variable in time or space, and essentially unpredictable or unmeasurable. For example, we may know something about the frequency and magnitude characteristics of rainfall but cannot predict the sequence of rainfall events over time spans of more than a few days. Similarly, we may have a good estimate of the average hydraulic conductivity of an aquifer but little or no information about its heterogeneity. Likewise, a hallmark of many nonlinear systems (including some geomorphologic systems) is sensitivity to initial conditions: a small difference in the initial state of a system can lead to markedly different outcomes (see Gleick 1988). *Stochastic* models are designed to address such uncertainties by including an element of random variability. Such models typically use a random number generator to create a series of alternative inputs (e.g. rainstorms) or to trigger discrete events (e.g. landslides). Discussion and examples of stochastic models are given by Kirkby *et al.* (1992, Chapter 5).

At the heart of most geomorphic process models, both deterministic and stochastic, lies the continuity of mass equation

$$\frac{\partial \eta}{\partial t} = -\nabla q_s$$

where η is the height of the surface, t is time, q_s is the bulk volume rate of mass transport (rock, sediment or solute) per unit width and ∇ denotes a gradient in two dimensions. The continuity equation simply states mathematically that matter can neither be created nor destroyed (short of nuclear reactions). This particular form of the continuity equation is not universally applicable to all geomorphic problems; a slightly different form would be needed to describe horizontal (as opposed to vertical) retreat of a cliff face, the evolution of a fault block undergoing horizontal motion, or surface change due to changes in density rather than in mass, for example. Nonetheless, the continuity law in its various guises is arguably one of a handful of fundamental principles in geomorphological modelling. When combined with a suitable expression for q_s that represents a particular process or processes, the continuity equation can, subject to certain simplifying assumptions, be solved in order to predict landform features such as the shape of hillslope profiles, river profile geometry and soil-depth profiles.

An obvious advantage of mathematical process models over conceptual models (such as the influential concepts of G.K. Gilbert) is that they allow one to make statements not only about 'how' or 'why' but also 'how much' – for example, a mathematical slope model allows one to predict, based on processes, the degree to which the relief in a mountain drainage basin would ultimately change if the rate of uplift were to double (e.g. Snyder *et al.* 2000).

Central to geomorphic process models is the concept of 'geomorphic transport laws' (Dietrich *et al.* 2003). A geomorphic transport law is a mathematical statement about rates of mass transport averaged over a suitably long period of time. The definition of 'suitably long' depends on the process in question, but in general is much longer than the recurrence interval of discrete transport events such as floods, raindrop impacts, landslides, and so on. One of the current research frontiers in geomorphic process research lies in understanding the relationship between short-term transport events and long-term average transport rates.

Solving the continuity equation to predict landform shape requires assuming idealized conditions – for example, in the case of landforms with uniform soil or sediment properties, uniform climate, and height variations in one direction only. Modelling three-dimensional landforms generally requires approximating the solution to an appropriate form of the continuity equation, usually through the use of numerical techniques such as finite differencing, finite volume or finite element methods, cellular automata, or (in some cases) a combination of methods (e.g. Press 2002; Slingerland *et al.* 1994). Typically, these methods produce an approximate solution by dividing up space into discrete elements. Fluxes of mass are then computed within or between these elements. Beginning with a specified initial landform configuration, the evolution of landforms over time is computed by iteratively calculating the transport rates at each point, extrapolating these rates forward in time over a discrete time increment, and then adjusting the topography accordingly. This in turn affects the transport rates at the next time increment, so that the landform emerges as the result of an interaction between its shape and the processes acting upon it.

The development and use of numerical models of landform evolution has grown considerably since the 1980s. Examples include models of rill erosion, river basin evolution, glacial valley formation, and many other coupled tectonic–geomorphic–sedimentary systems. Numerical models of landform and landscape evolution generally operate on what Schumm and Lichty (1965) termed 'cyclic time' – that is, time spans on which landforms can change significantly and which, apart from rapid processes like rill erosion, are generally much longer than a human lifetime. Alongside these 'cyclic time' models are 'event time' numerical models aimed at understanding process dynamics. Here, event time refers to the timescales of individual process events such as floods. This is the timescale on which direct experimentation, observation and application of Newtonian mechanics are most feasible. For example, computational fluid dynamics models have been used to great effect to examine phenomena such as river flooding, coastal sediment transport, soil erosion and debris flows. Often, such models are founded on basic theory in fluid dynamics or material rheology, and combine well-established physical principles (e.g. the Navier–Stokes equations for fluid flow) with empirical laws obtained from laboratory experiments (for background and examples, see Middleton and Wilcock, 1994).

Both event-time and cyclic-time models have had a tremendous impact on geomorphologists' ability to understand the dynamics of processes, and to link these with the landforms that they shape. The applications of models range quite widely, and include both pragmatic forecasting and investigative analysis. Event-time models are often used in an applied context, to make predictions for purposes of planning, land management and insurance assessment. Models of soil erosion, for example, are typically used in this way. In an applied, predictive mode of application, a given model is generally taken 'as read', usually calibrated with existing data, and used to forecast the outcomes of different scenarios.

Numerical models in geomorphology serve other important roles as well. Both event-time and cyclic-time models have been, and continue to be, used in a heuristic mode; that is, they are used as theoretical tools for developing general insight and understanding, rather than for making precise predictions in a particular case study. One of the most valuable roles of mathematical models in geomorphology, in fact, is to make testable predictions about process and form connections. For example, numerous river basin evolution models have been used in 'what if' mode to predict the morphological consequences of statements such as 'the long-term average incision rate of a stream channel is proportional to the rate of energy dissipation per unit bed area' (e.g. Whipple and Tucker 1999). This exploratory process of *forward modelling* makes it possible to reject some models in favour of others, based on their ability to reproduce observed landform characteristics given a plausible set of initial and boundary conditions.

As in other sciences, models in geomorphology both drive and are driven by the results of observational and experimental work. In some cases, a model is developed for the express purpose of explaining a set of data. In others, one or more models are proposed before any relevant data exist, and they stimulate the search for new observations. One example of the latter concerns the relationship between the thickness of soil and the rate of lowering of the soil–bedrock contact. Several models were proposed in the 1960s and 1970s (see Cox 1980). Of these, some predicted an exponential decline in regolith production rate with increasing soil depth, with the maximum production rate occurring at or near the surface. Others predicted a 'humped curve' with a maximum production rate at some optimal soil thickness, due to the added efficiency of water retention. These models remained essentially untested for many years, until cosmogenic nuclide analysis made it possible to infer rates of regolith production. Research beginning in the 1990s has provided evidence for an inverse dependence of regolith production rate on regolith thickness, in some cases with a near-surface maximum (e.g. Heimsath *et al.* 1997), in others with a maximum at depth (e.g. Small *et al.* 1999), depending on process and environment.

The example of regolith production models serves as a caution against the common myth that a model is of no value until and unless it has been validated. In fact, untested mathematical models in geomorphology – like the regolith production models when they were first proposed – have served the field well in two ways: first, by forcing rigour into our hypotheses, and second, by spurring the development of new efforts, ideas and technologies to test the models (for discussion see Bras *et al.* 2003).

A common limitation of models in geomorphology is that different models predict similar outcomes. For example, a range of different river process models predict that graded river profiles should be concave-upward in form – thereby providing multiple, competing explanations for the same observation. This classic problem of EQUIFINALITY, which is common across the Earth sciences, reflects a paucity of data about geomorphic systems. This limitation is part and parcel of the deep-time problem in the Earth sciences (and in other fields such as astronomy and astrophysics). The systems that geomorphologists study are often too big or too slow to allow for direct experiments. Furthermore, most geomorphic systems are dissipative in nature (Huggett 1988). Dissipative systems, by the 2nd law of thermodynamics, lose information as they evolve (consider, for example, trying to reconstruct a snow crystal from a drop of water). Geomorphologists are therefore forced to rely on inference, analogy and indirect evidence. It is no surprise that the problems of equifinality and deep time limit mathematical modelling in the same way that they limit geomorphic knowledge more generally. In principle, the solution to both problems is to obtain as much information as possible about denudation rates, boundary conditions (such as tectonic, climatic or sea-level variations), and the nature of changes in topography over the geologic past. Developing techniques to obtain such data arguably constitutes one of the foremost challenges in geomorphology.

The deep-time problem highlights the fact that, in building models of landform genesis, geomorphologists are forced to 'scale up' contemporary processes over geologic time, and over spatial scales relevant to the landforms in question. This approach is of course a natural outgrowth of Hutton's (1795) ideas. The impossibility of direct experiments makes it especially important to develop accurate constitutive process laws, and to pay careful attention to the role of natural variability in driving forces (such as weather and climate) and in materials (such as soil properties). This represents a considerable scaling challenge, because the formative processes often occur on timescales that are vastly smaller than the timescale required for significant landform change. For example, floods may last for minutes to days while the river basins they sculpt may take shape over hundreds of thousands of years.

Despite their limitations, the future of mathematical models in geomorphology looks bright. Continuing advances in computing power will make solving the scaling problem easier by allowing modellers to link together a wider range of time and space scales. While the deep-time problem will never go away, geomorphologists' ability to explore and test multiple working hypotheses will continue to grow. Likewise, continuing improvements in data describing the Earth's surface topography and in technologies for dating and estimating rates of change will make it possible to test models with increasing degrees of precision.

References

Bras, R.L., Tucker, G.E. and Teles, V. (2003) Six myths about mathematical modeling in geomorphology, in P.R. Wilcock and R. Iverson (eds) *Prediction in Geomorphology*, Geophysical Monograph, Washington, DC: American Geophysical Union.

Brown, H.I. (1996) The methodological roles of theory in science, in B.L. Rhodes and C. Thorn (eds) *The Scientific Nature of Geomorphology*, 3–20, Binghamton, NY: State University of New York.

Cane, M.A. and Molnar, P. (2001) Closing of the Indonesian seaway as a precursor to East African aridification around 3–4 million years ago, *Nature* 411(6,834), 157–162.

Cox, N.J. (1980) On the relationship between bedrock lowering and regolith thickness, *Earth Surface Processes and Landforms* 5(3), 271–274.

Dietrich, W.E., Bellugi, D., Sklar, L., Stock, J.D., Heimsath, A.M. and Roering, J.J. (2003) Geomorphic transport laws for predicting landscape form and dynamics, in P.R. Wilcock and R. Iverson (eds) *Prediction in Geomorphology*, Washington, DC: American Geophysical Union, 103–132.

Gilbert, G.K. (1914) The transportation of debris by running water, *US Geological Survey Professional Paper* 86.

Gleick, J. (1988) *Chaos: Making a New Science*, London: Heinemann.

Heimsath, A.M., Dietrich, W.E., Nishiizumi, K. and Finkel, R.C. (1997) The soil production function and landscape equilibrium, *Nature* 388(6,640), 358–361.
Huggett, R.J. (1988) Dissipative systems; implications for geomorphology, *Earth Surface Processes and Landforms* 13(1), 45–49.
Hutton, J. (1795) *Theory of the Earth*, Edinburgh.
Kirkby, M.J. (1971) Hillslope process-response models based on the continuity equation, slopes, form and process, *Transactions of the Institute of British Geographers, Special Publication* 3, 15–30.
Kirkby, M.J., Naden, P.S., Burt, T.P. and Butcher, D.P. (1992) *Computer Simulation in Physical Geography*, Chichester: Wiley.
Leopold, L.B. and Maddock, T., Jr (1953) The hydraulic geometry of stream channels and some physiographic implications, river morphology, *US Geological Survey Professional Paper* 252.
Major, J.J. and Iverson, R.M. (1999) Debris-flow deposition; effects of pore-fluid pressure and friction concentrated at flow margins, *Geological Society of America Bulletin* 111(10), 1,424–1,434.
Middleton, G.V. and Wilcock, P.R. (1994) *Mechanics in the Earth and Environmental Sciences*, Cambridge: Cambridge University Press.
Press, W.H. (2002) *Numerical Recipes in C++ : The Art of Scientific Computing*, Cambridge: Cambridge University Press.
Schumm, S.A. and Lichty, R.W. (1965) Time, space, and causality in geomorphology, *American Journal of Science* 263(2), 110–119.
Slingerland, R., Furlong, K. and Harbaugh, J.W. (1994) *Simulating Clastic Sedimentary Basins*, Englewood Cliffs, NJ: PTR Prentice Hall; Prentice-Hall International.
Small, E.E., Anderson, R.S. and Hancock, G.S. (1999) Estimates of the rate of regolith production using ^{10}Be and ^{26}Al from an alpine hillslope, *Geomorphology* 27(1–2), 131–150.
Snyder, N.P., Whipple, K.X., Tucker, G.E. and Merritts, D.J. (2000) Landscape response to tectonic forcing; digital elevation model analysis of stream profiles in the Mendocino triple junction region, Northern California, *Geological Society of America Bulletin* 112(8), 1,250–1,263.
von Neumann, J. (1963) The role of mathematics in the sciences and in society, and method in physical sciences, in *J. von Neumann – Collected Works Vol. VI*, ed. A.H. Taub, 477–498, New York: Macmillan.
Whipple, K.X. and Tucker, G.E. (1999) Dynamics of the stream-power river incision model; implications for height limits of mountain ranges, landscape response timescales, and research needs, *Journal of Geophysical Research*, B, Solid Earth and Planets 104(8), 17,661–17,674.

Further reading

Harmon, R.S. and Doe, W.W. III (eds) (2001) *Landscape Erosion and Evolution Modelling*, New York: Kluwer Academic/Plenum Publishers.
Rhodes, B.L. and Thorn, C.E. (ed.) (1996) *The Scientific Nature of Geomorphology*, Chichester: Wiley.
Wilcock, P.R. and Iverson, R. (ed.) (2003) *Prediction in Geomorphology, Geophysical Monograph*, Washington, DC: American Geophysical Union.

SEE ALSO: complexity in geomorphology; computational fluid dynamics; equifinality; laws, geomorphological; mathematics; mechanics of geological materials; non-linear dynamics

GREG TUCKER

MORAINE

A moraine is a glacial landform created by the deposition or deformation of sediment by glacier ice. Many different types of moraine exist, reflecting the many different processes by which glaciers deposit and deform sediment and the many locations and environments within the glacier system where deposition can occur. The material of which moraines are composed, which is generally referred to as till, is also highly variable, as its characteristics depend on the characteristics of the debris supplied by the glacier as well as on the processes and environment of GLACIAL DEPOSITION.

The term moraine has been used in a variety of different ways since it was originally introduced, and its definition remains controversial. Swiss naturalist Horace-Bénédict de Saussure originally introduced the term in 1779, and recognized that ancient moraines represented former extensions of existing glaciers. For the next two centuries the term was widely used to describe landforms created by glacial deposition, the sedimentary material of which the landforms were composed, and the debris in transport within, beneath or on the surface of glaciers. Although modern geomorphological definitions limit the term specifically to landforms, it is still sometimes applied more widely to glacial debris and glacially derived sediment. Many of the compound expressions that feature the term moraine, such as ground moraine and medial moraine, conflate elements of these different definitions, and so moraine continues to be used ambiguously in some geomorphological, glaciological and sedimentological literature. Dreimanis (1989) provides a useful review of the history of the term.

Moraines are classified both genetically according to the process by which they are created and geographically according to their position within the glacier system. There is a fundamental distinction between moraines that occur on the surface of the ice and moraines that occur on the ground surface beneath or at the margin of a glacier. Moraines on the ice surface, known as supraglacial moraines, are ephemeral features that move with the ice and are likely to be destroyed or redeposited on the ground surface when the ice beneath them ablates. They are not true landforms, and inherit the term moraine from now obsolete definitions that included debris in glacial transport.

Supraglacial moraines include lateral moraines (Small 1983), medial moraines (Vere and Benn 1989) and inner moraines (Weertman 1961). Lateral moraines occur at the edge of valley glaciers and comprise debris derived from the valley walls both above and below the ice. Medial moraines occur as longitudinal accumulations of debris downstream from junctions between confluent glaciers, and include debris derived from the lateral moraines of each tributary. Inner moraines are transverse accumulations of debris derived from the meltout of basal debris bands close to the glacier margin. All of these moraine types can develop into large ridges on the glacier surface as the debris cover protects the ice immediately beneath from melting while the surface of the surrounding, debris-free, ice is lowered by ablation. Thick and irregular accumulations of debris released onto the glacier surface by ablation have previously been referred to as ablation or disintegration moraine, forming part of a supraglacial land system, but these terms are increasingly being confined to terrestrial landforms that survive after ablation of the glacier. Supraglacial debris that does not form discrete topographic features on the glacier surface is not referred to as moraine.

Moraines can be formed on the ground surface subglacially or at the edge of the glacier, and by the lowering of supraglacial debris to the ground during deglaciation. They can be formed both by active (moving) ice and by stagnant ice. The principal processes by which moraines are created are the release of debris from ice by meltout and the deformation of proglacial or subglacial sediments by ice motion.

Moraines created by the lowering of supraglacial debris to the ground during deglaciation typically produce a chaotic topography and highly variable sedimentology as the landforms produced are strongly affected by resedimentation, water action and mass movement during their formation.

Subglacial moraines can occur parallel and perpendicular to ice flow or in irregular patterns. There is often a gradual transition between forms with different orientations such as Rogens (described below) and DRUMLINS. The origin of many of these features remains disputed. Areas of subglacial deposition without distinctive relief are sometimes referred to as ground moraine, but this term is falling into disuse and being replaced by non-topographic terms such as subglacial till.

Moraines parallel to ice flow include streamlined features within subglacial till, such as flutes and certain types of drumlins. The genesis of some of these features remains controversial. Whereas traditional analyses attribute them to subglacial deposition and the sculpting of subglacial deposits by moving ice, other interpretations based on subglacial meltwater processes (e.g. Shaw *et al.* 1989) imply that these features are not true moraines at all.

Transverse subglacial moraines include similar features in different locations that have been given various names and interpretations. The labels Rogen, De Geer, ribbed, washboard, corrugated, cyclic and cross-valley moraines have been applied to transverse features associated with subglacial processes. Rogen moraines are large ridges several tens of metres high, over 1km long and with crests several hundred metres apart, giving an irregular ribbed appearance to large areas of the landscape. They are often associated with flutes and drumlins, and are most commonly attributed either to thrusting of debris-rich basal ice into localized stacks beneath ice in compressive flow, or to deformation of subglacial sediment. The deformation hypothesis places Rogens at one end of a continuum of deformational forms that grades at the other end into longitudinal ridges such as flutes and drumlins. De Geer moraines are generally smaller in scale, and characterized by water-lain deposits within the moraine suggesting an origin beneath ice grounded in water.

Subglacial moraines lacking consistent orientation have been referred to as hummocky ground moraines. These are attributed either to the lowering of supraglacially released ablation moraines or to the release of debris beneath stagnant ice. The subglacial hypothesis places hummocky moraine at one end of a spectrum of forms that incorporates washboard moraines (weakly oriented hummocks) and drumlins (streamlined hummocks reflecting deposition beneath moving ice)

(Eyles *et al.* 1999). Some areas of hummocky moraine have recently been reinterpreted as complex assemblages of cross-cutting and discontinuous subglacial, supraglacial and ice-marginal moraine ridges.

Ice marginal moraines occur around the edges of glaciers and are defined by their position as either lateral or frontal moraines. Moraines marking the maximum extent of a glacial advance are referred to as terminal moraines. A terminal frontal moraine is called an end moraine. Moraines deposited at successive positions of the margin during a period of progressive retreat are referred to as recessional moraines. Moraines deposited at successive positions of the margin during periods of advance are usually destroyed by the advancing ice and are not preserved in the landscape, except for the terminal moraine.

Marginal moraines at existing glaciers are typically ridges of sediment resting partially on the edge of the glacier and partially on ice-free ground beyond the margin. Upon deglaciation, ice-cored moraines lose their ice support and therefore tend to shrink in size and may become structurally unstable (Bennett *et al.* 2000). Marginal moraines may be several tens of metres in height, tens or hundreds of metres across, and may stretch for hundreds of kilometres around the margins of large ice sheets. The main processes for the formation of marginal moraines are dumping of supraglacial, englacial or basal debris transported through the glacier, and pushing of sediments previously deposited in front of the glacier.

Small push moraines can be formed by seasonal bulldozing of proglacial sediment where an ice margin oscillates with seasonally varying ablation. Larger push moraines can form by the superposition of several seasonal moraines or by a substantial advance of the margin into deformable materials. Other glacitectonic features include moraines formed by the squeezing out of deformable sediment such as saturated till from beneath the ice margin.

Meltout or dump moraines occur where englacial or supraglacial sediment is transported to the margin and dumped where the glacier ends. Dump moraines grow in size for as long as an ice margin remains *in situ* to supply sediment, their rate of growth depending on the rates of sediment supply and ablation.

The morphology and sedimentology of moraines can be used to reconstruct the characteristics of former glaciers. The distribution of moraines reflects the geography of former glaciers and glacial process environments. Dated terminal and recessional moraines reveal the history of decay of a glacier, and process-controlled moraines reveal the locations of specific process. For example subglacial crevasse-fill ridges, which are formed by the squeezing of subglacial sediment into crevasses in the base of a glacier have been cited as indicators of glacier surging (Sharp 1985). Sediment characteristics reflect the source location of the debris: supraglacial debris is characteristically angular, while basally derived debris is typically basal faceted, subrounded and striated. Knight *et al.* (2000) showed how the distribution of clay-sized particles in a moraine reflected the distribution of a particular type of debris within the glacier that only occurred in certain process environments. Complex structures within moraines can reveal seasonal and long-term variations in processes of sedimentation. Small *et al.* (1984) showed how lateral moraine ridges derived aspects of their internal structure from seasonal variations in debris supply.

Moraines are one stage in the glacier sediment transfer system, providing long-term storage and a supply of debris to the proglacial zone. Sediment flux within glaciated basins is very sensitive to the position of glaciers relative to their moraines. When glaciers lie behind marginal moraines the bulk of sediment produced at the margin can go into storage in the moraine belt and not reach the proglacial region. When glaciers have no marginal moraines, sediment passes directly into the proglacial system. When glaciers re-advance over ancient moraines, large amounts of sediment from the moraine can be released from storage and transported into the proglacial landscape. Moraines can also focus meltwater discharge, localizing fluvial processes and causing meltwater from the glacier to be ponded up to form moraine-dammed lakes. These lakes are potentially unstable and pose a serious threat of catastrophic flooding.

Moraines are significant features within glaciated landscapes, useful indicators of past glacial activity and important components of the glacial sediment transfer system.

References

Bennett, M.R., Hambrey, M.J., Huddart, D. and Glasser, N.F. (2000) Resedimentation of debris on an ice-cored lateral moraine in the high-Arctic (Kongsvegen, Svalbard) *Geomorphology* 35, 21–40.

Dreimanis, A. (1989) Tills: their genetic terminology and classification, in R.P. Goldthwaite and C.L. Matsch (eds) *Genetic Classification of Glacigenic Deposits*, 17–83, Rotterdam: A.A. Balkema.
Eyles, N., Boyce, J.I. and Barendregt, R.W. (1999) Hummocky moraine: sedimentary record of stagnant Laurentide Ice Sheet lobes resting on soft beds, *Sedimentary Geology* 123, 163–174.
Knight, P.G., Patterson, C.J., Waller, R.I., Jones, A.P. and Robinson, Z.P. (2000) Preservation of basal-ice sediment texture in ice sheet moraines, *Quaternary Science Reviews* 19, 1,255–1,258.
Sharp, M. (1985) Crevasse-fill ridges – a landform type characteristic of surging glaciers? *Geografiska Annaler* 67A, 213–220.
Shaw, J., Kvill, D. and Rains, B. (1989) Drumlins and catastrophic subglacial floods, *Sedimentary Geology* 62, 177–202.
Small, R.J. (1983) Lateral moraines of Glacier de Tsidjiore Nouve: form, development and implications, *Journal of Glaciology* 29, 250–259.
Small, R.J., Beecroft, I.R. and Stirling, D.M. (1984) Rates of deposition on lateral moraine embankments, Glacier de Tsidjiore Nouve, Valais, Switzerland, *Journal of Glaciology* 30, 275–281.
Vere, D.M. and Benn, D.I. (1989) Structure and debris characteristics of medial moraines in Jotunheimen, Norway: implications for moraine classification, *Journal of Glaciology* 35, 276–280.
Weertman, J. (1961) Mechanism for the formation of inner moraines found near the edge of cold ice caps and ice sheets, *Journal of Glaciology* 3, 965–978.

Further reading

Benn, D.I. and Evans, D.J.A. (1998) *Glaciers and Glaciation*, London: Arnold.
Bennett, M.R. and Glasser, N.F. (1996) *Glacial Geology*, Chichester: Wiley.
Goldthwaite, R.P. and Matsch, C.L. (eds) (1989) *Genetic Classification of Glacigenic Deposits*, Rotterdam: A.A. Balkema.
Hambrey, M.J. (1994) *Glacial Environments*, London: UCL Press.
Knight, P.G. (1999) *Glaciers*, Cheltenham: Nelson Thornes.

SEE ALSO: glacial deposition

PETER G. KNIGHT

MORPHOGENETIC REGION

A morphogenetic region is an area where landforms are, or have been shaped, by the same or similar processes, mainly those controlled by climate. In climatic geomorphology there are two spatial categories: in morphoclimatic zones typical processes are considered, whereas in climato-morphogenetic regions the distinctive morphogenesis of an area is investigated. These definitions are more or less followed in continental Europe. In Anglo-American geomorphology, on the other hand, the term morphogenetic is used differently as 'the extent to which different climatic regimes are potentially capable of exerting direct and indirect influences on geomorphic processes, and thereby of generating different "morphogenetic" landform assemblages' (Chorley *et al.* 1984: 466). This nearly corresponds in German terminology to 'klimamorphologische Zonen' (morphoclimatic zones), and in French to 'les zones morphoclimatiques'. In the English literature these German and French terms are sometimes wrongly translated as 'climato-morphogenetic regions'.

The terms 'arid, humid, and nival' were introduced in 1909 by A. Penck as names for zones with distinct climate, hydrology and geomorphology. He had already recognized that these zones had shifted during the warm and cold periods of the Pleistocene and in 1913 introduced the term 'pluvial'. In 1926, in a symposion at Düsseldorf on the 'Morphologie der Klimazonen' (morphology of climatic zones), nine geomorphologists gave an overview of their research in certain areas ranging from the arctic to the humid tropics. Each one compared his findings with central Europe to stress the peculiarities. In 1948 Büdel introduced 'Das System der klimatischen Geomorphologie' (the system of climatic geomorphology). He gave a description of the typical processes in each morphoclimatic zone. The most important aspect was the interrelation of the processes in one zone, e.g. the work of a river is dependent on the relief of the area, which next to precipitation controls the amount and time of discharge. The load which has to be transported is generated from slopes and small creeks. By their interrelationships the relative strength or influence of the processes shaping the landforms should become clear. The relation between fluvial erosion and denudation was especially weighted. Thus there was some estimation of erosion rates, too. Not only were the most spectacular landforms looked for, but also the most widely distributed ones. Not only the catastrophic events but also the slowly working processes were investigated. For each morphoclimatic zone the processes were recorded as they were observed from recent occurances, from the observation of the REGOLITH, and from a check with the landforms, whose shape was interpolated and extrapolated with the processes, a feedback. The concept of morphoclimatic zones is a very open one and may be varied, e.g. according to rock resistance (petrovariance) or tectonics (tectovariance). This is a rather broad approach and of course uncertainty or even mistakes are possible. This does not spoil the concept though. Comparison of similar regions and results from different research has increased our knowledge of the different morphoclimatic zones, though no new complete version has been made since the handbook of Büdel (1977, 1982). However, many detailed studies are founded on this concept.

For the interrelation of forming processes the terms 'Prozessgefüge' (process fabric) or 'Formungsmechanismus' (relief forming mechanism) came into use. For one of the morphoclimatic zones, the humid mid-latitudes 'zone of Holocene retarded valley building', Büdel (1982: 14) named the following components, which make up or control the 'forming mechanisms for the highly complex phenomena and processes: solution, mechanical weathering, chemical weathering, plant cover, soil development, surface denudation, linear erosion, transport, and deposition'. They are connected on 'highly complex integration levels'. In quoting 'highly complex' twice and adding 'occuring only in nature, not reproducable', he wanted to stress that on this level field measurements and laboratory experiments should be combined with the 'predominant qualitative relief analysis'. The main methods are field observations in 'natural test sites', where the phenomena are typical and which have to be searched for. Then follows the comparison with similar areas, where e.g. the influence of different rocks can be observed. Thus by comparison the petrovariance and the tectovariance can be abstracted and the processes controlled by climate become clear.

It is easier to link relief forming mechanisms to ecological factors for which climate is an abbreviation, as these comply to a zonal order, than to build a system on lithology. Of course there are distinct landforms in limestones, sandstones and granites and excellent relevant handbooks, but there is no systematic arrangement of forms due to differences in rock hardness or structure. Thus a morphogenetic region according to one of these rock groups would more or less coincide with a geological map. That would not be a new insight. It is possible to outline morphotectonic domains, but the connection to geomorphological processes is only very slowly developing, as detailed knowledge of the influence of tectonic movements on processes, except landsliding, is very small so far, and for cratons almost unknown. In both cases palaeoforms are hard to incorporate systematically, but this is easy in climatic geomorphology.

A morphoclimatic zone defined by relief forming mechanisms is a framework for detailed studies. These may be of megaforms, mesoforms or microforms and it is possible to apply many different methods. For instance, if landform facets are linked to the thickness and texture of the regolith and/or sediments, their relative age and evolution is investigated. This can be verified by laboratory research of the material, and by absolute datings. If the extent of the landforms is mapped or their changes are derived from sequences or monitoring, there is an estimation of the volume of transport possible. As this holds mainly for several hundreds or thousands of years, this provides a long-term check for short-term measurements of material transport. Thus it is possible to discriminate between natural and human-induced erosion rates. The concept of morphoclimatic zones is helpful to provide working hypotheses with regard to the full breadth of processes possible, their interdependence and relative strength. Especially in extrapolation of measured properties, an appraisal of relief forming mechanisms should be incorporated.

The essence of the concept of morphoclimatic zones is the interrelation of processes and there are almost no attempts in climatic geomorphology, as understood in Europe, to link landforms to climatic data. As in a morphoclimatic zone the interrelation of weathering, denudation and fluvial erosion is described, and it is obvious that only a general combination with climate is possible. Büdel (1977, 1982) himself delineated ten morphoclimatic zones. Originally (1948) there were twelve, and in 1963 they were reduced to five with an emphasis on the tropical semi-humid zone of excessive planation and the subpolar zone of excessive valley cutting. The names were changed too, though only slightly. This may show that zonation was not Büdel's foremost interest. He never tried to link the boundaries of the zones to climatic data. He rather insisted on the complexity of relief analysis, covering as many ecological factors as possible. There is one inconsistency, too. The zone of excessive planation should be shifted to the perhumid tropics, as only there is weathering intense enough for the concept of double planation, which is still very valid. The term climatic geomorphology is a misnomer, but the attempt to change to dynamic geomorphology was not successful as the term was introduced for a long time in contrast to tectonic geomorphology.

The difference of the Anglo-American approach to morphogenetic regions is twofold: the relevance of climatic data at the start and the broadness of the approach. The first attempt at delineating morphogenetic regions in the USA was the diagram by Peltier (1950). It was much cited but had little influence on detailed studies. Even the more sophisticated diagram of Chorley *et al.* (1984) has not been filled by regional or areal studies. Thus climatic regions as a starting point and the deduction of possible processes does not seem to be very fruitful. Instead there are single features like drainage densities connected with climatic data, or gradients of rivers or slopes are linked to sediment transport and rainfall variables. Polygenetic landforms are quite often approached from the knowledge of palaeoclimates. On the other hand there are excellent books on tropical, desert, periglacial, glacial geomorphology and karst, which describe and explain landforms and processes. But there are few interrelations and almost no connection to climatic data, though these handbooks often contain a chapter on the climate of the zone.

An extension of the morphogenetic regions was done by Brunsden in creating tectono-climatic regions. He proposed linking geotectonic domaines with morphoclimatic zones. For example, he entered into a map of the present conditions of the Indo-Australian plate the recent morphoclimatic zones and second the environmental conditions of 18,000 BP. Comparison of these two pictures gives areas of tectono-climatic stability. These are interesting hypotheses, but here, too, the starting point is the concept from facts outside geomorphology. Only later shall it be filled with field observations. It is a way that proceeds from the top downward, not from the base upwards.

It is always possible to concentrate on a special process but this should not be done in an isolated way but in the realm of the relief forming mechanism. Thus it is tied up in an analysis of interrelations of larger to smaller forms, of single processes to the process fabric. Thus the extrapolation of single processes and the interpretation of landforms becomes more secure. An example might be river terraces in mid-latitudes. Are they of climatic or tectonic origin? Not only the material of the terraces and their gradient is indicative but the origin of the pebbles and the mode of transport from the source area on a slope (e.g. by solifluction into the rivers). Such features as periglacial ice wedges casts and covers like loess are studied in relation to former climatic conditions and age. Are similar terraces developed in neighbouring areas? Which forms are incised in the older terraces? This for instance led to a detailed history of incision for the middle Rhine valley. This part of the valley is antecedent and developed during slow uplift, but the forms in detail are climate controlled. This is an example for a morphogenetic region in German understanding. There are similar regional studies in the English literature. The methods are more detailed in CLIMATO-GENETIC GEOMORPHOLOGY.

References

Brunsden, D. (1990) Tablets of stone: toward the ten commandments of geomorphology, *Zeitschrift für Geomorphologie, Supplementband* 79, 1–37.

Büdel, J. (1977) *Klima-Geomorphologie*, Borntraeger: Berlin. Translated by L. Fischer and D. Busche (1982) *Climatic Geomorphology*, Princeton: Princeton University Press.

Chorley, R., Schumm, S.A. and Sugden, D.E. (1984) *Geomorphology*, London: Methuen.

Peltier, L.C. (1950) The geographical cycle in periglacial regions as it is related to climatic geomorphology, *Annals of the Association of American Geographers* 40, 214–236.

HANNA BREMER

MORPHOMETRIC PROPERTIES

Morphometric properties of a DRAINAGE BASIN are quantitative attributes of the landscape that are derived from the terrain or elevation surface and drainage network within a drainage basin. GEOMORPHOMETRY is the measurement and analysis of morphometric properties. Traditionally morphometric properties were determined from topographic maps using manual methods, but with the advent of geographic information system (GIS) technology, many morphometric properties can be automatically computed.

Size properties

Size variables provide measures of scale that can be used to compare the magnitudes of two or more drainage basins. Size variables are derived from measurements of the basin outline as defined by the drainage divide or are obtained from the drainage network. Many size variables are strongly correlated with one another so can be used interchangeably.

Drainage area, the two-dimensional projection of area measured in the map plane, is the most important size measure and is specified as the area contained within the drainage divide. RUNOFF GENERATION and the frequency of FLOODS is directly correlated with drainage area in many environments.

Basin length indicates the distance from the basin outlet to a point on the drainage divide, but many different methods for measuring basin length have been devised. For example, the end-point of the length measure can be the highest point on the divide or the point on the divide that is equidistant from the outlet along the divide. Perimeter is a measure of distance around the drainage basin measured along the drainage divide.

Main channel length is the length from outlet to channel head along a subjectively defined main channel, or, more objectively, the length of the longest flow path to the drainage divide. Total channel length is the sum of lengths of all channels in a basin.

Stream order can also be used to indicate basin size (see STREAM ORDERING). The order of a basin is the order of its outlet stream. Stream magnitude is the number of FIRST-ORDER STREAMS in a basin. Magnitude is a more discerning measure of size than is order.

Surface properties

Surface properties are quantities depicted by fields comprising a value at each point within a domain (drainage basin). GIS technology provides the capability to derive surface properties from a DIGITAL ELEVATION MODEL (DEM) which is the numerical representation of an elevation surface. The elevation surface is the most fundamental surface property field, and quantifies the ground surface elevation at each point (neglecting cave and overhang special cases). DEM types include square or rectangular digital elevation grids, triangular irregular networks, sets of digital line graph contours or random points (Wilson and Gallant 2000).

The flow direction field is the direction that water flows over a surface under the action of gravity. This may be defined by the horizontal component of the surface normal. The flow direction field is represented numerically by a flow direction grid. The simplest flow direction grid is the D8 flow direction grid in which flow direction is represented by one of eight values. The value depends on which of the eight neighbouring cells (four on the main axes, four on the diagonals) is in the direction of steepest descent and thus receives its drainage. Other numerical flow direction fields can be derived using finite difference or local polynomial or surface fits to elevations of grid cells in the neighbourhood of each point (Tarboton 1997).

Terrain slope is a field giving the slope of the terrain in the direction of the flow direction field at each point. This is evaluated numerically by taking elevation differences from the elevation field over a short distance centred on each point.

Contributing area is a field representing drainage area upslope of each point. It is defined by tracing flow paths up slope from each point along the flow direction field to the drainage divide and measuring the area enclosed. Within a grid-based GIS, contributing area is evaluated by counting the number of grid cells draining to each grid cell. Contributing area is also referred to as catchment area or flow accumulation area.

Specific catchment area is a field representing contributing area per unit contour length. On a smooth surface, the contributing area to a point may be a line that has zero area. Specific catchment area is quantified using the measurable area contributing to a small length of contour (Moore *et al.* 1991: 12). Specific catchment area has units of length. On a planar surface with parallel flow, specific catchment area is equal to the upslope distance to the drainage divide.

Shape properties

Drainage basin shape is a difficult morphometric property to characterize simply, and there have been numerous attempts at defining shape variables. The simplest shape measures employ area, length, width or perimeter of the drainage basin or of a shape with area equivalent to that of the basin. More complex functions of drainage basin or drainage network shape are best portrayed using two-dimensional graphical plots.

The cumulative area distribution function is defined as the proportion of a drainage basin that has a drainage area greater than or equal to a specified area. It is typically represented by plotting cumulative area versus area on a log-log line chart.

The distance area diagram depicts the area of the basin as a function of distance along flow paths to the outlet. The channel network width function is the number of channels at a given distance from the drainage basin outlet, as measured along the drainage network, and is typically plotted as a line or bar chart. The distance area diagram and channel network width function both give an indication of basin hydrological response and are related to the instantaneous unit hydrograph.

Relief properties

RELIEF properties bring the dimension of height into morphometric analysis. Because many landscape processes are driven by gravity, relief properties are frequently used as indicators of EROSION potential and DENUDATION rates.

Total basin relief is the difference in height between the outlet and the highest point on the drainage divide. Relief ratio removes the size effect by dividing total relief by basin length. Sediment yield (see SEDIMENT LOAD AND YIELD) in small drainage basins has been shown to be exponentially related to relief ratio (Hadley and Schumm 1961: 172).

A more complex representation of basin relief is the area-elevation relationship or hypsometric curve. The hypsometric curve is a plot of the area of a basin (on the x-axis) above each elevation value (on the y-axis). The axes are commonly normalized to range between zero and one. The hypsometric curve is equivalent to one minus the cumulative distribution of elevation within a drainage basin. Davisian model evolutionary stage can be inferred from the shape of a basin's hypsometric curve.

Texture properties

Texture indicates the amount of landscape dissection by a channel network. The contours on a map of a highly textured landscape will have many small crenulations (wiggles) indicating the presence of numerous channels.

DRAINAGE DENSITY (Horton 1945: 283), the best-known texture indicator, is defined as the lengths of all stream channels in a drainage basin divided by drainage area and has units of 1/length. Drainage density ranges from less than $1\,km^{-1}$ to over $800\,km^{-1}$, attaining maximum values in semi-arid areas (Gregory 1976: 291). High drainage densities indicate highly textured landscapes, short hillslopes and domination by OVERLAND FLOW runoff typical of BADLANDS.

The area–slope relationship quantifies the area draining through a point versus the terrain slope at that point, typically plotted on a log-log scale graph. The scatter when all points or grid cells are used is removed by binning (e.g. using a moving average) to reveal a characteristic area–slope relationship with two distinct regions. For small areas, slope increases with drainage area and for large areas, slope decreases with area. The turnover point in the relationship has been interpreted as the drainage area at which diffusive hillslope processes (see HILLSLOPE, PROCESS) are overtaken by fluvial processes and channels are initiated (Tarboton *et al.* 1992: 73).

References

Gregory, K.J. (1976) Drainage networks and climate, in E. Derbyshire (ed.) *Geomorphology and Climate*, 289–315, London: Wiley.

Hadley, R.F. and Schumm, S.A. (1961) *Sediment Sources and Drainage-Basin Characteristics in the Upper Cheyenne River Basin*, Washington: US Geological Survey Water Supply Paper 1,531.

Horton, R.E. (1945) Erosional development of streams and their drainage basins; hydrophysical approach to quantitative morphology, *Geological Society of America Bulletin* 56, 275–370.

Moore, I.D., Grayson, R.B. and Ladson, A.R. (1991) Digital terrain modelling: a review of hydrological, geomorphological, and biological applications, *Hydrological Processes* 5, 3–30.

Tarboton, D.G. (1997) A new method for the determination of flow directions and contributing areas in grid digital elevation models, *Water Resources Research* 33, 309–319.

Tarboton, D.G., Bra, R.L. and Rodriguez-Iturbe, I. (1992) A physical basis for drainage density, *Geomorphology* 5, 59–76.

Wilson, J.P. and Gallant, J.C. (2000) *Terrain Analysis: Principles and Applications*, New York: Wiley.

Further reading

Gardiner, V. (1975) *Drainage Basin Morphometry*, British Geomorphological Research Group Technical Bulletin No. 14, Norwich: GeoAbstracts.

SEE ALSO: Horton's Laws

CRAIG N. GOODWIN AND DAVID G. TARBOTON

MORPHOTECTONICS

Morphotectonics is the term pertinent to links between geomorphology and tectonics, although individual authors apparently understand the exact nature of these links in slightly different ways. Most often, morphotectonics is considered synonymous with TECTONIC GEOMORPHOLOGY and defined simply as the study of the interaction of tectonics and geomorphology. Embleton (1987) lists four main lines of interest in morphotectonic research: (1) study of landforms indicative of contemporary or recent tectonic movement, (2) study of deformation of PLANATION SURFACES, (3) study of geomorphological effects of earthquakes (see SEISMOTECTONIC GEOMORPHOLOGY), (4) use of geomorphological evidence to predict earthquakes. It needs to be emphasized that in some countries morphotectonics is a term of very limited usage. For example, two recent American textbooks about tectonic geomorphology (Burbank and Anderson 2001; Keller and Pinter 2002) do not mention morphotectonics, although they evidently deal with this kind of phenomenon.

Fairbridge (1968) offers a different explanation and understands morphotectonics as a means to classify major landforms of the globe rather than any landforms related to tectonic processes. Accordingly, he distinguishes morphotectonic units of first and second order. In the first order these are continents and oceanic basins, in the second one there are shields, younger mountain belts, older mountain massifs, basin-and-range areas, rift zones and basins. This global context of morphotectonics is also evident in the study of great ESCARPMENTs (Ollier 1985).

In practice, the morphotectonic approach frequently means using landforms or any other surface features (e.g. drainage patterns) as a key to infer the existence of tectonic features, especially in relatively stable areas where seismicity and present-day rates of uplift and subsidence are negligible. They acquire the status of geomorphic markers of tectonics. Geomorphological maps, drainage pattern maps, digital elevation models and their various derivatives are analysed with the aim of locating anomalies in landform distribution, river courses, channel form, terrace profiles, local relief or specific landforms such as slope breaks. These anomalies in turn, if no other explanation for their occurrence is available, are considered to reflect the presence of tectonically active zones or areas. Detailed analysis of river patterns can be a particularly valuable tool in morphotectonic research in lowland areas, where hardly any other evidence is at hand.

References

Burbank, D.W. and Anderson, R.S. (2001) *Tectonic Geomorphology*, Malden: Blackwell.
Embleton, C. (1987) Neotectonic and morphotectonic research, *Zeitschrift für Geomorphologie N.F., Supplementband* 63, 1–7.
Fairbridge, R.W. (1968) Morphotectonics, in R.W. Fairbridge (ed.) *Encyclopedia of Geomorphology*, 733–736, New York: Reinhold.
Keller, E.A. and Pinter, N. (2002) *Active Tectonics*, Englewood Cliffs, NJ: Prentice Hall.
Ollier, C.D. (ed.) (1985) Morphotectonics of passive continental margins, *Zeitschrift für Geomorphologie N.F., Supplementband* 54.

Further reading

Morisawa, M. and Hack, J.T. (eds) (1985) *Tectonic Geomorphology*, Boston: Allen and Unwin.
Ollier, C.D. (1981) *Tectonics and Landforms*, London: Longman.
Schumm, S.A., Dumont, J.F. and Holbrook, J.M. (2000) *Active Tectonics and Alluvial Rivers*, Cambridge: Cambridge University Press.
Summerfield, M.A. (ed.) (2000) *Geomorphology and Global Tectonics*, Chichester: Wiley.

SEE ALSO: active and capable fault; active margin; fault and fault scarp; global geomorphology; neotectonics; passive margin

PIOTR MIGOŃ

MOULIN

Moulins, or glacier mills, are sink holes that owe their name to the roaring noise of water that engulfs itself in them. They form in the ablation zone of GLACIERs (Paterson 1994), where meltwater (see MELTWATER AND MELTWATER CHANNEL) manages to cut stream channels into the ice, generally parallel to glacier slope. These channels are eventually intercepted by crevasses, which form perpendicular or oblique to glacier slope in response to ICE flow related to bedrock surface irregularities. Moulins are the result of meltwater flowing into crevasses (Rothlisberger and Lang 1987).

Moulins are characterized by a vertical shaft up to 100 m tall, developing along the planes of single or cross-cutting crevasses and prolonging into a downflow dipping gallery that follows structures related to glacial flow. The gallery dips approximately 45° and forms a succession of pools on an irregular floor, but the gallery sometimes dips approximately parallel to glacier slope when the shaft is less than 50 m tall. Shafts are circular or elliptical in horizontal cross section and range between less than 1 m to over 20 m in their long axis, but detail of their morphology is controlled by the dynamics of the water that flows into it (Holmlund 1988).

The bottom of moulins is often submerged. Water level can vary within a few hours and from one season to the next in relation to meteorological conditions, glacial flow, ice plasticity and according to the facility with which water can flow along the base of the glacier.

During the summer, moulins provide the main inputs of glacial aquifers. During the winter, they are separated from the surface by a snow bridge. However, water level in the moulins usually increases during the winter and then decreases in a jerky fashion, the moulins functioning like surge tanks (Schroeder 1998). This implies that drainage in the glacier tends to clog in the front first during the beginning of the cold season, while the water column that then remains stocked within the moulins prevents glacial flow from closing it. With the onset of the warm season, this water that was stocked within the moulins helps in reinitiating subglacial drainage.

The life expectancy of moulins can reach several dozens of years. Moving along with the glacier, they eventually lose their connection to glacier surface drainage at the profit of new moulins forming upflow from them. In dead ice, moulins often reach down through the entire glacier. Megapotholes (diameter > 50 m) developed in rock bars of regions that were glaciated during the Quaternary are thought to be the result of extended water circulation at the base of moulins in stagnant ice.

References

Holmlund, P. (1988) Internal geometry and evolution of moulins, Storglaciären, Sweden, *Journal of Glaciology* 34, 242–248.

Paterson, W.S.B. (1994) *The Physics of Glaciers*, New York: Pergamon.

Rothlisberger, H. and Lang, H. (1987) Glacial hydrology, in A.M. Gurnell and M.J. Clark (eds) *Glaciofluvial Sediment Transfer*, 209–284, New York: Wiley.

Schroeder, J. (1998) Hans glacier moulins observed from 1988 to 1992, Svalbard, *Norsk Geografisk Tidsskrift* 52, 79–88.

JACQUES SCHROEDER

MOUND SPRING

Small mounds formed preferentially along fault lines by artesian springs. Solutes and colloids are precipitated to form travertines or tufas (see TUFA AND TRAVERTINE) of calcium carbonate, together with various siliceous and ferruginous deposits. Wind-blown sand and accumulated plant debris, together with mud and sand carried up with the spring water, assist in their formation. Where springs display high rates of water flow there tends to be little or no mound formation – they are too erosive. However, springs with low discharge rates and laminar flow experience high rates of evaporation (especially in arid environments) and have a greater possibility of accumulating chemical precipitates.

Major examples of such features are known from the Great Artesian Basin of Central Australia (Ponder 1986) and from the depressions of the Western Desert in Egypt, where much accretion has occurred when vegetated fields, irrigated by the springs, have trapped aeolian sediment (Brookes 1989).

References

Brookes, I.A. (1989) Above the salt: sediment accretion and irrigation agriculture in an Egyptian oasis, *Journal of Arid Environments* 17, 335–348.

Ponder, W.F. (1986) Mound springs of the Great Artesian Basin, in P. de Deckker and W.D. Williams (eds) *Limnology in Australia*, 403–420, Dordrecht: Junk.

A.S. GOUDIE

Plate 82 A silty mound deposit associated with spring activity in the Farafra Oasis of the Western Desert of Egypt

MOUNTAIN GEOMORPHOLOGY

Mountain geomorphology is a 'regional component within geomorphology' (Barsch and Caine 1984). The region in this case is the world's mountains defined by absolute elevation (>600 m above sea level), available relief (>200 m km^{-2}) and topographic slopes (>10°). There is no international standard definition, but other elements which are frequently incorporated are high spatial variability, presence of ice and snow and evidence of late Pleistocene glaciation. Carl Troll (1973) who was the modern creator of mountain geomorphology, defined mountain systems as those which encompass more than one vegetation belt, but do not reach alpine elevation by contrast with high mountain systems (hochgebirge) which extend above the timberline.

Fairbridge (1968) rehearses the classification of mountains by scale and continuity: (a) mountain is a singular, isolated feature or a feature outstanding within a mountain mass; (b) a mountain range is a linear topographic feature of high relief, usually in the form of a single ridge; (c) a mountain chain is a term applied to linear topographic features of high relief, but usually given to major features that persist for thousands of kilometres; (d) a mountain mass, massif, block or group is a term applied to irregular regions of mountain terrain, not characterized by simple linear trends; and (e) a mountain system is reserved for the greatest continent-spanning features.

A simple genetic system of mountain types, which was developed before global plate tectonics was understood, is nevertheless useful in local-scale understanding. There are two broad categories: (1) structural, tectonic or constructional forms and (2) denudational or destructional forms. Under the first category can be identified: (a) volcanic; (b) fold and nappe; (c) block; (d) dome; (e) erosional uplift or outlier; (f) structural outlier or klippe; (g) polycyclic tectonic; and (h) epigene mountains. Under the second category are defined: (a) differential erosion; (b) exhumed; (c) plutonic and metamorphic complex; and (d) polycyclic denudational mountains (Fairbridge 1968). In the structural mountain categories it is the tectonic process that has played a primary role; in the denudational categories it is the denudational processes that are primary. The lithology and the climatic history are both extremely important with respect to the detailed modification of these mountain types. Indeed, much of the science of geomorphology is centred on the discrimination of these second-order effects.

The simplest typology of mountain geomorphology makes use of the tripartite division into historical, functional and applied mountain geomorphology. Historical mountain geomorphology focuses on the evolution of mountains and mountain systems over both long and medium timescales. It is common, at the largest scale, to distinguish between young active mountain belts which have evolved throughout the Cenozoic and are still associated with active plate margins and mountains on passive continental margins. Nearly all the literature on mountain building in the past forty years has concentrated on active margins where collision and subduction may explain both mountains and the structures within them. Most exciting in recent years is the trend towards quantifying rates of uplift and denudation with the development of new geochemical, geochronological and geodetic methods. But, in reality, there are mountains on PASSIVE MARGINS too (Ollier and Pain 2000). The evolution of these older mountain belts is intrinsically more complex as they do not easily fit into the simple plate tectonic story of mountain building at collisional sites and include the history of the Earth since the breakup of Gondwanaland during the Mesozoic. A major difference of opinion has emerged between those who place greatest emphasis on the data from FISSION TRACK ANALYSIS and those who use whatever landform, stratigraphic and geological data that can be found to constrain the interpretation. Whereas geomorphic models of denudation history are difficult to validate, interpretation of fission track data in terms of denudation history is complex.

Functional geomorphology of mountains includes the assessment of processes, rates and spatial and temporal patterns of mountain belt erosion. The process framework should ideally involve a consideration of the coupling of uplift and erosion; many geomorphic models have failed to include realistic models of this coupling. In mountain belts, such a consideration is obligatory as both uplift rates and erosion rates achieve maximum values and the coupling of the processes is even more critical than in lowland regions. Improvements in understanding of fluvial bedrock incision processes, hillslope mass wasting, glacial valley lowering and SEDIMENT ROUTING are leading to the development of improved mountain landscape evolution models.

Feedbacks between tectonic, climatic and geomorphic processes have been explored in geodynamic models and solid, solute and organic fluxes from mountain belts have been constrained and considered within a global geochemical context. The topographic evolution of mountain belts can be modelled with increasing realism, but the issue of equilibrium conditions versus transience is still far from resolved.

APPLIED GEOMORPHOLOGY of mountains: mountain habitats create or magnify natural hazards in the form of dangerous geomorphic processes. The interaction of geomorphic processes with mountain societies, their land uses and their response capabilities determines risk. Recent social and environmental changes in the mountains has led to the modernization of the natural hazard problematique. As a result, planned responses, including mitigation strategies for specific hazards and mountain disasters, must be developed to reduce the vulnerability of mountain peoples. The applied mountain geomorphologist has a distinctive role to play.

There are three formal or semi-formal attempts to define the field in the literature: Hewitt (1972) Barsch and Caine (1984) and Slaymaker (1991).

Hewitt addressed two issues: the idea of a high energy condition and the relation of distinctive morphological features to clima-geomorphic conditions and denudation history. These he said express the distinctiveness of mountain geomorphology. Under the high energy condition, he dealt with regional rates of net erosion, magnitude and frequency of erosional events and energy in the mountain geomorphic system. Under distinctive morphological features, he picked out accordant erosion surfaces, valley asymmetry and threshold slopes for detailed treatment. The beauty of Hewitt's vigorous statement is that he foreshadows the increasingly heavy emphasis on the operation of geomorphic processes in mountain regions, but also warns of the danger of not relating these observations to the larger questions of mountain landscapes and neglecting to either solve them or to restate them in better terms.

Barsch and Caine (1984) divide mountain geomorphology into studies of mountain form and morphodynamics in mountains. These two categories they further subdivide into (a) morphometry and structure, (b) relief generation and history, (c) morphoclimatic models and (d) process dynamics and activity. Morphometry and structure depend heavily on plate tectonic setting of the mountains. There are four convergent plate settings in which some of the most rapidly evolving mountain systems of the world are located. These are: oceanic to oceanic plate convergence (e.g. Japanese Alps and the Aleutian Arc, Alaska); oceanic to continental plate convergence (e.g. South Island, New Zealand and Cascade Ranges, Pacific North-West); continental to continental plate convergence (e.g. Himalayas); and displaced terranes along accreted margins (e.g. British Columbia). Divergent plate settings include sites of oceanic spreading, such as Iceland and the Galapagos Islands, and intra-continental rifts, such as the Gulf of Aqaba and the Scottish Highlands. Transform plate settings are threefold: ridge past ridge (e.g. Coast Ranges of California); trench past trench (e.g. Anatolia, Turkey) and ridge past trench (e.g. Pakistan–Afghanistan). It is not difficult to understand why mountains are preferentially located in all of the above plate marginal locations. But mountains are also found in plate interior settings, such as the following: hot spots (e.g. Hawaii and Yellowstone National Park); continental flood basalts (e.g. Deccan, India and Columbia Plateau, Pacific North-West); shields (e.g. Ahaggar Mountains, Sahara); intracratonic uplift sites (e.g. the San Rafael Swell, Utah); post-tectonic magmatic intrusion sites (Air Mountains, Nigeria); and evaporite diapers (e.g. Zagros Mountains, Iran). They note that in most mountain regions, the balance between denudation and tectonic uplift is resolved in favour of the latter. They fail to differentiate between high mountains and mountain systems on the basis of morphometry alone, but they make the case that there are four distinctive 'relief types' within high mountain systems, namely the Alp type, the Rocky Mountain type, polar mountains and desert mountains. The Alp type is associated with an overriding impact of glacial ice and glacial erosion; the Rocky Mountain type has a less pervasive impact of glacial erosion and includes areas of low relief on flat summits and rounded interfluves; polar mountains give evidence of intense glaciation, but are frequently with a local relief of less than 1,000 m; and desert mountains are high mountains in the 'true' sense even though they do not reach timberline and were only lightly glaciated during the Pleistocene.

Relief development in high mountains revolves around questions of accordant surfaces and valley benches as indicators of mode of valley dissection. Attention is directed to (a) the alpine summit accordance or 'gipfelflur', which has been explained as a remnant of an old erosion surface; (b) the alpine crest and summit accordance, explained as the product of regular patterns of dissection which constrain summits to approximately the same elevation; (c) the timberline and alp slope accordance, explained as an inter-glacial alp slope associated with a higher timberline than the present one; and (d) benches along the sides of major valleys, variously explained as Tertiary, Pleistocene glacial and inter-glacial timberline effects. Ford *et al.* (1981) have suggested that the age of the present relief of the southern Canadian Rockies is Pliocene, considerably older than had previously been thought.

Building on Caine (1974), Barsch and Caine (1984) distinguish four mountain geomorphic process systems: (1) the glacial system; (2) the coarse debris system; (3) the fine clastic sediment system; and (4) the geochemical system. Of the four, the glacial and the coarse debris systems are most characteristic of high mountain terrain. The final section of their paper summarizes contemporary geomorphic activity in high mountain areas using calculations of sediment flux (in $\mathrm{J\,km^{-2}\,yr^{-1}}$) from Sweden, Switzerland and the United Sates (Rapp 1960; Jackli 1957; Caine 1976). Most interesting was the observation that talus shift, solifluction, soil creep and other processes of slow mass wasting accounted for no more than 15 per cent of the geomorphic work done in all three areas and their relative importance decreased with increasing size of basin. The authors suggest that there are two urgent needs for mountain geomorphology: (1) linking process and form in a meaningful way and (2) identifying anthropogenic influences and ways in which they may be propagated through the mountain system.

Slaymaker (1991) suggests that the meso and macro-scales are the only spatial scales at which a distinctive mountain geomorphology signal is likely to be apparent. He then adopts a slightly modified version of the Chorley and Kennedy (1971) open systems framework to identify five mountain systems: (1) morphological; (2) morphologic evolutionary; (3) cascading; (4) process-response; and (5) control systems. Each of these mountain systems is examined at meso- and macro-scales in search of characteristic mountain geomorphology forms and processes. He claims that this typology is useful in that different measurement programmes are appropriate within each of the ten mountain systems identified.

In fact, these ten mountain geomorphic systems serve to underline the huge variety of forms and processes that characterize mountain geomorphology and support the contention that mountain geomorphology is characterized by its extreme gradients, not only of topography, but also of energy and mass balances and ecological responses. High vertical and horizontal rates of change over space of landforms and processes and rapid rates of change over time distinguish mountain geomorphic systems from other regions. Hence the validation of mountain geomorphology as a regional component within geomorphology.

References

Barsch, D. and Caine, N. (1984) The nature of mountain geomorphology, *Mountain Research and Development* 4, 287–298.

Caine, N. (1974) The geomorphic processes of the alpine environment, in J.D. Ives and R.G. Barry (eds) *Arctic and Alpine Environments*, 721–748, London: Methuen.

——(1976) A uniform measure of sub-aerial erosion, *Geological Society of America Bulletin* 87, 137–140.

Chorley, R.J. and Kennedy, B.A. (1971) *Physical Geography: A Systems Approach*, London: Prentice Hall.

Fairbridge, R.W. (1968) *The Encyclopedia of Geomorphology*, New York: Reinhold.

Ford, D.C., Schwarcz, H.P., Drake, J.J., Gascoyne, M., Harmon, R.S. and Latham, A.G. (1981) Estimations of the age of the existing relief within the southern Rocky Mountains of Canada, *Arctic and Alpine Research* 13, 1–10.

Hewitt, K. (1972) The mountain environment and geomorphic processes, in O. Slaymaker and H.J. McPherson (eds) *Mountain Geomorphology*, 17–34, Vancouver: Tantalus.

Jackli, H. (1957) Gegenwartsgeologie des bundnerischen Rheingebietes: ein Beitrag zur exogenen Dynamik alpiner Gebirgslandschaften, *Beiträge zur Geologie der Schweiz*, Geotechnische Serie No. 36.

Ollier, C.D. and Pain, C.F. (2000) *The Origin of Mountains*, London: Routledge.

Rapp, A. (1960) Recent development of mountain slopes in Karkevagge and surroundings, northern Scandinavia, *Geografiska Annaler* 42-A, 73–200.

Slaymaker, O. (1991) Mountain geomorphology: a theoretical framework for measurement programmes, *Catena* 18, 427–437.

Troll, C. (1973) High mountain belts between the polar caps and the equator: their definition and lower limit, *Arctic and Alpine Research* 5, 19–28.

SEE ALSO: plate tectonics

OLAV SLAYMAKER

MUD FLAT AND MUDDY COAST

The term mud is used to refer to sediments comprised chiefly of silts (size range 4 to 63 μm) and clays (finer than 4 μm). Such fine material is readily maintained in suspension and can be transported over long distances by coastal currents. Unlike coarser sands and gravels, muddy sediments tend to be cohesive. The electrochemical properties of clay mineral particles mean that these can bind together to form larger composite particles in a process known as flocculation. Flocculation is influenced by a variety of factors, notably salinity, fluid shear and suspended sediment concentration (Lick and Huang 1993). The effect of these processes may vary over quite short spatial and temporal scales, especially in estuaries, where mixing of freshwater and saltwater occurs and where marked variation in flow intensity occurs at tidal timescales. The cohesive nature of muddy sediments makes their behaviour far more complex than that of non-cohesive sands. Flocculated sediments typically settle from suspension far more rapidly than their constituent mineral particles, and the stability of natural muddy deposits is governed not only by physical processes but also by the activity of a rich and diverse biota including macroscopic and microscopic algae, invertebrates and bacteria (Paterson 1997).

Muddy coasts typically occur along low energy shorelines that are well supplied with silt and clay-sized sediments. They include many estuarine margins, delta shorelines, and areas of open coast subject to low wave energy. Such settings are usually dominated by tidal processes, and the characteristic landforms of muddy coasts – SALTMARSHes, MANGROVE SWAMPs and tidal flats – are often very well developed under macro-tidal conditions (Hayes 1975). Enormous quantities of muddy sediment are supplied by some of the world's major rivers, and their estuaries and deltas often feature extensive shore-attached mud banks. Open coast mud banks occur downdrift of major fluvial sediment sources, notably in the Gulf of Mexico (associated with the Mississippi River); more than 850 km of the Jiangsu coastline of China, supplied by the Huanghe and Changjiang Rivers (Ren 1987); and along the south-west coast of India. Both estuarine and open coast mud banks are highly dynamic landforms, which exhibit both seasonal and decadal style variability in response to variations in river flow and wave energy. Their deposits often have a high water content and include highly mobile 'fluid muds' that are highly effective in dissipating incident wave energy (Mehta and Kirby 2001). In other environmental settings, coastal and marine sediment sources are more important. In the North Sea, for example, erosion of unconsolidated Quaternary cliffs provides a major source of muddy sediments along the coast of eastern England (Ke *et al.* 1996).

The intertidal zone of muddy coasts typically comprises: a lower zone, characterized by sandy tidal flats; a middle zone of muddy tidal flats; and an upper intertidal of vegetated saltmarsh or mangrove. In some localities, the upper intertidal grades into a high supratidal plain or flat, inundated only by extreme water levels (e.g. during storm surges). The low topography is dominated by low gradient surfaces, crossed by shallow tidal channels (or 'creeks'). These channels vary in complexity from single 'rills' to intricate networks, and are generally best developed within the mud flat and saltmarsh sub-environments. Surface sediments generally decrease in grain size in a landward direction and the vertical stratigraphic sequence generally exhibits a fining upward sequence, usually attributable to transitions between tidal flat and saltmarsh as sedimentation proceeds.

The physical processes of mud flat sedimentation have been extensively studied, mainly from the perspective of sediment transport and deposition, with rather less emphasis being placed on processes of deposit consolidation and erosion (Amos 1995). A reduction in tidal current velocities in a landwards direction leads to the deposition of sediment suspended during the flood tide: the diminution in the competence of flows to transport material also explains the landward reduction in grain size. Although a portion of the newly deposited material is resuspended on the ebb tide, vertical and horizontal accretion of muddy intertidal sediments implies the dominance of flood-tide deposition (Evans 1965).

In the absence of any net (or 'residual') landward water transport, the accumulation of mud is further explained by reference to the concepts of 'settling lag' and 'scour lag'. Both these concepts were developed in the 1950s to account for tidal flat sedimentation in the Dutch Wadden Sea (see Amos 1995 for a recent review and evaluation of this work). Settling lag refers to the time elapsed between the slackening of tidal current intensity below the threshold of suspension for a given sediment and the deposition of the particle on the bottom. This means that particles are deposited some distance landwards of the point at which settling from suspension commences. Scour lag is a consequence of the higher flow intensity required to re-entrain a particle once it has been deposited. This is especially important for cohesive sediments, and means that ebb-directed transport occurs over a shorter duration than that of the flood tide. Both mechanisms tend to encourage the landward transport of mud and its accumulation in shallow intertidal areas.

Rates of mud flat sedimentation may be initially rapid (of the order of several centimetres a year), but diminish as the build-up of elevation reduces the frequency of inundation. Colonization by halophytic vegetation (and a transition to saltmarsh or mangrove) may be associated with a further increase in sedimentation rate owing to increased sediment retention under an energy-dissipative plant cover. However, this rapidly diminishes as vertical accretion further reduces inundation and wetland surface elevations tend towards a state of equilibrium between further sedimentation, the compaction of earlier deposits and sea level.

Mud flat topography arises from the dynamic interaction of tidal and wave-related hydrodynamics, sedimentation and morphology itself. Recent work has shown wave action to be more important than previously thought and has also drawn attention to the importance of biological processes in mediating sediment stability. Pethick (1996) draws an analogy between the morphological adjustment of mud flats to variations in wave energy and the morphodynamics of non-cohesive sandy beaches. The influence of waves differs between inner estuary sites, subject to small (fetch-limited) waves and outer estuary or open coastal sites, which experience a greater range of wave heights. At fetch-limited sites, waves may still exert oscillatory shear-stresses which exceed those generated by tidal currents and which are capable of resuspending mud flat sediments. A zone of resuspension migrates up and down the mud flat profile with the tidal variation in water level. Over time, the mud flat profile adjusts towards a form that is in equilibrium with wave induced stresses. The resulting profile is typically concave, a finding supported by numerical modelling experiments undertaken by Roberts *et al.* (2000). At more exposed sites, mud flats may undergo more episodic erosional adjustments in response to high wave energy conditions. In this case, the balance between individual erosion events and depositional recovery in the intervening periods determines longer term mud flat morphology.

Predominantly accretional mud flats tend to have a high and convex profile, whilst erosional mud flats are typified by a lower and concave profile. Mehta and Kirby (2001) attribute the contrasting stability of these mud flat morphologies to differences in their dissipative characteristics. In the case of high, convex mud flats, flexing of water-sediment mixture substantially dissipates tidal and wave-induced stresses, especially where thin surficial fluid mud layers are present. In low, concave mud flats, however, deposits are normally overconsolidated, such that hydrodynamic stresses are dissipated in overcoming interparticle cohesion, and in entraining sediment. Such systems are likely to be erosional.

The surficial sediments of mud flats support a variety of organisms, some of which act to stabilize the sediment and some of which act to increase the likelihood of erosion. Most mud flats support dense communities of benthic diatoms, which excrete large quantities of extracellular polymeric substances (EPS). EPS consist mainly of polysaccharides compounds and are a major component of surface films, which increase the stability of the sediment surface (Paterson 1997). Meso- and macro-fauna are active over a greater depth and may variously stabilize sediment (e.g. through the construction of EPS-coated tubular burrows) or reduce stability (e.g. by grazing on the micro-algae which helps bind sediment particles, or by reworking sediments through burrowing). Biological processes are extremely variable, both spatially and temporally, and are extremely important in determining the threshold stress at which erosion occurs. Once this threshold is exceeded, however, erosion may proceed more rapidly at a rate more closely controlled by bulk sediment properties.

Mud flats are increasingly valued as a habitat for large invertebrate populations which, in turn, provide a vital food source for wading birds. As landforms they are also of engineering significance as naturally dissipative systems which, allied to fixed defences, can provide an important component of integrated and more sustainable strategies for coastal protection. From both these perspectives, high and convex mud flats are preferable to low and concave ones (Kirby 2000). In the former case, waves are progressively attenuated as they approach the shore, a process which is further assisted by any saltmarsh fringe. Convex mud flats tend to have a larger invertebrate fauna, concentrated at a higher elevation within the tidal range, and capable of sustaining greater bird and fish populations. In contrast, erosional concave mud flats are less effective in dissipating wave energy and, in their upper portions, prone to rotational failure and slumping, with adverse consequences for the stability of sea defences.

References

Amos, C.L. (1995) Siliclastic tidal flats, in G.M.E. Perillo (ed.) *Geomorphology and Sedimentology of Estuaries*, 273–306, Amsterdam: Elsevier.

Evans, G. (1965) Intertidal flat sediments and their environment of deposition in The Wash, *Quarterly Journal of the Geological Society of London*, 121, 209–245.

Hayes, M.O. (1975) Morphology of sand accumulations in estuaries', in L.E. Cronin (ed.) *Estuarine Research, Volume II*, 3–22, New York: Academic Press.

Ke, X., Evans, G. and Collins, M. (1996) Hydrodynamics and sediment dynamics of The Wash embayment, eastern England, *Sedimentology* 43, 157–174.

Kirby, R. (2000) Practical implications of tidal flat shape, *Continental Shelf Research* 20, 1,061–1,077.

Lick, W. and Huang, H. (1993) Flocculation and the physical properties of flocs, in A.J. Mehta (ed.) *Nearshore and Estuarine Cohesive Sediment Transport*, 21–39, Washington, DC: American Geophysical Union.

Mehta, A.J. and Kirby, R. (2001) Muddy coast dynamics and stability, *Journal of Coastal Research, Special Issue* 27, 121–136.

Paterson, D.M. (1997) Biological mediation of sediment erodibility: ecology and physical dynamics, in N. Burt, R. Parker and J. Watts (eds) *Cohesive sediments*, 215–229, Chichester: Wiley.

Pethick, J.S. (1996) The geomorphology of mudflats, in K.F. Nordstrom and C.T. Roman (eds) *Estuarine Shores: Evolution, Environment and Human Alterations*, 185–211, Chichester: Wiley.

Ren, M. (ed.) (1987) *Modern Sedimentation in the Coastal and Nearshore Zones of China*, New York: Springer-Verlag.

Roberts, W., Le Hir, P. and Whitehouse, R.J.S. (2000) Investigation using simple mathematical models on the effect of tidal currents and waves on the profile shape of intertidal mudflats, *Continental Shelf Research* 20, 1,079–1,097.

Further reading

Healy, T., Ying Wang and Healy, J.A. (eds) (2002) *Muddy Coasts of the World: Processes, Deposits and Function*, Amsterdam: Elsevier Science.

SEE ALSO: mangrove swamp; saltmarsh; tidal creek; tidal delta

J.R. FRENCH

MUD VOLCANO

Mud volcanoes are positive topographic features formed by periodic venting of fluid mud, water and hydrocarbons (Kopf 2002). Individual mud volcanoes are elliptical mounds up to 2,000 m in diameter and 100 m in height. Cones and pools are often concentrated near the summit, and active portions of mud volcanoes are hummocky, unvegetated and covered by mud flows. Although clay and silt dominate mud-volcano deposits, pebble- to boulder-size clasts are common.

Mud volcanoes are known from approximately thirty regions worldwide. Most examples occur in compressional tectonic settings such as convergent plate margins. However, mud volcanoes also occur along passive margins and continental interiors. Excellent examples of subaerial mud volcanoes are present in Azerbaijan, Burma, Colombia, Indonesia, Iran, Italy, Mexico, Pakistan, Panama, Trinidad and Venezuela (Higgins and Saunders 1974). Subaqueous mud volcanoes occur in the Gulf of Mexico and the Barbados Ridge accretionary prism.

Mud volcanoes are commonly underlain by thick sequences of organic and clay-rich sediments (Hedberg 1974). Rapid sedimentation combined with methane generation, clay mineral diagenesis and tectonic compression produces high pore-fluid pressures, which mobilize fluid mud. Mud, water and hydrocarbons, migrate upward along fractures and faults that are typically associated with mud diapir-cored anticlines. If pressures are sufficient, fluid mud erupts at the seafloor or on the land surface to form mud volcanoes. In some instances, violent eruptions are accompanied by gas flares.

References

Hedberg, H.D. (1974) Relation of methane generation to undercompacted shales, shale diapirs, and mud volcanoes, *American Association of Petroleum Geologists Bulletin* 58, 661–673.

Higgins, G.E. and Saunders, J.B. (1974) Mud volcanoes – their nature and origin, in P. Jung (ed.) *Contributions to the Geology and Paleobiology of the Caribbean and Adjacent Areas*, 84, 101–152, Basel: Verhandlungen der Natureforschenden Gesellschaft.
Kopf, A.J. (2002) Significance of mud volcanism, *Reviews of Geophysics* 40(2), 1–52.

SEE ALSO: diapir; liquefaction; mudlump

ANDRES ASLAN

MUDLUMP

A diapiric structure composed of fine-textured material, especially clay, formed near the mouth of a delta's distributary. Mudlumps range in size from pinnacles to small, elongated islands. They are both subaqueous and subaerial with the subaerial forms often subject to rapid erosion by waves. The surface of mudlumps is usually irregular and most have gas (methane) and mud vents. Although several theories have been proposed for their formation, it is now generally accepted that they are the result of the intrusion of plastic, prodelta clay through overlying sand layers. They develop in sequence as distributary mouths advance seaward. Originally thought to have been unique to the distributaries of the Mississippi River, mudlumps are now known to exist in a few other deltaic areas.

Further reading

Lyell, C. (1889) *Principles of Geology*, Vol. 1, 11th edition, New York: Appleton.
Morgan, J.P., Coleman, J.M. and Gagliano, S.M. (1968) Mudlumps: diapiric structures in Mississippi delta sediment, in *Diapirism and Diapirs*, American Association of Petroleum Geologists, Memoir 8, 145–161.
Walker, H.J. and Grabau, W.E. (1992) Mudlumps, in D.G. Janelle (ed.) *Geographical Snapshots of North America*, 211–214, New York: Guilford.

H. JESSE WALKER

N

NATURAL BRIDGE

Remnant arch-shaped formation developed through erosion of the surrounding bedrock. Natural bridges, or stone arches, are unusual features that predominantly develop in horizontally bedded sedimentary rocks such as sandstone and limestone, though they hardly ever occur in metamorphosed or igneous rocks. They may form in a variety of ways, though all are ephemeral and will eventually collapse. The most common mode of formation is by water erosion, forming in deep valleys with highly sinuous rivers. Eventually, the river will cut across the neck of the entrenched meander by eroding a route through the obstructing rock outcrop. Often this can be accomplished without the arch collapsing, thus forming the natural bridge. Natural Bridge, Virginia, USA, has an uncertain evolutionary history, though meander cutting by the James River is a strong possibility (Malott and Shrock 1930). The other possible mode of formation is by the collapse of an underground drainage tunnel, leaving a remnant of the tunnel ceiling. Natural Bridge spans 30 m across Ceder Creek and is one of the few remaining natural bridges that is used as a transport bridge, at 60 m high.

Natural bridges may also be formed by the near complete collapse of underground tunnels. Such formations are common on the Hawaiian Islands, where recent lava tunnels roofed by a solidified crust may collapse leaving all but a small arch-shaped portion. Other origins of natural bridges include those cut by the sea resulting in coastal wave-cut arches, while a more unusual natural bridge can be found in Petrified Forest National Park, Arizona, USA, where a silicified tree trunk, known as the Onyx Bridge, spans a canyon 15 m wide.

Reference

Malott, C.A. and Shrock, R.R. (1930) Origin and development of Natural Bridge, Virginia, *American Journal of Science* 19, 257–273.

Further reading

Vokes, H.E. (1942) Rainbows of rock; how a natural bridge is carved (Utah), *Natural History* 50, 148–152.

SEE ALSO: arch, natural

STEVE WARD

NEBKHA

Nebkha, or nabkha, is an Arabic term given to mounds of wind-borne sediment (sand, silt of pelletized clay) that have accumulated to a height of some metres around shrubs or other types of vegetation. They are sometimes called shrub-coppice dunes. They may occur on bigger dunes, in interdune areas, on pan surfaces, near wadis and on or behind beaches. Morphometric data are provided by Tengberg and Chen (1998). The largest nebkhas (mega-nebkhas) accumulate around clumps of trees. In the Wahiba Sands of Oman these can be 10 m high and up to 1 km long (Warren 1988).

References

Tengberg, A. and Chen, D. (1998) A comparative analysis of nebkhas in central Tunisia and northern Burkina Faso, *Geomorphology* 22, 181–192.

Warren, A. (1988) A note on vegetation and sand movement in the Wahiba Sands, *Journal of Oman Studies Special Report* 3, 251–255.

A.S. GOUDIE

NEEDLE-ICE

Needle-ice (synonymous to 'pipkrake' or 'kammeis') is the accumulation of ice crystal growths in the direction of heat loss at, or directly beneath, the ground surface. Although needle-ice usually grows perpendicular to the ground surface, curved ice-filaments are sometimes observed owing to wind and gravity effects. Needle-ice may also connect normal to plant stalks that have drawn sufficient ground moisture. Needle-ice is common to areas of diurnal freeze–thaw, ranging from tropical alpine to subarctic environments.

Needle-ice best develops in moist, fine-textured sediment with at least 10 per cent clay/silt. Precise soil and near-surface thermal dynamics affecting needle-ice growth and decay are complex, thus making it difficult to predict annual frequency. Generally, needle-ice develops within the first hour of ground temperatures dropping below 0 °C. Further conditions necessary for needle-ice development include a relatively low soil water tension to enable ice segregation to take place and adequately rapid movement of unfrozen moisture to the freezing front, so that it corresponds with the rate of latent heat loss, and thus preventing the *in situ* freezing of pore water (Outcalt 1971). Very windy conditions may cause a rapid temperature drop through the soil pores, thus reducing the suction gradient and enhancing the development of pore ice rather than needle-ice. Typically, needle-ice phases will entail periods of growth, stagnation and ablation. The duration of freeze determines growth phases and consequently needle-ice length, which may vary from a few millimetres to several centimetres. Polycyclic or multilayered needle-ice, separated by thin veneers of sediment, occurs where there has been moisture stress. Alternatively, long-lasting growth phases over several days may produce multilayered needle-ice lengths exceeding 400 mm.

Needle-ice has been applied to studies examining SOIL CREEP, SOIL EROSION, the impacts on plant (particularly seedling) disruption and its function as a geomorphic process in miniature landform development. Needle-ice on stream banks or soil terraces commonly extrudes sediment, which is transferred by needle-ice induced direct particle fall, sliding and toppling failure and mini-mudflows. Geomorphological consequences of needle-ice as an erosion agent include notches and undercut fluvial banks, TURF EXFOLIATION and associated depositional microforms. Several soil structures including nubbin soils, gaps around stones and other varieties of PATTERNED GROUND, have been attributed to needle-ice. It is thought that soil stripes aligned parallel to the late morning sun may be a function of shadow and differential thaw effects during the ablation phase.

Reference

Outcalt, S.I. (1971) An algorithm for needle ice growth, *Water Resources Research* 7, 394–400.

Further reading

Lawler, D.M. (1993) Needle ice processes and sediment mobilization on river banks: the River Ilston, West Glamorgan, UK, *Journal of Hydrology* 150, 81–114.

Meentemeyer, V. and Zippin, J. (1981) Soil moisture and texture controls of selected parameters of needle ice growth, *Earth Surface Processes and Landforms* 6, 113–125.

Washburn, A.L. (1979) *Geocryology: A Survey of Periglacial Processes and Environments*, London: Edward Arnold.

SEE ALSO: freeze–thaw cycle; frost heave; ice

STEFAN GRAB

NEOCATASTROPHISM

Neocatastrophism, as defined by Schindewolf (1963), refers to the explanation of sudden extinctions in the palaeontological record. In geomorphology, George Dury (1975, 1980) expressed the view that high magnitude, low frequency events were more important in an absolute sense than low magnitude, high frequency events in moulding the Earth's landscapes. Dury's statement expresses the essence of the issue. Neocatastrophism is a response to one hundred years of geomorphic thinking in which the predominant role of low magnitude, high frequency events in landform evolution had become the prevailing paradigm. A side issue, expressed in an exchange between Brunsden (1996) and Yatsu (1996), is whether the word catastrophism should be excised from the geomorphic vocabulary, and hence, by implication, also the word neocatastrophism. I am not convinced that we need to fear this word; but there is a need for unambiguous definition. By contrast with catastrophism, which is an outmoded, pre-twentieth century mode of thought, neocatastrophism is thought to be an increasingly relevant way of viewing the geomorphic world.

Circumstances which have favoured the emergence of neocatastrophism include the following:

- improved precision in geochronology has demonstrated unexpectedly rapid past changes;
- the exploration of mass extinctions in the past has intensified;
- some geomorphological features, such as the Channeled Scablands of eastern Washington, are more amenable to explanation by low frequency, high magnitude events than by gradual, semi-continuous processes;
- space exploration has generated a strong interest in galactic scale events;
- interest in global environmental change has provided evidence of rapid past changes, such as found in the polar ice caps and the oceanic deep sediments;
- the rise of non-linear dynamics and chaos theory is beginning to provide ways of synthesizing gradualism and catastrophism.

Within geomorphology, it was the paper by Wolman and Miller (1960) which provoked a critical evaluation of the question of magnitude and frequency (see MAGNITUDE–FREQUENCY CONCEPT) of the operation of geomorphic processes. The authors directed attention to the importance of medium size and medium frequency events as having the greatest cumulative influence on the landscape. This was an important insight, but did not prevail after notable discussions by Wolman and Gerson (1978), Gould (1984), Gretener (1984) and Baker (1994).

Wolman and Gerson (1978), in following up their findings on magnitude and frequency, expanded on effectiveness of climate and relative scales of time such that they were forced to modify their view about the importance of the intermediate magnitude and intermediate frequency event in landform history. Introduction of the idea of the length of time over which a landform survived suggested that, in many cases, it was the extreme events which were most important. The influence of this paper on geomorphic thinking cannot be overestimated as it has emphasized the importance of combining measures of process magnitude and frequency with duration or lifetime of landforms.

Gould's discussion on punctuational change was an equally influential paper for the whole of Earth science (Gould and Eldredge 1977). The essential concept was a recognition that many important changes in Earth history have proceeded by relatively rapid flips between more stable conditions. Systems often absorb stress and resist change until the stresses accumulate past a breaking point. Systems then flip to a new stable state. This hypothesis, known as punctuated equilibrium, has gained widespread acceptance in the palaeontological community and it is viewed by other Earth scientists as a model for processes of inorganic change. Gretener (1984) advances the example of isostatic rebound to illustrate the relativity of gradualism. Isostatic rebound has been active during the last 10,000 years and is still in progress in such places as northern Canada and Scandinavia. The process covers all of humanity's conscious history and is generally perceived as a gradual phenomenon. However, if one considers that the Earth's skin can completely recover from the unloading of 1–2 km of ice within a period of 15–20,000 yrs, this process is effectively instantaneous from an Earth history perspective. This leads to a consideration of what constitutes an event? Gretener suggests that the duration of an event occupies no more than 1/100 of the total time span being considered. On this basis, geological processes may have durations as great as 10 Ma and still qualify as events. Indeed Earth history 'reveals long periods of tranquility interrupted by moments of action' (Gretener 1984: 86). The rare event in geology is a punctuation with such a low rate of occurrence that it has taken place, at most, a few times through all of Earth history. It is unscientific to call such events 'impossible'. Punctuationism would possibly be a better term than neocatastrophism. Nevertheless, either term is preferred to uniformitarianism, which fails to do justice to such extreme events. Brunsden (1996) illustrates Gretener's point well in his Figure 2.3.

Baker (1994) provides the most powerful geomorphic justification for neocatastrophism in his interpretation of the resistance of the geological community to Harlen Bretz's (1923) theory of the origin of the Channeled SCABLANDS of eastern Washington. He explains that the community was blinkered by its slavish adherence to gradualism and its suspicion of the mention of cataclysmic flood events. Nevertheless, Bretz's interpretation was finally vindicated in many of its neocatastrophist details, in part as a result of the identification of a source of this exceptional flooding (glacial Lake Missoula) which Bretz himself had not recognized, but also in part because of the recognition of the erosional and hydraulic concomitants of an extreme flood.

There remains an urgent need for geomorphologists to accommodate our thinking to the new diastrophic ideas associated with global plate tectonics. The focus on short timescales relevant to process studies has been partly responsible for a neglect of the longer timescales. Modes of vertical motion, the onset of Ice Ages and the appearance of volcanism all need to be reappraised in a neocatastrophist framework.

Thorn (1988) points out that there is an important intellectual issue associated with the rise of neocatastrophism. If greater significance is being attached to large events in a series, this only forces an adjustment of magnitude–frequency concepts. If the new perspective is one that identifies unique events as paramount in geomorphological records, then there can be no science of geomorphology because there is no science of uniqueness. Most of us are busily adjusting our magnitude–frequency concepts.

References

Baker, V.R. (1994) Geomorphological understanding of floods, *Geomorphology* 10, 139–156.

Bretz, J.H. (1923) The channeled scabland of the Columbia Plateau, *Journal of Geology* 3, 617–649.

Brunsden, D. (1996) Geo-apologia, in S.B. McCann and D.C. Ford (eds) *Geomorphology Sans Frontières*, 82–90, Chichester: Wiley.

Dury, G.H. (1975) Neocatastrophism? *Annales Academiensis Ciencias Brasiliensis* 47, 135–151.

——(1980) Neocatastrophism? A further look, *Progress in Physical Geography* 4, 391–413.

Gould, S.J. (1984) Toward the vindication of punctuational change, in W.A. Berggren and J.A. Van Couvering (eds) *Catastrophes and Earth History*, 9–34, Princeton: Princeton University Press.

Gould, S.J. and Eldredge, N. (1977) Punctuated equilibria: the tempo and mode of evolution reconsidered, *Paleobiology* 3, 115–151.

Gretener, P.E. (1984) Reflections on the 'rare event' and related concepts in geology, in W.A. Berggren and J.A. Van Couvering (eds) *Catastrophes and Earth History*, 77–90, Princeton: Princeton University Press.

Schindewolf, O.H. (1963) Neokatastrophismus? *Zeitschrift der Deutschen Geologischen Gesellschaft* 114, 430–445.

Thorn, C.E. (1988) *Introduction to Theoretical Geomorphology*, London: Unwin Hyman.

Wolman, M.G. and Gerson, R. (1978) Relative scales of time and effectiveness of climate in watershed geomorphology, *Earth Surface Processes and Landforms* 3, 189–208.

Wolman, M.G. and Miller, J.P. (1960) Magnitude and frequency of forces in geomorphic processes, *Journal of Geology* 68, 54–57.

Yatsu, E. (1996) Graffiti on the wall of a geomorphology laboratory, in S.B. McCann and D.C. Ford (eds) *Geomorphology Sans Frontières*, 53–58, Chichester: Wiley.

SEE ALSO: catastrophism; magnitude–frequency concept

OLAV SLAYMAKER

NEOGLACIATION

Neoglaciation is a geological term, originating in North America, used to describe the period during the latter half of the Holocene when valley GLACIERS in many mountain areas readvanced to their maximum extent following Pleistocene DEGLACIATION. The term was first used by Moss (1951) and Nelson (1954) who attribute it to Matthes (though the term appears in none of his papers). Neoglaciation was formally defined by Porter and Denton (1967) as a 'cool geologic-climate unit ... indicating a probable world wide synchrony of glacier fluctuations in response to climatic change'. Their classic paper established the standard division of the North American Holocene into a warmer and drier early Holocene (the Hypsithermal) followed by a cooler Neoglacial Interval characterized by several periods of glacier advance. The related term 'little ice-age' was first used by Matthes (1939) to define the period when glaciers re-established in the Sierra Nevada of California following the post-glacial climatic optimum. Matthes's 'little ice-age' was, in fact, what is now termed Neoglaciation. Subsequently, the term Little Ice Age (LIA) has been almost universally adopted to describe the latest and most severe part of the Neoglacial during the past few centuries when glaciers in many areas of the world reached their maximum Holocene extent (Grove 2003).

This terminology was established at a time when there were few detailed chronologies of Holocene glacier fluctuations with little absolute dating control (radiocarbon dates were just becoming generally available to Quaternary scientists). Holocene climates were interpreted on the basis of limited evidence from studies of glacier fluctuations and the zonation of pollen diagrams in Europe and North America. As the maximum Neoglacial (Little Ice Age) extent of glaciers at most northern hemisphere sites was between AD 1600 and 1850, almost all morphological evidence of earlier glacier events was destroyed. Stratigraphic evidence from lateral MORAINES and sections within the Little Ice Age limits was fragmentary, difficult to find and the dating often poorly constrained.

Over the past thirty to forty years new information has led to the modification of our knowledge and understanding of these glacial events. Significant glacier recession during the late twentieth century has exposed many new moraine sections and buried forest sites that yield detailed evidence of earlier glacier fluctuations. The advent of AMS and calendar-adjusted radiocarbon dates, plus dendrochronological dating of sub-fossil wood using millennial-length tree-ring reference chronologies, and the development of proximal varve sequences have improved the available record of dated Neoglacial sequences (Plate 83).

Most evidence of Neoglacial glacier fluctuations comes from western North America and western Europe where, generally, the LIA glacier advances were the most extensive. However, in the southern hemisphere several authors have identified deposits of an early Neoglacial advance *c.*4,400–4,600 yr BP, downvalley of the LIA limits. This evidence is critically reviewed by Porter (2000) who cautions that this conclusion should remain provisional until a larger population of better dated sites are available.

Early work in the northern hemisphere (mainly in Alaska and Scandinavia) identified three phases of Neoglaciation: early (*c.*6,000–4,000 BP), middle (2,500–3,500 BP) and late (last 1,000 years or LIA) with a minor event *c.* AD 700–900 (Denton and Karlen 1973). Evidence for the earliest events is fragmentary. Most investigations

Plate 83 Late Neoglacial (Little Ice Age) lateral and terminal moraines, Bennington Glacier, British Columbia, July 1990

date initial Neoglacial advances between 4,000–5,000 ^{14}C yr BP and link them with other PALEOCLIMATE evidence of climate deterioration at this time. Although the preceding Hypsithermal was originally defined as a time stratigraphic unit (Porter and Denton 1967), dates for the Hypsithermal–Neoglacial transition are clearly time transgressive with evidence for some alpine glacier advances before 6,000 BP. Therefore the early Neoglacial is not well defined. There is widespread evidence for glacier advances between *c.*3,500–2,800 ^{14}C yr BP in the Canadian Rockies, Alaska, Switzerland, Patagonia and Scandinavia. There is also evidence from several areas of glacier advances *c.*1,300–1,500 ^{14}C yr BP and an 'early medieval advance' *c.*AD 600–800. However, the most detailed (and best dated) reconstructions of glacier fluctuations from the Alps (Holzhauser 1997) indicate that at least seven advances of the Aletsch Glacier occurred between 3,200 yr BP and AD 1000, plus three major LIA advances. It seems unlikely that the history of glacier fluctuations at less well-dated sites is any less complex than that shown by the Aletsch. Therefore the history of glacier fluctuations during the early and middle Neoglaciation remains incomplete but probably consists of multiple, relatively short-lived (50–200 years?) advances that appear to have been progressively more extensive over time and were separated by periods of glacier recession. In assessing the synchroneity of these events it is critically important to determine both the precision of the dating technique used and the precision of dating control (i.e. its stratigraphic or geographic relationships with the event being dated). In many cases the limiting dates are +/−50 years at best which is often inadequate to differentiate between synchronous and closely spaced events or determine whether the events are correlative and synchronous over large areas rather than simply locally significant records.

The beginning of LIA is traditionally placed at the end of the Medieval Warm Period. The MWP was initially defined from non-glacial evidence in Europe (Hughes and Diaz 1994) and encompasses the period AD 800–1200 when there is little evidence of extended glacier cover. The status of this period as a global interval of generally warmer conditions remains questionable until an adequate database of high resolution palaeoclimate records becomes available. Well-dated, early LIA glacier advances occurred in the twelfth to fourteenth centuries in Patagonia, Canadian Rockies, Alaska, Switzerland and Scandinavia. These early advances were followed by an interval with little evidence of glacier fluctuations until the main LIA advances, dated between the sixteenth and nineteenth centuries. In many areas glaciers reoccupied positions at or very close to their maxima several times during the LIA, e.g. the early 1700s and mid-1800s in the Canadian Cordillera and coastal Alaska or the 1350s, 1650s and 1850s in the Alps. Most glaciers have receded rapidly during the twentieth century. The exposure of old buried forest sites and the alpine iceman suggests that this twentieth-century recession is the most rapid and severe during the Holocene. However, minor advances of glaciers occurred in many alpine areas during the 1960s–1970s and glaciers in western Norway advanced significantly during the 1980s and early 1990s as a result of increased winter precipitation due to changes in atmospheric circulation. In summary, these records indicate several intervals of glacier expansion over the past 5,000 years. Some, such as the nineteenth century, appear to be globally synchronous (at least at the centennial scale) whereas others may reflect local or more regional glacier histories.

The development of independent, high-resolution proxy climate records spanning the late Holocene (using tree rings, ice cores and other techniques) has greatly expanded our knowledge of climate variability and climate history. This work has shown that climate varies continuously at several spatial and temporal scales; that the relationships between glacier fluctuations and climate are complex; and that climate variability is rarely synchronous at the global scale. Recent palaeoclimate work also provides superior records of climate forcing. Some forcing factors have globally synchronous effects (e.g. variations in solar output, sunspot minima, etc.) whereas the effects of others may differ between hemispheres (orbital effects, volcanic eruptions) or between regions (e.g. circulation changes). Global climate variability reflects the interaction of all these factors as do climatically dependent fluctuations of glaciers. However, despite these strong links to climate, glacier behaviour and response times are also influenced by many factors unrelated to climate. The use of glacier-defined terms such as Neoglaciation and Little Ice Age to identify distinct, global, climate-geologic periods is inappropriate and misleading in the context of current knowledge of Holocene climates. Usage of these terms should be confined to describing the glacial advances of the late Holocene after *c.*5,000 BP and between *c.*AD 1000–1900, respectively.

References

Denton, G.H. and Karlen, W. (1973) Holocene climatic variations – their patterns and possible causes, *Quaternary Research* 3, 155–205.

Grove, J.M. (1988) *The Little Ice Age*, London: Methuen.

——(2003) *Little Ice Ages: Ancient and Modern*, London: Routledge.

Holzhauser, H. (1997) Fluctuations of the Grosser Aletsch Glacier and Gorner Glacier during the last 3200 years – new results, *Paleoklimaforschung* 24, 36–58.

Hughes, M.K. and Diaz, H.F. (1994) *The Medieval Warm Period*, Kluwer: Dordrecht.

Matthes, F.E. (1939) Report of the Committee on Glaciers, *Transactions of the American Geophysical Union* 20, 518–523.

Moss, J.H. (1951) *Early Man in the Eden Valley*, University of Pennsylvania, University Museum Monograph, 9–92, Philadelphia.

Nelson, R.L. (1954) Glacial geology of the Frying Pan River drainage, Colorado, *Journal of Geology* 62, 325–343.

Porter, S.C. (2000) Onset of Neoglaciation in the Southern Hemisphere, *Journal of Quaternary Science* 15, 395–408.

Porter, S.C. and Denton, G.H. (1967) Chronology of Neoglaciation in the North American Cordillera, *American Journal of Science* 265, 177–210.

Further reading

Calkin, P.E., Wiles, C.C. and Barclay, D.J. (2001) Holocene coastal glaciation in Alaska, *Quaternary Science Reviews* 20, 449–461.

Luckman, B.H. and Villalba, R. (2001) Assessing synchroneity of glacier fluctuations in the western cordillera of the Americas during the last millennium, in V. Markgraf (ed.) *Interhemispheric Climate Linkages*, 119–140, New York: Academic Press.

SEE ALSO: dating methods; dendrochronology; Holocene geomorphology; palaeoclimate

BRIAN LUCKMAN

NEOTECTONICS

Neotectonics concerns the study of horizontal and vertical crustal movements that have occurred in the geologically recent past and which may be ongoing today. While most crustal movements arise directly or indirectly from global plate motions (i.e. tectonic deformation), neotectonic studies themselves make no presumption about the mechanisms driving deformation. Consequently here 'movements' is a vague catch-all term that encompasses a myriad of competing deformation processes, such as the gradual pervasive creep of tectonic plates, discrete (seismic) displacements on individual faults and folds, and distributed tilting and warping through isostatic readjustment or volcanic upheaval. The phrase 'geologically recent past' is also intentionally vague. Early attempts to define the discipline by arbitrary time windows such as Late Cenozoic, Neogene or Quaternary have given ground to a more flexible notion that envisages neotectonism starting at different times in different regions. The onset of the neotectonic period, or the 'current tectonic regime', depends on when the contemporary stress field of a region was first imposed. For instance, in the Apennines of central Italy the 'current tectonic regime' began in the Middle Quaternary (~700,000 years ago) and it is even younger (< 500,000 years) in California; in contrast, in eastern North America the present-day stress regime has been in existence for at least the last 15 million years.

Typically then, neotectonic movements have been in operation in most regions for the last few million years or so. Over such prolonged intervals, neotectonic actions are revealed by the stratigraphical build-up of sediments in inland and marine basins, the burial or exhumation histories of rocks and the geomorphological development of landscapes. Geological studies of palaeobotany and palaeoclimate, numerical models of landscape evolution and techniques such as fission track analysis and cosmogenic dating are among the disparate tools unravelling this long-term tectonic activity. Over periods of many tens of, to several hundred thousand years, the actions of individual tectonic structures (faults and folds) can be determined, unmasked by their deformation of geomorphic markers, such as marine and fluvial terraces, and tracked with reference to the late Pleistocene glacial-eustatic time frame. The apparently smooth deformation rates discerned over intermediate timescales are revealed to be episodic and irregular when faults and folds are examined over Holocene (10,000 years) timescales. Over millennial timescales, secular variations in the activity of tectonic structures can be gleaned from a diverse set of *palaeoseismological* approaches, from interpreting the stratigraphy of beds that have been affected by faulting to detecting disturbances in the growth record of trees or coral atolls.

Although neotectonic movements continue up to the present day, the term *active tectonics* is typically used to describe those movements that have occurred over the timespan of human history. Active tectonics deals with the societal implications of neotectonic deformation (such as seismic-hazard assessment, future sea-level rise, etc.), since it focuses on crustal movements that can be expected to recur within a future interval of concern to society. Even contemporary crustal movements may reveal themselves in Earth surface processes and landforms, such as in the sensitivity of alluvial rivers to crustal tilting. In addition, geomorphological and geological studies are important in recording the surface expression of Earth movements such as earthquake ground ruptures which, due to their subtle, ephemeral or reversible nature, are unlikely to have been preserved in the geological record. However active tectonics also employs an array of high-tech investigative practices; prominent among these are the monitoring of ongoing Earth surface deformation using space-based or terrestrial geodetic methods (tectonic geodesy), radar imaging (interferometry) of ground deformation patterns produced by individual earthquakes and volcanic unrest, and the seismological detection and measurement of earthquakes (seismotectonics), both globally via the World-Wide Standardised Seismograph Network and regionally via local seismographic coverage. These modern snapshots of tectonism can be pushed back beyond the twentieth century through the analysis of historical accounts and maps to infer past land surface changes or deduce the parameters of past seismic events (historical seismology). In addition, earthquakes can leave their mark in the mythical practices and literary accounts of ancient peoples, the stratigraphy of their site histories, and the damage to their buildings (archaeoseismology). The time covered by such human records varies markedly, ranging from many thousands of years in the Mediterranean, Near East and Asia to a few centuries across much of North America. Generally they confirm that regions that are active today have been consistently active for millennia, thereby demonstrating the long-term nature of crustal deformation, but occasionally they reveal that some regions that appear remarkably quiet from the viewpoint of modern seismicity (such as the Jordan rift valley) are capable of generating large earthquakes.

In reality, the distinction between neotectonics and active tectonics is artificial; they simply describe different time slices of a continuum of crustal movement. This continuum is maintained by the persistence of the contemporary stress field, which means that inferences of past rates and directions of crustal movement from geological observations can be compared directly with those measured by modern geodetic and geophysical methods. Although the terms 'neotectonic' or 'active' are somewhat blurred and are often used interchangeably, societal demands (for instance, regulatory authorities for seismic hazard, nuclear safety, etc.) often require the incidence of tectonic movements to be strictly defined. For instance, the present definition in Californian law of an 'active fault' is one that has had surface-rupturing earthquakes in the last 11,000 years (established when the Holocene was considered to have begun at that time) (see ACTIVE AND CAPABLE FAULT). Other regulatory bodies recognize a sliding scale of fault activity: Holocene (moved in the last 10,000 years), Late Quaternary (moved in the last 130,000 years) and Quaternary (moved in the last 1.6 million years). Neotectonic faults, by comparison, are simply those that formed during the imposition of the current tectonic regime. 'Real' structures, of course, are unconstrained by such legislative concerns. Many modern earthquakes rupture along older (i.e. palaeotectonic) basement faults. Indeed, it is important to recognize that any fault that is favourably oriented with respect to the stress currently being imposed on it has the potential to be activated in the future, regardless of whether it has moved in the geologically recent past.

A more meaningful way to differentiate styles and degrees of neotectonic activity is in terms of tectonic strain rate, which is a measure of the velocity of regional crustal motions and, in turn, of the consequent tectonic strain build-up. Crustal movements are most vigorous, and therefore most readily discernible, where plate boundaries are narrow and discrete. In these domains of high tectonic strain, frequent earthquakes on fast-moving (>10 mm/yr) faults ensure that a century or two of historical earthquakes and a few years of precise geodetic measurements are sufficient to capture a consistent picture of the active tectonic behaviour. Intermediate tectonic strain rates characterize those regions where plate-boundary motion is distributed across a network of slower moving faults (0.1–10 mm/yr). Examples of such broad deforming belts are the Basin and Range Province of western USA or the Himalayan collision zone, where earthquake faults rupture every few hundred or thousand years, ensuring that the Holocene period is a reasonable time window over which to witness the typical crustal deformation cycle. In contrast, low-strain rates ensure that intraplate regions, often referred to as 'stable continental interiors', are low-seismicity areas with slow-moving (< 0.1 mm/yr) faults that rupture every few tens (or even hundreds) of thousands of years, making the snapshot of human history an unreliable guide to the future incidence of tectonic activity.

The global pattern of present-day crustal motions can be accounted for by PLATE TECTONICS theory, that elegant kinematic framework in which rigid plates variously collide, split apart and slide along their actively deforming boundaries. Closer inspection, however, reveals that the basic rules that govern global plate motions (i.e. rigid blocks separated by narrow deforming boundaries) break down at the regional and local scale. This is particularly so on the continents, where a patchwork of pre-existing geology and structure ensures that tectonic stresses are not applied in a uniform, straightforward fashion. Studies of how the contemporary stress field varies across the Earth's surface (Figure 108) distinguish between first- and second-order stress provinces. First-order provinces have stresses generally uniformly oriented across several thousands of kilometres. The largest of these are the midplate regions of North America and western Europe, where the stress fields are largely the far-field product of ridge push and continental collision. In contrast, first-order stress provinces in tectonically active areas are dominated by the downgoing pull of subducting slabs and the resistance to subduction. Second-order stress provinces are smaller, typically less than 1,000 km across, and are related to crustal flexure induced by thick sequences of sediments and postglacial rebound, and to deep-seated rheological contrasts. Although the bulk of the Earth's crust is in compression, significant regions of extension occur. In both the continents and oceans, these extensional domains are long and narrow and correspond to topographically high areas, though notable exceptions are the Basin and Range province and the Aegean region of the eastern Mediterranean. Most first-order stress provinces, and many second-order stress provinces coincide with distinct physiographic provinces.

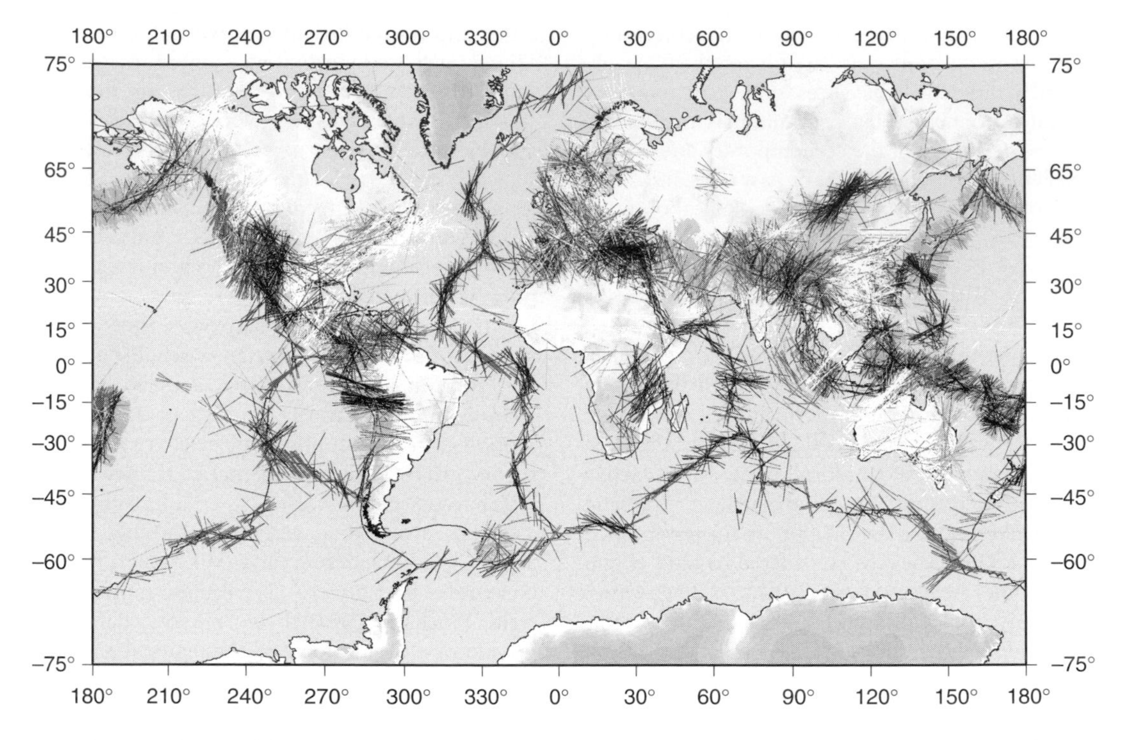

Figure 108 The World Stress Map with lines showing the directions of maximum horizontal compression. Black lines denote normal faulting (extension), dark grey lines denote strike-slip faulting, and light grey lines denote thrust faulting (compression); white lines show an uncertain tectonic regime. The longer the line length, the better the quality of the data. Around two-thirds of the stress data come from earthquakes and so highlights where the bulk of tectonic deformation is occurring; most of the remaining third comes from borehole stress measurements that are concentrated in petroleum-producing provinces. From Mueller *et al.* (2000)

Plate driving forces may exert the dominant control on the contemporary stress field, but another process contributes to crustal deformation at a global scale. That process is glacial isostatic adjustment (GIA), the physical response of the Earth's viscoelastic mantle to surface loads imposed and removed by the cycles of glaciation and deglaciation to which the planet has been subjected for the past 900,000 years (see GLACIAL ISOSTASY). Because large ice-mass fluctuations induce the sub-crustal flow of material, measurable crustal deformation extends for thousands of kilometres beyond the limits of the former ice margins (Figure 109); in short, the effects of GIA are felt globally. In addition, while the crust's elastic response to ice-sheet decay is geologically immediate, the delayed viscoelastic response of the mantle ensures that GIA persists long after the ice has gone. Although the effects of GIA can now be detected from space geodesy, its legacy is most clearly visible in the worldwide pattern of postglacial sea-level changes. Regions that were ice covered at the Last Glacial Maximum are uplifting (i.e. relative sea level is currently falling) as a consequence of postglacial rebound of the crust. Likewise the regions peripheral to the former ice sheets are subsiding (i.e. relative sea levels are rising) due to collapse of the 'glacial forebulge'. The effect of this subsidence outside the area of forebulge collapse is to draw in water from the central ocean basins, which is compensated by uplift in the ocean basin interiors in the far-field of the ice sheets. The final GIA component is the hydroisostatic tilting of continental coastlines due to the weight applied to the Earth's surface by the returning meltwater load, which produces a 'halo' of weak crustal subsidence around the world's major land masses. For the most part, geological studies of Holocene relative sea-level changes are consistent with the uplift/subsidence

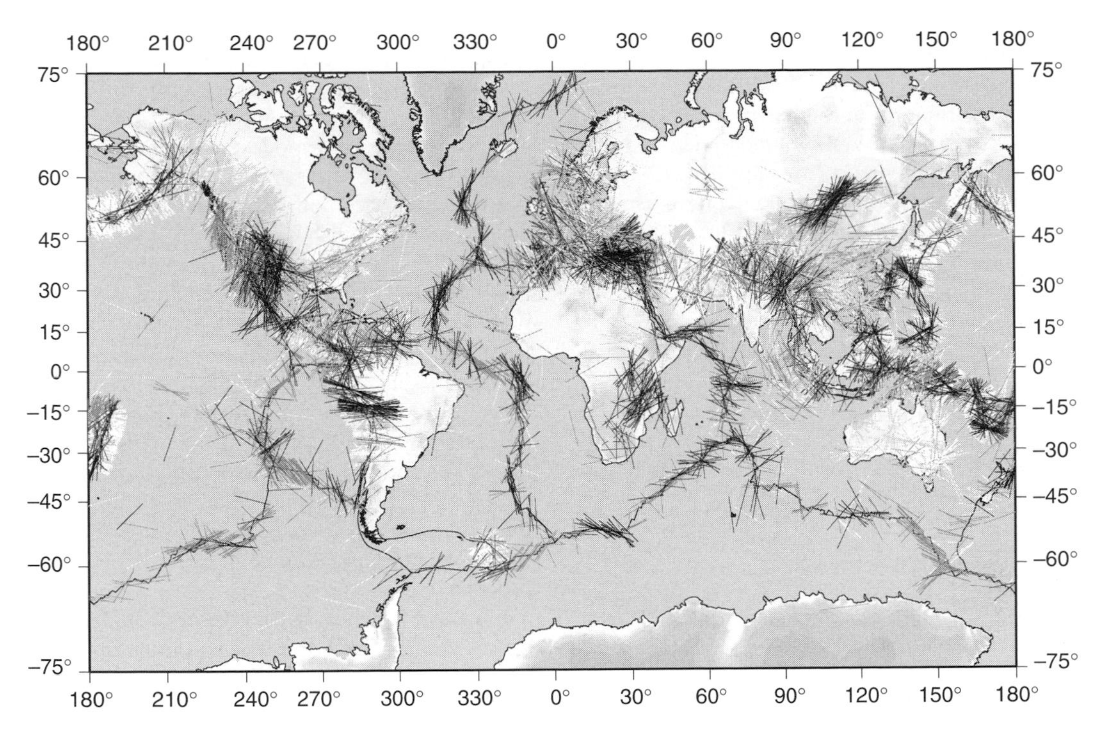

Figure 109 Map showing the outward radial motion of eastern North America (predicted by Peltier's (1999) postglacial rebound model) due to glacial isostatic adjustment to removal of the Laurentide ice sheet, and highlighting the concentration of contemporary seismicity along the former ice margins. From Stewart *et al.* (2000)

pattern predicted by global viscoelastic theory. The key areas of misfit are along plate boundary seaboards (especially subduction zones), where tectonic deformation dominates, and those areas 'contaminated' by local anthropogenic effects (groundwater extraction etc.).

The neotectonic implications of GIA are not confined to the coastline. Glacial rebound is now widely considered as an effective mechanism for exerting both vertical and horizontal stresses not only within the limits of the former ice sheets but for several hundred kilometres outside. Within the former glaciated parts of eastern North America and northern Europe both tectonic and rebound stresses are required to explain the distribution and style of both postglacial and contemporary seismotectonics. Outside in the ice-free forelands, predicted glacial strain rates are still likely to be one to three orders of magnitude higher than tectonic strain rates typical of continental interiors. Consequently, some workers argue that an apparent 'switching on' of Holocene earthquake activity in eastern USA and the occurrence of atypically large seismic events such as the great ($M > 8$) earthquakes that struck the Mississippi valley area of New Madrid in 1811–1812 may be associated with areas where glacial strains are particularly high. Glacial loading and unloading may also disturb the build-up of tectonic strain at glaciated plate boundaries, such as today in Alaska or previously when the Cordilleran ice sheet capped part of the Cascadia subduction zone. More recently, the isostatic component of glacier erosion in the mountain-building process is becoming appreciated.

In summary, the worldwide pattern of vertical and horizontal crustal movements arise from the global effects of plate motions and glacial isostatic adjustment. Regionally and locally, this is augmented by flexure from eustatic or sediment loading, volcanic deformation or anthropogenic change (dam impoundment). While many neotectonic investigations seek to disentangle movements arising from the imposition of tectonic strains from those augmented by non-tectonic processes, this is often a fruitless holy grail; because deformation of the Earth's crust typically induces compensatory flow underlying mantle, neotectonic movements are applied globally. Nevertheless, these disparate contributory mechanisms, coupled with the varying timescales over which their actions can de discerned, ensure that neotectonics encompasses a remarkable breadth of research disciplines. Few other fields easily blend topics as disparate as space science, seismology, Quaternary science, geochronology, structural geology, geomorphology, geodesy, archaeology and history. It is this interdisciplinary marriage that makes neotectonics particularly exciting and especially challenging.

References

Mueller, B., Reinecker, J., Heidbach, O. and Fuchs, K. (2000) The 2000 release of the World Stress Map (available online at www.world-stress-map.org)

Peltier, W.R. (1999) Global sea-level rise and glacial isostatic adjustment, *Global and Planetary Change* 20, 93–123.

Stewart, I.S., Sauber, J. and Rose, J. (2000) Glacio-seismotectonics: ice sheets, crustal deformation and seismicity, *Quaternary Science Reviews* 14/15, 1,367–1,390.

Further reading

Burbank, D.W. and Anderson, R.S. (2001) *Tectonic Geomorphology*, Oxford: Blackwell.

Stewart, I.S. and Hancock, P.L. (1994) Neotectonics, in P.L Hancock (ed.) *Continental Deformation*, 370–409, Oxford: Pergamon Press.

Vita-Finzi, C. (2002) *Monitoring The Earth*, Harpenden: Terra Publishing.

IAIN S. STEWART

NIVATION

Nivation is a morphogenetic term introduced by Matthes (1900) to describe and explain the processes associated with late-lying seasonal snow patches and landforms derived from them (nivation benches or terraces, and nivation hollows). The term became entrenched in periglacial geomorphology with little attention to process measurements until recently. One important vein of thinking envisages nivation hollows as precursors of glacial cirques.

While Matthes (1900) fails to produce a sharp definition of nivation, he exhibits a sophisticated grasp of snowpack accumulation dynamics. He invokes static snowpacks with intensified freeze–thaw around snowpatch peripheries, but assigns nivation only modest powers of landscape modification. Furthermore, while Matthes identifies a form continuum from nivation hollow to cirque, he distinguishes sharply between nivation and glacial effects and does not claim that nivation hollows enlarge into cirques. Nivation was soon adorned by others with bedrock freeze–thaw weathering, nivation hollows as precursors of cirques, the mobility of snowpacks, and solifluction as the primary mass wasting process. Thorn (1988) provides a comprehensive review of the development of nivation into the 1980s. The fundamental issue is to appreciate that nivation is a concept of weathering and transport intensification that invokes no unique processes.

Nivation benches or terraces are idealized as a gentle sloping flat or tread mantled in debris, unvegetated where snow is especially late-lying, with a steeper riser at the upslope end. Expansion is by headward incision promoted by the presence of late-lying snow at the inflection of slope. When suitably oriented such landforms, whatever their origin, are highly likely to become snow accumulation sites and it becomes tempting, if not irresistible, to move from correlation to causation, a step fraught with problems in the absence of process measurements. Nevertheless, available evidence does suggest that nivation is a likely mechanism for expansion in poorly consolidated materials (Thorn 1988; Berrisford 1992; Christiansen 1998). Where such forms occur in bedrock, headward incision becomes dependent upon more contentious weathering processes, rather than merely upon excavation by mass wasting.

While it is easy to envisage that the additional water supply associated with late-lying snow has considerable geomorphic potential, detailed field measurements were not undertaken until the 1970s (Thorn 1988). Important subsequent work includes Berrisford (1991, 1992) and Christiansen (1996, 1998). A reasonable summary of present knowledge of the mass wasting component of nivation is to state that late-lying snow does indeed accelerate or intensify periglacial mass wasting processes (e.g. solifluction, surface wash) by several factors, even by orders of magnitude, in comparison to nearby snow-free (or thinly snow-covered) surfaces.

However, the literature is not adequate to specify a consistent pattern of process or process rate intensification; indeed, considerable variability appears likely. On unconsolidated surfaces the elimination of vegetation cover by late-lying snow (not always the case) appears to represent an important process threshold. As rainfall inputs decrease and snowfall inputs increase proportionally, snowpatch meltwater influences emerge more starkly, especially in poorly consolidated materials (Christiansen 1998). In the presence of permafrost snowpatches may have important impacts on near-surface water flow (Ballantyne 1978), particularly by raising shallow subsurface flow to the surface where a snowpatch sustains a frozen subsurface.

While the role of nivation within the periglacial transport suite is generally non-problematic, and increasingly emphasizes meltwater impacts, the weathering role of nivation is problematic. For much of its history nivation was generally assigned no chemical weathering role. Intensification of chemical weathering processes beneath and around snowpatches is now documented (e.g. Thorn 1988), with a spatial pattern strongly dependent upon meltwater pathways that may even shift the impact downslope of the snowpatch itself. Knowledge of the role of nivation as a modifier of freeze–thaw weathering is largely constrained by the uncertainties associated with freeze–thaw weathering itself (Hall *et al.* 2002). Relevant ground climates (as opposed to largely irrelevant generalized air climates) are poorly known, laboratory studies do not necessarily mimic field conditions adequately, nor do they effectively isolate freeze–thaw from other possible mechanisms. Within the immediate context of nivation the critical issues rest with the interaction between snowpack insulation modifying, and perhaps eliminating, sufficient thermal regimes versus the obvious addition of abundant and necessary moisture through snowmelt. Berrisford (1991) found morphological evidence in the form of angular clasts beneath some portions of snowpatches to suggest enhanced mechanical weathering of coarse debris. However, he also emphasized the geomorphic importance of the annual temperature cycle, as opposed to shorter cycles, and views perennial snowpatches as protective.

Unlike CRYOPLANATION, of which it is a critical component, nivation research has been reinvigorated in recent years. While field data is increasingly available, definitional problems continue. Thorn (1988) suggested the term is so broad that it will always defy definition and should be abandoned, while Christiansen (1998) would like to expand it to embrace all snow-related processes making it equivalent to 'glaciation' in generality. Perhaps neither path is advisable, but the sharp contrast serves to highlight the problems presently associated with the term.

Snow-derived process will always be central to periglacial geomorphology and lead to some broad issues (Thorn 1978). Most nivation researchers appreciate that the wind-derived nature of a snowpatch means that there is potential for snow-bearing winter winds to orient landscape development through nivation. In fact, such orientation passes through a second filter, namely, available, suitable topographic traps because deep, late-lying snow cannot accumulate on a flat surface. Such concepts lead to ideas focused upon the landforms produced by snow-dominated regimes as opposed to those of full glaciation. Nelson (1989) goes so far as to suggest that not only is nivation central to cryoplanation, but that cryoplanation terraces are periglacial analogues of glacial cirques. His thesis invokes the presence of cryoplanation terraces where snowfall and temperature regimes are inadequate to generate cirque glaciation. Yet another view of nivation juxtaposes it with cold-based, that is non-erosive, glaciation. In such a context Rapp (1983) has suggested that interglacial nivation may represent the erosive, landforming regime and glaciation the quiescent, protective one.

The spatial extent of seasonal snowcover and lengthy interglacial periods, perhaps even more importantly relatively short pleniglacial periods, suggest that geomorphic processes derived from late-lying snow merit considerable attention. Clearly, nivation represents a core concept in such an appreciation, albeit not an exclusive one. Intensification of surficial mass wasting processes by nivation is now well established, but determination of systematic trends and rates must await generation of considerably more data. The weathering regime associated with nivation remains uncertain. Concentration of snowpatch meltwaters promotes chemical weathering; however, the location and degree of mechanical weathering with respect to a fluctuating seasonal snowpatch remains questionable, despite widespread willingness to invoke it. The extent to which nivation is able to shape a landscape is simply unknown.

Nivation cannot seemingly initiate topographic lows, but it can certainly modify them – but to what extent, in what fashion, and at what rates? Christiansen (1998) demonstrates headward expansion of nivation hollows in soft materials infused with permafrost, Berrisford (1992) suggests downslope expansion of nivation hollows. Thorn (1976) calculated nivation excavation rates in colluvium that would not produce a cirque from a nivation hollow in a feasible period of time. Quantitative research into such topics is urgently needed, but is immediately confronted by scale-linkage issues. In particular, periglacial process studies conducted on mesoscale phenomena must contend with the possibility that PARAGLACIAL conditions, rather than prevailing ones, hold the key.

References

Ballantyne, C.K. (1978) The hydrologic significance of nivation features in permafrost areas, *Geografiska Annaler* 60A, 51–54.

Berrisford, M.S. (1991) Evidence for enhanced mechanical weathering associated with seasonally late-lying and perennial snow patches, Jotunheimen, Norway, *Permafrost and Periglacial Processes* 2, 331–340.

——(1992) The geomorphic significance of seasonally late-lying and perennial snowpatches, Jotunheimen, Norway, Unpublished Ph.D. dissertation, University of Wales, Cardiff.

Christiansen, H.H. (1996) Nivation forms, processes and sediments in recent and former periglacial areas, *Geographica Hafniensia* A4.

——(1998) Nivation forms and processes in unconsolidated sediments, NE Greenland, *Earth Surface Processes and Landforms* 23, 751–760.

Hall, K., Thorn, C.E., Matsuoka, N. and Prick, A. (2002) Weathering in cold regions: some thoughts and perspectives, *Progress in Physical Geography* 26, 577–603.

Matthes, F.E. (1900) Glacial sculpture of the Bighorn Mountains, Wyoming, *United States Geological Survey, 21st Annual Report 1899–1900*, 167–190.

Nelson, F.E. (1989) Cryoplanation terraces: periglacial cirque analogs, *Geografiska Annaler* 71A, 31–41.

Rapp, A. (1983) Impact of nivation in steep slopes in Lappland and Scania, Sweden, in H. Poser and E. Schunke (eds) *Mesoformen des Reliefs im heutigen Periglazialraum*, Abhandlungen der Akademie der Wissenschaft in Göttingen, Mathematisch-Physikalische Klasse 3(35), 97–115.

Thorn, C.E. (1976) Quantitative evaluation of nivation in the Colorado Front Range, *Geological Society of America Bulletin* 87, 1,169–1,178.

——(1978) The geomorphic role of snow, *Annals of the Association of American Geogaphers* 68, 414–425.

——(1988) Nivation: a geomorphic chimera, in M.J. Clark (ed.) *Advances in Periglacial Geomorphology*, 3–31, Chichester: Wiley.

Futher reading

Thorn, C.E. and Hall, K. (2002) Nivation and cryoplanation: the case for scrutiny and integration, *Progress in Physical Geography* 26, 553–560.

SEE ALSO: cryoplanation; freeze–thaw cycle

COLIN E. THORN

NIVEO-AEOLIAN ACTIVITY

Niveo-aeolian processes involve the entrainment, transport and deposition of fine (mainly sand-sized) particles by wind in seasonally snow-covered areas, and modification of such sediments during snowmelt. The defining characteristic of niveo-aeolian activity is the deposition of wind-blown sediment on snowcover. This involves either simultaneous deposition of mixed sediment and drifting snow, or deposition of windblown sediment alone over earlier snowcover. Sediments deposited by wind on snowcover are referred to as niveo-aeolian deposits. The term *denivation* is used to refer to the processes, microforms and sedimentary structures associated with melting of the underlying snowpack.

Niveo-aeolian activity occurs in polar deserts, in subarctic environments, on alpine plateaux (Ahlbrandt and Andrews 1978) and on maritime mountains (Ballantyne and Whittington 1987). Most niveo-aeolian deposits are annual (associated with complete melting of snow in summer), though in exceptionally cold arid areas of Antarctica perennial deposits consisting of interstratified sediment and snow occur. Elsewhere, buried snow may persist under niveo-aeolian deposits for one or more summers (Bélanger and Filion 1991).

Sources of niveo-aeolian sediments include unvegetated or partly vegetated floodplains or outwash plains, raised beaches and deltas, aeolian sandsheets and dunes, glacial deposits, and sandy regolith. Sediments are entrained by wind in autumn or spring when snowcover is incomplete, or during winter when strong winds strip snow from the crests of ridges or hummocks. Sublimation of pore ice and abrasion by blowing sand are important in releasing sand particles from frozen surfaces. Most sand-sized particles travel by saltation or creep over ice or crusted snow, with fine sand and silt particles travelling in suspension. During violent storms coarse granules may travel up to 4 m above the surface and pebbles may be blown over ice (McKenna Neuman 1990).

Niveo-aeolian deposits have poor to moderate sorting and a wide range of modal grain sizes, but most are dominated by medium and coarse sand (0.2–2.0 mm). Fresh deposits often reveal concentrations of sediment at the top and base of the snowpack, layers of mixed snow and sediment, and sediment-rich layers separated by clean snow. During melt, sediment becomes concentrated at the snow surface. Such supranival deposits tend to be thickest on the lee of obstacles, notably on the slip faces of dunes. Melt of the underlying snow produces a range of denivation features including dimpled surfaces, snow hummocks, contorted bedding, sinkholes, cavities, tension cracks and faulted slipface strata (Koster and Dijkmans 1988; Dijkmans 1990). Meltwater fans may accumulate at the lower edge of niveo-aeolian beds (Lewkowicz and Young 1991).

Unless rapidly buried under prograding slipface beds, most denivation structures dry out and are destroyed by summer winds. Niveo-aeolian deposition therefore rarely leaves any distinctive sedimentological or structural signature in cold-climate aeolian sequences. For this reason the role of niveo-aeolian activity in the formation of the extensive Late Pleistocene COVERSANDS and dunefields (see SAND SEA AND DUNEFIELD) of Europe and North America remains contentious.

References

Ahlbrandt, T.S. and Andrews, S. (1978) Distinctive sedimentary features of cold climate eolian deposits, North Park, Colorado, *Palaeogeography, Palaeoclimatology, Palaeoecology* 25, 327–351.

Ballantyne, C.K. and Whittington, G.W. (1987) Niveo-aeolian sand deposits on An Teallach, Wester Ross, Scotland, *Transactions of the Royal Society of Edinburgh: Earth Sciences* 78, 51–63.

Bélanger, S. and Filion, L. (1991) Niveo-aeolian sand deposition in subarctic dunes, eastern coast of Hudson Bay, Québec, Canada, *Journal of Quaternary Science* 6, 27–37.

Dijkmans, J.W.A. (1990) Niveo-aeolian sedimentation and resulting sedimentary structures: Søndre Strømfjord area, western Greenland', *Permafrost and Periglacial Processes* 1, 83–96.

Koster, E.A. and Dijkmans, J.W.A. (1988) Niveo-aeolian deposits and denivation forms, with special reference to the Great Kobuk sand dunes, northwestern Alaska, *Earth Surface Processes and Landforms* 13, 153–170.

Lewkowicz, A.G. and Young, K.L. (1991) Observations of aeolian transport and niveo-aeolian deposition at three lowland sites, Canadian arctic archipelago, *Permafrost and Periglacial Processes* 2, 197–210.

McKenna Neuman, C. (1990) Observations of winter aeolian transport and niveo-aeolian deposition at Crater Lake, Pangnirtung Pass, NWT, Canada, *Permafrost and Periglacial Processes* 1, 235–247.

SEE ALSO: aeolian processes; nivation; periglacial geomorphology

COLIN K. BALLANTYNE

NON-LINEAR DYNAMICS

Geomorphologists have long recognized that landforms are the result of a complex set of interactions operating over different scales of space and time (Schumm and Lichty 1965; Schumm 1979; Brunsden and Thornes 1979). More recently, these ideas have been variously strengthened, confronted and extended by incorporating findings and analytical tools from the subject of non-linear system dynamics (NSD) developed in the mathematical and physical sciences. The term 'non-linear' expresses an unequal relationship between the driving force or the stress and the geomorphic response, most simply described as where *outputs are disproportionate to inputs*. A good example is the classic Hjulström curve of velocity of water plotted against the size of entrained sediment particles. It is a non-linear curve showing that the output (sediment size entrained) is not proportional to the input (water velocity); factors such as particle density and inter-particle cohesion are also important. The geomorphology of a landscape is governed by the interaction of a vast array of such processes operating in different parts of a landscape and over different timescales. Hence, the term 'non-linear dynamics' is used to describe the behaviour of the system rather than the behaviour of discrete process interactions. In the example of an unstable slope system, this means that the relationship between the intensity of precipitation falling on a slope and the size and timing of landslides may be complex and non-linear, rather than a simple cause and effect. It is widely believed that all complex systems consisting of interacting components behave non-linearly. The attractions of NSD lie with the possibilities of finding generic insights about geomorphic system behaviour and mapping well-studied model behaviours onto real systems that would normally be non-observable by conventional field methods. In practice, the demonstrable existence of model phenomena in real systems and hence the usefulness of NSD ideas are contested.

NSD may provide useful insights about (1) the predictability and unpredictability of geomorphic phenomena; (2) distinguishing between spontaneous and forced geomorphic change; (3) the sensitivity and resilience of landscapes to impact; and (4) the use of appropriate conceptual and modelling scales and methodologies.

A useful division in the discussion of non-linear system behaviour is to identify intrinsic or extrinsic changes. Intrinsic changes are those that spontaneously occur through self-organization as part of the system's own dynamic without any direct and proportional external forcing, analogous to the idea of biological evolution. One explanation is provided by the so-called 'arrow of time' implicit in the second law of thermodynamics, which predicts that a system will develop towards an equilibrium point where the free energy is minimized and thermodynamic entropy (disorder) is maximized. Thus, free energy in the form of flowing water drives the organization of a river network, especially its drainage density, so that the total potential of the flowing water to erode and carry sediment is minimized (or diffused) across the landscape. As in the classic Davisian cycle of landform evolution, thermodynamic equilibrium is reached when the relief is progressively lowered to a peneplain. However, since the Earth's surface is not dominated by peneplains but rather by an array of other geomorphic forms, it is clear that most landscapes have had their 'arrow of time' development arrested. Therefore, embedded within the idea of 'arrow of time' are other system behaviours that explain geomorphic systems set at some point far from equilibrium, known as dissipative systems (Prigogine 1996).

One of these, emergent complexity, describes the conditions found in open, and often partitioned systems that show ordered states maintained by flows of energy across the system boundary. Within human timescales, they may appear unchanging in their underlying forms but over longer timescales these systems evolve or emerge to become increasingly ordered and complex, as in the example of progressive weathering leading to soil horizons supporting a complex terrestrial ecosystem. Another, chaotic behaviour, may explain the local variability of many geomorphic systems. Chaos is most easily seen in mathematical equations that describe processes such as turbulent water flow, but the implications of chaos theory for geomorphology are large. Chaotic behaviour means that the exact pathway of a set of interacting processes over time is crucially sensitive to initial conditions and to even the smallest external perturbations. Over a long period of time, an observer of a chaotic landscape would see small initial differences in relief, drainage and soils amplified over time: divergence, rather than convergence. One result might be a mosaic of soil types overlying a fairly uniform parent material that initially differed from place to place only in small differences in texture. The variability would be amplified by subsequent vegetation succession and positive feedback controls. In one sense this means that a chaotic system, as we know for weather forecasts, is an unstable system becoming progressively unpredictable as the system moves from its starting point. However, model chaotic systems also evolve to lie within well-defined ranges, known as attractors, which translates to the natural environment in terms of the degree of variability at a higher scale being constrained and therefore predictable in probabilistic terms. Thus the overall range of soil types encountered under the emergent complexity displayed in a temperate woodland is fairly easy to predict from knowledge of the tree species, climate and geology but the local scale variability of soil properties, driven by chaos, may appear almost random.

Embedded system behaviours may also explain the remarkable fact that many emergent geomorphic patterns are the same, irrespective of the spatial scale. Aerial photographs of ripple beds and dunefields may be indistinguishable without an absolute distance scale. Stream networks, as measured by a stream-ordering parameter, such as Horton's bifurcation ratio, are often statistically similar whether viewed at the scale of the whole drainage basin or a small sub-basin. Many other geomorphic features show this so-called scale-invariance or fractal geometry (Figure 110a), which is easily identified by a power law (straight-line) relationship in a log–log plot of the variable and spatial scale. The behaviour of scale-invariant phenomena has been studied under the heading of self-organized criticality (SOC). This term was used to explain the response of a model sand pile to continuous additions of sand from above. The pile maintains a characteristic form (at the macrolevel) through losses of sand by avalanches (negative feedback at the microlevel), whose distribution over time (either size or frequency) conforms to a power law. The power law distribution means that it is possible to predict the overall

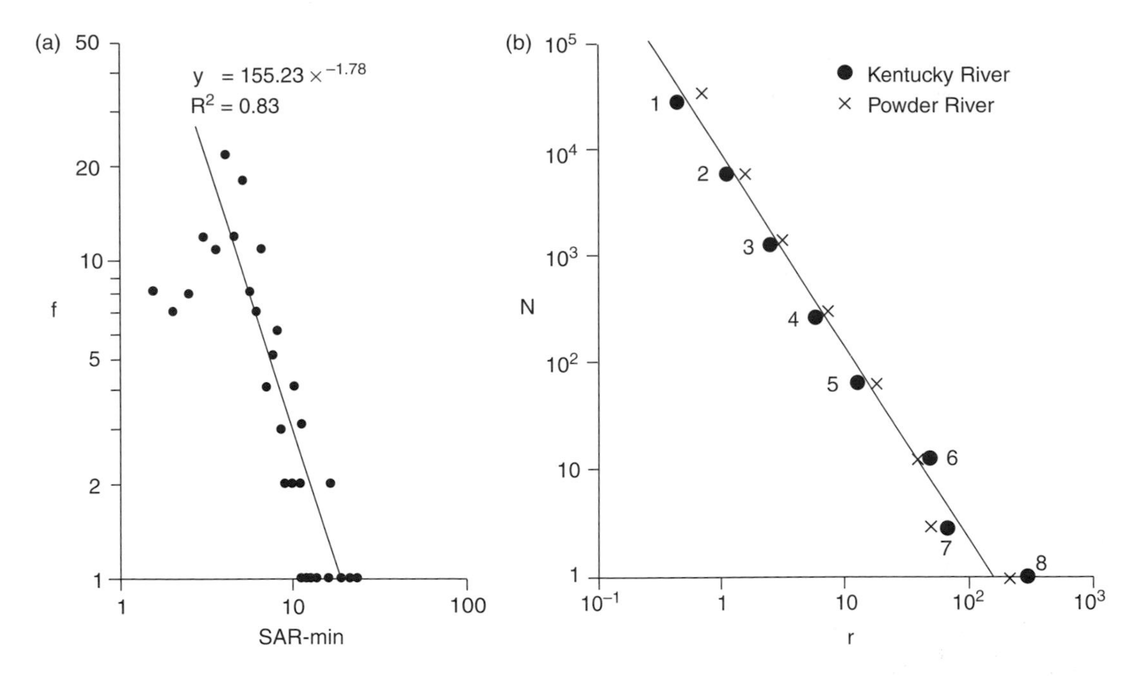

Figure 110 Scale-invariant phenomena in geomorphic patterns and processes. (a) Dependence of the number of streams N of various orders 1–8 on their mean length r for the Kentucky River basin in Kentucky and the Powder River basin in Wyoming (Turcotte 1997); (b) inverse power law relationship between the minerogenic sediment accumulation rate SAR-min and its frequency f for a mid-Holocene lake sediment record of catchment erosion at Holzmaar, Germany, indicative of self-organized criticality (Dearing and Zolitschka 1999)

properties of the system, such that there will be fewer large avalanches than small ones, but the details of timing and size of the next avalanche are unpredictable. A theory has developed that argues for systems of all kinds to evolve to critical states where they may similarly respond to small perturbations in a disproportionate manner, but importantly will evolve back to the original state (Bak 1996). Evidence for SOC now exists for many real spatial patterns, such as river networks and forest fires, and also real time series (Figure 110b), such as earthquakes, landslides and river sediment transport (e.g. Dearing and Zolitschka 1999). Thus, many geomorphic landscapes and landforms may be viewed at one scale as complex and ordered non-linear systems, emerging from the evolution of chaotic processes at another. Chaotic processes may produce highly variable or apparently random forms at a small scale, but through constraints imposed by the nature of the environment (e.g. geology, climate) may produce identifiable and self-similar patterns at large spatial scales, and these may exist in critical states.

Geomorphic systems also respond to external forces, like climate and human activities. The nature of the response depends on the force and the condition of the system and these may vary from direct and reversible to time-lagged and irreversible (Figure 111). Knowing the NSD dynamics that underlie a system may help to define the resilience or sensitivity of a geomorphic system. For example, how close a non-linear landscape system lies to a bifurcation point, as defined by the mathematical representation of that system, helps define its sensitivity to external impacts: the closer it is, the more likely is the system to be driven along a new irreversible trajectory towards an alternative steady state (Thornes 1983).

The implications of NSD in geomorphology are profound. We may have to accept that some geomorphic outcomes are unpredictable and strongly contingent on a landscape's history. It may be that reductionism as a methodology is unlikely to afford extrapolation of mathematical rules to large scales (Lane and Richards 1997). There may be a strong case for adopting some form of 'scientific

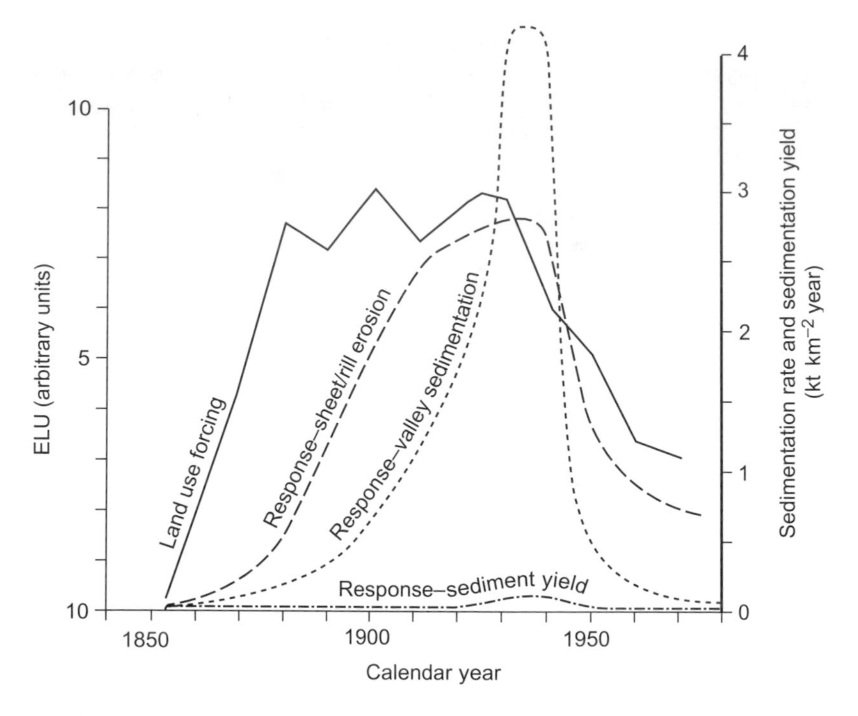

Figure 111 Non-linear relationships between forcings and geomorphic responses in space and time. Erosive land use change ELU in the Coon Creek basin, Wyoming, forces lagged temporal responses in hillslope sheet and rill erosion, main valley sedimentation, and catchment sediment yield (Trimble and Lund 1982; Wasson and Sidorchuk 2000)

realism' as the correct methodology, where we accept the existence of non-observable phenomena, structured and stratified systems with emergent properties, contingent relationships and prediction based on probabilities, while rejecting a belief in direct and enduring relationships between cause and effect (Richards 1990). Paradoxically, overcoming the difficulties of mapping NSD concepts and ideas onto real landscapes may be achieved through the use of simple computational models with simple rules. This 'new kind of science' (Wolfram 2002) uses cellular automata, where grids of interacting cells each containing a set of simple rules are updated at subsequent time steps in order to simulate the evolution of the system. Landscape models based on cellular grids and simple equations for sediment and water flows (e.g. Coulthard *et al.* 2000) show much promise in being able to simulate the spatial and temporal development of complex geomorphic forms with realistic non-linear behaviour. These may prove to be the most efficient means for simulating how non-linear landscapes will really respond to future combinations of climate and human activities.

References

Bak, P. (1996) *How Nature Works. The Science of Self-Organized Criticality*, New York: Springer.

Brunsden, D. and Thornes, J.B. (1979) Landscape sensitivity and change, *Transactions of the Institute of British Geographers* 4, 463–484.

Coulthard, T.J., Kirkby, M.J. and Macklin, M.G. (2000) Modelling geomorphic response to environmental change in an upland catchment, *Hydrological Processes* 14, 2,031–2,045.

Dearing, J.A. and Zolitschka, B. (1999) System dynamics and environmental change: an explanatory study of Holocene lake sediments at Holzmaar, Germany, *Holocene* 9, 531–540.

Lane, S.N. and Richards, K.S. (1997) Linking river channel form and process: time, space and causality revisited, *Earth Surface Processes and Landforms* 22, 249–260.

Prigogine, I. (1996) *The End of Certainty*, New York: The Free Press.

Richards, K.S. (1990) Editorial: real geomorphology, *Earth Surface Processes and Landforms* 15, 195–197.

Schumm, S.A. (1979) Geomorphic thresholds: the concept and its applications, *Transactions of the Institute of British Geographers* 4, 485–515.

Schumm, S.A. and Lichty, R.W. (1965) Time, space and causality in geomorphology, *American Journal of Science* 263, 110–119.

Thornes, J.B. (1983) Evolutionary geomorphology, *Geography* 68, 225–235.
Trimble, S.W. and Lund, S.W. (1982) Soil conservation and the reduction of erosion and sedimentation in the Coon Creek basin, Wisconsin, *US Geological Survey Professional Paper* 1,234, 1–35.
Turcotte, D.L. (1997) *Fractals and Chaos in Geology and Geophysics*, Cambridge: Cambridge University Press.
Wasson, R.J. and Sidorchuk, A. (2000) History for soil conservation and catchment management, in S. Dovers (ed.) *Environmental History and Policy: Still Settling in Australia*, Oxford: Oxford University Press.
Wolfram, S. (2002) *A New Kind of Science*, Champaign, IL: Wolfram Media.

Futher Reading

Brunsden, D. (2001) A critical assessment of the sensitivity concept in geomorphology, *Catena* 42, 99–123.
Phillips, J.D. (1999) *Earth Surface Systems*, Oxford: Blackwell.

JOHN DEARING

NOTCH, COASTAL

Notches can be cut at the cliff foot by waves, but they are generally poorly defined in fairly homogeneous rocks, and locally restricted to geologically favourable locations in more variable rocks. Notches, typically between 1 and 5 m in depth, are most common and best developed on tropical limestone coasts, where low tidal range concentrates the erosional processes. The formation of notches throughout the tropics is generally attributed to chemical or biochemical CORROSION, or to biological grazing and boring (see BORING ORGANISM), especially in sheltered locations. Nevertheless, ABRASION and other forms of mechanical wave erosion contribute to their formation in some areas. As there is little agreement over the level, or levels, that tropical notches are developing today, the occurrence of double or multiple notches in some places has been ascribed to changes in relative sea level, intermittent tectonic activity, variable rock structure and lithology, and the effect of organisms and other notch-forming mechanisms operating most efficiently at different elevations.

Further reading

Trenhaile, A.S. (1987) *The Geomorphology of Rock Coasts*, Oxford: Oxford University Press.
Trudgill, S.T. (1976) The marine erosion of limestone on Aldabra Atoll, Indian Ocean, *Zeitschrift für Geomorphologie Supplementband* 26, 64–200.

ALAN TRENHAILE

NUÉE ARDENTE

A French term most frequently translated into English as a 'glowing cloud', to refer to a pyroclastic flow (see PYROCLASTIC FLOW DEPOSIT) or surge from a volcano. First used by Lacroix (1904) to describe the pyroclastic clouds erupted from Mt Pelée, Martinique, on 8 May 1902 and subsequently. They destroyed the town of St Pierre killing most of its 27–28,000 inhabitants in the worst volcanic disaster of the twentieth century, as measured by loss of life. The term 'ardent' was originally used strictly to refer to 'very hot, burning or scorching' rather than 'glowing', and specified that such clouds were not incandescent at night, except close to the crater (Tanguy 1994).

Nuée ardente has come to be used as a general term for a hot, turbulent, self-expanding gaseous cloud with its entrained particles of rocks and ash, which may be incandescent, that descends the flank of a volcano at high or exceptionally high velocities. In the high velocity flows the denser, lower part hugs the ground surface and becomes strongly controlled by the pre-existing topography. This portion usually forms the bulk of a resultant deposit. Such deposits show massive coarse bouldery facies in channels or the axis of flow, and usually grade to sandy deposits both laterally and upwards. There is evidence of searing heat for only a few brief minutes, but if objects are buried in the deposit they show signs of being cooked. The lighter, upper part of the flow, comprising hot gases and ash particles, rapidly expands upwards as a dark, towering cloud to many kilometres in height. This cloud may spread over a much wider area than the dense lower flow, resulting in a widespread coeval ash.

In exceptionally high velocity surges, a blast of gases and entrained particles travels outward from the volcano, usually independent of the topography, removing most structures and trees in its path. Vertical surfaces become 'sandblasted', the bark of trees may be stripped, trees may be snapped and removed laterally for large distances, rocks may become embedded within materials they impact against, buildings may be razed with their contents twisted or carried from the scene. Resultant deposits show a high degree of fluidity (a low concentration fluid) of the gaseous mixture with cross-stratification, antidunes, highly irregular erosion breaks, considerable lateral variability in lithology and thickness, and frequently charcoal.

Nuée ardente has been used in a broad sense to encompass all pyroclastic flows and surges and in a restricted sense to refer specifically to small volume monolithologic block-and-ash flows generated by the collapse of actively growing lava domes or lava flows on steep terrain (Pelean eruptions).

Nuées ardentes are one of the most feared volcanic hazards. It is not simply the destructive energy of the cloud but the searing heat of gases and ash particles that make them so lethal to life and property.

References

Lacroix, A. (1904) *La Montagne Pelée et ses éruptions*, Paris: Masson.

Tanguy, J-C. (1994) The 1902–1905 eruptions of Montagne Pelée, Martinique: anatomy and retrospection, *Journal of Volcanology and Geothermal Research* 60, 87–107.

VINCENT E. NEALL

NUNATAK

Nunatak is an Inuit word referring to a mountain top that protrudes above the surface of a GLACIER or ice sheet. Such summits are subject to intense frost weathering but escape glacial erosion. On relatively flat surfaces this often results in the formation of autochthonous blockfields (see BLOCKFIELD AND BLOCKSTREAM), whilst sharper peaks and ridges can become broken along joints to produce sharp ARÊTEs and pinnacles. In areas that were glaciated in the past, the difference in weathering and erosion can be identified as a 'periglacial trimline' that separates glacially eroded terrain from higher ground that retains evidence of prolonged weathering.

A range of simple measures have been used to quantify the difference in degree of rock surface weathering above and below proposed trimlines, including rock surface roughness, joint depth and surface hardness as recorded using a SCHMIDT HAMMER (McCarroll *et al.* 1995). One of the most reliable indicators of past nunataks is the presence of the clay mineral gibbsite at the base of the soils. Gibbsite, an aluminium oxide, is an end product of the weathering of silicate minerals and in mountainous environments in the extra-tropics is thought to represent a long period of *in situ* weathering. Cosmogenic isotope exposure age dating (see COSMOGENIC DATING) can also be used to test the hypothesis that summits escaped glacial erosion and to date the exposure of glaciated bedrock and ERRATICS (Stone *et al.* 1998).

The identification of 'palaeonunataks' using geomorphological and dating evidence provides field evidence with which to test and improve models of past ice sheets in formerly glaciated areas such as the British Isles, northern Europe and North America (Ballantyne *et al.* 1998; McCarroll and Ballantyne 2000). It has also been argued that nunataks in these areas may have provided refugia for plant species to survive glaciations and then re-populate when the glaciers receded. In some cases this would help to explain the rather patchy occurrence of some species, though conditions would have been extremely harsh (Birks 1993).

References

Ballantyne, C.K., McCarroll, D., Nesje, A., Dahl, S-O., Stone, J.O. and Fifield, L.K. (1998) The last ice sheet in north-west Scotland: reconstruction and implications, *Quaternary Science Reviews* 17, 1,149–1,184.

Birks, H.J.B. (1993) Is the hypothesis of survival on glacial nunataks necessary to explain the present-day distribution of Norwegian mountain plants?, *Phytocoenologia* 23, 399–426.

McCarroll, D., Ballantyne, C.K., Nesje, A. and Dahl, S O. (1995) Nunataks of the last ice sheet in northwest Scotland, *Boreas* 24, 305–323.

——and Ballantyne, C.K. (2000) The last ice sheet in Snowdonia, *Journal of Quaternary Science* 15, 765–778.

Stone, J.O., Ballantyne, C.K. and Fifield, L.K. (1998) Exposure dating and validation of periglacial limits, NW Scotland, *Geology* 26, 587–590.

DANNY McCARROLL

O

ORGANIC WEATHERING

Organisms and organic compounds play a wide range of roles in both enhancing and retarding rock and mineral weathering processes in most environments. Indeed, as Reiche (1950: 5) recognized, weathering involves 'the response of materials which were at equilibrium within the lithosphere to conditions at or near its contact with the atmosphere, the hydrosphere, and perhaps still more importantly, the biosphere'. Thus, by definition, weathering in most places operates in a non-sterile environment and biological processes and influences need to be taken into account if we are to understand weathering fully. The term organic weathering is not widely used by geomorphologists, with biological weathering more commonly found in the literature, but both have a similar broad definition as the suite of biological weathering processes and indirect biological influences on weathering. Bioerosion is an allied term, referring to the erosive activity of organisms especially on bare rock surfaces, and there is a spectrum of such organic influences on weathering and erosion. Many books refer to a three-fold classification of weathering into physical, chemical and biological processes, but it is perhaps simpler to keep the classic division into physical and chemical weathering process groups, acknowledging that many processes within these groups can be seen to be biophysical or biochemical in nature.

Although there was a flourishing of interest in biological weathering in the late twentieth century, as analytical and microscopical techniques became available which permitted closer study of the rock:organism interface, interest in the possible roles of organisms and organic compounds in weathering is nothing new. Several late nineteenth-century workers carried out pioneering experiments on the role of plant seedlings (Sachs 1865), lichens (Sollas 1880) and organic acids (Julien 1879) in weathering common rocks and rock-forming minerals. However, there was little general assessment of the overall nature, rate and importance of biological weathering.

Research into organic weathering has tended to focus on a number of key types of organism or organic influence. First, there has been a huge concentration of effort into understanding the role played by organic acids and other organic compounds within soils on the weathering of minerals (Drever and Vance 1994). Second, there have been a number of studies on the role of plant roots in weathering both rocks and minerals, illustrating the important uptake of many elements by such processes (Kelly *et al.* 1998; Hinsinger *et al.* 2001). Third, a few studies have been made of the role of animals in weathering, especially their contribution to the decomposition of organic material leading to the production of organic acids. Fourth, there have been many studies on the role of lichens in the weathering of rock surfaces. These studies illustrate the complexity of roles played by organisms in weathering, with biophysical and biochemical lichen activity recognized, as well as a bioprotective role in some cases where they retard the action of other weathering processes and bind the rock surface together. Different lichen species play different roles, with some species being highly biodeteriorative and others not. For example, a recent study by Robinson and Williams (2000) in the Moroccan High Atlas show accelerated weathering of sandstone by *Aspicilia caesiocinerea* agg. producing notable scars on the rock surface. Finally, there have been many studies of the roles played by micro-organisms (notably algae, fungi and bacteria) in weathering. In most environments, combinations of different processes and organisms are involved producing a complex biological imprint on weathering. Thus, for example, the large percentage of the terrestrial landsurface covered by soils will experience weathering conditioned by soil organic acids, plant roots, animal decomposition and the many micro-organisms that inhabit soils (e.g. symbiotic mycorrhizal fungi have been shown by Jongmans *et al.* 1997 to bore into feldspars in soils, thereby allowing uptake of Ca and Mg by plant roots). Similarly, most bare rock surfaces are not actually bare at all but are coated with a diverse community of micro-organisms and lower plants which play a range of roles in weathering.

Biophysical weathering often involves the creation of stresses by the expansion and contraction of organisms or parts of organisms. Lichens, for example, can absorb a vast amount of water leading to huge expansion on wetting and a concomitant shrinkage on drying. Crustose lichens, which grow very closely attached to rock surfaces, can cause much weathering to the underlying rock. Fry (1927) and Moses and Smith (1993) provide experimental evidence of the physical weathering caused by such wetting and drying on a range of rock types. Natural processes of growth and decay of rock-dwelling organisms or parts of organisms can also cause biophysical weathering. Some crustose lichen thalli, for example, start to peel away from the rock surface at the centre as they senesce, producing patches of intensive weathering. Plant roots may also force their way into cracks and joints, with large pressures exerted during growth sufficient to induce weathering. Grazing and mechanical burrowing of animals can also produce physical weathering, although this is usually categorized as a form of bioerosion.

Biochemical weathering includes a host of individual processes, with the production of carbon dioxide, organic acids and chelation (often involving organic acids) being particularly important. Within soils, organic by-products can provide a key control on weathering processes, especially within the rhizosphere. Many organic acids, such as humic, fulvic, citric, malic and gluconic, have been shown to be capable of weathering a range of substrates. Several types of micro-organism have been found to be capable of boring into suitable substrates (such as rock, corals and mineral grains) through chemical means. The exact mechanism by which different micro-organisms bore has been much debated, but extracellular acids and other substances probably promote chemical decomposition of minerals. The production of a network of near-surface boreholes can encourage further weathering by other (often inorganic) processes as they weaken the surface and increase the reactive surface area. Biochemical weathering by lichens involves the production of carbon dioxide, oxalic acid and the complexing action of a range of sparingly soluble lichen substances. Lichens such as the highly biodeteriorating *Dirina massiliensis* forma *sorediata* produce oxalic acid which reacts with calcium carbonate within rocks to produce calcium oxalate (Seaward 1997).

Another important aspect of organic influences on weathering is bioprotection. In this case, organisms do not play an active role in encouraging weathering, but rather play a passive role. The very presence of a cover of lichens or biofilm, for example, protects the underlying rock surface from extremes of temperature (thus reducing the potential for damaging freeze–thaw weathering for example). Furthermore, the lichen or biofilm can also act as a sponge soaking up incident rainwater and providing a chemical buffer for the underlying surface, thus reducing the potential for chemical weathering.

Identification of the various organic weathering processes and influences is a challenge; an even greater challenge revolves around trying to quantify their effect and identify their contribution to overall weathering rates. Some measurements of biological weathering rates are presented in Goudie (1995) derived from detailed empirical studies in the field as well as experimental studies. Many biological processes have proved very difficult to measure in the field with currently available techniques (e.g. physical weathering by plant roots) and experimentation is also challenging where growing organisms are involved. Although some organic weathering processes operate at a fast rate and are quite dramatic, their action is often highly localized in time and space. Thus, for example, biodeteriorating lichens can act at a very fast rate, but they are often patchily distributed over a rock surface, so their net effect may be reduced. Furthermore, over time community dynamics will influence the species present on any surface, limiting the net contribution of biodeteriorating species.

Despite the wide range of studies on a variety of organic weathering processes and influences, there is still uncertainty about their general importance and contribution to weathering in different environments. Viles (1995) hypothesized that on bare rock surfaces biological weathering would increase in 'hostile' environments, and that less bioprotection would take place. The reasoning behind this assertion is that in hostile environments organisms extract more nutrients, water and shelter from the rock itself, producing a weathering effect as they do so. In more benign environments organisms tend to have a less close contact with the surface, and their net role will be a protective one. Thus, hot desert, cold tundra and coastal environments should be characterized by high rates of biological weathering, in comparison with humid temperate and tropical ones. However, many harsh environments are also characterized by slower biological growth rates which may complicate this pattern.

Even if biological weathering on bare rock surfaces can be seen to be more intense in some environments rather than others, this does not necessarily imply that it will be the dominant series of processes there. In hot arid areas, for example, although lichen weathering can be spectacular, heating and cooling and salt weathering can also be highly effective and operate at a much faster rate than can slow-growing lichens. Considering the soil-covered landscape it is probable that the reverse applies, with higher rainfall and temperatures encouraging organic growths within the whole ecosystem, thus enhancing the production of organic acids and increasing both the overall rate of weathering and the contribution of organisms to weathering.

In some instances, recognizable small-scale landforms can be produced by biological weathering processes. For example, lichen fruiting bodies create small pinhead-sized pits in rock surfaces and various authors have identified larger pitting and grooving as being organically produced (e.g. Danin and Garty 1983; Robinson and Williams 2000). On limestone surfaces such features may be called BIOKARST. It has proved difficult to establish convincing process-form links in many cases, as there are a multiplicity of ways in which similar landforms at this scale can be produced.

Future challenges for geomorphologists in understanding organic weathering include the need to provide quantitative comparisons of sterile vs. organically mediated weathering rates in different environments, and the need to provide a broader assessment of the overall role of organisms and organic by-products in weathering and sedimentation. Finally, scientists such as Robert Berner have suggested that biotically enhanced weathering has played a major role in the global carbon cycle over long time spans (Berner 1992), and geomorphologists can play a key role in providing reliable empirical data on biological weathering rates and their spatial and temporal variability in order to test such models.

References

Berner, R.A. (1992) Weathering, plants and the long-term carbon cycle, *Geochimica et Cosmochimica Acta* 56, 3,225–3,231.

Danin, A. and Garty, J. (1983) Distribution of cyanobacteria and lichens on hillslopes of the Negev Highlands and their impact on biogenic weathering, *Zeitschrift für Geomorphologie* 27, 413–421.

Drever, J.I. and Vance, G.F. (1994) Role of soil organic acids in mineral weathering processes, in M.D. Lewan and E.D. Pittman (eds) *The Role of Organic Acids in Geological Processes*, 491–512, New York: Springer-Verlag.

Fry, E.J. (1927) The mechanical action of crustaceous lichens on substrata of shale, schist, gneiss, limestone and obsidian, *Annals of Botany* 41, 437–460.

Goudie, A.S. (1995) *The Changing Earth*, Oxford: Blackwell.

Hinsinger, P., Fernandes Barros, O.N., Benedeth, M.F., Noack, Y. and Callot, G. (2001) Plant-induced weathering of a basaltic rock: experimental evidence, *Geochimica et Cosmochimica Acta* 65, 137–152.

Jongmans, A.G., van Breeman, N., Lundstrom, U., van Hees, P.A.W., Finlay, R.D., Srinivasan, M., Unestam, T., Giesler, R., Melkerud, P.A. and Olsson, M. (1997) Rock-eating fungi, *Nature* 389, 682–683.

Julien, A.A. (1879) The geologic action of humus acids, *Proceedings, American Association for the Advancement of Science* 28, 311–327.

Kelly, E.F., Chadwick, O. and Hilinski, T.E. (1998) The effects of plants on mineral weathering, *Biogeochemistry* 42, 21–53.

Moses, C.A. and Smith, B.J. (1993) A note on the role of *Collema auriforma* in solution basin development on a Carboniferous limestone substrate, *Earth Surface Processes and Landforms* 18, 363–368.

Reiche, P. (1950) *A Survey of Weathering Processes and Products*, New Mexico University Publication in Geology 3, Albuquerque.

Robinson, D.A. and Williams, R.B.G. (2000) Accelerated weathering of a sandstone in the High Atlas Mountains of Morocco by an epilithic lichen, *Zeitschrift für Geomorphologie* 44, 513–528.

Sachs, J. (1865) *Handuch der experimental-Physiologie den Pflanzen*, Leipzig: Wilhelm Englemann.

Seaward, M.R.D. (1997) Major impacts made by lichens in biodeterioration processes, *International Biodeterioration and Biodegradation* 40, 269–273.

Sollas, W.J. (1880) On the action of a lichen on a limestone, *Report, British Association for the Advancement of Science* 586.
Viles, H.A. (1995) Ecological perspectives on rock surface weathering: towards a conceptual model, *Geomorphology* 13, 21–35.

HEATHER A. VILES

ORIENTED LAKE

Oriented lakes are subparallel, elongated lakes, which commonly occur in extensive clusters. Many of these clusters, especially in bedrock, are the result of processes antecedent to lake formation. For instance, orientation on the Canadian Shield is commonly associated with glacial fluting. However, there are belts of oriented lakes covering thousands of km^2 in unglaciated areas of the sandy Arctic coastal plains of Russia, northern Alaska and north-west Canada. They also occur in non-permafrost areas, such as the Atlantic coastlands of Maryland, the Carolinas and Georgia, USA. Two forms of oriented lake have been recognized: the most common are elliptical, but rectangular shapes occur in the Beni Basin of northeastern Bolivia, and the Old Crow and Bluefish Basins of northern Yukon Territory, Canada.

In North America, the elliptical lakes occur outside the glacial limits, at sites where there is no evidence of glacial deposition. Some occur in postglacial marine terraces. Well-known examples are the Carolina Bays of the southern Atlantic coast, and the lakes near Liverpool Bay, NWT, and Point Barrow, Alaska, on the western Arctic coast. In permafrost environments, the lakes are in depressions formed by thermokarst processes, which have been elongated during their development. The lakes range in size over several orders of magnitude, from ponds with long axes of less than 30 m, to water bodies of over 1,500 ha. The lake-size distributions are skewed, with the mean being less than 250 ha. Along the western Arctic coast, the mean length to width ratio of the lakes is about 1:7. The major axes of the lakes are aligned perpendicular to the prevailing winds, and, in the case of the lakes near Liverpool Bay, the standard error of mean orientation is less than 3°.

Several theories have been advanced for the causes of lake orientation. Many of these, including bombardment by meteorite showers, effects from upwelling of artesian springs or the action of fishes hovering while spawning, have been discredited. However, consideration of the effects of wind action perpendicular to the long axes of the lakes has been supported by field data and laboratory experiment.

The hydrodynamic theory proposes that winds blowing across a lake establish a two-cell current circulation within the water body, with water returning to the windward shore around the ends of the lake (Rex 1961). The maximum littoral drift, and associated erosion, occurs at the ends of the lakes, where the angle between waves propagating in deep water and a line perpendicular to the shore is 50°. A similar circulation has been measured in large lakes of the Alaskan coastal plain (Carson and Hussey 1962), and has been reproduced at laboratory scale by blowing air across a 2 m square box containing a scale model pond. The model pond, initially circular, became elongated in the process. The equilibrium form of a finite shoreline in readily transported, unconsolidated material is cycloidal, which corresponds to the approximate shape of the oriented lakes along the western Arctic coast. The orientation of the lakes in the western Arctic is consistent with the hydrodynamic theory, when applied using the prevailing wind regime of the region (Carson and Hussey 1962; Mackay 1963).

The rectangular form of the oriented lakes in the Old Crow and Beni basins has been associated with aspects of bedrock structure propagating through the overlying sediments (Allenby 1989), but the explanation has not been verified. In northern Yukon, the lakes are THERMOKARST features that have developed in sediments up to at least 150 m thick, in which permafrost is present in the upper 60 m. The permafrost impounds the lakes. In Bolivia, the sediments are similarly thick, but lake levels are maintained by a high water table close to the surface.

References

Allenby, R.J. (1989) Clustered, rectangular lakes of the Canadian Old Crow Basin, *Tectonophysics* 170, 43–56.
Carson, C.E. and Hussey, K.M. (1962) The oriented lakes of arctic Alaska, *Journal of Geology* 70, 417–439.
Mackay, J.R. (1963) *The Mackenzie Delta Area, N.W.T.*, Geographical Branch Memoir 8, Ottawa: Department of Mines and Technical Surveys.
Rex, R.W. (1961) Hydrodynamic analysis of circulation and orientation of lakes in northern Alaska, in G.O. Raasch (ed.) *Geology of the Arctic*, 1,021–1,043, Toronto: University of Toronto Press.

C.R. BURN

OROGENESIS

Orogenesis is the building of mountains by the forces of PLATE TECTONICS. Driven by the internal heat of the Earth, motions of lithospheric plates produce changes in crustal thickness structure that result in vertical motion of the topographic surface. It is this motion that is responsible for creating impressive mountain landscapes that have been the inspiration of so much geomorphology.

Mountains are built slowly over geologic time as an accumulation of CRUSTAL DEFORMATION. Rates of mountain building are quantified as the rates of vertical motion of rock with respect to the geoid (rock uplift), the surface with respect to the geoid (surface uplift), or rock with respect to the surface (exhumation, see also DENUDATION) (England and Molnar 1990). The relationship, rock uplift = surface uplift + exhumation, is widely adopted. Surface uplift describes topographic growth and creation of the positive landforms of mountains. Long-term surface uplift is difficult to measure, but is approximated using the altitude dependence of fossil organisms or displacement of features with respect to eustatic sea level (Abbott *et al.* 1997). Short-term surface uplift can be measured using geodetic techniques (e.g. GPS and synthetic radar interferometry). Relative surface uplift can be constrained using geomorphic markers, such as river terraces (see TERRACE, RIVER) or erosion surfaces. Exhumation relates to erosion, which is critical for accommodating plate motion by transferring mass from thickened mountain belts to adjacent basins. It is generally inferred from rock cooling histories (see DENUDATION CHRONOLOGY and FISSION TRACK ANALYSIS), basin sedimentation records or COSMOGENIC DATING.

Active orogens may form at rates of rock uplift of 0.01–10 mm yr^{-1}, and a rate of ~1 mm yr^{-1} is representative of many actively growing mountains around the world. Mountain building tends to be stable over millions to tens of millions of years, such that major episodes of orogenesis are commonly given formal names (e.g. the Alpine Orogeny). The timescale of mountain building is short enough, however, that changes in plate motion and global climate throughout the latest Tertiary and Quaternary have had noticeable effects on most orogens.

Orogenic belts are most common along ACTIVE MARGINS, such as the arcuate mountain belts of the 'Ring of Fire' along the continental rim that surrounds the Pacific plate. The characteristics of these mountain ranges depend on the type of plate boundary. The majority of mountain uplift is produced by convergent tectonic motion, where two or more plates collide and increase crustal thickness. One setting in which this occurs is Cordilleran-type orogens, well represented by the Cascades of northwestern North America or the Andes. These mountains stretch above subduction zones and are produced by permanent convergent deformation of the overriding continental plate and thermally driven buoyancy of a magmatic arc. Their anatomy includes coastal ranges near the accretionary complex of the subduction zone, lines of volcanoes that are separated from the coast by elongate valleys, and highlands that rise above foreland fold and thrust belts that verge towards the continent interior. Width of these highlands is largely dependent on the dip of the subducting oceanic lithosphere.

A second setting of convergent mountain building is continental collision, typified by the active collision of India and Asia. India has underthrust beneath Asia over the past 50 Ma, resulting in uplift of the High Himalaya, Tibetan Plateau and associated mountains that penetrate thousands of kilometres into the interior of Asia (Hodges 2000). Several thousand kilometres of plate convergence has been accommodated by a combination of orogenic crustal thickening and lateral escape of microplates via strike-slip faults. That continental collision is the most effective mechanism of mountain building is evident. Half of all mountain peaks worldwide that rise above 7.5 km elevation occur in the Himalaya, while all of the remaining half are associated directly with the India–Asia collision. The crystalline core of the Himalaya has also experienced ~10–20 km of DENUDATION in the Neogene at rates locally as high as 1 cm yr^{-1}, leading to rapid deposition in the Bengal and Indus fans (Searle 1996). In Tibet, arid conditions and internal drainage basin geometry have hindered erosional exhumation, leading to formation of an orogenic plateau above the under-thrust Indian plate. Geodynamic processes limit the plateau's elevation to ~5 km, via lower crustal flow where strength is exceeded by gravitational load (Royden *et al.* 1997).

Mountains are also built at divergent and transform plate boundaries and within continental interiors. Crustal extension via normal faulting leads to tilting and uplift of large hanging-wall blocks and results in characteristic basin and range topography (e.g. the western USA). Strike-slip plate boundaries may produce narrow zones of orogenesis where plate motions are somewhat oblique in a convergent (transpressional) or divergent (transtensional) sense. The components of non-strike-slip motion along such boundaries may be accommodated by reverse or normal faulting, so that similar but laterally confined mountain systems result (e.g. the Transverse Ranges along the San Andreas fault). Mountains within continental interiors may represent earlier stages of continental deformation or the effect of geodynamic processes. Dynamic topography on cratons can occur where compositionally or thermally induced density contrasts occur in the sub-lithospheric mantle.

Although a mountain may owe its origin to tectonic construction, mountain landscapes are dominated by erosional processes. Mountain topography consists of valleys and hillslopes shaped by erosional agents (e.g. mass wasting, glacial erosion, fluvial erosion). Agents of erosion are poised to reduce the terrestrial surface to low RELIEF, such that tectonic orogenesis is critical for maintaining topographic variation above the mean elevation of the continents. The act of mountain building hence has important effects on the dynamic processes of erosion itself, both directly (e.g. changes in BASE LEVEL) and indirectly (e.g. the effect of mountains on local climate). Because of this erosional character of mountains, topographic character does not always discriminate orogens that are actively forming from those formed by prior plate motion.

Mountainous topography may linger for more than one hundred million years after cessation of active tectonic deformation (e.g. the Appalachian Mountains). Rock uplift and erosion continue as long as an orogen contains a thickened crustal root, that must be removed by DENUDATION. Topography in ancient mountains is maintained, even after erosion has removed many kilometres of rock, because of ISOSTASY. The ductile nature of the upper mantle permits adjustment to gravitational loads over timescales of ~10^3 yr. The increased thickness of crust beneath orogens is more buoyant than the mantle rocks that lie beneath adjacent crust, such that the topographic surface is higher than the surrounding region. As erosion removes mass from the mountains, the crust rebounds to remain in gravitational equilibrium. The magnitude of rebound is proportional to the ratio of crust and mantle densities, such that mean elevation decreases much more slowly than the rate of regional denudation. Isostatic rebound can even produce the uplift of peaks where mean elevation is decreasing, where valleys are incised faster than interfluves erode.

Although inactive mountain ranges are still referred to as 'orogens', they experienced their tectonic orogenesis in the geologic past and are thus distinct from mountains presently rising along active plate margins. Erosion rates of ancient mountain belts may become quite slow, such as average rates in the Appalachian Mountains of 0.02–0.04 mm yr^{-1} (Mills *et al.* 1987). Denudation this slow is essentially weathering-rate limited, yet it is enough to reduce crustal thickness over long periods. Variations in the geomorphic system, such as climate change, may impose upon the stagnant erosional setting of ancient mountains and force readjustment of erosional processes. This often leads to incision and REJUVENATION of topography.

The character of topography itself has traditionally been used to interpret the surface uplift and exhumation history of mountain belts. This is one goal of landscape evolution studies, which seek to define cyclic changes in landforms or topographic 'maturity' through stages of orogenic and post-orogenic development (see CYCLE OF EROSION). However, this approach requires tenuous assumptions about the relationships between topographic parameters and uplift and exhumation. Studies have shown that many aspects of topography, such as RUGGEDNESS, DRAINAGE DENSITY and hypsometry (see HYPSOMETRIC ANALYSIS), are dependent on the erosional resistance of rocks, the nature of local climate and the individual processes of the dominant erosional agent (e.g. Willgoose and Hancock 1998), such that direct interpretation of orogenic history from topographic parameterization is dubious. One parameter that has been demonstrated to correlate with exhumation rate is slope or short-wavelength relief. Slope is almost synonymous with erosion rate, as rates of all erosional processes increase with slope and slope must increase in areas of rapid surface uplift due to ever-growing gravitational instability. Nonetheless, the clues to orogenic evolution do not lie entirely in topographic parameterization using DIGITAL ELEVATION MODELS.

Because the geomorphic and geodynamic processes that shape orogenic belts are complex and occur over very long scales of time and space, the topographic evolution of mountain landscapes is difficult to study comprehensively with physical experimentation or direct observation. For this reason, the approach of numerical modelling has become important (see MODELS). Models can quantitatively test how elements of landscapes should evolve under a given set of conditions, based on the use of erosion laws derived from previous research. Models can be constructed as grids of cells with set boundary conditions, such as the distribution of tectonic uplift or base-level change, frequency and magnitude of precipitation events, and rheology of eroding materials. Erosion laws, such as stream power bedrock incision or diffusive hillslope transport, can be run under different conditions. The result is a representation of the long-term landscape evolution of a hypothetical setting, and can be instructive with respect to the role of boundary conditions or specific processes (Burbank and Anderson 2001). However, these models are limited by present degree of understanding of individual erosional processes and the difficulty of capturing real-world complexity.

The significance of erosion goes beyond shaping the landscape of orogens. Erosion can itself influence tectonic processes. For example, erosion modulates surface slope in deforming thrust wedges that can in turn affect deformation (e.g. critical taper wedges; Dahlen *et al.* 1984). Concentrated erosion in parts of an orogen also control trajectories of crustal motion, and thus influence deformation partitioning (e.g. the Southern Alps, New Zealand; Koons 1989). This in turn causes faster tectonic uplift and increases gravitational potential energy, leading to more rapid erosion (i.e. positive feedback loop). Eventually, erosion and tectonic rock uplift rates may become balanced in a steady state of mass flux. In the north-west Himalaya, for example, a steady state has been achieved where topography is sufficiently rugged as to result in bedrock landsliding that keeps pace with base-level induced incision driven by tectonic uplift (Burbank *et al.* 1996).

The most important external influence on erosion landscape evolution and mountain building may be climate. Rates of erosion tend to correlate with precipitation. Spatial concentrations of precipitation due to the orographic effect of topography can lead to concentrated erosion (e.g. Irian Jaya fold belt; Weiland and Cloos 1996). This is a core element of models of coupled erosion and tectonic evolution of orogens (Willett 1999). Latitudinal, altitudinal and temporal variations in climate also determine where glacial and fluvial erosion dominate. Glacial erosion is thought to be more effective, and can force relief production via valley incision and intense erosion at equilibrium line (see EQUILIBRIUM LINE OF GLACIERS) altitudes that can 'buzz-saw' mountain landscapes (Brozovic *et al.* 1997). The effects of climate may be so important that increases in erosion and sediment production in many mountain systems worldwide may have been caused by the onset of glacial climate ~4 Myr ago (Molnar and England 1990).

References

Abbott, L.D., Silver, E.A., Anderson, R.S., Smith, R., Ingle, J.C., Kling, S.A., Haig, D., Small, E., Galewsky, J. and Sliter, W. (1997) Measurement of tectonic surface uplift rate in a young collisional mountain belt, *Nature* 385, 501–507.

Brozovic, N., Burbank, D.W. and Meigs, A.J. (1997) Climatic limits on landscape development in the northwestern Himalaya, *Science* 276, 571–574.

Burbank, D.W. and Anderson, R.S. (2001) *Tectonic Geomorphology*, Malden, MA: Blackwell Science.

Burbank, D.W., Leland, J., Fielding, E., Anderson, R.S., Brozovic, N., Reid, M.R. and Duncan, C. (1996) Bedrock incision, rock uplift, and threshold hillslopes in the northwestern Himalaya, *Nature* 379, 505–510.

Dahlen, F.A., Suppe, J. and Davis, D. (1984) Mechanics of fold-and-thrust belts and accretionary wedges; cohesive Coulomb theory, *Journal of Geophysical Research* 89, 10,087–10,101.

England, P. and Molnar, P. (1990) Surface uplift, uplift of rocks, and exhumation of rocks, *Geology* 18, 1,173–1,177.

Hodges, K.V. (2000) Tectonics of the Himalaya and southern Tibet from two perspectives, *Geological Society of America Bulletin* 112, 324–350.

Koons, P.O. (1989) The topographic evolution of collisional mountain belts: a numerical look at the Southern Alps, New Zealand, *American Journal of Science* 289, 1,041–1,069.

Mills, H.H., Brackenridge, G.R., Jacobson, R.B., Newell, W.L., Pavich, M.J. and Pomeroy, J.S. (1987) Appalachian mountains and plateaus, in W.L. Graf (ed.) *Geomorphic Systems of North America: Centennial Special Volume 2*, 5–50, Boulder, CO: Geological Society of America.

Molnar, P. and England, P. (1990) Late Cenozoic uplift of mountain ranges and global climate change: chicken or egg? *Nature* 346, 29–34.

Royden, L.H., Burchfiel, B.C., King, R.W., Wang, E., Chen, Z., Shen, F. and Liu, Y. (1997) Surface deformation and lower crustal flow in eastern Tibet, *Science* 276, 788–790.

Searle, M.P. (1996) Cooling history, erosion, exhumation, and kinematics of the Himalaya–Karakorum–Tibet orogenic belt, in A. Yin and M. Harrison (eds) *The Tectonic Evolution of Asia*, 110–137, Cambridge: Cambridge University Press.

Weiland, R.J. and Cloos, M. (1996) Pliocene-Pleistocene asymmetric unroofing of the Irian fold belt, Irian Jaya, Indonesia: apatite fission-track thermochronology, *Geological Society of America Bulletin* 108, 1,438–1,449.
Willgoose, G. and Hancock, G. (1998) Revisiting the hypsometric curve as an indicator of form and process in transport-limited catchments, *Earth Surface Processes and Landforms* 23, 611–623.
Willett, S.D. (1999) Orogeny and orography: the effects of erosion on the structure of mountain belts, *Journal of Geophysical Research* 104, 28,957–28,982.

JAMES A. SPOTILA

OUTBURST FLOOD

Outburst floods or jökulhlaups, are high magnitude, low frequency catastrophic (see CATASTROPHISM) events involving the sudden release of glacial meltwater (see MELTWATER AND MELTWATER CHANNEL) stored in subglacial (see SUBGLACIAL GEOMORPHOLOGY) reservoirs or ice-dammed lakes. The volume of water discharged is usually orders of magnitude greater than normal flow, with modern outburst floods estimated up to 2,000,000 $m^3 s^{-1}$ and Pleistocene-aged flood peaks estimated at 21,000,000 $m^3 s^{-1}$. However, most discharges usually measure hundreds to thousands $m^3 s^{-1}$. Importantly, outbursts leave characteristic erosional and sedimentary signatures, which enables palaeo-outburst flood reconstructions.

Hydrographs with a slowly rising limb followed by a rapidly falling limb characterize outburst floods through glaciers. The steadily increasing discharge reflects a greater efficiency in subglacial water routing, aided by a positive feedback interaction of channel enlargement caused by melting and abrasion. The increasing discharge may overwhelm the subglacial drainage network, and it is not uncommon for water to burst from SUPRAGLACIAL positions at the ice margin. The rapid decrease in discharge is a reflection of either a drained reservoir, or rapid tunnel closure caused by ice deformation or collapse.

The water source is variable, dependent upon the location of the glacier and the reservoir. In volcanic regions, such as Iceland, high geothermal gradients and subglacial volcanic eruptions lead to the rapid production of meltwater and cyclic outburst floods. For example, Grimsvötn, beneath the Vatnajökull ice cap in Iceland, drains approximately every six years with discharges up to 50,000 $m^3 s^{-1}$. In non-volcanic areas, the collection and storage of water often takes much longer. In these areas, water production is a by-product of lower geothermal gradients, precipitation, insolation (see INSOLATION WEATHERING) and frictional heat from sliding and deforming ice. Water may be stored in supraglacial, englacial, subglacial or ice marginal positions. Supraglacial drainage is dependent upon connections with englacial or subglacial conduits such as crevasses or MOULINs. The largest known possible reservoirs are subglacial. About seventy subglacial lakes have been identified with radio-echo sounding beneath the Antarctic Ice Sheets. The lakes vary in size from a few kms^2 to 14,000 km^2 with between 4,000 and 12,000 km^3 of stored water.

Outburst floods also develop where proglacial (see PROGLACIAL LANDFORM) dams fail. Commonly in mountainous terrain and during glacial recession, proglacial lakes develop behind moraines or in ice-dammed valleys. Dam failure may be initiated by sudden inputs of water or iceberg calving, and is usually the result of rapid fluvial incision initiated by overflow, internal loss of support, or sapping processes, especially with dams composed of sediment or ice. These hydrographs show a rapid increase in discharge with a slowly falling limb.

Outburst floods may deeply cut canyons into bedrock or sediment, and form extensive outwash plains (sandurs) or discrete gravel bars. Deposits may consist of clast-supported, boulder-gravel sequences greater than 10 m thick, which coarsen upward, and are capped with a fining-upward sequence of gravels, sand and silt. However, boulder-gravel deposits described from Pleistocene-aged subglacial outburst floods tend not to show this fining-upward sequence. In backwater areas, rhythmically deposited couplets up to 15 m thick of fine gravel and sand with rip-up clasts and boulders (also called eddy bars), indicate pulsed flow with high sediment concentrations. HYPERCONCENTRATED FLOWs and DEBRIS FLOWs are commonly associated with outburst floods. Giant current ripples, deposited by the Pleistocene-age Lake Missoula floods that scoured out the Channeled SCABLANDs in central Washington, USA, have wavelengths up to 125 m and are up to 7 m high. These BEDFORMs were also instrumental in the development of expansion and pendant bars composed of foreset-bedded gravel. Such bars are also described from the Interior Plains of North America where outburst floods scoured channels across the prairie surface with discharges estimated at 10^5 m^{3}s^{-1}. The spillway geometry consists of an inner channel 25–100 m deep and 1–3 km wide, and an upper-scoured zone as wide as 10 km. On the upper scoured zone, channels have an anastomosing (see ANABRANCHING AND ANASTOMOSING RIVER) pattern with residual streamlined hills that resemble DRUMLINs, and boulder lags are common. Within the inner channel, streamlined hills are rare, gravel bars may be found in landslide alcoves or as point bars, and large fans (expansion bars) are found where the outburst flood entered a basin.

Outburst floods have also been used to explain Pleistocene-age subglacial bedforms. In this hypothesis, large subglacial reservoirs are released as sheet flows scouring the subglacial landscape, leaving erosional remnants such as drumlins, fluting, scabland and hummocky terrain up to 100 km wide. Some drumlins and Rogen moraines are explained as sediment moulds from cavities cut into the ice by the outburst floods. Tunnel channels, that often closely resemble spillway channels, are associated with the subglacial bedforms, and develop as the sheet flow collapsed into discrete channels.

Further reading

Baker, V.R. (1973) Paleohydrology and sedimentology of Lake Missoula flooding in eastern Washington, *Geological Society of America Special Paper* 144.

Clague, J.J. and Evans, S.G. (2000) A review of catastrophic drainage of moraine-dammed lakes in British Columbia, *Quaternary Science Reviews* 19, 1,763–1,783.

Dowdeswell, J.A. and Siegert, M.J. (1999) The dimensions and topographic setting of Antarctic subglacial lakes and implications for large-scale water storage beneath continental ice sheets, *Geological Society of America Bulletin* 111, 254–263.

Fisher, T.G., Clague, J.J. and Teller, J.T. (eds) (2002) The role of outburst floods and glacial meltwater in subglacial and proglacial landform genesis, *Quaternary International* 90, 1–115.

Kehew, A.K. and Lord, M.L. (1987) Glacial-lake outbursts along the mid-continent margins of the Laurentide Ice Sheet, in L. Mayer and D. Nash (eds) *Catastrophic Flooding*, 95–120, Binghamton Symposia in Geomorphology, Boston, MA: Allen and Unwin.

Maizels, J. (1989) Sedimentology, paleoflow dynamics and flood history of jökulhlaup deposits; paleohydrology of Holocene sediment sequences in southern Iceland sandur deposits, *Journal of Sedimentary Petrology* 59, 204–223.

Russell, A.J. and Knudsen, Ó. (1999) Controls on the sedimentology of the November 1996 jökulhlaup deposits, Skeiðarársandur, Iceland, in N.D. Smith and J. Rogers (eds) *Fluvial Sedimentology VI*, 315–324, International Association of Sedimentologists Special Publication, 28.

Shaw, J. (1996) A meltwater model for Laurentide subglacial landscapes, in S.B. McCann and D.C. Ford (eds) *Geomorphology Sans Frontières*, 181–236, New York: Wiley.

SEE ALSO: geomorphologial hazard; glacier; glacifluvial; palaeoflood; palaeohydrology

TIMOTHY G. FISHER

OVERCONSOLIDATED CLAY

Overconsolidated clays are those that have been highly compressed by burial. The burial may be by superincumbent sedimentation or by short-term loading, but glacial ice is the usual agent. Water is expelled from the clay as it assumes a denser packing. If the clay is subsequently unloaded, perhaps by the ice melting, the clay may become fissured and jointed. An example can be given from the glacial sediments of middle England.

Fissure studies in glacial lake clay and tills at Happisburgh and Cromer show well-defined fissure patterns which can be related to the glacial history of the area and the subsequent erosion history. The presence of fissures influences the strength, consolidation, and permeability characteristics of the clay; the strength along a striated fissure plane is almost reduced to its residual value. The coefficients of consolidation and permeability are significantly increased in the presence of fissures. Attention is drawn to the well-developed fissure systems in tills which have commonly been regarded as non-fissured materials.

Such fissured clay presents a lowered strength as a total mass than would be observed by a triaxial test on a small, unfissured sample. Such materials should be analysed in a manner similar to those used in rock mechanics, which consider the state of discontinuities in lowering the bulk strength. Another consideration in overconsolidated clays is that they have generally been sheared past their peak strength by an earlier loading phase. The subsequent performance of the material will thus be dictated by the residual strength, even though laboratory tests might indicate higher peak values.

During the formation of a sedimentary soil the total stress at any given elevation continues to build up as the height of the soil over that point increases.

The removal of soil overburden, perhaps by erosion (perhaps by bulldozer) results in a reduction of stress. A soil element that is at equilibrium under the maximum stress it has ever experienced is normally consolidated, whereas a soil at equilibrium under a stress less than that to which it was once consolidated is overconsolidated. This means that a clay soil whose *in situ* stress is less than the preconsolidation pressure is, regardless of the cause, called overconsolidated. Various geological and landscape factors responsible for causing preconsolidation stress have been recognized. Mechanisms that cause a preconsolidation pressure:

- Changes in total stress due to: (1) removal of overburden, (2) past structures, (3) glaciation.
- Changes in pore-water pressure due to: (1) change in water table elevation, (2) artesian pressures, (3) deep pumping, (4) desiccation due to drying, (5) desiccation due to plants.
- Changes in soil structure due to: (1) secondary compression, (2) environmental changes, such as pH, temperature, salt concentration, (3) chemical alteration due to: weathering, precipitation of cementing agents, ion exchange.
- Changes in strain rate on loading.

Further reading

Bell, F.G. (2000) *Engineering Properties of Soils and Rocks*, Oxford: Blackwell.
Costa, J.E. and Baker, V.R. (1981) *Surficial Geology: Building with the Earth*, New York: Wiley.

IAN SMALLEY

OVERFLOW CHANNEL

Ice-dammed lakes are often found in ice-free valleys tributary to glaciers. Meltwaters from snow and ice may be impounded between advancing glaciers and rock walls, between a main valley glacier and a tributary glacier, or by the advance of a glacier in a tributary valley across a valley not occupied by a glacier. Unless water drains beneath the glacier blocking the valley, meltwater accumulates and water levels rise, until the height of the lowest col is reached. At that stage, water overflows either to adjacent proglacial lakes, or to areas free of ice, creating overflow channels. The large volume of water that escapes ensures that the overflow channel is deepened in a comparatively short period of time.

When compared to normal regional drainage patterns, overflow channels are anomalous in terms of position, morphology and size. Characteristically overflow channels are trough-shaped, with flat floor and steep sides, forming an abrupt angle with higher ground above. For many, the longitudinal profile is undulating. Tributary valleys are rare or absent, but if present they display normal river valley morphology. Shorter overflow channels tend to be straight, or nearly so, while larger overflow channels may display a sinuous planform. Overflow channels are sometimes called spillways, although this term is also used for channels created by catastrophic OUTBURST FLOODs following the failure of an ice dam. Most of these channels are now dry, having been abandoned once the ice withdrew and the lakes they drained disappeared. However, on occasions they may be so deepened that the overflow channels retain drainage even after the ice melted and UNDERFIT STREAMs now flow in these channels. Not all anomalous drainage patterns can be explained by overflow, however. RIVER CAPTURE or subglacial drainage, for example, many provide alternative explanations.

Today, proglacial lakes and overflow channels are found in Norway, the Himalayan region, the Rocky and Andes mountains, Baffin Island, Iceland, and particularly Alaska. At the end of the last glaciation, significant volumes of water were impounded around the margin of continental ice sheets. Water levels in these lakes frequently changed configuration, depth and volume because of the interplay between ice margin position, subglacial topography, isostatic rebound and outlet erosion. As water levels changed, often abruptly, water was released into newly opened overflow channels. Thus, complex and extensive overflow channels resulting from late-Pleistocene glacial lakes are found in northern Europe and North America. Sparks (1960) provides detailed descriptions of overflow channels in northern England. Twidale (1968) describes the overflow channel that runs through central Sweden, from Stockholm, through two lakes Mälaren and Vänern and the Göta River, to Göteborg. Teller *et al.* (2002) describe the overflow channels of the largest late-Pleistocene lake, glacial Lake Agassiz, into the Mississippi and Hudson systems.

The term overflow channel has been extended to refer to abandoned channels in a floodplain that may carry water during periods of high flow, or at a dam site to refer to the spillway that can be opened to release lake water when water levels get high enough to threaten the safety of a dam.

References

Sparks, B.W. (1960) *Geomorphology*, London: Longman.
Teller, J.T., Leverington, D.W. and Mann, J. (2002) Freshwater outbursts to the oceans from glacial Lake Agassiz and their role in climate change during the last glaciation, *Quaternary Science Reviews* 21, 879–887.
Twidale, C.R. (1968) Glacial spillways and proglacial lakes, in R.W. Fairbridge (ed.) *Encyclopedia of Geomorphology*, 460–467, New York: Reinhold.

SEE ALSO: outburst flood

CATHERINE SOUCH

OVERLAND FLOW

Overland flow is the term used for water that flows over the surface of hillslopes. It is important because this route provides the fastest means by which rain falling on hillslopes can reach stream channels. Hence overland flow contributes significantly to the shape of a catchment storm hydrograph. Equally, it may be responsible for high erosion rates on hillslopes. Other terms that are used in this context are sheet wash (see SHEET EROSION, SHEET FLOW, SHEET WASH) and interrill flow, both of which denote unchannelled flow, and RILL flow, which is used to describe overland flow where it becomes concentrated into definable channels on the hillslope. Though interrill flow does not exhibit definable channels, it is common to observe that the flow is not of uniform depth. Instead, the flow converges and diverges around microtopographic obstacles forming anastomosing threads of deeper and faster flow within a layer of water that covers most of the surface. It is generally accepted that the presence of rills indicates that the flow is able to detach and transport sediment, whereas interrill flow, though capable of transporting sediment, does not have the erosive power to detach sediment. Instead, the sediment load of interrill flow is supplied to it by the detaching force of impacting raindrops. Three types of overland flow may be recognized. The first is that which is due to rain falling at an intensity in excess of the rate of infiltration into the soil. This type of overland flow is also termed Hortonian overland flow after R.E. Horton, who first described the process (Horton 1933). The second type of overland flow is termed saturation-excess overland flow, and the third is termed return flow.

Hortonian overland flow

According to Horton's description of the generation of overland flow, rain reaching the soil can be separated into two parts: one infiltrates into the soil, the other remains on the surface. The rate at which rain infiltrates and its relationship to the rainfall intensity is the basis of Horton's model for the generation of overland flow. Horton developed the equation

$$f = f_c + (f_o - f_c)e^{-kt}$$

to describe the way infiltration would change during a storm, in which f is the maximum instantaneous infiltration rate, f_c is minimum infiltration rate (infiltration capacity), assumed to be a constant for a given soil, f_0 is the initial (maximum) infiltration rate (at $t = 0$), k is a constant that varies with soil type, and t is time since the onset of rain.

In general, f_0 will be in excess of all but the highest rainfall intensities, so that initially all rain infiltrates. As the rainstorm proceeds, the pore spaces of the soil fill with infiltrated rainwater, cracks in the soil close and fine particles wash into the surface of the soil so that the instantaneous infiltration rate declines through time. Eventually, it may fall to the point where it is below the rainfall intensity, at which time some of the falling rain remains on the surface. The time taken for this to occur is known as the time to ponding and its completion can be recognized by glistening of the surface and the appearance of ponds of water in small depressions on an irregular ground surface. As these depressions fill with water they begin to overtop and interconnect, until the ground surface is covered by a connected series of these pools. The amount of water that is required for this to occur will vary with the surface irregularity of the hillslope, and is known as the depression storage. Once this stage is reached, flow from pool to pool begins to occur throughout the hillslope and water is discharged from the hillslope as overland flow. As the rainstorm proceeds, the rate of infiltration into the soil continues to decline so that the amount of water remaining on the surface increases. As the volume of water at the ground surface increases, so does the discharge from the hillslope. Once the infiltration capacity f_c of the soil has been approximated (assuming rainfall rate is constant), the discharge of water from the hillslope will also begin to level out towards an equilibrium value. The layer of moving water is known as the surface detention. The thickness of this layer and the time delay between the approximation of the soil's infiltration capacity and the achievement of equilibrium runoff will be a function of the runoff hydraulics (depth and velocity) and the length of the hillsope. Runoff hydraulics are, in turn, primarily controlled by the surface roughness of the ground, which determines the frictional resistance it affords to the flowing water. The relationship between water depth and velocity, on the one hand, and frictional resistance, on the other, can be expressed through the Darcy–Weisbach equation:

$$ff = \frac{8gds}{v^2}$$

in which ff is the dimensionless Darcy–Weisbach friction factor, g is the gravitational constant (m s^{-2}), d is water depth (m), s is slope (m m^{-1}) and v is water velocity (m s^{-1}).

Under the Hortonian model for the generation of overland flow it is assumed that flow will be generated more or less simultaneously over entire hillslopes. This is most likely to be the case where soils have very low initial soil moisture at the start of rainfall and/or the rainfall is intense and/or the soil has a very low infiltration rate. These conditions are most commonly met on bare soils (such as cleared agricultural land), in arid and semi-arid environments, and during convective thunderstorms during which peak rainfall intensities of 300 mm hr^{-1} can be attained and many minutes of rainfall at intensities exceeding 50 mm hr^{-1} are not uncommon. Bare soils may have quite low infiltration capacities because of crusting (either biological or mechanical) (see CRUSTING OF SOIL) of the surface. Kidron and Yair (2001) report infiltration capacities as low as 9 mm hr^{-1} on crusted dune soils in Israel, so that Hortonian overland flow can be generated at even quite low rainfall intensities.

Because Hortonian overland flow depends on the relationship between rainfall rate and infiltration rate discharge increases downslope. The ways in which this increase in discharge downslope affects flow width, depth and velocity (the HYDRAULIC GEOMETRY) is highly variable and depend primarily on the characteristics of the hillslope surface (Parsons *et al.* 1996). Both laminar and turbulent flow conditions are present and the flow may vary from laminar to turbulent both spatially and temporally. Horton (1945) used the term mixed flow to characterize this condition.

The downslope increase in discharge may be accompanied by a change from wholly unchannelled flow on the upper part of the hillslope, to a mix of channelled (rill) and unchannelled (interrill) flow on the lower part. The emergence of eroded channels under the operation of overland flow led Horton to term the upper part of hillslopes without rills the belt of no erosion. It is, however, not correct that no erosion takes place in this zone: simply that detachment in this zone is accomplished by raindrops and is spatially diffuse, as it is everywhere in interrill flow. Soil detachment by falling raindrops is controlled by the kinetic energy of the rainfall, but is diminished as the depth of interrill flow increases because the water dissipates some of the energy of the rainfall. The relationship may be expressed by the equation (Morgan *et al.* 1998)

$$D = \frac{k}{\rho_s} KEe^{-bd}$$

where D is the detachment rate, k is an index that varies with soil type, ρ_s is particle bulk density, KE is rainfall kinetic energy and b is a constant that varies with soil texture.

In rills, detachment is achieved by the shear stress exerted by the flow. Although there have been several attempts to quantify the threshold conditions for rill initiation by overland flow (e.g. Slattery and Bryan 1992), Nearing (1994) has pointed out that the shear stress exerted by shallow flow is of the order of a few pascals, whereas the shear strength of soils is of the order of a few kilopascals.

Saturation-excess overland flow

The acceptance of Horton's model for the generation of overland flow is at odds with the fact that it is very seldom observed in many environments, particularly those in which there is appreciable vegetation cover and/or rainfall is cyclonic, rather than convective. Kirkby and Chorley (1967) argued that rain falling onto an already saturated soil will also remain at the surface and become overland flow. Generation of this type of overland flow depends not so much on the relationship between rainfall intensity and soil infiltrability as on the amount of water that is already in the soil at the onset of rainfall, known as the antecedent moisture content, and the water-storage capacity of the soil. These amounts are both spatially and temporally variable. Antecedent moisture is likely to be highest on footslopes and in concavities, and areas with thin soils (such as spurs) have low total water storage capacity. Both of these areas will preferentially generate saturation-excess overland flow. Antecedent moisture will also depend on rainfall record prior to an individual storm event. Taken together, these two factors mean that, in contrast to Hortonian overland flow, saturation-excess overland flow is likely to be generated on only some parts of hillslopes (the concept of partial area contribution to overland flow – see Betson and Marius 1969), and be variable for two storms of similar characteristics (the concept of variable source area – see Dunne and Black 1970). Because saturation-excess overland flow is generated locally, particularly in areas close to rivers, it is an important control on catchment hydrographs. Conversely, because much is generated on low-angle footslopes, it is of much less importance for soil erosion on hillslopes.

Return flow

Water that infiltrates into the soil and moves downslope through the soil as throughflow or in pipes (see PIPE AND PIPING) may encounter saturated soil, thereby having its further downslope movement through the soil blocked. This water may be forced to the surface, where it is known as return flow, and travel further downslope as overland flow. Like saturation-excess overland flow, return flow is generated locally, often on footslopes, so its significance lies in its impact of catchment hydrographs.

References

Betson, R.P. and Marius, J.B. (1969) Source areas of storm runoff, *Water Resources Research* 5, 574–582.

Dunne, T. and Black, R.D. (1970) Partial area contributions to storm runoff in a small New England watershed, *Water Resources Research* 6, 1,296–1,311.

Horton, R.E. (1933) The role of infiltration in the hydrological cycle, *Transactions of the American Geophysical Union* 14, 446–460.

——(1945) Erosional development of streams and their drainage basins; hydrophysical approach to quantitative morphology, *Geological Society of America Bulletin* 56, 275–370.

Kidron, G.J. and Yair, A. (2001) Runoff-induced sediment yield over dune slopes in the Negev Desert. 1: quantity and variability, *Earth Surface Processes and Landforms* 26, 461–474.

Kirkby, M.J. and Chorley, R.J. (1967) Throughflow, overland flow and erosion, *Bulletin of the International Association for Scientific Hydrology* 12, 5–21.

Morgan, R.P.C., Quinton, J.N., Smith, R.E., Govers, G., Poesen, J.W.A., Auerswald, K., Chisci, G., Torri, D. and Styczen, M.E. (1998) The European soil erosion model (EUROSEM): a dynamic approach for predicting sediment transport from fields and small catchments, *Earth Surface Processes and Landforms* 23, 527–544.

Nearing, M.A. (1994) Detachment of soil by flowing water under turbulent and laminar conditions, *Soil Science Society of America Journal* 58, 1,612–1,614.

Parsons, A.J., Wainwright, J. and Abrahams, A.D. (1996) Runoff and erosion on semi-arid hillslopes, in M.G. Anderson and S.M. Brookes (eds) *Advances in Hillslope Processes*, 1,061–1,078, Chichester: Wiley.

Slattery, M.C. and Bryan, R.B. (1992) Hydraulic conditions for rill incision under simulated rainfall: a laboratory experiment, *Earth Surface Processes and Landforms* 17, 127–146.

Further reading

Emmett, W.W. (1970) The hydraulics of overland flow on hillslopes, *US Geological Survey Professional Paper* 662-A.

Parsons, A.J. and Abrahams, A.D. (eds) (1992) *Overland Flow: Hydraulics and Erosion Mechanics*, London: UCL Press.

SEE ALSO: soil erosion

A.J. PARSONS

OVERWASHING

Overwashing is generally regarded as the process of sediment transport across a BEACH RIDGE or barrier beach (see BARRIER AND BARRIER ISLAND), with deposition as a washover deposit on the back slope of the ridge or in the lagoon landward of the ridge occurring during storms. Morton *et al.* (2000) have, however, reported frequent overwash events occurring during non-storm periods. The term 'overwash' is generally applied to the process, and 'washover' to the resulting despositional landform. Overwashing is a major process by which the back slope of a barrier beach is renewed with the barrier building to landward. Overwash-dominated barriers tend to be relatively narrow and flat, with low and unstable dune systems. Where sediment supply is abundant, overwash barriers may grow and stabilize, but where sediment supply is limited, overwash causes barriers to roll over with sediment being moved from the seaward slope to the landward slope, thereby causing the barrier to migrate landward. Overwashing is an important process on both sand-dominated barriers (where AEOLIAN PROCESSES may also be important) and gravel barrier systems.

Overwash can occur along a significant length of barrier crest producing a washover ramp on the landward side, but more commonly flow is concentrated in channels called overwash throats. In some circumstances and in the long term an overwash throat may lead to the development of a tidal inlet. Washover fans develop on the landward side of the barrier where flow is no longer constrained. Orford and Carter (1984) relate the spacing of overwash throats and washover fans to beach rhythmic morphology (see BEACH; BEACH CUSP; WAVE).

Overwashing (which generally leads to the lowering of a beach barrier) must be differentiated from overtopping which is the process by which barrier systems are built up due to swash transport of sediments to the top of a berm or barrier crest, thereby causing the barrier to build in height and width. Overtopping and overwashing may occur at the same locations, dependent upon wave energy conditions.

Although often considered as a somewhat different process, overwash sedimentation also occurs on CORAL REEF islands, and can be important for the maintenance and development of reef beaches and cays.

References

Morton, R.A., Gonzales, J.L., Lopez, G.I. and Correa, I.D. (2000) Frequent non-storm washover of barrier islands, Pacific coast of Colombia, *Journal of Coastal Research* 16(1), 82–87.

Orford, J.D. and Carter, R.W.G. (1984) Mechanisms to account for the longshore spacing of overwash throats on a course clastic barrier in Southeast Ireland, *Marine Geology* 56, 207–226.

Further reading

Letherman, S.P. (ed.) (1981) *Overwash Processes*, Stroudsburg, PA: Hutchinson Ross.

SEE ALSO: beach ridge; storm surge

KEVIN PARNELL

OXBOW

An abandoned meander (see MEANDERING) loop along an alluvial river. It is the most common type of lake of fluvial origin. As a logical consequence of the lateral shifting and the general downstream migration of meanders, it develops from the interplay of channel erosion and accumulation separated in space. Oxbows are produced by meander cutoffs, which occur in two different ways. If a meander neck becomes narrow enough, streamflow is directed along the shortest route of greatest slope, instead of following the whole perimeter of the meander loop (neck cutoff). Alternatively, a new channel may develop along a swale between POINT BARS (chute cutoff). Natural LEVEE formation and FLOODPLAIN deposition soon build a silt or clay plug between the oxbow and the main channel, although a narrow batture (watercourse) may provide a connection. The American name emphasizes the crescent shape, while in many other languages an oxbow is called a 'dead arm' (e.g. in French: *bras mort*). Both types of cutoff cause channel shortening and scour upstream and deposition downstream. Thus the meandering river maintains its average SINUOSITY since each cutoff triggers the formation of further cutoffs on the long term (self-organization – Stolum 1996). The tranquil freshwater makes an oxbow a valuable aquatic habitat. Meander scars are oxbows completely filled up with mineral and organic matter. They remain discernible in the landscape for a long time.

Classic examples are found along several major rivers of the world, including the Mississippi, the Amazon and the large rivers of Siberia. Floodplains of some minor (regulated) rivers also abound in oxbow lakes.

Reference

Stolum, H.-H. (1996) River meandering as a self-organization process, *Science* 271(22), 1,710–1,713.

DÉNES LÓCZY

OXIDATION

Oxidation is the loss of a negative electron so an element becomes more positively charged, for example ferrous iron, Fe^{2+} (or Iron II) becomes oxidized to ferric iron Fe^{3+} or Iron III. This process commonly occurs in the presence of oxygen:

$$4FeO + O_2 \rightarrow 2Fe_2O_3$$

Iron II   Iron III

Oxidation is thus a common weathering reaction when a mineral formed in an anoxic environment becomes exposed to air at the surface of the Earth. The process is often combined with HYDROLYSIS, for example the weathering of olivine:

$$\underset{\text{Olivine}}{3MgFeSiO_4} + H_2O \rightarrow \underset{\text{Serpentine}}{H_4Mg_3Si_2O_9} + SiO_2 + \underset{\text{Ferrous oxide}}{3FeO}$$

the FeO then becoming oxidized to iron oxide, as above.

During oxidation, the strength of the minerals is reduced which also makes mechanical breakdown much easier.

The loss of a negative electron can equally occur when iron in an acid solution becomes less acid. The latter process accounts for the deposition of reddish iron oxides in the less acid lower parts of soil profiles where there can be, in fact, less oxygen than nearer the surface.

Iron oxides are an important constituent of tropical soils, being produced as a residual mineral through prolonged, intense weathering. The simple iron oxide, haematite (Fe_2O_3) is bright red and leads to the distinctive colour of soils in tropical and subtropical areas. If haematite is subject to HYDRATION then limonite ($2Fe_2O_3 \cdot 3H_2O$) forms which is yellow in colour. The content of iron also appears to be a key factor in the formation and hardening of laterite (Thomas 1974). Many of the theories of laterite formation involve the movement of iron in solution by mobile groundwater, or by upward diffusion, and their subsequent oxidation and mobilization near the surface; the harder laterites having a higher iron content.

Oxidation is not only a weathering process in itself, it can produce further weathering agents. The oxidation of iron sulphides (pyrites) can produce sulphuric acid:

$$\underset{\text{Pyrite}}{2FeS_2} + 15(O) + H_2O \rightarrow \underbrace{2Fe^{3+} + 4SO_4^{2-} + 2H^+}_{\text{Sulphuric acid}}$$

There is a notable example of this at Mam Tor in Derbyshire where pyrite oxidation and the production of sulphuric acid leads to the intense weathering of the illite and kaolinite present in the shale in which the pyrite occurs, leading to a marked rise in porosity and facilitating slope instability, contributing to the collapse of a road (Vear and Curtis 1981). The authors calculate that for 1.5 g of pyrite oxidized by a litre of acid-sulphate water, $0.0125\,gl^{-1}$ H^+ is produced.

While geomorphologists often focus on the oxidation of iron, other compounds can also be oxidized, for example manganese, and a key process in the nitrogen cycle involves the oxidation of ammonium produced by the decomposition of organic matter to nitrate:

$$\underset{\text{Ammonium}}{4NH_4^+} + 6O_2 \rightarrow \underset{\text{Nitrite}}{4NO_2^-} + 8H^+ + 4H_2O$$

and

$$\underset{\text{Nitrite}}{4NO_2^-} + 2O_2 \rightarrow \underset{\text{Nitrate}}{4NO_3^-}$$

References

Thomas, M.F. (1974) *Tropical Geomorphology*, London: Macmillan.

Vear, A. and Curtis, C.D. (1981) Quantitative evaluation of pyrite weathering, *Earth Surface Processes and Landforms* 6, 191–198.

STEVE TRUDGILL

OYSTER REEF

'Among organic reefs, those of the geologically young oysters are now second in size and distribution only to the coralline reefs' (Price 1968: 799). The reefs are built by the modern estuarine *Ostrea* and *Crassostrea* and the marine *Pycnodonte*. The tops of the reefs range from the intertidal zone to depths of as much as 12 m below sea level. Optimum temperatures are from 15–25 °C and optimum salinities occur in the central parts of bays and other estuaries midway between stream mouth and oceanic opening. In addition to forming reefs and visors, they can help to stabilize spits and other constructional features. Oyster reefs are widespread, and major locations include the Gulf of Mexico (*Crassostrea virginica*) and parts of China (*Crassostrea gigas*) (Wang Hong *et al.* 1995), but in some parts of the world they are being destroyed by human activities (including gastronomy and fishing); this has led to attempts to restore them or to provide artificial substances for their colonization (Coen and Luckenbach 2000).

References

Coen, L.D and Luckenbach, M.W. (2000) Developing success criteria and goals for evaluating oyster reef restoration: ecological function or resource exploitation? *Ecological Engineering* 15, 323–343.
Price, W.A. (1968) Oyster reefs, in R.W. Fairbridge (ed.) *Encyclopedia of Geomorphology*, New York: Reinhold.
Wang Hong, Keppens, E., Nielsen, P. and van Riet, A. (1995) Oxygen and carbon isotope study of the Holocene oyster reefs and paleoenvironmental reconstruction on the northwest coast of Bohai Bay, China, *Marine Geology* 124, 289–302.

A.S. GOUDIE

P

PALAEOCHANNEL

When a channel ceases to be part of an active river system it becomes a palaeochannel. Palaeochannels vary greatly in age. Those from the recent geological past (perhaps tens to hundreds of years old) include meander cut-offs and longer reaches abandoned by AVULSION. Although these channels are now isolated from the active river flow, except during periods of floodplain inundation, they are scaled to present flow regime. Many of the more ancient palaeochannels indicate discharges greatly in excess of those occurring at present. The oldest palaeochannels known are found on Mars where huge flows of surface water carved a complex pattern of channels more than 3.5 billion years ago. Clearly, the Martian channels could not have formed in the planet's present waterless environment.

Where palaeochannels are well preserved they provide valuable information about past flow regimes. The basis of discharge reconstruction is given by established statistical relationships between channel forming (bankfull) discharge and aspects of channel morphology including cross-sectional area and meander wavelength as documented by the US Geological Survey in the 1950s and 1960s and amply confirmed since. Meandering rivers are particularly suited to this work because of their potential to preserve planform and, sometimes, cross-sectional geometry. The relationship between meander wavelength and stream discharge (Dury 1965) provides a useful, although often imprecise, approximation of palaeoflow. Rotnicki (1983) argued, on the basis of fieldwork on the Prosna River in Poland, that channel cross sections in meander neck cut-offs provided more reliable estimates because of the excellent preservation of channel dimensions at such locations.

The high degree of channel preservation in neck cut-offs results from their mode of formation. Upon initiation of a short circuit, sedimentation rapidly seals the cut-off ends to form an OXBOW lake. In the low energy environment of the lake, fine-grained sediments from suspension form a drape over the old riverbed. The infill sediment cast effectively preserves the former channel cross section, which can be revealed subsequently in a series of auger holes.

In Rotnicki's (1983) comparative study, estimates of BANKFULL DISCHARGE based on fourteen equations linking meander wavelength and discharge were compared with estimates based on preserved cross-sectional dimensions at cut-offs. For a measured bankfull discharge of $22.5\,m^3\,s^{-1}$ the estimates based on wavelength ranged from 0.2 to $34.1\,m^3\,s^{-1}$, or more than two orders of magnitude. The frequently used equations of Dury and Carlston gave errors of 36 per cent and 78 per cent respectively. Estimates of bankfull discharge based on cut-off cross sections and the Manning Equation reduced the error to 10 per cent.

Many palaeochannels also indicate past regimes of markedly different channel pattern and sediment load. The transition from late glacial conditions to those of the Holocene produced a strong channel response globally. In regions directly affected by ice, many large braided (see BRAIDED RIVER) and bedload-dominated proglacial channels gave way to meandering mixed and suspended load channels. In North America the Mississippi provides an excellent example. At the same time, in regions far removed from the great Quaternary ice sheets parallel changes occurred. For example, on the Riverine Plain in southeastern Australia, low sinuosity, aggraded sand-bed channels (PRIOR

STREAMS) were converted to highly sinuous systems dominated by fine-grained sediments. Here, changing upper catchment conditions including rising temperatures and treelines, and shrinkage of the winter snowpack, produced the channel response. Selected examples of palaeochannels reported in the scientific literature are presented below.

Underfit streams

From the late nineteenth century it was recognized that some sinuous valleys in Europe contain floodplains on which present-day rivers describe smaller wavelengths than the enclosing valley (Davis 1899). Such streams were described as being manifestly (obviously) underfit. The inference being that a reduction in discharge had resulted in a reduction in meander wavelength. George Dury's (1964a,b, 1965) detailed studies in Europe and North America demonstrated the fluvial origin of large meandering valleys and the widespread regional distribution of UNDERFIT STREAMS. Following the elimination of other possible causes, including headwater capture and the loss of glacial meltwater, Dury deduced that regional climatic change had been responsible for the observed reduction in discharge. Radiocarbon dating of valley fills indicated that the last major discharge shrinkage occurred between 10,000 and 12,000 years ago, at the beginning of the Holocene.

Dury estimated the discharges of the large palaeochannels largely on the basis of the statistical relationship between meander wavelength and discharge. Although the computed discharges exceeded those of the present rivers by up to a factor of 60, Dury argued that they could have been produced by glacial climates characterized by reduced evapo-transpiration and a 50–100 per cent increase in precipitation. Many of the reconstructed discharges approach the largest discharges ever recorded on Earth for catchments of equal drainage area and thus were considered by many workers to be excessive, especially in areas of low relief unsuited to the production of extreme flows. In particular, there was concern about the influence of parameters other than discharge on meander wavelength. Various studies have subsequently shown that meander wavelength alone does not provide reliable estimates of bankfull discharge.

Superflood palaeochannels

In the 1920s, J. Harlen Bretz (1923) first described superflood palaeochannels in the Columbia Plateau region of the northwestern United States. Here, space imagery reveals a complex network of large anastomosing (see ANABRANCHING AND ANASTOMOSING RIVER) channels carved into basalt bedrock and overlying LOESS and other sediments. Fluvial features include great waterfalls, potholes, longitudinal grooves carved into the bedrock, and boulders that were transported and deposited in flood bars and giant current ripples. Bretz attributed these features to a cataclysmic flow he called the Spokane Flood.

The eventual verification of Bretz's catastrophic flood hypothesis, despite vehement opposition at the time of its proposal, is considered by Baker (1978) to be one of the most fascinating episodes of modern science. On the basis of flow reconstructions by Shaw *et al.* (1999) it now appears that the giant scabland floods arose from a combination of sources including late Pleistocene ice-dammed Lake Missoula and large subglacial reservoirs that extended over much of British Columbia. An estimated total volume of $10^5\,km^3$ flowed across the Scablands and achieved a peak discharge of some $17 \times 10^6\,m^3\,s^{-1}$. The power per unit area of streambed generated by these flows was up to 30,000 times greater than that produced in the present Amazon.

Similar outburst floods (jökulhlaups) have been documented for spillways marginal to the Laurentide Ice Sheet and in Swedish Lapland. Cataclysmic flows generated by Pleistocene ice-dammed lake failures in the Chuja Valley in the Altas Mountains of south-central Siberia (Baker *et al.* 1993), which exceeded $18 \times 10^6\,m^3\,s^{-1}$, are comparable to the largest of the Channeled Scabland flows. Spectacular palaeochannel landforms in the Chuja Valley include scoured channels, giant bars and gravel wave trains. The impressive hydraulic parameters associated with the Chuja Valley floods include flow depths of 400–500 m, supercritical flow velocities of $45\,ms^{-1}$ and stream powers approaching $10^6\,W\,m^{-2}$. The outburst floods of the Channeled Scabland and Chuja Valley are Earth's greatest known terrestrial discharges of freshwater.

Landform assemblages characteristic of cataclysmic flooding are also present on Mars (Baker *et al.* 1993). The Martian outflow channels, which were first recognized on the basis of Mariner 9 and Viking space mission imagery, are much larger than those of the Channeled Scabland and may have experienced discharges as great as $10^9\,m^3\,s^{-1}$. The Martian palaeochannel systems are therefore not only the largest known, but also the oldest, dating from before 3.5 billion years ago.

Palaeochannels in Australia

Australia's modest stream discharges, compared to those of other continents, result from its predominantly subtropical location and low average relief. Not surprisingly, the presence of large palaeochannel systems in Australia's two inland drainage basins with areas exceeding 1,000,000 km^2, the Murray–Darling and Lake Eyre (Figure 112), has been of particular interest to geomorphologists seeking to reconstruct Late Quaternary hydrological regimes.

In the Murray–Darling Basin, palaeochannels of the Murrumbidgee River have been studied extensively since the late 1940s. Previously described as prior streams, large palaeochannels here form an impressive distributary system in a region now characterized by small meandering rivers. Research before 1970 subdivided the palaeochannels into two genetically different categories: older prior streams and younger ancestral rivers. Channels described as prior streams were aggraded bedload systems characterized by low sinuosity, high width to depth levees and source bordering sand dunes. The sinuous ancestral channels were characterized by floodplains of lateral migration and discharges much larger than the present rivers in this region (Plate 84).

Thermoluminescence (TL) dating by Page *et al.* (1996) resulted in a major revision of the prior/ancestral model. It was shown that four major surface palaeochannel systems (Coleambally, Kerarbury, Gum Creek and Yanco) operated between 100,000 and 12,000 years ago with frequent alternations between prior and ancestral modes of channel behaviour. Stratigraphic investigations showed that bedload aggraded channels (prior streams) typically were bordered by fining-upwards deposits associated with laterally migrating channels (ancestral rivers). Clearly, there was a need to revise the existing model in which prior steams preceded ancestral rivers. Page and Nanson (1996) proposed that the first three phases of Murrumbidgee palaeochannel activity were characterized by alternations from laterally migrating to vertically aggrading channel behaviour with each phase terminating in vertical aggradation and the formation of source-bordering sand dunes (Figure 113). Only the final (Yanco) phase failed to terminate in bedload aggradation, probably because the onset of Holocene climates reduced the size of flood peaks, greatly diminished the supply of bedload from the upper catchments, and resulted in streams evolving into their present highly sinuous suspended load morphology.

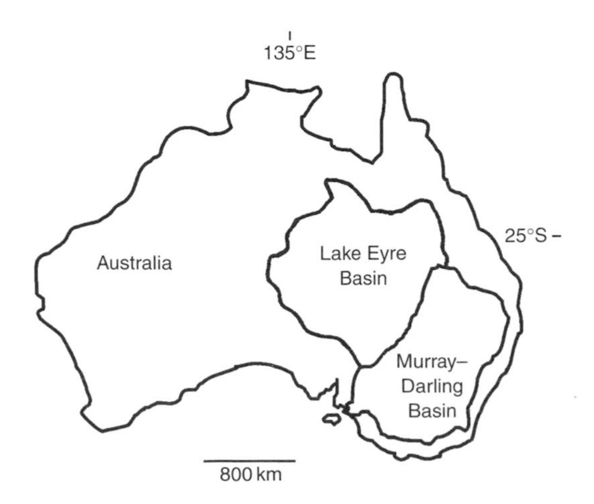

Figure 112 Map of Australia showing locations of Lake Eyre and Murray–Darling Basins

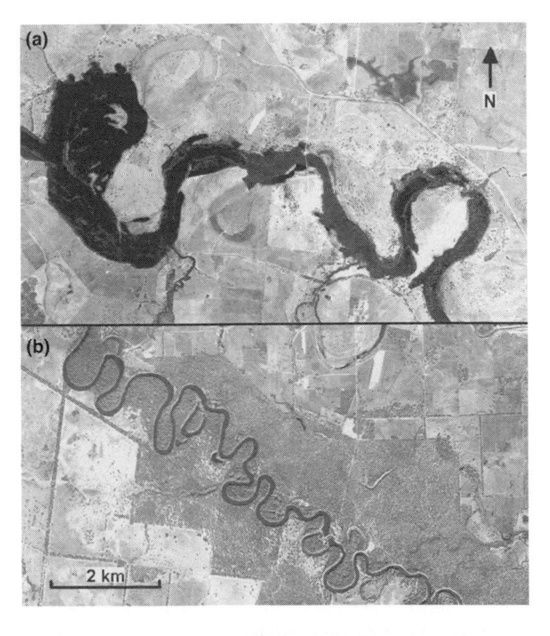

Plate 84 (a) Ancestral Green Gully palaeochannel, and (b) present channel of Murray River in southern Australia at same scale and drainage area

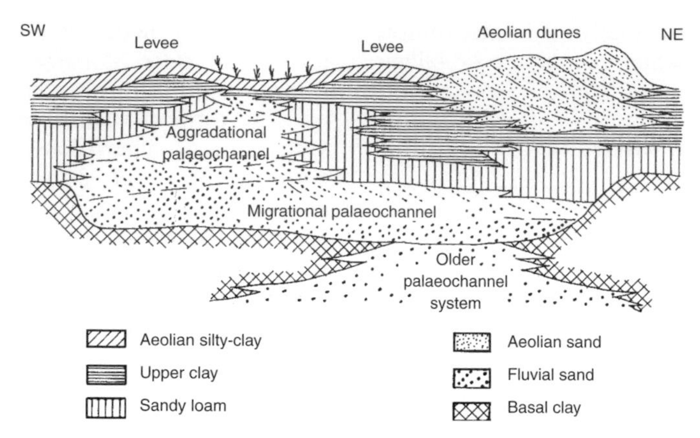

Figure 113 Stratigraphic model of Murrumbidgee River palaeochannels (Page and Nanson 1996: 943)

Discharge reconstructions at preserved channel cross sections of Murrumbidgee palaeochannels suggest that bankfull flows were between four and eight times greater than those of the present rivers.

The Channel Country of the Lake Eyre Basin (Figure 112), which includes Cooper Creek and the Diamantina River, comprises a vast system of low-gradient anastomosing channels dominated by fine-grained suspended load. The anastomosing channels, which date from about 80,000 years ago, are mud-lined, laterally very stable and underlain by extensive muddy floodplains. However, along the middle and lower reaches of Cooper Creek aerial photographs and subsurface exploration have revealed remnant scroll-bars and palaeochannels beneath the mud unit. The scrolls, which are scaled to river meanders larger than any present in the system today, were formed by mixed-load, laterally migrating rivers that deposited extensive sandy units with abundant flow structures (Katipiri Formation). TL dating by Nanson *et al.* (1988) showed that the Katipiri sands date from at least 250,000 years ago.

References

Baker, V.R. (1978) The Spokane Flood Controversy and the Martian Outflow Channels, *Science* 202, 1,249–1,256.

Baker, V.R., Benito, G. and Rudoy, A.N. (1993) Paleohydrology of Late Pleistocene Superflooding, Altay Mountains, Siberia, *Science* 259, 348–350.

Bretz, J.H. (1923) The Channeled Scabland of the Columbia Plateau, *Journal of Geology* 31, 617–649.

Davis, W.M. (1899) The drainage of cuestas, *Proceedings of Geological Society London* 16, 75–93.

Dury, G.H. (1964a) *Principles of Underfit Streams*, United States Geological Survey Professional Paper 452-A, Washington.

Dury, G.H. (1964b) *Subsurface Exploration and Chronology of Underfit Streams*, United States Geological Survey Professional Paper 452-A, Washington.

——(1965) *Theoretical Implications of Underfit Streams*, United States Geological Survey Professional Paper 452-C, Washington.

Nanson, G.C., Young, R.W., Price, D.M. and Rust, B.R. (1988) Stratigraphy, sedimentology and Late-Quaternary chronology of the Channel Country of western Queensland, in R.F. Warner (ed.) *Fluvial Geomorphology of Australia*, 151–175, Sydney: Academic Press.

Page, K.J. and Nanson, G.C. (1996) *Stratigraphic architecture resulting from Late Quaternary evolution of the Riverine Plain, southeastern Australia, Sedimentology* 43, 927–945.

Page, K.J., Nanson, G.C. and Price, D.M. (1996) Chronology of Murrumbidgee River palaeochannels on the Riverine Plain, southeastern Australia, *Journal of Quaternary Science* 11, 311–326.

Rotnicki, K. (1983) Modelling past discharges of meandering rivers, in G.K. Gregory (ed.) *Background to Palaeohydrology*, 321–354, London: Wiley.

Shaw, J., Munro-Stasiuk, M., Sawyer, B., Beaney, C., Lesemann, J-E., Musacchio, A., Rains, B. and Young, R.R. (1999) The Channeled Scabland: back to Bretz? *Geology* 27, 605–608.

KEN PAGE

PALAEOCLIMATE

The climate of the past: palaeo, from the Latin word meaning ancient or old; climate, refers to the interconnected group of Earth systems that control weather conditions (temperature, moisture, wind, etc. and the spatial/temporal variation in these factors) at the surface over an extended period of time. The study of past climates is referred to as palaeoclimatology. More specific definitions of palaeoclimate exist (e.g. climate prior to instrumental records) but in the present case a broad definition of palaeoclimate is taken as that meaning all climates prior to the present day. Further divisions of specific geological time periods may then be defined (e.g. Holocene, Quaternary, Permian) and some of these are discussed below briefly.

Climate has a significant influence on most geomorphic processes and an understanding of landscape systems cannot be achieved without knowledge of both present day climate and climatic history. Antecedence in geomorphic systems is often controlled to a large degree by palaeoclimate and interpretation of geomorphic records (e.g. through sedimentological records) is dependent upon an appreciation of climatic variations.

Climate varies on all temporal and spatial scales and the notion of scale is particularly important both in geomorphology and in climate studies. The nature of climate change at each spatial/temporal scale is determined to some extent by the factors forcing the change. Over long 'geological' timescales, the movement of tectonic plates, and associated volcanism, effect far-reaching but gradual changes in global climate. Over periods of hundreds of thousands of years, the influence on insolation of orbital variations (as described by the 'Milankovitch theory') may have driven the large-scale, high amplitude, changes in global climate of the Quaternary period (see below). Both tectonic and orbital forcing are examples of factors 'external' to the climate system. At shorter timescales (e.g. at the millennial scale), variations are probably due largely to 'internal' factors, such as the differences in response time of components within the climate system, and the subsequent chaotic dynamics of these strongly coupled (non-linear) systems. Internal controls often give rise to abrupt changes in climate as thresholds are reached and feedback systems evolve. Over these shorter timescales, the climate system is highly complex, meaning that cause and effect are often difficult to separate.

Information on past climate conditions can be obtained, in some cases, from historical records (e.g. farming records, instrumental data). However, such records are often sporadic, of questionable accuracy and limited in extent, both spatially and temporally. A second method is to employ the use of climate-dependent natural phenomena that leave traces in the geological record. Once calibrated, these traces can then be used as proxies for past climates. Calibration involves defining the dependence of each proxy on climate, sometimes using modern analogues, sometimes using theoretical calculations (the principal of uniformitarianism then applied). Broadly, there are two kinds of proxy data: (1) episodic/discontinuous records (e.g. flood deposits, glacial advances), which result from the integration of climatic conditions prior to the event; and (2) continuous/incremental records (e.g. constant accumulation of marine mud) which preserve a quasi-continuous record of environmental/climatic conditions. Some examples of proxy records are:

1 *Glaciological*: composition and macro-structure of ice cored from both large ice sheets (e.g. Antarctica, Greenland) and from mountain glaciers – provides information on both regional air temperatures and on global/regional atmospheric composition;
2 *Geological*: ocean sediment cores (geochemistry – particularly oxygen isotope profiles – and species composition of micro fauna), glacial features, sedimentary deposits (e.g. loess, sand dunes), chemistry of speleothems;
3 *Biological*: pollen recovered from terrestrial/marine sediments, remains of insects and micro fauna, composition and structure of tree rings.

The degree to which these proxies are affected by global/regional/local influences depends upon many factors (some of which are context specific) and, to some extent, what we see depends upon how we choose to look at the climate record. Broadly, globally integrated climate signals can be obtained from marine sediment geochemistry, loess/palaeosol deposits and ice sheet cores. However, all proxies are affected by both random fluctuations (noise), non-climate related processes and lags in response times, and all have different sensitivities to climatic conditions. These factors underline the importance of multi-proxy datasets in palaeoclimate research.

As with other geographical matters, it is not easy to discuss palaeoclimates as sets of primary data or even calibrated indices (e.g. temperatures) and further interpretations are required (sometimes quantitative, sometimes qualitative, e.g. 'wetter' or 'warmer'). For conceptual ease, palaeoclimates are commonly discussed in terms of simplifications such as by warmer/colder, drier/wetter or in other even more generalized terms such as reference to glacial/interglacial conditions. Care is needed when using such terms as they often have different meanings in different parts of the world. However, on the regional scale, these terms are indeed useful and are often retained in the palaeoclimate literature.

Climate has varied widely over the history of the Earth. Geological evidence from Precambrian times (approx. 860–550 Ma) points to large-scale glaciations, although the extent of ice cover during this period is debated. Some evidence suggests the Earth was almost completely ice covered, with glaciers reaching to the tropics, while other evidence puts the effected land masses at higher latitudes during these times. Since the Cambrian, the Earth's mean temperature has varied considerably, between the icy conditions of the Permian ice age to the more tropical temperatures of the late Cretaceous. Over the last ~55 Ma, there is strong evidence for a cooling trend at both poles and across the lower latitudes, culminating in the glacial conditions of the last ~2 Ma, the Quaternary period. While the deep geological past sets the wider stage on which contemporary processes operate, it is perhaps the Quaternary period, with its high amplitude, high frequency climatic changes, that is of most significance in understanding the geomorphic setting of most present-day landscapes.

Quaternary palaeoclimate

During the Quaternary period (~2 Ma) the Earth's climate has been through many oscillations in climate, that have operated over a range of temporal and spatial scales. Perhaps the most characteristic feature of the Quaternary has been the oscillation between glacial and interglacial conditions. During glacial conditions, large continental ice sheets grew on Northern Europe/Eurasia and Canada/North America, causing SEA LEVELS to fall by up to 120 m compared to the present day. During the warmer interglacials, ice sheets were restricted to the poles and to Greenland and temperatures were similar to those experienced at present. Insolation forcing due to orbital variation (the Milankovitch theory) at periods of roughly 100, 40 and 20 ka are believed to play a key role in these climate oscillations, although other factors cannot be ruled out. At lower latitudes, the effects of insolation forcing were different, affecting, for example, the intensity and distribution of precipitation (particularly the monsoon systems).

Other factors such as heat distribution due to the ocean current circulation and atmospheric gas composition are likely also to have played key roles. Some lines of evidence (e.g. ice and marine core data) suggest rapid high magnitude changes also occurred during the Quaternary, with mean regional temperatures changing by as much as 10 °C over decades or centuries. Such changes would have significant effects on geomorphic processes and these effects are often recorded in the sedimentological and morphological record.

Further reading

Bradley, R.S. (1996) *Palaeoclimatology: Reconstructing Climates of the Quaternary*, San Diego: Academic Press.

Kutzbach, J. (1976) The nature of climate and climatic variations, *Quaternary Research* 6, 471–480.

Lowe, J.J. and Walker, M.J.C. (1997) *Reconstructing Quaternary Environments*, 2nd edition, London and New York: Addison-Wesley-Longman.

Ruddiman W.F. (2001) *Earth's Climate: Past and Future*, San Francisco: W.H. Freeman.

SEE ALSO: El Niño effects; Holocene geomorphology; ice ages

RICHARD BAILEY

PALAEOFLOOD

Palaeoflood literally means 'ancient flood', but the word does not necessarily connote a specific age, and is often used for any flood not systematically gauged. The characteristics of ungauged floods may be inferred using historical, botanical or geological evidence (Wohl and Enzel 1995). Historical evidence comes from qualitative flood records kept by humans. High-water marks on buildings or canyon walls, diary entries, newspaper reports or damage reports for insurance purposes may all be used to estimate the magnitude and date of occurrence of floods. Such records may extend back 2,000 years in countries such as China.

Botanical evidence of past floods comes from vegetation growing along the riparian corridor (Hupp 1988). Flood-borne debris may impact riverside trees hard enough to destroy a portion of the tree's cambium and leave a corrasion scar that can be dated using annual growth rings in some tree species. Maximum scar heights may be used to estimate minimum peak stage. Flood debris may also bend or break trees near the channel. Many tree species can survive such damage and develop adventitious sprouts, usually within a year of the damage, which can also be dated using tree rings. Anomalies in the width or symmetry of annual growth rings result from changes in water availability, tilting of the tree or stripping of the tree's leaves, all of which may be associated with floods. The age structure of riparian vegetation indicates minimum time since initial deposition or scour of alluvial surfaces. Each of these botanical indicators may provide chronologically precise information on a range of flows for species with annual rings, but the use of these indicators is limited by the presence and age of the vegetation.

Geological evidence of ungauged floods may come from dimensions of relict channels, the size of fluvially transported sediment, or erosional and depositional features that indicate maximum flood stage. Palaeochannels may be preserved as exposed cross sections, abandoned channels on the surface or exhumed channels (Williams 1988). Form parameters including drainage density, terraces, channel pattern, meander wavelength and channel cross-sectional dimensions have been used to infer flow parameters including mean velocity and discharge by means of empirical equations developed for active channels. Many of these form parameters are best preserved along low gradient, unconfined alluvial rivers where continued lateral movement of the channel has left abandoned channels relatively well preserved. Along these rivers, form parameters commonly record low magnitude floods such as average discharge or mean annual discharge. Estimates of flow parameters using channel characteristics may be inaccurate because of deficient regression equations based on limited data; misapplication of available equations caused by extrapolation to different conditions of channel pattern and climate; inadequate preservation of abandoned channels; improper algebraic manipulation of the empirical equations; and uncertainty in the definition of some variables (Williams 1988).

Sediment characteristics may be related to flow parameters by first relating particle size to some index of local transport capability, such as STREAM POWER, and then transforming the transport variable into a discharge estimate using a hydraulic flow equation such as the Manning equation. Gravel and finer sediments may be used in the aggregate to reflect average flow. This approach is commonly used for lower gradient alluvial channels. Coarser sediments are often treated as individual particles, with a focus on the flow competence necessary to transport the largest particles present (O'Connor 1993). This approach is more commonly used for higher gradient confined channels such as bedrock canyons. Use of both finer and coarser sediments to estimate flow parameters relies on empirical relations developed between transported particles and observed, calculated or inferred flow conditions.

Erosional and depositional features may provide palaeostage indicators that record the maximum stage of individual flows. Erosional features include lines scoured into valley-wall soil and colluvium; truncation of landforms such as debris flow fans impinging on the channel; or vegetation limits below which individual species of vegetation are absent (Jarrett and Malde 1987). Depositional features include silt lines of very fine sediment and organics adhering to the channel banks; accumulations of organic debris from fine particles to logs; and slackwater deposits of sediment settling from suspension in areas of flow separation such as tributary mouths or channel-margin alcoves or caves (Kochel and Baker 1982). Palaeostage indicators are best preserved along confined channels with resistant boundaries where an increase in discharge produces a large increase in stage, and changes in channel geometry during and between floods are minimized; and in drier climates where the indicators are less likely to be weathered or obliterated by non-fluvial processes. Flood chronologies may be established from palaeostage indicators using both absolute geochronologic methods such as radiocarbon or thermoluminescence, and relative methods such as stratigraphic position or soil development. Combined with surveyed channel geometry, the stage indicators can be used to estimate flood magnitude (Webb and Jarrett 2002). Palaeostage indicators are commonly used to estimate the largest floods along a channel.

The majority of palaeoflood studies address floods that occurred during the late Pleistocene and Holocene. The late Pleistocene was characterized by immense outburst floods such as those in the Channeled SCABLAND and Siberia produced by the release of meltwater ponded along the margins of the continental ice sheets.

Geological methods used to estimate palaeoflood magnitude on Earth have also been applied to channels on Mars (Baker 1982).

Palaeoflood studies are distinguished from other types of fluvial palaeohydrology in that they usually focus on maximum flows along a channel rather than the entire range of flows. Palaeoflood studies may be a part of studies focusing on channel change as recorded in terraces (see TERRACE, RIVER), ARROYO formation or COMPLEX RESPONSE, or palaeoflood data may be used to examine issues of flood-frequency analysis, flood hydroclimatology and the geomorphic effectiveness of floods.

Flood-frequency analysis is largely based on the measured or extrapolated recurrence interval between discharges of a given magnitude. Measured recurrence intervals are limited by the time span of systematic discharge measurements, which is rarely longer than a hundred years. Extrapolated recurrence intervals may come from extending an existing flood-frequency curve beyond the time span of measurement, or from combining records from neighbouring regions and using the cumulative record length. Both approaches assume that the statistical properties of the hydrologic time series do not change with time, a condition known as stationarity. However, changes through time in the type or frequency of flood-producing storms, or changes in rainfall-runoff generation resulting from land use, are widespread (Hirschboeck 1988). Extending the systematic flood record with palaeoflood information avoids the problem of nonstationarity in the past because palaeoflood indicators record actual rather than hypothetical past floods (Baker *et al.* 2002). Palaeoflood records can also help to constrain the estimate of the probable maximum flood, the largest probable flood that could theoretically occur in a drainage basin. Statistical incorporation of palaeoflood data into systematic data relies on recognition of differences in the two types of data. For example, systematic data may include all floods above a fixed magnitude threshold, whereas the magnitude threshold for palaeoflood data may have varied through time (Blainey *et al.* 2002).

Palaeoflood indicators that record changes in flood frequency through time can also indicate changes in climatic circulation patterns (Redmond *et al.* 2002). And records of the magnitude and frequency of large floods may be used to infer rates of geomorphic change for channels dominated by floods (Wohl 2002) (see FLOOD).

References

Baker, V.R. (1982) *The Channels of Mars*, Austin, TX: University of Texas Press.

Baker, V.R., Webb, R.H. and House, P.K. (2002) The scientific and societal value of paleoflood hydrology, in P.K. House, R.H. Webb, V.R. Baker and D.R. Levish (eds) *Ancient Floods, Modern Hazards*, 1–19, Washington, DC: AGU Press.

Blainey, J.B., Webb, R.H., Moss, M.E. and Baker, V.R. (2002) Bias and information content of paleoflood data in flood-frequency analysis, in P.K. House, R.H. Webb, V.R. Baker and D.R. Levish (eds) *Ancient Floods, Modern Hazards*, 161–174, Washington, DC: AGU Press.

Hirschboeck, K.K. (1988) Flood hydroclimatology, in V.R. Baker, R.C. Kochel and P.C. Patton (eds) *Flood Geomorphology*, 27–49, New York: Wiley.

Hupp, C.R. (1988) Plant ecological aspects of geomorphology and paleoflood history, in V.R. Baker, R.C. Kochel and P.C. Patton (eds) *Flood Geomorphology*, 335–356, New York: Wiley.

Jarrett, R.D. and Malde, H.E. (1987) Paleodischarge of the late Pleistocene Bonneville Flood, Snake River, Idaho, computed from new evidence, *Geological Society of America Bulletin* 99, 127–134.

Kochel, R.C. and Baker, V.R. (1982) Paleoflood hydrology, *Science* 215, 353–361.

O'Connor, J.E. (1993) Hydrology, hydraulics, and geomorphology of the Bonneville Flood, *Geological Society of America Special Paper* 274.

Redmond, K.T., Enzel, Y., House, P.K. and Biondi, F. (2002) Climate variability and flood frequency at decadal to millennial time scales, in P.K. House, R.H. Webb, V.R. Baker and D.R. Levish (eds) *Ancient Floods, Modern Hazards*, 21–45, Washington, DC: AGU Press.

Webb, R.H. and Jarrett, R.D. (2002) One-dimensional estimation techniques for discharges of paleofloods and historical floods, in P.K. House, R.H. Webb, V.R. Baker and D.R. Levish (eds) *Ancient Floods, Modern Hazards*, 111–125, Washington, DC: AGU Press.

Williams, G.P. (1988) Paleofluvial estimates from dimensions of former channels and meanders, in V.R. Baker, R.C. Kochel and P.C. Patton (eds) *Flood Geomorphology*, 321–334, New York: Wiley.

Wohl, E. (2002) Modeled paleoflood hydraulics as a tool for interpreting bedrock channel morphology, in P.K. House, R.H. Webb, V.R. Baker and D.R. Levish (eds) *Ancient Floods, Modern Hazards*, 345–358, Washington, DC: AGU Press.

Wohl, E.E. and Enzel, Y. (1995) Data for palaeohydrology, in K.J. Gregory, L. Starkel and V.R. Baker (eds) *Global Continental Palaeohydrology*, 23–59, Chichester: Wiley.

SEE ALSO: flood

ELLEN E. WOHL

PALAEOHYDROLOGY

Palaeohydrology is the study of past occurrences, distributions and movements of continental waters. It is the highly interdisciplinary linkage of scientific hydrology with the sciences of Earth history and past environments (Schumm 1967). The linkage extends in both directions in that modern hydrological data can be used to create the means of reconstructing past environments (Schumm 1965), while data from past hydrological processes can be used to calibrate and test modern hydrological models (Baker 1998).

The term *palaeohydrology* was first used by Leopold and Miller (1954) in their study of past hydrological conditions associated with a sequence of late Quaternary alluvial terraces in Wyoming. Nevertheless, it is applicable to all elements of the hydrological cycle. Thus, many aspects of cave development in karst aquifers preserve indicators of paleohydrology for those aquifers. Similarly, past changes in lake levels can be documented in terms of a hydrological balance. All these branches of palaeohydrology derive from long traditions in geology and related Earth sciences. For example, Patton (1987) documents the interest by nineteenth and early twentieth-century geologists in past changes in river processes, as evidenced in deposits, terraces and other landforms. Particularly important was the example of Bretz (1923), who discovered the catastrophic flood origin of the Channeled Scabland region in the northwestern United States. Subsequent palaeohydrological quantification (Baker 1973) showed that immense catastrophic flood discharges generated the scabland features during the late Pleistocene bursting of ice-dammed glacial Lake Missoula.

Modes of palaeohydrological inference

There are three general modes of reasoning in palaeohydrology. In one mode general theories of hydrology are used to infer specific effects that can then be discerned in evidence of past hydrological processes. This is the classical deductive mode of rational inquiry. An example would be the problem of the catastrophic flooding associated with the failure of ice-dammed glacial lakes. The palaeohydrologist can use an existing theoretical model for how such a dam fails. Of course, the effective use of this model requires that the correct mode of dam failure be matched with the model (Figure 114). With this condition satisfied, the model may be capable of predicting the hydrograph of the resulting flood. Matching the predicted hydrograph properties to preserved field evidence then constitutes a kind of reconstruction of the past hydrological process.

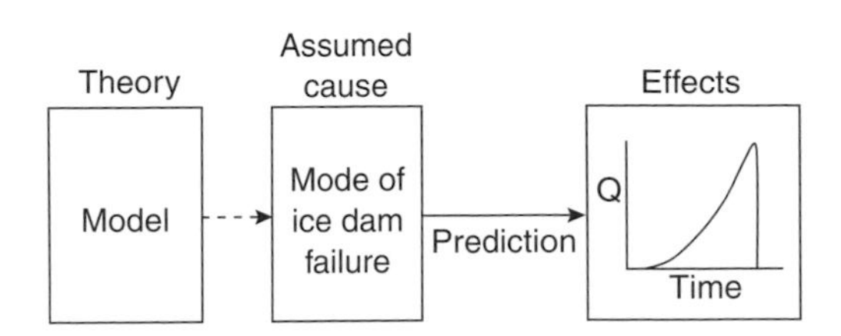

Figure 114 Schematic representation of the deductive mode of palaeohydrological inference applied to the problem of predicting an outburst flood hydrograph from a general theory for the failure of an ice-dammed lake

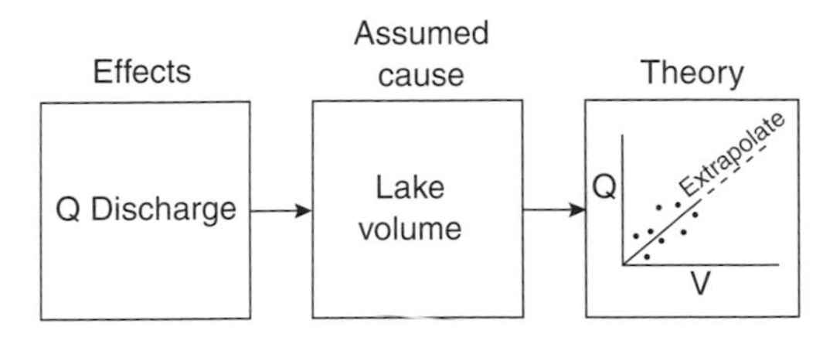

Figure 115 Schematic representation of the inductive mode of palaeohydrological inference applied to the problem of estimating the relationship between lake volume and peak outflow discharge for the failure of ice-dammed glacial lakes

Another common mode of palaeohydrological inference uses empirical relationships that are developed from numerous observations of related hydrological phenomena. This mode of scientific reasoning is inductive. Returning to the problem of ice dam failure, one can collect data on modern glacial lakes. By relating the peak outburst discharges to the associated lake volumes, one can derive an empirical relationship between these two variables (Figure 115). This relationship, extrapolated to the evidence for past lake volumes (or peak discharges) can then be used to estimate the associated discharge (or lake volume). Of course, this exercise must presume that the past phenomena fall in the same class as the data set on modern outburst floods. This is a limitation on all inductive reasoning, because nature is not constrained to behave as we presume it should from our limited set of observation.

Finally, a third mode of reasoning that is used extensively in palaeohydrology is retroductive, or abductive inference (Baker 1996, 1998). For the flood problem, retroductive inference can be accomplished by studying evidence or signs of the

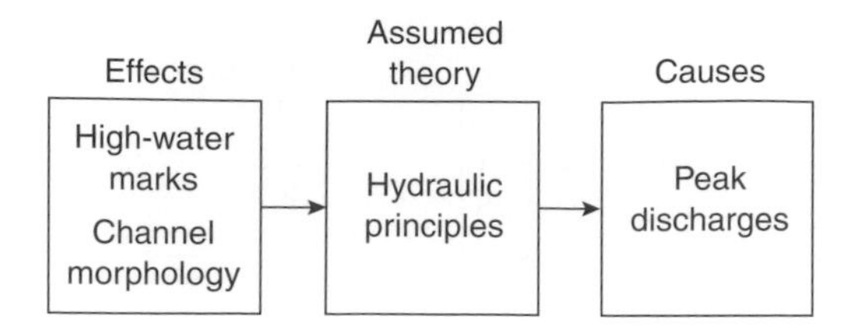

Figure 116 Schematic representation of the retroductive or abductive mode of palaeohydrological inference applied to the problem of estimating peak palaeoflood discharges from various evidence of palaeoflood stages, utilizing hydraulic theory

past floods. These might include the slackwater deposits emplaced marginal to flood channels, or other high-water marks for the past flow stages, as done in *palaeoflood hydrology* (Baker 1987). Then, using a hydraulic model, it is possible to associate the past flood effects with the causative discharges (Figure 116). Thus, retroductive reasoning proceeds from effect to cause, in contrast to the deductive reasoning that proceeds from cause to effect.

Fluvial palaeohydrology

The most basic relationships between river morphology and hydrology involve the supply of water and sediment from upstream of a channel reach of interest. The important dependencies are summarized in the following relationships:

$$Q_W \propto \frac{w, d, \lambda}{S}$$

$$Q_S \propto \frac{w, d, S}{d, P}$$

where Q_W is a measure of the mean annual water discharge and Q_S is a measure of the type of sediment given by the proportion of bedload (usually sand and gravel) to the total sediment load (which may include considerable clay and silt). Q_W and Q_S are the controlling, independent variables. The dependent variables include the channel width w and depth d, the slope or gradient of the channel S, the sinuosity of the reach P, and the meander wavelength λ. For palaeohydrological applications the above relationships are usually quantified by empirical equations, using the inductive approach.

An example of the foregoing reasoning is in regard to the phenomenon of *underfit streams*. These are streams for which some practical measure of the modern river, usually the meander wavelength λ, is too small in relation to the valley that contains the stream. Long recognized as being caused by stream capture, underfitness was also recognized in the context of climatic change by Dury (1954, 1965). Dury (1965) reasoned that, because meander wavelength is directly proportional to bankfull channel width, and because bankfull width is a function of discharge to the 0.5 power (Leopold and Maddock 1953), the wavelength of modern river meanders λ must be proportional to modern bankfull discharges q_b. Applying the same arguments to the enlarged valley meanders of wavelength L formed by ancient discharges Q, Dury (1965) finds

$$Q/q_b = L^2/l^2$$

Dury's study of many rivers in the United States and Europe showed that the ratio L/l varies from 5 to 10, which implies that the ancient discharges Q were 25 to 100 times larger. The immense climatic implications of such large changes led many to question Dury's estimates. In subsequent work it was discovered that the discharges responsible for valley meanders, which are often developed in bedrock, may have very different relationships to channel size than the empirical relationships that apply to modern alluvial rivers.

A significant discovery in fluvial palaeohydrology came when attempts were made to apply Dury's theory to underfit streams on the Riverine Plain of southeastern Australia. The modern Murrumbidgee River is underfit relative to the very large meanders of an ancestral channel. It has a much narrower channel and a much smaller meander wavelength. In addition, there are *prior channels*, which constitute an older system of paleochannels filled with sediment that is much coarser than that conveyed by either the modern Murrumbidgee or by its ancestral stream. Because the prior channels are much wider and have much greater meander wavelengths than the modern river, Dury's theory predicts that they should have experienced much larger bankfull discharges. However, Australian soil scientists insisted that the conditions at the time of prior stream activity were extremely arid. The apparent paradox was resolved when Schumm (1968) showed that the prior channels were formed by relatively high-gradient, low-sinuosity, coarse-sediment-transporting streams. From the proportional expressions above it is clear that the discharge factor relates to parameters other than meander wavelength and channel width, as presumed in Dury's theory. Slope, sediment size, sinuosity and the width-to-depth ratio are all factors, and these combine to produce the result of prior channel development during a drier climatic period.

While much of the foregoing concerned a regime approach to fluvial palaeohydrology, there are other procedures. The sizes of bedload particles moved during past flow events can be related to various measures of the event magnitude, including flow velocity, bed shear stress and power per unit area of bed (Costa 1983). One can also determine the past stages of flow events from a variety of palaeostage indicators, including the study of flood slackwater deposits (Baker 1987). These various techniques have now achieved global application both for the practical study of flood risk assessment, and for the academic study of extreme river processes that defy direct measurement in the field.

Lacustrine palaeohydrology

Ideally, the water balance for a closed basin can be described by the expression

$$dV/dt = d(P_L+R+U)/dt - d(E+O)/dt$$

where V is the water volume in the lake, t is time, P_L is precipitation input to the lake, R is runoff from the tributary basins that feed the lake, U is subsurface (groundwater) flow into the lake, E is evaporation from the lake, and O is the subsurface flow out of the lake. For any given lake stage, the hydrological balance can be considered in equilibrium, so that

$$dV/dt = 0$$

Subsurface inflow and outflow are generally rather small for many lakes, or they may be very difficult to estimate. By ignoring these factors, the equilibrium water balance equation can be simplified in relation to the area of the lake A_L and the area of the tributary catchment A_C from which water drains into the lake, as follows:

$$A_C\, P_L + A_C\, (P_C k) = A_L\, E_L$$

where P_C is the mean precipitation per unit area over the catchment, k is a runoff coefficient such that $P_C k$ will equal the runoff per unit area from the catchment, P_L is the mean precipitation per unit area over the lake, and E_L is the evaporation per unit area from the lake. Usually only A_L and A_C are known for ancient lakes, leaving a problem in estimating the relative influences of evaporation versus precipitation on the overall lake balance, as follows:

$$A_L/A_C = P_C k/E_L - P_L$$

Note that the area of the lake can expand if the evaporation E_L is reduced, if the runoff from the catchment $P_C k$ increases, if the precipitation over the lake P_L increases, or if some combination of these changes occurs. Because evaporation depends on temperature and other climatic factors, its determination may require some independent means of estimating the past climate. Additional complexities occur for precipitation. Thus, the relative simple appearance of expressions for lake palaeohydrology can be misleading in regard to the problem of actually estimating ancient lake balances.

References

Baker, V.R. (1973) Paleohydrology and sedimentology of Lake Missoula flooding in eastern Washington, *Geological Society of America Special Paper* 144, 1–79.

——(1987) Paleoflood hydrology of extraordinary flood events, *Journal of Hydrology* 96, 79–99.

——(1996) Discovering Earth's future in its past, in J. Branson, A.G. Brown and K.J. Gregory (eds) *Global Continental Changes: The Context of Palaeohydrology*, 73–83, London: Geological Society Special Publication 115.

——(1998) Paleohydrology and the hydrological sciences, in G. Benito, V.R. Baker and K.J. Gregory (eds) *Palaeohydrology and Environmental Change*, 1–10, Chichester, Wiley.

Bretz, J. H. (1923) Channeled Scabland of the Columbia Plateau, *Journal of Geology* 31, 617–649.

Costa, J.E. (1983) Paleohydrologic reconstruction of flash-flood peaks from boulder deposits in the Colorado Front Range, *Geological Society of America Bulletin* 94, 986–1,004.

Dury, G.H. (1954) Contribution to a general theory of meandering valleys, *American Journal of Science* 252, 193–224.

——(1965) Theoretical implications of underfit streams, *US Geological Survey Professional Paper* 452-C, 1–43.

Leopold, L.B. and Maddock, T. (1953) The hydraulic geometry of stream channels and some physiographic implications, *US Geological Survey Professional Paper* 252, 1–57.

Leopold, L.B. and Miller, J.P. (1954) Postglacial chronology for alluvial valleys in Wyoming, *US Geological Survey Water-Supply Paper* 1,261, 61–85.

Patton, P.C. (1987) Measuring the rivers of the past: a history of fluvial paleohydrology, in E.R. Landa and

S. Ince (eds) *The History of Hydrology*, 55–67, Washington, DC: American Geophysical Union History of Geophysics, Volume 3.
Schumm, S.A. (1965) Quaternary paleohydrology, in H.E. Wright and D.G. Frey (eds) *The Quaternary of the United States*, 783–794, Princeton: Princeton University Press.
——(1967) Palaeohydrology: application of modern hydrologic data to problems of the ancient past, in *International Hydrology Symposium, Proceedings Volume 1*, 161–180, Fort Collins, CO.
——(1968) River adjustment to altered hydrologic regimen, Murrumbidgee River and paleochannels, Australia, *US Geological Survey Professional Paper* 596, 1–65.

SEE ALSO: cave; palaeochannel; palaeoflood; pluvial lake; prior stream; scabland; underfit stream

VICTOR R. BAKER

PALAEOKARST AND RELICT KARST

Palaeokarst refers to KARST landforms that are completely decoupled from the hydrogeochemical system that formed them, as distinct from *relict karst* that is removed from the morphogenetic situation in which it was formed, but remains exposed to and may be modified by present geomorphic processes (Ford and Williams 1989). The terminology associated with palaeokarst can be complex and ambiguous, but a definitive discussion and explanation is provided by Bosak *et al.* (1989). Figure 117 illustrates the main geomorphic relationships encountered.

Palaeokarst is usually found buried unconformably beneath other rocks, the cover beds being younger than the karst. This is sometimes referred to as *buried karst*. When the burial is relatively recent, it tends to be by unconsolidated allochthonous clastic sediments such as alluvial, volcanic, marine or glacial deposits. Relict karst is still subject to modification by modern solution processes beneath the covering sediments and tends to be only partly buried.

Old and deeply buried palaeokarst arises from tectonic subsidence. It can also involve geological deformation. The caprock constitutes a confining formation, and the palaeokarst is interstratified between it and an underlying non-karst formation. This is a form of interstratal karst, but unlike currently active interstratal karst, the palaeokarst is older than the confining cover

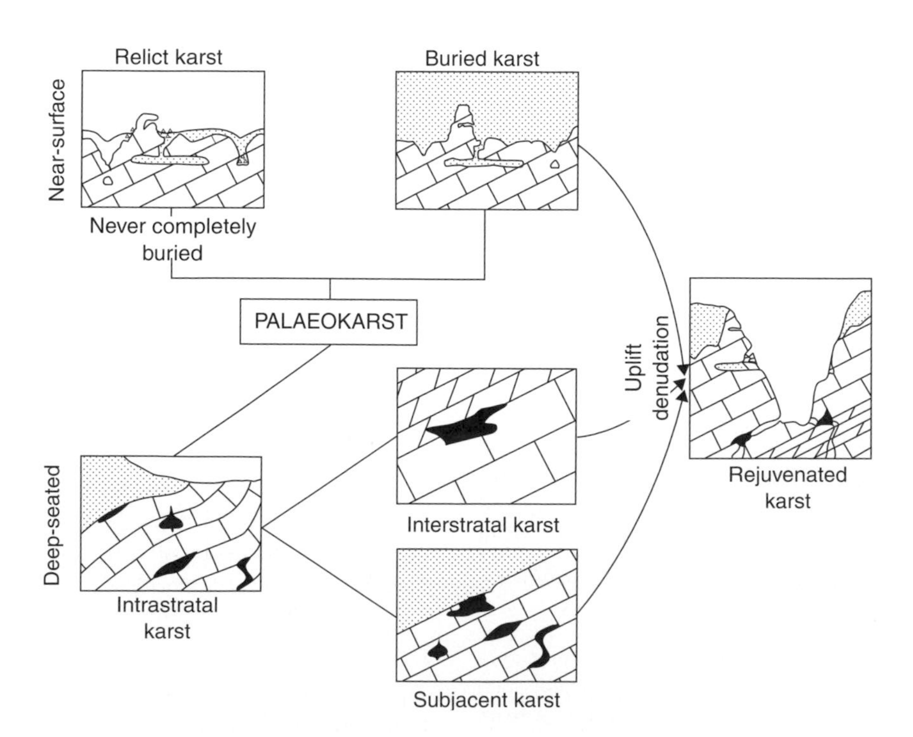

Figure 117 Main geomorphic relationships of palaeokarst

rocks and unconnected to the modern hydrogeological system. There is also an unconformity between the karstified rocks and the caprock. However, sometimes the palaeokarst is quite subtle and without major landforms, and is only recognizable by a disconformity in the carbonate sequence marked by a thin layer of insoluble residue. Such a situation represents a relatively brief interval of subaerial weathering followed by marine transgression.

A distinction is sometimes made between *interstratal* and *intrastratal* karst. The former develops along bedding planes and unconformities, whereas the latter is not restricted to such boundaries between strata. Karst beneath cover beds is sometimes referred to as *subjacent* karst, although if it is currently active it does not constitute palaeokarst. Deeply buried palaeokarsts serve as excellent traps for migrating hydrocarbons and contain some of the world's major oil and gas reserves. Later uplift may result in exposure of the cover beds to erosion and the exhumation of the palaeokarst. When this occurs it can sometimes be reintegrated into the modern hydrogeochemical system and therefore becomes rejuvenated.

Relict karst can arise in two ways: its hydrogeological context may change or its climatic (morphogenetic) situation may alter (Ford and Williams 1989). The first case is commonly found underground as a consequence of the incision of cave streams, because this leads to the de-watering and abandonment of high level cave passages, thus leaving them relict. They are not totally removed from the active hydrogeological system, because they remain in the vadose zone and receive percolation water and accumulate speleothems but, like river terraces in the case of surface rivers, they are removed from the streams that formed them.

The second case results from climate change on a timescale of 10^5 years or more. Climate change associated with major latitudinal shifts of climatic zones has resulted in landforms developed under one morphogenetic system (say humid subtropical) being exposed later to radically different process conditions (perhaps arid, cool temperate or even periglacial). This can arise from global changes to the Earth's climatic system, as experienced in the transition from the Tertiary to the Quaternary, or from continental-scale movement over millions of years arising from plate tectonics, which can result in latitudinal displacement and in wholesale uplift of very large tracts of land (including its karst), such as in Tibet. This leads to the karst being forced out of equilibrium with its process environment. Such landscapes in a different morphogenetic context than the one in which they were developed are sometimes referred to as fossil karst. Shorter term climatic changes, as experienced in glacial–interglacial cycles, can also have profound effects on landscapes, exposing them to polygenetic conditions without necessarily making them relict, but forcing frequent readjustments to new process regimes.

Although most karst is developed by processes associated with the circulation of cool meteoric waters, some is produced by dissolution by hydrothermal waters and some by hot hypogean solutions associated with the intrusion of magma bodies. Deep subjacent karst is formed where heated water is circulated in a confined aquifer. These karsts are often encountered during mineral exploration, because the cavities produced are often heavily mineralized (Bosak 1989) frequently with sulphide ores. When removed from the situation in which they were produced (which was often at depth and many millions of years ago), such karsts constitute *hypogene palaeokarst*.

Palaeokarst is widespread throughout the world and occurs in carbonate rocks to at least Cambrian age. Contributors to the book edited by Bosak (1989) provide the best international review currently available.

References

Bosak, P. (ed.) (1989) *Paleokarst: A Systematic and Regional Review*, Prague: Academia.

Bosak, P., Ford, D.C. and Glazek, J. (1989) Terminology, in P. Bosak (ed.) *Paleokarst: A Systematic and Regional Review*, 25–32, Prague: Academia.

Ford, D.C and Williams, P.W. (1989) *Karst Geomorphology and Hydrology*, London: Unwin Hyman.

PAUL W. WILLIAMS

PALAEOSOL

A palaeosol is a soil that formed on a landscape of the past (Retallack 2001). Soils are products of the physical, chemical and/or biologic weathering of sediments and rocks (see SOIL GEOMORPHOLOGY). Palaeosols typically occur at (1) major unconformities and (2) within basin-fill deposits representing aggradational systems. Although alluvial palaeosols are probably the most common type of palaeosol, they also occur in palustrine, aeolian, deltaic and coastal sediments, as well as in carbonate deposits (Kraus 1999). Palaeosols are especially abundant in Quaternary deposits and have been identified in rocks as old as 3.5 billion years (Retallack 2001).

Palaeosols are identified by a wide range of features including root traces, burrow fills, mottles, nodules, peds, clay films, cemented horizons (i.e. CALCRETE, SILCRETE, FERRICRETE), slickensides and matrix microfabrics. Concentrations of these features are used to identify palaeosol horizons, and individual profiles consist of vertically stacked horizons. Palaeosols should show vertical and lateral variations that mimic those observed in modern soils (see CATENA). One major difficulty in recognizing palaeosols is the effect of burial diagenesis. Diagenetic processes such as compaction, cementation and mineral transformations can significantly alter the texture, mineralogy and chemistry of palaeosols.

Palaeosols provide important records of past environments. Palaeosols at major unconformities are used to interpret past climates and changes in base level, and serve as important lithologic markers for correlating sedimentary deposits. In thick successions of sedimentary rocks, alluvial paleosols record the mode and tempo of basin filling (Kraus 1999). Weakly developed palaeosols are associated with rapid sediment accumulation rates and form close to ancient channel systems. In contrast, well-developed palaeosols reflect slow sediment accumulation rates and settings where sediment input is negligible. Rapid subsidence and sedimentation produce vertically stacked profiles whereas cumulative profiles reflect slow but steady rates of concurrent sedimentation and pedogenesis. Palaeosols also provide opportunities to study landscape development at a variety of spatial scales. At local scales, palaeosol properties vary according to changes in grain size and topography. At more regional scales, palaeosols reflect differences in climate, topography, tectonic setting and lithology.

One of the most promising applications of palaeosol research involves paleoclimate studies. Field-based and stable-isotope studies of iron- and carbonate-rich palaeosols have been used to document increases and decreases in atmospheric oxygen and carbon dioxide concentrations, patterns of global cooling and warming, and ancient mean annual temperature and precipitation (Retallack 2001). Mass balance studies of palaeosols have been used to quantify chemical weathering trends and ancient floodplain hydrology. Finally, palaeosols contribute to the record of ecosystem evolution. The colonization of land by plants and the development of forest and grassland ecosystems are recorded by the development of new palaeosol morphologies such as the mollic horizon (Retallack 2001).

References

Kraus, M.J. (1999) Paleosols in clastic sedimentary rocks: their geologic applications, *Earth Science Reviews* 47, 41–70.

Retallack, G.J. (2001) *Soils of the Past: An Introduction to Paleopedology*, Malden: Blackwell Science.

SEE ALSO: calcrete; catena; chemical weathering; chronosequence; climatic geomorphology; duricrust; ferricrete; silcrete; soil geomorphology

ANDRES ASLAN

PALI RIDGE

A Hawaiian term for a steep slope or large cliff. Palis are steep-faced scarp ridges between stream valleys, commonly composed of basalt, and typically over 1,000 m in height. Various mechanisms have been suggested for their origin, such as being the eroded wall of a dissected shield volcano, being shaped by higher past sea levels, extreme fluvial downcutting, and catastrophic landslides. It is likely that several of these processes are involved in the formation of a pali.

Further reading

Wierzorek, G.F., Wilson, R.C., Jibson, R.W. and Buchanan-Banks, J.M. (1981) Seismic slope instability; a consequence of sensitive volcanic ash? *Earthquake Notes* 52(1), 77.

STEVE WARD

PALSA

Palsas are small mounds of peat rising out of mires in the subarctic region characteristic to the discontinuous circumpolar permafrost zone provided that the peat layer is thick enough. They contain a permanently frozen core of peat and/or silt, small ice crystals and thin layers of segregated ice, which can survive the heat of summers. An insulating peat layer is important for preserving the frozen core during the summer. The peat should be dry during the summer, thus having a very low thermoconductivity, and wet in autumn, when freezing starts, giving a much higher thermoconductivity. This allows the cold to penetrate so deep into the peat layers that they do not thaw during the summer.

Palsas can be classified according to their morphology: dome-shaped, elongated string-form, longitudinal ridge-form, and extensive plateaux palsas as well as palsa complexes with many basins, hollows and ponds of thermokarst origin (Plate 85). The diameter of dome-shaped palsas ranges from 10 to 150 m and the heights from 0.5 up to 12 m. Longitudinal ridge-form palsas could be up to 0.5 km long and 6 m in height. Palsa plateaux rise 1–1.5 m above the surface of the surrounding peat surface and can cover areas of several square km.

Once a palsa hummock rises above the mire surface peat formation on its top ceases almost entirely. The surface peat on an old palsa is produced mainly by *Bryales* mosses, lichens and *Ericales* shrubs. It could also be by wind eroded old moss peat. Below the dry surface peat is the original mire peat formed by *Sphagnum, Carex* and *Eriophorum* remains. It is normally permanently frozen forming the permafrost core. In Finnish Lapland the summer thawing forms only a 50 to 60-cm thick active layer on the palsa surface. On the southern slopes of palsas the active layer gets deeper and on the edges the permafrost table is almost vertical. To date a palsa formation, samples should be collected from the contact of normal mire peat and of the dry peat formed on the palsa after its formation.

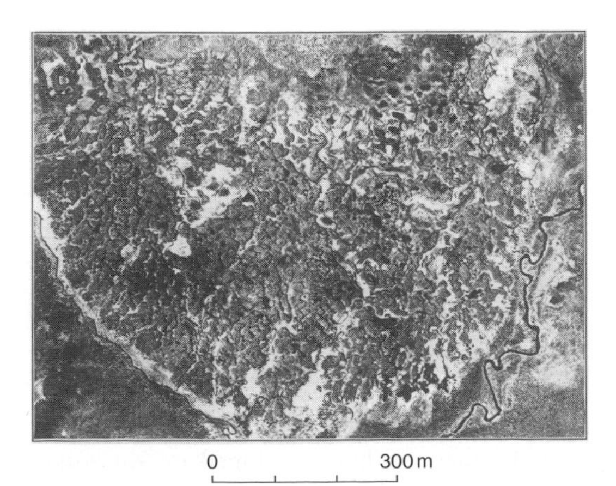

Plate 85 Aerial photograph (nr. 8634 17) of Linkinjeäggi palsa mire, Utsjoki, Finland. Published with the permission of Topografikunta

Low air temperatures together with low precipitation and a thin snow cover are found to be the most prominent factors for palsa formation. The hypothesis that palsas are formed in places with thin snow cover has been proved experimentally by cleaning the snow off the mire surface several times during three winters; a permafrost layer formed in the peat and a man-made small palsa.

Wind drift controls the thickness of snow cover on the mire surface. Thin snow cover allows the frost to penetrate deep into the peat, and in these places the frost fails to disappear completely during the seasonal thawing and part of it remains under the insulating peat. In the following winters the unthawed layer of frost becomes thicker and the mound starts to rise. The wind then carries away snow from the exposed hump more easily and the freezing process accelerates. The freezing front sucks moisture and segregated ice lenses are formed in the frozen core. This process increases the water content of the frozen core which can be 80–90 per cent of the volume.

The concept of cyclic palsa development is based on field observations and experimental studies in Finnish Lapland (Figure 118):

(A, B) The formation of a palsa begins when snow cover is locally so thin that winter frost penetrates sufficiently deeply to prevent summer heat from thawing it completely. The surface of the bog is then raised somewhat by frost processes.

(C) During succeeding winters the frost penetrates still deeper, the process of formation accelerates and the hump shows further upheaving due to freezing of pore water and ice segregation. As the surface rises, the wind becomes ever more effective in drying the surface peat and keeping it clear of snow.

(D, E) When the freezing of the palsa core reaches the till or silt layers at the base of the mire then the mature stage of palsa development begins. By this time the palsa stands well above the surface of the mire, typically displaying a relief of about 7 m in western Finnish Lapland.

(F) Degradation now starts, and peat blocks from the edges of the palsa collapse along open cracks into the pools which often surround the hummocks. During later stages,

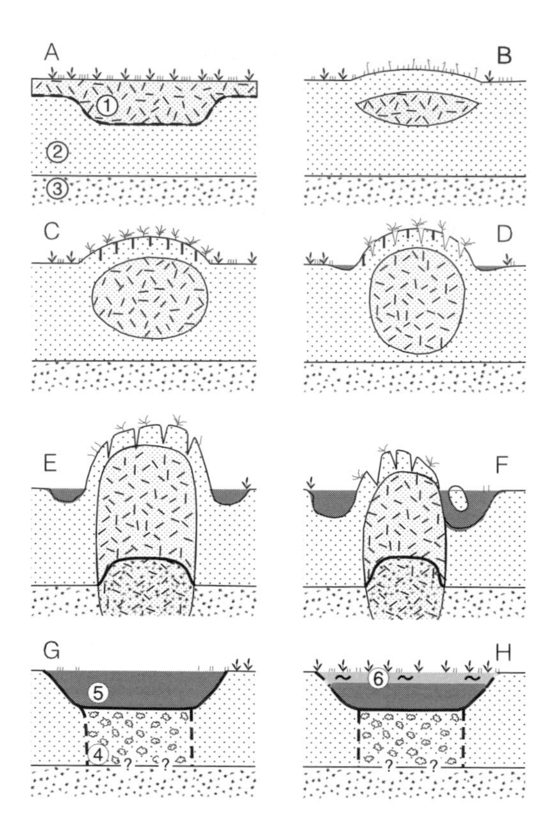

Figure 118 A general model of the formation of the frozen core (1) of a palsa in a mire (2) with a silty till substratum (3). A: the beginning of the thaw season. B: the end of the first thaw season. C: embryo palsa. D: young palsa. E: mature palsa. F: old collapsing palsa surrounded by a large water body. G: fully thawed palsa giving a circular pond on the mire (5). The thawed peat is decomposed (4). H: new peat (6) formation starts in the pond (Seppälä 1982, 1986, 1988)

the vegetation may be removed so that the palsa surface is exposed to deflation and rain erosion.

(G) Old palsas are partially destroyed by thermakarst, and become scarred by pits and collapse forms. Dead palsas are unfrozen remnants: either low (0.5 to 2 m high) circular rim ridges; or rounded open ponds and pond groups; or open peat surfaces without vegetation.

(H) From such pools a new palsa may ultimately emerge after a renewed phase of peat formation, and the cycle of palsa development recommences from the beginning.

References

Seppälä, M. (1982) An experimental study of the formation of palsas, *Proceedings 4th Canadian Permafrost Conference*, 36–42.

——(1986) The origin of palsas, *Geografiska Annaler* 68A, 141–147.

——(1988) Palsas and related forms, in M.J. Clark (ed.) *Advances in Periglacial Geomorphology*, 247–278, Chichester: Wiley.

Further reading

Åhman, R. (1977) Palsar i Nordnorge (Summary: Palsas in northern Norway), *Meddelanden Lunds Universitets Geografiska Institutionen, Avhandlingar* 78, 1–165.

Gurney, S.D. (2001) Aspects of the genesis, geomorphology and terminology of palsas: perennial cryogenic mounds, *Progress in Physical Geography* 25, 249–260.

Lundqvist, J. (1969) Earth and ice mounds: a terminological discussion, in T.L. Péwé (ed.) *The Periglacial Environment. Past and Present*, 203–215, Montreal: McGill-Queen's University Press.

Nelson, F.E., Hinkel, K.M. and Outcalt, S.I. (1992) Palsa-scale frost mounds, in J.C. Dixon and A.D. Abrahams (eds) *Periglacial Geomorphology*, 305–325, Chichester: Wiley.

Seppälä, M. (1994) Snow depth controls palsa growth, *Permafrost and Periglacial Processes* 5, 283–288.

——(1995) How to make a palsa: a field experiment on permafrost formation. *Zeitschrift für Geomorphologie N.F. Supplementband* 99, 91–96.

Zoltai, S.C. (1972) Palsas and peat plateaus in Central Manitoba and Saskatchewan, *Canadian Journal of Forest Research* 2, 291–302.

MATTI SEPPÄLÄ

PAN

Also called *playas, pfannen, sabkhas, chotts, kavirs*, etc. are closed topographic depressions that are features of low-angle surfaces in the world's drylands (Jaeger 1939). Their characteristic morphology has often been likened to a clam, a heart or a pork chop. They are especially well developed on the High Plains of the USA, in the Argentinian Pampas, Manchuria, the West Siberian steppes and Kazakhstan, western and southern Australia and the interior of southern Africa (Goudie and Wells 1995). Pans evolve on susceptible surfaces. In southern Africa, for example, they are best developed on the sandy Kalahari Beds and on fine-grained Ecca shales. They also occur in particular topographic situations – deflated lake floors, old drainage lines, in interdune swales, in the noses of parabolic dunes and on coastal plains (e.g. the Carolina Bays of the eastern seaboard of the USA). They are sometimes, though by no means invariably, associated with LUNETTES (Sabin and Holliday 1995). They are often oriented with respect to regional wind trends, and tend in many cases to have bulbous lee sides. In areas like the Pampas, the High Plains of the USA and the interior of South Africa there are literally tens of thousands of pans, and they may cover as much as a quarter of the ground surface.

The origin of pans has intrigued geomorphologists for over a century. Hypotheses have included deflation, excavation by animals, and karstic (see DAYAS) and pseudo-karstic solution. Arguments on this issue are recurrent and recent years have seen some important contributions to the debate (e.g. Gustavson *et al.* 1995). What is becoming clear is that a range of processes has been involved in the initiation and maintenance of pans and that no one hypothesis can explain all facets of their own long histories and their variable sizes and morphologies.

An integrated model of pan development is as follows (Goudie 1999). First, pans occur preferentially in areas of relatively low effective precipitation. This predisposing condition of low precipitation means that vegetation cover is sparse and that deflational activity can occur. Moreover, once a small initial depression has formed, and the water in it has evaporated to give a saline environment, the growth of vegetation is further retarded. This further encourages deflation. The role of deflation in the removal of material from a depression may be augmented by animals, who tend to concentrate at pans because of the availability of water, salt licks and a lack of cover for predators. Trampling and overgrazing expose the soil to deflation and the animals would also physically remove material on their skins and in their bladders. Aridity also promotes salt accumulation so that salt weathering could attack the bedrock in which the pan might be located. It is also important that any initial depression, once formed and by whatever means, should not be obliterated by the action of integrated or effective fluvial systems. Among the factors that can cause a lack of fluvial integration are low angle slopes, episodic desiccation and dune encroachment, the presence of dolerite intrusions and tectonic disturbance. This model of pan formation is similar to that developed for the USA High Plains by Gustavson *et al.* (1995).

In addition to their occurrence in deserts, various types of oriented lake are also a feature of some tundra areas (Carson and Hussey 1962).

References

Carson, C.E. and Hussey, K.M. (1962) The orientated lakes of Arctic Alaska, *Journal of Geology* 70, 419–439.

Goudie, A.S. (1999) Wind erosional landscapes: yardangs and pans, in A.S. Goudie, I. Livingstone and S. Stokes (eds) *Aeolian Environments, Sediments and Landforms*, 167–180, Chichester: Wiley.

Goudie, A.S. and Wells, G.L. (1995) The nature, distribution and formation of pans in arid zones, *Earth-Science Reviews* 38, 1–69.

Gustavson, T.C., Holliday, V.T and Hovorka, S.D. (1995) Origin and development of playa basins, sources of recharge to the Ogallala Aquifer, Southern High Plains, Texas and New Mexico, *Bureau of Economic Geology, University of Texas, Report of Investigations*, No. 229.

Jaeger, F. (1939) Die Trockenseen der Erde, *Petermanns Mitteilungen Ergänzungshelt* 236, 1–159.

Sabin, T.J. and Holliday, V.T. (1995) Playas and lunettes on the Southern High Plains: morphometric and spatial relationships, *Annals of the Association of American Geographers* 85, 286–305.

A.S. GOUDIE

PARAGLACIAL

Paraglacial is a term that was introduced by June Ryder (1971) to describe alluvial fans in the interior of British Columbia that had accumulated through the reworking of glacial sediment by rivers and debris flows following late Wisconsinan deglaciation. She mapped alluvial fan distribution throughout south-central British Columbia, noted that they were essentially inactive at present and concluded that they must have been dependent on the reworking of till, glacifluvial and glacilacustrine deposits by streams and debris flows in the earliest Holocene. She showed that fan accumulation was initiated soon after valley floors became ice free and continued until after the deposition of Mazama tephra (*c.*6,000 yrs BP).

The paraglacial concept was formalized by Church and Ryder (1972). They defined paraglacial as non-glacial processes that are directly conditioned by glaciation and added that 'it refers both to proglacial processes and to those occurring around and within the margins of a former glacier that are the direct result of the former presence of ice'. In this remarkable paper, they synthesized evidence from their field areas in

British Columbia (Ryder) and Baffin Island (Church) and they used the contemporary Baffin Island environment as an analogue for the early Holocene environment in south-central British Columbia. They concluded that, although fluvial sediment transport rates were likely to be greatest immediately after deglaciation, fluvial reworking of glacigenic sediments was likely to continue as long as such sediment was accessible to rivers. They identified three aspects of the influence of paraglacial sediment supply on fluvial transport: (a) the dominant component of reworked sediment may shift from till to secondary sources, such as alluvial fans and valley fills; (b) regional uplift will condition the timing of changes in the balance between fluvial deposition and erosion such that the cascade of sediment evacuation can be interrupted by sediment deposition; and (c) consequently, the total period of paraglacial effect is prolonged beyond the period of initial reworking of glacigenic sediments.

Slaymaker (1977) and Slaymaker and McPherson (1977) noted that in British Columbia primary upland denudation rates are low and that a large component of contemporary sediment load is derived from secondary remobilization of late Pleistocene and Holocene deposits. Slaymaker (1987) also showed that in the British Columbia and Yukon region, medium-scale (100–10,000 km^2) river systems exhibit the highest specific sediment yield, in contradiction to the conventional model of sediment yield vs basin area relations. Church *et al.* (1999) confirmed this result.

Church and Slaymaker (1989) emphasized the generality of the definition, specifying that it is applicable to all periods of glacier retreat and that a paraglacial period is not restricted to the closing phases of glaciation but may extend well into the ensuing non-glacial interval. The essence of the concept is that recently deglaciated terrain is often initially in an unstable or metastable state and thus vulnerable to rapid modification by subaerial agents. Effectively then the 'paraglacial period' is the period of readjustment from a glacial to a non-glacial condition. Different elements of paraglacial systems adjust at different rates from steep, sediment-mantled hillslopes (a few centuries) to large fluvial systems (>10,000 years). Increase in specific sediment yield with basin area for basins smaller than *c.*30,000 km^2 shows that specific sediment yield equals area raised to the power of 0.6. Isometry would dictate an exponent of 0.5 (because specific sediment yield can be reduced to a length dimension); hence sediment is recruited to streams at a rate greater than expected simply from an increase in area. The additional sediment is derived from erosion of both *in situ* and reworked glacigenic deposits along riverbanks. Effectively, these rivers are degrading through valley fill deposits forming entrenched trunk streams flanked by Holocene terraces. For basins greater than 30,000 km^2 specific sediment yield tends to decline as conventional models predict because non-alluvial riverbanks are protected from erosion.

These data demonstrate that the timescale of paraglacial sediment reworking in British Columbia includes the whole of Holocene time. Church and Slaymaker (1989) estimate that ultimate dissipation could take several tens of thousands of years. This implies that interglacial fluvial systems were still relaxing from the previous glacial period when the succeeding glacial period arrived. They also imply that there is no equilibrium between hydro-climate, denudation rate and sediment yield because all 'fluvial sediment yields at all scales above *c.*1 km^2 remain a consequence of the glacial events of the Quaternary'.

Owens and Slaymaker (1992) have examined sediment accumulation rates in three small lake-drained basins of less than 1 km^2 over the last 6,000 years and confirmed that these rates are 1–2 orders of magnitude lower than those of larger basins. Souch (1994) has traced the paraglacial signal downstream through a system of lakes progressively further from the glacial sources. Church *et al.* (1999) have expanded the analysis of suspended sediment yields across Canada and described seven Canadian regions with adequately monitored sediment data. Five of these regions were shown to have trends comparable with those of British Columbia; one, southern Ontario, is influenced by intensive land use disturbance and the data show no trend; one, the eastern Prairies, is a region of net fluvial aggradation and specific sediment yields decrease with basin area in accordance with conventional models. Evidently, paraglacial effects persist throughout the majority of Canada's regions.

Ballantyne (2002), in a magisterial summary and extension of the paraglacial concepts developed in British Columbia, points out that between 1971 and 1985, the paraglacial concept was largely ignored outside North America. Since 1985, he sees four trends: (a) an extension in the geomorphic contexts in which the paraglacial concept has been explicitly used; (b) a focusing of research on present-day paraglacial processes and land systems; (c) use of the paraglacial concept as a framework for research across a wide range of contrasting deglacial environments; and (d) a growing awareness of the palaeo-environmental significance of paraglacial facies in Quaternary stratigraphic facies. The working definition that he adopts for 'paraglacial' is 'non-glacial Earth-surface processes, sediment accumulations, landforms, land systems and landscapes that are directly conditioned by glaciation and deglaciation'.

The new perspective given by Ballantyne (2002) is most remarkable in its overview of the wide range of geomorphic contexts in which the paraglacial concept is already explicitly being used. These contexts are, in addition to the original debris cone, alluvial fan and valley fill deposits: (a) rock slopes; (b) sediment-mantled slopes; (c) glacier forefields; (d) glacilacustrine systems; and (e) coastal systems.

Wyrwoll (1977) was the first to identify rock slope response in a paraglacial context. Ice downwasting and retreat has resulted in the debuttressing of rockslopes and yields three responses: large-scale catastrophic rock slope failure; large-scale progressive rock mass deformation and discrete rock fall events.

The work of Ballantyne and Benn (1994) is significant in identifying sediment-mantled slopes in a paraglacial context. They note the processes of reworking sediment-mantled slopes yielding intersecting gullies, coalescing slope foot debris cones and valley floor deposits of reworked drift. Gully erosion and debris flow activity are the most obvious paraglacial processes invoked in this environment.

Matthews (1992) is credited with the first explicit identification of glacier forefields (forelands) in a paraglacial context. Effects conditioned by the former presence of a glacier include unconsolidated diamicton, steep slopes, unvegetated surfaces and the acceleration of mass movement, frost action, fluvial and aeolian processes.

Leonard (1985) was one of the early investigators of the paraglacial response of lake sediments. Such work accelerated in the 1990s and is now one of the most commonly used ways of assessing the changing rates of sediment production during the Holocene, specifically estimating the duration of the paraglacial effect in specific lake-drained basins.

The extension of the paraglacial concept to coastal systems is perhaps the most dramatic extension of the concept. Forbes and Syvitski (1994) defined paraglacial coasts as 'those on or adjacent to formerly glaciated terrain, where glacially excavated landforms or glacigenic sediments have a recognizable influence on the character and evolution of the coast and nearshore deposits'. They specifically exclude the effects of glacio-isostatic rebound and glacio-eustatic sea-level change on the grounds that these effects are more widely or even globally distributed.

It is clear from Ballantyne's discussion that the paraglacial concept has even wider significance than had previously been imagined. The data bring into question the possibility of any equilibrium or balanced condition in landscapes that have undergone Quaternary glaciation.

References

Ballantyne, C. (2002) Paraglacial geomorphology, *Quaternary Science Review* 21, 1,935–2,017.

Ballantyne, C.K. and Benn, D.I. (1994) Paraglacial slope adjustment and resedimentation following glacier retreat Fabergstolsdalen, Norway, *Arctic and Alpine Research* 26, 255–269.

Church, M. and Ryder, J.M. (1972) Paraglacial sedimentation: a consideration of fluvial processes conditioned by glaciation, *Geological Society of America Bulletin* 83, 3,059–3,071.

Church, M. and Slaymaker, O. (1989) Disequilibrium of Holocene sediment yield in glaciated British Columbia, *Nature* 337, 452–454.

Church, M., Ham, D., Hassan, M. and Slaymaker, O. (1999) Fluvial sediment yield in Canada: a scaled analysis, *Canadian Journal of Earth Sciences* 36, 1,267–1,280.

Forbes, D.I. and Syvitski, J.P.M. (1994) Paraglacial coasts, in R.W.G. Carter and C.D. Woodroffe (eds) *Coastal Evolution: Late Quaternary Shoreline Morphodynamics*, 373–424, Cambridge: Cambridge University Press.

Leonard, E.M. (1985) Glaciological and climatic controls on lake sedimentation, Canadian Rocky Mountains, *Zeitschrift für Gletscherkunde und Glazialgeologie* 21, 35–42.

Matthews, J.A. (1992) *The Ecology of Recently Deglaciated Terrain: A Geo-ecological Approach to Glacier Forelands and Primary Succession*, Cambridge: Cambridge University Press.

Owens, P. and Slaymaker, O. (1992) Late Holocene sediment yields in British Columbia, *International Association of Hydrological Sciences* 209, 147–154.

Ryder, J.M. (1971) The stratigraphy and morphology of paraglacial alluvial fans in south central British Columbia, *Canadian Journal of Earth Sciences* 8, 279–298.

Slaymaker, O. (1977) Estimation of sediment yield in temperate alpine environments, *International Association of Hydrological Sciences* 122, 109–117.

——(1987) Sediment and solute yields in British Columbia and Yukon: their geomorphic significance

re-examined, in V. Gardiner (ed.) *Geomorphology '86*, 925–945, Chichester: Wiley.
Slaymaker, O. and McPherson, H.J. (1977) An overview of geomorphic processes in the Canadian Cordillera, *Zeitschrift für Geomorphologie* 21, 169–186.
Souch, C. (1994) A methodology to interpret down valley sediments in records of Neoglacial activity, Coast Mountains, *Geografiska Annaler* 76A, 169–185.
Wyrwoll, K.-H. (1977) Causes of rock slope failure in cold area, Labrador Ungava, *Geological Society of America Reviews of Engineering Geology* 3, 59–67.

SEE ALSO: alluvial fan; glacifluvial; glacilacustrine

OLAV SLAYMAKER

PARALIC

Term referring to environments by the sea where shallow waters predominate, though nonmarine. Paralic environments are particularly associated with intertongued marine and continental deposits situated on the landward side of a coast. This includes lagoonal, littoral, alluvial and shallow neritic environments. Paralic sedimentation incorporates basins, swamps (paralic swamps), deltaic zones, heavily alluviated shelves and platform marshes. The word paralic is derived from the Greek word *paralia* meaning sea coast.

Paralic environments typically exhibit localized, abruptly changing facies tracts with a large variety of lithologies. The deposits are distinct by their thick terrigenous accumulations of clays, sands and silts (orthoquartzite to subgreywacke), intimately mixed with estuarine, marine and continental deposits. Paralic sediments can offer important stratigraphical information concerning the long-term changing coastal environment. Often the deposits are zones of subsequent coal formation (termed paralic coal), while petroleum accumulation is frequent within paralic basins. Paralic ecosystems are characterized by a large spectrum of biological species that are strictly bound to the particular environment, and are able to remain stable despite changing environmental conditions (Guelorget and Perthuisot 1992).

Reference

Guelorget, O. and Perthuisot, J.-P. (1992) Paralic ecosystems: biological organisation and functioning, *Vie et Milieu* 42, 215–251.

STEVE WARD

PARNA

A deposit of dust (suspended wind-blown mineral material) differentiated from LOESS by its higher clay content. The term was coined for deposits found in the interior of southeastern Australia and is attributed to an aboriginal word meaning 'sandy and dusty ground' (Butler 1956: 147). Parna can be regarded as synonymous with desert loess. High clay content (30–70 per cent) lead to differentiation of parna from the glacial loess of Europe, but is a feature of desert loess in Africa and elsewhere. High clay content, and particularly the inferred or observed presence of clay in the form of detrital aggregates, remains the main criterion for recognition of parna although the quartz fine sand or silt component is the most readily recognizable feature. These, and other, properties are inferred to arise from the origin of parna from the deflation of soils which have already experienced considerable weathering. Thus other deposits of clay-rich aeolian sediment, derived from deflation of lakes (see LUNETTE), for example, are not considered to be parna, although there is inconsistency in the application of the term. Other properties of parna, such as colour, calcium carbonate, salts, texture and structure vary with soil drainage in a catenary relationship. The depth and number of parna layers are variable and relate strongly to local topography and post-depositional erosion as well as proximity to the source areas. Parna layers were deposited during arid climate phases of the late Quaternary.

Reference

Butler, B.E. (1956) Parna: An aeolian clay, *Australian Journal of Science* 18, 145–151.

Further reading

Dare-Edwards, A.J. (1984) Aeolian clay deposits of south-eastern Australia: parna or loessic clay?, *Transactions of the Institute of British Geographers* N.S. 9, 337–344.
Hesse, P.P. and McTainsh, G.H. (2003) Australian dust deposits: modern processes and the Quaternary record, *Quaternary Science Reviews*, 22, 2007–2035.

PAUL HESSE

PASSIVE MARGIN

In plate tectonic theory, oceans are spreading from mid-ocean ridges, and being consumed by subduction at active margins. Passive continental margins are those that are not also edges of plates. They are also known as 'trailing edges' and 'Atlantic-type margins'.

They are presumed to be initiated as rift valleys, and when the rifts turn into oceans by seafloor spreading they become continental margins. The new margins may undergo some changes, but may also inherit landforms from pre-breakup times. In contrast to active margins which have many volcanoes, volcanicity is rare in passive margins: only east Australia has abundant volcanoes. In India the vast flows of the Deccan traps accompanied creation of the passive margin.

Based on morphotectonics there are two main types of passive margin: (1) passive margins without significant vertical deformation; and (2) passive margins with a marginal swell and Great Escarpment. We have no good explanation for the two types, or their distribution. Why does eastern Australia have a marginal swell-type margin, but most of the south coast is without vertical deformation? Why does most of southern Africa have Great Escarpments, while East Africa does not?

Passive margins without significant vertical deformation are formed by simple pull-apart of a continent. The Red Sea is an example of the early stage of the process. The Great Australian Bight exemplifies a later stage. Horizontal Tertiary limestones underlie the flat Nullarbor Plain, which is almost an old seafloor. In Patagonia (Argentina) the Atlantic is bordered by an extensive plain cut across ancient rocks. The offshore zone is characterized by many listric faults.

Margins with a marginal swell are the dominant type of passive margin (Ollier 2003), and include the Drakensberg, the Western Ghats, the Appalachians, parts of Greenland, Brazil, Antarctica, and elsewhere. The basic geomorphology of such margins is shown in Figure 119.

Plateau are upland areas with relatively flat topography and most are erosion surfaces. They may be extensive or dissected until only fragments are left. They occur on a wide range of rock types including horizontal strata, metamorphic rocks, granite and massive lava flow sequences.

The *marginal swell* is a widespread swell or bulge along the edges of a continent (*Randschwellen* in German; *bourrelets marginaux* in French). The whole land surface has been warped into an asymmetrical bulge, with the steeper slope

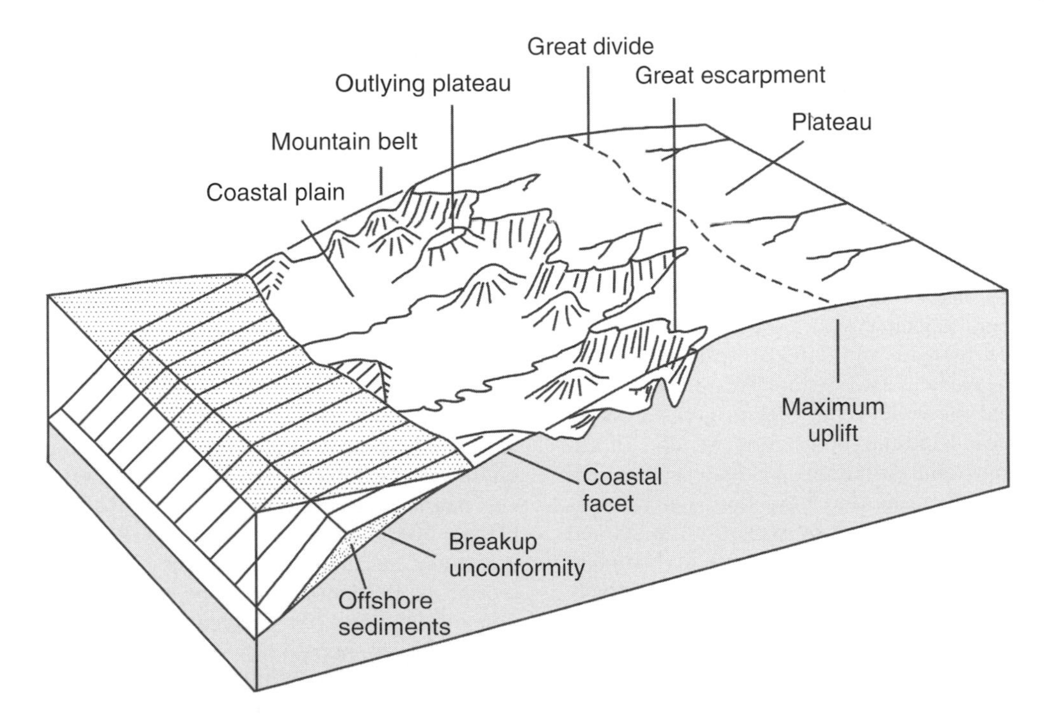

Figure 119 Morphotectonic features of a passive continental margin with marginal swell

to the coast (though the 'steep' slope is still only a few degrees). The marginal swell formed after the planation surface of the plateau, and after formation of major valleys.

Great Escarpments are scarps hundreds or thousands of kilometres long, and up to a thousand metres high. They occur on all sorts of rocks and are not structurally controlled. This is demonstrated especially in the Western Ghats of India, where the Great Escarpment in the north is cut across horizontal basalt, and continues with no change of form into the Precambrian gneisses and granites of southern India. Great Escarpments run roughly parallel to the coast, and they separate a high plateau from a coastal plain. The top of the Great Escarpment can be very abrupt. They are undoubtedly erosional. In places they are rather straight, but elsewhere can be highly convoluted. In some instances the top of the escarpment is the drainage divide between coastal and inland drainage (Brazil, Namibia); in other places the major drainage divide of the continental margin may be hundreds of kilometres inland of the Great Escarpment (east Australia). Many large waterfalls are found where rivers cross the Great Escarpment.

Mountainous areas, often quite rugged, form below the Great Escarpment, where the old plateau surface has been largely destroyed. Occasionally a patch of plateau is isolated to form a peninsula or isolated tableland.

Coastal plains lie between the mountains and the sea.

The landforms on such margins depend to some extent on present and past climates. In southern Norway the landscape is dominated by fjords and glaciated valleys, but the major features of plateau and Great Escarpment were present earlier. Glaciation straightened and deepened the valleys, but they originated before glaciation (Lidmar-Bergström *et al.* 2000). Greenland and Antarctica have some marginal swell-type margins that have been much modified but not obliterated by glaciation. In marked contrast is the Great Escarpment of Namibia, created by fluvial erosion but now in a largely desert environment.

A wedge of sediments is deposited offshore from the continental margin above an unconformity called the breakup unconformity (meaning related to the breakup of a supercontinent). The sediments record the history of uplift in their hinterland. In Scandinavia and southern Africa interpretation of the offshore sediments suggests that there were two main uplift phases – Palaeogene and Neogene. Individual river sources of sediment may be indicated: a delta developed after 103 Ma near the mouth of the present Orange River, South Africa.

There are two main models of passive margin evolution. One school, placing emphasis on fission track data and similar methods, believes there was a continuing uplift towards the margin, where the continental rim ended at a massive fault. Slope retreat moved from the initial faulted margin to the present Great Escarpment. The alternative is warping of the palaeoplain to below sea level. Valleys eroding the steeper, coastal side coalesced to form a Great Escarpment, which then retreats. This model equates the breakup unconformity with the plateau surface, and equates the marginal swell with the raised shoulder of present-day rift valleys (e.g. the Lake Albert rift, Uganda). This implies that the marginal swell dates back to the earliest days of continental breakup.

Some passive margins have simple drainage patterns with streams flowing in opposite directions away from a ridge at the top of the Great Escarpment (Brazil, Western Ghats). On some marginal swells major rivers were in existence before continental breakup and can still be traced in the modern landscape (Australia, South Africa). Drainage may be modified and even reversed. Original drainage divides may relate to the original tectonic movement that made a marginal swell, with the crest of the swell being the drainage divide. The location of divides can be modified by drainage evolution, especially headward erosion of coastward flowing rivers.

What caused uplift of the marginal swell is not known, though a wide range of proposals have been made, and it may not even be the correct question to ask. Some passive continental margins might have been high originally, like the high plateau bounding many present-day rift valleys. A secondary mechanism is isostatic adjustment to erosion of the land, loading by offshore deposition and (in places like Greenland and Antarctica) loading by ice sheets.

The geomorphic evolution of some margins has been traced back to the Mesozoic, as in eastern Australia. Elsewhere landscape evolution is thought to be Miocene and younger, as in the Piedmont of the USA. There may be more than one period of movement. Several passive margins are now thought to have Mesozoic beginnings modified by further movement in the Neogene.

Several passive margins do not fall into the two groups outlined above. Some margins are dominated by deposition of sediment as the crust sinks. The Gulf Coast of the United States, and the coastal part of the Chaco-Pampean Plain in Argentina are examples. The country east of Perth in Western Australia is dominated by the north–south Darling Fault. This makes a topographic fault scarp, bounding an erosion surface cut across Precambrian rocks. West of the fault, the Perth Basin has about 11 km of Silurian to Cretaceous sediments showing long-continued downfaulting. The basin is not a rift valley as no fault further west has been located, so this is a faulted passive margin. The southern coast of Western Australia, west of the Great Australian Bight, exhibits a simple warp, bending the West Australian planation surface to sea level. Ancient valleys that flowed across the land from Antarctica before breakup can be traced across the warp as lines of salt lakes, and the drainage is reversed to form the present south flowing rivers (Clarke 1994). Despite the long time available (Antarctica started to separate about 55 Ma) there is no sign of initiation of a Great Escarpment, suggesting that it requires a greater relief than this margin offers.

Considering the relationship of big rivers, deltas and global tectonics, Potter (1978) pointed out that the twenty-eight biggest rivers in the world all drain to passive margins. Twenty-five of the world's largest deltas are also found on passive margins.

References

Clarke, J.D.A. (1994) Evolution of the Lefroy and Cowan palaeodrainage channels, *Australian Journal of Earth Sciences* 41, 55–68.

Lidmar-Bergström, K., Ollier, C.D. and Sulebak, J.R. (2000) Landforms and uplift history of southern Norway, *Global and Planetary Change* 24, 211–231.

Ollier, C.D. (2003) Evolution of mountains on passive continental margins, in O. Slaymaker and P. Owens (eds) *Mountain Geomorphology*, London: Edward Arnold.

Potter, P.E. (1978) Significance and origin of big rivers, *Journal of Geology* 86, 13–33.

Further reading

Ollier, C.D. (ed.) (1985) Morphotectonics of passive continental margins, *Zeitschrift für Geomorphologie Supplementband, N.F.* 54.

Summerfield, M.A. (ed.) (2000) *Geomorphology and Global Tectonics*, Chichester: Wiley.

SEE ALSO: active margin; isostasy

CLIFF OLLIER

PATERNOSTER LAKE

A body of water set in a formerly glaciated environment, often divided by either moraine deposits and/or rock bars, and aligned with similar neighbouring lakes. They are often linked by a stream, rapids or waterfalls running through the valley so that it resembles, when viewed in plan, a string of beads. The term paternoster lake is derived from this pattern, with each lake resembling a paternoster (bead) in a rosary. Paternoster lakes are formed by the plucking and scouring out of a valley bed by a glacier, though they may also form through the damming effects of glacial deposits (by moraines, rock bars or by riegels). The varying rock resistance means that the glacier will erode away the weaker rock more quickly, forming depressions in the valley floor. Water then accumulates in these depressions upon glacial retreat, leaving a series of usually elongated lakes, reflecting the direction of scour from which they developed. The number, size and shape of paternoster lakes varies as a function of weakness, jointing and lithology of the underlying rock, alongside the varying characteristics of the glacier, valley steepness, and extending or compressive flow. Paternoster lakes are common in Sweden, draining into the Baltic. Also, Llyn Dinas and Llyn Gwynant, Snowdonia (UK), are examples of paternoster lakes.

STEVE WARD

PATTERNED GROUND

Patterned ground consists of a range of phenomena – circles, nets, polygons, steps and stripes – developed in surface materials. Such phenomena occur in a wide range of environments and have a large number of causes.

Particularly in the seasonal tropics, swelling clay and texture-contrast soils develop micro-relief consisting of mounds and depressions arranged in random to ordered patterns. These are normally termed GILGAI. Most mechanisms of gilgai development involve swelling and shrinking of clay subsoils under a severe seasonal climate.

In some arid and periglacial regions patterned ground is in the form of DESICCATION CRACKS AND POLYGONS. These result because of the volume reduction that takes place in fine-grained, cohesive sediments as they dry out by evaporation of water. This creates sufficient tensional stress for rupture to occur and cracks to be formed.

Elsewhere in dry regions, patterned ground can be associated with the presence of salt, particularly on the floors of playas and on SABKHAS (Hunt and Washburn 1960). Thrusting structures can develop which are called tepees, because of their resemblance to the shape of the hide dwellings of early American Indians (Warren 1983).

In other dryland regions patterns can be produced by vegetation banding. From the air many dryland surfaces can be seen to be characterized by alternating light and dark bands called BROUSSE TIGRÉE (tiger bush). The banding reflects differences in the proportions of grasses and shrubs. This in turn is related to the action of sheet flow on low angle surfaces (0.2–2 per cent) in areas with 50–750 mm mean annual rainfall (Mabbutt and Fanning 1987).

Organic processes also create patterns through the building of mounds by such organisms as communal rodents and termites (see TERMITES AND TERMITARIA). In the case of the MIMA MOUNDS of the USA and the Heuweltjies of southern Africa their mode of origin is uncertain (Reider *et al.* 1996).

However, patterned ground (Plate 86) is especially prevalent in periglacial regions (see PERIGLACIAL GEOMORPHOLOGY; ICE WEDGE AND RELATED STRUCTURES) and in areas underlain by PERMAFROST. A great diversity of forms and processes are involved (Washburn 1956), including thermal contraction cracking, seasonal frost cracking and desiccation cracking. Circular forms are produced by FROST HEAVE (cryoturbation). Periglacial areas also show the development of Earth hummocks (Thufur) (Schunke and Zoltai 1988), and cryoturbation plays a role in their formation. Relict periglacial patterned ground phenomena developed during former cold phases are widespread in mid-latitude areas (Boardman 1987).

Plate 86 Late Pleistocene patterned ground developed under periglacial conditions in the Thetford region of eastern England. The stripes, which have analogues in present-day Alaska, are formed by alternations of heather (*Calluna vulgaris*) and grass

References

Boardman, J. (ed.) (1987) *Periglacial Processes and Landforms in Britain and Ireland*, Cambridge: Cambridge University Press.

Hunt, C.B. and Washburn, A.L. (1960) Salt features that simulate ground patterns found in cold climates, *US Geological Survey Professional Paper* 400B.

Mabbutt, J.A. and Fanning, P.C. (1987) Vegetation bandings in arid western Australia, *Journal of Arid Environments* 12, 41–59.

Reider, R.G., Hugg, J.M. and Miller, T.W. (1996) A groundwater vortex hypothesis for Mima-like mounts, Laramie Basin, Wyoming, *Geomorphology* 16, 295–317.

Schunke, E. and Zoltai, S.C. (1988) Earth hummocks (Thufur), in M.J. Clark (ed.) *Advances in Periglacial Geomorphology*, 231–245, Chichester: Wiley.

Warren, J.K. (1983) Tepees, modern (southern Australia) and ancient (Permian-Texas and New Mexico) – a comparison, *Sedimentary Geology* 34, 1–19.

Washburn, A.L. (1956) Classification of patterned ground and review of suggested origins, *Geological Society of America Bulletin* 67, 823–865.

A.S. GOUDIE

PEAT EROSION

Peatlands cover large tracts of the microthermal northern hemisphere, in countries like Canada, Russia and Finland, but mostly these are low-lying and the peat remains largely intact. Upland blanket mire is much more rare and because of higher rainfall and greater slope angles, erosion is more likely. Some 8 per cent of the land surface of the British Isles is covered by blanket peat, mainly in the north and west. These blanket bogs form the largest single contribution (10–15 per cent) to a globally scarce resource. These areas of blanket peat are important for many reasons: water catchments, hill farming, shooting, recreation and landscape.

The blanket peat of the southern Pennines is unquestionably the most degraded in Britain with gullying affecting three-quarters of the blanket peat. These peatlands lie close to large urban areas (Manchester, Leeds, Sheffield), important sources of air pollution, and, compared to other areas of blanket bog, are climatically marginal, being more southerly than most and in areas receiving barely 1500 mm annual rainfall (Tallis 1997). Peat erosion has been studied over the last century, but it was largely Margaret Bower (1960, 1961) who stimulated recent work. She identified two types of gully system: a dense network of freely and intricately branching gullies on very flat ground (less than 3°); and, more linear gullies with much less tendency to branch on sloping land. Erosion rates from the heavily gullied blanket peat are high for the UK. Labadz *et al.* (1991) used reservoir sedimentation surveys to establish the long-term sediment yield: in total over 200 t $km^{-2} yr^{-1}$ including an organic fraction of almost 40 t $km^{-2} yr^{-1}$. These high sediment yields mean that many of the small reservoirs built in the nineteenth century are now largely full up with sediment and effectively useless for water storage. Whilst the peat erosion rates seem relatively small, given the low bulk density of peat, these do in fact represent large volumetric losses, implying that most of the gullies may have developed during the past three centuries.

John Tallis, in particular, has studied the history of peat erosion in the southern Pennines. His analysis of pollen profiles shows that there was drying out of the mire surface during the 'Early Mediaeval Warm Period' in the twelfth and thirteenth centuries. This was followed by a cooler, wetter period, and it seems possible therefore that climate change could have triggered gully erosion at that time (and perhaps in earlier dry phases too). More recently, human-induced pressures on the blanket peat have probably been more important, sometimes working in tandem with climate change. Fire (accidental or deliberate) and overgrazing by sheep are the most important direct pressures leading to erosion; the loss of pollution-sensitive mosses, particularly *Sphagnum*, is also likely to have been significant. Complete loss of *Sphagnum* soon after the start of the Industrial Revolution in the eighteenth century may well have initiated more widespread gully erosion than might have developed because of climatic change alone.

In intact peat, the water table remains close to the surface except during severe droughts. Most storm runoff is produced by saturation-excess overland flow therefore, although locally pipeflow may be important (Holden and Burt 2002). On flat ground a hummock and pool micro-topography often develops. If the peat dries out, gullies begin to form, further lowering the water table, especially during summer. As the peat re-wets in autumn, there is an increased tendency for leaching of dissolved organic carbon (DOC) discolouring local water supplies and leading to significantly increased costs of water treatment (Worrall *et al.* 2002). Together, enhanced export of particulate and dissolved carbon means that the blanket peat no longer continues to build up a store of carbon and increasingly becomes a source of carbon export instead. From the 1950s, many areas of blanket peat were drained (using narrow slot drains or 'grips') in an attempt to increase productivity. More recently, landowners are beginning to fill in the grips, in an attempt to restore habitat and reduce DOC export.

Further north in the Pennine Hills, there are clear signs today (2002) of revegetation of previously heavily eroded blanket peat. This may indicate that previous pressures leading to erosion have been reduced. In the southern Pennines, some gullies have begun to infill over the past twenty years but generally there is little sign of revegetation, perhaps showing that the combined influences of sheep grazing and air pollution continue to hinder recovery there.

References

Bower, M.M. (1960) The erosion of blanket peat in the Southern Pennines, *East Midland Geographer* 13, 22–33.

——(1961) The distribution of erosion in blanket peat bogs in the Pennines, *Transactions of the Institute of British Geographers* 29, 17–30.

Holden, J. and Burt, T.P. (2002) Infiltration, runoff and sediment production in blanket peat catchments: implications of field rainfall simulation experiments, *Hydrological Processes* 16, 2,537–2,557.

Labadz, J.C., Burt, T.P. and Potter, A.W.R. (1991) Sediment yield and delivery in the blanket peat moorlands of the southern Pennines, *Earth Surface Processes and Landforms* 16, 225–271.

Tallis, J.H. (1997) The Southern Pennine experience: an overview of blanket mire degradation, in J.H. Tallis,

R. Meade and P.D. Hulme (eds) *Blanket Mire Degradation: Causes, Consequences and Challenges*, Aberdeen: Macaulay Land Use Research Institute.
Worrall, F., Burt, T.P., Jaeban, R.Y., Warburton, J. and Shedden, R. (2002) Release of dissolved organic carbon from upland peat, *Hydrological Processes* 16, 3,487–3,504.

TIM BURT

PEDESTAL ROCK

A pedestal rock is an isolated erosional rock mass comprising a slender stem, neck or column supporting a wider cap. Also known as mushroom rocks balanced rocks and perched blocks, by local names such as *loganstones* (south-west England) and *hoodoo rocks* (North America), and by non-English equivalents such as *rocas fungiformas, roches champignons, Pilzfelsen*, pedestal rocks are developed in various climatic and lithological contexts; but especially well in sandstone, granite and limestone.

They are due to differential weathering and erosion of the cap and stem. Some are structural, the caprock being inherently more resistant than that of the stem. Pedestal rocks have been attributed to various epigene effects, and certainly, some standing in rivers and on the coast, especially where limestone is exposed, are due to physical, biotic and biochemical attack around water level. The occurrence of pedestal bedrock shapes just below the surface, however, shows that moisture attack there produces incipient indents, alcoves and concave shapes in the bedrock surface. Subsurface weathering all around the base of a block or boulder followed by lowering of the surrounding area produces a mushroom form.

Epigene effects such as differential wetting and drying on different aspects contribute to development and maintenance of the form after exposure. Sand blasting may be responsible for pedestal rocks in an immediate sense, but may exploit bedrock already weakened by weathering. Pedestal rocks are convergent forms, but most are of two-stage or etch origin.

Further reading

Twidale, C.R. and Campbell, E.M. (1992) On the origin of pedestal rocks, *Zeitschrift für Geomorphologie* 36, 1–13.

C.R. TWIDALE

PEDIMENT

A pediment is a gently inclined slope of transportation and/or erosion that truncates rock and connects eroding slopes or scarps to areas of sediment deposition at lower levels (Oberlander 1989). They have been reported from polar, humid and arid zones (Whitaker 1979), but they are most widely reported and studied in dryland environments, and are generally perceived as a phenomenon of DESERT GEOMORPHOLOGY.

Pediments are part of a family of landforms developed in the piedmont zone, an area of diverse geomorphology juxtaposed between uplands dominated by sediment erosion and lowlands dominated by sediment transport and deposition. Piedmonts may, therefore, be subjected to EROSION, transport or depositional process domains. These domains can vary spatially and temporally, giving rise to a complex variety of piedmont landforms, including ALLUVIAL FANS, BAJADAS and pediments.

Pediments have been variously defined in the literature. These definitions range from very general, such as 'a terrestrial erosional surface inclined at a low angle and lacking significant relief' (Whitaker 1979), to more specific, such as 'surfaces eroded across bedrock or alluvium, usually discordant to structure, with longitudinal profiles usually either concave upward or rectilinear, slope at less than 11°, and thinly and discontinuously veneered with rock debris' (Cooke 1970). Although gradients can range between 0.5° and 11°, pediments steeper than 6° are rare in the natural environment (Dohrenwend 1994). Where two pediments meet across a divide, breaking up the continuity of a mountain mass, a pediment pass is formed, often creating useful trafficable routes through upland regions. A pediplain is formed by the coalescence of numerous pediments. It should be distinguished from a PENEPLAIN, where slope decline is thought to be the major process, rather than slope retreat.

Pediments were first described by Gilbert (1877), but the term 'pediment' was coined by McGee (1897), and is derived from the architectural term for a low-pitched gable, especially the triangular form used extensively in classical architecture. Subsequently, the term has been applied to a variety of geomorphological forms, giving rise to considerable confusion and problems of definition (Whitaker 1979).

Difficulties are encountered when trying to define the outlines of pediments for the purpose of GEOMORPHOLOGICAL MAPPING. This also makes it difficult to derive their MORPHOMETRIC PROPERTIES, such as length, area, mean gradient, etc. (Cooke 1970). The upper margin is generally agreed to be at the piedmont angle (the junction between pediment and mountain front, usually defined as the line of maximum change in gradient in the slope profile), or at the watershed if a tributary upland area is absent. However, the downslope margin is more difficult to define. Cooke (1970) suggests this boundary should be placed where alluvial cover becomes continuous; Howard (1942) and Tator (1952, 1953) suggest it should be placed where the depth of alluvial cover equals the depth of stream scour (15 m). Other researchers have placed this boundary where the thickness of alluvial cover exceeds a small proportion (e.g. 1 per cent) of total pediment length (Dohrenwend 1994).

A variety of pediment types have been recognized and classified. Three different pediment forms can be identified using simple geomorphological criteria; an apron pediment is the common form that extends between an upland source area and a lowland depositional area; a pediment dome is formed by coalescing pediments, when the upland area has been removed; a terrace pediment is developed adjacent to a relatively stable base level such as a through-flowing stream. Other classifications distinguish between covered and exposed forms: a mantled pediment is one where crystalline bedrock is veneered by a residual weathering mantle and which is inferred to have been formed by subsurface weathering of the crystalline bedrock and wash removal of the resulting debris; a rock pediment is thought to be formed by removal of the overlying debris from a mantled pediment; and a covered pediment is developed discordantly across sedimentary rocks, having a veneer of coarse debris.

Oberlander (1989) makes an important distinction between two fundamental types of pediment; those that truncate softer rocks adjacent to a more resistant upland, and those where there is no change in lithology between upland and pediment. The first form has been widely reported, most notably along the northern margin of the Sahara Desert. These landforms have been widely studied by French geomorphologists, who term them GLACIS D'ÉROSION. These landforms truncate weak materials, and tend to be veneered by alluvial gravels, indicating the importance of fluvial processes in their creation. The second form, which has proved much more difficult to explain, is referred to by Oberlander (1989) as a 'true' pediment. Here, the pedimented surface has been cut across a lithology of similar resistance to the adjacent upland, usually a relatively resistant igneous or metamorphic rock. They typically lack the alluvial cover of the glacis-type, and show little clear evidence of fluvial processes in their formation. In the absence of an obvious mechanism a number of theories have been proposed for their formation and development, but they remain ill-defined and controversial.

At a basic level, pediments are normally viewed as a result of erosion of upland areas. This material is transported across the pediment into the lowland depositional area, and the retreating upland leaves behind an enlarging transportational pediment surface. However, problems arise when attempts are made to identify specific pediment-forming processes. Numerous processes have been proposed, but their significance is very difficult to demonstrate in practice. These proposed pediment-forming processes can be considered under three headings; surface WEATHERING, subsurface weathering and fluvial processes.

Surface weathering processes cover a wide range of SUBAERIAL processes leading to the breakdown of bedrock and REGOLITH. However, these processes do not explain the formation of a distinct piedmont angle on rocks with the same resistance to weathering. This has led Mabbutt (1966), and others, to emphasize the importance of subsurface weathering in the formation of pediments. This is largely based on the observation that pediments are widely developed on granitic bedrock, a lithology particularly susceptible to subsurface weathering. Perhaps most important here is the nature of the material produced by deep weathering of granite. The well-sorted, sand-sized GRUS forms non-cohesive channel bank material that are highly susceptible to lateral channel shifting and planation, resulting in a limited amount of channel incision (Dohrenwend 1994). The fine-grained grus can also be transported down the low pediment slopes. Mabbutt (1966) attributes the formation of a piedmont angle to slope foot notching (weathering in the subsurface layer at the base of the mountain front). However, much of this model of formation is based on assumptions based on form and occurrence, rather than on observed and well-characterized processes that can be easily validated.

Fluvial processes have been widely implicated in pediment formation, with the major emphasis being given to lateral planation. Streams debouching from the upland drainage basins are thought to erode back the mountain front by lateral channel migration. Channel incision is thought to be limited due to the high sediment load of these streams. Other research has focused on the importance of sheet floods, but this process occurs rarely in the natural environment, and its significance in pediment formation must be questioned. Sheet flooding cannot produce a planar surface, because a planar surface is necessary for sheet flooding to occur (Cooke *et al.* 1993). This vital distinction between pediment-forming and pediment-modifying processes was emphasized by Lustig (1969), who suggested that contemporary pediments were the wrong place to look for an explanation of how they were formed, since the pediments would already have to exist for these processes to operate. He suggested that geomorphologists should instead concentrate on studying the erosional processes operating in the adjacent upland drainage basins, as this is where erosion is most active. Other workers have suggested that much of the erosion leading to formation of pediments takes place in embayments, formed where streams debouch from the upland area (Parsons and Abrahams 1984). As with the subsurface weathering model detailed above, the fluvial models must be treated with caution due to the difficulties of linking observed forms with clearly defined physical processes.

The main difficulty in explaining development of pediments is the problem of maintaining parallel rectilinear retreat of permeable slopes in a SAPROLITE-mantled landscape. Oberlander (1989) proposes that rectilinear retreat occurs because sediment transport processes are limited by deep permeability of grus, eluviation of fines by throughflow, and accelerated subsurface weathering by soil moisture, concentrated at the base of slopes. Twidale (1978) suggests that lithological and structural features within granitic massifs (petrological variations and differences in joint density) are important controls on pediment morphology, but other work has failed to demonstrate any clear relationships. The importance of tectonics in pediment formation is also uncertain, although, in general, pediments appear to be best developed in areas of long-term stability (Dohrenwend 1994).

With improvements in dating techniques, there is a growing amount of evidence indicating that some pediments are extremely ancient. In the Sahara and Mojave desert (Cooke and Reeves 1972; Plate 87), lava flows can be seen to bury existing pediment surfaces. This raises the possibility that they may be relict landforms formed under different climatic conditions pertaining in Tertiary or even Late Mesozoic times (Oberlander 1989). Specifically, the arid-zone processes acting on contemporary desert pediments may not be appropriate to explain landforms that developed over timescales that embraced humid as well as arid phases. A variety of conditions have been proposed as being optimal for pediment development of crystalline rocks; these include seasonally wet, low-latitude forest, savanna and cold-winter deserts affected by cryogenic processes. Oberlander (1989) suggests that pedimentation is currently active in parts of central Arizona, which appears to replicate conditions in the Mojave Desert in the Miocene. It seems likely that many pediments must be regarded to some extent as relict landforms, currently being modified under very different environmental conditions from those that pertained during their initial formation.

Plate 87 Pediment in the Mojave Desert, southwest USA

References

Cooke, R.U. (1970) Morphometric analysis of pediments and associated landforms in the Western Mojave Desert, *American Journal of Science* 269, 26–38.

Cooke, R.U. and Reeves, R.W. (1972) Relations between debris size and slope of mountain fronts and pediments in the Mojave Desert, California, *Zeitschrift für Geomorphologie* 16, 76–82.

Cooke, R.U., Warren, A. and Goudie, A.S. (1993) *Desert Geomorphology*, London: UCL Press.

Dohrenwend, J.C. (1994) Pediments in Arid Environments, in A.D. Abrahams and A.J. Parsons (eds) *Geomorphology of Desert Environments*, 321–353, London: Chapman and Hall.

Gilbert, G.K. (1877) *Report on the Geology of the Henry Mountains: United States Geological Survey of the Rocky Mountain Region*, Washington, DC: Department of the Interior.

Howard, A.D. (1942) Pediment passes and the pediment, problem, *Journal of Geomorphology* 5, 3–31, 95–136.
Lustig, L.K. (1969) Trend surface analysis of the Basin and Range Province and some geomorphic implications, *US Geological Survey Professional Paper* 500–D.
Mabbutt, J.A. (1966) Mantle-controlled planation of pediments, *American Journal of Science* 264, 79–91.
McGee, W.J. (1897) Sheetflood erosion, *Geological Society of America Bulletin* 8, 87–112.
Oberlander, T.M. (1989) Slope and pediment systems, in D.S.G. Thomas (ed.) *Arid Zone Geomorphology*, 56–84, London: Belhaven.
Parsons, A.J. and Abrahams, A.D. (1984) Mountain mass denudation and piedmont formation in the Mojave and Sonoran Deserts, *American Journal of Science* 284, 255–271.
Tator, B.A. (1952) Pediment characteristics and terminology (part 1), *Annals of the Association of American Geographers* 42, 295–317.
——(1953) Pediment characteristics and terminology (part 2), *Annals of the Association of American Geographers* 43, 47–53.
Twidale, C.R. (1978) On the origin of pediments in different structural settings, *American Journal of Science* 278, 1,138–1,176.
Whitaker, C.R. (1979) The use of the term 'pediment' and related terminology, *Zeitschrift für Geomorphologie* 23, 427–439.

SEE ALSO: alluvial fan; desert geomorphology; glacis d'érosion

KEVIN WHITE

PENEPLAIN

Peneplain is the term coined by W.M. Davis to mean a surface of low relief worn down to near sea level and formed through erosion over protracted spans of time. His own words are:

> Given sufficient time for the action of denuding forces on a mass of land standing fixed with reference to a constant base-level, and it must be worn down so low and so smooth, that it would fully deserve the name of a plain. But it is very unusual for a mass of land to maintain a fixed position as long as is here assumed.... I have therefore elsewhere suggested that an old region, nearly base-levelled, should be called an almost-plain; that is a peneplain.
>
> (Davis in Chorley *et al.* 1973: 190)

The peneplain is thus not the end-product of a cycle of erosion and, if keeping with Davis's way of thinking, it should not be confused with an endless and featureless plain as is often implied. Rather, a peneplain is a regional landscape at the penultimate stage of development which is yet to be eroded down to a true plain. In another place Davis himself says:

> At a less advanced stage of degradation, the land will still possess low, unconsumed hills along the divides and subdivides between the broad-floored rivers. It will then be almost-a-plain, or a peneplain. A peneplain will be hardly above sealevel at its base, but if the area is large it may attain altitudes of 2,000, 3,000, or 4,000 feet far inland near the river heads, and its residual mounts and hills may rise still higher, although with gentle slopes.
>
> (Davis in King and Schumm 1980: 8)

The processes leading to a peneplain would be mainly subaerial, chiefly fluvial and gravity-driven hillslope processes. They ought to be in action long enough to obliterate the effect of unequal rock resistance, so only the hardest rocks would form bedrock-built hills rising above the peneplain, the monadnocks. Otherwise, gentle slopes would be underlain by deeply weathered rock, with the thickness of weathering mantle being in excess of 10 m. As far as the relative relief of a peneplain is concerned, Davis seemed to be rather vague in defining any critical hill heights or slope angles. Therefore, there are two crucial characteristics of a peneplain. One is its temporal context within the cycle of subaerial erosion. To be a peneplain, the surface of low relief must have formed in the course of protracted denudation. The second prerequisite is grading down to sea level.

Much of the substantial confusion around the term results from the fact that subsequent workers did not always keep with Davis's original definition and used the term in various contexts. For example, peneplains were often equated with PLANATION SURFACES, or a particular mode of slope evolution was implied for a peneplain. Many geomorphology textbooks contrast peneplains formed mainly through slope downwearing and consequent relief reduction, with pediplains formed by slope backwearing and relative relief maintained high until a rather late stage of development. In other cases the condition of being located close to, or graded to, sea level was ignored. In consequence, flattened summit surfaces within mountain ranges were frequently called peneplains despite the fact that neither their origin nor the age were known sufficiently to warrant the use of the term in its strict, Davisian sense. Davis himself suggested the term 'pastplain' to describe a peneplain which has been uplifted and now shows the initial stage of dissection.

Free usage of the term and its obvious connection with the Davisian model of cyclic landform development and the DENUDATION CHRONOLOGY approach, themselves strongly criticized since the 1960s, had eventually led to its declining popularity and gradual abandonment. Preference was given to more neutral 'planation surfaces' in describing landscapes, whereas in the field of theory a search for non-cyclic models of geomorphic development was pursued strongly.

Nonetheless, Fairbridge and Finkl (1980) proposed to return to 'peneplains', but realizing the potential for confusion and misuse they suggested disassociation from restrictions implied by the Davisian definition. Instead, they preferred to give the term a non-genetic meaning, simply to describe a near flat surface regardless of its origin, setting and evolutionary stage. This point has apparently been taken forward by Twidale (1983) who describes peneplains as 'rolling or undulating surfaces of low relief', without referring to their position in respect to BASE LEVEL. Moreover, he argues that there is no means to decipher the mode of past slope evolution leading to the present-day peneplain; hence the argument focused on the backwearing-or-downwearing issue is by and large pointless. On the other hand, he firmly adheres to the Davisian understanding of the peneplain as a landscape at the penultimate stage of evolution and introduces 'ultiplains' as the true end-products of relief development. By contrast, Phillips (2002) in his most recent review offers a broader definition of the peneplain in which the condition of being at any certain stage of a cycle has been made redundant. Given all these divergent views, the term is very difficult to recommend for routine application in describing and explaining landforms.

There has been much debate as to whether peneplains and peneplanation, in the truly Davisian sense, really occur. Phillips (2002), following his many predecessors, claims that contemporary peneplains eroded to near sea level are almost non-existent and seeks the reasons in constant variations in tectonic forcings, climate and base level, especially in the Quaternary. All these changes would not allow peneplanation to last for long, and induce surface dissection rather than planation. On the other hand, Twidale (1983) gives a number of examples of almost perfect rock-cut plains, but demonstrates their antiquity at the same time. Many of these plains date back to the Cretaceous or even beyond. In Fennoscandia, a peneplain of subcontinental extent undoubtedly existed at the end of the Precambrian (Lidmar-Bergström 1995). Further examples of past plains, or palaeoplains, have been reviewed by Ollier (1991). It appears that reconciling the evidence for little or no peneplanation at present and widespread planation in the geological past is one of the challenges for evolutionary geomorphology.

References

Chorley, R.J., Beckinsale, R.P. and Dunn, A.J. (1973) *The History of the Study of Landforms. Vol. 2: The Life and Work of William Morris Davis*, London: Methuen.

Fairbridge, R.W. and Finkl, C.W. Jr (1980) Cratonic erosional unconformities and peneplains, *Journal of Geology* 88, 69–86.

King, P.B. and Schumm, S.A. (eds) (1980) *The Physical Geography (Geomorphology) of William Morris Davis*, Norwich: GeoBooks.

Lidmar-Bergström, K. (1995) Relief and saprolites through time on the Baltic Shield, *Geomorphology* 12, 45–61.

Ollier, C.D. (1991) *Ancient Landforms*, London: Belhaven.

Phillips, J.D. (2002) Erosion, isostatic response, and the missing peneplains, *Geomorphology* 45, 225–241.

Twidale, C.R. (1983) Pediments, peneplains and ultiplains, *Revue de Géomorphologie Dynanique* 32, 1–35.

Further reading

Adams, G. (ed.) (1975) *Planation Surfaces*, Benchmark Papers in Geology 22, Stroudsburg, PA: Dowden, Hutchinson and Ross.

Melhorn, W.N. and Flemal, R.C. (eds) (1975) *Theories of Landform Development*, London: George Allen and Unwin.

SEE ALSO: Cycle of Erosion; denudation chronology

PIOTR MIGOŃ

PERIGLACIAL GEOMORPHOLOGY

Perhaps surprisingly, there is no agreement as to what exactly constitutes terrain which can be regarded as periglacial for there are no quantitative defining parameters which have gained universal acceptance. However, most would accept the proposition that there are two possible approaches to the demarcation of what is periglacial, and that both can be justified.

One would emphasize the requirement for intense frost action in the form of frequent FREEZE–THAW CYCLES and deep seasonal freezing. If these criteria are used to delimit the distribution of the periglacial domain, it follows from this that some 35 per cent of the Earth's continental surface (mainly in the northern hemisphere) falls into the periglacial category. The other approach would stipulate that the presence of perennially frozen ground, i.e. PERMAFROST, is paramount. Permafrost may be defined as any earth material that has maintained a temperature at or below 0 °C for a minimum period of two years. Note that there is no reference to water content or lithology in this definition. If permafrost is the fundamental attribute, then a more rigorous climate than that needed for frost action alone is required to qualify for periglacial status, as permafrost demands a mean annual temperature of below 0 °C. As a result the global periglacial area would be substantially less at around 20 per cent of the total terrestial surface. Nevertheless, from both perspectives the total area regarded as periglacial remains a considerable part of the Earth's terrestrial environment. By way of comparison, the area of permanent snow and glacial ice is only 3 per cent.

Relief also has an influential role in determining the distribution of periglacial regions. Both freeze–thaw cycles and permafrost are related to climate and this is influenced by both latitude and altitude. The outcome is that the most widespread periglacial areas are mainly lowlands in northern Eurasia and North America and these incorporate both tundra and boreal forest landscapes. Mountain temperatures are influenced by elevation which is sensitive to the lapse rate. This can produce sufficient cooling at lower latitudes to over print the more temperate conditions in the adjacent lower areas. As a consequence alpine periglacial areas can occur even at the equator. Usually they reveal a tundra zone with the lower limit roughly corresponding with the upper treeline.

Basic periglacial processes

Periglacial geomorphology focuses primarily upon those terrestrial surface processes, sediments and resultant landforms which characterize the cold non-glacial areas of the Earth's surface. Basic to comprehension of these is a knowledge of the somewhat anomalous physical behaviour of water substance. First, there is a 9 per cent increase in volume as the phase changes from liquid to solid and conversely by a decrease from solid to liquid which occurs during freezing and thawing. This is accompanied by a large latent heat of fusion (84 calories per gram) not much less than the heat required (100 calories per gram) in changing the temperature of the liquid phase from the solid to gas transitions. The net result is that the rate of both freezing and melting are delayed more than might be expected. Second, volumetric changes occur during temperature variations in the solid (frozen) state when cooling produces contraction and vice versa. This in itself is not unusual but to avoid confusion it has to be seen as being totally independent on the 9 per cent shift at the liquid–solid–liquid phase change. Third, maximum density is achieved in the liquid phase, some 3.98 °C above the freezing point, ensuring that the solid phase floats on the liquid. This has profound implications for life as it means that water is present beneath lakes and rivers with ice covers. Even in the harshest climates water bodies over 3 m deep do not freeze to their beds as annually developed ice rarely exceeds 2 m in thickness. Fourth, within the sediment pore the freezing point of water can be lowered down to −22 °C in extreme cases. This is especially effective in fine-grained sediments (clays and silts) where the movement of thin films of water occur even though the ground is technically frozen. This facilitates the aggradation of ice masses with volumes well in excess of the pore capacity. All these factors contribute to the landscape-forming processes associated with periglacial environments and collectively these determine the nature of periglacial processes. Driving these processes is temperature change and in its turn weather and climate.

Palaeoperiglacial activity

Apart from the unusual physical behaviour of water substance, a further complicating factor in understanding periglacial features is the temporal dimension, particularly that allied to climatic change. This can be illustrated by the example of Britain. If delimiting areas subject to periglaciation on the basis of freeze–thaw cycles is accepted, then the higher summits of Britain remain within the ambit of periglacial activity. This is the viewpoint taken by Ballantyne and Harris (1994) in their major regionally based synthesis. However, adoption of the alternative less permissive approach based on the presence or absence of permafrost, inevitably means that the British Isles cannot be regarded as periglacial since the last permafrost finally dissipated some 11,500 years ago towards the end of the Last Glacial stage. Since then through the ensuing Flandrian interglacial (postglacial) even the climatically most extreme locations have been unable to sustain any permafrost. This was the position taken by Worsley (1977) in reviewing British periglaciation when it was concluded that all the periglacial evidence in Britain was effectively relict.

Prior to 11,500 years ago, during the Last Glacial stage, most of Britain had experienced episodic extensive permafrost. Indeed, in many areas of the world the Last Glacial stage witnessed the dramatic expansion of the periglacial realm by up to 50 per cent. But this was probably never so dramatic as in western Europe where the mean annual temperature dropped by some 20 °C during the time when the Gulf Stream was inoperative. An underemphasized aspect of the global glacial stages is that if sea level is taken as a proxy for palaeoclimate then the durations of very low sea levels (maximum glacial ice volumes) were relatively short. Similarly Quaternary marine oxygen isotope ratio records are interpreted as reflecting in large part the degree of global glaciation and the negative ratio peaks in the curves are taken to correspond with the periods of most extensive glacial ice cover (corresponding to the lowest sea levels). In many areas of the world covered by glaciers in the Last Glacial stage, the stratigraphic evidence indicates that the glacial advance which culminated in the most extensive ice cover occurred late in the stage. These data imply that for much of the glacial stage those areas which were to become glaciated were cold but non-glacial in character, i.e. periglacial processes rather than glacial processes prevailed. Hence many of the glaciated landscapes bear a partial periglacial imprint. Naturally those areas immediately outside the maximum ice extent limits witnessed a periglacial regime for much of the glacial stage and hence the effects of periglaciation are clearer. Finally, the earlier phases of ice retreat from the maximum limits were primarily the result of reduced snowfall rather than a temperature increase. This enabled the perglacial environment to extend into the areas recently vacated by the ice. PARAGLACIAL conditions follow ice withdrawal and to an extent these might be regarded as part of periglaciation.

Historical development of the periglacial concept

The term periglacial was first coined in 1909 by the Polish geologist and pedologist W. Łoziński in his account of the mechanical weathering of sandstones and blockfield (see BLOCKFIELD AND BLOCKSTREAM) production under inferred cold climates in the Carpathian Mountains. Three years later Łoziński introduced the concept of a periglacial facies produced by mechanical weathering, although he did not give it any quantitative climatic parameters. However, it is clear that Łoziński was proposing the periglacial concept in an attempt to reconstruct the palaeoenvironmental context of his facies. He envisioned it as diagnostic of the former processes operative in terrain immediately adjacent to glaciers and ice sheets of Pleistocene age rather than as a function of contemporary activity. His second paper was published as part of the proceedings of an international geological congress held in Stockholm in 1910 and this ensured that the term periglacial was widely disseminated. Other activities at the congress included a field excursion to Spitzbergen where periglacial facies could be related to contemporary environmental processes, and thereby gave further impetus to the scientific study. Strictly therefore, periglacial, as originally envisaged by Łoziński, should refer to the area or zone formerly subject to arctic-type climatic conditions peripheral to a glacier.

From the standpoint of modern usage, an erroneous impression might be given that periglacial features are exclusively associated with the area around glacial margins. On the contrary, some of the major areas of permafrost today have never been glaciated or indeed been peripheral to former ice sheets. The prime example of this is east Siberia where the permafrost can exceed over 1 km in depth. This is probably explained by the fact that it has experienced the longest history of sustained permafrost development anywhere on the Earth. A further disadvantage is that there might be the assumption that areas peripheral to glaciers experience a rather less severe climate and that a climatic deterioration would necessarily lead from the periglacial to glacial, with glaciation representing the ultimate severe climatic state. Despite a number of workers having argued for the term periglacial to be abandoned because of its imprecision, its usage is now widespread and a degree of permissiveness in its definition is automatically accepted by its users. It is interesting to note that there was a change in the title between the first and second editions of A.L. Washburn's synthesis of periglacial environments (Washburn 1973, 1979). In the latter the term GEOCRYOLOGY was introduced (i.e. the investigation of frozen Earth materials). The word is derived from the Russian equivalent and although it can include glaciers it is usually directed towards frozen ground.

Although Łoziński formalized the notion of a periglacial zone early in the twentieth century, as with many concepts in Earth science earlier workers anticipated later formulations. For example, the commencement of government-sponsored geological mapping in England in the 1830s soon led to the identification of a relict mantle of rock debris by De la Beche overlying the slopes of Cornwall and Devon in the south-west peninsula. Field relationships demonstrated to him that this 'head' blanket of angular fragments had been derived from mechanical weathering of the bedrock cropping out on the slopes above and that there had been an ubiquitous downslope movement which had tended to even out any terrain irregularities, thereby anticipating something akin to SOLIFLUCTION.

Unique periglacial processes and landforms

The research literature arising from investigations of periglacial geomorphology has given rise to a wide range of specialized terms and names of unique landforms. These have come from a number of language sources and some duplication and confusion have arisen because of inconsistent usage.

This was exemplified in the planning of this encyclopedia since the draft list of topics specified both altiplanation and CRYOPLANATION as separate entries. In reality there is no difference between the two. Altiplanation terraces were first described by H.M. Eakin in 1918 following field mapping in part of eastern Alaska where he encountered benches and summit surfaces cut in bedrock largely independent of lithology and structure. Similar relationships had earlier been identified in Russia and the term 'goletz' terrace applied to them. Other terms which have been used include equiplanation and nivation terraces. Bryan (1946) undertook a wide-ranging review of the then existing periglacial terminology and amongst others proposed a new term – cryoplanation – to express the unified concept of frost action and frost-related downslope movement of debris to produce a degradation system eroding and lowering hillslopes. This is now the internationally agreed term for such features.

Fortunately the confused nomenclature has been subject to clarification in a comprehensive glossary produced by a very experienced interdisciplinary team of Canadian permafrost workers (Harris *et al.* 1988). This is an excellent source of current usage, definitions and synonyms used in periglacial geomorphology. It also has the additional merit of thoughtfully discussing many of them and the authors are not reticent in recommending the abandonment of some cherished terms!

Periglacial geomorphogy, like any morphogenetic geomorphology, should consider all the geomorphological agents which contribute to the landscape character. Naturally there is a tendency to concentrate upon those agents which are either unique to or are readily associated with it at the expense of those which are common to a range of environments. To illustrate this point there are over fifty entries in this encyclopedia which are relevant to periglacial geomorphology. Yet only half of these are likely to be discussed or referenced to the periglacial realm. Examination of most periglacial environments, in the field or from maps and air photographs, normally reveals an essentially fluvial landscape displaying a 'normal' drainage network. There are some exceptions and significant parts of the periglacial regions display desert landscapes. This is not surprising considering some of the most arid areas of the world are underlain by permafrost. But these deserts are cold. There is tendency to regard deserts as hot places for this is where the vast majority of desert geomorphologists work.

Applied periglacial geomorphology

Wherever there is ice within the subsurface there is always the possibility that it might melt. Under natural conditions this is an ongoing process and can occur through a range of incidents such as forest fires, coastal and riverbank erosion, or climatic amelioration. Indeed the prospect of global warming carries severe implications for the entire permafrost world.

Over the last century there has been progressive settlement of the permafrost terrain by people from more southerly regions who had an expectation of a similar range of facilities to those south of the permafrost region. This movement was given a particular impetus by the Second World War and operation of defence facilities during the Cold War. Later economic exploitation of mineral resources and hydrocarbon exploration placed further demands for transport, urbanization and allied installations.

Construction of all kinds on permafrost terrain is potentially hazardous if the natural ground thermal equilibrium is disturbed as this will induce melting of the ground ice and cause thaw consolidation. Under the stress of war a number of mistakes were made in road and pipeline construction but experience was gained from tackling the challenges presented by permafrost. A landmark publication was the compilation by Muller (1947) of the then state of the art understanding of permafrost and its allied engineering problems. This drew extensively upon Russian experience. It led to the founding of the US Army's Cold Regions Research and Engineering Laboratory which has subsequently been one of the leading institutions engaged in periglacial research. Similar laboratories were established in Yakutia and Canada with primarily civilian missions.

In Canada in the 1950s, Aklavik, the pre-existing administrative centre in the Mackenzie delta region, was suffering from annual breakup floodings. A decision was taken by the Federal Government to construct a new town to replace it which would incorporate the 'best practice' in permafrost construction (Johnston 1981). A number of sites were assessed in terms of their periglacial geomorphological attributes with that at Inuvik selected for development. Inuvik has since become the show piece of how a small town offering the facilities of the non-periglacial world can be created without significant environmental damage. There all the buildings are well insulated and usually placed on piles which penetrate pads of non frost-susceptible materials carefully placed on the original vegetation. A 1 m high air gap through the tops of the piles enables the maintenance of the natural ground thermal regime. Using the same approach, a system of water, sewage and heating pipes were installed in a duct network (utilidor). In some instances, such as power generating units, piles were not feasible and thick pads of granular materials, through which ventilation pipes were inserted, have succeeded in achieving the same objectives.

A vastly improved appreciation of periglacial LANDSCAPE SENSITIVITY has largely ensured that land use activities can be undertaken without major disastrous consequences. Even so construction has to be closely monitored by environmental managers versed in the basics of periglacial geomorphology and in the field of hydrocarbon exploration a number of drill sites have been closed in the summer for fear of excessive disturbance to the ground ice within the permafrost.

References

Ballantyne, C.K. and Harris, C. (1994) *The Periglaciation of Great Britain*, Cambridge: Cambridge University Press.

Bryan, K. (1946) Cryopedology – the study of frozen ground and intensive frost-action with suggestions on nomenclature, *American Journal of Science* 244, 622–642.

Harris, S.A., French, H.M., Heginbottom, J.A., Johnston, G.H., Ladanyi, B., Sego, D.C. and Everdingen, R.O. (1988) *Glossary of Permafrost and Related Ground-ice Terms*, Ottawa: National Research Council of Canada Technical Memorandum 142.

Johnston, G.H. (ed.) (1981) *Permafrost Engineering Design and Construction*, Toronto: Wiley.

Muller, S.W. (1947) *Permafrost or Permanently Frozen Ground and Related Engineering Problems*, Ann Arbor, MI: J.W. Edwards.

Washburn, A.L. (1973) *Periglacial Processes and Environments*, London: Arnold.

——(1979) *Geocryology*, London: Arnold.

Worsley, P. (1977) Periglaciation, in F.W. Shotton (ed.) *British Quaternary Studies Recent Advances*, 203–219, Oxford: Clarendon Press.

Further reading

Clark, M.J. (ed.) (1988) *Advances in Periglacial Geomorphology*, Chichester: Wiley.

French, H.M. (1996) *The Periglacial Environment*, 2nd edition, Harlow: Longman.

Harris, S.A. (1986) *The Permafrost Environment*, Beckenham: Croom Helm.

Jahn, A. (1975) *Problems of the Periglacial Zone*, Warszawa: Polish Scientific Publishers.

King, C.A.M. (ed.) (1976) Periglacial processes, Benchmark Papers in Geology 27, Stroudsburg, PA: Dowden, Hutchinson and Ross.

Williams, P.J. and Smith, M.W. (1989) *The Frozen Earth: Fundamentals of Geocryology*, Cambridge: Cambridge University Press.

SEE ALSO: active layer; alas; cryostatic pressure; frost and frost weathering; frost heave; hummock; ice wedge and related structures; icing; loess; needle-ice; nivation; niveo-aeolian activity; oriented lake; palsa; patterned ground; pingo; protalus rampart; rock glacier; thermokarst

PETER WORSLEY

PERMAFROST

Permafrost is defined as ground (soil or rock) that remains below 0 °C for at least two years, and the term is defined purely in terms of temperature rather than the presence of frozen water (Permafrost Subcommittee 1988). Permafrost may, therefore, not contain ice, or may contain both ice and unfrozen water. In many cases, however, ground ice forms a significant component of permafrost, particularly where the substrate comprises fine-grained unconsolidated sediments. The geothermal gradient below the ground surface averages around 30 °C km^{-1} (Williams and Smith 1989) and this increase in temperature with depth determines the thickness of the permafrost (Figure 120). Seasonal temperature fluctuations lead to above zero ground surface temperatures in summer, and the downward penetration of a thawing front. The surface layer that freezes and thaws seasonally is called the active layer, and its thickness depends on the ground thermal properties and on the ratio of the summer thawing index (the accumulated degree-days above freezing) to the winter freezing index (accumulated degree days below freezing). The annual cycle of winter cold and summer warmth is propagated downwards into the permafrost, but rapidly attenuated, so that it becomes undetectable below around 15 m (Figure 120). This is termed the depth of zero amplitude (Brown and Péwé 1973). Longer term changes in ground surface temperatures cause downward propagation of a thermal perturbation, and in many permafrost sites today the geothermal gradient is non-linear, with warm-side deviation that increases towards the surface (Lachenbruch and Marshall 1986), indicating warming over the past century or more (Figure 120).

In northern Canada, permafrost is up to 600 m thick (Figure 121) and its thickness decreases southwards as the climate becomes warmer. Eventually, local variation in ground conditions leads to breaks in the continuity of permafrost, and a complex pattern of discontinuous permafrost results. Under still warmer climatic conditions

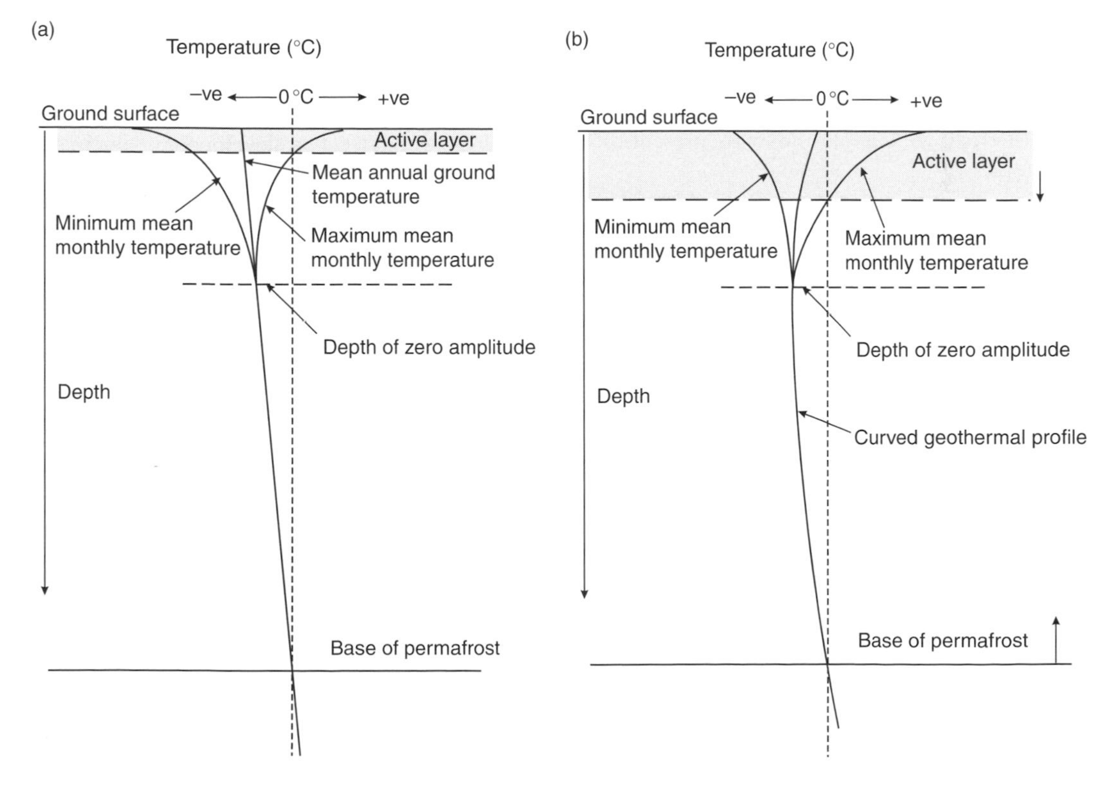

Figure 120 Thermal profiles in permafrost: (a) equilibrium and (b) during thermal adjustment to surface warming

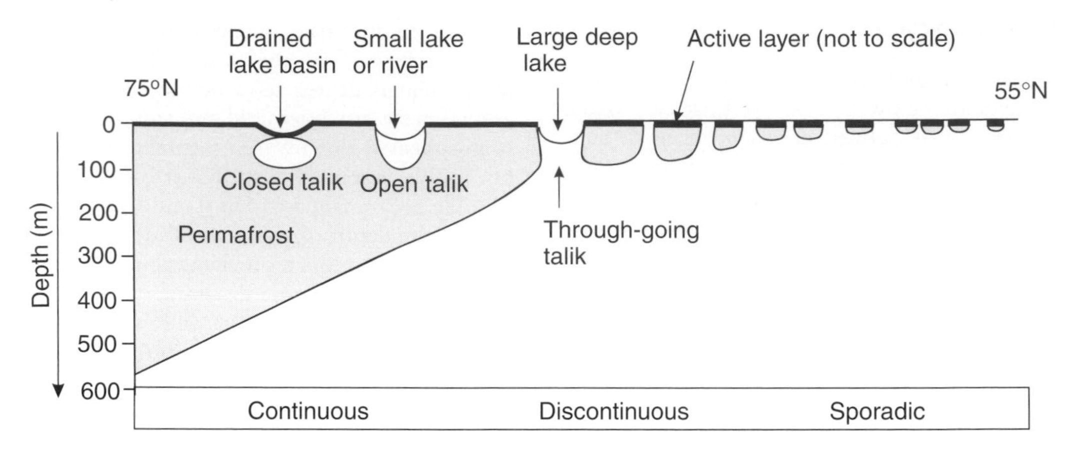

Figure 121 Typical permafrost characteristics along north–south transect, north-west Canada (after Lewkowicz 1989)

permafrost may form only isolated patches (often related to areas with peat cover), and is described as sporadic. Siberian permafrost is generally colder than North American and in places is in excess of 1,000 m thick (Williams and Smith 1989). However, it takes many millennia for thick permafrost to adjust to surface warming, and it is likely that Siberian permafrost remains chilled by the severity of the last Quaternary cold stage, and is not in thermal equilibrium with present-day conditions.

Taliks are unfrozen zones within permafrost terrain that generally occur beneath large bodies of water such as lakes or rivers that do not freeze to their beds in winter. The unfrozen lake or river water is warmer than 0 °C and therefore constitutes a heat source causing a thaw bulb to develop in the underlying permafrost. Drainage of lakes causes downward advance of permafrost, creating a closed talik, entirely surrounded by permafrost (Figure 121). Hydrochemical taliks may be cryotic (below 0 °C), but remain unfrozen due to the flow of mineralized ground water, while hydrothermal taliks may remain non-cryotic due to heat supplied by groundwater flow.

Mean annual permafrost ground surface temperatures are usually higher than the corresponding mean air temperature by a few degrees, so that defining permafrost distribution on the basis of air temperature can be misleading. However, Brown *et al.* (1981) used a mean annual air temperature of approximately −8 °C to delimit the boundary between continuous and discontinuous permafrost in North America, and −1 °C to define the southern limit of discontinuous permafrost. Williams and Smith (1989) stress the multitude of factors that influence the development and survival of permafrost, and point to a gradual southward transition from continuous to discontinuous to sporadic permafrost, with local factors leading to wide variations in associated mean air temperatures.

Where permafrost is developed in unconsolidated sediments, it commonly contains ground ice. Mackay (1972) has provided a classification of ground ice, identifying four categories: pore ice, segregation ice, vein ice and intrusive ice. Both pore ice and segregation ice occur in seasonally frozen soils, but vein ice and intrusive ice occur only in permafrost. Pore ice refers to the ice occupying the pore space in ice-cemented permafrost, and is particularly important in sands and gravels. In fine-grained soils (silts and clays) and porous rocks, much of the pore water occurs as thin films within which capillary and adsorption effects lower the freezing point by several degree celsius (Burt and Williams 1976; Williams and Smith 1989). Progressive freezing of such water results in development of cryosuction, causing water to migrate towards the freezing front. Here it freezes to form lenses of clear ice (segregation ice), increasing ice contents to well in excess of the natural saturated moisture content. Ice segregation during freezing of fine soils causes a significant increase in soil volume and upward frost heaving of the ground surface. Vein ice is the ice that accumulates within permafrost as ice wedges as a result of thermal contraction cracking (see ICE WEDGE AND RELATED STRUCTURES).

Finally, intrusive ice may form layers up to several metres in thickness as a result of pressurized water flow towards the freezing zone. The pressurized water may be derived from groundwater flow beneath permafrost (open system), or arise from porewater expulsion ahead of a penetrating freezing front in saturated coarse sands and gravels (closed system). Expulsion of water results from the expansion that occurs as pore water changes phase from water to ice. Freezing of pressurized water close to the ground surface in both open and closed systems is responsible for the formation of distinctive conical hills, or PINGOs, the pingo ice being a common form of intrusive massive ground ice (Mackay 1998). Not all massive ice bodies within continuous permafrost originated as intrusive ice, however. In Siberia and parts of northern Canada, ice bodies are considered by some to represent buried glacier ice (Astakhov *et al.* 1996; French and Harry 1990).

The presence of ice-rich permafrost results in high terrain sensitivity to surface thermal disturbance. Permafrost degradation caused by climate warming leads to slope instability and differential settlement as ground ice thaws (French and Egginton 1973). The resulting irregular surface relief is termed THERMOKARST. Widespread thaw settlement in the Arctic has been predicted due to twenty-first-century global warming (Nelson *et al.* 2001). In high altitude mountains, such as the Rockies, Himalayas and the European Alps, discontinuous permafrost is commonly present. Mountain permafrost distribution is generally complex, reflecting altitude, aspect and ground cover, particularly snow cover in winter. Terrain sensitivity to atmospheric warming is again high, and the presence of steep mountainsides increases potential hazards from landslides, debris flows and rockfall (Harris *et al.* 2001).

References

Astakhov, V.I., Kaplyanskaya, F.A. and Tarnogradsky, V.D. (1996) Pleistocene permafrost of West Siberia as a deformable glacier bed, *Permafrost and Periglacial Processes* 7, 165–192.

Brown, R.J.E. and Péwé, T.L. (1973) Distribution of permafrost in North America and its relationship to the environment: a review, 1963–1973, in *North American Contribution, Permafrost Second International Conference, Yakutsk*, 13–28 July, 71–100, Washington, DC: National Academy of Sciences.

Brown, R.J.E., Johnston, G.H., Mackay, R.J., Morgenstern, N.R. and Smith, W.W. (1981) Permafrost distibution and terrain characteristics, in G.H. Johnson (ed.) *Permafrost Engineering* 31–72, Toronto: Wiley.

Burt, T.P. and Williams, P.J. (1976) Hydraulic conductivity in frozen soils, *Earth Surface Processes and Landforms* 1, 349–360.

French, H.M. and Egginton, P. (1973) Thermokarst development, Banks Island, Western Canadian Arctic, in *North American Contribution, Permafrost Second International Conference, Yakutsk*, 13–28 July, 203–212, Washington, DC: National Academy of Sciences.

French, H.M. and Harry, D.G. (1990) Observations on buried glacier ice and massive segregated ice, western Arctic coast, Canada, *Permafrost and Periglacial Processes* 1, 31–43.

Harris C., Davies M.C.R. and Etzelmüller, B. (2001) The assessment of potential geotechnical hazards associated with mountain permafrost in a warming global climate, *Permafrost and Periglacial Processes* 12, 145–156.

Lachenbruch, A.H. and Marshall, B.V. (1986) Changing climate: geothermal evidence from permafrost in the Alaskan Arctic, *Science* 234, 689–696.

Lewkowicz, A.G. (1989) Periglacial systems, in D. Briggs, P. Smithson and T. Ball (eds) *Fundamentals of Physical Geography* (Canadian Edition), 363–397, Toronto: Copp, Clark, Pitman.

Mackay, J.R. (1972) The world of underground ice, *Annals of the Association of American Geographers* 62, 1–22.

——(1998) Pingo growth and collapse, Tuktoyaktuk Peninsula area, Western Arctic Coast, Canada: a long-term field study, *Géographie physique et Quaternaire* 52, 271–323.

Nelson, F.E., Anisimov, O.E. and Shiklomonov, O.I. (2001) Subsidence risk from thawing permafrost, *Nature* 410, 889–890.

Permafrost Subcommittee (1988) Glossary of permafrost and related ground-ice terms, *National Research Council Canada* Technical memorandum 142.

Williams, P.J. and Smith, M.W. (1989) *The Frozen Earth*, Cambridge: Cambridge University Press.

CHARLES HARRIS

PHYSICAL INTEGRITY OF RIVERS

Human activities have dramatically altered the forms and processes of the Earth's river systems. In the northern third of the globe, almost 80 per cent of the rivers are segmented by dams (Dynesius and Nilsson 1994), while in technologically advanced countries such as the United States more than 98 per cent of rivers are significantly impacted by human activities (Echeverria *et al.* 1989). The recognition that the far-reaching effects of channelization, levee building and dam construction have affected biodiversity of aquatic and riparian environments has produced governmental policies

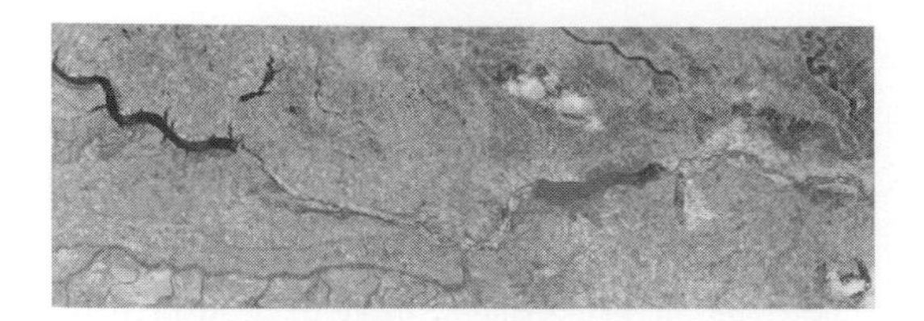

Plate 88 US Geological Survey LANDSAT image of the middle Missouri River in north-central United States. North is at the top of the image, which extends about 175 km east–west and about 88 km north–south. The Missouri River, largely lacking physical integrity, flows from the left (west) side of the image to the right (east) side. The dark, wide areas are reservoirs behind large dams, while the remaining connections between the dams and the next reservoir downstream are channels where fluvial processes are controlled by the dams. The Niobrara River enters the view from the lower left (south-west) corner of the image. It retains some of its physical integrity and does not include large dams

in many countries to restore rivers and their environments to more natural conditions. Although it is rarely possible to restore rivers to primeval anatural conditions, many nations have policies to promote the physical integrity of rivers, a term that often appears in legislation. Thus, the term integrity has its origins in legal language and usage.

From a scientific and engineering perspective, a physical integrity for rivers refers to a set of active fluvial processes and landforms wherein the channel, floodplains, sediments and overall spatial configuration maintain a dynamic equilibrium, with adjustments not exceeding limits of change defined by societal values (Plate 88). Rivers possess physical integrity when their processes and forms maintain active connections with each other in the present hydrologic regime (Graf 2001). Each term in this definition has particular meaning for the geomorphologist:

- streams and rivers: those parts of the landscape with confined water flow;
- fluvial processes and forms: those features related only to the fluvial domain;
- channel, near-channel landforms, sediments: channel area that is active in the present regime of the river (having a return interval of interaction with flow of 100 years or less), near-channel landforms include the functional surfaces that interact with fluvial processes and channels in the present regime, sediments that are active in the present regime;
- configuration: planimetric and cross-sectional arrangement of functional surfaces, landforms and sediments;
- dynamic equilibrium: the tendency for parameters describing the river to change annually about mean values which also change over periods of decades or centuries;
- limits of change defined by societal values: dimensional and spatial changes in forms and processes within ranges that are acceptable for economic, social or cultural reasons; changes greater than limits imposed by society result in re-engineering the channel to protect lives and property;
- present hydrologic regime: decade or century-long behaviour of daily stream-flow values for magnitude, frequency, duration, seasonality and rates of change.

Measurement of physical integrity for rivers depends on use of a few easily defined, readily assessed indicator parameters. These parameters must have strong roots in the geomorphological literature so that researchers may take advantage of existing knowledge and theory, but the parameters must also be understandable by decision-makers and the public. The parameters must be few in number, and be readily available or easily measured by non-specialists, because river management entails contributions from a variety of observers. Although the range of choices for such parameters in the literature is large (Leopold 1994), the following are the most commonly used and are measured at cross sections: daily water discharge, active channel width, sinuosity, pattern and particle size of bed material.

Of these indicator parameters, daily water discharge is the most important. These data, often collected and made widely available by governmental agencies, provide insight in to the primary driving forces and masses that control the river system. Human interactions with rivers often directly impact the water discharge, with subsequent effects rippling through the geomorphic system. Active channel width is the most easily assessed morphological variable for rivers, and it is the variable that is most sensitive to changes imposed by human activities through discharge adjustments. Classic hydraulic geometry usually shows that width adjusts more than depth or velocity with changes in discharge (Knighton 1998).

SINUOSITY and channel pattern are easily measured on aerial photography for small or medium-sized rivers, or on satellite imagery for large rivers. These parameters reflect upstream impacts of human activities that alter the delicate balance among water, sediment and channel form. Bed material size is readily assessed in field measurements and is sensitive to sediment supply as well as transport capacity (total amount of sediment the river is capable of transporting) and competence (the maximum size of particle that can be transported). Sediment discharge data are also informative, but such data are often not available because they are expensive to measure.

Physical integrity for rivers is important because it underpins biological integrity of river environments. Biological integrity, often characterized by biodiversity and sustainability of ecosystems, depends on the physical substrate of water, sediment and landforms. Efforts to restore rivers to more natural conditions are often thought of in biological terms by planners and managers, but restoration of the underlying physical system must occur first before the biological components of the system can assume more natural conditions.

The most significant human activity in reducing the physical integrity of rivers is the installation of dams. Dams segment the original river system and at least partially control the flow of water and sediment in downstream reaches. Dams reduce peak flows for flood protection, but high flows are important in activating functional surfaces near the channel. As a result, the floodplains downstream from many dams become inactive, causing substantial ecosystem changes that include wholesale vegetation changes. Birds and animals dependent upon the pre-dam vegetation face a loss of habitat under post-dam conditions when the physical and biological integrity of the river decline. In dryland settings, dams often divert the entire flow of the river, resulting in the dessication of reaches that once supported riparian habitat for diverse flora and fauna. In some urban areas, the original stream is replaced by a totally artificial channel without floodplains or other active features, and engineering works often seek to replace braided channels with single thread channels. The restoration of rivers with physical integrity in such cases is a scientific and engineering challenge (Brookes and Shields 1996; Petts and Carlow 1996) that must balance competing objectives subject to social valuation.

References

Brooks, A. and Shields, F.D. Jr (eds) (1996) *River Channel Restoration: Guiding Principles for Sustainable Projects*, New York: Wiley.

Dynesius, M. and Nilsson, C. (1994) Fragmentation and flow regulation of river systems in the northern third of the world, *Science* 266, 753–762.

Echeverria, J.D., Barrow, P. and Roon-Collins, R. (1989) *Rivers at Risk*, Washington, DC: Island Press.

Graf, W.L. (2001) Damage control: restoring the physical integrity of America's rivers, *Annals of the Association of American Geographers* 91(1), 1–27.

Knighton, D. (1998) *Fluvial Forms and Processes: A New Perspective*, London: Arnold.

Leopold, L.B. (1994) *A View of the River*, Cambridge, MA: Harvard University Press.

Petts, G. and Carlow, P. (eds) (1996) *River Restoration*, London: Blackwell.

SEE ALSO: floodplain

WILLIAM L. GRAF

PHYSIOGRAPHY

A word that has obscure origins, although it was in common currency in eighteenth-century Scandinavia, and in regular usage in the English-speaking world in the nineteenth century (see Stoddart 1975, for a historical analysis). Dana defined it in 1863:

> Physiography, which begins where geology ends – that is, with the adult or finished earth – and treats (1) of the earth's final surface arrangements (as to its features, climates, magnetism, life, etc.) and (2) its systems of physical movements and changes (as atmospheric and oceanic currents, and other secular variations in heat, moisture, magnetism, etc.).

One of the most notable exponents of physiography was the British naturalist T.H. Huxley, who published a highly successful text, *Physiography*, in 1877. Huxley's *Physiography* has some geomorphological content including chapters on 'the work of rain and rivers', 'ice and its work', 'the sea and its work', 'slow movements of the land' and 'the formation of land by Animal Agencies'. In the USA W.M. Davis preferred the term to GEOMORPHOLOGY, but he used it without the catholicity of meaning that it had for Huxley. Various other American geomorphologists, including J.W. Powell and N. Fenneman, divided up the USA into what they termed Physiographic Regions, Provinces or Divisions (see Atwood 1940).

References

Atwood, W.W. (1940) *The Physiographic Provinces of North America*, Boston: Ginn.
Dana, J.D. (1863) *Manual of Geology: Treating on the Principles of the Science*, Philadelphia: Bliss.
Huxley, T.H. (1877) *Physiography: An Introduction to the Study of Nature*, London: Macmillan.
Stoddart, D.R. (1975) 'That Victorian science': Huxley's *Physiography* and its impact on geography, *Transactions of the Institute of British Geographers* 66, 17–40.

A.S. GOUDIE

PIEZOMETRIC

A term used in the study of groundwater hydrology referring to the underground water pressure, though the term has particular relevance to underground aquifers and stability analysis of slopes (and PORE-WATER PRESSURE). Underground flow is largely unseen and so instruments called piezometers are used to provide an indication of pressure potential in the water table at particular depths. There are several types of piezometers, including well piezometers, standpoint piezometers and hydraulic piezometers, the commonest being a standpoint piezometer. A typical standpoint piezometer consists of a generally imperforated tube (typically 1–3 m length) inserted into the layer or horizon of interest, with a porous screen at one end to permit flow (about 25–50 mm diameter). Clean sand is placed around the screen, and the borehole surrounding the piezometer tube is filled with a seal (e.g. cement) to ensure that the pressure value given reflects only that on the screen tip. More comprehensive descriptions of piezometer instrumentation is provided in Goudie (1994: 237).

The piezometer gives the potential pressure in a soil or rock by measuring piezometric head, referring to the energy possessed by the water (also termed potentiometric head, hydraulic head and pressure head). Thus, the piezometric head at a point on the water table is the level above an appropriate datum (for instance sea level) that the water table reaches. Differences in head between points in the same water table will result in the transportation of energy (and underground flow) from the point of high piezometric head to that of lesser piezometric head. The velocity of flow between the two points should be directly proportional to the difference in head between, as long as all else remains equal (known as the piezometric gradient). The piezometric gradient can be influenced by external factors, such as precipitation (high amounts of precipitation resulting in a greater piezometric gradient).

By interpolating measurements of piezometric head, an imaginary surface, termed a piezometric surface (though the term potentiometric surface is preferable) can be formed. This represents the distribution of potential energy within the water body. A piezometric surface that lies above ground level will result in flowing water on land. This is termed an artesian well, and is typical of synclinal structures. Insufficient piezometric pressure to reach above ground level is termed subartesian. In unconfined aquifers, the slope of the piezometric surface defines the hydraulic gradient.

Maps of the piezometric surface can be developed, joining together points of equal piezometric head (using contours known as equipotential lines). Flow takes place perpendicular to these lines (down the piezometric gradient), and largely parallel to the overlying surface (Jones 1997: 93). Contour maps provide indications of the piezometric gradient and the pattern of the subsurface flow, and can also be used in stability analysis of soils and rocks.

References

Goudie, A.S. (ed.) (1994) *Geomorphological Techniques*, British Geomorphological Research Group, London: Unwin Hyman.
Jones, J.A.A. (1997) *Global Hydrology: Processes, Resources and Environmental Management*, Harlow: Longman.

STEVE WARD

PINGO

A pingo is a perennial PERMAFROST mound or hill formed through the growth of a body of ice in the subsurface. The term 'pingo' has been taken from a local Inuktitut word (meaning conical hill) used in the Mackenzie Delta region of the western Canadian Arctic. The term is now used globally to refer to this particular type of ice-cored mound, although in Siberia the Yakutian term *bulgannyakh* is sometimes used.

There are estimated to be some 5,000 or more pingos worldwide with the highest concentration occurring in the Tuktoyaktuk Peninsula area of the western Canadian Arctic, where there are around 1,350 pingos (Mackay 1998). Other areas with considerable numbers of pingos include the Yukon Territory (Canada), the Canadian Arctic Islands, northern Québec (Canada), northern Alaska (USA), northern areas of the former Soviet Union, the Svalbard archipelago (Norway) and Greenland. A few examples have been noted from Mongolia and the Tibetan Plateau.

Pingos can attain heights in excess of 50 m, with basal diameters of over 600 m. The ice volume of large pingos can exceed 1 million m^3. Whilst pingos can form almost conical hills, they may have more irregular forms and can be oval or elongate rather than circular in plan.

In order for a pingo to form and grow, water under pressure must be delivered to a position beneath the surface within a continuous or discontinuous permafrost environment. This water is frozen to form the ice core which is often described as intrusive or injection ice. As the pingo ice core grows the material overlying it (the pingo 'skin' or overburden) is forced upward forming the mound or hill.

Pingos can be classified as either hydraulic (previously termed 'open system' or 'east Greenland' type) or hydrostatic ('closed system' or 'Mackenzie Delta' type). This classification is based on the origin of the groundwater feeding the growth of the pingo ice core.

Hydraulic pingos are initiated by water under pressure of a hydraulic head/potential coming towards the surface in a valley bottom or lower valley side position and they are, therefore, features of high relief environments (for example, east Greenland, Alaska, Svalbard). The position of the upwelling of the ground water may shift over time and in this way a group or complex of pingos may develop within a relatively small area. The largest documented group is the 'Zurich Pingo group' in the Karup Valley of Traill Island, east Greenland (Worsley and Gurney 1996) which has some eleven pingos in various states of growth and decay.

Hydrostatic pingos are initiated by the drainage of a deep lake in a continuous permafrost environment. Following lake drainage the unfrozen saturated sediments that were beneath the lake are aggraded by permafrost and the pore water is progressively squeezed out. It is this water which feeds the growth of the ice core. In general the size of the pingo will be governed by the size of the lake basin from which it grows and usually one pingo will grow in the centre of the former lake basin. If the lake which drains has an irregular form and has two or more basins then one pingo may form in each of the basins.

Studies of the internal structure of hydrostatic pingos in the Tuktoyaktuk Peninsula have shown that beneath the pingo ice core there may be found a pressurized sub-pingo water lens and it is this water which feeds further growth of the pingo. Once such a pingo has become established there is little increase in its diameter and all subsequent growth is upwards, increasing its height. Similar details have not been proven for hydraulic pingos.

One of the longest surveys of a growing pingo was conducted on Ibyuk Pingo in the Tuktoyaktuk Peninsula (Mackay 1998). Although this large pingo was already some 47 m high at the initiation of the survey, the pingo was still seen to grow higher at an average rate of 2.7 cm per year during the survey period 1973–1994. Using this growth rate, along with other data concerning the geomorphological evolution of the area, suggests that Ibyuk may be of the order of 1,000 or more years old (Mackay 1998).

Growth of the pingo ice core leads to the progressive stretching of the overburden causing it to fail through cracking (the generation of dilation cracks) and slumping. The sediment cover at the summit will be thinnest due to this stretching and slumping and pronounced radial cracks may form here. The thinning and rupture of the thermally protective overburden will lead to the decay of the ice core and this will often result in the development of a crater at or near the summit which may contain a pond in summer. When pingos ultimately collapse, whether in a permafrost environment or due to climate change which sees the decay of the permafrost, they do so from the top down and invariably leave a circular or oval rampart surrounding a depression which may contain a pond or marshy area.

Since pingos only form in a permafrost environment, evidence of their previous existence can be used to infer the former presence of permafrost and hence they are extremely useful for palaeoclimatic reconstruction. The remains of pingos of Pleistocene age are often referred to as 'relict pingos' and such features have been documented from North America and western Europe. That these features have always been correctly interpreted, however, is still a matter of some dispute.

References

Mackay, J.R. (1978) Sub-pingo water lenses, Tuktoyaktuk Peninsula area, Northwest Territories, *Canadian Journal of Earth Sciences* 15, 1,219–1,227.

——(1998) Pingo growth and collapse, Tuktoyaktuk Peninsula area, western arctic coast, Canada: a long-term field study, *Géographie physique et Quaternaire* 52, 271–323.

Worsley, P. and Gurney, S.D. (1996) Geomorphology and hydrogeological significance of the Holocene pingos in the Karup Valley, Traill Island, northern east Greenland, *Journal of Quaternary Science* 11, 249–262.

Further reading

Gurney, S.D. (1998) Aspects of the genesis and geomorphology of pingos: perennial permafrost mounds, *Progress in Physical Geography* 22, 307–324.

SEE ALSO: ice wedge and related structures; palsa

STEPHEN D. GURNEY

PINNING POINT

Pinning points are topographic constrictions at which glaciers halt during advances or retreats. They are places where troughs shallow and/or narrow, bifurcate, join another valley or bend sharply. They operate at a range of scales; the fluctuations of CALVING GLACIERS in FJORD systems and ice-contact lakes are sensitive to pinning points, and the topography of land masses and continental shelves affects the behaviour of the floating extensions of ICE SHEETS (ice streams and ice shelves). At non-calving GLACIERS ice is lost primarily through melting so that ice losses are closely tied to climate change. At calving glaciers, however, calving may represent a significant proportion of total ablation, yet calving is only indirectly affected by climate. Calving rates increase with water depth and with the cross-sectional area of the calving terminus, so the mass of ice lost through calving is determined primarily by these non-climatic factors. This means that the fluctuations of calving glaciers can be largely controlled by the topographic geometry of the valley. At pinning points calving rates are reduced, and they therefore represent places of enhanced stability where ice losses are balanced by ice supply. If a glacier retreats from a pinning point into deep water it must continue to retreat until calving rates decrease to match ice supply at a pinning point upstream. Equally, a calving glacier may be unable to advance during periods of regional glacier growth if such an advance would take its terminus into deep, open water. Glacier stillstands are common at pinning points and large MORAINES may be constructed at these locations. Because these halts are determined by topography and not by climate, such moraine systems may have limited palaeoclimatic significance.

Further reading

Vieli, A., Funk, M. and Blatter, H. (2001) Flow dynamics of tidewater glaciers: a numerical modelling approach, *Journal of Glaciology* 47, 595–606.

SEE ALSO: mass balance of glaciers

CHARLES WARREN

PIPE AND PIPING

Natural soil pipes are linear voids formed by flowing water in soils or unconsolidated deposits (Plate 89). They occur throughout the world and vary from a few millimetres to several metres in diameter (Figure 122). Attempts have been made to provide a quantitative distinction between pipes and other soil macropores based on size, but none has been entirely satisfactory. The most fundamental property of soil pipes is that they actively drain water through the soil, which means that 'connectivity' and a drainage outlet are generally more critical than size in defining a pipe.

As pipes develop, they tend to create subsurface drainage networks akin to surface streams. Horizontal networks up to 750 metres long with

Plate 89 Pipe outlets in a riverbank in the English Peak District

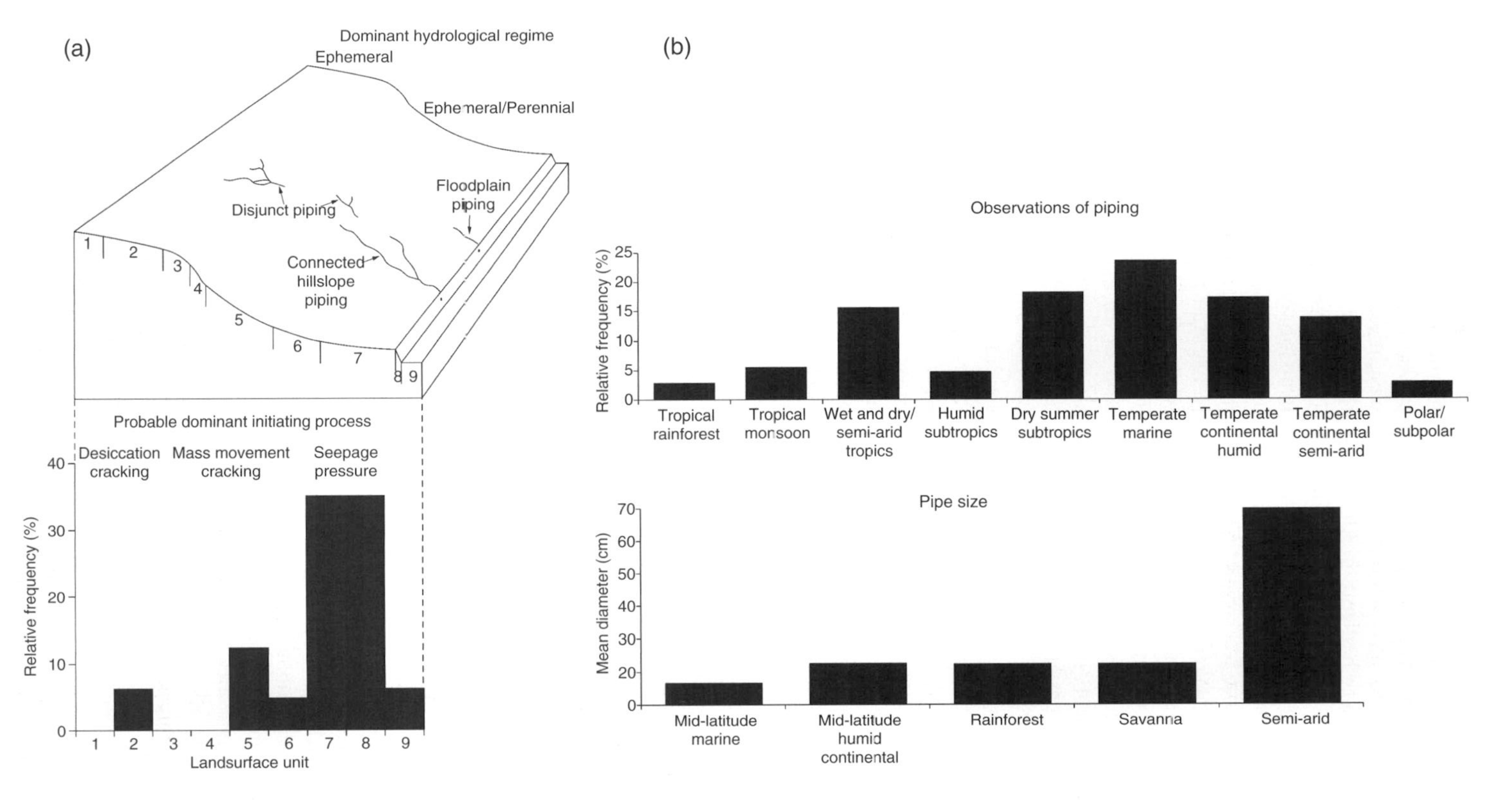

Figure 122 Geomorphic and climatic distribution of pipes: (a) frequency of piping in different landsurface units (Conacher and Dalrymple's (1977) NULM classification); (b) frequency and mean size of pipes in different climatic zones

many branching tributaries, reminiscent of dendritic stream networks, have been reported in shallow upland soils in Britain. In contrast, in badlands or deep loess deposits the networks tend to be more three-dimensional, as in the Loess Plateau in northern China or badlands in Alberta, Arizona, South Africa and Spain. Sometimes these consist of a number of horizontal networks formed above impeding layers with lower permeability and greater resistance to erosion linked by pipes eroding through the layers.

Pipes may be both a cause and a product of GULLYing. Gullies or channels may be formed when pipe roofs collapse. Pipe collapse provides an alternative to the 'classical' theory of channel extension based on headcutting by OVERLAND FLOW entering open channels (Jones 1987a). Conversely, gullies may trigger pipe formation by increasing the surrounding hydraulic gradient. American engineers have been concerned about pipes causing riverbank collapse (Hagerty 1991a,b). In such cases, traditional tests of bank stability based on the properties of the bank material may not give a good indication of risk, as the triggering mechanism may owe more to the source of pipeflow many tens of metres away.

The relationship between piping and LANDSLIDES is equally ambivalent. Pipes may initiate slides if they become blocked or water pressures exceed a critical threshold. Conversely, many peat bogs in Ireland that have well-developed piping do not display the periodic 'bog burst' phenomenon found in those lacking pipes. The pipes may prevent the buildup of water pressure. Again, pipes can develop in mass movement cracks following landslides.

These relationships indicate that shallow subsurface erosion processes can be significant agents in landscape development, sometimes with cyclical collapse and renewal. The term 'pseudo-karst' has been used for such landscapes, because the KARST-like features are predominantly formed by mechanical erosion rather than solution. The term SUFFOSION aptly evokes the mechanical winnowing and scouring. Clearly, the role of subsurface erosion is still undervalued in geomorphology, particularly in modelling.

The mechanics of pipe initiation were first defined by Karl Terzaghi in the founding years of the science of soil mechanics, because piping was the cause of many failures of earth dams. The process begins when seeping water produces sufficient force to entrain material at the seepage outlet point, which may be on a hillslope, a cliff face, riverbank or dam toe slope. The essential feature of Terzaghi's mechanism is that the *water pressure* renders the soil particles or aggregates weightless by counterbalancing gravity, so that the soil seems to 'boil' away (Terzaghi and Peck 1966). Seepage is increasingly focused on the head of the hole as it eats back into the bank. This may be called 'true' piping in the engineering sense. It is quite possible, however, for piping to begin at the upslope end, even on earth dams where it has been called 'rainfall erosion tunnels'. This second process involves progressive expansion of an existing conduit mainly through the shear stress exerted by the flowing water. Cracks caused by desiccation are commonly exploited by rainwater, which preferentially selects and erodes those cracks that run downslope and keeps them open underground even after re-expansion closes the cracks at the surface. Mass movement cracks, tree roots and animal holes can also be exploited. This process has often been distinguished as 'tunnel erosion'. However, the exact mode of initiation may not always be apparent from looking at the resulting landforms in the field. Indeed, in many cases the two processes interact in a very complex manner, as even those who have advocated distinguishing between them admit (Dunne 1990).

Reports of pipes and tunnels from almost every climatic region of the world clearly demonstrate that there is no single, unique set of initiating factors. One of the few truly universal contributory factors is a sufficient water supply to create the necessary pressure or shear stress. This water supply may be very spasmodic and some of the larger pipes occur in drylands, where short, intense rainstorms or the occasional rapid snowmelt event are the main cause.

Steep hydraulic gradients are also commonly needed, typically generated by a combination of water surplus and local relief. Piping is therefore most common where there is a high local relative relief, be it on upland hillslopes, in deeply incised badlands or simply in a riverbank.

Impeding layers within or beneath the soil, which concentrate vertical seepage and divert flow horizontally downslope, are also amongst the most universal preconditions. These are most effective where the layer conducting the flow is more erodible and/or contains pre-existing macropores, especially desiccation cracks, to speed the flow. There is a clear distinction here between the most common initiating factors in humid lands and those in the drylands. Whilst prior cracking aids pipe development in both climatic realms, most dryland piping is linked with dispersible soils, which are not common in humid lands.

Individual soil particles are more easily washed away than soil aggregates. Soil dispersal is predominantly a chemical process. It generally depends on the three-way balance between the total concentration of dissolved cations in the water flowing through the soil, the percentage of soluble or exchangeable sodium in the cation content of the soil itself and the type of clay minerals. In soils bonded by clays, concentrated salt solutions or weakly soluble salts like gypsum, dispersion may be accomplished by the dilution of the bonding salts by seepage water with low salinity. Alternatively, it may be produced by the chemical exchange of cations between the water and the soil, especially the replacement of divalent cations like calcium with monovalent cations like sodium in the soil. Sodium increases the repulsive forces between mineral particles and is the main dispersing or deflocculating agent. However, deflocculation does not necessarily increase erodibility and pipe development. Deflocculated soils may be stable if the sodium content of the soil is high enough to so thoroughly disperse the soil that permeability is reduced below a critical level, and to so reduce the structural stability of the soil that any voids that do develop quickly collapse and fill in. Experimental studies have identified an area of potential instability, swelling clays and deflocculation where piping is likely to be initiated in the transition zone between stable dispersion (high soil sodium and low salinity water) and stable flocculation (low soil sodium and saline seepage water). The exact boundaries between these zones vary according to the predominant clay mineral species. Montmorillonite is generally the least stable, with the broadest zone of instability. Nevertheless, evidence from earth dam engineers in America has revealed that the situation can be rather more complex, and many cases of piping have been observed in earth dams that fall well within the so-called stable zones (Sherard and Decker 1977).

Clay minerals affect the mechanical properties of the soil as well as dispersibility. Minerals of the smectite group like montmorillonite are noted for their higher rates of expansion and contraction. This increases susceptibility to desiccation cracking and infiltration rates. However, the expansive and dispersive properties of montmorillonites depend on the variety of the mineral. Sodium montmorillonite is more expansive than calcium or aluminium montmorillonite, and montmorillonites with cation substitution in their tetrahedral layer rather than their octahedral layer will not display the usual dispersive properties.

The minimum rate of seepage required to initiate piping depends on many properties of the soil, e.g. a threshold of $0.1\,mm\ s^{-1}$ in non-dispersive silts against only 0.001 in dispersive clays. Soils that contain impeding horizons, are subjected to periodic desiccation and intense rainstorms, have high susceptibility to cracking, especially clayey or organic soils, and/or contain highly erodible, especially dispersible, layers are the most prone to piping.

Human interference has often increased the incidence of piping. Deforestation and devegetation decrease evapotranspirational losses, increasing water surplus. They expose soils to more desiccation and reduce the stabilizing effect of roots. In these situations, piping generally combines with other processes of accelerated erosion causing land degradation and increasing flood hazard (Jones 1981).

Considerable research has been undertaken into methods of rehabilitating farmland damaged by piping in Australia and New Zealand (Crouch *et al.* 1986). These have successfully included planting trees and grasses with high evapotranspiration rates and deep, stabilizing root systems, as well as adding soil conditioners to improve crumb structure. Even so, agricultural land is still being lost to piping, especially in drylands and through over-irrigation, e.g. in Arizona and Spain. Nearly half the farmland in the San Pedro valley, Arizona, has been lost to piping (Masannat 1980).

Research on the hydrology of piping has shown that pipes can contribute significant amounts of water, especially to upland rivers. Nearly half of floodflow and baseflow in one Welsh headwater tributary is derived from ephemeral or perennially flowing pipes (Jones 1987b). The pipes generally flow when the water table rises to pipe level and their discharge is a variable mix of new rainfall and 'old' water pushed out by the new. Monitoring in Canada, Japan, China, India and Britain generally confirms their role in speeding runoff and increasing floodflows. Most contributions fall between 20 and 50 per cent of streamflow, though amounts vary considerably in both time and space, even within the same basin, and in some cases are insignificant (Jones and Connelly 2002). Comparison with the response characteristics of other hillslope drainage processes suggests that pipeflow falls between diffuse throughflow and saturation overland flow in both timing and volume.

References

Conacher, A. and Dalrymple, J.B. (1977) The nine unit landsurface model: an approach to pedogeomorphic research, *Geoderma* 18(1/2), 1–154.

Crouch, R.J., McGarity, J.W. and Storrier, R.R. (1986) Tunnel formation processes in the Riverina area of N.S.W., Australia, *Earth Surface Processes and Landforms* 11, 157–168.

Dunne, T. (1990) Hydrology, mechanics, and geomorphic implications of erosion by subsurface flow, in C.G. Higgins and D.R. Coates (eds) *Groundwater Geomorphology: The Role of Subsurface Water in Earth-surface Processes and Landforms*, 1–28, Boulder, CO: Geological Society of America Special Paper 252.

Hagerty, D.J. (1991a) Piping/sapping erosion: I Basic considerations, *Journal of Hydraulic Engineering* 117(8), 991–1,008.

——(1991b) Piping/sapping erosion: II Identification-diagnosis, *Journal of Hydraulic Engineering* 117(8), 1,009–1,025.

Jones, J.A.A. (1981) *The Nature of Soil Piping: A Review of Research*, Norwich: GeoBooks.

——(1987a) The initiation of natural drainage networks, *Progress in Physical Geography* 11(2), 207–245.

——(1987b) The effects of soil piping on contributing areas and erosion patterns, *Earth Surface Processes and Landforms* 12(3), 229–248.

Jones, J.A.A. and Connelly, L.J. (2002) A semi-distributed simulation model for natural pipeflow, *Journal of Hydrology* 262, 28–49.

Masannat, Y.M. (1980) Development of piping erosion conditions in the Benson area, Arizona, U.S.A, *Quarterly Journal of Engineering Geology* 13, 53–61.

Sherard, J.L. and Decker, R.S. (eds) (1977) *Dispersive Clays, Related Piping, and Erosion in Geotechnical Projects*, American Society for Testing Materials, Special Technical Publication No. 623.

Terzaghi, K. and Peck, R.B. (1966) Soil Mechanics in Engineering Practice, 3rd edition, New York: Wiley.

Further reading

Higgins, C.G. and Coates, D.R. (eds) (1990) *Groundwater Geomorphology: The Role of Subsurface Water in Earth-surface Processes and Landforms*, Boulder, CO: Geological Society of America Special Paper 252.

Jones, J.A.A. (1997) Subsurface flow and subsurface erosion: further evidence on forms and controls, in D.R. Stoddart (ed.) *Process and Form in Geomorphology*, 74–120, London: Routledge.

Jones, J.A.A. and Bryan, R.B. (eds) (1997) *Piping Erosion*, Special Issue, *Geomorphology*, 20 (3–4).

J. ANTHONY A. JONES

PLANATION SURFACE

Topographical surfaces which are nearly flat over longer distances are called in geomorphology 'planation surfaces'. Some relief is allowed, especially in the form of isolated residual hills, but otherwise slope gradients should be very low and drainage lines should not be incised. Ideally, planation surfaces should cut across bedrock structures.

Thus, the descriptive definition is a simple one, but the issue of planation surfaces is one of the more controversial in geomorphology. As the name implies, the state of low relief has been achieved in the course of planation of a formerly higher relief by means of various exogenic agents of destruction (Plate 90). There are at least a few points persistently disputed, including the nature of process, or processes, leading to planation, the meaning of planation surfaces in long-term landscape evolution, and the possibility of producing and maintaining flat surfaces without recourse to erosional processes acting over protracted time spans (Figure 123).

Several mechanisms have been proposed to account for the origin of near-level surfaces. Accordingly, specific types of planation surfaces are distinguished. These include PENEPLAINS formed by peneplanation (Davis 1899),

Plate 90 Planation surfaces ideally should truncate geological structures, as it is in the case of a coastal surface in the Gower Peninsula in south Wales. The actual origin of the surface, whether wave-cut or subaerial, is the subject of debate

pediplains formed by pediplanation (King 1953), etchplains formed by etchplanation (see ETCHING, ETCHPLAIN AND ETCHPLANATION) (Büdel 1957). Other, more localized modes of formation are planation by sea waves in the coastal zone, by frost processes in the periglacial environment (see CRYOPLANATION), by areal glacial erosion, or by ubiquitous salt weathering in some desert and coastal situations.

The discussion between protagonists of peneplanation and pediplanation is now to much extent historical. In short, the principal difference between the two models resides in the way of slope development. Peneplains develop primarily through downwearing, i.e. slope lowering. In the course of peneplanation divides are lowered, slopes become gentler with time, and the landscape is progressively graded towards BASE LEVEL. By contrast, in the pediplanation model slope retreat away from drainage lines, i.e. backwearing, plays a crucial part. Higher ground may persist for much longer than the peneplanation model would imply, but their areal extent diminishes in time as bounding escarpments retreat. In front of scarps gently inclined surfaces of PEDIMENTs form and then coalesce to form a regional, ever-growing planation surface of the pediplain type. Another difference is that pediplains are not necessarily graded towards base level and they may form stepped landscapes and develop simultaneously at different altitudes. In both theories, planation surfaces are the ultimate products of long-term landform development and need a long time to form, perhaps of the order of $>10^7$ my. Neither peneplanation nor pediplanation are geomorphic processes per se; rather, they include a variety of superficial processes, including fluvial erosion, surface wash and various categories of mass movement.

Etchplanation was initially seen as a specific variant of peneplanation, applicable to low latitudes, where deep chemical weathering is ubiquitous. However, it was shown later that the mechanism of planation is fundamentally different. Etchplains form in the subsurface, through rock decay which is intense enough to overcome local differences in rock resistance against weathering. This leads to the development of a planar boundary between weathered material and solid rock beneath (see WEATHERING FRONT). Subsequent removal of weathering products exposes the planar 'etched' topography, which now forms an etchplain. Etchplains are thus two-stage features, as opposed to peneplains and pediplains. Although almost featureless etchplains, fulfilling the descriptive criteria for a planation surface, have been described from several areas, the view seems to prevail now that long-term etching leads to diversification rather than to planation of relief. Many low-latitude surfaces of low relief are cut across the weathering mantle, but the hidden topography of the weathering front is much more varied.

As the residual topography rarely gives a clue to the mode of formation of surfaces of low relief, it is preferable to call them simply 'planation surfaces', without genetic connotations. Moreover, it is likely that plains of long geomorphic history have been shaped by various processes, alternating over time, hence they would be 'polygenetic' surfaces rather than any 'monogenetic' peneplains, pediplains, or etchplains (Fairbridge and Finkl 1980). For example, pedimentation may be a means of stripping products of deep weathering to expose an etched surface.

Marine action used to be a favoured mode of planation, and in the nineteenth and early twentieth centuries many flat surfaces, especially in Britain, were identified as 'abrasion surfaces', even if no marine sediments could have been demonstrated. Later studies have shown that wave-trimmed surfaces of regional extent are unlikely to exist, and abrasion platform would have only limited extent (King 1963). Demonstration of a subaerial origin for many surfaces previously claimed as of marine origin, undermined the concept even further.

Periglacial planation is supposed to be achieved by means of simultaneous action of frost weathering of bedrock, rock cliff development and retreat, and mass movement, chiefly solifluction. The resultant cryoplanation is essentially a variant of pediplanation, applied to high latitudes and the Pleistocene. As with marine planation, cryoplanation is now generally seen as unlikely to account for the origin of more extensive surfaces, which are mostly inherited from pre-Pleistocene times.

Extensive level terrains in the Canadian and Fennoscandian Shield were long believed to have been shaped by powerful glacial erosion exerted by consecutive ice sheets during the Quaternary. Later research has shown that areal glacial erosion is not as common as formerly thought (Sugden and John 1976). By contrast, remarkable flatness of basement surfaces is inherited from protracted pre-Quaternary subaerial development, whilst much of these surfaces are exhumed Precambrian features (Lidmar-Bergström 1997).

The recognition of the powerful role of salt weathering in low-lying desert environments has led to the proposal that this process may have been crucial in producing flat topography of some coastal plains or closed depressions such as Quattara in Egypt. The term 'haloplanation' has been suggested (Goudie and Viles 1997).

Planation surfaces occupied a central position in geomorphology in the days, when establishing DENUDATION CHRONOLOGY of a given area was considered the main objective of geomorphology and cyclic development of landforms served as a paradigm. The first step in geomorphic research was to identify planation surfaces in the present-day landscape, or more often their remnants surviving on divides after the landscape had been dissected. It was followed by recognition of

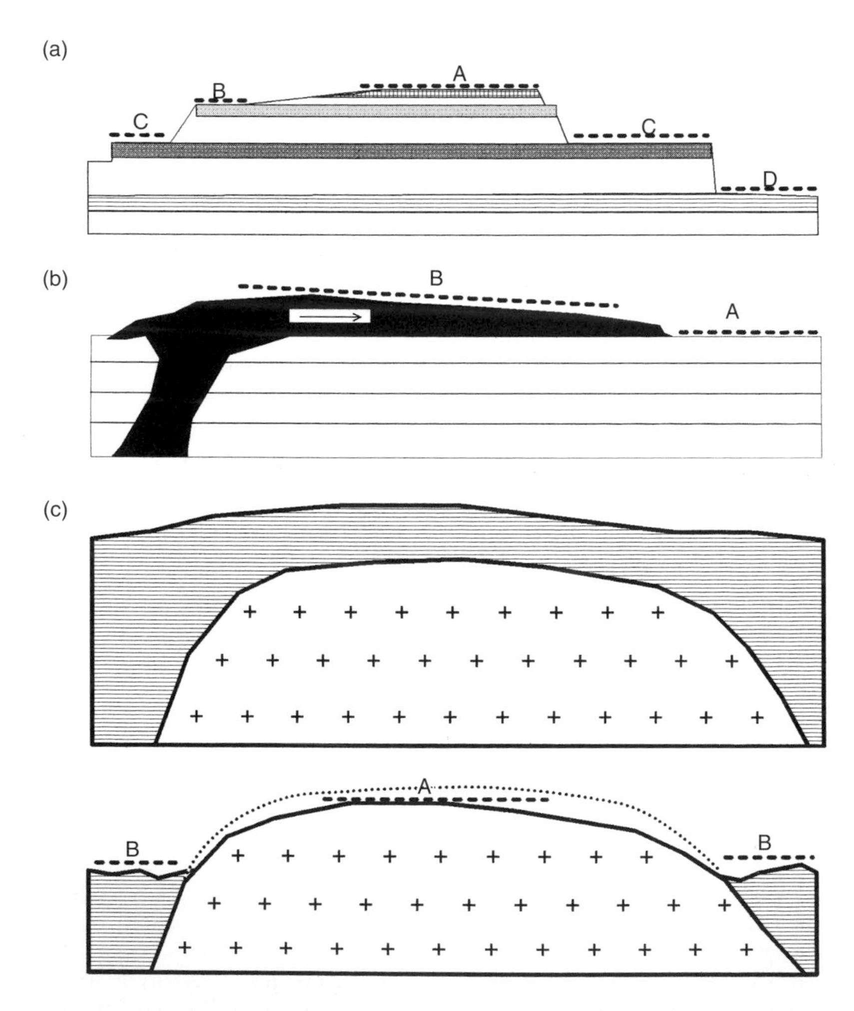

Figure 123 A diagram to show that not all plains are products of protracted planation. In areas built of flat-lying sedimentary rocks benches are distinctively controlled by structure (a). Flatness of surfaces underlain by lava flows may be largely primary and relate to the low viscosity of lava (b). The surface 'B', although located at higher altitude, is actually younger than the true planation surface 'A'. Surfaces of low relief common for many granite batholiths may be related to the original dome form assumed during emplacement (c). Different altitudes of surfaces cut in granite ('A') and country rock ('B') reflect different resistance of each rock complex rather than being indicators of different ages

altitude range of their occurrence, correlation of surfaces over wider areas and classification by height. The number of planation surfaces indicated the number of erosional cycles experienced by the landscape. As this approach frequently ignored influences differential tectonics may have had on landform development, relied on highly uncertain correlation procedures, and suffered from deficiency of accurate dating of surfaces, it began to be severely criticized in the 1960s and lost much of its popularity. Thus, whereas the issue of planation surfaces, their origin and chronology, features prominently in older regional studies, its importance has been greatly reduced in modern geomorphology. In addition, preoccupation with planation surfaces in historical geomorphology created an incorrect view that a search for ancient landscapes is essentially a search for planated relief. Recent work indicates that many palaeosurfaces preserved in geological and geomorphic record had a very complex topography, far from any state of advanced planation.

However, planation surfaces have retained significance in morphotectonic studies and are widely used as indicators of uplift and subsidence histories (Ollier and Pain 2000). By analysing spatial and altitude patterns of their distribution one can infer magnitudes of surface uplift, recognize direction of tilting and amount of warping, or locate fault zones in areas where conventional geological evidence is not at hand. In this context, the origin of a planation surface is usually not critical to the argument.

It is important to remember that the surface morphology alone may not give the clue to the reasons of flatness. Examination of geological structure underlying a flat topography can suggest origins alternative to the one implied by the name 'planation surface'. Moreover, if geological control can be demonstrated, tectonic quiescence is not a prerequisite any longer as inherent in the 'classic' modes of formation.

In many platform areas sedimentary strata lie horizontally over very long distances and topography may be adjusted to this negligible dip, especially if resistant rock layers occur at the surface. Likewise, backslopes of CUESTA ridges may show close adjustment to the dip of strata. In elevated plateaux there may exist several levels of 'planation surfaces' separated by escarpments, but in reality these are structural surfaces, each following a more resistant layer.

In formerly volcanic areas, topography may be adjusted to the geometry of lava flows. Basaltic lava, because of its low viscosity, may extrude in sheets of more or less uniform thickness over large areas. Flatness of the top surface of a flow will be then inherited from the time of extrusion and cooling. In case of multiple flows, denudation may expose top parts of each major lava flow and produce stepped topography, reminiscent of a generation of planation surfaces of various ages. In fact, all benches have structural foundation and may all have similar ages.

Remnants of ancient planation surfaces have also been sought in areas underlain by igneous rocks, especially by granite, which indeed often display a gently rolling topography or resemble laterally extensive, low-radius domes. A variety of structural controls is possible, including adjustment of form to the original roof of the intrusion, or to flat-lying joints.

That planation surfaces exist on Earth is indisputable. Examples are known from throughout the geological record, from Precambrian up to the present. In places such as the Fennoscandian Shield, Laurentide Shield or the Middle East, extensive surfaces of extreme flatness truncating various bedrock structures and disregarding differential rock resistance existed by the end of Precambrian. They were subsequently buried by Cambrian sediments and form sub-Cambrian planation surfaces, nowadays partially exhumed. Another generation of extensive planation surfaces evolved in the Mesozoic. Many upland areas are typified by the occurrence of surfaces of low relief at different altitudes, which probably have formed during the Cenozoic. It is the meaning, the mode of origin and age range of their formation which remain contentious and, paradoxically, underresearched issues.

References

Büdel, J. (1957) Die 'Doppelten Einebnungsflächen' in den feuchten Tropen, *Zeitschrift für Geomorphologie N.F.* 1, 201–228.

Davis, W.M. (1899) The Geographical Cycle, *Geographical Journal* 14, 481–504.

Fairbridge, R.W. and Finkl, C.W. Jr (1980) Cratonic erosional unconformities and peneplains, *Journal of Geology* 88, 69–86.

Goudie, A. and Viles, H. (1997) *Salt Weathering Hazards*, Chichester: Wiley.

King, C.A.M. (1963) Some problems concerning marine planation and the formation of erosion surfaces, *Institute of British Geographers Transactions* 33, 29–43.

King, L.C. (1953) Canons of landscape evolution, *Geological Society of America Bulletin* 64, 721–752.
Lidmar-Bergström, K. (1997) A long-term perspective on glacial erosion, *Earth Surface Processes and Landforms* 22, 297–306.
Ollier, C.D. and Pain, C.F. (2000) *The Origin of Mountains*, London: Routledge.
Sugden, D.E. and John, B.S. (1976) *Glaciers and Landscape. A Geomorphological Approach*, London: Edward Arnold.

Further reading

Phillips, J.D. (2002) Erosion, isostatic response, and the missing peneplains, *Geomorphology* 45, 225–241.
Twidale, C.R. (1983) Pediments, peneplains and ultiplains, *Revue de Géomorphologie Dynamique*. 32, 1–35.
Widdowson, M. (ed.) (1997) *Palaeosurfaces: Recognition, Reconstruction, and Palaeoenvironmental Interpretation*, London: Geology Society Special Publication 120.

PIOTR MIGOŃ

PLATE TECTONICS

Plate tectonics is a unifying theory that explains many of the major features of the Earth's lithosphere such as VOLCANOes, rift (see RIFT VALLEY AND RIFTING) zones and mountain belts. The theory asserts that the lithosphere – the crust and uppermost mantle – is divided into rigid bodies or 'plates' that move horizontally and interact at their boundaries to produce these features. The theory evolved from the earlier concepts of continental drift and SEAFLOOR SPREADING.

In the eighteenth century it was noted that the coastlines of the Atlantic Ocean fit together like a jigsaw puzzle. In 1915 Alfred Wegener published several geological arguments to hypothesize that the continents had drifted apart from a supercontinent named Pangaea. He pointed to features that could be aligned by closing the Atlantic including Palaeozoic fold belts like the Appalachian and Caledonian mountains (Figure 124), metamorphic shields like northern Scotland and Labrador, major faults, palaeoclimatic indicators such as Carboniferous subtropical coal beds and tropical evaporites and desert sandstones, tillites from a Carboniferous ice cap with radiating striations and glacial erratics, and unique fossils of ferns and reptiles. Shortly thereafter he noted that seismic velocities in oceanic rocks were faster than in continental rocks, indicating that the less dense continents would float on top of oceanic rocks.

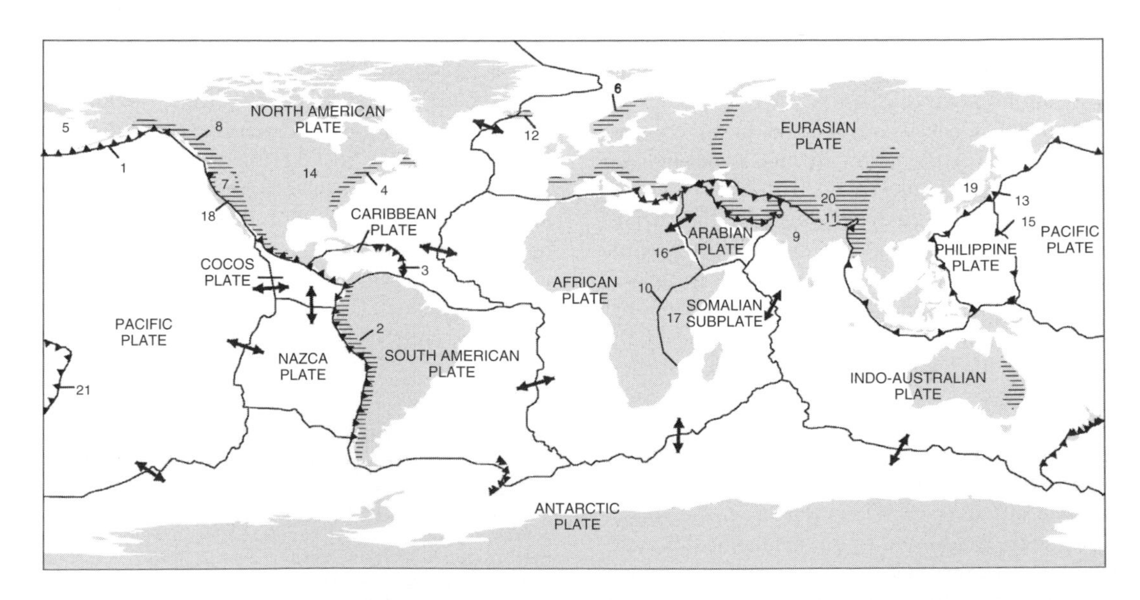

Figure 124 Plate boundaries of the Earth's crust. Features mentioned in the text: 1: Aleutian island arc; 2: Andes mountains; 3: Antilles island arc; 4: Appalachian mountains; 5: Bering Sea; 6: Caledonian mountains; 7: Columbia River basalts; 8: Cordilleran mountains; 9: Deccan basalts; 10: East African rift; 11: Himalayan mountains; 12: Iceland; 13: Japanese island arc; 14: Keweenawan aulacogen; 15: Mariana trench; 16: Mount Kilimanjaro; 17: Red Sea; 18: San Andreas fault; 19: Sea of Japan; 20: Tibetan plateau; 21: Tonga trench

Wegener's arguments were widely discredited for several decades. One exception was Alexander Du Toit who observed in 1937 that Pangaea split into the supercontinents of Laurasia (North America and Eurasia) and Gondwanaland (South America, Africa, India, Australia and Antarctica) in Triassic time, and that they split into the present continents in Jurassic time.

Geologic mapping increased dramatically worldwide in the 1950s, adding abundant evidence in support of continental drift. Critical geophysical arguments were added. For example, the advent of computers and related statistics enabled Bullard *et al.* (1965) to show the high probability of fit along the Atlantic continental margins. K-Ar radiometric age DATING METHODS and aeromagnetic anomaly maps confirmed the match of exposed and buried lithologies across the South Atlantic. Seafloor exploration confirmed that seafloor and continental rocks differed. Palaeomagnetic studies proved most persuasive. By measuring a rock unit's fossilized remnant magnetization in many samples, its location can be calculated relative to the Earth's palaeomagnetic pole with a known error. By comparing poles from coeval rock units on two continents, the relative rotation and displacement between them can be determined. Such studies statistically proved continental drift (Irving 1958).

Arthur Holmes in 1928 realized that radioactive decay heat would cause the Earth's mantle to convect and push continents along the Earth's surface. Dietz (1961) suggested that seafloor spreading occurred when convecting mantle magma intruded into the seafloor crust at mid-oceanic ridges, causing ACCRETION of new seafloor and carrying older seafloor away in both directions. Tests of the hypothesis followed, relying heavily on sonar maps of the submarine geomorphology and aeromagnetic maps of the seafloor that had been produced to track nuclear submarines. Quickly it was shown, on going perpendicularly away from a mid-oceanic ridge, that the youngest rock ages on oceanic islands increases linearly, that the K-Ar radiometric and fission track (see FISSION TRACK ANALYSIS) ages of seafloor basalts increase linearly, that the seafloor cools, and that the microfossils and tuffs in seafloor sediments record increasing ages – all supporting seafloor spreading. Most critical was the recognition that reversals of the Earth's magnetic field every 10^4 to 10^7 Ma were recorded in new lavas at the mid-oceanic ridges, which then spread away equally in both directions to create linear rift-parallel anomalies of increased and decreased magnetic field intensity. This analysis both confirmed the seafloor spreading hypothesis and provided a way to easily map and date the seafloor, which represents about two-thirds of the Earth's surface (Heirtzler *et al.* 1968). The plate tectonic theory evolved to explain how the Earth's crust is moving everywhere systematically.

The Earth's surface has seven large plates – the Pacific, Eurasian, North American, South American, African, Indo-Australian and Antarctic plates, five medium-sized plates (Figure 124), and an uncertain number of small plates in between. Plates move at rates up to about 20 cm yr^{-1} although the norm is only 5–10 cm yr^{-1}. This motion can be directly measured using very long baseline interferometry, satellite laser ranging, and satellite radiopositioning using the Global Positioning System's satellites. Past rates are measured using the spacing of seafloor magnetic anomalies and palaeomagnetic methods. Further, because crustal plates are fitted around a sphere, each plate actually rotates about its own Euler pole with a varying rate of motion across it. Plate boundaries have three types of ACTIVE MARGINS: rift zones, subduction zones and transform faults. Each may involve oceanic, continental or both types of crust and each combination of boundary and crust has its own typical topographic expression, dynamics and geologic characteristics.

Rift zones and ridges

Rift zones are linear zones of extension where new crust is added to the Earth's surface and where the motion is perpendicularly away from the lineament. Topographically, they have a central depression, typically 20 to 50 km wide and 1½; to 3 km deep, between uplifted plateaux that extend outwards for hundreds of kilometres on either side. Below the depression, convecting upwelling magma splits at the crust–mantle interface, pushing older crust upward and outward on inward-dipping normal faults to form the plateaux. Decreased lithostatic loading in the depression causes melting of the uppermost mantle to form a mafic magma. It intrudes along the rift lineament, particularly along its central axis, to form sheeted gabbroic intrusions at depth and basaltic lavas on the depression's floor, creating a new crust about 5 km thick. Because the crust is thin and the forces tensional, rift zones generate numerous relatively weak, shallow earthquakes.

On the seafloor, the rift zones are called 'ridges'. They form a great continuous submarine mountain system through the world's oceans that extends for about 60,000 km (Figure 124), but is exposed in only a few places such as Iceland. Typical mid-oceanic ridge basalts (MORB) are olivine tholeiites with only minor compositional variations. Erosion is minimal in the oceans, so that only a thin veneer of mostly chemical and biochemical sediments precipitate on the seafloor volcanics as siliceous and calcareous oozes. These sediments increase slowly in thickness progressively away from the ridge, typically at a rate of about 1 mm/10^2 yr. In the ridge depression, high heat flow from the Earth causes abundant geothermal activity. The hot hydrothermal fluids can precipitate elegant chimney structures around their vents that are rich in metal sulphides to form ore deposits. Also the fluids chemically alter the basalts and gabbros as they pass through and they create a marine microenvironment with a diverse range of exotic flora and fauna. Heat flow, seismic and gravity studies show that the seafloor crust progressively cools, increases in density and sinks, and thickens by underplating as magma freezes on its lower surface as it is pushed further away from the rift.

The East African rift zone is about 5,000 km long and exemplifies how the continental crust of a CRATON is split (Figure 124). Such rift zones are topographically and seismically similar to oceanic ridges. Because the bounding plateaux are exposed to WEATHERING, minor chemical sediments occur around hot springs only but EROSION produces abundant terrestrial clastic sediments that are deposited in the rift valley and the volcanic plateaux are progressively attacked by surficial weathering away from the valley. Tholeiite basalts sometimes fill and overflow the rift, producing large plateau lava sequences such as the Columbia River or Deccan basalts. Also, alkali basalt volcanism may occur in the adjacent plateaux through partial melting of continental crust, creating intrusive plugs, extrusive cinder cones and spectacular stratovolcanoes such as Mt. Kilimanjaro with its equatorial GLACIER peak. If the rift zone is truly active, the continental crust is entirely split in 10 to 20 Ma, the intervening crust becomes a mid-oceanic ridge, the sea invades as has happened in the Red Sea, and the continental edges become PASSIVE MARGINS.

Subduction zones

Subduction zones are convergent margins where the Earth destroys old crust to make room for the new crust created in rift zones. As two plates converge, one plate bends downward and is driven back into the Earth's mantle at an angle to be remelted. A subducted slab can reach depth extents of 1,400 km long and reach a depth of 700 km. The plates may converge head on or at a substantial angle, resulting in strong compressive forces that produce numerous powerful earthquakes with foci along the depth extent. In fact, about 90 per cent of all earthquake energy is released in subduction zones so that their locations are well known from seismology studies. Three types of crustal collisions are possible: ocean-to-continent, ocean-to-ocean, and continent-to-continent.

When oceanic crust collides with continental crust, as is happening around most of the Pacific Ocean (Figure 124), the thin (5–10 km) denser (~2.9 g/cc) oceanic plate dives under the thicker (30–50 km) less dense (~2.7 g/cc) continental plate at a 30° to 70° angle. At the surface a trench is formed that may be several thousand kilometres long in plan with typical widths of 50 to 100 km and depths of 7 to 9 km. Thus the epicentres for shallow earthquakes occur at the trench. Most trenches are arcuate with their concave side facing the continent like the Aleutian trench, but some are straight like the Tonga trench. Any sediment from erosion on the exposed continent that reaches the trench's bottom is subducted down into the Earth, so the trench never fills. However, such sediments do accumulate on the CONTINENTAL SHELF as TURBIDITY CURRENTs carry them into a fore-arc basin between the continental margin and the continent's shoreline. Commonly the turbidites are compressed, deformed and metamorphosed in the plates' collision to form a flysch belt along the coastline. Sometimes seafloor sediments and crust are obducted onto the shelf and preserved in the flysch. Inbound of the flysch, an ISLAND ARC forms. Heat, generated mostly by friction in the subduction zone, melts the overlying sialic and simatic rocks of the upper and lower continental crust and the mafic to ultramafic uppermost mantle, causing large diorite to granite intrusions to be emplaced into the old continental crust and mafic to felsic volcanics to be extruded onto it. Thus the world's most spectacular volcanoes are found mostly around the Pacific Ocean. Where there is a wide continental shelf, this intrusive–extrusive rock complex forms a mountainous island arc with a submarine back-arc basin behind, like the Aleutian arc and Bering Sea or the Japanese arc and Sea of Japan. Alternatively, if the shelf is narrow, the complex forms a range of high coastal mountains. Further compression deforms and stacks the rocks behind the arc into sheets on thrust faults, forming high mountains like the Andes and Cordillera with their photogenic MOUNTAIN GEOMORPHOLOGY.

Collisions between oceanic plates are less common than oceanic–continental collisions but the process is similar except that both crusts are relatively thin and dense. Consequently their subduction zones descend a steeper angle approaching 90° and go deepest into the Earth, producing most of the world's deepest earthquakes. Also their trenches, like the Mariana trench, are deeper and reach depths of about 11 km. Further, only the peaks of the arc volcanoes reach the surface as a chain of basaltic islands, such as the Antilles, on the margin of the non-subducting plate. With so little land above sea level, very little clastic sedimentation occurs, but chemical sediments and CORAL REEFs often form around such islands in tropical climates to form barrier (see BARRIER AND BARRIER ISLAND) reefs and atolls.

Collisions between continental plates create the world's greatest mountains with spectacular TECTONIC GEOMORPHOLOGY, like the 10-km high Himalayan Mountains where the Indian subcontinent rides on the Indo-Australian plate and butts against the Eurasian continent and plate. The mountains are high because both crusts are relatively thick but less dense than the Earth's mantle and so they float to high elevation. Their combined 70-km thick crust sinks deeper into the Earth's mantle because of isostasy. Prior to collision, the continents had oceanic crust between them with either a passive margin like most of the Atlantic coastline or more commonly with an oceanic–continental collisional margin as described above. As the continents collide, some seafloor sediments, volcanics and gabbroic intrusions are often squeezed up and trapped between them where they are complexly deformed, and metamorphosed to serpentinites. Finding serpentinites defines the suture between the plates. On both sides of the suture zone are the deformed and metamorphosed remains of island arcs and back-arc basins first, then very high thrust-faulted mountains of stacked sedimentary sequences, followed by high foothills of folded sedimentary rocks, and then high plateaux such as the Tibetan plateau, with gently deformed strata that are deeply dissected by canyons. The suture zone and thrust-fault belts are tectonically active, producing strong shallow earthquakes mainly. Below the suture, subduction stops on closure. However, it takes tens of millions of years for the relict slab of descending crust to melt so that some minor earthquakes still occur at intermediate depths. Closure also terminates most volcanism because frictional heating ceases in the subduction zone.

Transform faults and triple junctions

J. T. Wilson's (1965) description of a new type of fault, a transform fault (Figure 125), was the key to understanding the dynamics of plate tectonics. The motions of rift and subduction zones were well understood, but the seafloor scarps were thought to be transcurrent faults where the two sides slid in opposite directions to offset a marker. Noting that these scarps cut the ridges at right angles and offset them, Wilson reasoned that the seafloor was moving away from both offset ridge segments in both directions so that the fault's sides were moving: (a) towards each other between the offset ridges as an active plate boundary, and (b) in the same direction and speed outside of the two ridges where they are tectonically

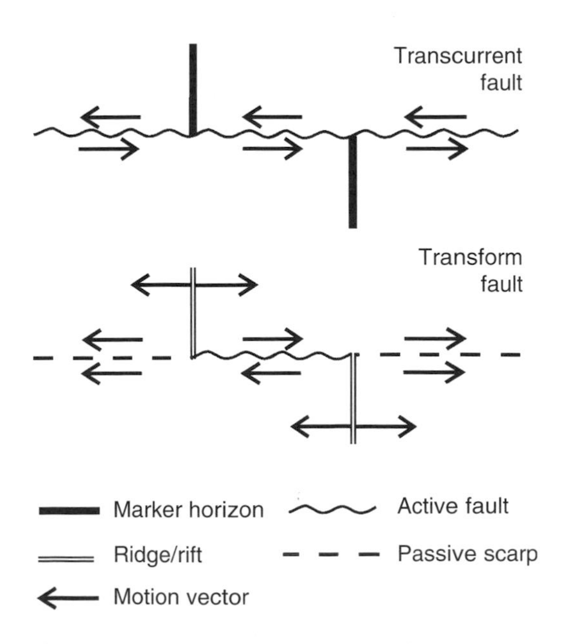

Figure 125 Dynamic comparison of transcurrent and transform faults

passive and inside the plates (Figure 125). He further explained how the transform fault motions interacted with ridge and subduction zone motions to create plate boundaries.

On the seafloor, transform fault lineaments can be over 3,000 km long and 3 km in height, being highest on the side of the closest ridge segment. Excluding very minor amounts of active volcanism, the Earth's crust is neither created nor destroyed. Between the ridges, horizontal shear generates weak to moderate, shallow earthquakes that are rarely destructive. In a few places, such as the San Andreas fault, the active part of a transform fault cuts much thicker continental crust, causing periodic powerful and highly destructive shallow earthquakes as the two sides grind past each other. Minor alkalic volcanics, sag ponds and other geomorphic features may occur along such faults.

Active plate boundaries form triple junctions where three plates meet, separated by three boundaries or arms. Each arm may be a rift, subduction zone or transform fault to make sixteen possible combinations. For example, the Pacific–Cocos–Nazca triple junction has three radiating rifts as arms whereas the Pacific–Cocos–North American triple junction has a rift, a subduction zone and a transform fault as its arms. The geometrically important fact is that the velocity and direction of each plate at a triple junction must form a spherical vector triangle. Thus, if the rotational motions of two plates are known from seafloor magnetic anomalies or palaeomagnetism, the rotational motion and Euler pole of the third plate can be calculated.

Importance of the theory

Understanding plate tectonics has led to a revolution in the Earth sciences. Virtually all of its supporting evidence relates to the last 200 Ma or so but, invoking UNIFORMITARIANISM, it has provided the basis for geographers to understand how the Earth's landscapes developed and for geologists to interpret the preceding 4,000 Ma of the Earth's rock record. For example, the Caledonian and Appalachian mountain belts are now recognized to be the deformation product of three Palaeozoic collisional events with oceans opening and closing in between in WILSON CYCLES. Similarly, geologists and geophysicists are using the theory to fit Precambrian supercontinents together such as Rhodinia, to identify failed rift valleys such as the 1,100 Ma Keweenawan aulacogen, to explain how geologic terranes that formed in radically different tectonic settings are now abutting, and to explain epeirogenic vertical motions in the continental interiors to form basins and plateaux. Finally, knowing how plates have moved over the past 200 Ma is enabling geophysicists to constrain models to investigate the Earth's magnetic field and to understand convective motions in its interior.

References

Bullard, E.C., Everett, J.E. and Smith, A.G. (1965) The fit of the continents around the Atlantic, *Philosophical Transactions of the Royal Society*, London 258A, 41–51.

Dietz, R.S. (1961) Continental and ocean basin evolution by spreading of the seafloor, *Nature* 190, 854–857.

Heirtzler, J.R., Dickson, G.O., Herron, E.M., Pitman, W.C. and LePichon, W.C. (1968) Marine magnetic anomalies, geomagnetic field reversals, and motions of the ocean floor and continents, *Journal of Geophysical Research* 73, 2,119–2,135.

Irving, E. (1958) Rock magnetism: a new approach to the problems of polar wandering and continental drift, in S.W. Carey (ed.) *Continental Drift: A Symposium*, 24–61, Hobart: University of Tasmania.

Wilson, J.T. (1965) A new class of faults and their bearing on continental drift, *Nature* 207, 343–347.

Further reading

Condie, K.C. (1989) *Plate Tectonics and Crustal Evolution*, 3rd. edition, Oxford: Pergamon Press.

Cox, A. and Hart, R.B. (1986) *Plate Tectonics: How It Works*, Palo Alto: Blackwell.

Kearey, P. and Vine, F.J. (1990) *Global Tectonics*, Oxford: Blackwell.

Moores, E.M. and Twiss, R.J. (1995) *Tectonics*, New York: W.H. Freeman.

D.T.A. SYMONS

PLOUGHING BLOCK AND BOULDER

Ploughing blocks and boulders are individual boulders that move downslope faster than their surrounding material by processes related to seasonal frost. Their movement ranges from millimetres to a few centimetres a year and is restricted to the annual freeze–thaw cycle (Ballantyne 2001; Berthling *et al.* 2001a). Due to the differential movement, the boulder pushes up a mound against its downslope side while leaving a depression along its upslope track. Ploughing boulders belong to the area of PERIGLACIAL GEOMORPHOLOGY. They are developed on slopes in the warmer part of the periglacial belt, commonly together with solifluction lobes. Only a few detailed process studies exist, but these indicate that boulder movements are caused by the same processes that occur in SOLIFLUCTION. During autumn and winter, the boulders protrude above the snow cover for some time. Combined with differences in thermal conductivity, this causes more intensive heat loss through the boulder than through ground elsewhere. This results in favourable conditions for ice segregation (Ballantyne 2001) and causes FROST HEAVE of the boulders. A heave of up to 7.5 cm has been demonstrated from southern Norway (Berthling *et al.* 2001b). It was shown that boulder heave stopped in midwinter regardless of snow conditions. Depletion of soil moisture might explain this behaviour. During spring and early summer, the boulders melt out of the snow and the soil beneath the boulder starts to melt earlier than the surrounding ground. Consolidation rates of up to 4.2 mm/day through a six-day period were measured by Berthling *et al.* (2001b). If melting of ice is more rapid than the ability of the released water to escape, water is trapped beneath the boulder so that the porewater pressure rises. This causes the soil beneath the boulder to lose strength and downslope deformations may occur. The process is referred to as gelifluction, and is the main cause for ploughing boulder movement. A second process that has been invoked is frost creep. Frost creep results from frost heave normal to the freezing plane and settlement along the vertical. This results in a net downslope movement in the cases where the freezing plane is essentially parallel to the sloping ground surface. The frost creep model in its simple form should be abandoned, as the boulders heave and tilt in directions determined by variations in heat removal, frost susceptibility and soil water content beneath the boulder, not slope. Yet, instability caused by tilting during frost heave might induce some displacements during thaw.

References

Ballantyne, C.K. (2001) Measurement and theory of ploughing boulder movement, *Permafrost and Periglacial Processes* 12, 267–288.

Berthling, I., Eiken, T., Madsen, H. and Sollid, J.L. (2001a) Downslope displacement rates of ploughing boulders in a mid-alpine environment: Finse, southern Norway, *Geografiska Annaler* 83A, 103–116.

Berthling, I., Eiken, T. and Sollid, J.L. (2001b) Frost heave and thaw consolidation of ploughing boulders in a mid-alpine environment, Finse, southern Norway, *Permafrost and Periglacial Processes* 12, 165–177.

SEE ALSO: frost heave; periglacial geomorphology; solifluction

IVAR BERTHLING

PLUVIAL LAKE

Bodies of water that accumulate in basins as a result of former greater moisture availability resulting from changes in temperature and/or precipitation. The study of pluvial lakes developed in the second half of the nineteenth century. Jamieson (1863) called attention to the former greater extent of the great saline lakes of Asia: The Caspian, Aral, Balkhash and Lop-Nor and Lartet (1865) pointed to the expansion of the Dead Sea. The term pluvial appears to have first been applied to an expanded lake by Hull (1885), but was originally applied by Tylor (1868) to valley fills in England and France. A major advance in the study of pluvial lakes came in the western USA with the work of Russell (1885) on Lake Lahontan and of Gilbert (1890) on Lake Bonneville. A discussion of these early studies and their bibliographic details is given in Flint (1971: Chapter 2).

The Great Basin of the USA held some eighty pluvial lakes during the Pleistocene, and they occupied an area at least eleven times greater than the area they cover today. Lake Bonneville (Plate 91), was roughly the size of present-day Lake Michigan, about 370 m deep and covered 51,640

Plate 91 A group of high shorelines that developed in the Late Pleistocene around pluvial Lake Bonneville, Utah, USA

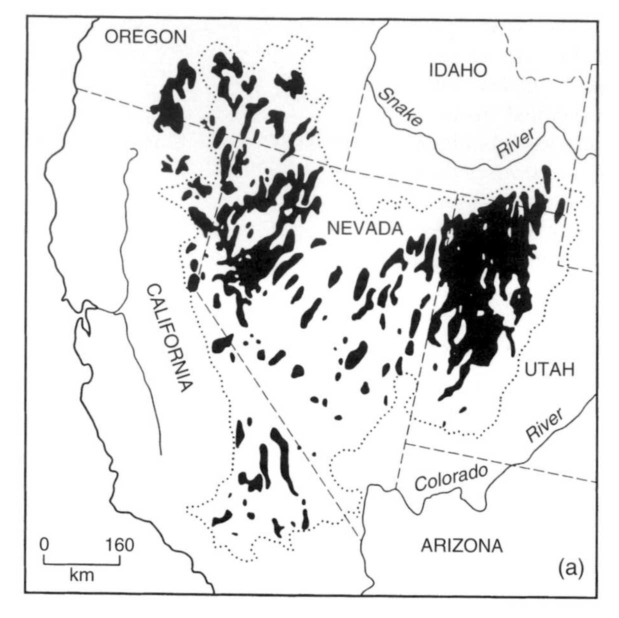

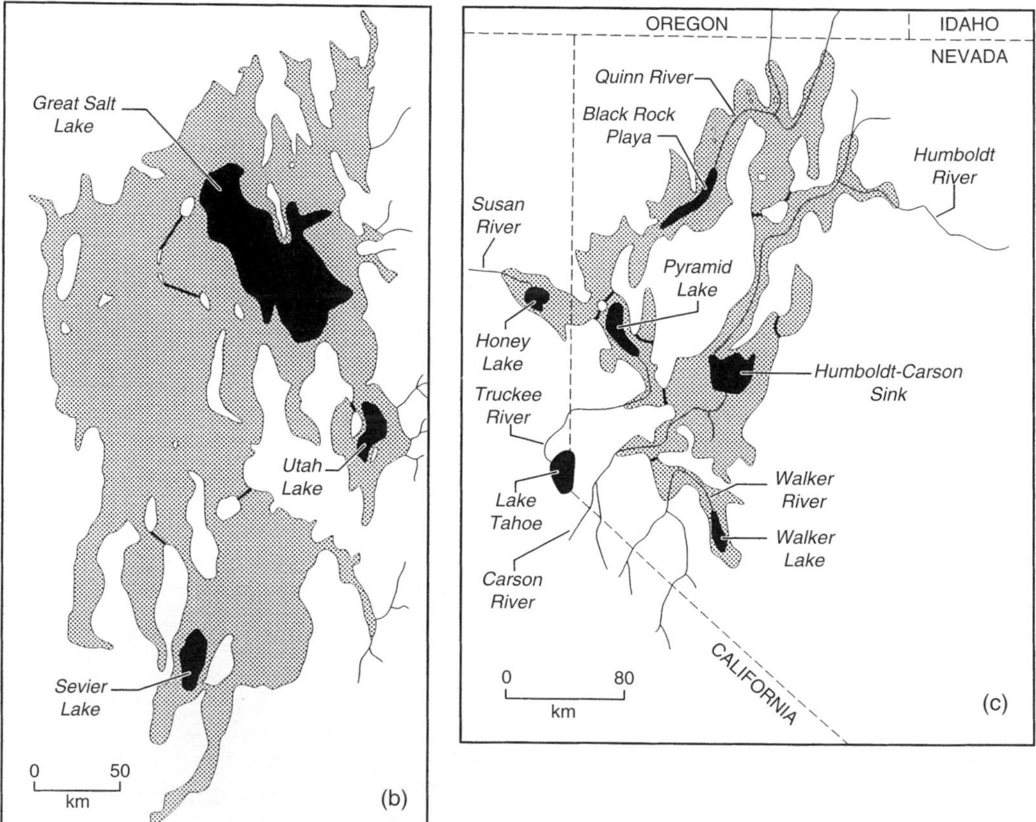

Figure 126 (a) The distribution of Pleistocene pluvial lakes in the southwestern USA; (b) Lake Bonneville; (c) Lake Lahontan

km². Lake Lahontan was rather more complicated in form, covered 23,000 km², and reached a depth of about 280 m at the site of today's Pyramid Lake (Figure 126). It covered an area nearly as great as present-day Lake Erie. River courses became integrated and lakes overflowed from one sub-basin to another. For example, the Mojave River drainage, the largest arid fluvial system in the Mojave Desert, fed at least four basins and their lakes in pluvial times: Lake Mojave (including present-day Soda and Silver lakes), the Cronese basin and the Manix basin (which includes the Afton, Troy, Coyote and Harper sub-basins; Tchakerian and Lancaster 2002). Also important was the Owens Lake–Death Valley system.

A very large amount of work has been done to date and correlate the fluctuations in the levels of the pluvial lakes. Much of the early work is reviewed by Smith and Street-Perrott (1983). They demonstrated that many basins had particularly high stands during the period that spanned the Late Glacial Maximum, between about 25,000 and 10,000 years ago. More recently there have been studies of the longer term evolution of some of the basins, facilitated by the study of sediment cores, as for example from Owens Lake, the Bonneville Basin, Mono Lake, Searles Lake and Death Valley.

The high lake levels during the Last Glacial Maximum may well be the result of a combination of factors, including lower temperatures and evaporation rates, and reduced precipitation levels. Pacific storms associated with the southerly branch of the polar jet stream were deflected southwards compared to the situation today.

Other major pluvial lakes occurred in the Atacama and Altiplano of South America (Lavenu *et al.* 1984). The morphological evidence for high lake stands is impressive and this is particularly true with regard to the presence of algal accumulations at high levels (as much as 100 m) above the present saline crusts of depressions like Uyuni (Rouchy *et al.* 1996). There is a great deal of variability and confusion about climatic trends in the Late Quaternary in this region, not least with respect to the situation at the Late Glacial Maximum and in the mid-Holocene (Placzek *et al.* 2001). Nonetheless, various estimates have been made of the degree of precipitation change that the high lake stands imply. Pluvial Laguna Lejíca, which was 15–25 m higher than today at 13.5 to 11.3 Kyr BP and covered an area of 9–11 km² compared to its present extent of 2 km², had an annual rainfall of 400–500 mm, whereas today it has only around 200 mm. Pluvial Lake Tauca, which incorporates present Lake Poopo, the Salar de Coipasa and the Salar de Uyuni and which had a high stillstand between 15 and 13.5 Kyr BP, had an annual rainfall of 600 mm compared with 200–400 mm today.

In the Sahara there are huge numbers of pluvial lakes both in the Chotts of the north, in the middle (Petit-Maire *et al.* 1999) and in the south (e.g. Mega-Chad). In the Western Desert there are many closed depressions or playas, relict river systems and abundant evidence of prehistoric human activity (Hoelzmann *et al.* 2001). Playa sediments contained within basins such as Nabta Playa indicate that they once contained substantial bodies of water, which attracted Neolithic settlers. Many of these sediments have now been subjected to radiocarbon dating and they indicate the ubiquity of an early to mid-Holocene pluvial phase, which has often been termed the Neolithic pluvial. A large lake formed in the far north-west of Sudan, and this has been called 'The West Nubian Palaeolake' (Hoelzmann *et al.* 2001). It was especially extensive between 9,500 and 4,000 years BP, and may have covered as much as 7,000 km². If it was indeed that big, then a large amount of precipitation would have been needed to maintain it – possibly as much as 900 mm compared to the less than 15 mm it receives today.

In the Kalahari of southern Africa, Lake Palaeo-Makgadikgadi encompassed a substantial part of the Okavango Delta, parts of the Chobe–Zambezi confluence, the Caprivi Strip, and the Ngami, Mababe and Makgadikgadi basins. At its greatest extent it was over 50 m deep and covered 120,000 km². This is vastly greater than the present area of Lake Victoria (68,800 km²) and makes Palaeo-Makgadikgadi second in size in Africa to Lake Chad at its Quaternary maximum. Dating it, however, is problematic (Thomas and Shaw 1991) as is its source of water. Some of the water may have been derived when the now DRY VALLEYS of the Kalahari (the *mekgacha*) were active and much could have been derived from the Angolan Highlands via the Okavango. However, tectonic changes may also have played a role and led to major inputs from the Zambezi.

In the Middle East expanded lakes occurred in the currently arid Rub-Al-Khali and also in Anatolia (Roberts 1983). In Central Asia the

Aral–Caspian system was hugely expanded. At several times during the late Pleistocene (Late Valdai) the level of the lake rose to around 0 m (present global sea level) compared to −27 m today and it inundated a huge area, particularly to its north. In the early Valdai glaciation it was even more extensive, rising to about +50 m above sea level, linking up to the Aral, extending some 1,300 km up the Volga River from its present mouth and covering an area in excess of 1.1 million km^2 (compared to 400,000 km^2 today). At its highest it may have overflowed through the Manych depression into the Black Sea. In general, transgressions have been associated with warming and large-scale influxes of meltwater (Mamedov 1997), but they are also a feature of the glacial phases when there was also a decrease in evaporation and a blocking of ground water by permafrost. Regressions occurred during interglacials and so, for example, in the Early Holocene the level of the Caspian dropped to −50 to −60 m below sea level.

Large pluvial lakes also occur in the drylands and highlands of China and Tibet and levels appear to have been high from 40,000 to 25,000 BP (Li and Zhu 2001). Similarly the interior basins of Australia, including Lake Eyre, have shown major expansion and contractions, with a tendency for high stands in interglacials (Harrison and Dodson 1993).

As can be seen from these regional examples, pluvial lakes are widespread (even in hyper-arid areas), reached enormous dimensions, and had different histories in different areas. Pluvials were not in phase in all regions and in both hemispheres (Spaulding 1991). In general, however, dry conditions during and just after the Late Glacial Maximum and humid conditions during part of the Early to Mid Holocene appear to have been characteristic of tropical deserts, though not of the south-west USA.

References

Flint, R.F. (1971) *Glacial and Quaternary Geology*, New York: Wiley.

Harrison, S.P. and Dodson, J. (1993) Climates of Australia and New Zealand since 18000 yr BP, in H.E. Wright, J.E. Kutzbach, T. Webb, W.F. Ruddiman, F.A. Street-Perrott and P.J. Bartlein (eds) *Global Climates since the Last Glacial Maximum*, 265–293, Minneapolis: University of Minnesota Press.

Hoelzmann, P., Keding, B., Berke, H., Kröpelin, S. and Kruse, H-J. (2001) Environmental change and archaeology: lake evolution and human occupation in the Eastern Sahara during the Holocene, *Palaeogeography, Palaeoclimatology, Palaeoecology* 169, 193–217.

Lavenu, A., Fournier, M. and Sebrier, M. (1984) Existence de deux nouveaux episodes lacustres quaternaries dans l'Altiplano peruvo–bolivien, *Cahiers ORSTOM ser Géologie* 14, 103–114.

Li, B.Y. and Zhu, L.P. (2001) 'Greatest lake period' and its palaeo-environment on the Tibetan Plateau, *Acta Geographica Sinica* 11, 34–42.

Mamedov, A.V. (1997) The Late Pleistocene–Holocene history of the Caspian Sea, *Quaternary International* 41/42, 161–166.

Petit-Maire, N., Burollet, P.F., Ballais, J-L., Fontugne, M., Rosso, J-C. and Lazaar, A. (1999) Paléoclimats du Sahara septentionale. Dépôts lacustres et terrasses alluviales en bordure du Grand Erg Oriental à l'extreme-Sud de la Tunisie, *Comptes Rendus Académie des Sciences*, Series 2, 312, 1,661–1,666.

Placzek, C., Quade, J. and Betancourt, J.L. (2001) Holocene lake-level fluctuations of Lake Aricota, southern Peru, *Quaternary Research* 56, 181–190.

Roberts, N. (1983) Age, paleoenvironments, and climatic significance of Late Pleistocene Konya Lake, Turkey, *Quaternary Research* 19, 154–171.

Rouchy, J.M., Servant, M., Fournier, M. and Causse, C. (1996) Extensive carbonate algal bioherms in Upper Pleistocene saline lakes of the central Altiplano of Bolivia, *Sedimentology* 43, 973–993.

Smith, G.I. and Street-Perrott, F.A. (1983) Pluvial lakes of the western United States, in S.C. Porter (ed.) *Late Quaternary Environments of the United States. Vol. 1. The Late Pleistocene*, 190–212, London: Longman.

Spaulding, W.G. (1991) Pluvial climatic episodes in North America and North Africa: types of correlation with global climate, *Palaeogeography, Palaeoclimatology, Palaeoecology* 84, 217–229.

Tchakerian, V. and Lancaster, N. (2002) Late Quaternary arid/humid cycles in the Mojave Desert and Western Great Basin of North America, *Quaternary Science Reviews* 21(7), 799–810.

Thomas, D.S.G. and Shaw, P. (1991) *The Kalahari Environment*, Cambridge: Cambridge University Press.

A.S. GOUDIE

POINT BAR

Point bars are a form of river bar. They are located along the convex banks of river bends. They typically have an arcuate shape that reflects the radius of curvature of the bend. The cross-sectional slope of the bar is inclined towards the centre of the channel, reflecting the asymmetrical channel geometry at the bend apex. Textural attributes of the bar reflect patterns of secondary helical flow over the bar surface as the thalweg shifts to the outside of the bend at high flow stage.

Point bars are most commonly found along meandering rivers where there are clear genetic links between instream processes that form and maintain pool–riffle sequences, the channel morphology that results, the formation of point bars on the insides of bends, and resulting channel planform attributes (the meandering behaviour of the channel).

At bankfull stages, helical flow in bends carries sediment up the convex slope of point bars, while the concave bank is scoured. Sand or gravel bedload material is moved by traction towards the inner sides of channel bends via this helical flow. In this lateral accretion process, bedload materials are deposited on point bar surfaces. Lateral accretion deposits are detectable in the sedimentology of point bars by their oblique structures dipping towards the channel. Differing patterns of sedimentation are imposed by the radius of bend curvature (bend tightness) as well as the flow regime and sediment load.

Grain size typically fines down-bar (around the bend) and laterally (away from the channel). This produces a longitudinal 'around the bend' set of sedimentary structures comprising bedload material at the head of the bar, where the thalweg is aligned adjacent to the convex bank (at the entrance to the bend). As the thalweg moves away from the bend down-bar, lower energy suspended load materials are deposited. A mix of bedload and suspended load is generally evident at the tail of the bar. The most recently accumulated deposits are laid down as bar platform deposits at the bend apex. Typically these unit bar forms are largely unvegetated.

In many instances, point bars are compound features. These bank attached compound point bars comprise a mosaic of geomorphic units. In gravel situations, the bar platform is a relatively flat, coarse feature atop which a range of features are deposited or scoured. In the centre of the point bar is a gravel lobe. This likely represents the position of the shear zone during high flow stage. At high flow stage, compound point bars are dissected by chute channels, often with associated ramp deposits (McGowen and Garner 1970). Flow around vegetation produces a series of depositional ridges that have distinct grain size distributions (Brierley and Cunial 1998). The bar apex is a shallower feature inclined towards the channel.

References

Brierley, G.J. and Cunial, S. (1998) Vegetation distribution on a gravel point bar on the Wilson River, NSW: a fluvial disturbance model, *Proceedings of the Linnean Society (NSW)*, 120, 87–103.

McGowen, J.H. and Garner, L.E. (1970) Physiographic features and stratification types of coarse-grained point bars: modern and ancient examples, *Sedimentology* 14, 77–111.

SEE ALSO: bar, river; channel, alluvial; fluvial geomorphology

KIRSTIE FRYIRS

POLJE

A form of large, flat-floored closed depression formed in KARST regions. They are a distinctive feature of limestone geomorphology. The term comes from the Slav word for field. Poljes are often covered with alluvium, are subject to periodical flooding, and they may have sinks, called *ponors*, into which streams may disappear. There is much debate as to the origin of poljes. They are probably polygenetic, with solution and tectonics playing important roles.

Further reading

Gams, I. (1978) The polje: the problem of definition, *Zeitschrift für Geomorphologie* 22, 170–181.

A.S. GOUDIE

POOL AND RIFFLE

Pools and riffles are one of the most common and recognizable bedform sequences in channels. In the most basic sense, pools are classified as deep areas with low velocities at low stage, while riffles exhibit higher water-surface slopes and faster velocities. Pools and riffles exist in both alluvial and bedrock channels, but are best developed in gravel-bed substrates and meandering channels. Pools and riffles occur in moderate gradient channels with a transition to step-pool morphologies at higher gradients. These undulations in bed topography provide the primary framework for aquatic habitat in channels, and are of great interest because of their importance for macro invertebrates and fish species. The features also create tremendous form drag and flow resistance that may help rivers achieve equilibrium.

Because basic flow characteristics in pools and riffles change with stage (Richards 1976), limitations in the definition of pools and riffles exist. The zero-crossing method, where pools and riffles are recognized as residuals above a calculated mean bed profile, and spectral analysis provide robust methods for identifying the bedforms provided minimum size criteria are specified (Carling and Orr 2000). The residual-depth criteria or control-point method is used to define a pool based on the idea that riffles pond water in pools if flow ceases (Lisle and Hilton 1992). The residual pool is easily recognized in the field as the area that would be inundated, but it is still necessary to use minimum depth and width criteria to avoid subdividing a channel into extreme numbers of small morphologic units. Complications can arise because the residual depth of a pool is also influenced by sediment deposition, which can fill much of the pool at low flow. However, this fine sediment infilling can be used to estimate the sediment supply in a particular channel based on the idea that the residual volume of a pool prior to low-flow deposition is represented by the elevation of the coarse substrate below the overlying fines (Lisle and Hilton 1992). This assessment of pool sedimentation is particularly useful to evaluate land-use impacts on channels. It is also worth recognizing an important distinction between different types of pools where pools formed by some obstruction to flow are called forced pools and the remaining pools are termed free-formed pools (Montgomery *et al.* 1995).

The depth of pools and height of riffles is clearly maintained by some process, but the nature of the process is still debated. One idea that receives considerable attention is the velocity-reversal hypothesis. Keller (1971) used near-bed velocity measurements to show that velocities in pools are initially lower than those in riffles but increase at a faster rate and may exceed riffle velocities near bankfull stage. The water-surface elevations also change so that water-surface slopes equalized at high stage and may be steeper over riffles above bankfull level. The idea is closely related to the concept of two-phased bedload transport where low-flow deposition occurs in pools and high-flow deposition occurs in riffles (Jackson and Beschta 1982). Although the velocity-reversal hypothesis is often cited, it is frequently criticized based on continuity of mass concerns. However, recirculating eddies can form in some channels, increase velocities in pools and maintain continuity (Thompson *et al.* 1999). Meanwhile, the constrictions create backwater and locally elevated water-surface slopes in pools that can exceed water-surface slopes in riffles at bankfull stage. Evidence for recirculating-eddy enhanced velocity reversals exists, but flow routing of sediment around pools may dominate in other locations (Brooker *et al.* 2001). The combination of sedimentological properties and turbulence characteristics also is used to explain pool formation and maintenance. According to this idea, an obstacle to flow temporarily creates turbulence fluctuations, the perturbations to the flow generate the pool-riffle morphology, and differences in turbulence intensities and sediment characteristics between pools and riffles help to maintain disparities in sediment movement along the sequence (Clifford 1993). Much of the remaining work on pool formation and maintenance focuses on meandering channels and draws links between meandering and pool formation because pools tend to form at bends. Yang (1971) used this linkage and the idea that channels would adjust to minimize unit stream power along a channel in a theoretical approach with an equalization of water-surface slopes over pools and riffles at high stage. He concluded that the resulting formation process was a combination of dispersion and sorting of sediments. Studies also draw links between helical flow development and pool and riffle formation, but these processes are so clearly linked it is difficult to determine a casual relationship between them.

Pools and riffles exert an important influence on sediment sorting along a channel, especially the downstream end of pools, which create an uphill climb for particles moving downstream (Thompson *et al.* 1999). Sediment size, packing density and relative protrusion can all differ between pools and riffles. However, the combination of low-flow and high-flow sediment deposition can create a large variability in the size of bed sediments within a small area, and make it difficult to recognize distinct differences between pools and riffles (Richards 1976). Although channel-bed sediments in pools are often reported to be smaller than those in riffles (Clifford 1993), the opposite trend is reported in sediment supply-limited channels (Thompson *et al.* 1999). The disagreement in the general sorting trend probably results from two-phase bedload transport. During low flows, fine sediments can cover coarse substrate in pools along channels with high sediment loads, while supply-limited channels generally preserve the sediment-sorting patterns established at high flow.

Another fundamental characteristic of pools and riffles is the distance between successive pools or riffles, a measure termed the pool and riffle spacing. Values between five and seven average bankfull widths are often reported for pool and riffle spacing (Keller and Melhorn 1978), and this spacing has been attributed to reach-scale influences related to meander wavelengths in sinuous channels (Carling and Orr 2000). For example, variations in bed topography follow second-order autoregressive models as a result of a combination of periodic and random effects that may be related to meander wavelength (Richards 1976). Variation in spacing occurs because the channel-bed slope exerts a control on average spacing between pools (Wohl *et al.* 1993). Average spacing also varies due to the influence of obstructions to flow and variations in how pools are defined (Montgomery *et al.* 1995). In channels dominated by forced pools, local-scaling effects related to recirculating eddies behind randomly spaced channel constrictions can build morphologies with spacing values that agree with published values for a range of channel conditions (Thompson 2001). Therefore, it is unclear if reach-length or local-scaling effects create the semi-rhythmic spacing reported in natural channels.

Channel slope, channel-bed resistance and drainage area influence pool and riffle dimensions. Pool length and depth both tend to decrease with increased channel-bed slope (Wohl *et al.* 1993), and riffles become smaller in deeper water (Carling and Orr 2000). Given the fact that stream power increases with slope, the inverse relation between pool size and slope reflect changes in channel-bed resistance with more resistant beds associated with higher slopes. Pools also increase in size on larger channels, presumably because of the simultaneous increase in stream power with an increase in discharge on these larger systems. The relative magnitude of the bed undulations also tends to decrease with increased sediment supply because pools begin to fill and lose their distinct characteristics (Lisle and Hilton 1992).

As demonstrated by past research, pools and riffles will continue to be a central focus of research in fluvial geomorphology because of their important influence on both the physical and biological characteristics of natural channels.

References

Brooker, D.J., Sear, D.A. and Payne, A.J. (2001) Modelling three-dimensional flow structures and patterns of boundary shear stress in a natural pool-riffle sequence, *Earth Surface Processes and Landforms* 26, 553–576.

Carling, P.A. and Orr, H.G. (2000) Morphology of riffle-pool sequences in the River Severn, England, *Earth Surface Processes and Landforms* 25, 369–384.

Clifford, N.J. (1993) Differential bed sedimentology and the maintenance of a riffle-pool sequence, *Catena* 20, 447–468.

Jackson, W.L. and Beschta, R.L. (1982) A Model of Two-Phase Bedload Transport in an Oregon Coast Range Stream, *Earth Surface Processes and Landforms* 7, 517–527.

Keller, E.A. (1971) Areal sorting of bed-load material: the hypothesis of velocity reversal, *Geological Society of America Bulletin* 82, 753–756.

Keller, E.A. and Melhorn, W.N. (1978) Rhythmic spacing and origin of pools and riffles, *Geological Society of America Bulletin* 89, 723–730.

Lisle, T.E. and Hilton, S. (1992) The volume of fine sediment in pools: an index of sediment supply in gravel-bed streams, *Water Resources Bulletin* 28, 371–383.

Montgomery, D.R., Buffington, J.M., Smith, R.D., Schmidt, K.M. and Pess, G. (1995) Pool spacing in forest channels, *Water Resources Research* 31, 1,097–1,105.

Richards, K.S. (1976) The morphology of riffle-pool sequences, *Earth Surface Processes and Landforms* 1, 71–88.

Thompson, D.M., Wohl, E.E. and Jarrett, R.D. (1999) Pool sediment sorting processes and the velocity-reversal hypothesis, *Geomorphology* 27, 142–156.

Thompson, D.M. (2001) Random controls on semi-rhythmic spacing of pools and riffles in constriction-dominated rivers, *Earth Surface Processes and Landforms* 26, 1,195–1,212.

Wohl, E.E., Vincent, K.R. and Merritts, D.J. (1993) Pool and riffle characteristics in relation to channel gradient, *Geomorphology* 6, 99–110.

Yang, C.T. (1971) Formation of Riffles and Pools, *Water Resource Research* 7, 1,567–1,574.

Further reading

Knighton, D. (1998) *Fluvial Forms and Processes*, London: Arnold.

Wohl, E.E. (2000) *Mountain Rivers*, Washington, DC: American Geophysical Union.

SEE ALSO: bedform; meandering; point bar; rapids; step-pool system

DOUGLAS M. THOMPSON

PORE-WATER PRESSURE

REGOLITH and highly fractured rock at Earth's surface (here termed soil) contain voids (pores) that are variously wetted or filled with water (pore water). Forces acting on pore water establish gradients of fluid potential, the work required to move a unit quantity of fluid from a datum to a specified position, and pore-water flows in response to these gradients. The concept of hydraulic head usefully describes pore-water potential. Total hydraulic head, or potential per unit weight of fluid, is usefully described in terms of gravitational, pressure and kinetic energy potential. The total hydraulic head (h) for an incompressible fluid (fluid having a constant density; ρ_w for water) is given by (Hubbert 1940):

$$h = \xi + \frac{p}{\rho_w g} + \frac{u^2}{2g}$$

where ξ is the gravitational, or elevation, potential; $p/\rho_w g$ is the pressure potential, in which p is the gauge pressure of the water relative to atmospheric pressure and g is gravitational acceleration in the co-ordinate direction; and $u^2/2g$ is the kinetic energy, or velocity, potential, where u is water velocity. Flow velocity in soil is usually very small, so calculations of hydraulic head in soils typically neglect velocity head. Pore-water pressure therefore constitutes one of two dominant components of the fluid potential in many soils.

Pore-water pressure is isotropic, but it varies with position relative to the water table (the depth horizon where pore-water pressure is atmospheric, which defines the zero-pressure datum) and with the proportion of soil weight carried by intergranular contacts. Below the water table, pore-water pressure is greater than atmospheric and positive; above the water table pore-water pressure is less than atmospheric and negative owing to tensional capillary forces exerted on pore water (e.g. Remson and Randolph 1962). If intergranular contacts carry all of the soil weight and water statically fills pore space, then the hydrostatic pore-water pressure (p_h) at a depth z normal to the soil surface (Figure 127) is given by

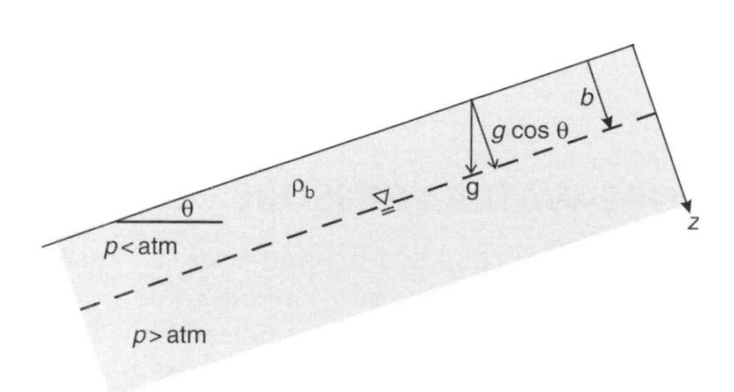

Figure 127 Schematic profile and definition of geometric parameters in an infinite slope having a water table at depth. Above the water table pore-water pressure is less than atmospheric; below it is greater than atmospheric

$$p_h = \rho_w g \cos\theta\,(z - b)$$

where b is depth to the water table and θ the slope of the soil surface. Pore-water pressure can exceed or fall short of hydrostatic under hydrodynamic conditions or if a soil collapses or dilates under load. If a soil collapses, it compacts. Below the water table this will cause a transient increase in pore-water pressure, the duration and magnitude of which are governed mainly by the rate of collapse and the permeability of the soil. The gauge pressure (p) at depth z can then be written as $p = p_h + p_e$, where p_e is a nonequilibrium pressure in excess of hydrostatic. If collapse thoroughly disrupts intergranular contacts, then the pore fluid may bear the entire weight of the solid grains, and the soil will liquefy. In that case, gauge pressure can be written as

$$p = \rho_w g \cos\theta(z - b) + \rho_b g \cos\theta\, b + (\rho_s - \rho_w)(1 - \phi)g\cos\theta(z - b)$$

where ρ_b is the moist bulk density of soil above the water table, ρ_s is grain density and φ is soil porosity. When a soil liquefies, the excess water pressure equals the sum of the unit weight of soil above the water table and the buoyant unit weight of the soil below the water table:

$$p_e = \rho_b g \cos\theta b + (\rho_s - \rho_w)(1 - \phi)g \cos\theta(z - b)$$

In unsaturated soil, such as occurs above the water table or (occasionally) when a saturated soil dilates (expands) under load, water does not fill pore space completely, and pore-water pressure locally is less than atmospheric. In that case capillary and electrostatic forces cause water to adhere to solid particles. As soil water content decreases, tensional forces increase and negative pore-water pressure bonds solid particles, increasing soil strength. The magnitude of negative pore-water pressure depends on soil texture and physical properties as well as on water content. Fine soils have a broader pore-size distribution and larger particle-surface area than do coarse soils. As a result fine soils have a greater range of negative-pressure potential because they hold more water than coarse soils, and the water bonds more tightly to particle surfaces. Piezometers are used to measure positive pore-water pressure; tensiometers commonly are used to measure negative pore-water pressure (e.g. Reeve 1986).

References

Hubbert, M.K. (1940) The theory of ground-water motion, *Journal of Geology* 48, 785–944.

Reeve, R.C. (1986) Water potential: piezometry, in A. Klute (ed.) *Methods of Soil Analysis Part 1 – Physical and Mineralogical Methods*, 545–561, Madison: American Society of Agronomy, Inc.

Remson, I. and Randolph, J.R. (1962) Review of some elements of soil-moisture theory, *US Geological Survey Professional Paper* 411-D, D1-D38.

SEE ALSO: debris flow; effective stress; landslide; liquefaction; mass movement; piezometric; undrained loading

JON J. MAJOR

POSTGLACIAL TRANSGRESSION

The postglacial transgression (see TRANSGRESSION) has had many names, depending from the area and the time period studied. The name of Flandrian stage was first proposed in the late nineteenth century to indicate the 'Campinian sands' of Flanders and of the Anvers' Campine. Dubois (1924) included in the Flandrian stage all sediments, marine and continental, that characterize the displacement of the shoreline from the sea-level minimum, corresponding to the last glacial maximum, to the present situation.

In the British Isles, on the other hand, the deposits called Flandrian correspond more or less to those of the Holocene period, i.e. to the last 10 ka (Shotton 1973). An Irish variant for the Flandrian is the Littletonian, the base of which is placed near the base of a peat dated about 10,130 BP in a core, or at the maximum of *Juniperus* pollen in another core.

In the Mediterranean, Blanc (1936) described in the Versilia Plain the stratigraphy of deposits, *c.*90 m deep, covering all epiglacial and postglacial times, that he considered equivalent to the Flandrian transgression. The term Versilian was subsequently applied to other Mediterranean deposits by several authors, following the chronostratigraphic meaning proposed initially by Dubois (1924).

In Japan the postglacial transgression has received several local names: Yurakucho transgression, after the name of a marine Holocene bed in the Tokyo area, Numa transgression, after the raised coral bed of the Numa Terrace in the Boso Peninsula (Naruse and Ota 1984), or Umeda transgression, corresponding to deposits found in the Kinki area. In uplifted Japanese regions, the higher part of the marine deposits are also called Jomon transgression, after the age of the older stage of the Japanese Neolithic culture (between *c.*9,400 and 2,800 BP) (Takai *et al.* 1963).

In west Africa, the Nouakchottian episode, defined by Elouard (1966), consists of shell deposits, a few decimetres to over 2 m thick, which are found along most of the coast of Sénégal and Mauritania, at elevations close or slightly higher than the present sea level (between −2 m and +2 or +3 m). These RAISED BEACH deposits date from 5,500 to 1,700 BP and follow a shoreline that formed several gulfs. In the Ndrhamcha Sebkha area, where the coastline is now rectilinear, the Nouakchottian gulf extended into the continent about 90 km. Other similar transgression stories, each with a different name, could be added from other coastal regions of the world.

The use of many names to identify the same phenomenon in various areas, in order to attempt correlations, may have been useful in the past, when precise dating tools were rare or unavailable, but seems not to be justified today. Radiochronology has shown that the last maximal glacial peak occurred about 18,000 radiocarbon years BP, i.e. after calibration, *c.*21 ka ago (Bard *et al.* 1990). A more general use of the terms 'postglacial transgression' (for the last 21 ka) or 'Holocene transgression' (for the last 10 ka) would certainly contribute to clarify and unify the international terminology.

A marine transgression may occur, however, not only with a rising sea level, but even with a falling sea level if sediment supply is depleted and erosion can occur; conversely, a regression of the sea often results from a sea-level fall, but may also occur with a rising sea level in the case of high sediment supply and coastal progradation. Transgression–regression sequences, i.e. lithostratigraphic evidence of marine deposits inter-fingered with freshwater or terrestrial sediments, usually correspond to major changes in sea level. However, when the inter-fingered layers are not continuous in space and time, interpretation should be careful, especially in the case of Late Holocene sediments deposited near river mouths or near plate boundaries where tectonic displacements may have taken place.

In postglacial times, especially in the mid- to late Holocene, interpretation of transgression

–regression sequences has been reported from several coastal localities, again with various names. This was the case, e.g. on the southern coasts of the North Sea, where within the upper part of the Holocene transgression, three Dunkerquian transgressions were distinguished above Calais deposits (Tavernier and Moorman 1954). For each transgression a former sea-level position was deduced from the present elevation of the deposits. Correlations between the Flemish Dunkerquian stratigraphic sequences and other local deposits were subsequently attempted by various workers in France, the Netherlands and Germany, giving rise to the reconstruction of sea-level histories showing oscillations of varying amplitudes. Nevertheless, some fifty years later, the precise amount of these Dunkerquian sea-level oscillations, as well as their existence, remains to be demonstrated.

As SEA LEVEL is concerned, the last deglaciation seems to have occurred mainly in two eustatic (see EUSTASY) steps, with a first warming period that peaked at the Bölling (about 13–12 ka BP) and a second warming period after about 11.6 ka BP, separated by a temporary cooling (Younger Dryas). The melting histories of the various ICE SHEETs were non-synchronous and the last deglaciation ended in each place when the former ice sheet had completely melted. This seems to have occurred around 10 ka BP in Scotland and for the Cordilleran ice sheet, close to 9 ka BP in the Russian Arctic, around 7.5 ka BP in Scandinavia and around 6 ka BP for the Laurentide ice sheet. In Antarctica and Greenland, only the outer part of the ice domes melted.

The melting of the ice sheets caused considerable glacio-isostatic (see ISOSTASY) vertical movements of the Earth's crust: mainly uplift in unloaded areas, and subsidence in wide peripheral areas around the former ice sheets. The load of melted water on the ocean floor caused the latter to subside (hydro-isostasy). This is expected to have caused a flow of deep material from beneath the oceans to beneath the continents, with possible reactivation of seaward flexuring at the continental edges. Part of the above isostatic movements were elastic, i.e. contemporaneous to loading and unloading. Part continued, at gradually decreasing rates, for several thousand years after loading or unloading had stopped, because of the viscosity of the Earth's material. Part of such isostatic vertical displacements is still going on.

When the subsidence in areas peripheral to former ice sheets took place under the sea, it increased locally the volume of the oceanic container. This caused hydrostatic imbalance, with an indraught of water from other areas, decreasing the sea-level rise or even producing a slight sea-level fall in oceanic regions far away from the influence of former ice sheets. It is generally towards 7 to 6 ka BP, with the ending of ice melting, that slight emergence of isostatic or tectonic origin started to occur in many areas. Combination of the above processes produced a variety of relative sea-level and postglacial transgression histories along the world's coastlines. According to Mörner (1996), during the last 5 ka relative sea-level changes have been affected also by the redistribution of water masses due to changes in oceanic circulation systems.

In former ice-sheet areas, where marks of past shorelines on ice vanished with ice flowing and melting, local sea-level histories can be reconstructed only after deglaciation. They usually show continuing regression/sea-level fall, though at decreasing rates, up to the present.

In peripheral areas to the former ice sheets, the maximum postglacial sea-level peak has not yet been reached and the postglacial transgression is more or less still going on. Finally, in areas remote from the ice sheets and in most uplifting regions, a sea-level maximum, now emerged at variable elevations, has generally been reached towards the mid-Holocene.

References

Bard, E., Hamelin, B., Fairbanks, R.G. and Zindler, A. (1990) Calibration of the 14C timescale over the past 30,000 years using mass spectrometric U-Th ages from Barbados corals, *Nature* 345, 405–410.

Blanc, A.C. (1936) La stratigraphie de la plaine côtière de la basse Versilia (Italie) et la transgression flandrienne en Méditerranée, *Revue de Géographie Physique et Géologie Dynamique* 9, 129–160.

Dubois, G. (1924) Recherches sur les terrains quaternaires du Nord de la France, *Mémoires de la Société géologique du Nord*, 8, 16,360.

Elouard, P. (1966) Eléments pour une definition des principaux niveaux du Quaternaire Sénégalo–Mauritanien, I. Plage à Arca senilis, *Bulletin de l'Association Sénégalaise pour l'Etude du Quaternaire. Ouest-Africain* 9, 6–20.

Mörner, N.A. (1996) Sea level variability, *Zeitschrift für Geomorphologie, Supplementband* 102, 223–232.

Naruse, Y. and Ota, Y. (1984) Sea level changes in the Quaternary in Japan, in S. Horie (ed.) *Lake Biwa*, 461–473, Dordrecht: Junk Publishers.

Shotton, F.W. (1973) General principles governing the subdivision of the Quaternary System, in G.F. Mitchell, L.F. Penny, F.W. Shotton and R.G. West (eds) *A Correlation of Quaternary Deposits in the British Isles*, 1–7, London: Geological Society, Special Report No. 4.
Takai, F., Matsumoto, T. and Toriyama, R. (1963) *Geology of Japan*, Tokyo: University of Tokyo Press.
Tavernier, R. and Moorman, F. (1954) Les changements du niveau de la mer dans la plaine maritime flamande perdant l'Holocéne, *Geologie en Mijnbouw* 16, 201–206.

Further reading

Pirazzoli, P.A. (1991) *World Atlas of Holocene Sea-Level Changes*, Amsterdam, Elsevier.
——(1996) *Sea-Level Changes: The Last 20 000 Years*, Chichester: Wiley.

P.A. PIRAZZOLI

POT-HOLE

Pot-holes are vertical, circular and cylindrical erosion features in consolidated rock of various lithologies. They are common in fluvial, fluviglacial and shore environments. Their sizes (diameter and depth) range from a few dm to a few m; but mega pot-holes many metres deep and wide also occur in formerly glaciated areas. Pot-holes are produced by abrasion, CAVITATION, dissolution and/or corrosion. A tool (pebble, sand) is necessary for abrasion whereas cavitation is a mechanical process of wearing created by a turbulent flow. Dissolution is active in carbonate rocks whereas corrosion, a more complex process, is manifested in most rock types, particularly in warmer climates. Many pot-holes have a complex origin. Shallow cavities made by cavitation or dissolution are often subsequently eroded through abrasion. Coastal pot-holes are less common than fluvial and fluviglacial. A few anthropogenic pot-holes on shore platforms have been reported in Brittany. The use of the term pot-hole for kettle and moulin should be avoided.

Further reading

Tschang, H. (1974) An annotated bibliography of pot-hole forms, *Chun Chi Journal* 12, 15–53.

JEAN CLAUDE DIONNE

PRESSURE MELTING POINT

The melting point is the temperature at which a solid changes into a liquid. The application of pressure to a solid depresses the melting point, and the melting point under a given pressure is referred to as the pressure melting point. For ice, the lapse rate of melting point with pressure is $-0.072\,°C\,MPa^{-1}$, the effect of which, for example, is to depress the melting point to $-1.28\,°C$ under 2,000 m thickness of ice.

The pressure melting point is significant in glacial geomorphology because of its effect on the interaction of glacier ice with its substrate. Ice exists on the surface of the Earth at temperatures very close to, and sometimes at, its melting point. The application of pressure to Earth surface ice, as a result of either the hydrostatic pressure of ice overburden or the interaction of moving ice with undulations in the substrate, can lead to the depression of the melting point sufficient to allow melting. The process of pressure melting on the upstream over-pressured side of bedrock bumps, and refreezing of water on the downstream under-pressured side, plays a major role in glacier motion, and is also significant in the entrainment and transport of subglacially eroded debris. This REGELATION process allows material generated by subglacial abrasion and quarrying to be entrained in re-frozen layers, and is one process by which debris-laden ice can be added to the basal layer of a glacier or ice sheet.

WENDY LAWSON

PRESSURE RELEASE

Many rocky outcrops show sets of horizontal or curvilinear fractures (SHEETING or EXFOLIATION structures) that are roughly parallel to the topographical surface. They are known with different names, some of them equivalent, such as pressure release, relief of load or offloading. It is obvious that all rock fractures are an expression of erosional offloading because only through the release of vertical and/or lateral pressure can the closed discontinuities become opened. But the gist of the pressure release concept is that rocks which cool and solidify deep in the Earth's crust (e.g. magmatic rocks), do so under conditions of high lithostatic pressure, i.e. loading by overlying materials (either rocks, sediments or even water or ice). So, many people suppose that when the rock outcrops in the Earth's surface it suffers a pressure release and this causes the development of stress and subsequent fractures parallel to the land surface. That is why the form of the land surface in broad terms could determine the geometry of the so-called pressure release fractures. According to this the fractures so generated would be secondary features (i.e. developed after the topography). But that is not always true because it is generally accepted that many plutons have been emplaced at shallow depths and the related structure was generated by the stresses imposed on magmas during injection or emplacement and, hence, so was the shape of the original pluton. Moreover, the so-called pressure release structures (i.e. sheet jointing) are well developed in rocks such as sedimentary and volcanic, which have never been emplaced, and even in granites, the magnetic orientation contemporaneous with the emplacement is clearly discordant to the pressure release structures. The pressure release theory may be questioned on several other grounds. Unloading appears to be mechanically incapable of producing fractures because if expansive stress developed during erosion, it would be accommodated along pre-existing lines of weakness and does not need to generate new ones, namely the sheet fractures. Another reason is that in fact several morphological and structural features developed on and in granitic rocks are incompatible with the tensional or expansive conditions implied by offloading. It is the case of structural domes, wedges and overthrusting associated with sheeting and is impossible to explain in terms of an extensional regime. Furthermore, evidence of dislocation and mylonitization along sheet fractures suggests that they are true tensional faults. Thus the pressure release structures may be better interpreted as primary features of the rock and accordingly the joints (see JOINTNG) were first developed in the bedrock and the shape of the land surface is a response to this previous internal structure.

Further reading

Gilbert, G.K. (1904) Domes and dome structure of the High Sierra, *Geological Society of America Bulletin* 5(15), 29–36.

Twidale, C.R., Vidal-Romani, J.R., Campbell, E.M. and Centeno, J.D. (1996) Sheet fractures: response to erosional offloading or to tectonic stress? *Zeitschrift für Geomorphologie, Supplementband* 106, 1–24.

Vidal-Romani, J.R. and Twidale, C.R. (1998) *Formas y paisajes graníticos*, Servicio de Publicaciones de la Universidad de Coruña, Serie Monografías 55.

JUAN RAMON VIDAL-ROMANI

PRIOR STREAM

Prior streams are Late Quaternary PALAEOCHANNELS of the semi-arid Riverine Plain in southeastern Australia. These ancient rivers were first described in the scientific literature by Butler (1958), who was given the task of producing soil maps in a region set aside for expanded irrigated agriculture in the period following the Second World War.

The 77,000 km^2 Riverine Plain consists of the coalescing floodplains of westward-flowing rivers of the southern Murray–Darling system. Despite its exceedingly subdued topography and low surface elevation (the great majority is less than 100 m above sea level), the Plain displays a complex pattern of sediments, soils and micro-topography. At first, the apparently featureless nature of the landscape frustrated Butler's attempts to make sense of his field observations. However, with the aid of aerial photographs, it became clear that well-drained sandy linear depressions that stood a little above the adjacent plain marked the locations of ancient aggraded palaeochannels that Butler called prior streams.

Soil variation on the Plain was controlled by proximity to a prior stream. Well-drained calcareous soils on prior stream levees graded laterally into heavy clays on the distal floodplain. Beneath the prior stream channels were thick beds of pebbly sand. As Butler mapped the regional soil landscape in more detail he discovered that the prior streams formed a complex distributary pattern (Plate 92) that petered out to the

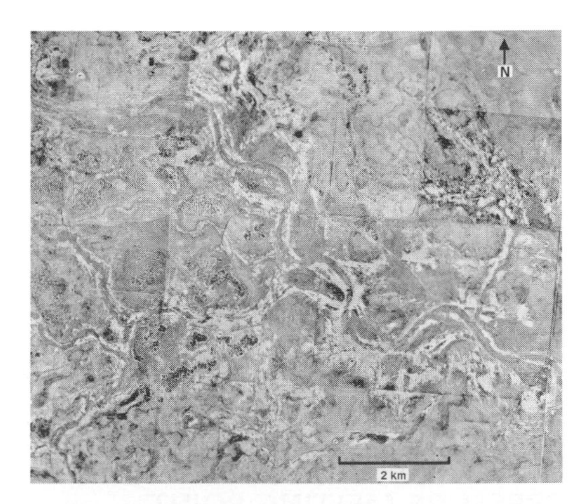

Plate 92 Air photograph mosaic of distributary prior stream channels on the Riverine Plain, Australia

west. Because the channels often intersected one another it was clear that there had been more than one period of prior stream activity. Butler invoked a cyclic model to interpret the different prior stream phases. Channel incision was thought to occur during more humid conditions when an absence of deposition permitted the development of well-organized soil profiles. The stable surfaces on which soils developed were called *groundsurfaces*. Channel aggradation occurred during more arid conditions when copious amounts of bedload sediment from upland catchment regions resulted in channel aggradation and extensive deposition.

Phases of prior stream activity

The oldest groundsurface described by Butler was the Katandra. It was thought to represent a long period of soil formation under more humid conditions than exist at present. A switch to more arid conditions resulted in a new phase of fluvial deposition (Quiamong) that progressively buried the Katandra surface. As aridity intensified vegetation breakdown in the region to the west led to widespread clay deflation by westerly winds and the deposition of an extensive blanket of pelletal clay (Widgelli PARNA) which mantled the earlier Quiamong deposits. As the peak of aridity passed parna deposition waned and renewed deposition by Mayrung prior streams occurred. This final phase of prior stream deposition was followed by a long humid phase when stream incision occurred and well-developed soils formed across the surface of the Plain. Soils of the Mayrung groundsurface developed on fluvial and aeolian parent materials of Quiamong, Widgelli and Mayrung age. A generalized summary of Butler's stratigraphic units is shown in Figure 128.

The modern channels of the Riverine Plain developed in the post-Mayrung period. They occupy narrow floodplain trenches incised two to three metres below the Mayrung surface and are characterized by high sinuosity, low width to depth ratio and a dominance of suspended sediment. According to Butler, these younger Coonambidgal deposits display very weak soil organization.

Post Butler

Not all workers agreed with Butler's interpretation of the prior stream deposits. The geomorphologist Langford-Smith (1960) argued that large meander wavelengths of the prior stream channels demanded greater discharges associated with late glacial pluvial, rather than arid, conditions. However, the absence of absolute dates on the prior streams (they all appeared to be beyond the radiocarbon limit of about 30,000 years) precluded any secure correlation with the glacial and interglacial episodes of the Late Quaternary.

Butler's early ideas were extensively revised during the latter part of the twentieth century. In the 1960s, Pels (1971) concluded that Butler's youngest stratigraphic unit, the Coonambidgal, was more complex than previously supposed. The early Coonambidgal phase was characterized by distinctive ancestral rivers that post-dated the prior streams and were the immediate precursors

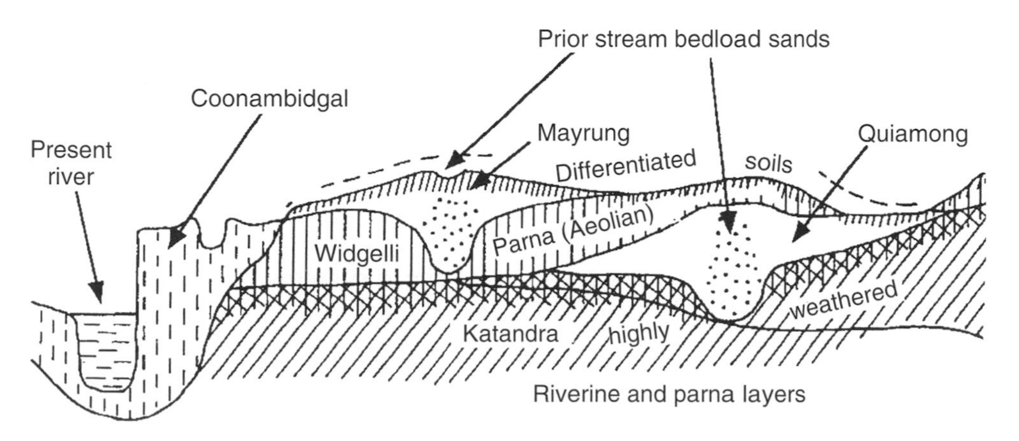

Figure 128 Generalized cross section of Riverine Plain showing Butler's (1958) prior stream sediments and soils (groundsurfaces)

of the modern drainage. The ancestral rivers were deep, sinuous, without levees and dominated by suspended load. They were much larger than the present rivers and maintained their courses across the Riverine Plain. Pels' model of Riverine Plain ancestral rivers and prior streams gained wide acceptance in the 1970s.

However, Bowler (1978), questioned the classification of ancient channels into two exclusive sequential types. He noted that both prior and ancestral attributes sometimes occur in different reaches of the same palaeochannel. In addition, Bowler found that some ancestral channels carried appreciable quantities of bedload, were bordered by sand dunes and mantled by well-developed soils similar to those of Butler's Mayrung groundsurface. Bowler concluded that Pels' separation of palaeochannels into two genetically different categories was unjustified.

Despite Bowler's misgivings, further progress on the nature and chronology of fluvial deposition awaited the development of thermoluminescence dating (Page *et al.* 1996). In brief, it was shown that four phases of palaeochannel activity occurred between approximately 100,000 and 12,000 years ago. Channel activity was characterized by alternations between sinuous, laterally migrating, mixed-load and straighter, vertically aggrading, bedload channel modes. Although the laterally migrating and vertically aggrading channel modes respectively approximate ancestral and prior streams, the sequence of channel activity was more complex than envisaged by Pels.

References

Bowler, J.M. (1978) Quaternary climate and tectonics in the evolution of the Riverine Plain, southeastern Australia, in J.L. Davies and M.A.J. Williams (eds) *Landform Evolution in Australia*, 70–112, Canberra: Australian National University Press.

Butler, B.E. (1958) *Depositional Systems of the Riverine Plain of south-eastern Australia in Relation to Soils.* Soil Publication No. 10, Australia: CSIRO.

Langford-Smith, T. (1960) The dead river systems of the Murrumbidgee, *Geographical Review* 50, 368–389.

Page, K.J., Nanson, G.C. and Price, D.M. (1996) Chronology of Murrumbidgee River palaeochannels on the Riverine Plain, southeastern Australia, *Journal of Quaternary Science* 11, 311–326.

Pels, S. (1971) River systems and climatic changes in southeastern Australia, in D.J. Mulvaney and J. Golson (eds) *Aboriginal Man and Environment in Australia*, 38–46, Canberra: Australian National University Press.

KEN PAGE

PROGLACIAL LANDFORM

The literal meaning of 'proglacial' is 'in front of the glacier'; it is the area that receives the products of glaciation. The proglacial environment is complicated, especially where warmer glaciers produce more meltwater. It can include terrestrial environments, streams, lakes and the ocean. The deposits include moraines, large outwash fans, deltas, marine fans and thick packages of sediment deposited in the marine realm. Other than where streams rework and erode previously deposited material, this is a depositional environment. The proglacial zone moves with the ice edge. As the ice advances, the zone of proglacial deposition moves forward as well. As the ice retreats, the proglacial setting follows the retreating margin 'backwards' and former subglacially deposited sediment-landform assemblages (see SUBGLACIAL GEOMORPHOLOGY) will be partially eroded, redeposited and/or buried.

The use of the term 'proglacial' has not been consistent in the literature. It is clear that there is a transition between ice-marginal and proglacial fluvial or lacustrine/marine processes. Some have suggested a further transition between proglacial fluvial processes and paraglacial processes, defined as any non-glacial processes conditioned by glaciation (Ryder, 1971). On this definition PARAGLACIAL processes strictly subsume proglacial fluvial processes. Thus the terms 'ice-marginal', 'proglacial' and 'paraglacial' have been used in overlapping senses and there is no universally agreed set of definitions. Table 37 classifies proglacial environments according to distance from the ice margin and lists the associated landforms. As most of these landforms are covered by separate entries (see cross references in the fourth column of the table) proglacial landform assemblages in the following will be viewed at the larger scale of ice sheet systems, mountain valley systems and subaquatic landsystems. A detailed review of these three landsystems can be found in Benn and Evans (1998).

Ice sheet systems

The processes and patterns of proglacial deposition in front of ice sheets and ice caps are strongly influenced by glacier thermal regime. Temperate glacier margins are wet-based for at least part of the year. Meltstreams are the main agents of sediment transport and deposition. Ice-marginal forms are produced by alternating glacifluvial

Table 37 Classification of proglacial landforms according to different land-forming environments

Environment	Process	Landform	See also:
Terrestrial ice-marginal	Meltwater erosion	Ice-marginal meltwater channels	Meltwater and meltwater channel Urstromtäler
	Mass movement/ meltwater deposition	Ice-marginal ramps and fans Dump and push moraines Recessional moraines	Moraine Glacitectonics Kame Kettle and kettle-hole
	Glacitectonics	Composite ridges and thrust block moraines Hill-hole pairs and cupola hills	 Ice stagnantion topography
	Meltwater deposition	Kame and kettle topography	
Subaquatic ice-marginal	Mass movement/ meltwater deposition	Morainal banks De Geer moraines	Glacilacustrine Glacimarine Moraine
	Meltwater deposition	Grounding-line fans Ice-contact (kame-) deltas	
	Debris flows	Grunding-line wedge	
Transitional from ice-marginal to fluvial	Meltwater erosion	Scabland topography Spillways	Meltwater and meltwater channel Outburst flood
	Meltwater deposition	Outwash plain (sandur) Outwash fan Valley train Pitted outwash Kettle hole/pond	Glacifluvial Kettle and kettle-hole
Transitional from ice-marginal to lacustrine and marine	Meltwater deposition/mass movement	Deltas	Glacideltaic Glacilacustrine Glacimarine
	Deposition from suspension settling and iceberg activity	Cyclopels, cyclopsams, varves Dropstone mud and diamicton Iceberg dump mounds Iceberg scour marks	

and gravitational processes, locally modified by glacitectonic deformation. In addition, extensive proglacial rivers may rework glacigenic sediments. In some cases, virtually all the evidence of a temperate ice lobe is in the form of glacifluvial sediments. At temperate glacier margins with moderate debris cover deposition typically produces small dump moraines, or push and squeeze moraines derived from sediment exposed on the glacier foreland. During deglaciation suites of recessional moraines are commonly formed. The areas between recessional moraines often exhibit well-preserved subglacial landforms. At temperate glacier margins covered by considerable quantities of glacifluvial sediment uneven ablation of sediment-covered ice can lead to the development of a karst-like topography with a relative relief of up to several tens of metres. Sediment deposited in supraglacial outwash fans and lakes produces a complex assemblage of landforms including ESKER systems, kame ridges and plateaux, and pitted outwash. Distinctive spatial associations of glacifluvial landforms deposited during ice wastage can be recognized: proglacial outwash passes upvalley into kame and kettle topography. Kame and kettle topography is locally refashioned into suites of river terraces by proglacial and postglacial streams.

Subpolar margins are characterized by cold-based conditions near the snout and an upglacier wet-based zone. Subpolar margins are commonly affected by glacitectonic processes. Meltwater is available on the glacier surface but not on the bed. It can have some impact on ice-marginal landforms, but in general these are only locally reworked into outwash deposits. Where thick accumulations of unconsolidated sediments are present, terminal moraines in the form of composite ridges are constructed by proglacial tectonics. Where proglacial thrusting does not occur, ice margin positions are recorded by frontal aprons built up from fallen debris. Upvalley end moraines or frontal aprons pass into ice-cored moraines of chaotic hummocky or transverse ridges and/or kames. In the cases of totally cold-based glaciers and where bedrock is close to the surface the amount of available debris reaches a minimum. Former glacier margins in such areas are often marked by lateral meltwater channels cut into bedrock.

Entirely cold-based polar-continental glacier margins leave very little imprint on the landscape. Features that appear to be terminal and hummocky moraines often constitute a thin veneer of debris overlying buried ice.

The margins of surging glaciers typically have a very high debris content. Surging glaciers also tend to be associated with widespread subglacial and proglacial glacitectonic deformation. Near the margin there is often extensive tectonic thrusting, particularly if the substratum has been weakened by high pore-water pressures. In addition, large discharges of meltwater and sediment associated with glacier surges are responsible for major changes in deposition rates in proglacial lakes and sandar. Sharp (1988) described the geomorphic effects of a glacier surge cycle. During the advance (surge) phase, fluted tills and thrust moraines are formed. At the termination of the surge, crevasse-fill ridges form at the bed, then hummocky moraine and outwash are deposited during glacier recession.

Under circumstances where glacially dammed lakes are breached, exceptionally high magnitude discharges are generated and the proglacial environment will be characterized by extensive erosional landforms as well as outwash deposits. The Channeled Scablands region of Washington State for example derives its name from the dramatic erosional forms generated by the draining of glacial lake Missoula.

Mountain valley systems

The majority of the debris transported by valley glaciers is derived from mass wasting of valley walls. Valley glaciers in high relief settings typically have extensive covers of supraglacial debris. Following ice retreat, high-relief mountain environments are subject to paraglacial reworking of ice-marginal sediments and landforms.

The margins of valley glaciers in mountain areas are commonly delimited by latero-frontal dump and push moraines. Meltwater deposition produces ice-marginal ramps and fans. In low relief mountain areas (e.g. the Scottish Highlands or the Norwegian mountains) Neoglacial end moraines are typically 2–5 m high. In debris-rich high relief settings the latero-frontal moraines can be much higher, so that repeated glacier advances may terminate at the same location and contribute to moraine-building, resulting in large landforms which can exceed 100 m in height.

Glacier retreat in mountain environments is normally recorded by recessional moraines. However, debris-mantled glaciers tend to remain at the limits imposed by their latero-frontal moraines until advance or retreat is triggered by significant climatic change. The landform record of a retreating debris-mantled glacier often consists of major moraine complexes, separated by extensive tracts of hummocky moraine deposited during episodes of ice-margin wastage and stagnation.

Glacial lakes are common features in mountainous environments. In low relief mountains they are mainly ice dammed as a result of the blocking of side or trunk valleys by expanded glacier tongues. Good examples of landform and sediment assemblages formed in Pleistocene lakes are found in the Highlands of Scotland and the water levels of former Glen Roy lake, for instance, are recorded by very prominent shorelines known locally as the 'Parallel Roads'. In high-relief mountain environments proglacial lakes dammed by moraines or by rockfalls, landslides and debris flow fans are more important than ice-dammed lakes. Some of the modern proglacial lakes impounded by latero-frontal moraines in the Andes and the Himalaya present a high risk of outburst floods to downvalley settlements.

Glacifluvial deposits are commonly well preserved in low to moderate-relief glaciated valleys. The focusing of meltwater flow by valley sides results in the erosion of gorges or the deposition of ribbon-like valley trains along valley axes. Staircases of terraces occur along the floors of many glaciated valleys, and the highest members may show signs of ice-marginal kame deposition. In valleys of high mountain environments mass movement features, lacustrine and fluvial sediment accumulations, and river terraces may be much more widespread than glacigenic landforms soon after their deglaciation. Paraglacial reworking of glacial landforms and sediments is less effective where glaciers advanced from high-relief mountainous regions to the foothills, and in such settings the preservation potential of the substantial ice-marginal landforms is greater. The margins of the Pleistocene piedmont glaciers in the northern foothills of the European Alps are delimited by high semicircular terminal moraines surrounding excavational basins. The ice proximal flanks of the moraines are characterized by kame and kettle topography and the central parts of the basins are either occupied by lakes or exhibit well-preserved DRUMLIN fields. The outer flanks of the moraines are skirted by large outwash fans, in which flights of several terraces were incised by postglacial rivers. For this spatial arrangement of landforms the term 'glacial sequence' was coined by Penck and Brückner (1909).

Subaquatic systems

Given the extent of water-terminating glacier margins today and during the past, the subaquatic proglacial environment is an important one. More than 90 per cent of the Antarctic ice sheet margins terminate in the sea. The Pleistocene northern hemisphere ice sheets were bordered in many places by lakes hundreds of kilometres across, ponded between the ice and topographic barriers, or by epicontinental seas. In Europe, the largest proglacial lake was the Baltic ice lake which, around 10,500 years BP, stretched for some 1,200 km along the southern margin of the Scandinavian ice sheet. The most extensive of the North American lakes, inundating 2,000,000 km^2, has been named proglacial Lake Agassiz (Teller 1995).

The type of sediment-landform association deposited adjacent to the glacier grounding line depends on ice velocity and calving rate, sediment supply, input of meltwater from the ice, and water depth and salinity.

Sediment supply and subglacial discharges of meltwater are highest at temperate glaciers with a tidewater front. Large amounts of coarse debris are deposited in morainal banks and grounding-line fans, and fine-grained sediments are carried away in turbid plumes to form a distal zone of laminated mud deposits. At the grounding line of temperate glaciers ending as an ice shelf meltwater-related processes tend to be less important and grounding-line fans are less common. Grounding-line deposits pass into drapes of dropstone muds, released by meltout of sediments embedded in the basal zone of the ice shelf and further out into dropstone muds derived from icebergs. Iceberg-rafted debris is generally more important in the vicinity of ice shelves where icebergs are often trapped close to the ice margin by sea ice, whereas in the forefront of grounded temperate ice margins sea ice does not restrict iceberg drift.

In contrast to wet-based ice bodies, all glaciers and ice sheets, which are frozen to their beds, provide little debris and little meltwater. Grounding lines below ice shelves are associated with grounding-line wedges, composed of mass flow deposits. In the case of ice margins ending with a tidewater front hardly any sediment will be released to the lacustrine/marine environment and proglacial landforms are rare.

References

Benn, D.I. and Evans, D.J.A. (1998) *Glaciers and Glaciation*, London: Arnold.

Penck, A. and Brückner, E. (1909) *Die Alpen im Eiszeitalter*, Leipzig: Tauchnitz.

Ryder, J.M. (1971) The stratigraphy and morphology of paraglacial alluvial fans in south central British Columbia, *Canadian Journal of Earth Sciences* 8, 279–298.

Sharp, M. (1988) Surging glaciers: geomorphic effects, *Progress in Physical Geography* 12, 533–559.

Teller, J.T. (1995) History and drainage of large ice dammed lakes along the Laurentide Ice Sheet, *Quaternary International* 28, 83–92.

CHRISTINE EMBLETON-HAMANN

PROTALUS RAMPART

A ridge, series of ridges or a ramp of debris formed at the downslope margin of a perennial or semi-permanent snowbed, which is located typically near the base of a steep bedrock slope in a periglacial environment. Observations of active examples indicate that constituent sediments range from diamicton to accumulations of coarse rock fragments. Roundness of rock fragments can vary from subangular to very angular, depending on the source of sediment supply. In planform, ramparts range from curved, to sinuous and complex. Typically, they have thicknesses (measured perpendicular to the slope) of up to 10 m. Examples on relatively steep slopes tend to have short proximal (i.e. adjacent to the snowbed) and long distal slopes.

The term 'pronival rampart' was preferred by Shakesby *et al.* (1995) on the basis that all examples lie at the foot of a snowbed (as 'pronival' indicates) but not all lie at the foot of a TALUS slope (as 'protalus' indicates). Until the early 1980s, there had been few observations of active forms. Suggested origins were based almost entirely on circular reasoning linking logical but hypothetical processes to supposed fossil 'ramparts', which might easily have been mistaken for other landforms (e.g. ROCK GLACIER, LANDSLIDE, MORAINE, avalanche impact ridge). It was reasoned that ramparts were formed entirely by coarse rockfall debris rolling, bouncing or sliding down a snowbed surface with very little if any fine debris reaching the rampart (White 1981). Any fine sediment in the ramparts, it was suggested, had been derived by *in situ* weathering or by the impacts of transported rock fragments. During the mid- to late 1980s, other processes (AVALANCHES and DEBRIS FLOWS) were found to be supplying fine as well as coarse material across snowbed surfaces to actively forming ramparts (Ono and Watanabe 1986; Ballantyne 1987). During the 1990s, the range of processes was expanded to include those operating beneath snowbeds, both as regards sediment supply (snowmelt, debris flow, SOLIFLUCTION) and modification of pre-existing sediment (bulldozing by a moving snowbed) (Shakesby *et al.* 1995, 1999). Because of confusion with other upland depositional landforms, there have been a number of attempts to identify diagnostic criteria, which have included morphological and sedimentological characteristics. In particular, attention has focused on the distinction between ramparts and moraines formed by small GLACIERS. Since, however, many 'rampart' characteristics have been based on (1) conjectural fossil examples, and (2) assumed formation at the bases of static snowbeds (although a rampart origin by a mobile snowbed has been demonstrated (Shakesby *et al.* 1999)), such diagnostic criteria must be viewed with extreme caution.

References

Ballantyne, C.K. (1987) Some observations on the morphology and sedimentology of two active protalus ramparts, *Journal of Glaciology* 33, 246–247.

Ono, Y. and Watanabe, T. (1986) A protalus rampart related to alpine debris flows in the Kuranosuke Cirque, northern Japanese Alps, *Geografiska Annaler* 86A, 213–223.

Shakesby, R.A., Matthews, J.A. and McCarroll, D. (1995) Pronival ('protalus') ramparts in the Romsdalsalpane, southern Norway: forms, terms, subnival processes, and alternative mechanisms of formation, *Arctic and Alpine Research* 27, 271–282.

Shakesby, R.A., Matthews, J.A., McEwen, L. and Berrisford, M.S. (1999) Snow-push processes in pronival (protalus) rampart formation: geomorphological evidence from southern Norway, *Geografiska Annaler* 81A, 31–45.

White, S.E. (1981) Alpine mass-movement forms (non-catastrophic): classification, description and significance, *Arctic and Alpine Research* 13, 127–137.

Further reading

Shakesby, R.A. (1997) Pronival (protalus) ramparts: a review of forms, processes, diagnostic criteria and palaeoenvironmental implications, *Progress in Physical Geography* 21, 394–418.

SEE ALSO: nivation; periglacial geomorphology

RICHARD A. SHAKESBY

PSEUDOKARST

A term first employed by von Knebel in 1906 (Bates and Jackson 1980), which has been widely used to describe topography, landform assemblages or features developed on non-carbonate rocks which exhibit a morphology similar to those characteristic of carbonate KARST terrain.

Such a lithology based definition excludes landforms in non-carbonate rocks from genuine karst. This classification is still in use by some geomorphologists and speleologists. However a more recent, all-encompassing definition of karst is now becoming increasingly widely accepted (Jennings 1983). Jennings's designation is less restrictive and he argued that karstic processes and landforms may be found on any rock type where the 'process of solution is critical, although not necessarily dominant'. Pseudokarst landforms should therefore be considered as those that morphologically resemble karst, but have formed through processes that are not dominated by solutional weathering or solution-induced subsidence and collapse.

Pseudokarst includes landforms morphologically similar to those commonly associated with carbonate or GYPSUM KARST landscapes, and include subterranean drainage, CAVES, DOLINES, BLIND VALLEYS, grikes, SPELEOTHEMS and surface KARREN.

Examples of pseudokarst fulfilling the conditions of Jennings definition, that is, where solution is not a critical formative process, include (1) caves in glaciers, or topographic depressions in permafrost regions (thermokarst), caused by a change in phase (i.e. from solid to liquid water) rather than dissolution (Otvos 1976); (2) VOLCANIC KARST, comprising tunnels within lava, formed where molten lava continued to flow inside an already partially solidified lava bed (i.e. caves are formed at the same time as the host rock) (Anderson 1930), and also depressions associated with the mechanical collapse of such caves; and (3) caverns and karst-like features caused by predominantly mechanical erosion of rock by animals such as abrasion caused by molluscs in the tidal zone of limestone outcrops on tropical and temperate coasts (Sunamura 1992), or moving water, wind or ice. Some workers also class as pseudokarst depressions and pipes (see PIPE AND PIPING) formed in soils or other unconsolidated sediments by the mechanical erosion of unconsolidated material (piping) (Otvos 1976) as, for example, often found within loess deposits.

In view of its original definition and long-term usage, the term pseudokarst can also be widely found in the literature referring to any karst-like features in rocks other than limestone (or gypsum) (including rocks such as basalt, granite or diorite) regardless of their mode of formation. Examples of these so-called pseudokarst features include basins, runnels, caves, underground drainage and even small speleothems. Provided these features can be ascribed to a range of physical or chemical weathering and erosive processes that do not rely on solution to any significant extent, the use of the term pseudokarst is appropriate.

The term pseudokarst has, however, also often been applied to landforms on rocks of relatively low solubility such as quartzites or highly siliceous sandstones, which consist almost entirely of silica (SiO_2) (e.g. Pouyllau and Seurin 1985). Such usage has been based on the widely held but incorrect assumption that quartzose rocks are practically immune to chemical weathering (Tricart 1972). This belief is based on the fact that the equilibrium solubility of many carbonate rocks ranges between 250 and 350 $mg\,l^{-1}$ at normal temperatures, whilst under the same conditions the solubility of crystalline silica (quartz) does not exceed 15 $mg\,l^{-1}$, and even that of amorphous silica is less than half that of many carbonates. Quartzose rocks were thus generally considered not to develop solutional and therefore 'genuine' karst, but rather pseudokarst (e.g. Pouyllau and Seurin 1985).

However, during the past few decades features of considerable dimensions and striking morphological similarity to dissolutional karst have been identified in quartzose rocks in Africa, South America and Australia. For example, in Africa solutional landforms and caves are found in the quartz sandstones and quartzites of Tchad, Nigeria, South Africa and the Transvaal; the great South American quartzite landscapes of Brazil and the Venezuelan Roraima display numerous large and small, remarkably carbonate-like, surface forms, silica speleothems and many cave systems with lengths exceeding 2.5 km and depths of 350 m; and the quartz sandstones of the Arnhem Land and Kimberley regions of northern Australia and even the Sydney region of south-eastern Australia displays many caves, tower karst, smaller surface karren and speleothems (see Wray 1997 for a detailed review).

A range of studies carried out in these highly quartzose regions has now either argued or directly shown that the prime process leading to these 'pseudokarst' features is the direct dissolution of silica. Where quartz grains are held together by amorphous silica cement, the dissolution of this comparatively soluble material (up to about 150 $mg\,l^{-1}$) may isolate individual quartz grains from the parent rock (arenization) (Jennings 1983), which may then be removed by flowing water. However, arenization also occurs in rocks with very little amorphous cement when individual quartz grains and crystalline overgrowths are dissolved despite their low solubility (see especially Jennings 1983 for northern Australia; Wray 1997 for south-eastern Australia; and Chalcraft and Pye 1984 for South America). In a study investigating cave passages in the well-developed quartzite karst in Venezuela, Doerr (1999) has even argued that, under specific conditions, such karstforms may develop largely through dissolution, with arenization playing only a minor role.

A range of workers conclude that in many areas with quartzose rocks, dissolution is the key process in the formation of karst-like features, and argue that genuine karst may develop in highly siliceous rocks, where very long periods of weathering offset slow rates of dissolution (e.g. Jennings 1983; Chalcraft and Pye 1984; Wray 1997; Doerr 1999). Following the earlier urgings of Jennings (1983), Wray (1997) argued in a wide-ranging and comprehensive analysis of the worldwide karst-like features in quartzites and quartz sandstones, that in these features, the critical role of solution clearly identifies these forms as true karst (i.e. quartzite or sandstone karst) and not pseudokarst.

References

Anderson, C.A. (1930) Opal stalactites and stalagmites from a lava tube in northern California, *American Journal of Science* 20, 22–26.

Bates, R.L., and Jackson, J.A. (eds) (1980) *Glossary of Geology*, 2nd edition, Falls Church, VA: American Geological Institute.

Chalcraft, D. and Pye, K. (1984) Humid tropical weathering of quartzite in Southeastern Venezuela, *Zeitschrift für Geomorphologie* 28, 321–332.

Doerr, S.H. (1999) Karst-like landforms and hydrology in quartzites of the Venezuelan Guyana shield: pseudokarst or 'real' karst?, *Zeitschrift für Geomorphologie* 43, 1–17.

Jennings, J.N. (1983) Sandstone pseudokarst or karst?, in R.W. Young and G.C. Nanson (eds) *Aspects of Australian Sandstone Landscapes*, Australian and New Zealand Geomorphology Group Special Publication No.1, University of Wollongong, Wollongong.

Otvos, E.G. (1976) 'Pseudokarst' and 'pseudokarst terrains': problems of terminology, *Geological Society of America Bulletin* 87, 1,021–1,027.

Pouyllau, M. and Seurin, M. (1985) Pseudo-karst dans des roches grèso-quartzitiques da la formation Roraima, *Karstologia* 5, 45–52.

Sunamura, T. (1992) *Geomorphology of Rocky Coasts*, Brisbane: Wiley.

Tricart, J. (1972) *The Landforms of the Humid Tropics, Forests and Savannas*, London: Longman.

Wray, R.A.L. (1997) A global review of solutional weathering forms on quartz sandstones, *Earth-Science Reviews* 42, 137–160.

Further reading

Ford, D. and Williams, P. (1989) *Karst Geomorphology and Hydrology*, London: Chapman and Hall.

Jennings, J.N. (1985) *Karst Geomorphology*, Oxford: Blackwell.

SEE ALSO: biokarst; chemical weathering

STEFAN H. DOERR AND ROBERT WRAY

PULL-APART AND PIGGY-BACK BASIN

Pull-apart and piggy-back sedimentary basins are typically associated with convergent plate tectonic settings (see PLATE TECTONICS). Pull-apart basins are topographic lows developed by rifting along strike-slip faults in areas of transtension (i.e. areas subjected to both transform and extension tectonics). The term 'pull-apart' was first used by Burchfiel and Stewart in 1966 to describe features in the Death Valley region of the USA and was later used by Crowell in 1974 to describe features along the San Andreas fault. Pull-apart basins have also been referred to as 'rhombochasm' and 'rhombograben' for the largest features (several kilometres by tens of kilometres in dimensions) and 'sag pond' for the smallest features (a scale of tens to hundreds of metres) (Seyfot 1987). The areas of transtension that develop pull-apart basins are typically associated with either (1) bends in the fault system (known as releasing bends) or (2) fault off-sets. These bends or off-sets need to step over to the left for a left lateral fault system or right for a right lateral fault system in order to generate the required transtension for basin development. The resulting basin is bounded on two sides by the strike-slip faults (which also have a significant normal component to the fault movement) and on the other two sides, approximately perpendicular to the main strike-slip faults, by normal faults. With continued extension of the basin the floor can be stretched and thinned to the extent that volcanism may occur and thus may cover the floor of the basin. Sedimentary fill of the basins may be developed in one main depocentre (area of maximum subsidence) in the central part of the basin or two depocentres, each adjacent to the bounding normal faults (Deng *et al.* 1986). Modelling by these authors suggests that the number and position of the depocentres is dependent on the geometry of the basin which in turn is dependent on three main factors (1) separation between the overlapping lateral fault strands; (2) degree of overlap between the main lateral faultstrands; and (3) the depth to the basement. Basins elongated parallel to the main lateral faults (overlap is more than separation) tend to have two depocentres, whereas 'shorter' basins where the separation is more than the overlap tend to have one depocentre. In most cases the depth of the basins typically tends to be greater than typical rift basins developed in divergent plate tectonic settings, and tends to be dominated by alluvial or lacustrine sedimentary fill.

Piggy-back basins, in contrast, are typically associated with thrusted terrain where basin development is complicated by deformation of earlier basin deposits by more recent thrusting. The term 'piggy-back basin' was first used by Ori and Friend in 1984 to describe minor sedimentary basins that rest on moving thrust sheets. Such basins have also been termed 'thrust-sheet-top basins' (Ori and Friend 1984)', 'satellite basins' (Ricci-Lucchi 1986) and 'detached basins' (Steidmann and Schmitt 1988). These basins are typically a few tens of kilometres across and are physically separated from the foredeep (the basin in front of all the active thrusts). Classic examples of piggy-back basins are found throughout the Alpine mountain chains of Europe. The basin fill comprises sediment sources from all basin margins, with a dominant provenance from the uplifted ramp of the older thrust behind the basins. Sedimentary environments range from coarse submarine fan and fan delta to fluvial deposits. Fluvial systems typically comprise a transverse drainage from the thrust ramps on both sides of the basin and a longitudinal drainage which enters the basin from the topographic lows that develop above lateral fault terminations (Miall 1999).

References

Burchfiel, B.C. and Stewart, J.H. (1966) Pull-apart origin of the central segment of Death Valley, California, *Geological Society of America Bulletin* 77, 439–442.

Crowell, J.C. (1974) Origin of late Cenozoic basins in southern California, in W.R. Dickinson (ed.) *Tectonics and Sedimentation*, Society of Economic Palaeontologists and Mineralogists Special Publication 22, 190–204.

Deng, Q., Zhang, P. and Chen, S. (1986) Structure and deformation character of strike-slip fault zones, *Pure and Applied Geophysics* 124, 203–223.

Miall, A.D. (1999) *Principles of Sedimentary Basin Analysis*, 3rd edition, New York: Springer Verlag.

Ori, G.G. and Friend, P.F. (1984) Sedimentary basins formed and carried piggyback on active thrust sheets, *Geology* 12, 475–478.

Ricci-Lucchi, F. (1986) The Oligocene to recent foreland basins of the northern Apennines, in P.A. Allen and P. Homewood (eds) *Foreland Basins*, International Association of Sedimentologists Special Publication 8, 105–139, London: Blackwell Science.

Seyfot, C.K. (1987) *Encyclopaedia of Structural Geology and Tectonics, Encyclopaedia of Earth Sciences Series*, Vol. 10, New York: Van Nostrand Reinhold.

Steidmann, J.R. and Schmitt, J.G. (1988) Provenance and dispersal of tectogenic sediments in thin-skinned, thrusted terrains, in K.L. Kleinsehn and C. Paola (eds) *New Perspectives in Basin Analysis*, 353–366, Berlin and New York: Springer Verlag.

Further reading

Burbank, D.W. and Anderson, R. (2001) *Tectonic Geomorphology*, Malden, MA: Blackwell Science.

Hatcher, R.D. Jr (1995) *Structural Geology Principles and Concepts*, 2nd edition, Upper Saddle River, NJ: Prentice Hall.

SEE ALSO: fault and fault scarp; plate tectonics; rift valley and rifting; tectonic geomorphology

ANNE E. MATHER

PUNCTUATED AGGRADATION

The theory that the long-term aggradation of sediment (through geological time) has been via episodic SEDIMENTATION. This is in contrast with the traditional concept of UNIFORMITARIANISM and the continual and gradual build-up of sediments through time. Early studies such as that by Barrell (1917) provided the initial challenge to the long-held paradigm of gradual aggradation. The theory of punctuated aggradation began to gather momentum once more in the early 1980s. Ager (1980: 43) fuelled the debate by referring to sediment stratigraphy as having 'more gaps than record', and argued that the large disparities between modern sediment deposition (for a specific environment) and ancient calculated deposition was a result of the episodic nature of aggradation. The theory of punctuated aggradation treats each bedding plane as a pause in sedimentation, whereas continual aggradation considers bedding planes as merely signifying a change in diagenesis or texture, and treats the formation as the basic stratigraphic unit, each one a product of a particular environment.

The term punctuated aggradational cycle (or PAC) was coined by Goodwin and Anderson (1985), within their hypothesis for episodic stratigraphic accumulation. The hypothesis argues that, allowing minor exceptions, the stratigraphic record consists of thin (1–5 m thick), basin-wide, shallowing-upward cycles. These are sharply defined by surfaces produced by geologically instantaneous relative BASE-LEVEL rises (termed punctuation events). Deposition occurs during intervening periods of base-level stability. A host of depositional environments can be included in the PAC hypothesis (e.g. fluvial, deltaic, shelf, slope, etc.), as PACs are assumed to exist in all depositional environments influenced by rapid base-level rises.

The PAC hypothesis proposes that allogenic processes such as sea-level change are responsible for changes in the stratigraphic record, rather than autogenic processes (e.g. channel migration, etc.) that are held as responsible in continuous aggradation. Autogenic processes are not dismissed entirely, but are treated as localized stratigraphic influences, superimposed on the allogenic processes. The bounding surfaces between the PACs are often traceable laterally for vast distances since they are formed by large-scale allogenic processes. This allows them to be accurate stratigraphic markers in the field. Base-level rise during a punctuation event can be rapid (reaching 1 m per 100 years) whereas stratigraphical analysis indicates that the recurrence of such punctuation events can be as frequent as 50,000 years, thus reflecting the rapidity of the base-level rise. Thickness of PACs, though generally thin, varies considerably though long-term aggradation rates remain similar. Goodwin and Anderson suggest that the most likely mechanisms responsible for PACs would include episodic crustal movement, episodic movement of the geoidal surface and global eustatic sea-level changes.

References

Ager, D.V. (1980) *The Nature of the Stratigraphical Record*, 2nd edition, New York: Wiley.

Barrell, J. (1917) Rhythms and the measurement of geologic time, *Geological Society of America Bulletin* 28, 745–904.

Goodwin, P.W. and Anderson, E.J. (1985) Punctuated aggradational cycles: a general hypothesis of episodic stratigraphic accumulation, *Journal of Geology* 93, 515–533.

Further reading

Dott, R.H. (1982) SEPM presidential address: episodic sedimentation – how normal is average? How rare is rare? Does it matter? *Journal of Sedimentary Petrology* 56, 601–613.

STEVE WARD

PYROCLASTIC FLOW DEPOSIT

Pyroclastic flow deposits are the products of fragmental material transported laterally by gas-charged, concentrated flows (sometimes called NUÉES ARDENTES). Pyroclastic flows are generated in many ways, with a spectrum from the 'passive' collapse of oversteepened lava-flow or dome margins, through the gravitational collapse of high eruption columns, to powerful overpressured blast-like events. In contrast to gravity-controlled lava flows (see LAVA LANDFORM) and water-charged LAHARS, pyroclastic flows may possess considerable momentum and cross substantial obstacles (sometimes >1-km high mountains). Pyroclastic flow deposits are so diverse that they are here described in terms of five spectra.

The first spectrum is in densities of the juvenile (newly erupted) component, which reflect the relative importance of expansion of dissolved volatiles in frothing and fragmenting the magma. Densities are higher (1.0–2.7 $Mg\,m^{-3}$) in many small-volume deposits, particularly those associated with composite VOLCANOes and/or the collapse of lava domes, where the magma is fragmented by external means such as crushing or shattering by interaction with water. In larger deposits, clast densities reduce to $<1\,Mg\,m^{-3}$ (i.e. pumice), commensurate with an increasing role for expansion of dissolved volatiles. Simultaneously, contents of fine ash (<1/16 mm) increase; deposits with dense juvenile clasts have low contents (typically <2–5 per cent), those containing pumice have higher contents (>10–15 wt per cent). Dense clast-rich deposits often contain abundant large (dm–m-sized) juvenile clasts, and are labelled as, e.g. 'block-and-ash flow deposits' or 'dome-collapse avalanche deposits'. Deposits where the juvenile component is pumice are termed 'ignimbrites' or 'ash-flow tuffs'; collectively they represent by far the greatest volume of pyroclastic flow deposits worldwide.

The second spectrum is size. Distances travelled range from a few hundred metres in lava-collapse flows, to >150 km for prehistoric large ignimbrites. Areas range from a few thousand square metres to $>30,000\,km^2$. Volumes range from about $1,000\,m^3$ for individual dome-collapse events to $>1,000\,km^3$ for large ignimbrites. Observed pyroclastic flow eruptions generated only relatively small examples, with distances travelled up to 30–40 km, areas up to $400\,km^2$, and volumes up to $\sim15\,km^3$. Small pyroclastic flow deposits ($<1\,km^3$) are generated from vents on composite volcanoes or from collapse of lava flows/domes. Intermediate-sized deposits (up to a few tens of km^3) can be generated from composite volcanoes or CALDERA volcanoes, often associated with caldera collapse. Larger deposits are associated with eruptions of gas-rich, evolved magmas (particularly rhyolite) from caldera volcanoes.

The third spectrum is deposit morphology. Individual, small-volume pyroclastic flows form tongue-like deposits, often with surface ridging, marginal levees and lobate flow fronts akin to those developed on DEBRIS FLOWs. However, most deposits form during many (tens to hundreds of) individual flow events, and so the gross deposit morphology then reflects the energetics of flow emplacement and deposit volume. The energetics are represented by the 'aspect ratio', which is the ratio of the average deposit thickness to the diameter of a circle with the same area as the deposit. Sluggishly emplaced deposits have a high aspect ratio (as high as 1 : 200), that is, the material is relatively thick for its extent. Energetically emplaced deposits have low aspect ratios ($>$ 1 : 10,000), that is, the material is very widespread for a given volume of material. The volume of the pyroclastic deposit coupled with its aspect ratio then yields three major morphologies: landscape-mantling, landscape-modifying, and landscape-forming (Figure 129). The largest deposits can create wholly new land surfaces over areas of $>$1,000 km^2, forming fan- or pediment-like surfaces around the source volcano.

The fourth spectrum is in the internal structure of the deposits. Single pyroclastic flows generate single flow units, that may be composed of a number of layers and facies that in turn reflect the mechanics of flow emplacement. Deposits of multiple flows should, in principle, show multiple flow units, but the clarity with which flow-unit boundaries can be discerned within such deposits is very variable. Thick stacks of ignimbrite may show no stratification, or only vague bedding or fluctuations in grain size, to suggest that they are the product of multiple flows. Grading structures within individual flow units vary widely also, and can reflect both migration of coarse clasts (regardless of density) under shearing forces, and

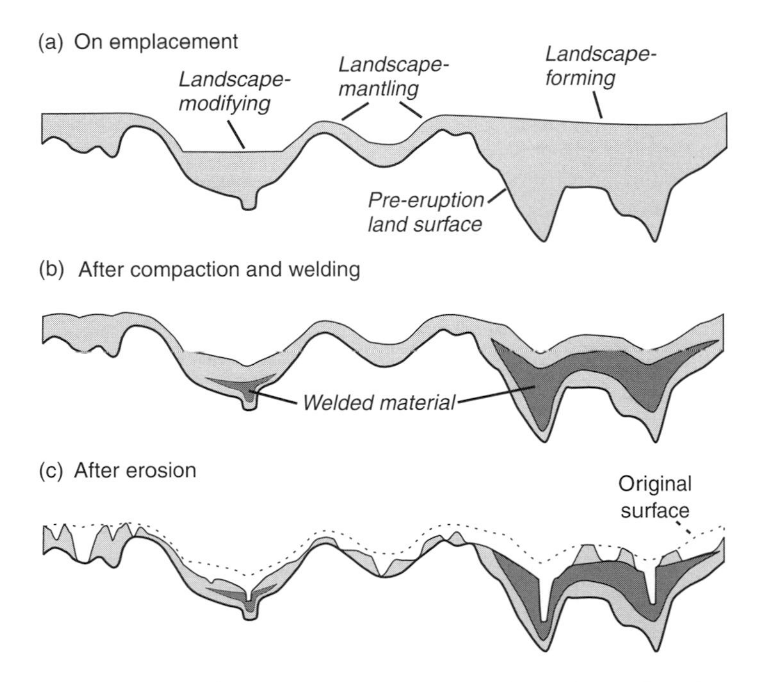

Figure 129 Schematic diagram to illustrate the morphologies of pyroclastic flow deposits. (a) On immediate emplacement, showing how the pre-eruptive landscape may be mantled, modified, or buried, depending on the thickness of the deposit, the topographic relief and the energetics of emplacement. (b) After consolidation and welding; note how compaction is greatest where the deposits are thickest, thus new valleys are generated along the line of pre-eruptive buried valleys. (c) After erosion; valleys are re-cut along their old courses. Non-welded deposits are preferentially removed, but summit heights may still be concordant, reflecting the original surface

flotation/sinking of lighter/denser coarse clasts, respectively, under buoyancy forces.

The fifth spectrum is in the lithologies of the deposits. Pyroclastic flows are efficient conservators of heat, and so many deposits are emplaced at temperatures above those at which the juvenile material can flow plastically (e.g. >550–600 °C for rhyolitic pumice). The combination of retained heat and load stresses imposed by overlying deposits causes the juvenile fragments to adhere and flatten (weld) to form a coherent rock. At its most extreme, welding can eliminate all initial pore space and the rock may be so hot as to continue to flow plastically as a kind of lava flow. Welding can only occur as long as the juvenile phase is glassy, but in most welded deposits the glass has subsequently devitrified. In addition, gases released from the juvenile material can cause further crystallization and vapour-phase alteration of the deposit, either along discrete pathways ('fossil fumaroles') or pervasively through the porous rock mass. Non-welded deposits show little or no JOINTING, but welding (and any other causes of induration) is generally accompanied by formation of jointing in the rock mass. The orientation and spacing of the joints can vary, but columnar joints, spaced at decimetres to metres apart, are characteristic of the interior of thick ignimbrites. Closer to the base, top or sides of the deposits, or in places where local fluxes of hot gases have occurred, the jointing can be more closely spaced and fan-like in disposition.

The morphologies of freshly emplaced pyroclastic flow deposits (Figure 129) are generally very rapidly modified by erosion, as loose pyroclastic-flow material is readily eroded, generating syn- and post-eruptive debris flows, lahars and HYPERCONCENTRATED FLOWS. Incision by streams often occurs so rapidly that interaction may occur between water and the still-hot interior of the deposits, leading to 'rootless' phreatic explosions. In non-welded deposits, incision rates of metres to tens of metres per rain event are known. Incision tends to recur along the lines of the pre-eruption valleys; the greatest thicknesses of deposits (and hence the greatest compaction) occur there and so the pre-eruptive topography is mirrored in subdued fashion on the surface of the deposits, controlling the paths of re-established streams. Erosion slows considerably when hard (welded) material is reached, or the non-welded deposits are stabilized by regrowth of vegetation.

Landscape morphologies seen in areas covered by pyroclastic flow deposits reflect a complex interplay between the initial depositional morphology, the presence or absence of welding or induration to create hard rock, and the local climate. A characteristic feature in dissected large ignimbrites is a concordance of ridge or summit heights, defining a surface parallel to the original deposit surface. Slopes in non-welded deposits are typically at or close to the angle of rest, except along streams or river where undercutting leads to vertical cliffs. Slopes in welded deposits are often cliffed, as the removal of material is controlled by vertical jointing that allows toppling of columnar masses as they are undermined by erosion.

Although pyroclastic flow deposits are volumetrically important in many volcanic terrains, the enormous variety of characteristics these deposits can display, and the hazards associated with flow emplacement, mean that there is still much to be discovered about the processes and products of pyroclastic flows.

Further reading

Cas, R.A.F. and Wright, J.V. (1987) *Volcanic Successions Modern and Ancient*, London: Allen and Unwin.

Druitt, T.H. (1998) Pyroclastic density currents, in J.S. Gilbert and R.S.J. Sparks (eds) *The Physics of Explosive Eruptions*, 145–182, London: Geological Society Special Publication 145.

Fisher, R.V. and Schmincke, H.-U. (1984) *Pyroclastic Rocks*, Berlin: Springer.

Freundt, A., Wilson, C.J.N. and Carey, S.N. (2000) Block-and-ash flows and ignimbrites, in H. Sigurdsson (ed.) *Encyclopedia of Volcanoes*, 581–599, San Diego: Academic.

Ross, C.S. and Smith, R.L. (1960) *Ash-flow Tuffs: Their Origin, Geologic Relations, and Identification*, US Geological Survey Professional Paper 366.

Walker, G.P.L. (1983) Ignimbrite types and ignimbrite problems, *Journal of Volcanology and Geothermal Research* 17, 65–88.

COLIN J.N. WILSON

Q

QUICK FLOW

Hydrologists generally separate streamflow into two operationally defined components: event flow, considered to be the direct response to a given water-input event (also called direct runoff, storm runoff or stormflow), and base flow, which is water that enters from persistent, slowly varying sources and maintains streamflow between water-input events (derived largely from groundwater circulation). Quick flow is simply another term for event flow. The mechanisms involved may be one, or a combination, of Hortonian overland flow, saturation overland flow, and near-stream subsurface storm flow via groundwater mounding. In the latter case, at least some of the water identified as quick flow is 'old water' that entered the basin in a previous event. Quick flow can also be 'delayed', which involves storm runoff from distal sources via predominantly subsurface routes.

SEE ALSO: runoff generation

MICHAEL SLATTERY

QUICKCLAY

The quickclays (quick clays, quick-clays, Swedish: *kvicklera*) are clay-sized postglacial marine sediments of very high sensitivity (see SENSITIVE CLAY). The term relates to the old Nordic *qveck*, meaning living. They are found in Norway, Sweden and Canada, and to a much lesser extent in Alaska, Finland and Russia, and they have been defined as having a sensitivity of greater than 50. The original definition was: a clay whose consistency changed by remoulding from a solid to a viscous fluid. Very high sensitivity values have been found – up to 200 for the Champlain clays of east Canada. The literature is dispersed; there are reviews by Bentley and Smalley (1984), Cabrera and Smalley (1973), Maerz and Smalley (1985), McKay (1979, 1982), Brand and Brenner (1981) and Locat (1995). The high sensitivity value means that the clays lose most of their strength on remoulding, and this can lead to catastrophic landslides, which progress rapidly as flowslides. Soderblom (1974) proposed that two types of quickclays should be recognized: rapid quickclays and slow quickclays. The rapid materials lose their strength very quickly on reworking; but the slow materials require the input of a fairly large amount of energy before they convert to a liquid. The strength parameters of the remoulded clays can be difficult to measure.

The classic quickclay explanation by I.Th. Rosenqvist (1953) depended on postglacial uplift, and leaching. The clay material was deposited in shallow salty seas in immediate postglacial times. As postglacial uplift occurred these deposits became dry land and were exposed to rainfall and groundwater flow. This had the effect of leaching out the salts and changing the electrochemical environment of the soil particles. The loss of the soil cations meant that the system became more metastable and responded to stress via soil structure collapse, LIQUEFACTION and flowsliding. The Rosenqvist theory appeared to work for the rapid Scandinavian clays, but not to be so suitable for the slower Canadian clays.

As mineralogical analysis became more sophisticated it became apparent that in many quickclays the actual clay mineral content was quite low and that they were perhaps better described as very fine silts. This fitted in rather well with their observed distribution on the fringes of glaciated regions. Glacial action could provide the very fine primary mineral material required to form the quickclay deposits. In fact the geomorphological observations led to a new approach to quickclays which has become known as the inactive-particle, short-range bond theory. This requires that the quickclay systems be cohesive (by virtue of the small particle size) but not plastic (because of the predominance of primary mineral particles, e.g. quartz, and the shortage of clay mineral particles). The fine blade-shaped primary mineral particles sediment in the shallow sea as Rosenqvist required, and form an open rigid structure; but the interparticle bonding is not the long-range clay mineral-type bonding but rather a short-range contact bond, enhanced by cementation.

References

Bentley, S.P. and Smalley, I.J. (1984) Landslips in sensitive clays, in D. Brunsden and D. Prior (eds) *Slope Instability*, 457–490, Chichester: Wiley.

Brand, E.W. and Brenner, R.P. (eds) (1981) *Soft Clay Engineering*, Amsterdam: Elsevier.

Cabrera, J.G. and Smalley, I.J. (1973) Quickclays as products of glacial action: a new approach to their nature, geology and geotechnical properties, *Engineering Geology* 7, 115–133.

Locat, J. (1995) On the development of microstructure in collapsible soils, in E. Derbyshire, T. Dijkstra and I.J. Smalley (eds) *Genesis and Properties of Collapsible Soils*, 93–128, Dordrecht: Kluwer.

Maerz, N.H. and Smalley, I.J. (1985) The nature and properties of very sensitive clays: a descriptive bibliography, Waterloo, Ontario: University of Waterloo Press.

McKay, A.E. (1979, 1982) *Compiled bibliography of Sensitive Clays*, Ottawa, Ontario Ministry of Natural Resources.

Rosenqvist, I.Th. (1953) Considerations on the sensitivity of Norwegian quick-clays, *Geotechnique* 3, 195–200.

Soderblom, R. (1974) New lines in quick clay research, Swedish Geotechnical Institute: Reprints and Preliminary Reports 55, 1–17.

Further reading

Ter-Stepanian, G. (2000) Quick clay landslides: their enigmatic features and mechanism, *Bulletin Engineering Geology Environment* 59, 47–57.

SEE ALSO: liquefaction; sensitive clay

IAN SMALLEY

QUICKSAND

Quicksand requires a flow of water. As the water flows through sands and silts and loses pressure its energy is transferred to the particles that it is flowing past, which in turn creates a drag effect on the particles. If the drag effect is in the same direction as the force of gravity, then the effective pressure is increased and the system is stable. In fact the soil/sediment tends to become denser. Conversely, if the water flows towards the surface, then the drag effect works against gravity, and reduces the effective pressure between the particles. If the velocity of the upward flow is sufficient it can buoy up the particles so that the effective pressure is reduced to zero. This represents a critical condition where the weight of the submerged soils is balanced by the upward-acting seepage force. This critical condition sometimes occurs in sands and silts. If the upward velocity of flow increases beyond the critical hydraulic gradient a quick condition develops.

Quicksands, if subjected to deformation or disturbance, can undergo a spontaneous loss of strength, which causes them to flow like viscous liquids. Karl Terzaghi, in 1925, explained the quicksand phenomenon as follows: first, the sand or silt concerned must be saturated and loosely packed. Second, on disturbance the constituent grains become more closely packed, which leads to an increase in pore-water pressure, reducing the forces acting between the grains. This brings about a reduction in strength. If the pore water can escape very rapidly the loss in strength is momentary. The third condition is that the pore water cannot escape readily. This occurs if the sand or silt has a low permeability or the seepage path is long, or both.

Casagrande, in 1936, demonstrated that a critical packing porosity existed above which a quick condition could be developed. He proposed that many coarse-grained sands, even when loosely packed, have porosities just about equal to the critical condition, while medium- and fine-grained sands, especially if uniformly graded (a narrow range of particle size), exist well above the critical porosity when loosely packed. Thus fine sands (say 60–150 μm) tend to be potentially more unstable than coarse-grained sands. The finer sands tend to have lower permeabilities.

Further reading

Bell, F.G. (1999) *Geological Hazards: Their Assessment, Avoidance and Mitigation*, London: Spon.

SEE ALSO: liquefaction; quickclay

IAN SMALLEY

R

RAINDROP IMPACT, SPLASH AND WASH

One of the most important driving forces in soil and hillslope EROSION is the kinetic energy of raindrops striking the soil surface. Raindrop impact contributes to soil erosion directly by splashing particles downslope, by entraining particles in OVERLAND FLOW which is below the threshold conditions necessary to pick up material. It can also affect erosion indirectly by disrupting soil aggregates, increasing ERODIBILITY, and by beating the surface into an almost impermeable seal or crust (see CRUSTING OF SOIL), which reduces infiltration and increases runoff (see RUNOFF GENERATION) discharge during rainstorms.

The kinetic energy of a moving object is expressed by $0.5\,MV^2$ where M = mass of the object and V = velocity. In the case of raindrops, the velocity is the terminal velocity which, in still air, reaches values around $9\,m\,s^{-1}$, for drops of 5 mm diameter (Laws 1941). During rainstorms, this value can be significantly affected by near-ground turbulence and wind. Raindrop mass is an even more critical control on the kinetic energy of raindrop impact. Raindrop size varies greatly from minute droplets a few microns in diameter, to an upper limit around 6.5 mm. As raindrop mass is directly proportional to diameter, there is a huge difference in the kinetic energy expended by small and large drops as they strike the surface. Comprehensive understanding of the relationship between raindrop impact and rainstorm characteristics is limited by the scarcity of accurate drop size measurements, particularly during rainstorms of very high intensity. However, Hudson (1981), amongst many others, has shown that rainstorms typically have a normally distributed spectrum of drop sizes, which can be expressed by the median drop diameter. This ranges from around 1.8 mm for a rainstorm of $12.7\,mm\,h^{-1}$ intensity to about 2.3 mm for a rainstorm of $65–115\,mm\,h^{-1}$ intensity. Information about characteristics of very high intensity rainfall is limited, because the most intense storms are usually of very limited duration and extent. It was thought that intensities above $150\,mm\,h^{-1}$ are very rare and largely limited to the tropics, but recent observations suggest that intensities as high as $400\,mm\,h^{-1}$ are by no means uncommon, particularly for very short periods, particularly at the beginning of thunderstorms.

Although information about raindrop size and rainfall intensities is still deficient, it is clear that there are major systematic differences between different types of rainfall and different parts of the world, which are reflected in the kinetic energy expended and the capacity of raindrop impact to generate erosion. The highest energy expenditure is certainly associated with the large drops and high intensities of severe thunderstorms or orographic rainfall and so the highest annual rainfall erosivities (see EROSIVITY) occur in areas like Assam or Hawaii, where such rainfall is combined with high annual totals. By comparison, the predominantly frontal rainfall of temperate areas produces very low kinetic energy, though occasional severe storms can, of course, cause much damage. The effect of raindrop impact is, however, strongly affected by vegetation. A dense vegetation cover can absorb virtually all the kinetic energy of raindrops, almost eliminating erosional hazard. However, although it takes about 30 m fall for drops to achieve full terminal velocity, they can achieve 60–70 per cent with a fall of some 3 m. Unless there is dense vegetation near or on the surface, raindrops can therefore regain much of their kinetic energy before hitting the surface. As a result, trees are not usually effective in controlling soil erosion in the absence of ground cover.

Raindrop impact affects the soil surface in several different ways. It may cause crusting by compacting the surface, increasing soil density and shear strength. It may also disrupt unstable soil aggregates, producing small fragments which can wash into pores and cracks, effectively sealing the surface. The resulting thin seal (often <1 mm in thickness) can make the soil surface almost entirely impermeable. The effectiveness of raindrop impact in causing compaction, disruption, crusting and sealing depends on rainfall characteristics, soil properties and on soil moisture content. Aggregate disruption by SLAKING is most effective on dry soils, while compaction is most effective on wet clay soils where cohesion drops close to zero. Although bursts of extremely high intensity rainfall, which cause most disruption, are usually very short-lived, they can strongly influence the subsequent effectiveness of erosional processes. This is particularly true in the case of intense summer thunderstorms where initial very high intensity rainfall often falls on a dry surface. These bursts usually last only a few minutes, but by initiating sealing, can result in almost instantaneous overland flow.

Raindrop impact may be entirely absorbed by soil and vegetation, but in intense storms there is usually sufficient energy available to generate some erosional processes as well. The exact processes depend on the balance between the amount of water (rainfall) arriving at the surface and the soil infiltration capacity. This will determine whether all the water can infiltrate or whether excess will be available to generate surface ponding and overland flow. Where no excess occurs, wash erosion processes are absent, but splash erosion can occur. On dry soils raindrop impact can produce miniature surface craters, but usually does not move soil particles. As the water content increases, however, soil strength drops rapidly and the surface can become fluidized. Raindrop impact is converted to an upward force which can entrain soil particles and transport them in a parabola away from the point of impact. The distance of movement depends on the mass of the particle, but is rarely more than 0.6 m above the surface, or more than 2 m in a horizontal direction, unless splash is carried by a strong wind. On a horizontal surface (in the absence of wind), movement is not significant, because the ultimate effect of many raindrops striking the surface is abundant movement, but no net transport in any direction. When the surface slopes, however, this changes as up to 60 per cent of entrained material is deposited downslope from the original impact point, so significant net transport can occur.

The relative vulnerability of soil particles and aggregates to entrainment by splash is an important component of soil ERODIBILITY. Poesen and Savat (1981) have shown in laboratory experiments that the relationship between particle size and the threshold impact energy necessary to cause entrainment is quite similar to the Hjulstrom Curve for flowing water. Entrainment of particles with diameters around 0.125 mm typically requires the lowest impact energy. Splash erosion on most slopes during most storms is therefore a selective process, which ultimately transforms the surface material, producing an *erosional lag deposit* which progressively protects the underlying soil from entrainment.

Pure splash erosion (in which material is both entrained and transported by splash) is comparatively rare, but De Ploey and Savat (1968), who originally identified the influence of slope gradient on the balance of upslope and downslope deposition of splashed material, also described the evolution of sandy hillslopes near Kinshasa, Congo, which is almost exclusively controlled by splash. Elsewhere the effects of splash erosion are subtle and often indistinguishable, but where parts of very erodible surfaces are protected by stones or bits of vegetation, the effect of splash is easily seen by the occurrence of miniature Earth pillars or hoodoos.

Splash erosion can occur without any surface water layer, but in the intense rainfall conditions which produce most splash, such a water layer usually forms quite swiftly. Initially this concentrates in micro-depressions, but ultimately it increases sufficiently in depth to overtop roughness elements and generate overland flow. Before reaching this point, however, it starts to influence the splash process. Initially, except on sandy soils, the water layer actually increases splash transport, up to a critical depth which, laboratory experiments suggest, ranges from about the diameter of the raindrops (Palmer 1963) to about one-fifth of that value (Torri *et al.* 1987). As drop size varies greatly in any rainstorm, the precise result is a very complex mixture of processes on the surface. Eventually, however, the increasingly deep water layer protects parts of the surface from splash erosion. As the first areas protected are microtopographic depressions, the overall effect of continued splash erosion is diffusion of soil particles from higher points to these depressions, progressively reducing the amplitude of the microtopography. Another important effect is the increasing heterogeneity of soil infiltration characteristics, as the *structural* crusts which form on the high points typically have infiltration capacities up to six times higher than the *depositional* crusts which form in depressions (Boiffin and Monnier 1985).

The interaction of spatially varied rainfall, splash and microtopography produces complex, heterogeneous conditions on most hillslopes, particularly with regard to transition from splash-dominated areas to those dominated by overland flow and wash processes. On simple, idealized, homogeneous hillslopes, it is possible to distinguish an upper splash-dominated zone from a lower wash-dominated zone, and finally, from a zone in which concentrated RILL erosion occurs. In practice, the boundaries between these zones are highly irregular and dynamic. However, a transition does occur downslope as surface water deepens progressively, ultimately protecting the surface from raindrop impact. The first stages of overland flow are, however, typically very shallow. Conceptually, on very smooth surface there may actually be a thin, continuous sheet of water, but in practice as most surfaces are quite irregular, this is very rare. The initial flow usually consists of irregular, tortuous concentrations in depressions, which vary significantly in depth and width, and are separated by microtopographic protuberances. Numerous field and laboratory studies have shown that flows of this sort are typically laminar or transitional, with Reynolds numbers often well below 2,500, and relatively smooth, with Froude numbers well below 1. The flows, whether as a sheet or as more or less concentrated streams, are slow and pulsatory, and typically do not exert sufficient shear stress to entrain soil particles. However, as the Hjulstrom curve shows, flow velocities necessary to transport fine silts and clays are significantly lower than those required for entrainment. In these circumstances, raindrop impact and splash are still important, as they may be able to entrain material which can then be transported by flow. Such flows are usually referred to as *rain-impacted flows*, and the erosional process as *rainflow* or *rainwash* erosion (De Ploey 1971). The particle transport distance and the effectiveness of rainflow erosion are governed largely by particle density and settling velocity (Kinnell 2001). Significant transport is typically limited to shallow flows no more than about 1.5 times the average raindrop diameter (Kinnell 1991). Because surfaces are irregular, and flow often discontinuous, the transport distance is frequently very short, resulting in small patches of sediment deposition on the hillslope. Nevertheless, in many areas, rainflow is the most effective and frequent erosional process on upper slopes and interrill areas and can ultimately result in highly significant movement of soil to the base of the slope. This is particularly true where loose soil aggregates are of low density or are water-repellent. In some cases, the patches of sediment deposited on the slope by intermittent flows progressively join to form quite extensive *sedimentary* or *depositional seals*. These are usually highly impermeable, and become preferred locations for runoff generation and wash erosion during subsequent rainstorms (Bryan *et al.* 1978).

Once overland flow is sufficiently deep to protect the surface from raindrop impact, rainflow erosion gives way to *wash erosion*. Surface irregularities ensure that most hillslopes will have patches of wash erosion intermixed with splash and rainflow. Once the surface is fully protected, the only force which can cause entrainment is the bed shear stress exerted by flow. Transport will then occur only if shear stress exceeds the threshold necessary to move the most erodible particle. This critical value depends on soil properties, but Moore and Burch (1986) found that it was equivalent to a unit stream power of $0.002\,\mathrm{m\,s^{-1}}$ for many soils. Once unit stream power exceeds values of $0.01\,\mathrm{m\,s^{-1}}$, transport increases rapidly, and wash erosion tends to be replaced by concentrated rill erosion.

References

Boiffin, J. and Monnier, G. (1985) Infiltration rate as affected by soil surface crusting caused by rainfall, in F. Callebaut, D. Gabriels and M. DeBoodt (eds) *Assessment of Soil Surface Crusting and Sealing*, 210–217, Ghent: State University.

Bryan, R.B., Yair, A. and Hodges, W.K. (1978) Factors controlling the initiation of runoff and piping in Dinosaur Provincial Park Badlands, Alberta, Canada, *Zeitschrift für Geomorphologie, Supplementband* 34, 48–62.

De Ploey, J. (1971) Liquefaction and rainwash erosion, *Zeitschrift für Geomorphologie, Supplementband* 15, 491–496.
De Ploey, J. and Savat, J. (1968) Contribution a l'étude de l'érosion par le splash, *Zeitschrift für Geomorphologie* 12, 174–193.
Hudson, N.W. (1981) *Soil Conservation*, London: Batsford.
Kinnell, P.I.A. (1991) The effect of flow depth on sediment transport induced by raindrops impacting shallow flow, *Transactions of the American Society of Agricultural Engineers* 34, 161–168.
——(2001) Particle travel distances and bed and sediment compositions associated with rain-impacted flows, *Earth Surface Processes and Landforms* 26, 749–768.
Laws, J.O. (1941) Measurement of fall velocity of water-drops and raindrops, *Transactions of the American Geophysical Union* 22, 709–721.
Moore, I.D. and Burch, G.J. (1986) Sediment transport capacity of sheet and rill flow: application of unit stream power theory, *Water Resources Research* 22, 1,350–1,360.
Palmer, R.S. (1963) The influence of thin water layer on water drop impact forces, *International Association of Scientific Hydrology Publication* 68, 141–148.
Poesen, J. and Savat, J. (1981) Detachment and transportation of loose sediments by raindrop splash. Part II Detachability and transportability measurements, *Catena* 8, 19–41.
Torri, D. Sfalanga, M. and Del Sette, M. (1987) Splash detachment: runoff depth and soil cohesion, *Catena* 14, 149–155.

Further reading

Morgan, R.P.C. (1995) *Soil Erosion and Conservation*, London: Longmans.

RORKE BRYAN

RAINFALL SIMULATION

The purpose of a rainfall simulator is to deliver rainfall to the soil surface in a controlled manner with realistic simulation of rainfall intensity and drop-size distribution. Rainfall simulators have been used widely over the past few decades, both in the field and the laboratory. Various factors influence the method of rainfall generation including the purpose of the experiment, the soil surface area to be studied, the drop-size distribution of the simulated rainfall, the need to reproduce realistic terminal velocities, and the need for precise replication of rainfall characteristics between experiments.

Broadly, rainfall simulators fall into three categories (Foster *et al.* 2000): sprays, rotating sprays and drip-screens. Because they eject raindrops relatively high above the ground surface, spray systems are capable of achieving rainfall delivery at terminal velocities approaching that of natural rainfall. However, rainfall intensities can be hard to control, because of variation in pumping rates, and rainfall intensity usually decreases with distance from the rotating nozzle. To overcome this latter problem, multiple rotating nozzles are employed, with the overlap distance between the nozzles being determined by the area over which the simulation is to be performed (Foster *et al.* 2000). Drip systems, using hypodermic needles or drop formers, are usually used over small surface areas (typically $< 1\,m^2$). They are less likely to achieve realistic terminal velocities, because of the difficulty of raising the drip screen high enough, but give much better control of rainfall intensity. Intensities as low as $3\,mm\,hr^{-1}$ can be maintained, and replication between experimental runs is good (Bowyer-Bower and Burt 1989).

Despite the widespread use of rainfall simulators in geomorphological research, until recently there has been little co-ordinated effort to collate all the available information regarding the design and purpose of such simulators, or to discuss future developments relating to the use of this technique. To this end, the British Geomorphological Research Group established a Rainfall Simulation Working Group in 1995 to address these issues. The work resulted in a special issue of *Earth Surface Processes and Landforms* (Volume 25, Number 7, 2000) and creation of a website: http://www.geog.le.ac.uk/bgrg/index.html which includes a database of simulators and a lengthy reference list. Lascelles *et al.* (2000) make the point that when rainfall simulation is used for explicitly spatial studies, some prior analysis of the simulator's inherent variability is vital.

References

Bowyer-Bower, T.A.S. and Burt, T.P. (1989) Rainfall simulators for investigating soil response to rainfall, *Soil Technology* 1–16.
Foster, I.D.L., Fullen, M.A., Brandsma, R.T. and Chapman, A.S. (2000) Drip-screen rainfall simulators for hydro- and pedo-geomorphological research: the Coventry experience, *Earth Surface Processes and Landforms* 25, 691–707.
Lascelles, B., Favis-Mortlock, D.T., Parsons, A.J. and Guerra, A.J.T. (2000) Spatial and temporal variation in two rainfall simulators: implications for spatially explicit rainfall simulation experiments, *Earth Surface Processes and Landforms* 25, 709–721.

TIM BURT

RAISED BEACH

A raised beach is a relict depositional landform comprising mostly wave-transported sedimentary material and preserved above and landward of the active shoreline. First described by Jamieson (1908), raised beaches can form along marine coasts or lake shorelines and are well recognized as indicators of a fall in relative sea (or lake) level. In certain situations, multiple raised beaches may form adjacent to one another, producing a BEACH RIDGE plain, or strandplain (Otvos 2000). Raised beaches are distinguished here from raised marine terraces on the basis that the former are solely the product of physical depositional mechanisms, whereas the latter have a broader genesis that may incorporate depositional, erosional and/or biogenic (i.e. reefal) processes.

The elevated position of a raised beach relative to active shoreline processes may be the product of one or more of the following mechanisms: (1) tectonic uplift associated with plate-margin convergence (e.g. New Zealand east coast; Garrick 1979); (2) isostatic rebound related to ice-unloading of a land mass (e.g. mainland Scotland; Smith *et al.* 2000); (2) depositional regression involving delivery of sediment to a shoreline at a rate sufficient to allow formation and stranding of successive beaches (e.g. east coast of Australia; Thom 1984), and; (3) forced regression whereby eustatic sea-level fall leads to abandonment of a shoreline (e.g. southern Australia coast; Murray-Wallace and Belperio 1991). In the case of depositional and forced regression, the beach remains at its original elevation, as is the case for many shoreline deposits formed during the Last Interglacial sea level high-stand *c.*125 ka BP. Thus the word 'raised' is applied to all stranded fossil beaches regardless of whether the associated landmass has undergone uplift or remained stable.

Clear identification of a raised beach deposit requires satisfying a range of criteria related to the morphology and sedimentology of that deposit. Doing so allows separation from similar coastal depositional landforms such as cheniers (see CHENIER RIDGE) and linear dune ridges. For ice-free coasts, Tanner (1995) identifies four depositional processes that lead to beach ridge formation: wave-swash action, settling lag, storm surge and aeolian action. Along coasts that experience annual freeze-over of the sea (or lake) surface, ice-push is an additional mechanism for beach ridge formation. Each of these five physical mechanisms produces a shoreline deposit with different morphology and sedimentology, as described below.

The most common form of raised beach is produced by wave-swash processes on sandy to gravelly shores. Onshore transport and sorting of sediment across a beach face produces a berm that accretes to maximum wave run-up under spring tidal conditions. Subtle variations in berm morphology exist, ranging from a linear, convex-up ridge with low-angle cross-bedding to a gently landward-sloping uniform surface with continuous subhorizontal bedding. Given alongshore variations in wave energy, both forms may be present along different parts of a shoreline at one time. Consequently, it is possible to find equally variable morphology and internal structure within a raised beach.

Formation of a beach ridge by settling-lag processes is comparatively rare, developing under fetch-limited shallow water conditions such as a small lagoon or pond. Deposition occurs by sand settling out of the water column to produce a low subaqueous flat-topped ridge or bar with discontinuous horizontal bedding. Because wave action is minimal, sediments are not as well sorted as on a swash-formed beach ridge and cross-bedding is characteristically absent. Preservation of a settling-lag ridge as a raised beach typically requires a relatively rapid and permanent lowering of relative sea (or lake) level.

Storm surge is known to result in deposits at elevations above mean high water spring tidal level, either at the beach–dune interface or as a strandline feature on supratidal flats to landward of the fair-weather beach. Grain size is more varied than for fair-weather swash deposits, incorporating the largest materials available. Sedimentary structures reflect higher wave energy, ranging from complex trough cross-bedded sands to imbricated cobbles. The distinction is drawn here between these truly raised storm deposits and an overwash (see OVERWASHING) fan that is also a product of storm surge and typically located on the lagoon side of a low-lying coastal barrier, but is not raised above the elevation range of active sedimentary processes. The role of storms as an agent in the formation of raised beaches is debated in the literature (Tanner 1995), with some authors arguing for storms as an agent of net beach erosion rather than deposition. Documented instances of storm ridge formation (e.g. hurricane ridges, Florida; Tanner

1995), record these as ephemeral features, lasting only until the next storm. Good examples of multiple raised storm beaches exist along the Ross Sea coast of Antarctica where glacio-isostasy during the Holocene has driven coastal uplift (e.g. Hall and Denton 1999).

Aeolian action may contribute to the formation of a raised beach to the extent that wind-blown sand is placed directly on top of a swash-built or settling-lag initiated ridge. A raised beach with aeolian decoration is characterized by an irregular hummocky morphology with low to high-angle cross-bedding that is multidirectional and discontinuous. If vegetated, the internal structure may be weakly bedded to massive in the root zone. Relict dune ridges that are oriented parallel to a shoreline but are solely the product of aeolian processes are excluded from the range of raised beach forms.

Ice-push may also lead to formation of a beach ridge along shorelines that undergo annual freezing of the sea (or lake) surface. An ice-push ridge is typically a discontinuous accumulation of poorly sorted sand- to boulder-sized sediment that forms along the margins of winter sea-ice sheets. Ridge height is a function of the available sediment size, with boulder ridges attaining elevations of ~5 m. Due to the lack of grain sorting, the internal structure of ice-push ridges is characteristically massive. Summer wave-reworking may produce some subsequent sorting of sediment and generation of low-angle cross-bedding of the sand fraction. However, these features are mostly ephemeral, being reworked by the next ice-push.

A raised beach can be used as a proxy for palaeo-sea (or lake) level, providing the range of diagnostic physical sedimentary structures and texture noted above are preserved in the deposits. In particular, a distinction between wave-formed and aeolian sedimentary units is necessary. Thus, a vertical transition from subhorizontal or low-angle cross-bedding in medium to coarse-grained sand to high-angle cross bedding in fine to medium sand, or massive rooted structure would allow this distinction between beach berm and foredune to be drawn. Where multiple raised beaches are preserved on a strandplain, mapping of the beach-foredune contact along a dip-oriented profile can provide for reconstruction of sea (or lake) level change. Examples of this application of the raised beach sedimentary record range from decadal scale fluctuations in shoreline position along Lake Michigan (Thompson and Baedke 1995) to inferred sea level fall along the New Zealand north-east coast toward the close of the Last Interglacial period (Nichol 2002).

Where material suitable for reliable age-dating is incorporated into a raised beach deposit, it is possible to construct a chronology of formation. This is particularly useful for calculating rates of isostatic uplift (e.g. Smith *et al.* 2000), or for estimating rates of shoreline progradation in relation to local sea level and sediment supply (e.g. Tanner 1993). Traditionally, chronological analysis of raised beaches has applied radiocarbon dating to the remains of marine organisms such as shallow water molluscs (Taylor and Stone 1996). Difficulties arise with this method, however, if the material used for dating is not *in situ*. Most of the organic material incorporated in a raised beach is typically reworked from offshore environments and may therefore be considerably older than the enclosing beach sediment. Alternative dating techniques, such as optical dating of beach and dune sands, offer a more reliable avenue for establishing a detailed and accurate chronology of raised beaches, thereby enhancing their utility as a landform that can be used as an indicator of regional geomorphological and geological processes.

References

Garrick, R.A. (1979) Late Holocene uplift at Te Araroa, East Cape, North Island, New Zealand, *New Zealand Journal of Geology and Geophysics* 22, 131–139.

Hall, B.L. and Denton, G.H. (1999) New relative sea-level curves for the southern Scott Coast, Antarctica: evidence for Holocene deglaciation of the western Ross Sea, *Journal of Quaternary Science* 14, 641–650.

Jamieson, T.F. (1908) On changes of level and the production of raised beaches, *Geological Magazine* 5, 22–25.

Murray-Wallace, C.V. and Belperio, A.P. (1991) The Last Interglacial shoreline in Australia – a review, *Quaternary Science Reviews* 10, 441–461.

Nichol, S.L. (2002) Morphology, stratigraphy and origin of last interglacial beach ridges at Bream Bay, New Zealand, *Journal of Coastal Research* 18, 149–160.

Otvos, E.G. (2000) Beach ridges – definitions and significance, *Geomorphology* 32, 83–108.

Smith, D.E., Cullingford, R.A. and Firth, C.R. (2000) Patterns of isostatic land uplift during the Holocene: evidence from mainland Scotland, *Holocene* 10, 489–501.

Tanner, W.F. (1993) An 8000-year record of sea level change: data from beach ridges in Denmark, *Holocene* 3, 220–231.

Tanner, W.F. (1995) Origin of beach ridges and swales, *Marine Geology* 129, 149–161.
Taylor, M. and Stone, G.W. (1996) Beach-ridges: a review, *Journal of Coastal Research* 12, 612–621.
Thom, B.G. (1984) Transgressive and regressive stratigraphies of coastal sand barriers in eastern Australia, *Marine Geology* 56, 137–158.
Thompson, T.A. and Baedke, S.J. (1995) Beach-ridge development in Lake Michigan – Shoreline behaviour in response to quasi-periodic lake-level events, *Marine Geology* 129, 163–174.

SEE ALSO: beach ridge; chenier ridge; sea level; strandflat

SCOTT NICHOL

RAMP, COASTAL

The term 'ramp' has been used by some workers to refer to gently sloping SHORE PLATFORMS, particularly to those in the north Atlantic, in order to distinguish them from the subhorizontal platforms which are more common in Australasia. Generally, however, the term is either used for sections of higher gradient at the rear of gently sloping shore platforms, or for steeply sloping rock surfaces (commonly 4° to 10°) that occupy the entire intertidal zone and may extend to elevations that are well above the high tidal level. Both types of ramp have been reported most frequently from the swell wave environments of Australasia and elsewhere around the Pacific, and less frequently from the storm wave environments of the mid-latitudes of the northern hemisphere. It has been suggested that ramp occurrence and morphology are related to the strength and frequency of the swash generated by storm waves, to waves of translation that sweep across the platforms, and to the presence of abrasive material at the cliff foot. In northeastern England, ABRASION accomplishes rapid erosion, ranging up to 30 mm yr^{-1}, on the steeply sloping ramp where there is a sand and pebble beach, whereas dessication of the shale is dominant on the more gently sloping platform (Robinson 1977). In some places the occurrence of ramps appears to reflect variations in rock structure and lithology. The presence of thick shale beds and other weak material near the high tidal level seems to be particularly suitable for the development of prominent ramps in eastern Canada and in northeastern England, and this is supported by mathematical modelling, which suggests that ramps are most common where rapid erosion produces wide intertidal platforms. Where contemporary rates of erosion are low, however, as in northwestern Spain, ramps extending up to several metres above the modern high tidal level are the result of higher SEA LEVEL during the last interglacial (see ICE AGES) (Trenhaile *et al.* 1999). In southern Australia, sloping ramps, which extend up to more than 10 m above present sea level, are probably polygenic, having developed under rising and falling sea level during the Cenozoic Era (Young and Bryant 1993).

References

Robinson, L.A. (1977) Erosive processes on the shore platform of northeast Yorkshire, England, *Marine Geology* 23, 339–361.
Trenhaile, A.S., Pérez Alberti, A., Martínez Cortizas, A., Costa Casais, M. and Blanco Chao, R. (1999) Rock coast inheritance: an example from Galicia, northwestern Spain, *Earth Surface Processes and Landforms* 24, 605–621.
Young, R.W. and Bryant, E.A. (1993) Coastal rock platforms and ramps of Pleistocene and Tertiary age in southern New South Wales, Australia, *Zeitschrift für Geomorphologie* 37, 257–272.

ALAN TRENHAILE

RAPIDS

Rapids in bedrock channels are not technically defined in fluvial literature, but imply steep reaches with rough water and very variable depth between lower gradient pools (Leopold 1969). Their origin is attributed to the erratic and episodic supply of boulders into the channel, both debris flows from tributaries and rock avalanches and rock fall from the valley sides (Howard and Dolan 1981; Webb *et al.* 1984). Subsequent accelerated flow through the constriction redistributes boulders downstream and partly reshapes the channel bed into quasi-stable form of boulder-strewn bars (Graf 1979; Kieffer 1987).

References

Graf, W.L. (1979) Rapids in canyon rivers, *Journal of Geology* 87, 533–551.
Howard, A.D. and Dolan, R. (1981) Geomorphology of the Colorado River in the Grand Canyon, *Journal of Geology* 89, 269–298.
Kieffer, S.W. (1987) The rapids and waves of the Colorado River, Grand Canyon, Arizona, Report 87–096, *United States Geological Survey*.
Leopold, L.B. (1969) The rapids and the pools – Grand Canyon, *United States Geological Survey Professional Paper* 669-D, 131–145.
Webb, R.H., Pringle, P.T., Reneau, S.L. and Rink, G.R. (1984) Monument Creek debris flow, 1984: implications for formation of rapids on the Colorado

River in Grand Canyon National Park, *Geology* 16, 50–54.

KEITH J. TINKLER

RASA AND CONSTRUCTED RASA

The term rasas, of Spanish derivation, refers to old and perched littoral levelling surfaces or planation surfaces. Their width can reach several kilometres. The erosion surfaces are bordered inland by steep relief and by cliffs towards the sea. They were described for the first time by Hernandez-Pacheco (1950) on the Cantabrian Coast, northern Spain. Guilcher (1974) made a remarkable synthesis. These forms were also observed in Galicia (Nonn 1966), northern Chile (Paskoff 1970), southern Morocco, Brittany and Cornwall in England (Guilcher 1974), and Sardinia (Ozer 1986). Guilcher distinguished three types of rasas. The first one was described above, the second is more complex and is constituted by a succession of levellings arranged in stairs, and the third is when the passage towards the inland is gradual.

Plate 93 Rasa: Coast of Gallura (north Sardinia)

Plate 94 Constructed rasa: Coast of Anglona (north Sardinia). Accumulation of aeolianites on the terrace of the last interglacial sea level

Many of these rasas are covered by marine deposits (sand and rounded pebbles). These sediments were brought at a later date, during tertiary transgressions which only slightly retouched these levelling surfaces. Evidence of this process is found through ancient reefs in Brittany (Guilcher 1974), northern Sardinia (Ozer 1986) and south of Tangier, Morocco (Ozer, 2001 observation).

However, a convergence of shapes can exist, which is then called constructed rasas. This is a littoral aeolian accumulation, generally indurated (aeolianites), often mixed with local deposits of torrential origin. These accumulations are cut again in a shelf shape, slightly sloping towards the sea subsequent to runoff erosion.

The most spectacular constructed rasas are developed on slopes preceded by a well-developed continental shelf exposed to dominant winds. During Quaternary regressions, winds transported abandoned sands from the continental shelf until the first relief was formed by ancient cliffs which developed during the Quaternary transgressions. These deposits, essentially aeolian, became consolidated and were later shaped into cliffs by the current sea level. They are bounded inland by strong relief which is a previous Quaternary dead cliff.

References

Guilcher, A. (1974) Les 'rasas': Un problème de morphologie littorale générale, *Annales de Géographie* 455, 1–32.

Hernandez-Pacheco, E. (1950) Las rasas litorales de la costa cantabrica en su segmento asturiano, *C.R. Congrès Internationnal Géographie de Lisbonne* 2, 29–86.

Nonn, H. (1966) *Les régions côtières de la Galice (Espagne), étude morphologique*, Paris, Strasbourg: Thèsè.

Ozer, A. (1986) Les niveaux marins au Pléistocène supérieur en Méditerranée occidentale, Atti del Convegno 'Evoluzione dei litorali', ENEA, Policoro (Italia), 241–261.

Paskoff, R. (1970) *Recherches géomorphologiques dans le Chili semi-aride*, Bordeaux: Thèse.

ANDRÉ OZER

RATES OF OPERATION

Rates of operation of geomorphic processes are determined in a number of different ways depending on the time and space scales of interest, and whether one is interested in rates of operation of individual processes or in the aggregate rates resulting from all processes combined.

The current dynamic tectonic conditions need to be considered alongside of the overall denudation rates in order to place measurement programmes conducted at site or watershed scale into proper perspective (Brunsden 1990). Brunsden notes that with respect to the major geotectonic provinces the Cenozoic orogenic regions, and especially the subduction areas on plate margins, can experience greater than 20 $mm yr^{-1}$ of vertical movement at the same time as the overall denudation rate rarely exceeds 1 $mm yr^{-1}$. At the other extreme, shields, platforms, cratonic regions and intracratonic basins experience less than 1 mm/1,000 yrs of vertical movement; nevertheless, overall denudation rates scarcely exceed 1 mm/10,000 yrs. Superimposed on these orogenic and epeirogenic movements are the isostatic readjustments which occur in regions recently emerged from under thick ice sheet cover. In the cratonic regions of the Baltic Sea and Hudson Bay, rates of isostatic readjustment were as high as 1–10 m/100 yrs at the close of the Wisconsinan glaciation and remain as high as 10 $mm yr^{-1}$ in the Gulf of Bothnia. The implication drawn from these data corresponds closely with that of Schumm (1963) namely that 'the style and location of landform change is determined by the type, location and rate of tectonic movements and their associated stress fields over the relevant time and space framework of the landform assemblage' (Brunsden 1990: 3). There is general agreement on the order of magnitude of these rates at global to regional scale; the extent to which they are relevant to site and watershed scale is open to debate.

Average rates of operation of processes can obscure the fact that many processes are episodic and that land surfaces may evolve in a series of step jumps, with periods of relative stability followed by brief periods of rapid erosion or accelerated uplift. Variations in rates of change through time are further complicated by variations in space. Even within geotectonic provinces, spatial variations can be large.

Fundamentally, landscape stability and rates of change depend upon the ratio of resistances to change to the forces promoting change. Where these forces are in balance, little change occurs; where resistance exceeds the forces of denudation, weathering processes permit the deepening of soil profiles. This condition is called transport-limited. Where the forces of denudation are greater than the landscape resistance, erosion removes soil and weathering products as quickly as they are formed. This condition is called weathering-limited.

It is apparent that erosion rates will depend in large measure on the availability of transportable soil and sediment. As soil can only accumulate to considerable depths under stable conditions and eroded sediment can only accumulate at regional scale under conditions of continental-scale glaciation, the most extreme erosion rates occur when there is a marked change from one set of processes to another. When a threshold between one set of processes and another is crossed, extremely high rates of denudation may occur. The time period over which these accelerated rates can last is limited by the supply of readily eroded soil and sediment. Paraglacial geomorphology is one striking example of accelerated erosion and sedimentation following threshold exceedance. Landscape sensitivity to change is therefore as effective in controlling the short-term denudation rate as is the energy of the processes of erosion and transport.

In a brief historical sketch of the development of interest in rates of operation of geomorphic processes, Archibald Geikie and Charles Darwin are two of the early researchers who attempted to determine the rates of operation of individual processes. Geikie estimated the rate of rock weathering by measuring changes on dated tombstones in Edinburgh churchyards and Darwin estimated the rate of soil movement on slopes caused by worm casting. The first spatially representative estimates of the overall rate of ground loss derived from a summary of river sediment loads in the United States. Early twentieth-century estimates of the rate of cliff retreat in Germany on sandstones and in Brazil on granites under rainforest found that the rates in Brazil were an order of magnitude greater than those in Germany. Seasonal rates of movement of stones on talus in the Alps and longer term integrations of postglacial creep of till (135 m in 30,000 years) and Lester King's estimates of the rate of retreat of the Drakensberg scarp in South Africa (240 km in 150 million years) were some of the few quantitative rates of erosion estimated before the 1950s. No one seems to have correlated these data as they were simply too scattered and lacking in formal methodology. One notable exception was the US Soil Conservation data. The first systematic programme to measure soil erosion came about in the United States during the 1930s when one of the New Deal programes of President Roosevelt, intended to stem the growth of unemployment, resulted in the construction of tens of thousands of small dams by the US Soil Conservation Service. Large data sets of volumes of sediment delivered to small reservoirs thereby became available. Accelerated erosion plots usually included an adjacent control plot to demonstrate the negative effects of poor land use practices. From a strictly geomorphic perspective, the control plot data gave indications of spatial variability of surface wash rates, but integrated analyses were not published until the late 1940s. One of the important theoretical contributions from the US Soil Conservation data was the formulation of the dynamic concept of sediment sources. There was a recognition of the difference between sediment sources and sediment delivery at the outlet of each basin and the sediment delivery ratio became a useful tool to determine sediment storage. The 1950s were a decade of pioneering studies on rates of geomorphic process, all the way from sediment budgeting (Jackli 1957; Rapp 1960; Leopold *et al.* 1966) to surface wash (Schumm 1956) and a variety of creep processes (Jahn 1961).

By 1983, Saunders and Young summarized (somewhat uncritically) literally thousands of reported data on rates of process operation. Data are no longer the problem but standardized data, both in terms of methods of collection and units of measurement, remain a serious problem. With respect to endogenic processes, England and Molnar (1990) summarized the major difficulties. Many reports of surface uplift in mountain ranges are based on mistaking exhumation of rocks or uplift of rocks for surface uplift and provide no information whatsoever on the rates of surface uplift. Some observations provide reliable measures of the uplift of rocks but, because erosion rates may be high, the mean surface elevation may be decreasing while the rocks are uplifting.

Standardization of data

How does one compare (a) the linear downslope movement of the uppermost layer of the regolith with (b) the volumetric downslope movement of the whole regolith with (c) the slope retreat or ground loss perpendicular to the ground surface with (d) the mass of sediment transported past a control section with (e) the bedrock mass uplifted above the geoid surface? These are all common ways of reporting the results of contemporary process measurements. Caine (1976) stated the problem coherently. Not only is there a problem of the use of disparate units and dimensions, but there is a need to define hillslope erosion and river channel erosion in terms that are mutually compatible, and storage effects within river systems should also be accounted for. His solution is the calculation of a unit of geomorphic work which incorporates the product of the mass of sediment, the change in elevation and the gravitational acceleration. The approach is logically compelling but has not been widely adopted.

An alternative solution has been to convert all data to a linear measure of denudation distributed evenly across the basin. The Bubnoff unit (Fischer 1969), which is equivalent to 1 mm of denudation per 1,000 years, has also encountered some resistance, partly on account of the somewhat arbitrary specific gravity and packing corrections that have to be made, but also because of the impression created of even denudation across a highly spatially variable surface. It seems fair to say that the prevailing attitude is to maintain different units of measurement for slope, channel and basin data.

Equilibria between hillslope erosion and sediment yield

A number of studies have engaged the question of the quantitative balance between hillslope erosion or contemporary uplift and sediment yield. Here we consider just two examples of apparent balance between measured rates. Adams (1980) examined the Southern Alps of New Zealand and compared rates of crustal shortening, tectonic uplift, river sediment and dissolved load, and offshore deposition. In billions of $kg\,yr^{-1}$, the rates were respectively of the order of 700, 600, 700 and 580. Data on crustal shortening derived from geophysical estimates of the rate of convergent plate motion across the Indian–Pacific plate boundary, amounted to about $22\,mm\,yr^{-1}$. This process would lead to a build-up of crustal lithosphere. Data on tectonic uplift were calculated by converting the shortening to uplift along the Alpine Fault. Data on river loads were taken from water analyses (dissolved load), estimates from formulae and field measurement (bedload) and monitored data supplemented by runoff vs sediment concentration relations (suspended load). The average amount removed was adequate (on an annual basis) to balance the build-up effect from tectonic uplift. Finally, data on offshore deposition showed similar order of magnitude effects, thereby removing sediment to the east and west to the converging plate margins. The model described by Adams is a steady-state mountain range with rapid uplift being balanced by rapid erosion. The details are contentious, but the example is instructive in that it demonstrates the extensive data demands placed on such an interpretation. The author is fully cognizant of the errors inherent in the calculations. He confirms his findings in an interesting appeal to the shapes of New Zealand's mountains. The Southern Alps are spiky mountains (suggesting a steady-state condition) whereas immediately adjacent, in Otago, the mountains are flat-topped and are the remains of a pre-uplift surface of low relief.

Reneau and Dietrich (1991) examined a part of the southern Oregon Coast Range and compared data on bedrock exfoliation rates, thicknesses and dates of accumulations of colluvial fill in topographic hollows and the size of the contributing source area with monitored suspended and dissolved load data from the region. The novelty of this approach derives from some premises with respect to the effectiveness of topographic hollows in trapping colluvium and the ability to satisfactorily date the colluvial fill at up to five stratigraphic levels. If it be admitted that colluvial transport rates down the axis of a hollow are dependent on gradient and are constant in the part being evaluated, then net deposition is entirely due to colluvium added from the adjacent side slope. Calculations of volumetric colluvial transport rates into each hollow involved using measures of local topographic convergence, average soil density and the mass depositional rate of colluvium. Calculated average erosion rates from dated hollows were equivalent to about 70 Bubnoffs (mm/1,000 yrs); calculated exfoliation rates were equivalent to about 90 B and calculated denudation rates varied from 50–80 B. Again, the authors carefully identify error bars on their data but conclude that because hillslope and basin-wide erosion rates are so similar hillslope sediment production and stream sediment yield in the Oregon Coast Range are roughly in balance. Net changes in sediment storage downstream are necessarily also minor. Again it should be noted that the data needs are onerous and creative field measurement programmes are necessary.

By contrast with rates of geomorphic process in apparently steady-state environments, relatively few measurement programmes on rates of bedrock incision have been reported.

Whipple *et al.* (2000) took advantage of the diversion of the upper Ukak River in Alaska by an ash flow in 1912 to measure rates of incision along a newly formed bedrock channel. Although the minimum rates of incision are high (10–100 $mm\,yr^{-1}$), they are within the range of previously published estimates (e.g. Stock and Montgomery 1999). In this branch of process geomorphology there are substantially more modelled rates of operation of process than confirmed field data.

There remains considerable ambiguity over the significance of measured rates of erosion at site scale and over short periods of time vis-à-vis the evolving shape of the landscape. During the 1960s there was optimism that measured rates might be extrapolated from site scale and from short-term measurements to larger landscapes and longer term rates. Such expectations have been shown to be naive and derived in part from assumptions about equilibrium, a balanced condition and the ignoring of contingent environmental constraints. Perhaps the central question now being engaged is that of how to link measurements of rates of operation of geomorphic process at one scale (whether temporal or spatial) to another scale. The information is urgently required in the context of concerns about global environmental change (at what scales are the effects of human activity clearly differentiable from the effects of climate change?) and also in the context of a better understanding of Earth history.

References

Adams, J. (1980) Contemporary uplift and erosion of the Southern Alps, New Zealand, *Geological Society of America Bulletin* 91, 1–114.

Brunsden, D. (1990) Tablets of stone: toward the ten commandments of geomorphology, *Zeitschrift für Geomorphologie Supplementband* 79, 1–37.

Caine, N. (1976) A uniform measure of subaerial erosion, *Geological Society of America Bulletin* 87, 137–140.

England, P. and Molnar, P. (1990) Surface uplift, uplift of rocks and exhumation of rocks, *Geology* 18, 1,173–1,177.

Fischer, A.G. (1969) Geological time-distance rates: the Bubnoff unit, *Geological Society of America Bulletin* 80, 549–552.

Jackli, H. (1957) Gegenwartsgeologie des bundnerischen Rheingebietes: ein beitrag zur exogenen Dynamik Alpiner Gebirgslandschaften, *Beiträge Geologie Schweiz Geotechnische Serie*, No. 36.

Jahn, A. (1961) Quantitative analysis of some periglacial processes in Spitzbergen, *Panstwowe Wydawnictno Naukowe*, Warsaw, Geophysics, Geography and Geology, IIB.
Leopold, L.B., Emmett, W.W. and Myrick, R.M. (1966) Channel and hill slope processes in a semi-arid area, *US Geological Survey Professional Paper* 352-G. Washington, DC: US Geological Survey.
Rapp, A. (1960) Recent development of mountain slopes in Karkevagge and surroundings, northern Scandinavia, *Geografiska Annaler* 42A, 65–200.
Reneau, S.L. and Dietrich, W.E. (1991) Erosion rates in the southern Oregon Coast Range: evidence for an equilibrium between hill slope erosion and sediment yield, *Earth Surface Processes and Landforms* 16, 307–322.
Saunders, I. and Young, A. (1983) Rates of surface processes on slopes, slope retreat and denudation, *Earth Surface Processes and Landforms* 8, 473–501.
Schumm, S.A. (1956) Evolution of drainage systems and slopes in badlands at Perth Amboy, New Jersey, *Geological Society of America Bulletin* 67, 597–646.
——(1963) The disparity between present rates of erosion and orogeny', *US Geological Survey Professional Paper* 454-H, Washington, DC: US Geological Survey.
Stock, J.D. and Montgomery, D.R. (1999) Geologic constraints on bedrock river incision using the stream power law, *Journal of Geophysical Research* 104, 4,983–4,993.
Whipple, K.X., Snyder, N.P. and Dollenmeyer, K. (2000) Rates and processes of bedrock incision by the Upper Ukak River since the 1912 Novarupta ash flow in the Valley of Ten Thousand Smokes, Alaska, *Geology* 28, 835–838.

Further reading

Burbank, D.W. and Beck, R.A. (1991) Rapid, long-term rates of denudation, *Geology* 19, 1,169–1,172.
Gage, M. (1970) The tempo of geomorphic change, *Journal of Geology* 78, 619–625.

SEE ALSO: Bubnoff unit; chemical denudation; denudation

OLAV SLAYMAKER

REDUCTION

Reduction is the gain of a negative electron so an element becomes less positively charged, for example ferric iron, Fe^{3+} (or Iron III) becomes reduced to ferrous iron Fe^{2+} or Iron II. This process commonly occurs in the absence of oxygen but can equally occur when iron is in an acid solution. The latter process accounts for the solubilization and loss of iron oxides in the upper parts of soil profiles where there is, in fact, oxygen available but where organic acids derived from the decomposition of plant material acidifies the soil.

In geomorphology, the focus is on the mobilization of iron under reducing conditions and its transport in anoxic/acid waters, often in deep ground water, and the redeposition of iron oxides in oxic conditions. Retallack (1992) proposes that a study of fossil soils shows how the Earth's atmosphere evolved with the gradual increase in oxygen due to the rise of the plants. Around 1,000 million years BP virtually all palaeosols contained oxidized iron, but palaeosols with reduced iron present occur before that date and 3,000 million years ago there were very few paleosols with oxidized iron present.

Reference

Retallack, G.J. (1992) *Soils of the Past*, London: Unwin Hyman.

STEVE TRUDGILL

REEF

Reefs can broadly be classified as spatially heterogeneous, three-dimensional structures which have morphological form that is different from that of the underlying substrata. Historically and currently the term reef has been used to classify a whole host of organic and inorganic structures including stone reefs, OYSTER REEFs, CORAL REEFs, ATOLLs, SERPULID REEFs, algal reefs and artificial reefs. Due to the range of disparate features being classified as reefs, there has been much debate in the literature over what does and does not constitute a reef.

Although many reef specialists, both biologists and geologists, have argued that reefs must be of biogenic origin to be classified as reefs, numerous applications of the term reef have been applied to inorganic structures. In the late nineteenth and early twentieth centuries, emminent natural scientists and geologists referred to inorganic structures, such as beachrock or the curious bar at Pernambuco, Brazil, as stone reefs (e.g. Branner 1905). More recently, there has been a resurgence of the use of the term reef for inorganic structures, as artificial reefs. In many coastal environments, artificial reefs are being built for a range of purposes including as offshore coastal defences, such as those at Sea Palling, Norfolk, England; or as subtidal structures designed to enhance biodiversity in inshore waters. From a geomorphological perspective, both organic and inorganic reef structures influence geomorphological processes and the morphology of coastal environments. As such, both organic and inorganic reef structures are classified as reefs for geomorphological research purposes.

Smaller reefs have often been termed bioherms and biostromes: bioherms are reef-like, mound-like or lens-like features of purely organic origin which are found embedded in rocks of different lithologies, while biostromes are organic layers, which are thinner and less developed structures than bioherms, such as oyster reefs (Cummings 1932).

Importantly, Cummings was one of the first authors to stipulate that reefs are organic forms which can be produced by several different species and they exhibit a variety of forms, ranging from reefs to bioherms and biostromes, where corals are only one type of reef form. Although this subdivision of reefs into more specialized categories had many merits, modern authors still preferentially use the term reef.

Reefs are found in temperate to tropical marine ecosystems, with the most prominent reef types, corals and atolls, being found in tropical and subtropical zones. Algal reefs and bioherms are commonly found in more moderate climatic zones, such as the Mediterranean, and include corniches, trottoirs and mini-atolls built primarily by calcarous algae, vermetids and serpulids. In temperate regions, reefs are often more like bioherms or biostromes in structure and include reef communities such as *Sabellaria*, oyster or skeletal carbonate reefs. Temperate reefs are found in the eulittoral to pelgaic zones and typically develop on a firm substrata. Reefs can enhance the growth and persistence of other species, by providing sheltered habitat or by providing a fixed substrata upon which cryptic communities can colonize and they can also influence sediment dynamics by trapping and storing sediment.

References

Branner, J.C. (1905) Stone reefs on the northeast coast of Brazil, *Geological Society of America Bulletin* 16, 1–12.

Cummings, E.R. (1932) Reefs or bioherms? *Geological Society of America Bulletin* 43, 331–352.

Further reading

Fagerstrom, J.A. (1987) *The Evolution of Reef Communities*, New York: Wiley.

Riding, R. (2002) Structure and composition of organic reefs and carbonate mud mounds: concepts and categories, *Earth-Science Reviews* 58 (1–2), 163–231.

Wood, R. (1993) Nutrients, predation and the history of reef-building, Palaios 8, 526–543.

LARISSA NAYLOR

REGELATION

Means to 'freeze again'. In the glacial context it refers to those processes which permit a glacier to slide over a rough bed by means of melting on the upglacier side of an obstacle and to refreeze on the downglacier side. Regelation occurs because the greatest resistance to glacier movement is on the upstream side of an obstacle. This results in locally high pressures and a consequent lowering of the pressure melting point. Thus melting of ice occurs immediately upglacier of the obstacle, and the resulting meltwater migrates to the lower pressure zone on the downglacier side of the obstacle. There it refreezes because the pressure melting point is higher. It is through this mechanism that the ice in effect overcomes the obstacle by temporarily turning to water and back again. It is, therefore, an important process in glacier sliding and has been confirmed by direct observation in subglacial cavities.

Further reading

Weertman, J. (1957) On the sliding of glaciers, *Journal of Glaciology* 3, 33–38.

A.S. GOUDIE

REGOLITH

The term was coined by Merrill (1897) to describe an 'incoherent mass of varying thickness composed of materials essentially the same as make up the rocks themselves, but in greatly varying conditions of mechanical aggradation and chemical combination'. He went on to point out regolith may be formed *in situ* or from sediments transported from another source.

Merrill derived the word from the Greek regos ($\rho\eta\gamma o\sigma$) meaning blanket or cover and lithos ($\lambda\iota\theta o\sigma$) meaning rock or stone. Jackson (1997) defines regolith as a term for 'the layer or mantle of fragmental and unconsolidated material, whether residual or transported and highly varied in character, that nearly everywhere forms the surface of the land and overlies the bedrock'. Another more simple definition is everything that lies between fresh rock and fresh air.

Regolith is restricted to terrestrial environments and is generally considered to comprise mechanically and chemically weathered rock debris whether *in situ* or transported. It includes rock weathered to varying degrees, sediments of colluvial, alluvial, aeolian, marginal marine and glacial origin as well as volcanic ash and lag gravels, pisolites and sand. It ranges from soft and loose to consolidated and/or cemented and very hard.

Regolith is the earth material usually called 'soil' by many scientists and engineers. Engineers tend to call any earth materials that can be moved with a bulldozer or mechanical digger soil. Forensic geologists call their sampling media soil. Agricultural scientists on the other hand think of soil as a growing medium from crops and pasture. To a regolith scientist soil is a part of the regolith at the uppermost part of the whole body of unconsolidated material they call regolith. Regolith also may contain buried soils that formed during periods when little accretion occurred in an accretionary (sedimentary or volcanic) landscape.

Both Merrill and Jackson consider regolith to be unconsolidated, but when duricrusts are considered this concept falls down. A silcrete for example is as tough a rock as one can find, but it is considered by most to be part of the regolith. Equally in many parts of the world lava flows are encapsulated by regolith (Figure 130). Does this mean that the lavas are part of the regolith or are there two different regolith units above and below the lava? In Figure 130 it is clear that at section A this is the case, but laterally in the section there is only one regolith, part a lateral equivalent of the one below the lava and one laterally equivalent to that above.

This dilemma raises the issue of regolith stratigraphy and dating. Within transported parts of regolith it is possible to apply the principles of lithostratigraphy remembering that this provides little in the way of chronological control on regolith materials. The lithostratigraphy in section A (Figure 130) is very different from that in section B. The age of weathering in the regolith unit below the lava may very well be very different from that above it. The age of weathering in section B will be complex because this section has been exposed to weathering for a longer time than the upper regolith unit in section A and probably has a complex weathering profile carrying components of pre- and post-lava weathering. Moreover because weathering occurs continuously, albeit at different rates, weathering overprints on regolith materials cannot be used for correlation unless dating of weathering demonstrates equivalence. Without dates on regolith materials this dilemma cannot be resolved except in a relative sense, and even then with some difficulty.

The age of regolith does not form part of its definition, but many would consider that Palaeozoic (or even older) materials now at the surface are not regolith but exhumed surfaces on which some regolith is preserved. In many parts of the world ancient regolith exists at or near the modern surface. In some cases it is unlikely that these surfaces and materials have ever been buried (Craig and Brown 1984). In other cases they were buried and have since been exhumed (Lidmar-Bergström 1995). Carboniferous weathering profiles have been dated by palaeomagnetic methods within 1–2 m of the surface in central

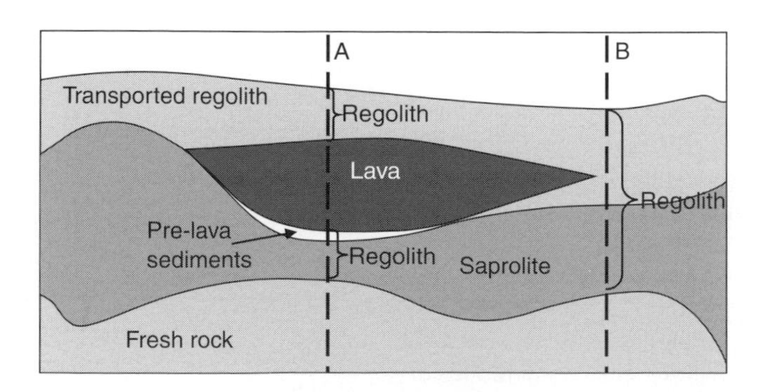

Figure 130 Is it logical to include the lava in with regolith or define it as detached and possibly part of the fresh rock even though it may be a different age and composition?

New South Wales, Australia (Pillans *et al.* 1999), but it has been suggested this profile is exhumed several times, 3.5 km during the Permo-Carboniferous and another 2.5 km during the Triassic to early (O'Sullivan *et al.* 2000). Most regolith however is very much younger than the Palaeozoic and it is still forming across the Earth's surface.

In situ regolith generally forms a weathering profile and these profiles often have a characteristic sequence of materials developed in them. Taylor and Eggleton (2001) provide detailed descriptions and interpretations of weathering profiles. Essentially the sequence is:

- soil
- ferruginous and/or aluminous lag
- collapsed saprolite (may be mottled by ferric oxihydroxides)
- saprolite mottled by ferric oxihydroxides
- bleached saprolite (composed of kaolinite and/or quartz grading downward into more complex clay minerals and quartz ± other primary minerals)
- saprock
- weathering front
- fresh rock.

Weathering profiles of this type are often considered to be the norm and if the upper parts of the profile (e.g. the lag and/or collapsed saprolite) are not present it is often inferred that there has been erosion. This is a misguided inference as there is often no evidence to suggest that the profile was completely developed or that it ever had all those components. Such inferences can lead to erroneous conclusions regarding landscape evolution and the formation of various regolith materials.

References

Craig, M.A. and Brown, M.C. (1984) Permian glacial pavements and ice movement near Moyhu, north-east Victoria, *Australian Journal of Earth Sciences* 31, 439–444.

Jackson, J.A. (1997) *Glossary of Geology*, 4th edition, Alexandria, VA: American Geological Institute.

Lidmar-Bergström, K. (1995) Relief and saprolites through time on the Baltic Shield, *Geomorphology* 12, 45–61.

Merrill, G.P. (1897) *A Treatise on Rocks, Rock Weathering and Soils*, New York: Macmillan.

O'Sullivan, P.B., Pain, C.F., Gibson, D.L. *et al.* (2000) Long-term landscape evolution of the Northparkes region of the Lachlan Fold Belt, Australia: constraints from fission track and paleomagnetic data, *Journal of Geology* 108, 1–16.

Pillans, B., Tonui, E. and Idnurm, M. (1999) Palaeomagnetic dating of weathered regolith at Northparkes Mine, N.S.W., in G. Taylor and C.F Pain (eds) *Regolith '98: New Approaches to an Old Continent*, 237–242, Perth: CRC LEME.

Taylor, G. and Eggleton, R.A. (2001) *Regolith Geology and Geomorphology*, Chichester: Wiley.

GRAHAM TAYLOR

REJUVENATION

Rejuvenation stems from *juvenis*, Latin for young. Thus rejuvenation is to make young again. The term has been applied to individual landforms such as a hillslope or a river channel, but it is most commonly and more appropriately applied in the context of the entire landscape. The term enjoys wide usage among physical geographers and historically based geomorphologists. Its origin and usage in geomorphology can be traced to the interpretation of several lengthy philosophical discourses in the late nineteenth century when some of the major paradigms of long-term landscape evolution were first established (Davis 1889, 1899).

The geographic cycle of Davis (1899) (see CYCLE OF EROSION) continues to influence modern thoughts on long-term landscape evolution. Davisian theory explains landscapes and their constituent landforms primarily in the context of the amount of time that they have been subjected to the forces of erosion. Landscapes are viewed as being born from impulsive rock uplift above sea level. This uplift is followed by a protracted period of erosion that lowers the mean elevation of the landscape by first incising deep, narrow valleys, then widening the valley bottoms and rounding the hillslopes, finally leading to the decline of interfluves to the point that the entire landscape has been reduced to a flat plain or PENEPLAIN. During the valley incision stage, the landscape is traditionally described as youthful, in the valley widening and hillslope rounding phase, the landscape is thought of as mature, and as a peneplain, the landscape is thought of as old. Davis (1899), as well as the subsequent generation of geomorphic thought, recognized that in reality, the geographic cycle almost never proceeded to completion creating a widespread peneplain. Rather, tectonism was understood to be frequent enough such that landscapes in various stages of maturity or old age were uplifted, increasing mean elevation, causing renewed stream incision, and effectively making the landscape appear young again. Such active tectonics has the effect of rejuvenating the landscape.

Rejuvenation is a useful concept when viewing landscape evolution over long (10^6–10^7 yrs) timescales, especially when the flux of sediment that is eroded from those landscapes is considered (Schumm and Rea 1995). Long-term sediment yield from landscape erosion tends to follow a decaying exponential relationship that records an initial, large erosion response in concert with the rock uplift, followed by a long period of time where the rate of erosion decreases as mean elevation and mean slope are reduced (Ahnert 1970; Pazzaglia and Brandon 1996). Impulsive increases in sediment yield over these timescales are probably correctly interpreted as some major change in the erosion processes and rates operating on a rejuvenated landscape imposed by renewed rock uplift, a change in climate, or both.

Unfortunately, use of the term rejuvenate has been extended to explain the forms and changes in individual components of a landscape over shorter timescales (10^0–10^5 yrs), but its applicability in this context is probably not correct. For example in the strict Davisian interpretation, a meandering river channel flowing in a wide river valley is a mature or even old landform whereas a steep river channel flowing in a narrow valley is a youthful landform. Individual landforms such as river channels are much better explained as a DYNAMIC EQUILIBRIUM expression between driving and resisting forces where form and process are mutually dependent. The meandering channel speaks more to the fact that the river has a stable discharge, primarily fine-grain size, gentle slopes, and stable, vegetated channel banks rather than its age in the geographic cycle. In fact, active meander channels in bedrock are known to exist in even the most rapidly uplifting landscapes such as Taiwan where there is no evidence that they have been superimposed or inherited from earlier forms (Hovius and Stark 2001; Hartshorn *et al.* 2002). Similarly, steep, narrow river valleys are common on the great ESCARPMENTs of the southern continents which are known to be among the most slowly eroding and changing landscapes on the planet (Bierman and Caffee 2001). The term rejuvenation is improperly used in these cases of attempting to explain relative landform age or changes in the landscape through an investigation of forms only, without consideration of process or tectonic setting.

References

Ahnert, F. (1970) Functional relationship between denudation, relief, and uplift in large mid-latitude drainage basins, *American Journal of Science* 268, 243–263.

Bierman, P.R. and Caffee, M. (2001) Slow rates of rock surface erosion and sediment production across the Namib desert and escarpment, southern Africa, *American Journal of Science* 301, 326–358.

Davis, W.M. (1889) The rivers and valleys of Pennsylvania, *National Geographic Magazine* 1, 183–253.

——(1899) The Geographical Cycle, *Geographical Journal* 14, 481–504.

Hartshorn, K., Hovius, N., Dade, W.B. and Slingerland, R.L. (2002) Climate-driven bedrock incision in an active mountain belt, *Science* 297, 2,036–2,038.

Hovius, N. and Stark, C.P. (2001) Actively meandering bedrock rivers, *EOS Transactions* 82(47), 506.

Pazzaglia, F.J. and Brandon, M.T. (1996) Macrogeomorphic evolution of the post-Triassic Appalachian mountains determined by deconvolution of the offshore basin sedimentary record, *Basin Research* 8, 255–278.

Schumm, S.A. and Rea, D.K. (1995) Sediment yield from disturbed earth systems, *Geology* 23, 391–394.

FRANK J. PAZZAGLIA

RELAXATION TIME

Geomorphological change may be envisaged as a set of responses to the varying frequencies and magnitudes of formative events at all scales (Graf 1977; Brunsden and Thornes 1979). The concept of LANDSCAPE SENSITIVITY to changes in the operation of controlling processes suggests three divisions of time (Brunsden 1980, 1990; Figure 131): the time taken to react to an impulse of change (lag or *reaction time*); the time taken to attain the characteristic state (*relaxation time*); and the time over which the form exists (*characteristic time* or *landform lifetime*) (McSaveney and Griffiths 1988). Relaxation time is an important measure because landforms can only reach a slowly changing (stable?) state if the interval between form-changing events is greater than the sum of reaction and relaxation times. If the interval is shorter then the landforms will be in a state of constant readjustment and strong flux. This state may be called *transient*.

A further application of the idea of relaxation is to define the term as '*recovery*'. After a severe or land forming event the more 'normal ' frequent events will seek to erase the landform or to modify the form until it is compatible with them.

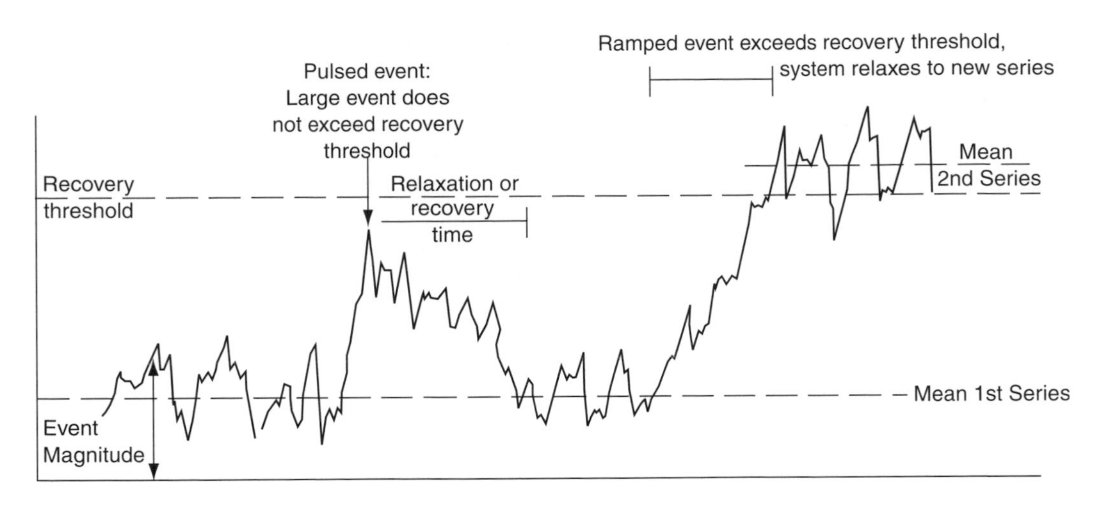

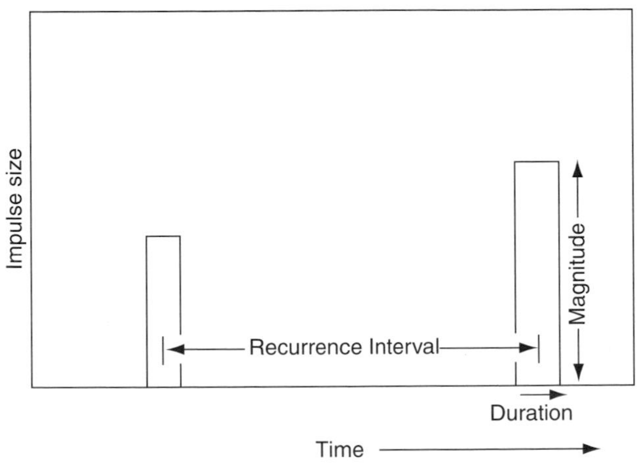

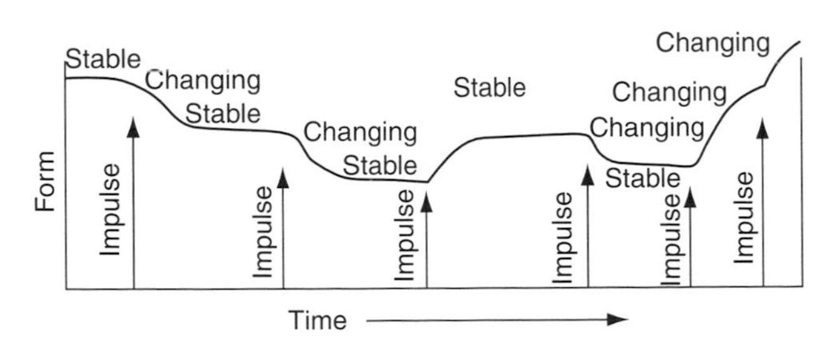

Figure 131 A schematic representation of the concepts of *reaction* (lag) time, *relaxation* (recovery, healing, form adjusting) time and *characteristic form* (form constant?) time

The process can also be an attempt to 'heal' the scars and to return the landscape to its former state. Crozier (1986; see also Crozier *et al.* 1990) regards this as a process of 'ripening' in which the landscape is again prepared for another effective event. This idea is usually applied to soil erosion and mass movement on hillslopes where hollows produced by these processes are weathered and infilled until critical depth is reached and failure can again take place (Deitrich and Dorn 1984, Deitrich *et al.* 1992).

References

Brunsden, D. (1980) Applicable models of long term landform evolution, *Zeitschrift für Geomorphologie N.F. Supplementband* 36, 16–26.
——(1990) Tablets of Stone: toward the ten commandments of geomorphology, *Zeitschrift für Geomorphologie N.F. Supplementband* 79, 1–37.
Brunsden, D. and Thornes, J.B. (1979) Landscape sensitivity and change, *Transactions Institute of British Geographers* NS4, 463–484.
Crozier, M.J. (1986) *Landslides: Causes, Consequences and Environment*, London and Dover: Croom Helm.
Crozier, M.J., Vaughan, E.E. and Tippett, J.M. (1990) Relative instability of colluvial-filled bedrock depressions, *Earth Surface Processes and Landforms* 15, 326–339.
Deitrich, W.E. and Dorn, R. (1984) Significance of thick deposits of colluvium on hillslopes, a case study involving the use of pollen analysis in the coastal mountains of southern California, *Journal of Geology* 92, 147–158.
Deitrich, W.E., Wilson, C.J., Montgomery, D.R., McKean, J. and Bauer, R. (1992) Erosion thresholds and land surface morphology, *Geology* 20, 675–679.
Graf, W.L. (1977) The rate law in fluvial geomorphology, *American Journal of Science* 277, 178–191.
McSaveney, M.J. and Griffiths, G.A. (1988) *A General Theory for Frequency Distribution of Age and Lifetime of Steepland Elements Formed by Physical Weathering*, New Zealand Geological Survey, Christchurch, NZ 1–10.

DENYS BRUNSDEN

RELIEF

Relief may be defined most generally as the elevation difference over a predetermined area or inferred length scale. This simple definition allows for specifying a number of particular types of relief. The relief of a mountain range, for example, may be considered as the difference in elevation between the highest peak and the base of the range front. Alternatively, the relief of a mountain range can refer to the absolute height of the highest peak, with the implicit reference to sea level as a datum. When defined over much shorter length scales, relief can be defined as the range in elevation spanned by a particular hillslope, ridge to valley transect, or physiographic feature such as an escarpment. Relief may also simply refer to topography in general, or more specifically to the collective elevations or their inequalities of a land surface. In other words, the term relief has a variety of possible meanings depending upon the context within which it is used. The most common use of the term, however, generally refers either to topography itself or to the elevation difference between the highest and lowest points within an area of interest.

Differences in elevation that produce relief arise from the interaction of spatial variations in rock uplift and erosion. Volcanic and tectonic processes that raise rocks above sea level are ultimately responsible for elevating mountain ranges, although normal faulting also may produce local relief in extensional settings. Erosional processes may limit the total relief maintained by rock uplift but also cut valleys and produce relief over shorter length scales. Fluvial and glacial processes that incise the landscape produce relief, whereas mass-wasting processes (such as soil creep and many types of landsliding) tend to reduce relief. The overall relief of a mountain range ultimately depends on the balance between uplift and erosion, unless accumulation of crustal material exceeds the mechanical limit supportable by crustal strength, leading to the growth of a high plateau.

Several kinds of relief can be used to describe different aspects of a drainage basin. Fluvial relief represents the elevation drop measured down the longitudinal profile of a river network, as given by the elevation difference between the channel head and the basin outlet. This portion of the total basin relief may be influenced by changes in fluvial processes and rates of river incision. Hillslope relief defined by the elevation difference from the channel head to the drainage divide at the head of the basin represents that portion of the total relief within a drainage basin beyond the immediate influence of fluvial processes and which is instead controlled by hillslope processes. The geophysical relief of a drainage basin has been described as the local elevation difference between the ridgetop and valley bottom, which consists of both hillslope and fluvial relief. In addition to these specific types of relief, local relief may be defined by the elevation difference between the highest and lowest point on the topography measured over an area of predetermined size or a proscribed length scale.

Local relief is inherently scale dependent. The larger the length scale over which it is measured, the larger the relief. Generally, local relief increases as a non-linear function of the diameter of the area over which it is measured, with an exponent <1 and typically about 0.7 to 0.8. In addition, mean local relief is strongly correlated with mean local slope. But in comparison to mean slopes, which have a strong grid-size dependence, mean local relief is less grid-size dependent when calculated from digital elevation models.

Fundamental relationships between relief and erosion rates have been posited since early workers argued that greater relief and steeper slopes lead to faster erosion. In one of the first modern studies of the influence of relief on erosion rates, Schumm (1963) reported a linear relation between erosion rate and drainage basin relief (the height above sea level of the highest point in the basin) for large North American drainage basins. Ahnert (1970) subsequently reported that erosion rates increase linearly with mean local relief (the difference in elevation measured over a specified length scale) for mid-latitude drainage basins. Later studies bolstered Ahnert's relation with data from other regions and showed that local relief and runoff are dominant controls on erosion rate for major world drainage basins (e.g. Summerfield and Hulton 1994). Different relations between erosion rates and mean elevation characterize tectonically active and inactive mountain ranges (Pinet and Souriau 1988), and Montgomery and Brandon (2002) recently reported evidence for a strongly non-linear relation between long-term erosion rates and mean local relief.

Until relatively recently, the relief of bedrock hillslopes was thought not to be strength limited because of the great cohesive strength of intact rock. But the development of discontinuities in rock strength at the scale of an entire hillslope, valley side, or mountain can limit relief development through large-scale bedrock landsliding (Schmidt and Montgomery 1995). The catastrophic 1991 failure of the crest of Mt Cook – the highest point in New Zealand – illustrates how bedrock landsliding can limit relief in steep, highly dissected terrain. Arguing for the generality of strength-limited hillslope relief, Burbank and others (1996) demonstrated that the gorge of the Indus River had strong gradients in incision rate through a region where mean hillslope gradients are independent of the local river incision rate. Hence, they concluded that the development of strength-limited hillslopes allowed bedrock landsliding to efficiently adjust slope profiles such that ridgetop lowering keeps pace with rapid bedrock river incision. This emerging view of the role of relief on erosion rates holds that in steep tectonically active regions erosion rates adjust to high rates of rock uplift primarily through changes in the frequency of landsliding rather than increased hillslope steepness or increased relief (Montgomery and Brandon 2002). In contrast, in lower gradient landscapes the steepness of hillslopes, and therefore local relief, may respond to changes in the controls on landscape-scale erosion rates.

Climate setting and variability constrain the total relief of mountain ranges. Highly orographic rainfall variability can either limit or increase the fluvial relief depending upon the nature of the feedback operating in a specific mountain range (Roe *et al.* 2002). Enhanced erosion by glaciers and periglacial processes can preclude development of relief substantially above the perennial snowline (Brozovic *et al.* 1997). The role of erosion in reducing mass accumulation in mountain ranges is perhaps best illustrated by the exceptional cases where lack of rainfall allows mass accumulation to engage the mechanical limit to crustal thickening and results in development of high plateaux like the Altiplano and Tibet. The position of Earth's high plateaux in the dry latitudes suggests that plateau formation reflects the coincidence of high rates of tectonically driven mass convergence and low rates of erosion due to an arid climate.

A new view of the coupling and feedback among climate, erosion and tectonic processes is coalescing from recent studies focused on their interactions. Geologists are recognizing that spatial gradients in the climate forcing that drives erosion can influence the development and evolution of geologic structures. Development of mountain ranges strongly influences patterns of precipitation and numerical simulations of evolving and steady-state orogens show that both the topography and the resulting metamorphic gradients exposed at the surface reflect the influence of spatial variability in erosion (Willett 1999). Gradients in climate and tectonic forcing strongly influence erosional intensity, and this interaction in turn governs the development and evolution of topography. Hence, the development of relief is strongly coupled to large-scale feedback involving the interplay of climate, erosion and tectonics. Whereas the geographical distribution of plate tectonic environments has changed over geologic time, the global pattern of climate variability exhibits robust latitudinal patterns characterized by abundant and intense rainfall in the equatorial tropics, a low latitude belt of deserts and stronger glacial influences toward the poles. In this context, the feedback between climate, tectonics and erosion implies large-scale climatic controls on the global distribution of topography; high plateaux are likely to form astride the desert latitudes, whereas high mountains are unlikely to form in the equatorial or polar regions where erosion rates are high due to either intense rainfall or glacial processes.

Substantial debate has centred on the relation of climate change to the relief of mountainous topography. An increase in the absolute relief of a mountain range, and consequent increase in the area of alpine environments, can increase rates of weathering through rapid mechanical breakdown of fresh rock by periglacial and glacial processes. Hence, increased relief in alpine areas potentially could influence global carbon cycles and large-scale climate. Wager (1933) noted the proximity of high Himalayan peaks to deep valleys and proposed that isostatic rebound in response to valley incision was responsible for elevating Himalayan peaks above the Tibetan Plateau. Molnar and England (1990) proposed that much of the evidence for substantial late Cenozoic uplift of mountain ranges may simply represent the effect of climatic deterioration on increased erosional exhumation of rocks or the uplift of mountain peaks in response to deepening and enlargement of valleys. Analyses of valley geometry show that such an effect could account for up to about a quarter of the elevation of mountain peaks, although the potential magnitude of such an effect depends on the strength of the crust and the nature of erosional processes (Gilchrist *et al.* 1994; Montgomery 1994). However, recent studies have concluded that there is minimal potential for valley incision to substantially influence local relief in tectonically active mountain ranges (Whipple *et al.* 1999; Montgomery and Brandon 2002).

In summary, relief is a simple concept with many variants of meaning that depend on the specific context in which it is used. Nonetheless, an understanding of the controls on relief generation is central to understanding the linkages between geomorphic processes, tectonics and climate that together shape the Earth's surface.

References

Ahnert, F. (1970) Functional relationship between denudation, relief, and uplift in large mid-latitude drainage basins, *American Journal of Science* 268, 243–263.

Brozovic, N., Burbank, D.W. and Meigs, A.J. (1997) Climatic limits on landscape development in the Northwestern Himalaya, *Science* 276, 571–574.

Burbank, D.W., Leland, J., Fielding, E., Anderson, R.S., Brozovic, N., Reid, M.R. and Duncan, C. (1996) Bedrock incision, rock uplift and threshold hillslopes in the northwestern Himalayas, *Nature* 379, 505–510.

Gilchrist, A.R., Summerfield, M.A. and Cockburn, H.A.P. (1994) Landscape dissection, isostatic uplift, and the morphologic development of orogens, *Geology* 22, 963–966.

Molnar, P. and England, P. (1990) Late Cenozoic uplift of mountain ranges and global climate change: chicken or egg? *Nature* 346, 29–34.

Montgomery, D.R. (1994) Valley incision and the uplift of mountain peaks, *Journal of Geophysical Research* 99, 13,913–13,921.

Montgomery, D.R. and Brandon, M.T. (2002) Non-linear controls on erosion rates in tectonically active mountain ranges, *Earth and Planetary Science Letters* 201, 481–489.

Pinet, P. and Souriau, M. (1988) Continental erosion and large-scale relief, *Tectonics* 7, 563–582.

Roe, G.H., Montgomery, D.R. and Hallet, B. (2002) Effects of orographic precipitation variations on the concavity of steady-state river profiles, *Geology* 30, 143–146.

Schmidt, K.M. and Montgomery, D.R. (1995) Limits to relief, *Science* 270, 617–620.

Schumm, S.A. (1963) *The Disparity between Present-day Denudation and Orogeny*, US Geological Survey Professional Paper 454-H.

Summerfield, M.A. and Hulton, N.J. (1994) Natural controls of fluvial denudation rates in major world drainage basins, *Journal of Geophysical Research* 99, 13,871–13,883.

Wager, L.R. (1933) The rise of the Himalaya, *Nature* 132, 28.

Whipple, K.X., Kirby, E. and Brocklehurst, S.H. (1999) Geomorphic limits to climate-induced increases in topographic relief, *Nature* 401, 39–43.

Willett, S.D. (1999) Orogeny and orography: the effects of erosion on the structure of mountain belts, *Journal of Geophysical Research* 104, 28,957–28,981.

DAVID R. MONTGOMERY

RELIEF GENERATION

Almost everywhere landforms are composed of elements evolved under different climates, i.e. shaped by different exogenetic forces. The relevant form assemblage is called relief generation. They are the constituents of CLIMATO-GENETIC GEOMORPHOLOGY. This concept has no relation to the stages of the Davisian CYCLE OF EROSION and its DENUDATION CHRONOLOGY, which are based on tectonics.

In central Europe, rumpfflächen (etchplains, plains cutting rocks of different hardness; see ETCHING, ETCHPLAIN AND ETCHPLANATION) came into existence in Tertiary time as is concluded from the form, deposits and relics of tropical weathering. Into these plains valleys are incised, starting with broad terraces of Pliocene/lower Pleistocene age with almost pure quartz gravels, sometimes with a few pisoliths. Obviously they are the eroded and transported result of tropical weathering. The flight of terraces in the middle and lower part of the valleys is of periglacial origin as is proved by pebbles from different rocks, syngenetic permafrost features, and LOESS or dunes (only Würm) on top. A solifluction cover on almost every slope, as strata in which the recent soils are developed, shows the overall small amount of Holocene erosion. This applies too for fluvial processes as the floodplain is only about 3 m below the Würm terrace and the incision was mainly in the late Würm, or early Holocene. The term relief generations was introduced by Büdel in 1955 in a paper about the Hoggar in the central Sahara. Here the rumpfflächen carry red loam, relics of tropical soils, under dated basalt flows. A loams terrace in the valleys does not correspond with the recent processes, and has very old artefacts on top. The recent river bed consists of sand.

These examples show the main methods to distinguish relief generations: (1) separation of landforms of different origin as the younger ones are nested or incised into the older; rarely the younger form is on top of the older ones as for instance dunes; (2) observation of the recent processes thus delineating the recent forms; (3) search for weathering relics and/or correlated sediments giving an indication of a different climate and linking these to the older relief generation. If a complete picture is derived by observations in the field, perhaps added to by laboratory analysis, one then tries to compare the forms of the relief generation with similar forms in a different climatic zone. Absolute datings are helpful as they provide an age, which, via the geologic timescale, gives an idea of paleoecological conditions.

The basis for the comparison is 'Klimatische Geomorphologie', which is quite different from CLIMATIC GEOMORPHOLOGY, which studies the landform assemblage and the relative importance of the recent processes in morphoclimatic zones. A regular assemblage of landforms presents the chance to classify relic forms that are only sporadically preserved and to search for additional forms. For example, if one observes overdeepening and glacial striations in rocks, a wall of mixed deposits in a certain position, then it is most likely a moraine. This can be backed up by looking for drumlins or other features of glacial erosion. The concept of relief generations is broader than investigating palaeoforms, for it asks for form assemblages and their relief forming mechanisms. After all these investigations one might ask for palaeoclimatic data from e.g. palaeobotany for comparison. It has almost never been tried to fix recent climatic data to the boundaries of regions with different relief generations.

The relief forming processes of different climates in Tertiary to recent times might also be seen at the small scale. Blocks in blockfields in the Harz Mts have a red rind from weathering in the Tertiary red loam. This is topped by a small white rind. On several blocks a triangular or square piece has been split off by frost weathering. Here the edges of the block have a white rind only. On the rim of the blockfield further blocks are uncovered by recent wash as most probably happened during warm periods of the Pleistocene, too. Thus these fields may be called 'Mehrzeitformen' (multitude forms of different climates). Ayers Rock has weathering forms in the hard rock, which are nested and which were formed in different climates.

For methodological discussions relief generations should be the basis for the distinction of relief elements of different sensitivity to recent geomorphological processes. Elements of older relief generations are more stable than younger ones, and they may be eroded mainly by valley incision. The strength and place of occurence of thresholds can be explained by relief generations, e.g. the edge of an old plain. The elements of older relief generations are certainly not in equilibrium with recent processes. They have not been formed by them nor are they considerably changed by them. Equilibrium is almost never provable even for the recent generation for two reasons: the influence of older generations on the discharge and water movement paths, and the different resistance of existing forms. The ergodic principle can only be applied to relief forms of one generation, preferably those which change fast. Thus the ergodic principle is limited in time.

Further reading

Büdel, J. (1977) *Klima-Geomorphologie*, Berlin. Translated by L. Fischer and D. Busche (1982) *Climatic Geomorphology*, Princeton: Princeton University Press.

Bremer, H. (1965) Ayers Rock, ein Beispiel für klimagenetische Morphologie, *Zeitschrift für Geomorphologie N.F. Supplementband* 9, 249–284.

HANNA BREMER

REMOTE SENSING IN GEOMORPHOLOGY

Remote sensing is the acquisition of information about an object without physical contact. In geomorphology, remote sensing often implies the collection of information from aerial platforms (e.g. airplanes, balloons or kites) or from spacecraft orbiting the Earth. The term remote sensing is credited to Evelyn L. Pruitt and her staff in the United States Office of Naval Research. It was coined during the early 1960s in recognition that instruments other than cameras and regions of the electromagnetic spectrum outside those visible to the human eye and to which photographic film is sensitive were increasingly being used to image the Earth. The current American Society of Photogrammetry and Remote Sensing's (ASPRS) definition of photogrammetry and remote sensing reads 'the art, science, and technology of obtaining reliable information about physical objects and the environment through the process of recording, measuring, and interpreting imagery and digital representations of energy patterns derived from noncontact sensor systems' (Colwell 1997: 3).

While remote sensing will not replace the traditional geomorphic field study, the value of remote sensing to provide a synoptic overview of a landscape cannot be overlooked. Historically, the use of remote sensing in geomorphology has been mainly interpretive, enabling geomorphologists to develop a 'mental picture' of the landscape and as a map-making aid (Hayden *et al.* 1986). However, the use of remote sensing for quantitative geomorphic study is growing rapidly.

Remote sensing provides unique global views at different spatial scales and in different regions of the electromagnetic spectrum. These global views are extremely useful for the subdiscipline of megageomorphology, which emphasizes the study of planetary surfaces at large scales (Baker 1986). The global views provided by remote sensing are not static, but are being continuously refreshed. This repeated global monitoring captures geomorphic events that might otherwise go unnoticed. For example, in 1983 astronauts aboard the STS-8 Space Shuttle mission photographed a dust storm over northwestern Argentina that was transporting material from exposed salt flats on the Puna Plateau of the South American Andes eastward toward the Argentine Pampas. These remote observations helped confirm the source of the Pampas loess and demonstrated that silt accumulation on the Pampas is an ongoing geomorphic process (Hayden *et al.* 1986). Remote sensing can also enable geomorphic study of areas that are inaccessible to field-based investigations.

Remote sensing provides a unique historical archive of geomorphic change. Aerial photography suitable for geomorphic analysis began to be collected as early as the 1920s. The satellite image archive suitable for geomorphic analysis began with the launch of the first of the Earth Resources Technology Satellite satellites ERTS-1, later renamed to Landsat-1, on 23 July 1972.

The history of remote sensing as a tool for geomorphic analysis is intimately tied to advances in photography and the acquisition of photographs from aerial platforms. Photography was born in 1839 when the photographic processes developed by Joseph Nicephore Niepce, Louis Jacques Mande Daguerre and William Henry Fox Talbot were publicly disclosed. One year later the use of photography to aid in the development of topographic maps was advocated by François Arago, Director of the Paris Observatory (Fischer 1975: 27). The first aerial photograph was taken by Gaspard Felix Tournachon, also known as Nadar, from a balloon outside Paris, France in 1858. However, recognition of photography's value to geomorphology also arose in part from terrestrial photographs of the landscape taken during the latter half of the nineteenth century. In 1890, the Geological Society of America formed a Committee on Photographs and its first report described photogeology (Fischer 1975: 34) which as a science took shape in the 1920s and 1930s. The basis of photogeologic analysis rests on the simple notion that landforms developed under similar geologic and geomorphic processes will appear similar in remotely sensed images (Way and Everett 1997: 117). By the early 1940s, photo interpreters were able to recognize the distinguishing surface features of approximately thirty-five major landforms and realized that aerial photographs provided important information on the origin, composition and history of landforms (Colwell 1997: 26). These landforms, and other features, can be identified in remote sensing images based on their location, size, shape, tone and/or colour, shadow, texture, pattern, height/depth as well as site characteristics and associations among features in the landscape (Jenson 2000: 121–132). The first systematic, although low resolution, observation of Earth from satellite began in 1960 from TIROS I, the world's first meteorological satellite. Since then, remote sensing of the Earth has expanded significantly as spaceborne imaging systems have grown in number and in sophistication.

Successful application of remote sensing images to geomorphic study requires careful matching of an instrument or image's spatial, temporal and spectral characteristics with the requirements of the geomorphic study at hand. A sensor's spatial resolution is the smallest angular or linear separation that it can resolve. The resolution of digital images acquired by non-photographic instruments is often described by the length of one dimension (in metres) of the individual elements (pixels) that comprise the two-dimensional image. In determining the required spatial resolution required for a particular application, a useful rule of thumb is that for an instrument to detect a feature of a certain size, its spatial resolution should be at most one-half of the feature's smallest dimension.

The temporal resolution of a remote sensing system refers to how frequently it can image a certain area. Most orbital sensors have fixed repeat cycles which control how often an area is imaged. They typically range from less than one day to one or two weeks. Aircraft overflights or manned space missions usually acquire images much more infrequently and at much more irregular intervals.

Various wavelength regions of the electromagnetic spectrum provide quite different information about the chemical, physical and biological properties of a landscape. The two most commonly used regions of the electromagnetic spectrum for geomorphic study are the optical, where the propagating electromagnetic energy has wavelengths of 0.3 to 14 micrometres (μm, 1 μm $= 10^{-6}$ m), and the microwave with millimetre to metre wavelengths.

The optical region, which historically has been the most widely used, can be divided into two subregions, a reflected optical region (0.3–3.0 μm) and a thermal infrared region (3.0–14.0 μm). Remote sensing instruments operating in these two wavelength regions are typically passive; the energy supplying the signal to the sensor comes from an external source. In the reflected optical region, solar energy reflected off the landscape provides the signal while in the thermal infrared region, energy emitted directly by the Earth itself as a function of its surface temperature is the primary energy source. These two spectral regions contain numerous atmospheric windows or wavelength intervals in which the atmosphere is fairly transparent to solar energy or emitted energy making remote observations possible. While the atmosphere may be fairly transparent in these windows, for some geomorphic applications correction for atmospheric effects still may be important. Cloud cover can also obscure the surface over all wavelengths in the optical.

The reflected optical wavelengths are the most commonly used in terrestrial remote sensing. The reflected optical is typically subdivided into three spectral regions: the visible (0.4–0.7 μm) to which the human eye is sensitive, the near infrared (0.7–1.1 μm) and mid or short-wave infrared (1.1–2.5 μm). As the name suggests, reflected optical images are formed from the energy reflected from the surface towards the sensor. The reflectance of surface materials varies as a function of wavelength making it often possible to discriminate between different surface materials based solely on their reflectance. While two materials may appear similar at one wavelength they may be quite easy to distinguish at another.

Remote sensing instruments can also provide a valuable three-dimensional view of a landscape through the use of stereopairs, which are two remote sensing images that when viewed together add the illusion of relief to a landscape. Stereoscopic measurements can be used to make topographic maps or digital representations of topography known as digital elevation models (DEMs). Stereopairs also are stellar in their support of one of the original stated purposes of photogeology which was 'to provide better illustrations for teaching geology' (Fischer 1975: 34). Two excellent modern examples of the educational value of stereopairs and satellite images for teaching about the Earth's landforms are the *Atlas of Landforms* (Curran *et al.* 1984) and *Geomorphology from Space* (Short and Blair 1986).

The microwave region (wavelengths ranging from one mm to one m) of the electromagnetic spectrum is also important for geomorphic remote sensing. Synthetic Aperture Radars (SARs) are active remote sensing instruments that illuminate the ground with their own electromagnetic signal and then record the amount of energy that is scattered from the target back to the sending antenna. Therefore, SAR images are sometimes referred to as backscatter images. SARs offer advantages over optical sensors as they can penetrate through clouds and obtain images at night making them ideal for studying cloudy regions and for capturing short-lived dynamical geomorphic processes like flooding. SAR images capture quite different characteristics of the landscape than do optical sensors. The backscatter signal received at a SAR antenna is affected by the roughness of the surface and the moisture present in the soil and in vegetation. This makes SAR useful for assessing such important landscape properties as soil moisture, melting conditions on the surface of glaciers, the biomass of plant communities and flooding or inundation.

One unique feature of SAR that has proved valuable for geomorphic research is the ability of SAR to penetrate into dry materials, such as sand or dry snow. The longer the SAR wavelength the deeper into the subsurface the microwave energy can penetrate. A classic geomorphic study demonstrating SAR's ability to provide subsurface information was the identification of an extensive drainage network under the Selima Sand Sheet covering portions of western Egypt and eastern Sudan not easily visible from field observation or optical sensors (Hayden *et al.* 1986).

In geomorphology, remote sensing is not limited to the collection of images of terrestrial surfaces from the air or space. Ground-based remote sensing techniques including ground penetrating radar (GPR) and seismic reflection profiling provide detailed two or three-dimensional images of the near subsurface useful for studying the internal structure of landforms and glaciers and for environmental site analysis. Since the 1960s, multibeam acoustical sounding instruments, often known as side scanning SONARs (sound, navigation and ranging), have been widely used in marine geomorphology and bathymetric mapping. Remotely operated vehicles (ROVs) are providing fascinating views of otherwise unseen marine environments. Remote sensing of landscapes is not limited to Earth. By necessity, planetary geologists have made extensive use of remote sensing to study other planets and even asteroids in our solar system (Hayden *et al.* 1986).

References

Baker, V.R. (1986) Introduction: regional landform analysis, in N.M. Short Sr and R. Blair Jr (eds) *Geomorphology from Space, NASA SP-486*, Washington, DC: National Aeronautics and Space Administration.

Colwell, R.N. (1997) History and place of photographic interpretation, in W.R. Philipson (ed.) *Manual of Photographic Interpretation*, 2nd edition, 3–47, Bethesda, MD: American Society for Photogrammetry and Remote Sensing.

Curran, H.A., Justus, P.S., Young, D.M. and Garver, J.B. (1984) *Atlas of Landforms*, 3rd edition, New York: Wiley.

Fischer, W.A. (1975) History of remote sensing, in R.G. Reeves (ed.) *Manual of Remote Sensing*, 1st edition, 27–50, Falls Church, MD: American Society of Photogrammetry.

Hayden, R.S., Blair, R.W. Jr, Garvin, J. and Short, N.M. Sr (1986) Future outlook, in N.M. Short Sr and R. Blair Jr (eds) *Geomorphology from Space*, NASA SP-486, Washington, DC: National Aeronautics and Space Administration.

Jenson, J.R. (2000) *Remote Sensing of the Environment: An Earth Resource Perspective*, Upper Saddle River, NJ: Prentice Hall.

Short, N.M. Sr and Blair, R.W. Jr (1986) *Geomorphology from Space. A Global Overview of Regional Landforms*, NASA SP-486, Washington, DC: National Aeronautics and Space Administration. Available online at: http://daac.gsfc.nasa.gov/DAAC_DOCS/geomorphology/GEO_HOME_PAGE.html

Way, D.S. and Everett, J.R. (1997) Landforms and geology, in W.R. Philipson (ed.) *Manual of Photographic Interpretation*, 2nd edition, 117–165, Bethesda, MD: American Society for Photogrammetry and Remote Sensing.

ANDREW KLEIN

REPOSE, ANGLE OF

The maximum angle at which a mass of debris under given conditions will remain stable. The angle of repose generally varies between 25° and 40°. For instance, the angle of repose for sand is between 30° and 35°, whereas for scree it is between 32° and 36°. The exact angle of repose depends upon slope conditions such as the size, shape, roughness and degree of interlocking, sorting, the height of fall, and density of the individual sediment grains. Also, the length of slope and the pore-water pressure of the sediment are important, as increased water content enhances structural integrity of the sediment due to surface tension between grains. A general understanding of the angle of repose is known, though studies concerning factors influencing the angle of repose have produced diverse results.

Further reading

Francis, S.C. (1986) The limitations and interpretation of the 'angle of repose' in terms of soil mechanics; a useful parameter? in A.B. Hawkins (ed.) *Engineering Geology Special Publication* 2, 235–240.

Kinya, M., Kenichi, M. and Shosuke, T. (1997) Method of measurement for the angle of repose of sands, *Soils and Foundations* 37(2), 89–96.

STEVE WARD

RESIDUAL STRENGTH

Also termed ultimate strength. It refers to the minimum remaining degree of strength (i.e. resistance to movement) in a soil or rock after loss of strength following significant displacement (relative to the material but typically >1 m). The term is thus linked with slope movements, and is extremely important in slope stability analysis in order to gauge the strength of a pre-existing active slope. Residual strength in sands is typically the same as the critical shear strength (a steady state subsequent to shearing in which the effective stresses remain constant and no volume changes occur), whereas materials with high levels of clay provide a residual value of about half the critical shear strength. Soils high in platy clay materials cause a considerable reduction in strength (from peak to residual) as they tend to align themselves parallel to the direction of displacement following movement.

Further reading

Carrubba, P. and Moraci, N. (1993) Residual strength parameters from a slope instability, in P. Shamsger (ed.) *Third International Conference on Case Histories of Geotechnical Engineering*, 1,481–1,486, Rolla: University of Missouri.

Spangler, M.G. and Handy, R.L. (1982) *Soil Engineering*, New York: Harper and Row.

STEVE WARD

REYNOLDS NUMBER

A dimensionless number used in fluid dynamics to determine the transition from laminar to turbulent flow through a pipe, developed by Osborne Reynolds in 1883. The parameter is based upon the fact that the ratio of kinetic energy to energy transferred by viscous forces was correlated with turbulent flow. The number is defined by the equation $Re = V L / v$, where Re is the Reynolds number, V is the velocity, L is the length, and v is the kinematic viscosity (viscosity/density). When this ratio is less that 1,000 laminar flow will be observed, whereas high Reynolds numbers represent turbulent flow. However, the actual definition of a high Reynolds number is determined by the shape of the system. Reynolds also studied the effects of flow resistance in pipes, demonstrating that the friction coefficient is a unique function of the Reynolds number at various surface roughnesses.

Further reading

Reynolds, O. (1883) An experimental investigation of the circumstances which determine whether the motion of water shall be direct or sinuous and of the law of resistance in parallel channel, *Philosophical Transactions of the Royal Society* 174, 935–982.

Rott, N. (1990) Note on the history of the Reynolds number, *Annual Review of Fluid Mechanics* 22, 1–11.

SEE ALSO: boundary layer

STEVE WARD

RIA

A coastal inlet resulting from the drowning of a former river valley system or estuary. Rias are formations originating during the postglacial glacioeustatic transgression of the seas across the continental shelf during the Flandrian TRANSGRESSION, following the melting of the ice sheets and glaciers. This resulted in the development of an extremely irregular, indented coastline, where only the pre-existing hill peaks remained above sea level. Rias are non-glaciated, having been formed originally by subaerial erosion, and are characteristically long, narrow, often funnel-shaped inlets, whose depth and width uniformly decreases inland. They are also shorter and shallower than a fjord. The term ria originates from the type locations of Galicia and Asturias, north-west Spain, where a series of long mountainous-sided estuaries exists, once drowned by postglacial eustatic sea-level rise. Other examples include the south-west of Ireland (Kerry and Bantry Bay). A less restricted use of the term ria exists, pertaining to any broad estuarine river mouth, including fjords. However, the original application of the term is preferred in geomorphology.

Further reading

Cotton, C. (1956) Rias *sensu stricto* and *sensu lato, Geographical Journal* 122, 360–364.

STEVE WARD

RICHTER DENUDATION SLOPE

A hillslope type, common in extreme environments such as alpine and polar regions, that develops through cliff retreat and forming a uniform (rectilinear) slope at the angle of rest of the accumulating talus. Richter denudation slopes were first noted by E. Richter in 1900, following studies in the Alps. The formation of such slopes was later expanded on by Bakker and Le Heux (1952), who modelled Richter slope formation. Richter denudation slopes develop essentially by rock fall, where the resulting talus is moved gradually by rolling and sliding, and forming a thin veneer over the basal slope. The cliff retreats steadily, often cutting across bedrock, while the basal slope may either accumulate talus, thus raising the foot of the slope, or be removed by weathering or abrasion of sliding talus. The free face will eventually be eliminated resulting in a smooth hillslope of uniform gradient. Examples of such slopes are common in the Transantarctic Mountians and Koettlitz Valley, Antarctica.

Reference

Bakker, J.P. and Le Heux, J.W.N. (1952) A remarkable new geomorpholgical law, *Koninklijke Nederlandsche Akademie von Wetenschappen* B55, 399–410, 554–571.

Further reading

Selby, M.J. (1982) *Hillslope Materials and Processes*, Oxford: Oxford University Press.

STEVE WARD

RIDGE AND RUNNEL TOPOGRAPHY

Ridge and runnel topography comprises a series of alternating intertidal bars and troughs and is typically found on sandy beaches in fetch-limited, macrotidal coastal environments. The number of ridges (and runnels) is 3–6, the height of the ridges ranges from 0.5 to 1 m and the distance between the ridges varies from 50 to 100 m. The intertidal gradient of ridge and runnel beaches is approximately 0.015, but the seaward slope of the ridges is significantly steeper and may be up to 0.05. Storm wave conditions result in a flattening, or even destruction, of the ridge morphology. Calm wave conditions, on the other hand, induce ridge build-up and promote onshore migration of the ridges. Ridge and runnel topography is relatively stable and rates of onshore bar migration rarely exceed 1 m per tide.

It was previously thought that the ridges develop as swash bars during stationary tide conditions. However, some doubt has been cast on a swash origin of the ridges and it seems more likely that the ridges are breaker bars. Whatever its origin, ridge and runnel topography is subjected to a range of hydrodynamic processes over a tidal cycle, including swash, surf and shoaling wave processes. Depending on the wave/tide conditions and the position on the beach profile, the ridges will be affected and controlled to varying degrees by each of these hydrodynamic processes.

In the American coastal literature, welded bar systems that develop following storm erosion (see BEACH) are also sometimes referred to as ridges and runnels. This usage of the term 'ridge and runnel topography' is considered inappropriate and should be avoided.

Further reading

Orford, J.D. and Wright, P. (1978) What's in a name? – descriptive or genetic implications of 'ridge and runnel' topography, *Marine Geology* 28, M1–M8.

GERHARD MASSELINK

RIEDEL SHEAR

Refers to a conjugate set of overlapping *en echelon* faults which develops during the early stages of shearing, usually at inclinations of ~15° (R Shears or synthetic fractures) and ~75° (R' Shears or antithetic fractures) to the principal displacement zone (PDZ) boundary. Riedel shear zones form in dip-slip fault regimes and are composed of fault and fracture elements marked by standard physical properties of brittle shear zones (e.g. slickenside surfaces, slickenlines, gouge and/or breccia and abundant fracturing), though others can be observed as deformation bands and zones of deformation (Davis *et al.* 2000). Second *en echelon* synthetic and antithetic shears, termed P-shears, may form through development of the Riedel shear system, although P-shears can sometimes develop before R-shears or at the same time. Lesser shears that can develop in relation to Riedel shear include Y-shears, and T fractures. Riedel shearing was first observed by Cloos (1928) and Riedel (1929) during studies on clay-cake deformation.

References

Cloos, H. (1928) Experimenten zur inneren Tektonic, *Centralblatt für Mineralogie, Geologie und Paleontologie* 1928B, 609.
Davis, G.H., Bump, A.P., Garcia, P.E. and Ahlgren, S.G. (2000) Conjugate Riedel deformation band

shear zones, *Journal of Structural Geology* 22(2), 169–190.
Riedel, W. (1929) Zur mechanic geologischer bruderscheinungen, *Centralblatt für Mineralogie, Geologie, und Paleontologie* 1929B, 354.

SEE ALSO: shear and shear surface

STEVE WARD

RIFT VALLEY AND RIFTING

Rift valleys are elongate depressions in the Earth's surface that are formed as a result of extension in the crust and upper mantle. Many are large enough to be easily visible from space from where they resemble large cracks that cut across continental landmasses (Plate 95). At ground level they are well defined and easily recognizable geomorphological features that have been the subject for exploration and research for more than a centrury. The term 'rift valley' was first used by Gregory in the nineteenth century to describe the East African Rift (alternatively called the Great Rift or the Gregory Rift) which runs from Afar in the north of Ethiopia to Blantyre to the south of Lake Malawi – a total length of 35,000 km. In Ethiopia this rift reaches its greatest depth of over 3 km. Other well-known rifts include the Baikal Rift of south-central Siberia, the Rhine Graben of Europe and the Rio Grande Rift of western USA.

The morphology of rifts is always similar with a central depression or rift valley flanked on both sides by uplifted areas. The uplifted shoulders of rifts are each associated with a staircase of (mostly) normal faults of varying magnitude which step the underlying rocks down towards the central depression.

The classical view of the structure of a rift was that it is symmetrical, with a flat bottom, flanked by two equally large border faults, a structure for which the German word 'graben' was coined. This interpretation was based almost entirely on an assumption that the surface expression of a rift is the same as the structure, ignoring the importance of erosion and deposition in modifying the landscape. The flat bottom observed e.g. in the East African Rift is due to deposition of lake and river sediments (Frostick and Reid 1989). Seismic evidence from rifts worldwide has

Plate 95 False colour satellite image of part of the Kenyan section of the East African Rift. Rift valley and other sediments show white

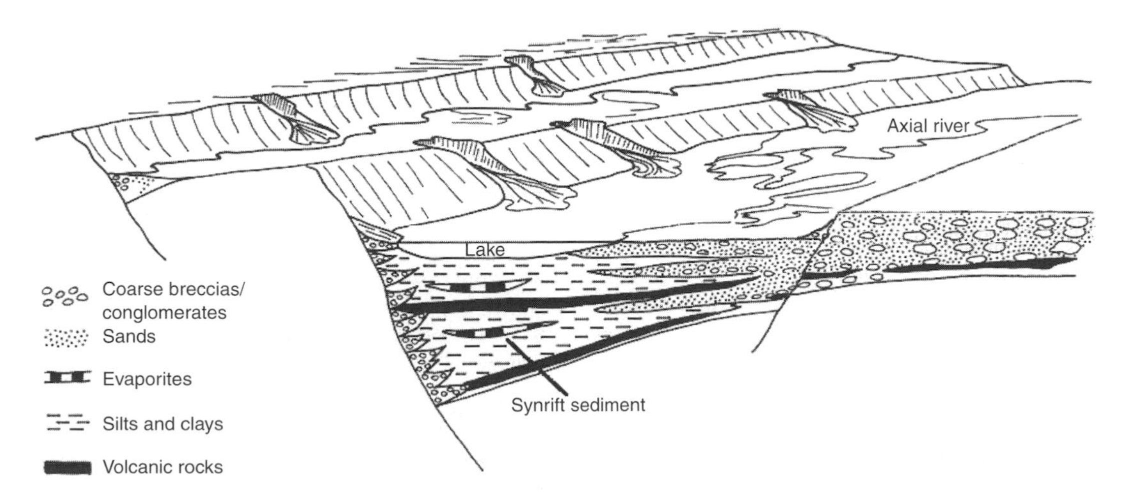

Figure 132 Schematic diagram of a cross section through rift structure. Note the asymmetry with a thicker package of sediments filling the basin close to the border fault

shown that the dominant structure is an asymmetrical half graben with one margin more intensely faulted than the other (Figure 132). The location and character of the main border faults is controlled by the structure and lines of weakness in the pre-rift rocks. In some areas there are a few large faults, with vertical displacement in excess of 2 km, and in others a plethora of smaller ones. The rocks between the faults are tilted away from the rift axis forming parallel valleys between tilted fault blocks. Another characteristic of the faulted margin is uplift, as the valley floor is displaced downwards the shoulder of the rift rises upwards to emphasize the topographic step. Rift valleys vary in width from less than 30 to over 200 km. Along the less faulted margin there are smaller faults inclined both towards and away from the rift axis (antithetic and synthetic faults). Most of these are also normal extensional faults that carve the pre-rift rocks into a series of small HORST blocks with intervening valleys.

Although continental rifts can be viewed overall as continuous elongate depressions, it is interesting to note that the underlying structure, and to some extent its topographic expression, is segmented into a series of smaller basins which vary in length from tens to hundreds of kilometres. At the divides between basins there are transverse or oblique structural elements, the nature of which is the matter of debate, named variously transfer zones, accommodation zones, relay ramps, relay zones and segment boundaries. Across these zones the basin margin occupied by the major fault can alter, giving a sinuous form to the deepest zone of the valley along rift axis.

The origin of rifts has been the subject of great debate over many decades. Rifts develop in a variety of plate tectonic (see PLATE TECTONICS) settings and can be formed anywhere the crust of the Earth is placed under tension. The most obvious circumstance in which this will occur is when a continent is splitting apart to form a new ocean (e.g. in the Red Sea–Gulf of Aden; see Girdler 1991), but it can also occur where plates are moving laterally past or even towards each other in a non-uniform way that places local areas of the crust under tension (e.g. the Dead Sea pull-apart basin). Some researchers favour classifying rifts into those associated with the constructive plate margins that lead to the development of oceans and those that are within a plate that is not splitting apart, so called intraplate settings. However, this division is difficult to justify given that continental rifts may be in an intraplate setting but still associated with oceanic development. This is the case with both the Benue trough in west Africa, which was formed during the opening of the Atlantic Ocean, and the East African Rift which is associated with the opening of the Red Sea–Gulf of Aden (Frostick 1997). Both are failed arms of oceans that had the potential to become mid-ocean ridges. Such failed oceanic rifts are called aulacogens.

The biggest rift systems in the world are on the ocean floor in the centres of mid-ocean ridges. The worldwide network of ocean ridges constitutes the most significant topographic feature on the Earth's surface, surpassing even the Himalayas in scale. A typical ridge is 1,000–2,000 km wide and 2–3 km high. The central rift is the focus of intense earthquake activity and volcanicity.

Although it is well known that rifts form as a consequence of crustal extension, the cause of the instability has been hotly debated. The cause might be convection cells in the mantle that pull apart areas of the crust or the crust might be placed under tension by other plate movements. Whatever the mechanism, the stretching of the Earth's crust to form a rift causes it to thin in a manner similar to the thinning of semi-solid toffee as it is pulled apart. Hot, low-density mantle material wells up close to the surface resulting in high heat flows in and around most rifts. The topographic expressions of hot material from lower in the Earth penetrating closer to the surface can be the development of large domes and extensive volcanic activity. The development of large uplifted domes is a feature of the early stages of ocean opening. For example, examination of the topography and drainage of the West African margin reveals a series of large domes approximately 1,000 km in diameter that predate the opening of the Atlantic Ocean (Summerfield 1991). Similar structures are associated with the East African Rift, centred on Robit in Ethiopia and Nakuru in Kenya.

Rifts are often the focus for volcanic activity which commences at an early stage of rift development and can be extensive, for example in East Africa where an area of over 500,000 km^2 is covered with rift-related volcanic rocks and many of the well-known mountains are volcanoes including Ol Doinyo Lengai, Kilimanjaro and Mount Kenya. The nature of the volcanic rocks in rifts is distinctive and contains high concentrations of so-called volatile elements (particularly carbon dioxide and halogens). Rock types include basalts, trachytes, tuffs and carbonatites. Salts leached from these rocks can contribute to the development of saline lakes, e.g. Lakes Natron and Magadi.

The new topography that results from the development of a rift valley in a continental landmass will impact on hydrology, climate and ecology in a variety of different ways. Uplift along the rift margins reduces the ambient temperature and tends to increase rainfall while the centre of the rift remains warmer and can be more arid, depending on the latitude of the rift. Rift flanks are often the sites of more lush and temperate vegetation which, in the tropics, can form rainforest. Both the topography and the contrast in habitat from flank to valley bottom act as barriers to the migration of animals and, to a lesser extent, plants. The relatively isolated environment of the rift valley bottom is one that has played a unique role in human evolution. It is now widely accepted that the hominids found in the East African Rift, largely by members of the Leakey family, show that critical stages in the evolution of hominids occured in this area prior to migration out of Africa.

The evolution of rift morphology will disrupt the pre-existing continental drainage patterns, reversing, diverting and beheading river systems in a systematic and effective way. Pre-rift, most continental drainage systems comprise a limited number of very large, long-lived rivers fed by a well-integrated network of smaller streams and draining towards the nearest ocean margin. The impact of the incipient rift will depend upon the orientation of the new structure to the existing river system. If the rift is aligned with the main drainage direction it might capture all or some of a local river system. In contrast, a rift which cuts across the pre-existing drainage often diverts and reverses sections of the drainage. Domed sections of a rift are particularly effective at drainage diversion and develop a radial stream pattern that diverts all but very local and small rivers away from the rift basin. As the structure develops further, and faults begin to carve the surface into a series of ridges, there are new adjustments. The uplift and tilting of fault blocks create new river systems which drain along the 'saddle' between adjacent fault blocks, bypassing the basin centre. Most of these bypass rivers finally gain access to the rift axis through transfer zones where the throw on the border fault reduces to zero (see e.g. the Kerio river of northern Kenya described in Frostick and Reid 1987).

In some rifts, topographic barriers pond the drainage, forming lakes. Examples are Lakes Baikal, Tanganyika and Malawi. These lakes vary in salinity from hypersaline to fresh water depending on the surrounding geology and volcanicity. In other rifts there are no lakes and axial rivers drain the length of the valley, for example in the Rhine and Benue rifts. The marginal fault scarps are cut by alluvial fans that feed water and sediment into the basinal rivers and lakes.

As the rift basin floor subsides the uplifted flanks will be progessively eroded and sediments will accumulate in the valley at a rate that largely depends on climate and hydrology. Sediments that accumulate during rifting are normally called 'synrift' sediments. The lowest areas of the valley fill first, generally with lacustrine and river sediment. In the later stages of development into an incipient ocean, sea water may penetrate into the rift valley and the whole area will become a large marine inlet with an uncertain connection to the open ocean. This can lead to the accumulation of thick salt sequences as the sea water evaporates. As the filling progresses wedge-shaped masses of synrift sediments develop and, if subsidence ceased, the valley would eventually lose its topographic identity. In the rifts we see today, subsidence is ongoing and successive wedges of sediment are superimposed on each other. Over geological time many kilometres of sediments can accumulate in rifts.

Continental rifts offer conditions favourable to the development of a number of economic deposits that are rare in other parts of the continents. Some rifts contain sediments that can produce and trap oil and gas in large quantities given the right burial history (e.g. the oil and gas of the North Sea is in a Jurassic rift). Salts that accumulate from both saline lakes and sea water are exploited in some areas for example the Dead Sea Works is situated in the Dead Sea Rift and supplies much of the world's bromine. In addition, the river sands and gravels that accumulate in these basins can be an important source of building materials if the rift is sufficiently close to a developing centre of population.

The spectacular scenery of rifts is, perhaps, their most striking feature and some rifts have therefore become attractive tourist centres. One good example of this is Death Valley in the western USA where the desert conditions reduce vegetation to a minimum and the striking geomorphology is evident even to the untrained eye.

References

Frostick, L.E. (1997) The East African Rift basins, in R.C. Selley (ed.) *African Basins. Sedimentary Basins of the World*, 3, 187–209, Amsterdam: Elsevier.
Frostick, L.E. and Reid, I. (1987) Tectonic controls of desert sediments in rift basins ancient and modern, in L.E. Frostick and I. Reid (eds) *Desert Sediments: Ancient and Modern*, Geological Society Special Publication 35, 53–68.
Frostick, L.E. and Reid, I. (1989) Is structure the main control on river drainage and sedimentation in rifts? *Journal of African Earth Sciences* 8, 165–182.
Girdler, R.W. (1991) The Afro-Arabian rift system – an overview, *Tectonophysics* 197, 139–153.
Summerfield, M.A. (1991) *Global Geomorphology*, 424–425, Harlow: Longman.

Further reading

Frostick, L.E., Renaut, R.W., Reid, I. and Tiercelin, J.J (eds) (1987) *Sedimentation in the African Rift*, Geological Society Special Publication 25, Oxford: Blackwell Scientific.
Hovius, N. and Leeder, M.R. (1998) Clastic sediment supply to basins, *Basin Research* 10, 1–5.
Miall, A.D. (1996) *The Geology of Fluvial Deposits: Sedimentary Facies, Basin Analysis and Petroleum Geology*, Berlin: Springer Verlag.
Selley, R.C. (1997) *African Basins. Sedimentary Basins of the World, 3*, Amsterdam: Elsevier.

LYNNE FROSTICK

RILL

At the start of a rainfall event, rainwater which has fallen upon a hillslope begins to 'pond', i.e. OVERLAND FLOW moves rather slowly under the influence of gravity into small closed depressions in the soil's irregular surface (its 'microtopography'). This 'detention storage' gradually fills, although some of the stored water is constantly lost to infiltration into the soil. Meanwhile, if the rain is of moderate or high intensity then each raindrop which impacts upon an unprotected area of soil will possess sufficient kinetic energy to detach soil particles (see RAINDROP IMPACT, SPLASH AND WASH), which are thus redistributed over the soil's surface. Soil in the ponded areas is however largely protected from raindrop impacts. As a result, rainsplash redistribution usually decreases over time within a storm as the area and depth of surface water increases. There is a net downslope movement of splashed soil but this is generally small.

If the rain continues, then provided precipitation rate exceeds infiltration rate the deepening ponds on the soil's surface will eventually overtop their depressions. Overland flow which is released from overtopped ponds is likely to flow downhill more quickly and in greater quantities (i.e. possess greater kinetic energy) than the more diffuse and shallow flow into depressions: it may therefore be sufficiently competent (see SEDIMENT RATING CURVE) to transport soil particles which are splashed into it. Such soil particles can be carried some distance, being deposited only when flow velocity decreases (due to, for example, a reduction in gradient or the presence of vegetation).

Flow with still greater kinetic energy will generate a shear stress which is sufficient to detach soil particles from the body of the soil. These particles will then be transported along with splashed-in sediment. At locations where such detachment occurs, the soil's surface is lowered slightly. Such lowered areas form preferential paths for subsequent flow, and will thus be eroded further. Rather quickly, this positive feedback (see SYSTEMS IN GEOMORPHOLOGY) results in small, well-defined linear concentrations of flow (Favis-Mortlock 1998), known as 'microrills' or 'traces', with a width and depth of a few millimetres.

Many microrills will eventually become ineffective due to deposition within the microrill itself. But a fortunately located subset may grow further to become rills, with a maximum width and depth of a few tens of centimetres. This process of competition between individual channels leads to the self-organized formation (see COMPLEXITY IN GEOMORPHOLOGY) of networks of microrills and rills. Rill networks tend to be dendritic (see DRAINAGE PATTERN) in form on natural soil surfaces, but are constrained by the direction of tillage on agricultural soils. Such networks form hydraulically efficient pathways for the removal of water from hillslopes. However, sediment which is being transported within the rill network may be redeposited after a short distance if the flow loses its competence (such sediment possibly being detached again later in the same rainfall event, if flow conditions change; or during a subsequent rainfall event). Or the sediment may be carried some distance, perhaps even off the field and into a GULLY and/or a permanently flowing channel (see FIRST-ORDER STREAM). Once the rain stops, however, flow in the rill network will gradually cease: all sediment which is being transported at that time will then be redeposited within the network itself.

Rill networks on agricultural land are regularly erased by tillage, and regularly reinitiated (see SHEET EROSION, SHEET FLOW, SHEET WASH). On natural landscapes, however, rill networks persist and may in time cause such serious dissection of the hillslope as to lead to the formation of BADLANDs.

An eroding hillslope, then, normally consists of a flow-dominated channel network in which rill erosion occurs, separated by interrill areas where the dominant processes are rainsplash and diffuse flow. Soil loss from these areas is known as interrill erosion. But it is rill erosion which is the more effective agent for detachment and removal of soil, and so in many parts of the world rill erosion is the dominant subprocess of SOIL EROSION by water on hillslopes (De Ploey 1983). Boundaries between rill and interrill areas of the hillslope are frequently ill-defined and are constantly shifting. Note that subsurface flow may, in some circumstances, rival hillslope topography in importance in determining where channel erosion will begin and develop, e.g. at the base of slopes, and in areas of very deep soils such as tropical saprolites (see GROUND WATER).

Mean flow velocities within individual rills are usually between one and ten centimetres per second. Interestingly, there is evidence that for actively eroding rills, flow velocity is not dependent on rill gradient: this may be due to some compensatory increase in within-rill roughness on steeper slopes (Nearing *et al.* 1997). Velocity-depth profiles and planform patterns of velocity in rills are qualitatively similar to those of larger channels. Velocities may vary noticeably along the rill, however, with increased flow rates and scouring at 'headcuts' (Slattery and Bryan 1992), i.e. breaks of along-channel slope (such as that which is often found at the upstream extremity of each rill). Headcuts tend to move slowly upstream as their headward facets are eroded.

References

De Ploey, J. (1983) Runoff and rill generation on sandy and loamy topsoils, *Zeitschrift für Geomorphologie N.F. Supplementband* 46, 15–23.

Favis-Mortlock, D.T. (1998) A self-organising dynamic systems approach to the simulation of rill initiation and development on hillslopes, *Computers and Geosciences* 24(4), 353–372.

Nearing, M.A., Norton, L.D., Bulgakov, D.A., Larionov, G.A., West, L.T. and Dontsova, K. (1997) Hydraulics and erosion in eroding rills, *Water Resources Research* 33(4), 865–876.

Slattery, M.C. and Bryan, R.B. (1992) Hydraulic conditions for rill incision under simulated rainfall: a laboratory experiment, *Earth Surface Processes and Landforms* 17, 127–146.

Further reading

Abrahams, A.D., Li, G. and Parsons, A.J. (1996) Rill hydraulics on a semiarid hillslope, southern Arizona *Earth Surface Processes and Landforms* 21(1), 35–47.

Brunton, D.A. and Bryan, R.B. (2000) Rill network development and sediment budgets, *Earth Surface Processes and Landforms* 25(7), 783–800.

Bryan, R.B. (1987) Processes and significance of rill development, *Catena Supplement* 8, 1–15.

Merritt, E. (1984) The identification of four stages during microrill development, *Earth Surface Processes and Landforms* 9, 493–496.

Rauws, G. and Govers, G. (1988) Hydraulic and soil mechanical aspects of rill generation on agricultural soils, *Journal of Soil Science* 39, 111–124.

SEE ALSO: erodibility; erosivity; runoff generation; sheet erosion, sheet flow, sheet wash; soil erosion; universal soil loss equation

DAVID FAVIS-MORTLOCK

RIND, WEATHERING

Weathering rinds are zones of chemical alteration on the outer portions of rocks. In some, but not all cases, a distinct colour difference highlights this zone of intense CHEMICAL WEATHERING. Weathering rinds are important in geomorphology for their role in weathering processes, their role in the development weathering forms such as CASE HARDENING, and in their use in dating landforms. Obsidian hydration rinds are a related phenomenon.

A weathering rind is not just a zone of chemical alteration at the outer edge of a clast; weathering rinds represent the redistribution of elements. Some rinds are dominated by an enrichment in iron, while others are depleted in such mobile cations as calcium and sodium. A variety of processes develop weathering rinds. Dissolution, for example, leaves void space in the rock and does not necessarily change the colour. Oxidation of iron, in contrast, leaves a band of discolouration. The appearance of the zone of discolouration varies by location and rock type. For instance, rinds can appear white on the upper slopes of Mauna Kea and appear orange on the lower slopes, all in a basalt lithology (Plate 96). Andesite in Japan can appear brown to pale grey (Matsukura *et al.* 1994), and sandstone

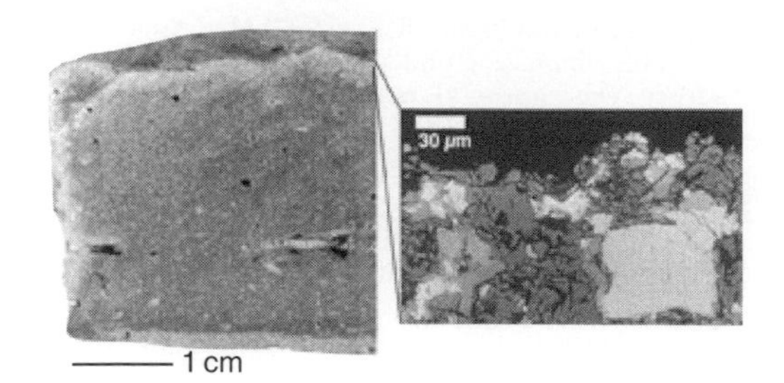

Plate 96 Weathering rind developed on a glacially polished basalt, Mauna Kea, Hawaii. This rind developed over a 16,000-year period. The left photograph shows an optical rind visible in a hand specimen. The right image shows an electron microscope (backscatter) image of a small section of the rind, illustrating three aspects of rind development. First, dissolution of minerals dominates rind formation, as exemplified by the pores (black areas). Second, the bright spots in the image are reprecipitated iron hydroxides, responsible for reddening. Third, rinds may not necessarily thicken over time. Often, they undergo erosion as pieces of weathered minerals progressively detach, that is if rinds are not protected by rock coatings

rinds in New Zealand can appear whitish (Knuepfer 1988).

Weathering rinds form on all three rock types: igneous (e.g. andesites, basalts, granitic), sedimentary (e.g. sandstones) and metamorphic (e.g. schists). Weathering rinds occur in a wide range of locations and in temperate, tropical, arctic and arid environments, for example, Hawaii (Jackson and Keller 1970), the coterminous United States (Colman and Pierce 1986), New Zealand (Chinn 1981), Japan (Matsukura *et al.* 1994) and northern Europe (Dixon *et al.* 2002). Weathering rinds are found in clasts at the surface and within the soil profile (Chinn 1981; Knuepfer 1988).

Weathering rinds are often used in geomorphology to estimate ages of landforms and landscape surfaces (Chinn 1981). This approach assumes that rinds begin to form soon after emplacement of the host rock, and that rinds grow thicker with time (Knuepfer 1988). Weathering rinds thus serve as a relative age indicator where thicker rinds occur on older landforms, and as a calibrated age indicator if accurate forms of age calibration are available in the study area. Prior to the use of cosmogenic nuclides (see COSMOGENIC DATING), use of weathering rinds was prevalent in Quaternary research where moraines, outwash sheets and other landforms correlated climatic changes (Colman and Pierce 1986). The thickness of the discoloured zone of a number of clasts in a deposit is measured normal to the surface, usually with a caliper. Statistical methods differentiate groups of thicknesses among different deposits or surfaces.

Because weathering rinds are so often felt to be synonymous with discolouration, we stress that the study of weathering rinds should not be limited to the measurement of colour changes in hand samples for several reasons. First, a weathering rind can occur without any noticeable colour change. Second, colour change provides only one indication of weathering; microscope studies reveal that the zone of chemical weathering continues into the rock well underneath the zone of colour change. Third, although weathering rinds are not ROCK COATINGS, a single clast may exhibit both a weathering rind and a rock coating (Matsukura *et al.* 1994), a distinction not always recognized in the field. Fourth, where weathering rinds are not protected by rock coatings, weathered mineral fragments readily spall off.

Research into weathering rinds is expanding into exciting new dimensions. Physical and chemical characteristics of weathering rinds are being used to help discern geochemical weathering processes in a given region or area (Dixon *et al.* 2002). The use of cosmogenic nuclides as a dating method has made weathering rind analysis more important than ever. A key uncertainty in cosmogenic dating surrounds the prior exposure history of a possible sample. With each cosmogenic measurement costing about US$2,000 in sample processing and analysis, weathering-rind measurements provide an inexpensive field check on the possibility that a particular sample might have a complex geomorphic history. In addition, *in situ* measurements of weathered minerals in rinds are providing new insight into quantitative rates of weathering; this method is being used, for example, to establish long-term rates of glass dissolution with the goal of understanding GEOMORPHOLOGICAL HAZARDS associated with nuclear waste storage (Gordon and Brady 2002).

References

Chinn, T. (1981) Use of rock weathering-rind thickness for Holocene absolute age-dating in New Zealand, *Arctic and Alpine Research* 13, 33–45.

Colman, S.M. and Pierce, K.L. (1986) Glacial sequence near McCall, Idaho: weathering rinds, soil development, morphology, and other relative-age criteria, *Quaternary Research* 25, 25–42.
Dixon, J.C., Thorn, C.E., Darmody, R.G. and Campbell, S.W. (2002) Weathering rinds and rock coatings from an Arctic alpine environment, northern Scandinavia, *Geological Society of America Bulletin* 114, 226–238.
Gordon, S.J. and Brady, P.V. (2002) In situ determination of long-term basaltic glass dissolution in the unsaturated zone, *Chemical Geology* 90, 115–124.
Jackson, T.A. and Keller, W.D. (1970) A comparative study of the role of lichens and 'inorganic' processes in the chemical weathering of recent Hawaiian lava flows, *American Journal of Science* 269, 446–466.
Knuepfer, R.L.K. (1988) Estimating ages of late Quaternary stream terraces from analysis of weathering rinds and soils, *Geological Society of America Bulletin* 100, 1,224–1,236.
Matsukura, Y., Kimata, M. and Yokoyama, S. (1994) Formation of weathering rinds on andesite blocks under the influence of volcanic gases around the active crater Aso Volcano, Japan, in D.A. Robinson and R.B.G. Williams (eds) *Rock Weathering and Landform Evolution*, 89–98, Chichester: Wiley.

SEE ALSO: case hardening; chemical weathering; rock coating

STEVEN J. GORDON AND RONALD I. DORN

RING COMPLEX OR STRUCTURE

> A petrologically variable but structurally distinctive group of hypabyssal or subvolcanic igneous intrusions that include ring dykes, partial ring dykes and cone sheets. Outcrop patterns are arcuate, annular, polygonal and elliptical with varying diameters ranging from less than 1 to 30 km or greater. The majority of ring complexes represent the eroded roots of volcanoes and their calderas.
>
> (Bowden 1985: 17)

Ring dykes are thick, approximately vertical igneous bodies that form concentric circles around a central intrusion. They are associated with a process called cauldron subsidence. Cone sheets tend to be thinner and have a general form as a set of inverted cones. They result from stresses set up in the Earth's crust as the magma body with which they are associated forced its way upwards. Other circular structures are associated with impact events.

Reference

Bowden, P. (1985) The geochemistry and mineralization of alkaline ring complexes in Africa (a review), *Journal of African Earth Sciences* 3, 17–39.

A.S. GOUDIE

RIP CURRENT

Many of the world's BEACHes are characterized by the presence of strong, concentrated seaward flows called rip currents. The term was introduced by Shepard (1936) to distinguish rips from the misnomers 'rip tide' and 'undertow', which are unfortunately often still used to describe rips today. Rips are an integral component of nearshore cell circulation and ideally consist of two converging longshore feeder currents which meet and turn seawards into a narrow, fast-flowing rip-neck that extends through the surf zone, decelerating and expanding into a rip-head past the line of breaking WAVEs. The circulation cell is completed by net onshore flow due to wave mass transport between adjacent rip systems (Figure 133a). Rip flows are often contained within distinct topographic channels between bars (see BAR, COASTAL) and are a major mechanism for the seaward transport of water, sediments and pollutants (Figure 133b). Rips are also a major hazard to swimmers and it is of concern that many aspects of rip occurrence, generation and behaviour remain poorly understood.

Rip currents are generally absent on pure dissipative and reflective beaches, but are a key component of sandy intermediate beach states in microtidal environments. Short (1985) identified three types: (1) accretion rips occur during decreasing or stable wave energy conditions and are often topographically arrested in position with mean velocities typically on the order of $0.5-1\,\mathrm{m\,s^{-1}}$; (2) erosion rips are hydrodynamically controlled and occur under rising wave energy conditions. They are transient in location, having mean flows in excess of $1\,\mathrm{m\,s^{-1}}$; and (3) mega-rips, which occur in embayments under extremely high waves, and can extend more than 1 km offshore with mean velocities greater than 2 $\mathrm{m\,s^{-1}}$. All are associated with localized erosion of the shoreline and often create rhythmic embayments termed mega-cusps. Relatively permanent rips located adjacent to headlands, reefs and coastal structures, such as GROYNEs, are referred to as topographically controlled rips.

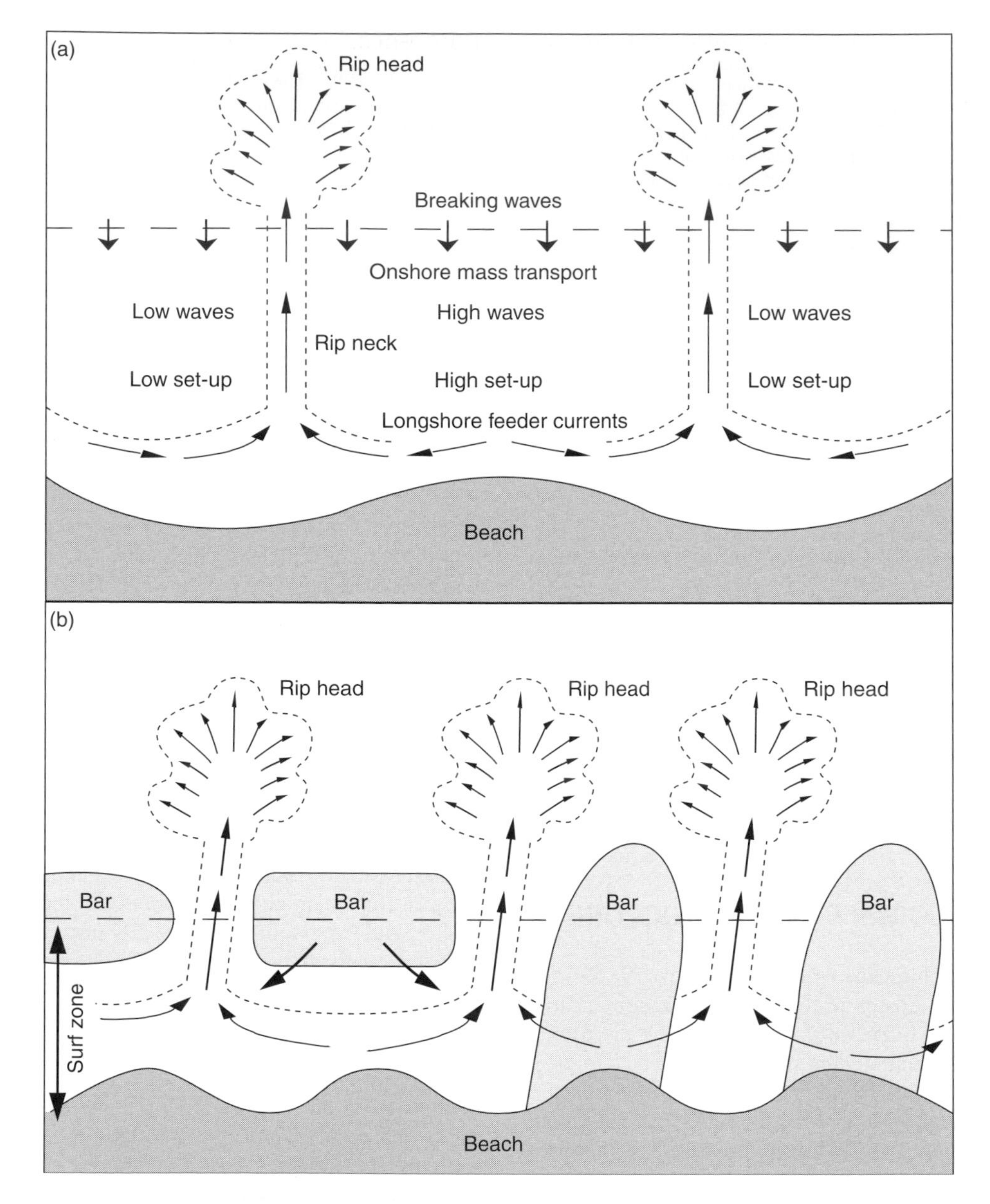

Figure 133 Idealized patterns of rip current flow and components in relation to: (a) nearshore cell circulation and wave set-up gradients; and (b) coastal bar topography

The primary limitation to our understanding of rips has been the difficulty obtaining quantitative field measurements from an energetic environment. Early attempts at describing rips (e.g. McKenzie 1958) were largely qualitative, but correctly identified that rips often display a periodic longshore spacing, increase in intensity and decrease in number as wave height increases, and flow fastest at low tide. Subsequent theoretical, laboratory and field studies have attempted to explain these characteristics with varying degrees of success, although it is generally accepted that rips exist as a response to an excess of water, termed wave set-up, built up on shore by breaking waves. The flow is forced by longshore variations in wave height, which produce gradients in the set-up that drive water alongshore from regions of high to low waves (Bowen 1969; see Figure 133a).

Existing models for the generation of rip cell circulation have thus incorporated various mechanisms to account for the existence of these longshore gradients and can be grouped into three main categories: (1) the wave–boundary interaction model involves wave modification by non-uniform topography and/or coastal structures. For example, wave refraction can produce regions of high and low waves, such that rips can occur in the lee of offshore submarine canyons (Shepard and Inman 1950), but more commonly adjacent to headlands and groynes; (2) wave–wave interaction models have shown theoretically and in laboratory experiments (Bowen and Inman 1969) that incident waves can generate synchronous edge waves that produce alternating patterns of high and low wave heights along the shoreline. Rips occur at every other antinode with a spacing equal to the edge wave length; and (3) instability models suggest that longshore uniformity in set-up is unstable to any small disturbance caused by hydrodynamic or topographic factors and rip spacing is predicted to equal four times the surf zone width. It should be emphasized that validation of these models has primarily been restricted to laboratory experiments and has not been adequately verified in the field. Short and Brander (1999) used a global field dataset to show that rip spacing is related to regional wave energy environments. Patterns of rip spacing (L_r) were consistent within west coast swell ($L_r \cong 500$ m), east coast swell ($L_r \cong 200$ m), and fetch-limited wind wave environments ($L_r \cong 50-100$ m).

The wave–boundary model is best supported by Sonu (1972) who found that on a beach consisting of alternating sandbars and topographic channels with uniform longshore wave height, constant and extensive wave energy dissipation across the bars and local and intense wave breaking over the channels created a set-up gradient towards the channels, which controlled rip flow. Set-up gradients generated in this manner support field data confirmation (e.g. Brander 1999) that rip flows are tidally modulated, since stronger flows at low tide would be expected with increased wave dissipation associated with shallower water depths over the bars. Field studies have also shown that rip velocities increase steadily from the feeders, attaining maximums in the middle of the rip-neck, are greater near the water surface and experience short duration and strong velocity pulses every few minutes, the forcing of which is likely related to infragravity motions such as shear waves or wave groups.

References

Bowen, A.J. (1969) Rip currents. 1. Theoretical investigations, *Journal of Geophysical Research* 74, 5,467–5,478.

Bowen, A.J. and Inman, D.L (1969) Rip currents. 2. Laboratory and field observations, *Journal of Geophysical Research* 74, 5,479–5,490.

Brander, R.W. (1999) Field observations on the morphodynamic evolution of a low-energy rip current system, *Marine Geology* 157, 199–217.

McKenzie, P. (1958) Rip-current systems, *Journal of Geology* 66, 103–111.

Shepard, F.P. (1936) Undertow, rip tide, or rip current, *Science* 84, 181–182.

Shepard, F.P. and Inman, D.L. (1950) Nearshore water circulation related to bottom topography and wave refraction, *Transactions of the American Geophysical Union* 31, 196–212.

Short, A.D. (1985) Rip current type, spacing and persistence, Narrabeen Beach, Australia, *Marine Geology* 65, 47–61.

Short, A.D. and Brander, R.W. (1999) Regional variations in rip density, *Journal of Coastal Research* 15(3), 813–822.

Sonu, C.J. (1972) Field observation of nearshore circulation and meandering currents, *Journal of Geophysical Research* 77, 3,232–3,247.

SEE ALSO: bar, coastal; beach; beach sediment transport; groyne; wave

ROBERT W. BRANDER

RIPARIAN GEOMORPHOLOGY

Riparian geomorphology is concerned with the dynamics, form and sedimentary structure of riparian zones. Riparian zones have been variously described as 'three-dimensional zones of direct interaction between terrestrial and aquatic ecosystems' (Gregory *et al.* 1991: 540); zones that extend 'from recently colonized fluvial landforms exposed at low flow to the limits of the area wherein biota are adapted to, or characteristic community structures are influenced by, flooding' (Dykaar and Wigington 2000: 88); and the 'part of the biosphere supported by, and including, recent fluvial landforms. . . inundated or saturated by the BANKFULL DISCHARGE' (Hupp and Osterkamp 1996: 280). From these and other definitions, it is apparent that the riparian geomorphology of a river reach is dependent upon: present and past flow magnitude and frequency (see MAGNITUDE–FREQUENCY CONCEPT); the amount and calibre of sediment transported by the river; and the slope and degree of confinement of the reach. Whilst the past and present flow and sediment transport regime govern the materials delivered to the reach for landform building, the local slope and confinement of the reach govern the river's energy and its ability to construct and erode landforms.

Nanson and Croke (1992) explored these controls to develop a genetic classification of FLOODPLAIN types that is explicitly linked to the river types that construct the floodplains. They defined three broad groups of floodplain (high-energy non-cohesive, medium-energy non-cohesive, low-energy cohesive) based upon the river's ability to do work and expressed by its specific STREAM POWER at bankfull discharge, and the erosional resistance of the floodplain materials (non-cohesive implies gravel or sand; cohesive implies silt and clay). They subdivided the three groups into thirteen different floodplain classes. These were discriminated by details of the sediment from which they are constructed, the river planform or pattern, its characteristic erosional and depositional processes, and thus the typical landforms present on the floodplain and within the river margins. Importantly, this classification links process and form in a dynamic way, illustrating that riparian zones may possess an enormous variety of landforms and that the nature and dynamism of the landforms varies between floodplain types. Thus, if there is a change in the controlling processes, riparian geomorphology also changes. For example, changes in climate, flow regulation and flood defence engineering affect river flow and sediment transport regimes and the ERODIBILITY of channel margin materials, and so can have far-reaching impacts on riparian zone character (e.g. Steiger and Gurnell 2002).

In most analyses of riparian zone form and process, vegetation has been seen to play a largely passive role, responding to present and past environmental conditions created by fluvial processes (e.g. Hupp 1988). Thus, floodplain vegetation patterns have been interpreted to depend on the type and age of the mosaic of riparian landforms. Migrating, MEANDERING rivers provide a simple illustration. As the river erodes the outer banks of meander bends, POINT BARs develop on the inner banks. Vegetation colonizes point bar surfaces and plant species are gradually replaced as sediment, moisture, light and disturbance on the bars change during their aggradation and incorporation into the floodplain.

Recently, more emphasis has been placed on the active role of vegetation in influencing riparian zone geomorphology. For example, Gurnell and Petts (2002) consider both biotic and abiotic ways in which vegetation can influence the form, sedimentary structure and dynamics of riparian zones. Abiotic influences include the impact of root systems on the erodibility of sediments and the flow resistance of the vegetation canopy. Roots can cause significant reinforcement of riparian sediments, making them more resistant to river erosion. When the riparian zone is flooded, the ROUGHNESS of the vegetation canopy can reduce flow velocities across the vegetated surface, reducing rates of erosion and increasing rates of sedimentation. These abiotic processes can significantly affect patterns of erosion and aggradation, and thus the form and sedimentary structure of riparian zones. The geomorphological significance of these abiotic influences depends on the species, age and density of the vegetation cover, which is related to several biotic processes. The degree to which riparian plants reproduce from seeds or by vegetative reproduction is important because, in general, riparian vegetation growth is more rapid when plants propagate vegetatively. The timing of seed or vegetative propagule release can greatly influence the likelihood of successful vegetation establishment, because many riparian plant propagules are transported and deposited by the river. For example, the timing of propagule release in relation to the climate and river flow regimes can influence whether suitable colonization sites are exposed or inundated by the river, and whether their moisture and temperature characteristics are appropriate to support the successful germination and growth of young plants.

Riparian tree species can be particularly important riparian zone engineers. Poplar and willow species can grow very rapidly, propagating through both seeds and vegetative reproduction. Rivers may erode, transport and deposit whole trees as well as fragments (branches, twigs, root boles) and seeds. Entire trees may survive rafting by floods, deposition and burial within river margins and on bars, and they can sprout to form patches of new sizeable shrubs within a year. The importance of these processes for riparian geomorphology varies with tree species and environmental conditions but also with riparian tree management. The pruning and felling of riparian trees to prevent LARGE WOODY DEBRIS entering rivers is often carried out to maintain the FLOOD conveyance of the river channel. Its impact on riparian geomorphology and ecology is far reaching, leaving little impression of the diverse geomorphological and ecological character and high dynamism of unimpacted riparian zones (Gurnell *et al.* 1995, 2002).

References

Dykaar, B.B. and Wigington, P.J. (2000) Floodplain formation and cottonwood colonization patterns of the Willamette River, Oregon, USA, *Environmental Management* 25, 87–104.

Gregory, S.V., Swanson, F.J., McKee, W.A. and Cummins, K.W. (1991) An ecosystem perspective of riparian zones *BioScience* 41, 540–551.

Gurnell, A.M., Gregory K.J. and Petts G.E. (1995) The role of coarse woody debris in forest aquatic habitats: implications for management, *Aquatic Conservation* 5, 143–166.

Gurnell, A.M. and Petts, G.E. (2002) Island-dominated landscapes of large floodplain rivers, a European perspective, *Freshwater Biology* 47, 581–600.

Gurnell, A.M., Piégay, H., Swanson, F.J. and Gregory, S.V. (2002) Large wood and fluvial processes, *Freshwater Biology* 47, 601–619.

Hupp, C.R. (1988) Plant ecological aspects of flood geomorphology, in V.R. Baker, R.C. Kochel and P.C. Patten (eds) *Flood Geomorphology*, 335–356, New York: Wiley.

Hupp, C.R. and Osterkamp, W.R. (1996) Riparian vegetation and fluvial geomorphic processes, *Geomorphology* 14, 277–295.

Nanson, G.C. and Croke, J.C. (1992) A genetic classification of floodplains, *Geomorphology* 4, 459–486.

Steiger, J. and Gurnell, A.M. (2002) Spatial hydrogeomorphological influences on sediment and nutrient deposition in riparian zones: observations from the Garonne River, France, *Geomorphology* 49(1), 1–23.

Further reading

Gurnell, A.M., Hupp, C.R. and Gregory, S.V. (eds) (2000) Linking hydrology and ecology, *Hydrological Processes*, Special Issue 14, 2,813–3,179.

Stanford, J.A. and Gonser, T. (eds) (1998) Rivers in the landscape: riparian and groundwater ecology, *Freshwater Biology*, Special Issue 40, 401–585.

Tockner, K., Ward, J.V., Kollmann, J. and Edwards, P.J. (eds) (2002) Riverine Landscapes, *Freshwater Biology*, Special Issue 47, 497–907.

ANGELA GURNELL

RIPPLE

Ripple is a general term applied to a range of normally unrelated, very small bedforms that occur in trains and record sediment mobilization and transport in various aqueous and aeolian environments (see BEDFORM; BEDLOAD; ROUGHNESS). The main kinds are current and rhomboid ripples, oscillation or 'symmetrical' ripples (see WAVE), ballistic or impact ripples (see AEOLIAN PROCESSES; SALTATION), adhesion ripples and warts (see AEOLIAN PROCESSES; SALTATION), and rain-impact ripples (see RAINDROP IMPACT, SPLASH AND WASH). With the exception of adhesion warts, and some complex oscillation types, ripples are characterized by crests that lie transversely to flow.

Trains of current ripples, restricted to the coarser silts and the finer sands, are typical of rivers, but also appear in tidal environments (estuaries, barred beaches) where flows can be unidirectional for several hours at a time. As equilibrium bedforms, current ripples have linguoid crests in plan, heights of up to about 0.02 m, wavelengths of 0.1–0.2 m, and strongly asymmetrical profiles, the short leeward face lying at the angle of repose (see REPOSE, ANGLE OF). When generated from a smooth bed, however, current ripples evolve toward a linguoid shape through a range of long-crested forms, the crests of which increasingly lose straightness. Internally, current ripples are cross-laminated, commonly in climbing sets, a testimony to high rates of sediment deposition on a scale of minutes or hours. Ripple dimensions are independent of flow depth but increase weakly with grain size. Diamond-shaped rhomboid ripples are developed where ripple-generating flows are sufficiently shallow as to be supercritical.

Other flows being absent, wind waves generate within the affected water-body symmetrical, oscillatory currents which are superimposed on a much weaker drift in the direction of wave-propagation. When sufficiently powerful, their combined effect on sand beds is to create trains of ripples with long, regular crests and steep, almost symmetrical, trochoidal profiles which, as revealed by internal cross-laminae, migrate very slowly in the direction of wave-propagation. Ripple scale depends in a complex manner on the properties of the waves and the sediment, the wavelength and height increasing markedly with grain size. Wavelengths are of the order of 0.01 m in silt, 0.1 m in fine sand and 1 m in coarse sands and fine gravels. Broadly, wavelength is about 500 times the median grain diameter. Wave ripples are most familiar from estuaries and beaches but, after storms, appear on continental shelves to water depths of 100–200 m. Complex forms of ripple occur where barriers reflect waves and where, especially on beaches and in estuaries, unrelated unidirectional and wave currents operate either simultaneously or sequentially. Wave ripples are valuable indicators of shallow water and of shoreline location and orientation.

The saltation of wind-driven grains over a dry bed is generally accompanied by the development of trains of ballistic ripples, resulting from an unstable interaction between the surface and the flow of sediment. These ripples are rather flat, asymmetrical structures which vary in form and scale with increasing grain size and the average length of the jumps made by the particles. Typically, ripples in the finer sands have crests that are long and regular in plan and wavelengths of about 0.05 m. Those in sediments of very coarse sand or granule grade take wavelengths of the order of 1 m and generally have short, irregular crests, along which the coarser particles conspicuously lie. Ballistic ripples are cross-laminated internally, but the structure is difficult to see in the well-sorted sands of which the smaller examples are formed. The ripples have long been reported from deserts and sandy coasts, wherever the wind is free to mobilize sufficiently coarse grains.

The capture of saltating particles by a damp or wet surface, such as a coastal sand beach, river bar or sabkha, gives rise to upwind-facing, centimetre-scale adhesion ripples (uniform wind-direction) or adhesion warts (wind-direction variable). These common and widespread structures have no particular climatic significance but are valuable proofs of surface exposure and aeolian activity. Advancing in the opposite direction to the wind, adhesion ripples create a steep internal bedding that dips downwind.

Rain-impact ripples are centimetre-scale, upwind-facing, transverse ridges shaped when heavy rain driven by a strong wind descends at a fine angle onto an exposed, water-saturated sand bed, such as a beach, tidal sand shoal or river bar. The ridges advance very slowly in the direction of the wind under the repeated impact of the drops. If rain-impact ripples have a fossil record, which is uncertain, they would afford a further proof of atmospheric exposure.

Further reading

Allen, J.R.L. (1979) A model for the interpretation of wave ripple-marks using their wavelength, textural composition and shape, *Journal of the Geological Society, London* 136, 673–682.

——(1982) *Sedimentary Structures*, Amsterdam: Elsevier.

Anderson, R.S. (1987) A theoretical model for aeolian impact ripples, *Sedimentology* 34, 943–956.

Baas, J.H. (1999) An empirical model for the development of and equilibrium morphology of current ripples in very fine sand, *Sedimentology* 46, 123–138.

Bagnold, R.H. (1946) Motion of waves in shallow water. Interaction between waves and sand bottom, *Proceedings of the Royal Society, London* A187, 1–16.

Clifton, H.E. (1977) Rain-impact ripples, *Journal of Sedimentary Petrology* 47, 678–679.

Doucette, J.S. (2002) Geometry and grains-size sorting of ripples on low-energy sandy beaches; field observations and model predictions, *Sedimentology* 49, 483–503.

Fryberger, S.G., Hesp, P. and Hastings, K. (1992) Aeolian granule ripple deposits, Namibia, *Sedimentology* 39, 319–331.

Kahle, C.F. and Livchak, C.J. (1996) Nature and significance of rhomboid ripples in a Silurian sabkha sequence, north-central Ohio, *Journal of Sedimentary Research* 66, 861–867.

Kocurek, G. and Fielder, G. (1982) Adhesion structures, *Journal of Sedimentary Petrology* 52, 1,229–1,241.

J.R.L. ALLEN

RIVER CAPTURE

River capture, sometimes called stream capture or stream piracy, refers to the occurrence of the seizure of the waters of a stream or drainage system by a neighbouring one. It is based on the difference in local BASE LEVEL heights, with the captured stream having a higher base level and for that reason with a low erosion potential. The predatory stream, with a lower base level, is capable of diverting in its favour the waters of the less active stream, and in this way enlarging its drainage net and catchment area. Integration of both drainage systems leads to a higher order network. It does not only occur because of a steeper gradient but also because the pirate stream is cutting its valley in softer rock.

Capture constitutes a common event in the erosional evolution of a drainage net of a region and is a traditional concept in geomorphology that can be found in classical authors. Gilbert (1877) described the process in relation to the role of unconsolidated materials in mine dumps, calling it abstraction, a term often applied to the simplest type of capture, which results from competition between adjacent consequent gullies and ravines. He was also aware that a stream flowing down the steeper slope of an asymmetrical ridge erodes its valley more rapidly than the one flowing down a more gentle slope, and as a result the divide migrates away from the more actively eroding stream. This principle has been refered to as the *law of unequal slopes* (Thornbury 1969). The same concept was integrated by Davis (1899) in his model of relief evolution by the geographical cycle, capture taking place in the young or early mature stages of development. Another classical author, Horton (1945), in his slope runoff model also takes into consideration the capture process and uses it in order to explain the development of a hierarchical drainage net, that is, the process by which the drainage lines become integrated into a few dominant stream courses. Unequal rainfall on two sides of a divide may contribute to divide migration, especially where winds are prevailing from one direction, as in the trade wind belts (Thornbury 1969).

At the point at which the capture takes place, the captured stream bends sharply, forming a right angle turn into the pirate stream, which is called the elbow of capture. The valley stretch in which the captured stream continues to flow after losing the upper part of its catchment becomes a beheaded valley. This valley is then too large for the stream that continues to flow in it and thus becomes an UNDERFIT STREAM, that is a stream too small to be hydrologically related to the valley in which it now flows. On the other hand, the captured part of the stream now has a lower local base level, which increases its erosion potential and makes it able to incise into its former alluvial valley floor, producing a terrace in its former floodplain.

The capture process mainly occurs in two different ways: by headward erosion and by lateral erosion. *Headward erosion* is the probable cause of most easily recognizable stream captures. It takes place when the tributaries of the high energy stream are working back towards its head, and eventually reach the neighbouring valley head and cut through the divide. Capture by *lateral erosion* occurs when two streams flow parallel at no great distance from each other. Progressive erosion produces a lateral shifting of the stream which can finally produce a planation of the water divide. If this continues at the cost of one of the neighbouring streams, it ends in the lateral capture of its waters. Capture by subterranean waters can also occur in soluble rocks when water from a stream at a higher level percolates and meets an underground stream flowing at a lower level.

Examples of river captures have been described in many regions of the world, both at large and at small scales. In the large scale, one of the classical examples is that of a tributary of the Indus captured by the Ganges which implied the transfer of the drainage of a large area of the Himalayas from Pakistan to India. In Yunnan Province, China, rivers flowing towards the Red river were captured by the middle Yangtze tributaries. In Queensland, eastern Australia, the Fitzroy River has reached an old divide at the Connors Range and captured several of the rivers flowing in this area. In New Zealand the capture of the Silver Stream by the Karori near Wellington is well known. In Europe waters were diverted from the Danube towards the Rhine by a small head-ward tributary. In North America, in the Appalachian region of the eastern United States, there are many captures which are controlled by differences in rock hardness.

Amongst the implications of river captures is their role and significance for the evolution of relief and the history of drainage patterns, providing an interesting geomorphic challenge. In that sense, a highly integrated stream system with a large main stream is usually an indication of a long period of development (Ahnert 1998). Another important issue mentioned by Schumm (1977) is the role of captures in the discovery of new placer deposits, which are alluvial deposits containing valuable minerals, because the regional distribution of placers can be strongly influenced by stream capture. The result of this process from an economic point of view is that the source of the valuable minerals may be abruptly isolated from the downstream depositional area.

References

Ahnert, F. (1998) *Introduction to Geomorphology*, London: Arnold.
Davis, W.M. (1899) The Geographical Cycle, *Geographical Journal* 14, 481–504.
Gilbert, G.K. (1877) Report on the geology of the Henry Mountains, 141, Washington, *US Geographical and Geological Survey of the Rocky Mountains Region*.
Horton, R.E. (1945) Erosional development of streams and their drainage basins: hydrological application of quantitative morphology, *Geological Society of America Bulletin* 56, 281–370.
Schumm, S.A. (1977) *The River System*, Chichester: Wiley.
Thornbury, W.D. (1969) *Principles of Geomorphology*, New York: Wiley.

SEE ALSO: base level; gully; underfit stream

MARIA SALA

RIVER CONTINUUM

The biological concept of a river continuum describes a regular downstream progression of such physical variables as channel width, diel temperature pulse and stream order, in relation to biotic adjustments (Vannote *et al.* 1980). The concept was originally developed for rivers in regions with deciduous forests. In these regions, headwater streams (orders 1–3) are narrow and shaded by riparian vegetation. The vegetation reduces instream or autotrophic production from algae by shading, and contributes large amounts of coarse organic detritus (>1 mm diameter), such as leaf litter. The ratio of photosynthesis/respiration (P/R) is less than 1. The diversity of soluble organic compounds is high, and the diel temperature pulse is low. Communities of aquatic insects in headwater streams are dominated by insects that shred coarse organic matter (shredders), and insects that filter finer organic matter from transport, or gather such material from sediments (collectors). Fish populations have cool-water species that feed mainly on invertebrates. Biotic diversity is low.

Medium-sized streams (orders 4–6) are sufficiently wide that sunlight reaches a greater portion of the stream channel. Algae and rooted plants in the stream are more plentiful, and the ratio of P/R exceeds 1. The diversity of soluble organic compounds drops sharply relative to headwater streams, and the diel temperature pulse reaches a maximum. Fine particulate organic matter (50 μm–1 mm diameter) becomes more important. Collectors remain important, shredders form a smaller percentage of insect communities, and grazers that shear attached algae from surfaces in the stream increase in abundance. Fish populations now have more warm-water species that feed on invertebrates and other fish. Biotic diversity reaches a maximum.

Large rivers (>6 order) are very broad and open to sunlight, but photosynthesis may be limited by depth and turbidity. Large quantities of fine particulate organic matter from processing of dead leaves and woody debris come from upstream, and the ratio of P/R again drops below 1. The diel temperature pulse is low. Aquatic insects are primarily collectors. Fish are warm-water species that feed on plankton, invertebrates and other fish. Biotic diversity drops off again.

The general pattern described above may differ in mountainous areas where headwater streams flow through alpine meadows, in dry regions where riparian vegetation is restricted, or along deeply incised channels where shading from valley walls limits photosynthesis. However, the river continuum does provide a conceptual model of spatial gradients in physical and biological variables. This conceptual model is one of the first holistic theories of a river as an ecosystem, rather than individual segments. The river continuum emphasizes the connections between the river and its terrestrial setting. The continuum also suggests that aquatic communities can be explained by the mean state of environmental variables and their degree of temporal variability and spatial heterogeneity (Minshall *et al.* 1985). Together with hypotheses of stream succession that predict changes in habitat and species following a disturbance such as a FLOOD, the river continuum concept facilitates predictions of reach-specific patterns in habitat, communities or life-history strategies.

References

Minshall, G.W., Cummins, K.W., Petersen, R.C., Cushing, C.E., Bruns, D.A., Sedell, J.R. and Vannote, R.L. (1985) Developments in stream ecosystem theory, *Canadian Journal of Fisheries and Aquatic Science* 42, 1,045–1,055.

Vannote, R.L., Minshall, G.W., Cummins, K.W., Sedell, J.R. and Cushing, C.E. (1980) The river continuum concept, *Canadian Journal of Fisheries and Aquatic Science* 37, 130–137.

SEE ALSO: fluvial geomorphology; large woody debris; stream ordering

ELLEN E. WOHL

RIVER DELTA

River deltas are coastal accumulations of terrestrial sediments that rivers have brought to the sea. The Greek historian Herodotus (*c.*450 BC) originally applied the term 'delta' to the triangular subaerial deposit surrounding the mouth of the Nile River. In modern usage, however, deltas can be either subaerial or subaqueous accumulations and may have a variety of geometries. Although deep sea fans may also be considered deltas, they are not discussed here. Here, the 'subaqueous delta' is assumed to be confined to deposits on the continental shelf. The prevailing shape of any given delta depends on the rates of sediment supply by the rivers and the patterns and rates of sediment dispersal by coastal ocean processes and by gravity. In many cases, the subaqueous deltaic deposits are much more extensive than the subaerial deposits and, in some cases such as that of Papua New Guinea's Sepik River which discharges directly into deep water, the subaerial delta may be missing altogether. Historically, deltas have played important socio-economic roles. Subaerial deltas were the sites of early agriculture and formative civilizations and presently support some of the world's largest urban centres (e.g. Shanghai, Bangkok, Cairo). Subaqueous deltas are sinks for terrestrial carbon and are sources of fossil fuel.

Deltas vary immensely in both area and volume. The size of a delta depends at the lowest order on the annual sediment discharge of the river but the most extensive deltas also tend to be developed where wide, low gradient continental shelves provide a platform for prolonged sediment accumulation and morphological progradation. Hence, the largest deltas are found on passive (as opposed to active) continental margins (Wright 1985). Despite this fact, active margins are probably equally or more important than passive margins in supplying river sediment to the sea; Milliman and Syvitski (1992) showed that the numerous small mountainous streams, particularly those of the humid tropics, are collectively the most important source of terrestrial sediment to the sea. However, since these rivers are spatially distributed and since much of this sediment is bypassed to deep water, large deltas typically do not result. Other factors that influence delta area and the relative sizes of subaerial vs subaqueous components include tectonic subsidence and the energy of waves and currents that resuspend or prevent shallow water deposition of sediments. Table 38 lists characteristics of five deltas.

Satellite images of the Changjiang (Yangtze) and Mississippi Deltas (Plates 97 and 98) illustrate the diversity of major deltas. In the case of the Changjiang (Plate 97), the river-supplied sediments have built a large lobate protrusion into the East China Sea that supports and surrounds Shanghai. More exaggerated seaward protrusion of this delta is constrained by strong currents and waves, which disperse newly discharged sediments over the shelf and into Hangzhou Bay. These sediments are accumulating on the shelf at a rate of 5 cm yr^{-1} (DeMaster *et al.* 1985) to form the subaqueous component of the delta. The Mississippi Delta, one of the world's most extensively studied deltas, is composed of sediments from a catchment that covers 60 per cent of the continental United States. Its elongated and narrow digitate shape is rare and is attributable to a combination of fine cohesive sediment, low wave

Table 38 Characteristics of five major river deltas

Property	River delta				
	Amazon	Ganges–Brahmaputra	Changjiang (Yangtze)	Huanghe (Yellow R.)	Mississippi
Drainage Basin Area ($km^2 \times 10^6$)	6.15	1.48	1.94	0.77	3.27
Water discharge ($km^3\,yr^{-1}$)	6,300	971	900	42	580
Sediment discharge (10^6 tonnes yr^{-1})	900	1,620	480	1,060	210
Sediment/water ratio ($kg\,m^{-3}$)	0.14	1.67	0.53	25.25	0.36
RMS wave height (m)	1.6	2.5	1.5	2.0	1.1
Spring tide range (m)	5.8	3.6	3.0	1.4	0.4
Total delta area ($km^2 \times 10^3$)	467	106	67	36	29
Ratio subaerial/ subaqueous area	6.4	2.4	1.7	3.3	5.3

Sources: Coleman and Wright (1975); Milliman and Meade (1983); Wright and Nittrouer (1995)

Plate 97 Satellite image of the Changjiang (Yangtze) delta and estuary showing the city of Shanghai, the turbid river effluent and Hangzhou Bay which serves as a sink for much of the sediment

Plate 98 Satellite image of the Mississippi Delta showing the active 'bird's foot' in the right portion of the image and the abandoned La Fourche delta lobe on the left

height and negligible tidal currents, a regime that allows sediments to accumulate near the river mouths without being widely dispersed by oceanographic forces (Wright and Nittrouer 1995). The modern Mississippi 'bird's foot' (Plate 98) now extends across the continental shelf creating a barrier to east–west currents. In recent geologic history, a series of such lobate overextensions have been followed by avulsions: at least sixteen such lobes have been created and abandoned in Holocene time (Kolb and Van Lopik 1966). Abandoned deltaic lobes make up most of Louisiana's coastal plain, which is experiencing a high rate of coastal land loss because of regional subsidence and erosion.

The processes that disperse, transport and deposit the sediment discharged by a river determine the configuration of the resulting delta. This is true not only for the subaqueous component but also for the subaerial delta, which must surmount the subaqueous deposits in order to prograde. Wright and Nittrouer (1995) argued that the fate of sediment seaward of river mouths involves at least four stages: (1) supply via river plumes; (2) initial deposition; (3) resuspension and onward transport by marine forces (e.g. waves and currents); and (4) long-term net accumulation. Different suites of processes dominate each stage. Immediately upon leaving the confines of a river mouth, a river effluent spreads as either a positively buoyant (lighter than seawater because of the salinity difference) or negatively buoyant (because of very high suspended sediment concentrations in the river water) plume while mixing and exchanging momentum with the ambient seawater. This is the first stage of dispersal. Most of the larger rivers that drain to passive continental margins have positively buoyant effluents because they carry low concentrations of suspended sediment and large volumes of low-density fresh water. Examples include the Mississippi, Amazon, Ganges–Brahmaputra and Changjiang among numerous others. The most prominent exception among the large rivers is the Huanghe (Table 38), which often transports suspended sediment in concentrations greater than $25\,kg\,m^{-3}$ creating an effluent bulk density greater than that of seawater (Wright *et al.* 1990). Such negatively buoyant effluents are referred to as hyperpycnal (excessively dense) and tend to move downslope within the near-bed layer under the influence of gravity. Hyperpycnal conditions are somewhat more common at the mouths of smaller rivers that drain mountainous catchments near the coasts of active margins. The Eel River of northern California is a prominent example (Geyer *et al.* 2000).

The second stage of sediment dispersal is represented by initial, but usually temporary, deposition from the expanding and decelerating effluents of stage 1. River-mouth bars of varying geometries are among the morphologic products of this deposition. The deposition is caused by sediment flux convergence produced by effluent deceleration and sediment particle settling. The more rapidly the effluent gives up its momentum through mixing and bed friction and the greater the particle setting velocity, the closer the one would expect initial deposition to be to the river mouth. On the other hand, high waves and strong coastal currents enhance mixing and effluent momentum exchange with the sea but may also act to resuspend sediment or to inhibit initial deposition. Along high-energy coasts, the initial deposition may be delayed until the sediment reaches a deeper, more quiescent environment such as the mid-shelf region (Ogston *et al.* 2000). Energetic oceanographic processes also disperse sediment parallel to the coast or isobaths, generally preventing the formation of digitate or protruding deposits.

The third, resuspension and transport, stage of dispersal may act concurrent with or subsequent to the initial deposition stage (stage 2). When the coastal regime is highly energetic throughout the year or when high energy coincides with high river discharge, deposition is delayed as explained above. However, in many cases (e.g. Huanghe; Wright *et al.* 1990), the maximum input of river sediments to the sea and the maximum agitation of the bed by waves occur in different seasons. In such cases, the initial deposition may take place near the river but be removed, partially or wholly, by wave-induced resuspension a few months later. Depending on the amount of time that elapses between initial deposition and eventual resuspension, sediments may undergo varying degrees of consolidation making them more resistant to erosion and more likely to remain at the initial deposition site.

In the fourth dispersal stage, the river sediments reach their 'final' resting place and the rate of net accumulation exceeds the rate of erosion and resuspension. It is the accumulated products of this final stage that leave the most lasting geologic record. In the case of the Amazon delta, Pb-210 analyses of cores (Kuehl *et al.* 1986) indicate that on century timescales, roughly half the river's sediment load is accumulating on the mid-shelf (depth 30–50 m) at an average rate of 10 cm yr^{-1}. The remainder of the sediment is sequestered within the subaerial delta. In contrast, the mild energy regime of the Mississippi Delta, together with a rather rapid rate of tectonic subsidence, has permitted the formation of thick accumulations at the mouths of prograding distributaries on the centuries timescale. On a longer timescale, episodic delta lobe switching has yielded multiple thick and elongate accumulations distributed along the Louisiana coast.

As deltas prograde seaward over a continental shelf and spread laterally along a coast, a subaerial delta plain usually surmounts the underlying subaqueous platform. Although this subaerial surface, which includes the intertidal environments, is typically quite thin vertically, at least in comparison to subaqueous deposits, it is this component that supports most human activities and with which people are most familiar. The suites of geomorphologic features that distinguish any particular deltaic surface are, as for the subaqueous delta, products of the coastal process regime that moulded the delta as well as of the regional climate and human land use practices. Other factors include the degree to which a delta is undergoing submergence because of rising sea level, tectonic subsidence or both. Wright (1985) describes some of the most common subaerial features.

References

Coleman, J.M. and Wright, L.D. (1975) Modern river deltas: variability of processes and sand bodies, in M.L. Broussard (ed.) *Deltas: Models for Exploration*, Houston Geological Society, 99–149.

DeMaster, D.J., McKee, B.A., Nittrouer, C.A., Qian, J. and Cheng, G. (1985) Rates of sediment accumulation and particle reworking based on radiochemical measurements from continental shelf deposits in the East China Sea, *Continental Shelf Research* 4, 153–158.

Geyer, W.R., Hill, P.S., Milligan, T. and Traykovski, P. (2000) The structure of the Eel River plume during floods, *Continental Shelf Research* 20, 2,067–2,093.

Kolb, C.R. and Van Lopik, J.R. (1966) Depositional environments of Mississippi River deltaic, southeastern Louisiana, in M.L. Shirley (ed.) *Deltas in their Geological Framework*, Houston Geological Society, 17–61.

Kuehl, S.A., DeMaster, D.J. and Nittrouer, C.A. (1986) Nature of sediment accumulation on the Amazon continental shelf, *Continental Shelf Research* 6, 209–226.

Milliman, J.D. and Meade, R.H. (1983) World-wide delivery of river sediment to the oceans, *Journal of Geology* 91, 1–21.

Milliman, J.D. and Syvitski, J.P.M. (1992) Geomorphic/tectonic control of sediment discharge to the ocean: the importance of small mountainous rivers, *Journal of Geology* 100, 525–544.

Ogston, A.S., Cacchione, D.A., Sternberg, R.W. and Kineke, G.C. (2000) Observations of storm and river flood-driven sediment transport on the northern California continental shelf, *Continental Shelf Research* 20, 2,141–2,162.

Wright. L.D. (1985) River deltas, in R.A. Davis (ed.) *Coastal Sedimentary Environments*, 1–76, New York: Springer-Verlag.
Wright, L.D. and Nittrouer, C.A. (1995) Dispersal of river sediments in coastal seas: six contrasting cases, *Estuaries* 18, 494–508.
Wright, L.D., Wiseman, W.J., Yang, Z.-S., Bornhold, B.D., Keller, G.H., Prior, D.B. and Suhayda, J.M. (1990) Processes of marine dispersal and deposition of suspended silts off the modern mouth of the Huanghe (Yellow River), *Continental Shelf Research* 10, 1–40.

L.D. WRIGHT

RIVER PLUME

A plume is a vertically or horizontally moving, rising or expanding fluid body, such as the contrails of a fighter jet, emissions from a stack, or river discharge into a lake. Under the strong influence of gravity, rivers can enter a lake or ocean as a fully turbulent jet (Baines and Chu 1996), such as from discharge across steep riverbeds (>0.5°), or under the influence of a flood wave. A river will reduce its velocity at its mouth, if it is moving too fast, by undergoing a hydraulic jump and thickening its flow (Bursik 1995). Many rivers discharge more slowly. The river's momentum and the hydraulic head at the river mouth carry the plume up to hundreds of kilometres into the lake or ocean, depending on the size and power of the river (Syvitski *et al.* 1998).

The plume's behaviour is dependent on the density contrast between the river water and the standing water. Compared to most lake water, the contrast in effluent density is small and controlled by the river's suspended load. Ocean water has a higher density, and the plumes often flow buoyantly on the surface (hypopycnal: Plate 99). The pathway that a hypopycnal plume will take, depends on a variety of factors:

1 Angle between the river course at the entry point and the coastline;
2 Strength and direction of the coastal current;
3 Wind direction and its influence on local upwelling or downwelling conditions;
4 Mixing (tidal) energy near the river mouth; and
5 latitude of the river mouth and thus the strength of the Coriolis effect.

Often there are strong interactions between these factors. For example if the angle of entry is in the direction of the Coriolis effect (i.e. move to the right in the northern hemisphere), then the plume will likely form a coast-hugging plume. Otherwise the plume will detach from the coast.

While hypopycnal plumes may form when river water enters a freshwater lake, they are just as likely to form hyperpycnal plumes (Plate 99). Sometimes referred to as underflows or turbidity currents, these dense flows penetrate the lake

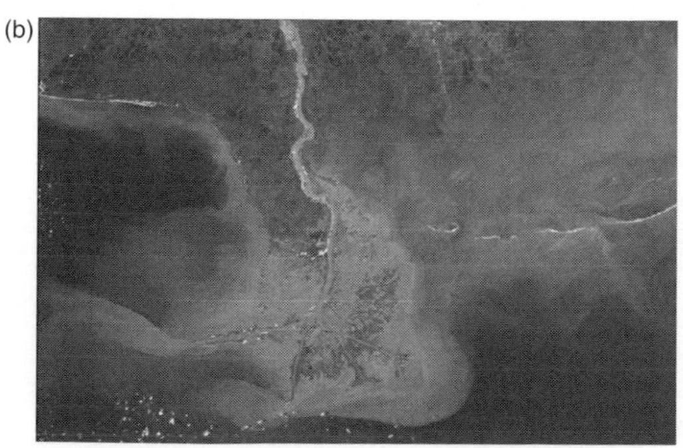

Plate 99 (a) A hyperpycnal plume forming seaward of Skeiđaràrsandur (Iceland) (1996 photograph by M.T. Gumundsson and F. Pálsson). The surface plume disappears at the plunging point and subsequently flows seaward along the seafloor. (b) SeaWiFs image showing hypopycnal plumes emanating from the Mississippi River

under the influence of gravity, remaining in contact with the lake floor (Kassem and Imran 2001). If a hyperpycnal plume accelerates, additional sediment can be resuspended into the flow from the lake floor. As the hyperpycnal plume spreads and thickens due to the entrainment of ambient fluid, velocity is reduced and sediment is deposited. Hyperpycnal plumes rarely occur in rivers that discharge to the ocean (Wright *et al.* 1986). Globally, only a few dozen rivers generate hyperpycnal plumes on an annual basis, and most rivers would see such events happen once every hundred or so years, if at all (Mulder and Syvitski 1995).

Another major difference in comparing sediment plumes flowing into oceans and lakes is in the dynamics of particle settling. In freshwater environments, finer particles settle out of the plume slowly and individually. In ocean environments, river particles quickly flocculate and settle out rather quickly (Syvitski *et al.* 1995). Flocculation is the process that sees particles come into contact with one another and stick, wherein the new larger particles (flocs) have greatly enhanced settling velocities.

River plumes may also enter a lake or ocean at depth: tidewater glaciers directly discharge their stream water at or near the base of the ice front.

References

Baines, W.D. and Chu, V.H. (1996) Jets and Plumes, in V.P. Singh and W.H. Hager (eds) *Environmental Hydraulics*, 7–61, Netherlands: Kluwer Academic.

Bursik, M. (1995) Theory of the sedimentation of suspended particles from fluvial plumes, *Sedimentology* 42, 831–838.

Kassem, A. and Imran, J. (2001) Simulation of turbid underflow generated by the plunging of a river, *Geology* 29, 655–658.

Mulder, T. and Syvitski, J.P.M. (1995) Turbidity currents generated at river mouths during exceptional discharges to the world oceans, *Journal of Geology* 103, 285–299.

Syvitski, J.P.M., Asprey, K.W. and LeBlanc, K.W.G. (1995) In-situ characteristics of particles settling within a deep-water estuary, *Deep-Sea Research II* 42, 223–256.

Syvitski, J.P.M., Nicholson, M., Skene, K. and Morehead, M.D. (1998) PLUME1.1: Deposition of sediment from a fluvial plume, *Computers and Geosciences* 24, 159–171.

Wright, L.D., Yang, Z.-S., Bornhold, B.D., Keller, G.H., Prior, D.B. and Wiseman, W.J. (1986) Hyperpycnal plumes and plume fronts over the Huanghe (Yellow River) delta front, *Geo-Marine Letters* 6, 97–105.

JAMES SYVITSKI

RIVER RESTORATION

River restoration is a term used to describe a wide range of approaches aimed at improving the environmental quality of engineered river systems (see CHANNELIZATION; DAM; STREAM RESTORATION). The objective may be to recreate the river's natural forms and processes (sometimes referred to as 'naturalization'), although the assumption that nature can be created has been criticized. Restoration has also been described as 'nudging nature'. This reflects the fact that, to date, much restoration work has been focused on lower energy streams which have less ability to recover naturally following river channelization and therefore require active intervention (see Brookes 1987, for a discussion of the recovery of channel sinuosity on straightened rivers in Denmark). Furthermore, improvements in urban rivers, sometimes undertaken for aesthetic reasons, have also been referred to as 'restoration'. And some restoration schemes may also involve the development of a resource, such as a riverside wetland, that did not previously exist at the site. 'Creation' may therefore be a more appropriate term in these situations. Consequently, there is no simple definition of river restoration. However, Brookes and Shields (1996: 4) make the following useful distinction between enhancement, rehabilitation and restoration.

Enhancement they define to be 'any improvement in environmental quality'. This would include, for example, the increased diversity of marginal river vegetation achieved following works to raise flood banks on the River Torne (UK). The enhancement works comprised bank re-profiling to create narrow wetland shelves (berms), shallow bays, channel margins of varying shape and depth and linear still ponds from borrow pits on the floodplain (Clarke and Wharton 2000). The opportunity for river enhancement, which was achieved at negligible cost, arose because contractors changed the usual practice of importing materials and obtained the spoil for the flood banks from the channel margins and the floodplain. A large number of case studies describing river enhancement techniques are also illustrated in *The New Rivers and Wildlife Handbook* (RSPB *et al.* 1994) and there is much scope for these small-scale river improvements.

Rehabilitation, as defined by Brookes and Shields (1996: 4), is 'the partial return to a pre-disturbance structure or function' (see Brookes and Shields 1996 for examples from northern Europe and the USA). An example from the UK of a small-scale rehabilitation project is the Redhill Brook. This is typical of many lowland streams in England which have undergone rehabilitation since the mid-1980s. A 100 m reach had been artificially straightened and a further realignment was planned in 1991 as part of a floodplain development project. However, in issuing a land drainage consent for this work the National Rivers Authority required the realigned section to be designed to reflect the characteristics of a natural lowland stream. This included creating a channel with varying channel cross sections incorporating pools, riffles and point bars. Sediment was also reinstated which would not erode in a bankfull flood event (Brookes and Shields 1996: 246–247). Much more extensive rehabilitation has been undertaken on the rivers Brede, Cole and Skerne comprising a joint Danish and British EU–LIFE demonstration project (see Holmes and Nielsen 1998 for information on the background to the project; and Kronvang *et al.* 1998 for details on the restoration of the channel morphology and hydrology).

Finally, 'restoration' in its strictest sense is the term employed by Brookes and Shields (1996: 4) to describe 'the complete return to a pre-disturbance structure or function'. There are, however, several constraints to full river restoration. First, there is likely to be disagreement about the most appropriate pre-disturbance state. Practically, there is a need to establish whether the baseline for restoration should be set immediately before the most recent channelization works, before the first evidence for channel modification or at some point in between. Second, there is the related problem of establishing pre-disturbance data. Rarely are data comprehensive and accurate enough for reconstruction to be fully informed. In this context, Tapsell (1995) has asked 'what are we restoring to?' and Graf (1996, 2001) has discussed the issue of what is natural and how closely restored systems can approximate natural conditions. And third, the desirability of river restoration must be questioned. If sustainable river management is the aim, then the river must be considered within its catchment context, with river forms and processes able to respond to controlling factors such as flow regimes and sediment transport rates, that in turn respond to changing drainage basin conditions. A river restored to some pre-disturbance state is unlikely to be in balance with present conditions. Erskine *et al.* (1999) also document how restoration of the pre-dam situation on the Snowy River (Australia) was neither possible nor desirable because the conditions below the dam had stabilized themselves to a new regime.

Thus, the term rehabilitation is more appropriate in reflecting the reality of river restoration. In the UK, the River Restoration Centre, a non-profit making organization working to restore and enhance rivers, views restoration as a visionary target 'of pristine rivers that are wholly returned to an undisturbed state requiring no management' (Holmes 1998: 139) while accepting rehabilitation as a more practical alternative.

The involvement of all stakeholders, including the participation of the public, is a key element in the success of river restoration schemes by helping to engender a sense of ownership by the local community. In the EU–LIFE demonstration project on the River Skerne (UK) a community liaison officer was a vital member of the team working with the experts and local residents to ensure effective public consultation and dialogue (Holmes and Nielsen 1998). And Waley (2000) describes the socio-cultural value of river restoration in Japan and how science and ethics combine in restoration programmes.

The restoration of rivers may be driven by many factors, including environmental, economic and political. Attempts to restore the geomorphology, hydrology, water quality and ecology of rivers may arise from a desire to redress the environmental impacts of past engineering schemes (see CHANNELIZATION; DAM). Thus most river restoration activity is concentrated in developed countries which have a long history of river engineering. And in the UK, improvements to the physical habitat and ecology have been shown to be the main drivers behind restoration initiatives. The removal of dams which no longer generate hydro-electric power at a competitive rate and the restoration of a more natural flow regime and river environment has been reported in the USA (Graf 1996). This can have benefits for wildlife and generate income from the recreational use of the river (e.g. fishing and canoeing). Changes in environmental legislation also have a significant influence on river restoration. For example, in Denmark the 1982 Watercourse Act provides powers for safeguarding the physical environment of streams by focusing on ecologically acceptable maintenance practices, and incorporates special provisions for stream restoration and the potential for financial support of such activities. Across Europe, the European Union Water Framework Directive (2000/60/EC) is providing further impetus to river restoration by requiring member states to protect, enhance and restore all bodies of surface water not designated as artificial or heavily modified.

Brookes (1988: 217–237) and Wharton (2000) describe procedures for restoring channel capacity, river-bed sediments, cross-sectional form and pattern. Clearly, however, these components should not be viewed separately in the process of restoring a river's geomorphology. Based on research in Sweden, Petersen *et al.* (1992) advocate a 'building block approach' for restoring river environments in a number of stages. By combining different elements such as the construction of pools and riffles, the re-meandering of reaches and the creation of buffer strips and wetlands, the design and implementation of the restoration scheme can be tailored to specific sites. Brookes and Shields (1996) have also published guiding principles on river restoration and the UK River Restoration Centre has produced a second edition of its *Manual of River Restoration Techniques* (RRC 2002). By maintaining a database on completed projects and river restoration practitioners and researchers, the RRC also plays a pivotal role in disseminating information on river restoration and forms part of the European Centre for River Restoration (ECRR).

As more river restoration projects are undertaken it becomes increasingly important to appraise these schemes so that failures as well as successes can be documented and evaluated (Kondolf 1998). Importantly, it should be recognized that river restoration can have impacts on the fluvial system similar to those reported for channelization. Information from immediate post-project appraisal and longer term monitoring will help to develop the science of restoration. Specifically, there is a need to improve predictions of river channel sensitivity to change and to incorporate this understanding in integrated catchment management planning. There is also a need for further evidence on the link between the restoration of geomorphology and physical habitat and subsequent improvements to river ecology. And the appraisal of river restoration schemes in terms of their management and implementation will help inform the development of future policy and practice.

References

Brookes, A. (1987) The distribution and management of channelized streams in Denmark, *Regulated Rivers* 1, 3–16.

——(1988) *Channelized Rivers: Perspectives for Environmental Management*, Chichester: Wiley.

Brookes, A. and Shields, F.D. Jr (eds) (1996) *River Channel Restoration, Guiding Principles for Sustainable Projects*, Chichester: Wiley.

Clarke, S.J. and Wharton, G. (2000) An investigation of marginal habitat and macrophyte community enhancement on the River Torne, UK, *Regulated Rivers: Research and Management* 16, 225–244.

Erskine, W.D., Terrazzolo, N. and Warner, R.F. (1999) River rehabilitation from the hydrogeomorphic impacts of a large hydro-electric power project: Snowy River, Australia, *Regulated Rivers: Research and Management* 15, 3–24.

Graf, W.L. (1996) Geomorphology and policy for restoration of impounded American rivers: what is 'natural'? in B.L. Rhoads and C.E. Thorn (eds) *The Scientific Nature of Geomorphology*, 443–473, Chichester: Wiley.

——(2001) Damage control: restoring the physical integrity of America's rivers, *Annals of the Association of American Geographers* 91, 1–27.

Holmes, N.T.H. (1998) The river restoration project and its demonstration sites, in L.C. De Waal, A.R.G. Large and P.M. Wade (eds) *Rehabilitation of Rivers: Principles and Implementation*, 133–148, Chichester: Wiley.

Holmes, N.T.H. and Nielsen, M.B. (1998) Restoration of the rivers Brede, Cole and Skerne: a joint Danish and British EU–LIFE demonstration project, I – Setting up and delivery of the project, *Aquatic Conservation: Marine and Freshwater Ecosystems* 8, 185–196.

Kondolf, G.M. (1998) Lessons learned from river restoration projects in California, *Aquatic Conservation: Marine and Freshwater Ecosystems* 8, 39–52.

Kronvang, B., Svendsen, L.M., Brookes, A., Fisher, K., Moller, B., Ottosen, O., Newson, M. and Sear, D. (1998) Restoration of the rivers Brede, Cole and Skerne: a joint Danish and British EU–LIFE demonstration project, III – Channel morphology, hydrodynamics and transport of sediment and nutrients, *Aquatic Conservation: Marine and Freshwater Ecosystems* 8, 209–222.

Petersen, R.C., Petersen, L.B.-M. and Lacoursiere, J. (1992) A building-block model for stream restoration, in P.J. Boon, P. Calow and G.E. Petts (eds) *River Conservation and Management*, 293–309, Chichester: Wiley.

RRC (2002) *Manual of River Restoration Techniques*, 2nd edition, Silsoe, Bedfordshire, UK: River Restoration Centre.

RSPB, NRA and RSNC (1994) *The New Rivers and Wildlife Handbook*, The Lodge, Sandy, Bedfordshire, UK: Royal Society for the Protection of Birds.

Tapsell, S.M. (1995) River restoration: what are we restoring to? A case study of the Ravensbourne River, London, *Landscape Research* 20, 98–111.

Waley, P. (2000) Following the flow of Japan's river culture, *Japan Forum* 12, 199–217.
Wharton, G. (2000) New developments in managing river environments, in A. Kent (ed.) *Reflective Practice in Geography Teaching*, 26–36, London: Paul Chapman.

SEE ALSO: anthropogeomorphology

GERALDENE WHARTON

ROCHE MOUTONNÉE

Roches moutonnées are asymmetric bedrock bumps or hills with one side ice-moulded and the other side steepened and often cliffed. They are widespread features in formerly glaciated hard-rock terrain and often to be found in clusters or fields. The name was first introduced by de Saussure (1786), based on a fancied resemblance to the wavy wigs of that period, which were called moutonnées after the mutton fat used to hold them in place. The term encompasses a wide range of feature sizes. Typically, roches moutonnées vary from 1 to 50 m in height and a few metres to hundreds of metres in length, but Sugden *et al.* (1992), for instance, describe large roches moutonnées hills in eastern Scotland with lee side cliffs up to 160 m high.

The morphology of roches moutonnées seems to reflect the contrast between ABRASION on the smoothed up-ice side and plucking on the lee side (see GLACIAL EROSION).

Abrasion acting on the stoss side is marked by STRIATIONS at a variety of scales together with polished facets and crescentic fractures on more gently sloping surfaces. As the glacier moves against the upstream side of an obstruction the ice overburden pressure increases. The basal ice may reach the PRESSURE MELTING POINT and partially melt, causing the glacier to slide. The direction of basal ice flow in this position is pointed towards the bed and the embedded clasts are dragged over the bedrock with some force, effectively abrading it.

On the lee side of the obstruction the ice overburden pressure is lower than average, encouraging the formation of a subglacial water-filled cavity. The presence of a cavity together with fluctuations of the water pressure within it strongly promotes the process of glacial plucking. Sugden *et al.* (1992) showed that block removal starts at the furthest point down ice in the cavity and from there extends successively up ice, thereby transforming the lee side of the bedrock bump into a staircase cliff. The detailed morphology of the plucked surface is also influenced by the properties of the parent bedrock, since plucking is encouraged by a favourable oriented system of joints.

Some roches moutonnées do appear to be only slightly modified preglacial hills, but in many areas initial bedrock eminences were clearly sculptured and reshaped by differential glacial erosion. Between these end-members there is likely to be a continuum of forms with varying degree of inherited topography. If the quarried and rough lee side as a distinctive feature of a roche moutonnée is little developed, it may also be difficult to distinguish roches moutonnées from whalebacks or rock DRUMLINs. Whalebacks are elongated and approximately symmetrical bedrock bumps whereas rock drumlins are asymmetrical with a steep upstream side and a gently inclined downstream side. Both are smoothed and rounded on all sides.

Reference

Sugden, D.E., Glasser, N. and Clapperton, C.M. (1992) Evolution of large roches moutonnées, *Geografiska Annaler* 74A, 253–264.

Further reading

Benn, D.I. and Evans, D.J.A. (1998) *Glaciers and Glaciation*, London: Arnold.

CHRISTINE EMBLETON-HAMANN

ROCK COATING

About 15 per cent of the Earth's landscape consists of bare rock surfaces. Yet the common phrase 'bare rock' is truly a misnomer, because paper-thin accretions coat almost all of these rock surfaces in all terrestrial environments. Studies on the physical and chemical characteristics, origin, geography and utility of these deposits has spawned over 3,000 scientific papers. Plate 100 illustrates a few examples.

Alexander von Humboldt (1812) initiated the scholarly study of rock coatings by studying the composition, origin, spatial distribution and environmental relations of coatings such as those found along tropical rivers. In the past two centuries, researchers have documented hundreds of different types of rock coatings found within the fourteen major categories listed in Table 39.

The three most common rock coatings are rock varnish (see DESERT VARNISH), silica glaze and iron films. Silica glazes occur in warm deserts, cold

Plate 100 Upper left: a vertical face at Canyon De Chelly, USA, is streaked with heavy metal skins, iron films, lithobiotic coatings, oxalate crusts, rock varnish and silica glaze. Upper right: alluvial fan in Death Valley deposits the same light-coloured rock types in active streams. But over time, rocks in abandoned stream courses are darkened by desert varnish. Lower row: the electron microscope images (backscatter detector) illustrate that rock coatings are external accretions, exemplified by an oxalate crust in the lower left image that is about 0.5 mm thick and desert varnish on the lower right that is about 0.1 mm thick

deserts like Antarctica, on dry tropical islands, along tropical rivers, mid-latitude humid temperate settings, and various archaeological settings. Silica glazes probably precipitate from soluble Al-Si complexes $[Al(OSi(OH)_3)^{2+}]$ that are released from the weathering of clay minerals. Rust-coloured iron films display a wide variety of characteristics in very different climates and microenvironments. For example, rocks in the Dry Valleys of Antarctica host iron hydroxides that both form a micron-scale accretion and a weathering rind (see RIND, WEATHERING) over a millimetre thick. In a very different setting, iron oxyhydroxides impregnate rocks in arctic streams (Dixon *et al.* 2002).

Geomorphologists have long used intuition to interpret rock coatings and their relationship to the geomorphic setting. For example, some have believed that PEDIMENTs are fossil landforms, in part because the presence of rock coatings must

Table 39 Major categories of rock coatings

General type	Description	Related terms
Carbonate skin	Coating composed primarily of carbonate, usually calcium carbonate, but could be combined with magnesium or other cations	Caliche, calcrete, patina, travertine, carbonate skin, dolocrete, dolomite
Case hardening agents	Addition of cementing agent to rock matrix material; the agent may be manganese, sulphate, carbonate, silica, iron, oxalate, organisms or anthropogenic	Sometimes called a particular type of rock coating
Dust film	Light powder of clay- and silt-sized particles attached to rough surfaces and in rock fractures	Gesetz der Wüstenbildung; clay skins; clay films; soiling
Heavy metal skins	Coatings of iron, manganese, copper, zinc, nickel, mercury, lead and other heavy metals on rocks in natural and human-altered settings	Described by chemical composition of the film
Iron film	Composed primarily of iron oxides or oxyhydroxides; unlike orange rock varnish because it does not have clay as a major constituent	Ground patina, ferric oxide coating, red staining, ferric hydroxides, iron staining, iron-rich rock varnish, red-brown coating
Lithobiontic coatings	Organic remains form the rock coating, e.g. lichens, moss, fungi, cyanobacteria, algae	Organic mat, biofilms,
Nitrate crust	Potassium and calcium nitrate coatings on rocks, often in caves and rock shelters in limestone areas	Saltpetre; nitre; icing
Oxalate crust	Mostly calcium oxalate and silica with variable concentrations of magnesium, aluminium, potassium, phosphorus, sulphur, barium and manganese. Often found forming near or with lichens. Usually dark in colour, but can be as light as ivory	Oxalate patina, lichen-produced crusts, patina, scialbatura
Phosphate skin	Various phosphate minerals (e.g. iron phosphates or apatite) that are mixed with clays and sometimes manganese	Organophosphate film; epilithic biofilm
Pigment	Human-manufactured material placed on rock surfaces by people	Pictograph, paint, sometimes described by the nature of the material
Rock varnish	Clay minerals, Mn and Fe oxides, and minor and trace elements; colour ranges from orange to black produced by variable concentrations of different manganese and iron oxides	Desert varnish, desert lacquer, patina, manteau protecteaur, Wüstenlack, Schutzrinden, cataract films
Salt crust	The precipitation of sodium chloride on rock surfaces	Halite crust, efflorescence, salcrete

Table 39 Continued

General type	Description	Related terms
Silica glaze	Usually clear white to orange shiny lustre, but can be darker in appearance, composed primarily of amorphous silica and aluminium, but often with iron	Desert glaze, turtle-skin patina, siliceous crusts, silica-alumina coating, silica skins
Sulphate crust	Composed of the superposition of sulphates (e.g. barite, gypsum) on rocks; not gypsum crusts that are sedimentary deposits	Gypsum crusts; sulphate skin

infer long-term stability. Others have guessed at the ages of such features as flooding events on ALLUVIAL FANs, based on an intuitive feeling about the appearance of rock coatings (see gradual darkening of alluvial fan surfaces in the Plate 100). The complexities associated with formative processes have made rock coatings extraordinarily difficult to use as geomorphological tools to indicate either age or infer palaeoclimate. Rock coatings will be getting increased attention in future years as they are identified on Mars and as planetary scientists attempt to use rock coatings to infer Martian geomorphic processes (Kraft and Greeley 2000).

Rock coatings have applied significance in a variety of contexts. Heavy metal skins assist in identifying metal pollution (Dong *et al.* 2002). Some believe that artificial rock coatings have potential to aid in the conservation of priceless stone monuments (Borgia *et al.* 2001). Construction and development in desert regions contrasts bright uncoated rocks and darker natural rock coatings; the desire to live in natural-appearing settings leads to the application of artificial rock coatings to mimic natural colouration (Henniger 1995). Rock coatings, called patina in archaeology, are also used in the study of surface artefacts, rock paintings and rock engravings.

References

Borgia, G.C., Bortolotti, V., Casmaiti, M., Cerri, F., Fantazzini, P. and Piacenti, F. (2001) Performance evolution of hydrophobic treatments for stone conservation investigated by MRI, *Magnetic Resonance Imaging* 19, 513–516.

Dixon, J.C., Thorn, C.E., Darmody, R.G. and Campbell, S.W. (2002) Weathering rinds and rock coatings from an Arctic alpine environment, northern Scandinavia, *Geological Society of America Bulletin* 114, 226–238.

Dong, D., Hua, X. and Zhonghua, L. (2002) Lead adsorption to metal oxides and organic material of freshwater surface coatings determined using a novel selective extraction method, *Environmental Pollution* 119, 317–321.

Henniger, J. (1995) Fooling mother nature with Permeon artificial desert varnish, *Rocky Mountain Construction* 76(8), 48–52.

Kraft, M.D. and Greeley, R. (2000) Rock coatings and aeolian abrasion on Mars: application to the Pathfinder landing site, *Journal of Geophysical Research – Planets* 105, 15,107–15,116.

von Humboldt, A. (1812) *Personal Narrative of Travels to the Equinoctial Regions of America During the Years 1799–1804 V. II*, translated and ed. T. Ross in 1907, London: George Bell & Sons.

Further reading

Dorn, R.I. (1998) *Rock Coatings*, Amsterdam: Elsevier.

SEE ALSO: alluvial fan; desert pavement; desert varnish; pediment

RONALD I. DORN

ROCK CONTROL

Rock control in geomorphology is defined as the influences of differences in earth materials on the development of landforms. Earth materials that form the Earth's surface or landforms are simply called landform materials, and include rocks, weathered materials and soil. The concept of *rock control* was first proposed explicitly and argued passionately by Yatsu (1966) and then expanded by him to a concept of *landform material science* in 1971. Yatsu stressed that to understand the formation of landforms, it is necessary to quantitatively evaluate the behaviours of landform materials in terms of their physical, mechanical, chemical and mineralogical properties in relation to the geomorphological processes concerned. His severe criticism of geomorphology is based on the fact that geologic structure and lithology have only been used qualitatively to explain the development of erosional landforms since the birth of modern geomorphology.

Typical examples often described in textbooks on geomorphology (e.g. Thornbury 1954; Sparks 1971) as structural landforms or landforms resulting from rock control include cuestas, hogbacks, mesas, structural benches, dyke ridges, knickpoints, karst and inversion topography. These landforms are relatively higher or steeper than their surroundings, and are generally composed of a relatively *resistant* or *hard* rock (e.g. sandstone, limestone, lava) that adjoins the relatively *less resistant* or *weak* rocks (e.g. mudstone, shale, tuff). However, rock control is not as simple as the vague terms *resistant* and *less resistant imply*. This is because the resistance and behaviour of landform materials vary markedly with geomorphological process and geomorphic setting.

For instance, the rocky coast of Arasaki, southwest of Tokyo, Japan, is underlain by steeply dipping, alternating beds of mudstone and scoria tuff. Differential erosion between the two rocks varies with altitude (Figure 134). On the vegetation-free sea cliffs behind the uplifted wave-cut benches, mudstone forms ridges and tuff forms shallow furrows. On the benches, mudstone forms the furrows and tuff forms the ridges, producing a washboard-like relief. On the shallow offshore sea bottom, there is no differential erosion. The mudstone is mechanically about two times as strong as the tuff. However, it is well jointed and forms fragments about 1 cm in size due to wet–dry slaking, whereas the tuff does not. The explanation for this differential erosion is that (1) both rocks are eroded at rates proportional to their mechanical strengths on the sea cliff above tidal zone, (2) the fragments of mudstone produced by wet–dry slaking are rapidly washed away by waves in the tidal zone, and (3) wave abrasion offshore erodes both rocks at the same rate (see Suzuki 2002).

Thus, the behaviour of landform materials generally does not merely depend on their geological structure and lithology, but also strongly depends on their physical and mechanical properties. This is because even lithologically similar rocks have wide ranges of physical and mechanical properties,

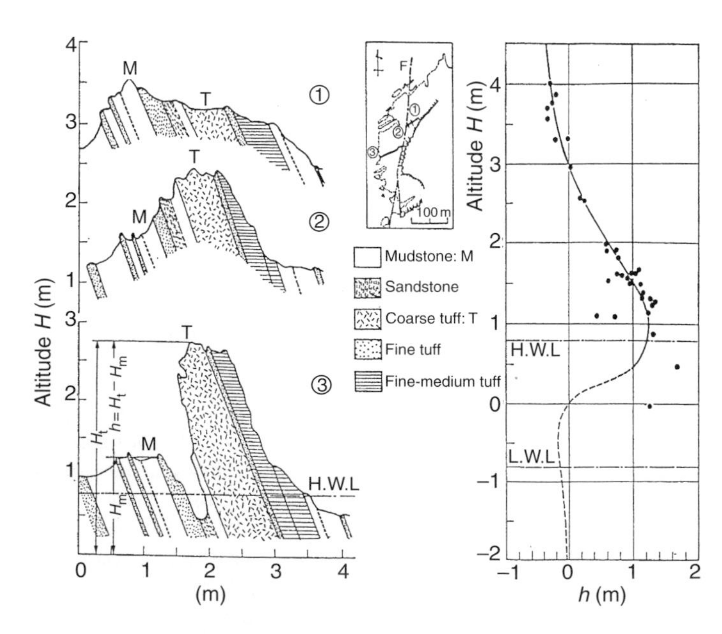

Figure 134 Change in relative relief between a tuff bed (T) and a mudstone bed (M) with altitude (H) on the Arasaki coast, Japan. *Left*: three geologic sections that are different in the altitude (H_m) of the mudstone surface (M). *Right*: relationship between relative relief (h) and H_m

reflecting their origin and history, such as diagenesis, tectonic deformation, unloading and weathering. Further, weathering results in changes in rock properties and hence is of importance as a preparatory process for erosion and mass movements.

Properties of landform materials are grouped into two major categories: geological properties and rock (material) properties. Geological properties are those described in terms of geology, and include lithology (such as grain size, mineral composition and texture), chemical composition, geological structure (such as stratification, joints, faults, unconformities, and their dip and strike), occurrence of rock masses (such as lava flow, dyke, batholith, etc.), weathering grades, and so on. Rock (material) properties, on the other hand, are those described in terms of physics and engineering (particularly rock and soil mechanics), and which can be further subdivided into physical properties and mechanical properties. Physical properties are the intrinsic characteristics (such as density, porosity and pore-size distribution) that do not depend on applied forces. Mechanical properties are the behaviours and responses of rocks against forces acting upon them, and hence vary with the kind of forces, the conditions of the rocks, such as water content, and the test methods used. Mechanical properties include strength (compressive, tensile, shear and bending strengths), hardness (e.g. abrasion, impact and scratch), deformation properties (e.g. deformation modulus, adhesive forces and internal friction), dynamic properties (elastic wave velocities), thermal properties (e.g. thermal conductivity and thermal expansion coefficient), permeability (e.g. permeability coefficient and infiltration capacity), behaviour in relation to water (such as swelling, slaking and solution) and so on.

These physical and mechanical properties are determined by the measurements of the rock mass in the field and for test pieces in the laboratory using precise instruments and equipment. Some practical test methods have also been applied to evaluate the mechanical properties of rocks, including standard penetration tests (N-value), rebound hardness with a Schmidt rock hammer, cone penetration hardness, needle penetration hardness for the rock mass, and rock quality designation for drilling cores.

Rock control problems are addressed by looking for the important rock properties in each process and quantitatively evaluating the roles of the properties with respect to the process. In the context of Yatsu's argument, therefore, physical, mechanical and chemical properties of landform materials have been intensively measured in both field and laboratory and in field and laboratory experiments since the 1970s, particularly by Japanese geomorphologists. Based on the measurements and experiments, much persuasive substantiation has been found for the modes and rates of various erosional processes and landforms (Suzuki 2002). Notable examples include formative processes along rocky coasts (Sunamura 1992), wind abrasion, lateral planation, slope evolution, hillslope morphology, valley development and some minor landforms such as tafoni. Processes and rates of bedrock weathering have also been studied actively in both field and laboratory.

Landform evolution is controlled not only by rock properties and geological properties but also by many other variables, such as geomorphological setting (initial landform), climate, geomorphic agents from which the various forces derive, and elapsed time. The research on rock control problems mentioned above, therefore, has been directed toward developing quantitative models of geomorphological processes that are capable of predicting types and rates of landform development. The models have been expressed as geomorphological equations including the geomorphic quantities concerned and the main controlling variables, i.e. site- and time-independent process equations with dimensionless constants. To develop the models, it is indispensable to study the rock control problems all over the world, because landforms are never changed unless the landform materials are moved. Thus, research on the rock control problems will be one of the core fields in process geomorphology in the twenty-first century.

References

Sparks, B.W. (1971) *Rocks and Relief*, London: Longman.

Sunamura, T. (1992) *Geomorphology of Rocky Coasts*, Chichester: Wiley.

Suzuki, T. (2002) Rock control in geomorphological processes: research history in Japan and perspective, *Transactions, Japanese Geomorphological Union* 23, 161–199.

Thornbury, W.D. (1954) *Principles of Geomorphology*, New York: Wiley.

Yatsu, E. (1966) *Rock Control in Geomorphology*, Tokyo: Sozosha.

——(1971) Landform material science – rock control in geomorphology, in E. Yatsu, F.A. Dahms, A. Falconer, A.J. Ward and J.S. Wolfe (eds) *Research Method in*

Geomorphology (Proceedings of the 1st Guelph Symposium on Geomorphology, 1969), 49–56, Ontario: Science Research Associates.

Further reading

Selby, M.J. (1993) *Hillslope Materials and Processes*, Oxford: Oxford University Press.
Yatsu, E. (1988) *The Nature of Weathering: An Introduction*, Tokyo: Sozosha.

SEE ALSO: rock mass strength; weathering

TAKASUKE SUZUKI

ROCK AND EARTH PINNACLE AND PILLAR

Within areas built of poorly consolidated sediments, subject to intensive linear erosion, sheet wash and susceptible to weathering, bedrock may be sculpted into groups of weirdly shaped erosional residuals in the form of pinnacles, pillars and cones. They are relatively common in semi-arid areas, where scarce vegetation provides little protection against surface erosion, hence pinnacles are typical components of BADLANDs. Steep slopes underlain by erodible sediments, for example of newly deposited MORAINEs, may also support pinnacle assemblages.

Two types of sediments yield to this type of erosional relief in particular, tills and pyroclastic deposits. Some tills and other glacial deposits contain boulders 'floating' in an otherwise fine-grained material. After exposure, boulders will protect an underlying softer mass against erosion, whereas the surrounding unprotected material will be eroded away, leaving the boulder-capped part standing as a residual pillar. Later, the boulder cap will provide a shield against the direct destructive impact of rain and the pillar may increase in height as long as the cap remains in place. Once the boulder falls from the top of the pinnacle, the column built of soft rock will rapidly be destroyed. Classic localities of this type of earth pinnacles have been described from the Tyrol in the Alps.

Plate 101 A group of rock pinnacles in Cappadocia, central Turkey. Remnants of resistant welded tuff act as a cap to the underlying softer sediment

In pyroclastic deposits, volcanic bombs within softer tuff play the similar protective role as boulders in tills do. In Cappadocia, central Turkey, bomb-capped pyramids reach up to 20 m high. In other cases, caps are provided by remnants of a welded tuff horizon overlying thicker and softer strata beneath.

Not all rock and earth pinnacles have a protective cap, and there are other reasons why they remain as isolated residuals. In the semi-arid badlands of Cappadocia, surfaces of tuff cones isolated by fluvial and sheet wash erosion are subject to case hardening, and it is the crust which protects the cones from further destruction. Owing to the presence of the crust, the earth pinnacles of Cappadocia could have grown up to 15–20 m high and were found stable enough for churches, hermitages and cave dwellings to be dug into them in early Christian times.

Tufa (see TUFA AND TRAVERTINE) deposits may form curiously shaped pinnacles too, but in these cases pinnacles are constructional, and not erosional features. At Mono Lake, California, and Searles Lake, California, tufa pyramids and pillars up to 15 m high formed around underwater springs and were exposed at the surface, when lake levels were lowered.

SEE ALSO: hoodoo

PIOTR MIGOŃ

ROCK GLACIER

Rock glaciers (German: *Blockgletscher, Blockstrom*; French: *Glacier rocheux*; Russian: *Kammeni gletscher*; Spanish: *glaciar rocoso*) occur in most alpine-mountainous regions and are distinct tongues of rock rubble which flow slowly downhill. Most features are elongate and are generally distinct from blockstreams (see BLOCKFIELD AND BLOCKSTREAM) which occur on very low angle slopes.

The substantial literature on these features is complex and often confusing, being hindered by difficulties of *in situ* investigation. Markedly different viewpoints have been taken to explain their origin, dynamic behaviour and environmental significance (Potter *et al.* 1998). It is important to avoid explicit designation of an origin so they are best defined by their morphology and appearance; a simple morphological definition, after Washburn (1979), is: 'a tongue-like body of angular boulders that resembles a small glacier, which generally occurs in high mountains and usually has ridges, furrows and sometimes lobes on its surface with a steep front or snout at the angle of repose'.

Plate 102 An active rock glacier (maximum velocity about $0.25\,\mathrm{m\,a^{-1}}$) in northern Iceland with a small corrie glacier at its head. There is a gradation from a thin cover of debris to much thicker debris (*c.*1 m) near the snout. The lateral margins are distinct from the sides of the valley and the longitudinal furrow is conspicuous. The whole rock glacier is about 800 m long and estimated to have formed in the past 200–300 years

Plate 103 Merging rock glaciers, Wrangell Mountains, Alaska. These are typical rock glacier forms, emerging from corries now containing little or no ice. That there is probably very little forward movement of the wrinkled and furrowed surfaces is indicated by the vegetated surface

Distribution maps and reviews can be found in Whalley and Martin (1992) and Barsch (1996). They may even occur on Mars (Whalley and Azizi 2003). Although originally thought to be confined to continental areas and to give way to glacier bodies in more maritime regions examples in the latter have been found. They were first recognized in North America and Greenland (Martin and Whalley 1987; Barsch 1996).

The surface velocity is generally $<1\,\mathrm{m\,a^{-1}}$, although some with velocities $>5\,\mathrm{m^{-1}}$ have been described (Gorbunov *et al.* 1992). If no movement can be detected they are generally referred to as 'relict' and, if highly vegetated with subdued features, as 'fossil' and are recognized by morphology alone. However, 'inactive' rock glaciers are sometimes recognized where creep rates are very low and may even have trees growing on them. The steep fronts (snouts) may advance over other features; valley floors, moraines and lakes. These characteristics, variable over time, make it difficult to show that they result from one origin or relate to a single set of environmental conditions.

Rock glaciers are generally about 1 km long but many shorter examples exist and some may be up to 3 km long; width is generally a few hundred metres. Typically, they have their heads in corries (see CIRQUE, GLACIAL) in which there may be a small glacier, although this may not always be visible. The elongate forms are regarded as rock glaciers proper, but have also been called 'tongue-shaped', 'valley floor' or 'debris rock glaciers'. Some forms may be 'spatulate' where they spread over a main valley floor. Rock glaciers are usually separated from valley sides by 'lateral furrows'. The term 'rock glacier' has also been used for features which are broader than long and which typically have their upper sections against cliffs or scree rather than emanating from corries. This form has been called a 'valley side rock glacier', 'lobate rock glacier' or 'talus rock glacier'. It has been suggested that the latter features are best termed 'protalus lobes' rather than rock glaciers because of the differences in form and location (Hamilton and Whalley 1995).

Flow-related features are commonly seen as ridges and furrows on the surface although it is not known if these extend to any depth. Some rock glaciers have mainly transverse ridges, especially near the snout, others have a predominantly longitudinal ridge pattern. Such flow features have been related to flow regime; compressing or extending (Whalley and Azizi 2003). Many rock glaciers have distinct longitudinal furrows and some show pools or small lakes developed in flatter areas ('thermokarst ponds').

There are four main theories of rock glacier origin. One suggests that they are glacially derived with a veneer of weathered rock debris (a few cm to >1 m thick) over a thin (<50 m) glacier ice core. The debris protects a thin body of ice which flows only slowly. This may be termed the 'glacial model'. It has been suggested that rock glaciers are nothing more than debris-covered glaciers. However, the subdued dynamic behaviour of rock glaciers indicates that ice volumes are small and that SUPRAGLACIAL debris has been supplied via the surface of the small glacier. Debris-covered glaciers gain debris in their lower reaches by ablation of ice releasing englacially transported debris but there is probably a transition between the two. What gives rise to rock glaciers is the relative abundance of debris to active glacier ice.

The 'permafrost model' explains the slow movement as creep of ice dispersed within weathered rock debris (derived from SCREE) and that a glacial derivation is not necessary to explain flow. It does require PERMAFROST conditions (mean annual air temperature <−1.5 °C) for the formation and continued creep of the ice. The ice needs to be above 'saturation', i.e. more than fills void spaces, or as ice lenses, for creep to be efficient. Ice-cemented debris (at or below saturation) colder than the PRESSURE MELTING POINT of ice is mechanically stronger than ice and will not flow unless at high shear stresses (high surface slope and/or thickness).

The third model suggests that some rock glaciers (or protalus lobes) are formed by sudden, perhaps catastrophic, failure of scree slopes (Johnson 1974) or as a single catastrophic rock avalanche (Bergsturz or STURZSTROM). This view is not widely held although there is evidence that *some* rock glacier forms might have been constructed in this way to produce topography similar to a rock glacier. Where the features are old there might be confusion with a fossil rock glacier.

A fourth model, a variant of the first (glacier-derived) and third (catastrophic), is that a rockfall covered a small or decaying glacier. The thin debris cover would thus insulate the thin glacier core but suddenly rather than progressively.

Plate 104 Complex protalus lobes, Alpes Maritimes, southern France. The inner ridges look a little like protalus ramparts and the feature lies below a cliffed area which probably had a thin glacier or large snowpatch at its foot. These features differ in form to rock glaciers per se

Rock glaciers have been used as indicators of permafrost (past permafrost for relict features) but only if the permafrost model is valid. This may be considered as being a 'zonal' model. The glacial model is thus 'azonal' as the contributing glacier may occur whether or not permafrost is present (Washburn 1979).

The origin and flow mechanism of rock glaciers is frequently attributed exclusively to creeping permafrost (Barsch 1996). Although traces of glacier ice have been seen in some rock glaciers, permafrost conditions were considered to be the only way in which the features could exist. Observations of glacier ice down the length of some rock glaciers have now been established and thus show that at least some rock glaciers have glacier ice cores. It is possible that modern dating and isotopic techniques will allow ice from such rock glaciers to provide a climatic record. The full implications for recognizing climate change through rock glaciers still needs investigation.

Geophysical measurements (seismic, gravity, resistivity and ground penetrating radar) have been used to investigate the structure of rock glaciers. Resistivity has been used to differentiate between the mode of ice formation. It is claimed that high resistivity (>10 MΩm) is indicative of glacier ice but that rather lower values are typical of ice of permafrost origin. This has been disputed by some authors who claim that the high resistivity is not typical of the small glaciers which provide glacier cores because such ice is contaminated by dust which lowers the value. The difficulty is of linking geophysical signals with an appropriate mixture (ice and debris) model (Whalley and Azizi 1994). The complexity is enhanced because there may be grades of mixture, from permafrost to glacier ice core, in one feature and is particularly significant near rock glacier snouts. This ambiguity of origin also suggests that using rock glaciers to identify past conditions may be difficult.

'Protalus lobes' are related to rock glaciers where permafrost conditions may be required to preserve ice but where a glacier is unlikely to have formed. These are distinctive enough to be given a separate name. PROTALUS RAMPARTs are long, rather narrow, ridges below cliffs and are thought to have a snow-bank (nival) origin. Suggestions have been made that they might be precursors to rock glaciers of permafrost origin (Barsch 1996).

References

Barsch, D. (1996) *Rock Glaciers*, Berlin: Springer.

Gorbunov, A.P., Titkov, S.N. and Polyakov, V.G. (1992) Dynamics of rock glaciers of the northern Tien Shan and the Djungar Ala Tau, Kazahkstan, *Permafrost and Periglacial Processes* 3, 29–39.

Hamilton, S.J. and Whalley, W.B. (1995) Rock glacier nomenclature: a re-assessment, *Geomorphology* 14, 73–80.

Johnson, P.G. (1974) Mass movement of ablation complexes and their relationship to rock glaciers, *Geografiska Annaler* 56A, 93–101.

Martin, H.E. and Whalley, W.B. (1987) Rock glaciers: Part 1: rock glacier morphology: classification and distribution, *Progress in Physical Geography* 11, 260–282.

Potter, N. Jr, Steig, E.J., Clark, D.H., Speece, M.A., Clark, G.M. and Updike, A.B. (1998) Galena Creek rock glacier revisited – new observations on an old controversy, *Geografiska Annaler* 80A, 251–265.

Washburn, A.L. (1979) *Geocryology: A Survey of Periglacial Processes and Environments*, London: Arnold.

Whalley, W.B. and Azizi, F. (1994) Models of flow of rock glaciers: analysis, critique and a possible test, *Permafrost and Periglacial Processes* 5, 37–51.

Whalley, W.B. and Azizi, F. (2003) Rock glaciers and protalus landforms: analogous forms and ice sources on Earth and Mars, *Journal of Geophysical Research, Planets* 108(E4), art.no. 8,032.

Whalley, W.B. and Martin, H.E. (1992) Rock glaciers: Part II: models and mechanisms, *Progress in Physical Geography* 16, 127–186.

BRIAN WHALLEY

ROCK MASS STRENGTH

Rock mass strength (RMS) refers to the specific properties of the rock mass that control its strength and subsequent rock slope stability. Importantly, it allows the prediction of the stable inclination of natural rock slopes, as well as the recognition of strength EQUILIBRIUM SLOPEs in the landscape adjusted to the prevailing SUBAERIAL processes. The standard method of RMS determination applied in geomorphology was developed by Selby (1980, 1982) and has been extensively tested over the past twenty years as a reliable method for the assessment of rock slope stability. This classification scheme is a modification of RMS classifications developed for engineering purposes (e.g. Bieniawski 1979) which have been extensively used to aid excavation design for tunnels, slopes and foundations. However, these classifications often do not incorporate a quantitative assessment of the influence of a reduction in rock mass strength due to WEATHERING. Furthermore, the engineering classifications contain different definitions of the rating classes so that the results derived from the various methods are not directly comparable. The method of Selby (1980), and modified by Moon (1984), explicitly and quantitatively incorporates and weights the influence of weathering on the strength of the rock mass in the field through evaluation of intact rock strength, estimation of state of rock weathering, joint spacing, continuity and infilling. Since weathered rock is the norm, the scheme developed by Selby (1980) is a more appropriate measure of the RMS of a natural rock slope than those developed for engineering purposes.

Although geomorphologists have long recognized that rock slope failure often occurs along discontinuities such as joints, bedding planes and faults (see FAULT AND FAULT SCARP) rather than through intact rock, logistical difficulties and frequent inability to access the appropriate equipment for the laboratory and field assessment of rock strength has often meant that studies of the strength of the rock mass tended to be qualitative in nature. It is in this context that the development of the rock mass strength classification has had important consequences for our understanding of the morphology and evolution of rock slopes as it provides a basis for understanding the features of the rock mass that provide resistance to weathering and EROSION, as well as the maintenance of slope stability. The only equipment required is a SCHMIDT HAMMER, tape measure and inclinometer, and it can be applied to any rock mass where there is enough exposure to allow measurement of the rock JOINTING (usually at least 10 m^2). If the slope contains more than one morphological element, it must be subdivided into zones with similar RMS properties, with the RMS assessment undertaken within each slope element.

The rock mass strength classification system developed by Selby (1980, 1982) was based on an examination of rock slopes in Antarctica and New Zealand, and has subsequently been applied in a range of settings (e.g. Augustinus 1992; Moon and Selby 1983; Allison and Goudie 1990). Slopes adjusted to their RMS (strength equilibrium slopes) are common in nature, and the recognition of over steepened slopes that have been undercut by erosion, as well as structurally controlled rock benches of lower slope angle and RICHTER DENUDATION SLOPES, indicates its utility in resistance-form studies. The RMS classification involves measurement of a range of properties of the rock mass: (1) Schmidt hammer impact as a measure of the strength of the intact rock; (2) state of rock weathering; (3) jointing characteristics of the rock mass: spacing of rock joints, joint width, joint continuity, joint infilling and orientation of the dominant joint set; and (4) water seepage from the rock face (Table 40). Since not all these parameters are of equal importance in controlling rock mass strength, each of these factors is weighted and given a rating value according to their perceived influence on stability of the rock slope using the scheme given in Selby (1980) or as modified by Moon (1984). However, the usefulness of the further subdivision of the classification proposed by Moon has been questioned, so that the simpler scheme of Selby (1980) is preferred. The sums of the individual weightings for the rock mass being evaluated is its RMS rating. A maximum value of 100 applies and the range is divided into five classes (Table 40). The higher the

Table 40 Geomorphic rock mass strength classification and ratings

Criteria	*(1)* Very strong	*(2)* Strong	*(3)* Moderate	*(4)* Weak	*(5)* Very weak
Intact rock# strength	100–60	60–50	50–40	40–35	35–10
Rating	*20*	*18*	*14*	*10*	*5*
Weathering	Unweathered	Slightly weathered	Moderately weathered	Highly weathered	Completely weathered
Rating	*10*	*9*	*7*	*5*	*3*
Joint spacing	>3 m	3–1 m	1–0.3 m	0.3–0.05 m	<0.05 m
Rating	*30*	*28*	*21*	*15*	*8*
Joint orientation	>30° into slope	<30° into slope	Horizontal and vertical	<30° out of slope	<30° out of slope
Rating	*20*	*18*	*14*	*9*	*5*
Joint width	<0.1 mm	0.1–1 mm	1–5 mm	5–20 mm	>20 mm
Rating	*7*	*6*	*5*	*4*	*2*
Joint continuity	None continuous	Few continuous	Continuous, no infill	Continuous, thin infill	Continuous, thick infill
Rating	*7*	*6*	*5*	*4*	*1*
Groundwater outflow	None	Trace	Slight	Moderate	Great
Rating	*6*	*5*	*4*	*3*	*1*
Total rating	100–91	90–71	70–51	50–26	<26

Source: Modified from Selby (1980), and Moon (1984)
Note: # N-type Schmidt hammer rebound values

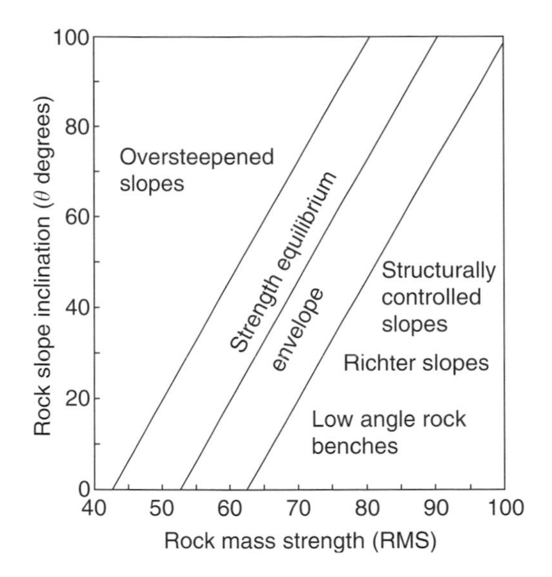

Figure 135 Plot of slope gradient and rock mass strength, with strength equilibrium envelope (after Moon 1984; Abrahams and Parsons 1987)

RMS rating, the higher mass strength and the steeper the slope inclination that can be sustained.

The graphical representation of the RMS classification involves plotting the total RMS rating against the slope inclination at each site (Figure 135). Note that the slope (θ) vs RMS graph is accompanied by an equilibrium line which relates the RMS rating to the stable slope angle, as defined by numerous measurements of slopes assessed to be in a stable equilibrium condition (Selby 1980, 1982). Superimposed on this plot is the RMS envelope as modified from that of Selby (1980) by Moon (1984), and further refined by Abrahams and Parsons (1987). Within this envelope there is a 95 per cent probability that the slopes are in strength equilibrium (Figure 135). Abrahams and Parsons (1987) re-evaluated the published RMS data for strength equilibrium slopes and produced a more statistically rigorous relationship between slope inclination and RMS. Using this plot and envelope, predictions of stable slope angles can be achieved, and it is possible to identify equilibrium or non-equilibrium slopes, with the latter either oversteepened or low angle, structurally controlled or Richter denudation slopes (Figure 135).

Strength equilibrium slopes have inclinations in balance with their RMS and are not controlled by other exo- or endogenic processes such as structural or tectonic factors. These slopes also require considerable time for this balance to develop (>10,000 years) so that young slopes are often not in strength equilibrium (Selby 1987). Nevertheless, many slopes have an inclination adjusted to their RMS, and oversteepened slopes undercut by processes such as GLACIAL EROSION can be easily recognized, although Augustinus (1995) demonstrated that equilibrium and structurally controlled slopes are more common in youthful, tectonically active mountains with deeply incised glacial valleys. Furthermore, the widespread recognition of strength equilibrium rock slopes suggests that many of them probably retreat whilst preserving strength equilibrium (Moon and Selby 1983; Selby 1987). Consequently, a change in the slope angle during retreat can occur where RMS changes as a consequence of progressive WEATHERING or rapid rupture of the rock mass as a consequence of external factors such as earthquake shaking. The tendency for slopes to equilibrate rapidly as a consequence of a change in RMS means that oversteepened slopes will have a short life span (on a geological timescale) before they evolve towards strength equilibrium forms as soon as fractures open, increase in continuity, widen or rotate.

The importance of rock mass control in geomorphology is exemplified by the application of the rock mass strength classification to the development of an understanding of rock slope form evolution and stability. However, rock mass strength and its resistance to EROSION processes may also be crucial to understanding the evolution of erosional landforms. For example, the development of features such as glacial valley longitudinal profiles (as well as the glacial valley cross-profile forms) will be dependent on the RMS of the rock being eroded as well as the EROSIVITY of the processes, since the intact rock strength, orientation of the rock joints and their spacing will influence the rock mass resistance to glacial erosion processes such as plucking. Clearly, in this situation the rock mass properties that control the stability and morphology of slopes will not be applicable to quantifying rock resistance to erosion and would require redefinition for this purpose.

References

Abrahams, A.D. and Parsons, A.J. (1987) Identification of strength equilibrium rock slopes: further statistical considerations, *Earth Surface Processes and Landforms* 12, 631–635.

Allison, R.J. and Goudie, A.S. (1990) The form of rock slopes in tropical limestone and their associations with rock mass strength, *Zeitscrift für Geomorphologie* 34, 129–148.
Augustinus, P.C. (1992) The influence of rock mass strength on glacial valley cross-profile morphometry: a case study from the Southern Alps, New Zealand, *Earth Surface Processes and Landforms* 17, 39–51.
——(1995) Rock mass strength and the stability of some glacial valley slopes, *Zeitschrift für Geomorphologie* 39, 55–68.
Bieniawski, Z.T. (1979) *Engineering Rock Mass Classifications*, New York: Wiley.
Moon, B.P. (1984) Refinement of a technique for determining rock mass strength for geomorphological purposes, *Earth Surface Processes and Landforms* 9, 189–193.
Moon, B.P. and Selby, M.J. (1983) Rock mass strength and scarp forms in Southern Africa, *Geografiska Annaler* 65A, 135–145.
Selby, M.J. (1980) Rock mass strength classification for geomorphic purposes, *Zeitschrift für Geomorphologie* 24, 31–51.
——(1982) Controls on the stability and inclinations of hill slopes formed on hard rock, *Earth Surface Processes and Landforms* 7, 449–467.
——(1987) Rock Slopes, in M.G. Anderson and K.S. Richards (eds) *Slope Stability*, 475–504, Chichester: Wiley.

Further reading

Selby, M.J. (1993) *Hillslope Materials and Processes*, 2nd edition, Chapter 6, Oxford: Oxford University Press.

PAUL AUGUSTINUS

ROCKFALL

Rockfall is the free or bounding fall of rock debris down steep slopes. Rockfalls vary from individual pebbles to catastrophic failures of several million cubic metres (STURZSTROM, rock avalanches). Smaller rockfalls ($<10^1$–$10^2\,m^{-3}$) are the primary process associated with the formation of SCREE (talus) slopes and may be classified into two types (Rapp 1960). Massive vertical cliffs are dominated by primary rockfalls where detachment is followed by direct transfer to the scree below. These are triggered mainly by pressure release or FREEZE–THAW CYCLE activity. However, debris may accumulate on irregularities in the cliff (benches, gullies, etc.) and subsequently be dislodged by other rockfalls, snow avalanches (see AVALANCHE, SNOW), surface flows, animals, etc. These secondary rockfalls have different magnitude–frequency characteristics than primary rockfalls. Triggering mechanisms for rockfalls are inferred from inventory studies that compare the pattern of rockfalls with simultaneous observations of temperature and precipitation. Most investigators identify diurnal maxima during times of solar illumination and seasonal maxima in spring and fall (Luckman 1976; Gardner 1980).

References

Gardner, J.S. (1980) Frequency, magnitude and spatial distribution of mountain rockfalls and rockslides in the Highwood Pass area, Alberta, Canada, in D.R. Coates and J. Vitek (eds) *Thresholds in Geomorphology*, London: Allen and Unwin.
Luckman, B.H. (1976) Rockfalls and rockfall inventory data: some observations from Surprise Valley, Jasper National Park, Canada, *Earth Surface Processes* 1, 287–298.
Rapp, A. (1960) Talus slopes and mountain walls at Templefjorden, Spitzbergen, *Norsk Polarinstitutt Skrifter* 119.

SEE ALSO: geomorphological hazard; hillslope, process; mass movement; pressure release; unloading

BRIAN LUCKMAN

ROCKPOOL

Rockpools (synonymous with tidepool, pool) can broadly be defined as depressions in eulittoral and supralittoral rocky SHORE PLATFORMs which store surface water and form as a result of dissection of rock material by a combination of chemical, physical and/or biological means. It is generally accepted that the presence or creation of an initial depression enables the commencement of a positive feedback loop – where surface water storage provides a suitable environment where weathering and erosion processes widen and deepen pits in rock surfaces to develop rockpools (Elston 1917).

Rockpool development can be divided into three phases: (1) pool initiation, (2) pool widening and deepening and (3) coalescing of smaller pools. Pool initiation is thought to be largely controlled by geological conditions such as rock hardness (with softer rocks such as sandstone and limestone being more prone to erosion), joint planes, irregular bedding and concretions which provide an initial depression from which rockpools gradually develop. Pool deepening and widening are often caused by a suite of biological, chemical and physical processes. Biological processes include bioerosion by boring and/or grazing species such as polychaete worms, sea urchins and limpets.

Dissolution is often caused by biochemical activity when respiration increases the CO_2 concentration in the pool. Chemical weathering also occurs due to evaporation and drying of seawater. The dominant physical erosion process is scouring and abrasion of pools by harder rock debris being moved or rotated by the action of waves. The third phase of pool formation is caused by the continual erosion of narrow walls separating adjacent pools. This leads to the coalescing of smaller pools into large irregular or elliptical pool forms and, in some instances, secondary, inset pools develop as another line of stratification is eroded in the base of pools.

Rockpools vary in size from small features a few centimetres in diameter to large, irregular forms which are up to 6 m in diameter and range from 0.1 to 2 m in depth. Two main types of rock pools have been defined in geomorphological literature: solution pools and pot-holes. Solution pools (synonymous with shallow pools) are typically defined as shallow, flat-bottomed depressions found on gently sloping shore platforms (Sunamura 1992). While solution pools have greater width than depth and vary in shape, pot-holes are typically cylindrical or bowl-shaped depressions that have more similar depth to width ratios. Pot-holes are thought to form primarily by abrasion. This classification is quite narrow in scope as rockpools typically form by a combination of biological, chemical and physical means rather than being dominated by one process over another.

References

Elston, E.D. (1917) Potholes: their variety, origin and significance, *Scientific Monthly* 5, 554–567.

Sunamura, T. (1992) *Geomorphology of Rocky Coasts*, Chichester: Wiley.

Further reading

Emery, K.O. (1946) Marine solution basins, *Journal of Geology* 54, 209–228.

Emery, K.O. and Kuhn, G.G. (1980) Erosion of rock shores at La Jolla, California, *Marine Geology* 37, 197–208.

Wentworth, C.K. (1944) Potholes, pits and pans: subaerial and marine, *Journal of Geology* 52, 117–130.

LARISSA NAYLOR

ROCKY DESERTIFICATION

Rocky DESERTIFICATION is the process that leads to KARST lands being turned into stony ecological deserts. It is a consequence of devegetation followed by intensive agriculture and extreme soil erosion. Soils on karsts are usually thin, because limestones are often composed of at least 90 per cent calcium carbonate and so there is little insoluble residue from their solution that can form the mineral basis of a soil. In the humid subtropical karst of China, it has been calculated that 0.25–0.85 million years are required to form 1 m of soil. Where thick soils are found on karst, it is usually because they are formed of foreign materials transported from beyond the boundary of the karst, such as loess, alluvium, volcanic ash or glacial drift. But even thick soils can be stripped from karst, although it takes longer.

It is well known in any environment that deforestation leads to accelerated SOIL EROSION. In karst this is exacerbated because of soil loss down countless voids opened by corrosion into caves, where it has a major deleterious impact on subterranean biota. It is eventually evacuated from cave systems via underground streams that discharge at springs, but lowered water quality results. The free draining nature of karst therefore contributes to the loss of its soil if the fragile hold accomplished by plant cover is disturbed.

In parts of the Mediterranean basin, the rocky nature of karst is so characteristic that it has come to be considered natural, rather than a consequence of millennia of human impact (Gams *et al.* 1993). The word *karst* itself can be traced back to pre-Indo-European origins, where it stems from *karra* meaning stone. But the landscapes involved were originally forested and have been cleared, tilled and overgrazed. The first evidence of forest clearance around the northern Mediterranean was in the Neolithic about 6,000 years ago. This continued through Greek, Roman and more recent times, and as population increased lands were subdivided and grazing became more intensive. This relentless impact contributed to the stripping of the hills; so that naked karst now seems the norm. But recent political and land use changes in Slovenia have seen rural migration and abandonment of farms, followed by natural regrowth and spread of forests, indicating that recovery is possible if human pressure is reduced.

A similar sequence of events has occurred in China (Yuan 1995), especially in Guizhou Province, where removal of subtropical monsoon forest and intense population pressure in recent centuries has seen the denudation of karstic hillsides. The expansion of rocky desertification in the area was at a rate of 933 square kilometres per year during the 1980s. Similar problems though of smaller scale are encountered in deforested and intensively farmed karstlands of the Gunung Sewu of Java and in parts of Central America and the Caribbean.

References

Gams, I., Nicod, J., Sauro, U., Julian, E. and Anthony, U. (1993) Environmental change and human impacts on the Mediterranean karsts of France, Italy and the Dinaric region, in P.W. Williams (ed.) *Karst Terrains: Environmental Changes and Human Impact*, Cremlingen-Destedt, *Catena Supplement* 25, 59–98.

Yuan Daoxian (1995) Rock desertification in the subtropical karst of South China, in P.W. Williams (ed.) Tropical and subtropical karst, *Zeitschrift für Geomorphologie, Supplementband* 108, 81–90.

PAUL W. WILLIAMS

ROUGHNESS

The term 'roughness' refers to the roughness of a channel bed, which is an important component of the overall resistance to water flow along the channel. Water flows along a channel under the influence largely of two forces: the downslope component of its own weight (which acts to propel it along the channel) and the resistance of the channel (which acts to hold it back). If the resistance is low, then a given flow has a high velocity and a low depth. If the resistance is high, the same flow has a low velocity and a high depth. Quantification of flow resistance is thus fundamental to the calculation of flow conditions in a channel.

Several relationships linking flow resistance, velocity and depth have been in use for a century or more, each quantifying the resistance with a single coefficient. They are the Darcy–Weisbach equation:

$$U = (8 g R S_f / f)^{1/2}$$

the Manning equation (in SI units):

$$U = R^{2/3} S_f^{1/2} / n$$

and the Chézy equation:

$$U = c (R S_f)^{1/2}$$

where U is mean flow velocity, R is hydraulic radius (flow cross-sectional area/channel wetted perimeter), S_f is friction slope (a measure of energy loss often approximated by water surface slope), g is acceleration due to gravity and f, n and c are respectively, the Darcy–Weisbach, Manning and Chézy coefficients. The central problem in flow resistance is therefore evaluation of the coefficient.

In a straight channel of uniform slope, uniform cross section and large width/depth ratio with no sediment transport or BEDFORMS, the resistance to flow is determined primarily by the frictional resistance of the bed. This varies with the roughness of the bed, itself dependent on the material of which the bed is composed, e.g. sand or gravel. In a popular approach, fluid mechanics and boundary layer theory are invoked to calculate resistance as a function of the logarithm of relative roughness, defined as the ratio of bed roughness height to flow depth. Roughness height is often evaluated as an equivalent sand grain size or as a selected percentile from the measured size distribution of the bed material. For example

$$(8/f)^{1/2} = 5.62 \log (d/D_{84}) + C$$

where D_{84} is the particle size for which 84 per cent of the particles are finer and C is a coefficient. Because theory cannot yet provide a full quantification of the resistance, the coefficient is determined empirically from experimental data.

Natural channels rarely conform to the ideal conditions described above and additional terms may be required in the resistance equation to account for deviations from these conditions. For example, sand beds develop bedforms such as ripples and dunes which increase the resistance above that of the roughness of the grains alone. Consequently there are a variety of formulae and methods for determining the resistance coefficient. Users should be careful to select a method which is appropriate for the conditions and data availability with which they are concerned.

Further reading

Bathurst, J.C. (1993) Flow resistance through the channel network, in K. Beven and M.J. Kirkby (eds) *Channel Network Hydrology*, 69–98, Chichester: Wiley.

Raudkivi, A.J. (1998) *Loose Boundary Hydraulics*, Rotterdam: Balkema.

JAMES C. BATHURST

RUGGEDNESS

A property of the landscape which describes the complexity of the topography and the roughness of the terrain. More rugged landscapes tend to exhibit a greater amount of complexity, having rough and uneven surfaces. Ruggedness is a naturally qualitative term, though several ruggedness indexes have been proposed (e.g. Riley *et al.* 1999) that provide a quantitative frame. Melton (1958) developed the ruggedness number to describe the ruggedness of land on a drainage-basin scale. This is a dimensionless number calculated from the formula $H / \sqrt{A}$ where H is the vertical relief above fan apex (miles^2), and A is basin area (miles^2). In general, the ruggedness number can be as high as 2.0 or 3.0 for a first or second-order basin, and rarely above 1.0 for a third or fourth-order basin.

References

Melton, M.A. (1958) Geometric properties of mature drainage basins and their representation in a E_4 phase space, *Journal of Geology* 66, 35–54.

Riley, S.J., DeGloria, S.D. and Elliot, R. (1999) A terrain ruggedness index that quantifies topographic heterogeneity, *Intermountain Journal of Sciences* 5, 23–27.

STEVE WARD

RUNOFF GENERATION

Runoff generation refers to a suite of processes that produce and route flow from landscape segments to stream channels in response to precipitation (i.e. rainfall and/or snowmelt). Runoff is generated by three different mechanisms: infiltration-excess overland flow (Horton-type), saturation-excess overland flow (Dunne-type) and subsurface stormflow. Infiltration-excess overland flow is overland flow that results from saturation from above (Horton 1933, 1945). This occurs where the water-input rate exceeds the infiltration capacity of the soil long enough for ponding to occur; the excess water then flows quickly over the surface to stream channels (see QUICK FLOW). Once the precipitation volume exceeds the moisture storage capacity of the soil, however, saturation-excess overland flow occurs. First described in detail by Dunne and Black (1970), this mechanism is controlled by the saturated hydraulic conductivity of the soil. The soil becomes saturated from below, either by (1) the presence of an impeding layer which causes a perched water table to develop that may gradually rise to the surface; (2) extension of the capillary fringe to the ground surface; or (3) the presence of a permanent water table at or near the ground surface. Saturation-excess overland flow consists of direct water input to the saturated area plus the return flow contributed by the exfiltration of ground water from upslope.

The third runoff mechanism, subsurface stormflow, consists mainly of the displacement of old pre-event water by new rainwater. This is due to near-stream ground water mounding (the same process that ultimately produces saturation overland flow) and/or via flow from perched saturated zones (saturated wedges). The process also includes throughflow or interflow, which is water that infiltrates into the soil and moves laterally within the soil matrix either in the unsaturated zone between the ground surface and a perched or regional water table or through macropores such as cracks, root and animal holes and pipes (Wilson *et al.* 1990; Jones 1997). The latter reaches the stream channel quickly and differs from other subsurface flow by the rapidity of its response and its relatively large magnitude.

It is widely accepted that Hortonian overland flow is the dominant response mechanism in semi-arid to arid regions, on impermeable zones and in human-disturbed areas. On the other hand, the response mechanism in humid regions is generally saturation-excess overland flow and/or some form of subsurface flow. Saturation-excess overland flow most frequently occurs near stream channels but it also can be generated in hillslope hollows, where subsurface flowlines converge in slope concavities, at concave slope breaks, or where soil layers conducting subsurface flow are locally thin. Subsurface storm flow dominates where soils are deep and permeable, especially under forest cover, and where slopes are steep. All three runoff mechanisms can occur simultaneously in a basin, even during a single water-input event.

References

Dunne, T. and Black, R.D. (1970) Partial area contributions to storm runoff in a small New England watershed, *Water Resources Research* 6, 1,296–1,311.

Horton, R.E. (1933) The role of infiltration in the hydrologic cycle, *Transactions of the American Geophysical Union* 14, 446–460.
Horton, R.E. (1945) Erosional development of streams and their drainage basins: hydrophysical approach to quantitative morphology, *Geological Society of America Bulletin* 56, 275–370.
Jones, J.A.A. (1997) Pipeflow contributing areas and runoff response, *Hydrological Processes* 11, 35–41.
Wilson, G.V., Jardine, P.M., Luxmoore, R.M. and Jones, J.R. (1990) Hydrology of a forested hillslope during storm events, *Geoderma* 46, 119–138.

SEE ALSO: quick flow

MICHAEL SLATTERY

S

SABKHA

A sabkha is the English form of the Arabic word *sebkha* which means 'salt flat'. Kinsman (1969) defines a sabkha as a surface of deflation down to the level of ground water or the zone of capillary evaporation. Neal (1975) describes a sabkha as a geomorphic surface the level of which is dictated by the water table. Warren (1989) and Briere (2000) describe a sabkha as a marginal marine and continental mudflat where evaporite minerals are forming in the capillary zone above the saline water table. Sabkhas were described subsequently in many other areas of the world, such as the coast of Baja California, Mexico and the coast of Sinai. Equivalent features are the Solonchak salt flats of the Caspian Sea and some kavir depressions of Iran. Certain salt pans in South Africa and playas in the southwestern United States may be similar but true sabkhas are not fed by streams or runoff.

In North America, the words 'playa' and 'salina' have both been applied to sabkha-like areas in the desert. Holm (1960) states that 'playa' is synonymous with 'mamlahah' (inland sabkha). Von Engeln (1942), on the other hand, says that if the percentage of salts in a playa is high enough for a salt crust to form when the flat is dry, it is then called a salina.

If the word sabkha means 'salt flat' then there are both coastal and continental sabkhas. Some coastal sabkhas, such as those along the Trucial Coast, pass laterally into continental sabkhas without any noticeable change in surface morphology on the sabkha (Kinsman 1969). The marine portion of the Abu Dhabi sabkha is characterized by a matrix of marine sediments soaking in largely marine-derived ground water and the continental portion by non-marine sediments and ground water.

Modern marine sabkhas are forming along the coasts of many stable land areas such as the western and southern coasts of the Arabian Gulf, the coasts of Australia and northern Africa. Some sabkhas are slightly above present sea level (0.5–3.0 m) and may have been inherited from short Mid-Holocene eustatic oscillations.

A review of the global distribution of sabkha indicates its extensive presence in the Middle East, including Egypt, Sudan, Libya, Tunisia, Algeria and Ethiopia. Sabkha also exists in India, Australia and southern Africa. Contrary to expectations, sabkha and sabkha-like sediments can occur also in relatively cold climates. Aridity, therefore, seems to play a more important role than hot weather in the formation of sabkha. Figure 136 shows the distribution of sabkha around the world.

Sabkhas are part of a landform sequence that extends from the shoreline, with barrier islands or dunes, through a lagoon then to the sabkha and perhaps into a dune system before truly terrestrial systems are reached. Sabkha surfaces are extremely flat and often extend for long distances. The sabkha depositional sequence can be divided into three units: the subtidal, the intertidal and the supratidal. When the sequence is prograding the three units are superimposed one on top of the other to form a shallowing upward sabkha cycle.

The subtidal unit is usually divided into open marine and restricted marine sedimentation. Restricted marine sedimentation occurs on the landward side of barrier islands. The lagoon sediments contain a diverse biota of many benthic species, molluscan sands occur in the more restricted areas and pelleted carbonate mudstones occur in the more open areas of the

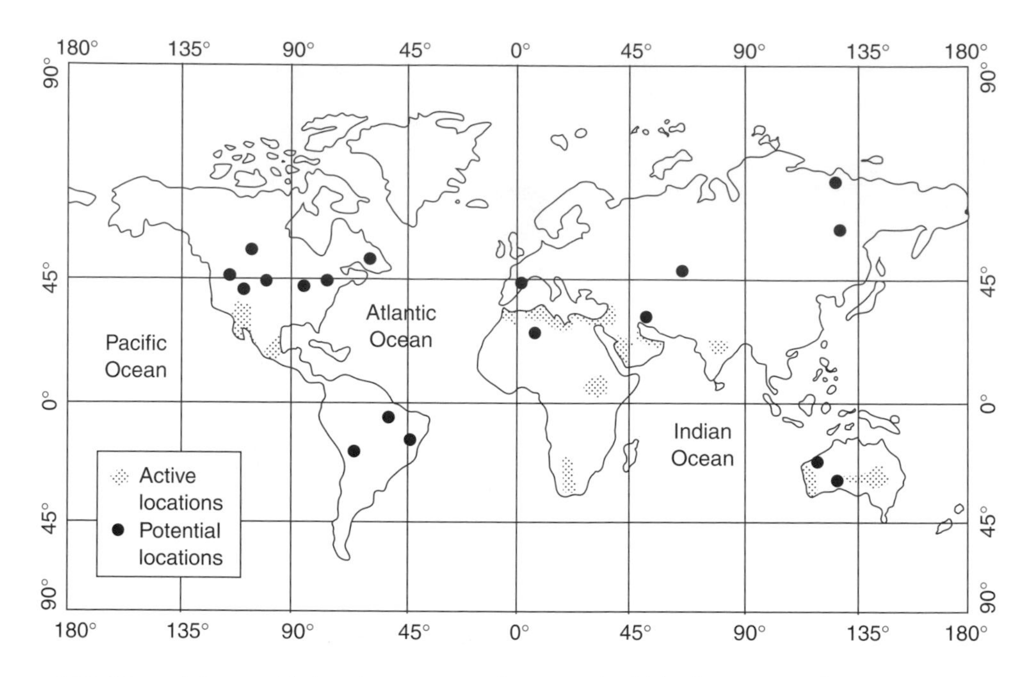

Figure 136 Map of the world showing active and potential sabkha locations (Al-Amoudi 1994)

lagoon. The intertidal zone can be divided into an upper and lower intertidal facies. However, the lower intertidal facies may be the same as the lagoonal facies as described by a number of workers. This facies is dominated by an algal peat composed of the bioturbated remnants of an algal mat. The upper intertidal facies is usually a laminated algal mat often cross-cut by desiccation cracks containing lenticular gypsum crystals. Aragonite, magnesite and dolomite may locally cement the surface sediments (Butler *et al.* 1982). The lower supratidal belt, which is flooded once or twice a month, is characterized by gypsum up to 30 cm thick. Diagenetic nodular anhydrite occurs in the deposits of the middle supratidal zone and there is often a surface crust of ephemeral halite. Such deposits are flooded on less than monthly intervals. In the upper supratidal zone, flooded once every four to five years, the gypsum has been replaced by coalesced nodules of anhydrite.

Sabkhas are broad coastal supratidal and intertidal flats developed along the margins of arid landmasses. Sediments that accumulate on sabkhas include: (1) siliciclastic detritus sediments that are eroded from adjacent land and washed onto the sabkha; (2) offshore deposits of sand and mud that periodic storms wash up and onto the sabkha; and (3) the indigenous sediments of the sabkha itself.

Much of the evaporite sediment produced in sabkhas precipitates as saline ground water seeps into and out of the sabkha (Figure 137). Much of this ground water is seawater that is continually recharged beneath the sabkha, but ground water from the adjacent landmass can also feed the system. Groundwater circulation is driven by capillary action and evaporative pumping. Intermittent flooding by the sea also occurs, and beach ridges can trap a reservoir of additional seawater.

Typical sabkha evaporite minerals are anhydrite, gypsum and dolomite. Much of the anhydrite occurs as irregularly shaped lumps or nodules. These nodules replace altered gypsum crystals originally formed within layers of interbedded carbonate mud or shale. The term chickenwire structure is used to refer to this mixture of elongate, irregular clumps of anhydrite separating thin stringers of carbonate and/or siliciclastic mud (Plate 105a,b). This structure is particularly common in sabkha evaporites.

Cyclicity is common in sabkha evaporite sequences. As deposition proceeds, sabkha deposits

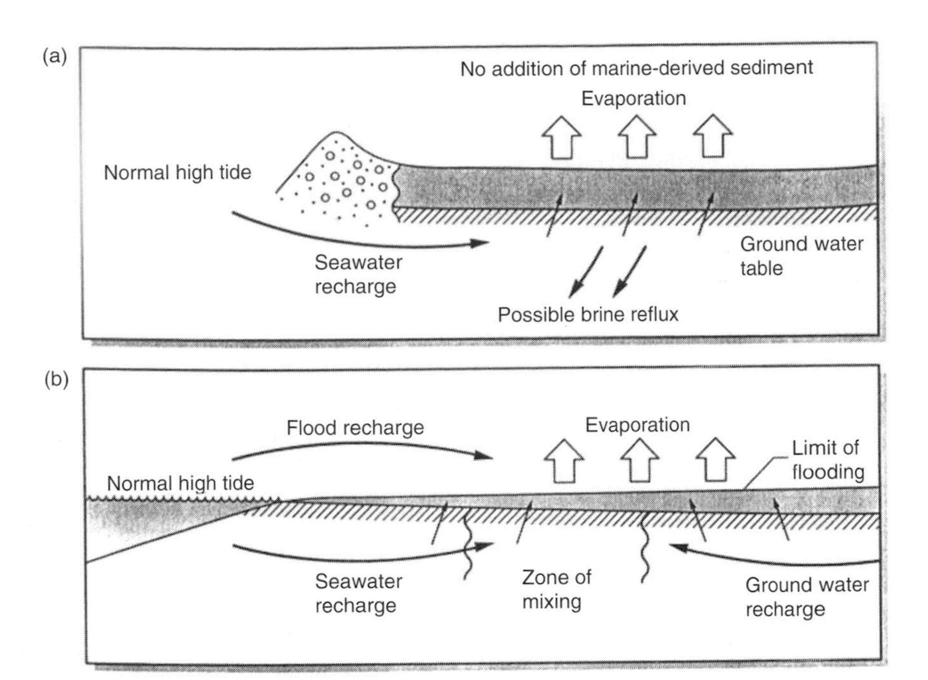

Figure 137 Sabkhas receive water from a variety of sources: (a) sabkha with seawater recharged through the subsurface and with relatively little groundwater influx; (b) sabkha groundwaters are recharged by a mixture of seawater and groundwater, plus seawater flooding from major storms (from Walker 1984)

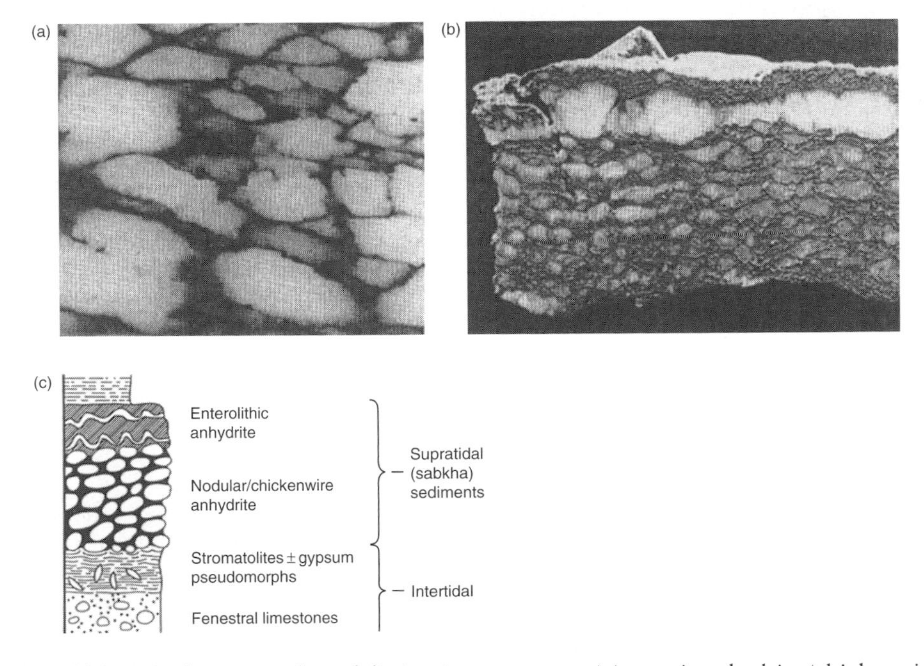

Plate 105 Sabkhas produce a number of distinctive structures: (a) mosaic anhydrite (chickenwire structure) commonly formed when anhydrite nodules coalesce, shown at actual size; (b) nodular gypsum is common just below the surface of the sabkha; (c) typical vertical cycle of sabkha sediments. Such cycles range from several metres to several tens of metres in thickness (after Tucker 1981)

naturally prograde oceanward and eventually lie upon intertidal sediments (STROMATOLITES, gypsum, fenestral birdseye pelleted carbonate mud). These in turn rest on oolitic and bioclastic subtidal carbonate rocks of the subtidal zone (Plate 105c).

The Arabian Gulf coastal sabkhas occur on the southern margin of the Gulf with the best-studied sequences being situated south-east of Abu Dhabi City, where they are now partially covered by urban and industrial development. If an idealized landward transect is traced, it passes in order through offshore open-marine skeletal sediments, belts of oolitic grainstones, belts of lagoonal and/or barrier sequences, and then crosses intertidal algal mats to terminate in a supratidal sabkha sequence (Butler *et al.* 1982). If the intertidal algal mats are included, the sabkhas constitute a zone 10–15 km wide, with an along-strike continuity of more than 150 km. The surface transect from the lagoonal sands and muds up onto the flat plain of sabkha first crosses the dark, black to grey, flat-laminated, algal mats of the intertidal zone (Figure 138).

In some arid areas the surface of the sabkha is encrusted by a veneer of salt and scattered discoidal crystals of gypsum. Dust and sand storms occur through most of the year, which results in the sabkha plain being covered by a layer of drifting quartz sand.

The landward boundary of the supratidal area is characterized by a zone of vegetation and is dominated by the halophyte *Halocnemon Strobilaceum*. The vegetation on the sabkha surface acts as a trap for the moving sand. Most of this drifting sand is usually washed off during storm tides and redistributed on the sabkha surface.

Inland sabkhas develop where water flowing in wadis intermittently floods low-lying depressions (see PAN) to leave behind damp, salt-encrusted sediments. They are also found in depressions where, for one reason or another, the water table reaches the surface.

In inland sabkhas, the salt crust forms as the result of the concentration of salts caused by evaporation of the water. Gypsum crystals are common in the sediments of inland sabkhas. Algae are known, but the algal mats so commonly associated with coastal sabkhas have not been recognized. A coastal sabkha, on the other hand, is characterized by marine flooding and evaporitic conditions. It is a diagenetic environment whose sediments are of continental and adjacent marine origin.

References

Al-Amoudi, O.S.B. (1994) A state-of-the-art report on the geotechnical problems associated with sabkha soils and methods of treatment, *Proceedings of the ASCE-SAS 1st Reg. Conference Exhibition*, Bahrain, 18–20 Sept. 53–77.

Briere, P.R. (2000) Playa, playa lake, sabkha: proposed definitions for old terms, *Journal of Arid Environments* 45, 1–7.

Butler, G.P., Harris, P.M. and Kendall, C.G. St.C. (1982) Recent evaporites from the Abu Dhabi coastal flats, in depositional and diagenetic spectra of evaporites, *A Core Workshop: SEPM Core Workshop 3*, Calgary, 33–64.

Holm, D.A. (1960) Desert geomorphology in the Arabian Peninsula, *Science* 132(3,427), 1,369–1,379.

Kinsman, D.J.J. (1969) Modes of formation, sedimentary associations and diagnostic features of shallow water and supratidal evaporites, *American Association Petroleum Geologists Bulletin* 53, 830–840.

Neal, J.T. (1975) Playa surface features as indicators of environment, in J.T. Neal (ed.) *Playas and Dried Lakes*, 363–380, Benchmark Papers in Geology, Stroudsburg, PA: Dowden, Hutchinson and Ross.

Tucker, M.E. (1981) *Sedimentary Petrology: An Introduction*, 161–173, Oxford: Blackwell Scientific.

Von Engeln, P.D. (1942) *Geomorphology*, New York: Macmillan.

Walker, R.G. (1984) *Facies Models*, 2nd edition, Toronto: Geological Association of Canada.

Warren, J.K. (1989) *Evaporite Sedimentology*, Upper Saddle River, NJ: Prentice-Hall.

SEE ALSO: deflation

ADEEBA E. AL-HURBAN

SACKUNG

A German term describing a type of slope-sagging, gravitational lateral spreading or deep-seated gravitational deformation in mountainous alpine landscapes. Sackungen (plural) display rounded morphology, commonly trend parallel to the contours of the slope, and form a characteristic ridge-top trench in the adjacent valley. They typically display a bulge at the toe

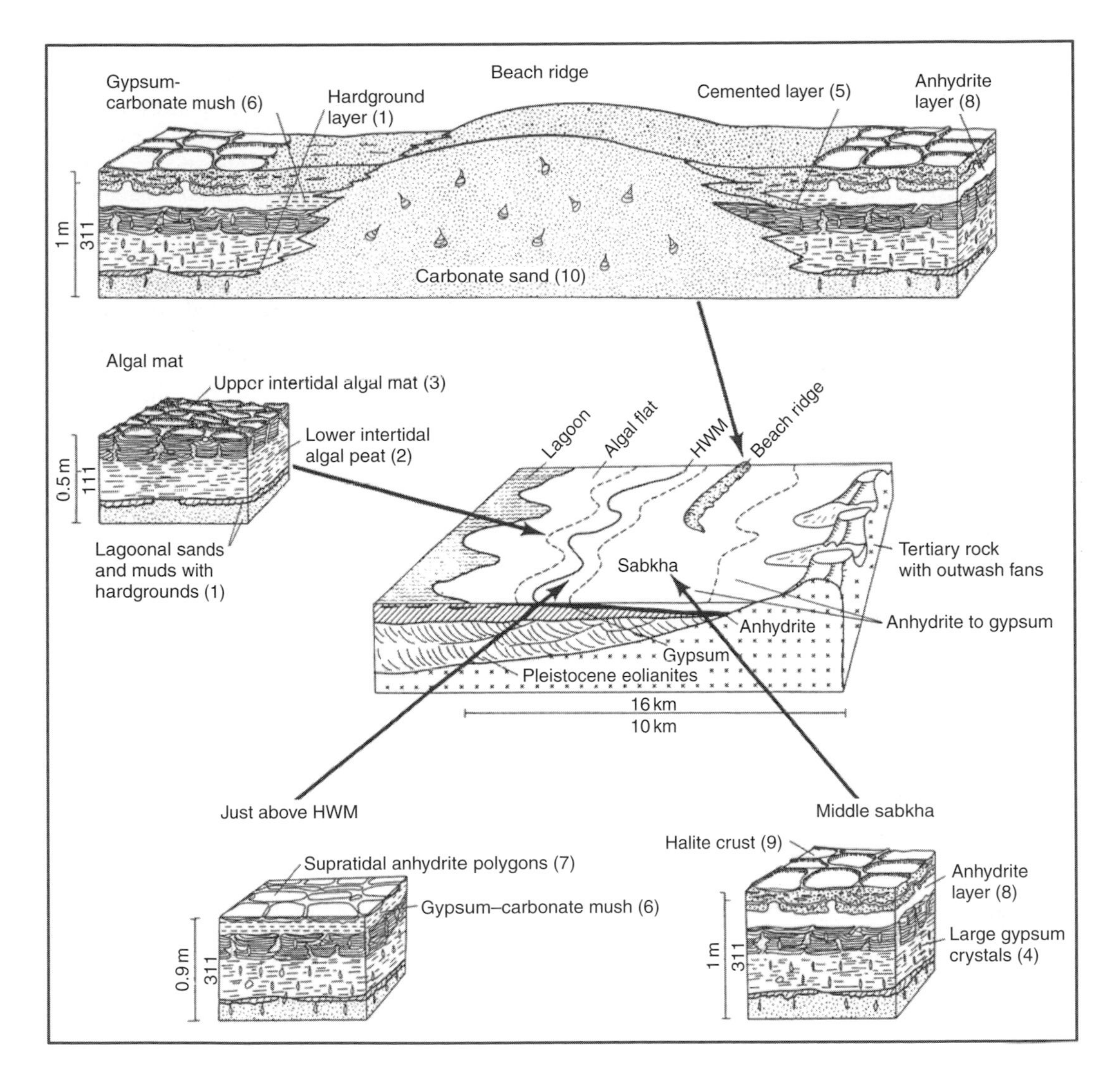

Figure 138 Schematic block diagrams showing sediment and evaporite distribution in Abu Dhabi sabkha, Arabian Gulf. All peripheral diagrams are keyed to central block of sabkha. HWM = high-water mark. (1) Lagoonal carbonate sands and/or muds with carbonate hardgrounds; (2) vaguely laminated lower tidal-flat carbonate-rich algal peat; (3) upper tidal-flat algal mat formed into polygons; (4) large gypsum crystals (many lenticular); (5) cemented carbonate layer; (6) high tidal-flat to supratidal mush of gypsum and carbonate; (7) supratidal anhydrite polygons with wind-blown carbonate and quartz; (8) anhydrite layer replacing gypsum mush and forming diapiric structures; (9) halite crust, formed into compressional polygons; (10) deflated beach ridge of cerithid coquina and carbonate sand. (After Butler *et al.* 1982)

of the slope, known as the 'Talzuschub', as well as tensile or normal faults near their crest, termed 'Bergzerreißung' (Zischinski 1969). Rates of material creep vary significantly (1 mm to several metres), with variations in activity corresponding to changes in precipitation and water table (increased creep activity with higher precipitation). Sackungen are indicative of gravitational spreading (Varnes *et al.* 1989), resulting from low mass strength in the underlying material (a product of high density jointing and faulting) to a substantial depth.

References

Butler, G.P., Kendall, C.G. St. C. and Harris, P.M. (1982) Abu Dhabi Coastal Flats, in C.R. Handford, R.E. Loucks and G.R. Davies (eds) *Depositional and Diagenetic Spectra of Evaporites a Core Workshop*. Society of Economic Palaeontologists Core Workshop, number 3, Calgary, Canada, 33–64.

Varnes, D.J., Radbruch-Hall, D.H. and Savage, W.Z. (1989) Topographic and structural conditions in areas of gravitational spreading of ridges in the western United States, *US Geological Survey Professional Paper* 1496, 28.

Zischinski, U. (1969) Uber sackungen, *Rock Mechanics* 1, 30–52.

Further reading

Savage, W.Z. and Varnes, D.J. (1987) A model for the plastic spreading of steep-sided ridges ('Sackung'), *Bulletin of the International Association for Engineering Geology* 35, 31–36.

SEE ALSO: mass movement

STEVE WARD

SALCRETE

A salty surface crust, primarily composed of sodium chloride, that cements a sand surface as a result of evaporation of moisture and the consequent chemical concentration of dissolved material. The term, which was coined by Yasso (1966), has been mainly used to describe crusts developed through evaporation of sea spray on beaches (e.g. Pye 1980).

References

Pye, K. (1980) Beach salcrete in North Queensland, *Journal of Sedimentary Petrology* 50, 257–261.

Yasso, W.E. (1966) Heavy mineral concentrations and sastrugi-like deflation furrows in a beach salcrete at Rockaway Point, *Journal of Sedimentology Petrology* 36, 836–838.

A.S. GOUDIE

SALT (EVAPORITE) KARST

Evaporites, including common salt (halite) and calcium sulphate (gypsum and anhydrite) are the most soluble of common rocks (see GYPSUM KARST). They are also widespread, being found, for example, in 32 of the 48 contiguous states of the USA (Johnson 1997) in rocks of every geologic system from the Precambrian through to the Quaternary. Karst-creating evaporites are also extensively developed in Canada (Ford 1997), and have been described from many parts of Europe, including the Ripon area of England and the Betic Cordillera and Ebro basins of Spain. Salt dome structures are often affected by solutional processes (see SALT RELATED LANDFORMS).

Evaporite outcrops display a full array of solutional features, including subsidence depressions, collapse breccias, sinkholes, vertical shafts and water-filled chimneys (Last 1993; Calaforra and Pulido-Bosch 1999; Gutierrez-Elorza and Gutierrez-Santollala, 1998). Because of the great solubility of evaporites, rates of karst denudation can be high, and even under the dry conditions of the arid parts of Israel, can approach 500–750 mm $1{,}000\,yr^{-1}$ (Frumkin 1994). Salt karst processes present a range of engineering problems and hazards (Paukstys *et al.* 1999) and these can be exacerbated as a result of human activities, including mining, ground water abstraction and other hydrological modifications.

References

Calaforra, J.M. and Pulido-Bosch, A. (1999) Gypsum karst features as evidence of diapiric processes in the Betic Cordillera, Southern Spain, *Geomorphology* 29, 251–264.

Ford, D.C. (1997) Principal features of evaporite karst in Canada, *Carbonates and Evaporites* 12, 15–23.

Frumkin, A. (1994) Hydrology and denudation rates of halite karst, *Journal of Hydrology* 16, 171–189.

Gutierrez-Elorza, M. and Gutierrez-Santollala, F. (1998) Geomorphology of the Tertiary gypsum formations in the Ebro Depression (Spain), *Geoderma* 87, 1–29.

Johnson, K.S. (1997) Evaporite karst in the United States, *Carbonates and Evaporites* 12, 2–14.

Last, W.M. (1993) Salt dissolution features in saline lakes of the northern Great-Plains, Western Canada, *Geomorphology* 8, 321–334.

Paukstys, B., Cooper, A.H. and Arustiene, J. (1999) Planning for gypsum geohazards in Lithuania and England, *Engineering Geology* 52, 93–103.

A.S. GOUDIE

SALT HEAVE OR HALOTURBATION

A cause of damage to engineering structures in desert areas as a result of the presence of soluble salts (including sodium chloride, magnesium sulphate and sodium sulphate). The process is akin to frost heave, in that the hydration and crystallization of salts plays a major role, but it is also akin to needle-ice (pipkrake) formation in that salt whiskers may grow vertically. The problem is especially severe when saline ground water approaches the ground surface. Possible techniques to deal with it have been developed (Horta 1985), including brooming, embankments, barriers and the use of thick, impervious surfacings. In the case of some gypsum areas, volume changes associated with solution and recrystallization can produce what are termed 'mega-tumuli' and dome-shaped hills (Ferrarese *et al.* 2002).

References

Ferrarese, F., Macaluso, T., Madonia, G., Plameri, A. and Sauro, U. (2002) Solution and recrystallisation processes and associated landforms in gypsum outcrops of Sicily, *Geomorphology* 49(1), 25–43.

Horta, J.C. de O.S. (1985) Salt heaving in the Sahara, *Géotechnique* 35, 329–337.

A.S. GOUDIE

SALT RELATED LANDFORMS

Evaporites, including halite (sodium chloride) are widespread both geographically and in the geological record. As much as one-quarter of the world's continental areas may be underlain by evaporites of one age or another. When evaporite beds are thick they can have many important geomorphological consequences: the development of diapiric structures (including salt domes), the production of folds and faults (see FAULT AND FAULT SCARP), tectonic uplift and associated drainage modification, the creep of salt as 'salt glaciers', and widespread solution, subsidence and karst formation. Salt is important also as a cause of weathering (see SALT WEATHERING).

Because it has a low density and low rheidity (the ease at which it flows as a viscous solid), salt flows readily under burial conditions when the surrounding sediment is still undeformed. Flow rates can increase in the presence of water (as brine) and at temperatures in excess of around 245 °C. The flowage of salt can transform relatively tabular evaporite bodies into a wide variety of structures that tend to evolve from concordant, low-amplitude features through to discordant high-amplitude intrusions (DIAPIRS), and thence to extrusions. The immature concordant structures include salt anticlines (which have approximately symmetrical cross sections, planar bases and arched roofs), salt rollers (which are also ridge-like, but are asymmetric with a faulted scarp) and salt pillows (periclinal, subsurface domes). Sediment covers tend to be thin over the crests of such structures, but the area above the pillow tends to be a topographic high surrounded by a topographic low or primary rim syncline. These form simultaneously with the accumulation of salt in the area of ongoing uplift, and result from the downwarping of the overlying strata into the space vacated by the salt flowing into the growing structure.

High amplitude diapiric intrusions, involving piercement, with uplift characteristically at 2–4 mm per year, develop in the next stage of structure growth. Among the forms described are salt walls. These are elongated like salt anticlines but are intrusive and of much greater amplitude. They tend to be 4 to 5 km in breadth, have a length of over 120 km and are generally 8 to 10 km apart. Another intrusive form is the salt stock. These vary in shape from squat to columnar and can be conical or barrel-shaped. The round varieties are 2 to 8 km in diametre in their upper parts, and in many places they are linked together in parallel-striking straight or winding lines that have been likened to strings of pearls. When the whole or part of the shallower portion of a diapir extends laterally beyond the cross-sectional area of the diapir roots, an overhang develops, producing balloon or mushroom shapes (see Jackson *et al.* 1990).

The final stage of salt structure evolution is the postdiapir stage (Warren 1989), during which the salt supply is dwindling as the volume of the salt mass decreases. Meteoric processes become important and solutional loss occurs. Large collapse depressions can form.

If the rate of diapiric growth is greater than the rate of salt solution, salt will be extruded at the ground surface. This may tend to be favoured in arid areas where low precipitation values cause low rates of dissolution. Because of its mechanical properties such extruded salt may begin to flow, and such flows are called 'salt glaciers'. These are termed 'namakiers' (from *namak* – the Farsi word for salt – and glacier) (Talbot 1979).

Some of the Zagros Mountains' salt glaciers in Iran are large. The example at Kuh-e-Namak is 2,000 m long, 3,500 m wide and up to 50 m thick, but perhaps the biggest example is Kuh-e-Gach, which is also 50 m thick, but attains a width of around 4,700 m and has an area of around 23.5 km^2. Their speed of movement is less than that of true glaciers, and the average speed is only a few metres per annum with movement tending to occur after rainfall events. They display complex folds where they flow over bedrock irregularities, and at their distal ends they may feather out in a mass of unbedded detritus analogous to a terminal moraine. Salt glaciers are also subject to retreat if the balance of wastage by solution should exceed that of outward flow of salt from the diapir, and this explains the presence of isolated exotic blocks, analogous to erratics, some kilometres beyond the plugs themselves.

The growth of salt structures can create characteristic drainage patterns and slope forms. A model of this has been proposed by Berger and Aghassy (1982) who envisage three development phases. In the 'positive relief stage' there is a central dome on which radial drainage by outbound consequent streams is dominant, and these dissect long isoclinal slopes. In the 'breached stage' the initial topographic high in the centre of the dome becomes lowered, an inversion of topography begins to occur and a major depression develops at the dome's centre. Inward-facing scarp slopes and inbound obsequents develop. In the 'obliterative stage' the inbound obsequents expand headwards and capture much of the consequent outward-bound drainage. Sedimentation, floodplain development and marsh formation occur in the core.

Finally, a large number of the world's 'fold and thrust' belts have developed over evaporites. Examples include the Jura, the Pyrenees, the Franklin Mountains of north-west Canada, the Canadian arctic fold mountains, the Salt Range of Pakistan, the Zagros of Iran, the Sierra Madre Oriental of Mexico, the Cordillera Oriental of Colombia, the Atlas of Tunisia and Algeria, the South Urals, the mountains of the Tadjik Republic, and the anticlinal province of the Amadeus basin in Australia. The presence of salt encourages what is termed 'thin-skinned deformation'.

References

Berger, Z. and Aghassy, J. (1982) Geomorphic manifestation of salt dome stability, in R.G. Craig and J.L. Craft (eds) *Applied Geomorphology*, 72–84, London: Allen and Unwin.

Jackson, M.P.A., Cornelius, R.R., Craig, C.H., Gansser, A., Stöcklin, J. and Talbot, C.J. (1990) Salt diapirs of the Great Kavir, Central Iran, *Geological Society of America Memoir* No. 177.

Talbot, C.J. (1979) Fold trains in a glacier of salt in southern Iran, *Journal of Structural Geology* 1, 5–18.

Warren, J.K. (1989) *Evaporite Sedimentology*, Englewood Cliffs, NJ: Prentice Hall.

A.S. GOUDIE

SALT WEATHERING

The weathering of rocks and building materials by salt. This is an important group of processes particularly in deserts, on coasts and in cities. Salt weathering probably plays an important role in the formation of PANs, TAFONI, STONE PAVEMENTs, SHORE PLATFORMs, rock flour and split cobbles. The build up of salt in rocks can also cause SALT HEAVE OR HALOTURBATION and SLAKING. Conventionally, salt weathering can be divided into mechanical and chemical mechanisms (Goudie and Viles 1997).

The most cited cause of salt weathering is generally the process of salt crystal growth from solutions in rock pores and cracks (Evans 1970). Various mechanisms can cause crystal growth to occur. For example, some salts rapidly decrease in solubility as temperatures fall. This is particularly true of sodium sulphate, sodium carbonate, magnesium sulphate and sodium nitrate. Thus nocturnal cooling could cause salt crystallization to occur. Such a crystallization of a salt solution on a temperature fall affects a much larger volume of salt per unit time than crystallization induced by evaporation, which is a more gradual process.

Nevertheless, evaporation helps to create saturated solutions from which crystallization can occur, and when this happens highly soluble salts will produce large volumes of crystals. In this context it is important to note that of the common salts, gypsum is very much less soluble than many of the others, and that less crystalline material will be available in a given volume of solution to cause rock disruption.

A salt's crystal habit may also affect its power to cause rock breakdown. For instance, the needle-shaped habit of sodium sulphate crystals (mirabilite) might tend to increase their disruptive capability.

Air humidity is an important control of the effectiveness of salt crystallization, for a salt can crystallize only when the ambient relative humidity is lower than the equilibrium relative humidity of the saturated salt solution. If that is the case on a rock surface then the salt will crystallize and cause decay. The equilibrium relative humidities of different salts vary considerably, and those with low values will be prone to dissolution in humid air (Plate 106). The equilibrium relative humidities of hydrated sodium carbonate and sodium sulphate are high, whereas those of sodium chloride, sodium nitrate and calcium chloride are relatively lower.

Another type of salt weathering process is salt hydration. Certain common salts hydrate and dehydrate relatively easily in response to changes in temperature and humidity. As a change of phase takes place to the hydrated form, water is absorbed. This increases the volume of the salt and thus develops pressure against pore walls. Sodium carbonate and sodium sulphate both undergo a volume change in excess of 300 per cent as they hydrate.

For some salts a change of phase may occur at the sorts of temperatures encountered widely in nature; sodium sulphate's transition temperature is 32.4 °C for a pure solution, and falls to 17.9 °C in a NaC1 saturated environment. Moreover for some salts the transition may be rapid. At 39 °C the transition from thenardite (Na_2SO_4) to mirabilite ($Na_2SO_4 \cdot 10H_2O$) may take no longer than twenty minutes (Mortensen 1933).

Winkler and Wilhelm (1970) have calculated the hydration pressures of some important common salts at different temperatures and relative humidities, and find that the greatest hydration pressures (maximum value 2,190 atm at 0 °C and 100 per cent relative humidity) occur when anhydrite changes into gypsum. This exceeds the crystallization pressure of ice at −22 °C, and is in excess of the pressure required to exceed the tensile strength of rocks.

The number of occasions upon which rock surface temperatures cycle across the temperature thresholds associated with the change of phase is probably substantial in many desert areas. If one assumes that an air temperature of *c*.17 °C translates into a rock surface temperature of *c*.32 °C (the transition temperature for sodium carbonate and sodium sulphate) then that value is crossed daily between 5 and 9 months of the year depending on the desert station selected. In other words, there may typically be around 150 to 270 days in the year in which rock temperature conditions are favourable to the salt hydration mechanism of rock decay.

A third possible mechanism of rock disruption through salt action has been proposed by Cooke and Smalley (1968), who argue that disruption of rock may occur because certain salts have higher coefficients of expansion than do the minerals of the rocks in whose pores they occur. Halite expands by around 0.9 per cent between 0 and 100 °C, whereas the volume expansion of quartz and granites is generally about one-third of that value. Gypsum and sodium nitrate are other common salts that have a relatively greater expansion potential compared to rock minerals.

It is difficult to assess the actual importance of this process, and while some early experimental simulations (e.g. Goudie 1974) suggested that it was not very effective, much more work is required on this mechanism before its potential can be dismissed.

In addition to these three main categories of mechanical effects, salt can cause chemical

Plate 106 Damaged water pipelines in central Namibia. They proved to be unable to withstand the corrosive conditions associated with the foggy, salty environment of the Namib, and had to be replaced after only a few years in service

Plate 107 Raised beach cobbles being split by halite and nitrate in the Atacama Desert near Iquique in Chile. (Beer can for scale)

weathering. Some saline solutions can have elevated pH levels. Why this is significant is that silica mobility tends to be greatly increased at pH values greater than 9. Indeed, according to various studies, silica solubility increases exponentially above pH9. The presence of sodium chloride may also effect the degree and velocity of quartz solution. At higher NaCl concentrations quartz solubility and the reaction velocity both increase. The growth of salt crystals may be able to cause pressure solution of silicate grains in rocks, for silica solubility increases as pressure is applied to silicate grains. This is a mechanism that has been identified as important in areas where calcite crystals grow, as for example in areas of calcrete formation.

Schiavon *et al.* (1995) have found petrographic evidence from granites in urban atmospheres that suggests chemical reactions occur between granite minerals and weathering solutions responsible for the precipitation of gypsum. They found feldspar minerals that were partially or totally replaced by sulphate crystalline salts while still retaining their primary mineral outline and texture.

The attraction of moisture into the pores of rocks or concrete by hygroscopic salts (e.g. sodium chloride) can accelerate the operation of chemical weathering processes and of frost action (MacInnis and Whiting 1979; Plate 108) and the disruptive action of moisture trapped in rock capillaries is well known.

Plate 108 Concrete buildings in Ras Al Khaimah, United Arab Emirates, illustrating salt weathering caused by migration of salt solutions up the pillars by the wick effect. The process is exacerbated by high ground water levels associated with the spread of irrigation

Salt can have a deleterious impact on iron and concrete. Many engineering structures are made of concrete containing iron reinforcements. The formation of the corrosion products of iron (i.e. rust) causes a volume expansion to occur. If one assumes that the prime composition of such corrosion products is $Fe(OH)_3$, then the volume increase over the uncorroded iron can be fourfold. Thus when rust is formed on the iron reinforcements, pressure is exerted on the surrounding concrete. This may cause the concrete cover over the reinforcements to crack, which in turn permits the ingress of oxygen and moisture which then aggravates the corrosion process. In due course, spalling of concrete takes place, the reinforcements become progressively weaker, and the whole structure may suffer deterioration.

Rates of corrosion are accelerated by chloride ions which may occur in a concrete because of the use of contaminated aggregates or because of penetration from a saline environment. However, the electro-chemical corrosion of metals can also be produced by sulphates for there are often sulphate-reducing bacteria in a saline soil containing sulphates, which can cause strong corrosion to metals.

Sulphates can cause severe damage to, and even complete deterioration of, Portland cement concrete. Although there is controversy as to the exact mechanism of sulphate attack (Cabrera and Plowman 1988), it is generally accepted that the sulphates react with the alumina-bearing phases of the hydrated cement to give a high sulphate form of calcium aluminate known as ettringite. Magnesium sulphate is particularly aggressive because in addition to reacting with the aluminate and calcium hydroxide as do the other sulphates, it decomposes the hydrated calcium silicates and, by continued action, also decomposes calcium sulphoaluminate. The formation of ettringite involves an increase in the volume of the reacting solids, a pressure build up, expansion and, in the most severe cases, cracking and deterioration. The volume change on formation of ettingite is very large, and is even greater than that produced by the hydration of sodium sulphate.

Another mineral formed by sulphates coming into contact with cement is Thaumasite. This causes both expansion and softening of cement and has been seen as a cause of disintegration of rendered brick work and of concrete lining in tunnels. Building materials may also be damaged by SULPHATION, which creates disruptive gypsum on rock surfaces.

Finally, one controversial issue in weathering studies is whether the presence of salts accelerates the rate at which frost action operates, and, if it does, why this should be. Some laboratory studies (see Williams and Robinson 1991) have indicated that some rocks disintegrate more rapidly when they are frozen after soaking in salt solution rather than in pure water, and various studies have shown that de-icing salts can promote the breakdown of concrete by freeze–thaw. However, in some laboratory simulations salts may reduce or even prevent frost weathering.

This whole issue is important in terms of understanding weathering in high latitude coastal situations (e.g. on shore platforms), to understand the effects of de-icing salts on road surfaces and engineering structures like bridges, and also because salts generated by acid rain could conceivably enhance frost action. Williams and Robinson (1991), in a useful review, have looked at some of the mechanisms that might explain why under some conditions salts could accelerate frost weathering.

Identification of areas where salt weathering is a hazard to engineering structures is an important task for applied geomorphologists and a detailed discussion of this in the context of ground water conditions in arid regions is provided by Cooke *et al.* (1982). As irrigation spreads, leading to ground water rise, areas subject to salt attack may increase. There are already many examples of the acceleration of salt weathering by human activities causing the decay of cultural treasures, such as the Sphinx in Cairo, Petra in Jordan and Monhenjo-Daro in Pakistan.

References

Cabrera, J.G. and Plowman, C. (1988) The mechanism and rate of attack of sodium sulphate on cement and cement/pfa pastes, *Advances in Cement Research* 1(3), 171–179.

Cooke, R.U. and Smalley, I.J. (1968) Salt weathering in deserts, *Nature* 220, 1,226–1,227.

Cooke, R.U., Brunsden, D., Doornkamp, J.C. and Jones, D.K.C. (1982) *Urban Geomorphology in Drylands*, Oxford: Oxford University Press.

Evans, I.S. (1970) Salt crystallisation and rock weathering: a review, *Revue de Géomorphologie Dynamique* 19, 153–177.

Goudie, A.S. (1974) Further experimental investigation of rock weathering by salt and other mechanical processes. *Zeithscrift für Geomorphologie Supplementband* 21, 1–12.

Goudie, A.S. and Viles, H.A. (1997) *Salt Weathering Hazards*, Chichester: Wiley.

MacInnis, C. and Whiting, J.D. (1979) The frost resistance of concrete subjected to a deicing agent, *Cement and Concrete Research* 9, 325–336.

Mortensen, H. (1933) Die Salzprengung und ihre Bedeutung für die regional klimatische Gliederung der Wüsten, *Petermanns Geographische Mitteilungen* 79, 130–135.

Schiavon, N., Chiavari, G., Schiavon, G. and Fabbri, D. (1995) Nature and decay effects of urban salting on granite building stones, *Science of the Total Environment* 167, 87–101.

Williams, R.B.G. and Robinson, D.A. (1991) Frost weathering of rocks in the presence of salts – a review, *Permafrost and Periglacial Processes* 2, 347–353.

Winkler, E.M. and Wilhelm, E.J. (1970) Saltburst by hydration pressures in architectural stone in urban atmosphere, *Geological Society of America Bulletin* 81, 567–572.

A.S. GOUDIE

SALTATION

Derived from the Latin *saltare*, saltation, coined by Gilbert (1914), refers to the hopping, jumping or leaping of sediment grains transported by a fluid, whether wind or water (Bagnold 1956). The grains are launched from their bed in a high angle trajectory by lift forces. In air this trajectory is between 20° and 40° downwind. Grains then accelerate in the flow direction as a result of fluid drag. Then, as a result of gravitational and drag forces they fall back to the bed on a more gentle trajectory (10°–15° in air). The entrainment of a grain in water is due to direct fluid lift and drag forces. In air, entrainment can also result from grain collision (Plate 109). The force at which a

Plate 109 Saltating sand grains leaping to a metre or so above the surface during a sand storm at Budha Pushkar in the Rajasthan Desert, India

grain is set in motion (the entrainment threshold), as well as the height and length that it attains in its subsequent jump, is proportional to the shear velocity of the fluid and to grain size. In air, saltation hop-lengths are about 12 to 15 times the height of bounce. Because a sand grain is only two to three times denser than water, the inertia of the rising grain only carries it to a height of about three grain diameters.

In air, on the other hand, saltating grains rise higher, sometimes to several metres, especially after bouncing impacts on rock or pebble surfaces. It is the mode of travel of most wind-blown sand. In addition, the impact of saltating grains can cause a slow downwind movement of grains by surface creep. There is a transition between surface creep, where grains do not lose contact with the bed, and saltation. Some grains may make very short trajectory paths where they barely leave the bed. This process is termed reptation (Rice *et al.* 1995).

Saltation also affects snow and contributes to the development of avalanches (Sato *et al.* 2001).

References

Bagnold, R.A. (1956) The flow of cohesionless grains in fluids, *Philosophical Transactions of the Royal Society of London* A249, 235–297.

Gilbert, G.K. (1914) Transportation of debris by running water, *United States Geological Survey Professional Paper* 85.

Rice, M., Willetts, B. and McEwan, I. (1995) An experimental study of multiple grain-size ejecta by collisions of saltating grains with a flat bed, *Sedimentology* 42, 595–706.

Sato, T., Kosugi, T. and Sato, A. (2001) Saltation-layer structure of drifting snow in wind tunnel, *Annals of Glaciology* 32, 203–208.

A.S. GOUDIE

SALTMARSH

Coastal and estuarine saltmarshes are depositional landforms situated within the upper part of the intertidal zone. Implicit in their definition is the presence of halophytic (salt-tolerant) vegetation. This distinguishes saltmarshes from tidal flats, from which they commonly develop. Inland areas characterized by alkaline soils may also develop a similar vegetation cover. Associated with aridity, these areas are more commonly referred to as salt flat, salt steppe or salt desert and are not considered further in this entry.

Saltmarshes have a wide geographic distribution along temperate and high latitude coasts, but are replaced by MANGROVES SWAMPS in the tropics. Locally, their occurrence is restricted to low wave energy environments which favour the accumulation of fine, generally muddy, sediments. The morphology of most saltmarshes is characterized by a seaward sloping vegetated platform, dissected by networks of tidal channels (also termed 'creeks' or 'sloughs'). The subtle topography of the marsh surface is often associated with a zonation in plant productivity and/or species composition, which results from complex linkages between factors such as salinity stress, nutrient availability, frequency of flooding (a function of elevation) and plant competition.

At a global scale, major differences in marsh character result from the interaction between ecological, climatic, edaphic and hydrographic influences. These are mediated at a regional scale by the nature and abundance of fine sediments and by the range of depositional settings afforded by particular coastal configurations. Saltmarshes are characterized by a particularly strong interplay of physical, biological and geochemical processes and, accordingly, have long been the subject of considerable scientific interest.

Early scientific studies were concerned mainly with the processes of halophyte colonization under the influence of various environmental factors (notably elevation, as the crucial factor determining the frequency of tidal inundation, salinity and soil aeration) and the importance of vegetation (especially dense swards of tall marsh grasses) in the trapping and binding of fine sediment. Research in Europe and North America emphasized the role of coastal halophytes as 'land-building agents', leading to a model of saltmarsh morphological development under the influence of an autogenic plant succession (Chapman 1974). Geographical variations in the plant succession provide one basis for the classification of marsh types (e.g. Adam 1990). Subsequent ecosystem studies have shown that saltmarshes are sites of high biological productivity, the cycling of which is governed by complex vegetation–substrate–fauna interactions and by tidal exchanges of water, sediments and nutrients with the marine environment. The so-called 'outwelling hypothesis' (Odum 1971) attributed much of the enhanced biological productivity of coastal waters to exports of organic material and nutrients from intertidal marshes and subtidal seagrass beds. Empirical studies provide only partial support for such nutrient exports, and highlight the importance of geomorphological controls on marsh configuration and processes, operating over a range of scales (Nixon 1980).

Geomorphologists have tended to assign a secondary, more opportunistic, role to the colonization of intertidal surfaces by halophytic vegetation. From this perspective, marsh ecological characteristics are viewed as contingent upon the provision of viable substrates for plant colonization. Four main sets of physical factors are implicated: sediment supply; tidal regime; wave climate; and relative sea-level movement (Allen and Pye 1992).

The configuration and extent of coastal margins exert a first-order control on both sediment supply and the 'accommodation space' for saltmarsh development. Frey and Basan (1985), for example, draw attention to physiographic (see PHYSIOGRAPHY) contrasts between the Pacific and Atlantic coasts of North America. On the tectonically active and sediment-deficient Pacific coast, saltmarshes are fragmented and are restricted to narrow fringes around protected embayments, estuaries (see ESTUARY) and (in the north) FJORDS. The more extensive Atlantic coastal plain marshes are continuous over larger areas. Regional variation in the width of the CONTINENTAL SHELF also exerts a control on tidal range which, in turn, defines a zone within which saltmarsh sedimentation can occur.

Saltmarshes are important sinks for fine sediment and play an important role in sediment exchange between estuarine and coastal waters. The nature of saltmarsh sediments varies markedly between *allochthonous* systems, characterized by the deposition of externally derived inorganic sediments, and *autochthonous* systems, dominated by the accumulation of internally produced organic material (Dijkema 1987). The relative importance of inorganic and organic matter accumulation determines the nature of saltmarsh morphodynamic development as well as the ability of both physical and ecological components of the system to adjust to changes in environmental boundary conditions (notably tidal range and mean sea level; French 1994).

Tidal hydrodynamic processes are especially important in controlling the rate and pattern of sedimentation within allochthonous marshes, some of the best-developed examples of which occur under large tidal ranges (e.g. Davidson-Arnott *et al.* 2002). In these systems, the introduction of muddy sediment is controlled by both elevation (which determines the frequency and duration of flooding, or 'hydroperiod') and by proximity to the tidal channels through which most of the tidal water movement occurs (French and Spencer 1993). The feedback between elevation, tidal flooding and sedimentation is a key determinant of long-term marsh morphodynamics (Allen 2000; Friedrichs and Perry 2001). Thus, newly formed marshes typically exhibit rapid rates of vertical sedimentation, whilst the sedimentation is much slower in older marshes, which are higher in elevation and therefore less frequently inundated. Non-tidal inundation, such as that resulting from occasional storm surge events, is of greater importance in introducing sediment to marshes with a very small tidal range.

Wave climate exerts an important local control on horizontal marsh extent, even in otherwise sheltered embayments and estuaries, where small geographical variations in fetch may give rise to significant differences in the character and energetics of the intertidal zone. Wave-induced stresses determine the viability of vegetation establishment, although the influence of waves on the stability of the underlying substrate seems to be of more importance than the mechanical strength of the plants. Wave climate also determines the morphology of the saltmarsh to tidal flat transition. Under moderate wave energies this may take the form of an erosional 'micro-cliff'.

The formation and development of saltmarshes is also related to sea-level change. SEA LEVEL provides a moving boundary condition which, along with tidal range, determines the vertical extent of saltmarsh growth. Modern saltmarshes formed in response to Holocene sea-level rise, and minor oscillations in sea level appear to have been associated with distinct episodes of saltmarsh expansion in areas conducive to fine sediment accumulation. This 'depositional paradigm' (Stevenson *et al.* 1986) has been challenged by the discovery of sedimentary deficits within subsiding deltaic marshes. This has focused attention on the extent to which saltmarshes are able to accumulate sufficient material to keep pace with the forecast rates of sea-level rise under global warming scenarios. In general, contemporary rates of sedimentation within non-deltaic marshes in both North America and Europe exceed present rates of sea-level rise. Furthermore, marsh elevations can adjust to higher rates of sea-level rise through the increased sedimentation which accompanies more frequent inundation (French 1994). This adjustment does, of course, depend on sediment supply. The effect of increased water levels on vegetation and soils is also important, especially in autochthonous marshes with limited inputs of inorganic sediment.

Saltmarshes in many parts of the world have experienced high rates of historical loss through reclamation and destructive industrial uses (such as salt production using evaporation ponds). In addition, large areas of estuarine and open coastal marsh have been lost through recent erosion. This erosion is widely attributed to a combination of contemporary sea-level rise and the presence of seawalls and other structures which restrict any natural landward migration of the intertidal zone. From an ecological perspective, remaining saltmarshes are not only important for the maintenance of estuarine and coastal food chains, but also provide valuable wetland habitats. They also act as a naturally dissipative landform that forms an important element of sustainable coastal defence and flood protection strategies. In particular, a number of studies have shown saltmarsh to be much more effective than unvegetated tidal flats in dissipating wave energy. This function translates into significant engineering cost savings when sea defences are constructed behind a strip of saltmarsh.

These ecological and engineering functions have stimulated efforts to restore saltmarsh, for example through the re-establishment of tidal conditions in reclaimed areas no longer required for agriculture. The success of such schemes has been mixed, and it is now clear that successful engineering of ecological and flood defence functions is crucially dependent upon a sound understanding of the geomorphological processes which act to shape the corresponding natural systems (French and Reed 2001).

References

Adam, P. (1990) *Saltmarsh Ecology*, Cambridge: Cambridge University Press.

Allen, J.R.L. (2000) Morphodynamics of Holocene saltmarshes: a review sketch from the Atlantic and Southern North Sea coasts of Europe, *Quaternary Science Reviews* 19, 1,155–1,231.

Allen, J.R.L. and Pye, K. (1992) Coastal saltmarshes: their nature and importance, in J.R.L. Allen and K. Pye (eds) *Saltmarshes: Morphodynamics, Conservation and Engineering Significance*, 1–18, Cambridge: Cambridge University Press.

Chapman, V.J. (1974) *Salt Marshes and Salt Deserts of the World*, 2nd edition, Lehere: Cramer.

Davidson-Arnott, R.G.D., van Proosdij, D., Ollerhead, J. and Schostak, L. (2002) Hydrodynamics and sedimentation in salt marshes: examples from a macrotidal marsh, Bay of Fundy, *Geomorphology* 48, 208–231.

Dijkema, K.S. (1987) The geography of salt marshes in Europe, *Zeitschrift für Geomorphologie* 31, 489–499.

French, J.R. (1994) Tide-dominated coastal wetlands and accelerated sea-level rise: a NW European perspective, *Journal of Coastal Research Special Issue* 12, 91–101.

French, J.R. and Reed, D.J. (2001) Physical contexts for saltmarsh conservation, In: A. Warren and J.R. French (eds) *Habitat conservation: managing the physical environment*, 179–228, Chichester: Wiley.

French, J.R. and Spencer, T. (1993) Dynamics of sedimentation in a tide-dominated backbarrier saltmarsh, Norfolk, UK, *Marine Geology* 110, 315–331.

Friedrichs, C.T. and Perry, J.E. (2001) Tidal salt-marsh morphodynamics: a synthesis, *Journal of Coastal Research Special Issue* 27, 7–37.

Frey, R.W. and Basan, P.B. (1985) Coastal salt marshes, in R.A. Davis (ed.) *Coastal Sedimentary Environments*, 2nd edition, 225–301 New York: Springer-Verlag.

Odum, E.P. (1971) *Fundamentals of Ecology*, 3rd edition, Philadelphia: Saunders.

Nixon, S.W. (1980) Between coastal marshes and coastal waters – a review of twenty years of speculation and research in the role of salt marshes in estuarine productivity and water chemistry, in R. Hamilton and K.B. McDonald (eds) *Estuarine and Wetland Processes*, 437–525, New York: Plenum.

Stevenson, J.C., Ward, L.G. and Kearney, M.S. (1986) Vertical accretion in marshes with varying rates of sea level rise, in D.A. Wolfe (ed.) *Estuarine Variability*, 241–260, Orlando: Academic Press.

Further reading

Allen, J.R.L. and Pye, K. (eds) (1992) *Saltmarshes: Morphodynamics, Conservation and Engineering Significance*, Cambridge: Cambridge University Press.

SEE ALSO: mangrove swamp; mud flat and muddy coast; tidal creek; tidal delta

J.R. FRENCH

SAND-BED RIVER

An alluvial river in which the bed material is predominantly sand-sized (0.0625–2 mm). Sediment size is a primary control on river form and process and sand-bed rivers therefore exhibit a number of characteristic process and morphological attributes that distinguish them from channels dominated by other sediment sizes. They are recognized by geomorphologists as a fundamental river type, distinct from GRAVEL-BED RIVERS, in which the bed material is coarser (>2 mm) and less well sorted.

Within the drainage network DOWNSTREAM FINING ensures that sand-sized sediments dominate the sediment load delivered to distal reaches. Gravel-bed rivers in upland and piedmont settings therefore give way to sand-bed channels in the lowlands, though sandy reaches will develop wherever sand is supplied in abundance, for example as a function of local lithology or land use.

The grain characteristics and relative importance of SUSPENDED LOAD and BEDLOAD transport vary in sand-bed rivers with sediment supply and flow characteristics, but suspended grains are typically finer than 0.2 mm and account for a greater proportion of total sediment yield. In the limit case, sand-bed rivers can be thought of as suspension-dominated (e.g. Parker in press). Examination of such channels reveals that the boundary shear stresses generated by modest flows (for example bankfull) are at least one order of magnitude larger than the stresses needed to entrain median sizes. This indicates that in sand-bed rivers mass sediment entrainment and transport occur at discharges well below discharges associated with rare floods, and that modest flows are responsible for maximum cumulative sediment yield. Transport conditions are fundamentally different in bedload-dominated, gravel-bed rivers where flows reach entrainment thresholds relatively infrequently. Sand-bed channels are therefore characterized by excess rather than threshold hydraulic stresses, have more mobile and responsive 'live' beds, and carry sediment loads that are limited by supply issues rather than the flow's competence to move available particle sizes.

When stresses are sufficient to entrain grains but turbulent eddying is insufficient to suspend them, grains move as bedload, primarily by SALTATION. Near their entrainment thresholds grains move randomly over a plane bed, but as flow intensity increases patterns of erosion and deposition become ordered in space and groups of sand grains move together as migrating BEDFORMS. Ripples and then dunes form, but a point is reached where structured transport collapses and dunes are replaced by an upper-regime plane bed. With further increases in flow the water surface may develop waves, beneath which antidunes grow. This bedform sequence correlates with both bedload and suspended sediment transport rates and increases of an order of magnitude have been observed between successive stages in flume experiments. The complex mutual adjustments between bedform generation, macroturbulence (burst cycles), flow resistance, and suspended sediment concentrations that ultimately determine hydraulic characteristics and sediment transport rates are incompletely understood, but sufficient has been learned to allow palaeohydraulic interpretation of bedform traces preserved in the modern and lithified deposits of sand-bed rivers.

Because bedform dimensions far exceed grain dimensions in sand-bed channels, a large proportion of total boundary flow resistance is due to bedforms and grain resistance is relatively unimportant. Drag increases as flat beds (grain roughness only) develop ripples then dunes. However, the transition from dunes back to a plane bed is accompanied by a significant drop in resistance that marks a shift from the so-called lower to upper flow regime and ensures that stage-velocity relations are non-linear. Energy losses rise again if the flow becomes supercritical and the water surface develops breaking waves. An additional control on flow velocity in sand-bed channels is suspended sediment concentration, which reduces resistance by dampening turbulence intensity.

The common transition from gravel to sand bed is often abrupt and accompanied by a distinct break of slope. This has been explained in terms of the gradient required to transport sand and gravel loads and the respective importance of channel capacity and competence in suspension and bedload-dominated systems. Slopes in sand-bed reaches are relatively small, typically between 0.002 and 0.0001 (2–0.1 m per kilometre) and exhibit less downstream adjustment than in gravel-bed reaches where large changes in grain size require consequent changes in flow competence.

Cross-section morphology may be regarded as the more adjustable morphological dimension in sand-bed rivers. An element of distinctiveness is apparent in HYDRAULIC GEOMETRY relations for sand-bed channels, but the multivariate nature of the problem makes widespread generalizations difficult. For example, many sand-bed channels transport significant quantities of fine-grained, cohesive sediment (silts and clays) that are deposited in over-bank positions during floods. This facilitates floodplain accretion and increases bank strength, producing channels that have lower width–depth ratios than gravel-bed rivers carrying similar discharges. However, sand-bed channels that do not carry significant quantities of fines have particularly weak banks and tend to be comparatively wide.

Straight, meandering, braided and anabranching rivers may all have sand or gravel beds. Although the patterns are common to both, Lewin and Brewer (2001) suggest that braid and meander formation are fundamentally different in gravel- and sand-bed channels as a function of differences in excess shear stress and thence sediment mobility and bedform persistence. Certainly a degree of distinctiveness is apparent in the controls on planform type. For example, numerous studies have identified threshold values of discharge and channel slope that discriminate meandering and braided forms, but detailed analysis indicates that the threshold slope for braiding also depends on bed material size, with sand-bed rivers tending to occur on more gentle gradients for a given discharge (Knighton 1998: 211). Similarly, examination of the controls on meander geometry suggest that channels carrying greater suspended loads and more cohesive materials (by extension, suspension-dominated sand-bed channels) tend to be more sinuous and have smaller wavelengths than meandering channels in bedload-dominated systems.

References

Knighton, D. (1998) *Fluvial Forms and Processes*, London: Arnold.

Lewin, J. and Brewer, P.A. (2001) Predicting channel patterns, *Geomorphology* 40, 329–339.

Parker, G. (in press) Transport of gravel and sediment mixtures, in *Sedimentation Engineering*, American Society of Civil Engineers, Manual 54.

Further reading

Simons, D.B. and Simons, R.K. (1987) Differences between gravel- and sand-bed rivers, in C.R. Thorne, J.C. Bathurst and R.D. Hey (eds) *Sediment Transport in Gravel-bed Rivers*, 3–15, Chichester: Wiley.

SEE ALSO: bedform; downstream fining; gravel-bed river; hydraulic geometry; suspended load

STEPHEN RICE

SAND RAMP

Topographically controlled aeolian deposits amalgamated with layers of fluvial, colluvial and talus sediments derived from local mountain sources, and palaeosols representing relatively stable geomorphological periods. Mountains act as barriers to sand transport and sand accumulates on piedmont slopes in their lee and to their windward sides. Where sections are present, multiple periods of sand accumulation and palaeosol development can be identified and such phases can be dated by techniques such as luminescence dating. They can provide a record of past episodes of aeolian activity and stabilization (see, for example, Allchin *et al.* 1978 on the Thar; and Lancaster and Tchakerian 1996 on the Mojave).

References

Allchin, B., Goudie, A.S. and Hegde, K.T.M. (1978) *The Prehistory and Palaeogeography of the Great Indian Desert*, London: Academic Press.

Lancaster, N. and Tchakerian, V.P. (1996) Geomorphology and sediments of sand ramps in the Mojave Desert, *Geomorphology* 17, 151–165.

A.S. GOUDIE

SAND SEA AND DUNEFIELD

Collections of dunes are best called by easily understood terms: 'dunefields', and, when large, 'sand seas'. Many geomorphologists use the term 'erg' for any large collection of dunes (see DUNE, AEOLIAN), but, apart from the indiscriminate use, there are problems with this term. First, 'erg' is an indigenous term from a small part of the northwestern Sahara, where it was used on the topographic maps of over a century ago for anything from a large dune to a very large sand sea. Second, there are many other indigenous terms for bodies of dunes in other parts of the world (even in the northwestern Sahara). And third, better mapping shows there to be two distinct groups of dunefield.

There is a sharp break in the size distribution of collections of dunes at about 32,000 km^2 and the larger group has a sharp peak in size at about 200,000 km^2 (Wilson 1973) (Figure 139). In other words, large dunefields, termed here 'sand seas', are a distinct population. A very similar size break also distinguishes 'seas' in common English usage from smaller bodies of water. The peaked distribution is a function of tectonics. The gentle folds in the basement rock of the African, Asian and Australian deserts are of this size order (Figure 140). Most dunefields and sand

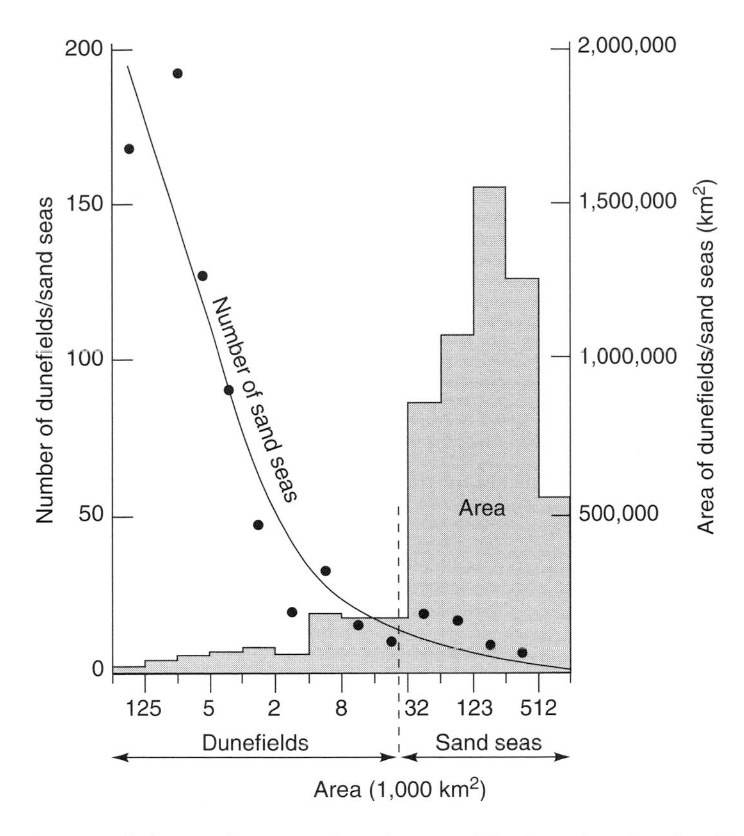

Figure 139 Distribution of the areal extent of aeolian sand bodies, showing the distinction between sand seas and dunefields (after Wilson 1973)

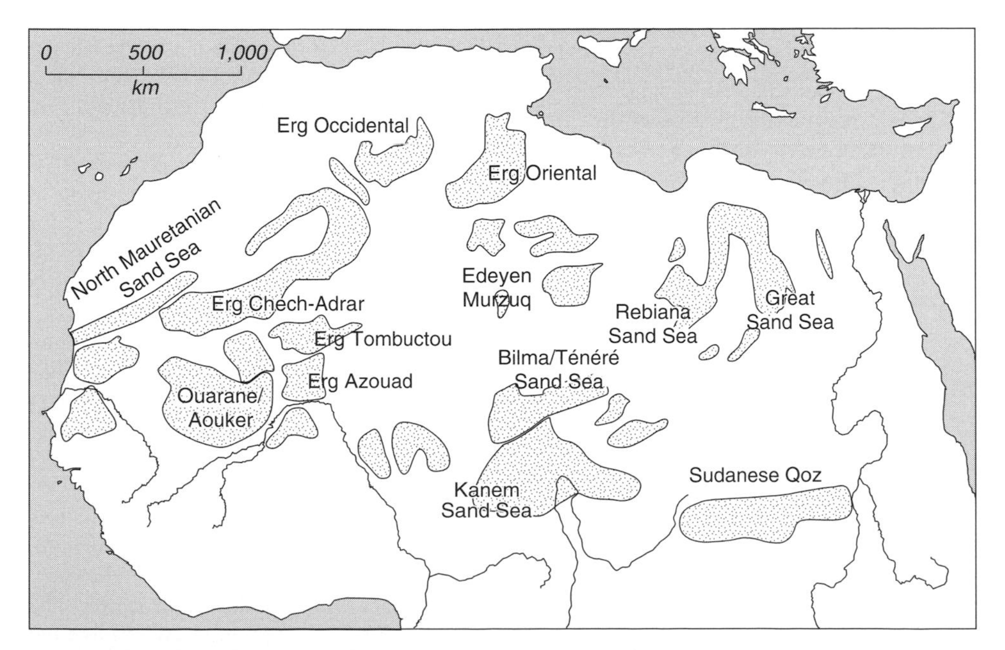

Figure 140 The principal sand seas of the Sahara and Sahel

seas occur in basins or lowlands, and many sand seas nearly fill their basins. Dunefields do so less commonly, although some, as in the Kelso dunefield in California and the Great Sand Dunes in Colorado, fill a large proportion of their small basins.

If the aeolian sand in a dunefield is broken by a patch of desert floor (pavement or bare rock) that is bigger than an interdune, then strictly speaking, this is the edge of the field, but this is a somewhat pedantic restriction. Judgements like this are just one example of the ambiguities in defining the extent of a sand sea (or dunefield), but even with this proviso, the largest active/semi-active sand seas include: the Rub' al Khali in Arabia (about 560,000 km^2) and the Great Eastern Erg in Algeria (about 192,000 km^2). Another somewhat vain definition is of the lower size limit of a dunefield: it could be two dunes.

Sources of sand

Apart from topographic control, the most important control on the distribution of bodies of dunes is proximate supplies of sand. These are of various kinds: weathered bedrock, fluvial deposits, coastal sedimentary cells, and earlier or other dunefields are the most common. Many large dunefields and sand seas derive their sand from a mixture of these sources. In many of the older sand seas, sand has accumulated and been reworked over many cycles. Many dunefields are parts of regional systems of sand movement, one field feeding sand to the next. In the Sahara the movement is generally from north-east to south-west. Regional movements also occur in the Mojave, in the Namib, and probably in Australia. The issue of source, which is best illuminated by studies of mineralogy, has been extensively debated in relation to individual sand seas (Muhs 2003).

Dune assemblages

Many sand seas, unlike most dunefields, contain a variety of dune types. There are several reasons. First, variety is a function of size, for different wind regimes can occur over such large areas, and wind regime is a major determinant of dune type. The northern parts of the Saharan and Arabian sand seas experience frontal winds in winter, and trade winds in summer. Further south in the same sand sea, the influence of the frontal winds fades. In the lower latitude sand seas of southern Africa and Australia, the trade wind systems dominate and dune form is less varied.

Second, variety of dune form is a function of age. Large bodies of sand can only accumulate over many thousands of years, lengths of time that have seen changes in climate. Older dunes may differ in type and orientation from younger ones. Moreover, many older dunes have been subdued by erosion, and have developed soils at times when the climate was wetter. Their gentle topography may be scarred by haphazard reactivation or it may be buried by younger sands to varying degree. This can be seen in the Great Western Erg in Algeria. The northern portion is dominated by ancient, mostly linear, and now subdued dunes, which have developed soils. Further south there are more active and much higher dunes in various formations (transverse, linear and star) (Callot 1988). Another example of how age creates variety is the Wahiba sand sea in Oman. The oldest dunes here have been lithified (see AEOLIANITE), and eroded to an almost level plain. A later generation of large linear dunes partially covers these lithified sands and has moved over them to the north. In the south, these linear dunes have in turn been covered by transverse and network dunes built with new Late Pleistocene sand. These sands have invaded the sand sea from the coast, and are themselves derived from the marine erosion of the ancient lithified aeolianites (Warren 1988). Age, in a sense, allows greater entropy and disorder in dunefields and sand seas.

Third, variety of dune type in large sand seas is a function of the movement of the dune body as a whole. Porter (1986) developed a model for ancient sand seas, now forming aeolian sandstones. It is also to be applicable to some Saharan sand seas. In the model, the finer sand and associated dunes moves forward more quickly than the coarse sand, which remains as a trailing edge of low zibars.

Non-aeolian features

Sand seas and dunefields have some distinctive non-aeolian features. The most striking of these is associated with the blockage of the preexisting and marginal drainage by the dunes. The drainage and dune water tables feed or have fed many thousands of lakes, some large. These fill up even after the rare storms of today, but were more often full in the wetter periods of the past. These ancient lakes have left deposits ranging from chara limestones to diatomites, sometimes in very thin (5–10 cm thick) drapes on the sand. Round these lakes, in many Saharan and Arabian sand seas, there are signs of human (Palaeolithic and Neolithic) settlement, in the form of tools and middens. Lakes like these are common in the Taklamakan in China and in the Nebraska Sand Sea.

Activity/planetary sand seas

Dunefields and sand seas are in a spectrum from fully active to lithified and largely buried. Some small dunefields are almost wholly active, in the sense that all the dunes are in movement, and most of the surface bare and blowable. But even in the hyper-arid central Sahara and Arabia, many sand seas have portions that are partly stabilized, being remnants of earlier phases bearing remnants of soils. As one moves to the margins of these deserts, as in northern Arabia, or either edge of the Saharan or Australian sand seas, dunes become progressively more stabilized by a cover of vegetation and a developed soil. The climatic limits of aeolian activity are debatable, if only for the reason that a migrating climatic boundary has left a complex legacy of stabilized, reactivated and new dunes. There is no dispute, however, that there are dunefields and sand seas that are now almost wholly inactive, some indeed covered by deep forest. Some of the largest surface sand seas, for example of the proto-Kalahari (2,500,000 km^2), or the Nebraska Sand Hills (57,000 km^2), and the Sudanese *qoz* (about 240,000 km^2) are now almost wholly stabilized. Stabilized dunefields and sand seas also occur in areas in which there are now no deserts, as in northern Canada, the North European Plain, Hungary and northern Tasmania.

Sand seas and dunefields slowly lithify under various influences, and become aeolian sandstones. The sediments of ancient 'ergs' occur from as far back as the Precambrian. The deposits of these sand seas generally show great complexity, as a result of changes in climate and tectonics (Blakey 1988). There are also sand seas and dunefields on Mars, some very similar to terrestrial features (Lancaster and Greeley 1990).

References

Blakey, R.C. (1988) Basin tectonics and erg response, *Sedimentary Geology* 56, 127–151.

Callot, Y. (1988) Evolution polyphasée d'un massif dunaire subtropical: Le Grand Erg occidental (Algérie), *Bulletin de la Société géologique de France*, Série 8, 4, 1,073–1,079.

Lancaster, N. and Greeley, R. (1990) Sediment volume in the North Polar Sand Seas of Mars, *Journal of Geophysical Research – Solid Earth and Planets* 95, 921–927.

Muhs, D.R. (2003) Mineralogical maturity in dune fields of North America, Africa and Australia, *Geomorphology* in press.

Porter, M.L. (1986) Sedimentary record of erg migrations, *Geology* 14, 497–500.

Warren, A. (1988) The dunes of the Wahiba Sands, *Journal of Oman Studies, Special Report* 3, 131–160.

Wilson, I.G. (1973) Ergs, *Sedimentary Geology* 10, 77–106.

Further reading

Cooke, R.U., Warren, A. and Goudie, A.S. (1993) *Desert Geomorphology*, London: University College Press.

Livingstone, I. and Warren, A. (1996) *Aeolian Geomorphology: An Introduction*, Harlow: Addison-Wesley Longman.

SEE ALSO: dune, aeolian

ANDREW WARREN

SANDSHEET

An area of predominantly aeolian sand where dunes with slipfaces are generally absent. Sandsheet surfaces can be rippled or unrippled, and range from flat to regularly undulatory to irregular (Kocurek and Nielson 1986). They form in ergs (sand seas) where conditions are not suitable for dunes, or particular factors act to interfere with dune formation. These factors include a high water table, periodic flooding, surface binding or cementation, the presence of vegetation, and a significant coarse grain-size component. These same factors are also effective in promoting sandsheet accumulation where otherwise only sand transport without deposition would occur.

The classic sandsheet is the Selima Sand Sheet of the Libyan Desert. It covers around 120,000 km^2 and is a largely featureless surface of lag gravels and fine sand broken only by widely separated dunefields and giant ripples. Maxwell and Haynes

(2001) have stressed the role of both fluvial deposition and aeolian modification in its development.

References

Kocurek, G. and Nielson, J. (1986) Conditions favourable for the formation of warm-climate aeolian sandsheets, *Sedimentology* 33, 795–816.

Maxwell, T.A. and Haynes, C.V. Jr (2001) Sandsheet dynamics and Quaternary landscape evolution of the Selima Sand Sheet, Southern Egypt, *Quaternary Science Reviews* 20, 1,623–1,647.

A.S. GOUDIE

SANDSTONE GEOMORPHOLOGY

Landforms developed on sandstone can be arranged in a hierarchical series increasing from the microscopic to the regional scale. Examples of features at these various levels are: (1) etch pits on quartz grains, silica skins; (2) TAFONI, tesselated surfaces, solution runnels; (3) cliffs, domes, towers, arches (see ARCH, NATURAL); (4) CUESTA and scarp assemblages; plateau and canyon assemblages; PSEUDOKARST assemblages, ruiniform assemblages. Explaining features at any of these scales requires an understanding of the variable properties of sandstones.

Sandstones can be most simply defined as clastic rocks in which sand-sized fragments are dominant, but there is considerable variation amongst them, and this variation is of geomorphological significance. The size of the dominant clasts is important, for a fine, silty sandstone responds to erosion differently to a coarse, pebbly one. The proportion of intergranular matrix also is significant, and 'clean' sands or arenites (<15 per cent matrix) need to be distinguished from 'dirty' sands or wackes (>15 per cent matrix). The composition of the grains can vary greatly, for arenites can be divided into quartz (<5 per cent feldspar or rock fragments), lithic (>25 per cent rock fragments), arkose (>25 per cent feldspar) and volcanic (>50 per cent volcanic fragments) subtypes. The amount and composition of intergranular cement, porosity and permeability, and the pattern of bedding and jointing exert important influences on the mechanical and chemical properties of sandstones.

The strength of sandstone depends greatly on its composition. For example, the uniaxial (or unconfined) compressive strength varies from 200 MPa (1 Megapascal = 145 lb/in^2) in a strongly cemented quartz arenite, 20 MPa in a weak sandstone, to 2 MPa in very weakly consolidated sands. Moreover, the strength of sandstones varies greatly with the type of stress applied. Their shearing strength is generally less than half, and their tensile strength only 5 to 15 per cent, of their compressive strength. Furthermore, the strength of saturated sandstone is from 90 to only 10 per cent of the strength of the same rock when dry; the difference between wet and dry strengths increases mainly with the clay content of sandstone. The clay content, together with the degree of cementation, also largely determines the deformability of sandstones. Those which are highly cemented and have little clay tend to fail by brittle fracture, whereas those which are poorly cemented and have substantial clay contents tend to deform elastically before rupturing.

The strength of sandstone mass depends not only on the strength of the intact pieces, but also on their freedom of movement, which, in turn, depends on the spacing, orientation and shearing strength of the discontinuities. Jointing and bedding provide major pathways along which water penetrates sandstones, and thereby influence both weathering of the rock, and the build up of pore-water pressure that may promote failure in the rock.

Strong sandstones generally form major cliffs, but the modes of failure on cliffs, and therefore the morphology of the cliffs themselves, varies with the characteristics of particular rock masses. Because sandstones are weakest when in tension, horizontal projections of rock from cliffs, unless in the form of supported arches, generally extend no more than 10 to 20 m. The patterns of scars caused by tensional rupturing indicate that about 50 per cent of the failures through intact sandstone are generated upwards, about 40 per cent laterally, and, contrary to the predictions of commonly used models, only about 10 per cent downwards. In well-jointed sandstones, the dominant mechanism of failure is the collapse of individual blocks that have been undercut. Once undercutting has penetrated beneath the central third of a block, the block will generally topple outwards, unless held in place by pressure exerted by adjoining blocks.

Undercutting of sandstone is mainly the result of the breakdown of less resistant, underlying rocks. While seepage promotes the plastic failure of clay-rich rocks such as shales, even in these rocks, failure frequently takes the form of brittle fracture along closely spaced beds and joints. Undercutting can also occur along prominent bedding planes in the sandstone itself, and in beds of conglomerate that commonly occur at the base of upward fining cycles of sandstone deposition. Seepage promotes undercutting, especially in very permeable sandstone, but undercutting is essentially the result of the very high concentration of stress generated at the base of a cliff by the weight of the rock above. As the rock at the base is compressed, it deforms laterally, and when the lateral stress pushing outwards into the undercut exceeds the tensional strength, the rock fractures. Tensional stresses generated in this fashion at the base of sandstone cliffs can be greatly augmented by active, or residual, tectonic stresses. For example, the measured horizontal stresses south of Sydney, Australia, are three times greater than the vertical stress. In these conditions, zones of tension, in which joints separating the blocks are opened, extend up the entire cliff face and onto the edge of the adjacent plateau surface. Where sandstones are tilted, joint-bounded blocks may topple or slide down the slope. High and narrow blocks tend to topple; wide and flat blocks tend to slide. Movement of blocks depends also on the steepness of the slope and the frictional resistance at the base of a block.

The form of cliffs thus depends not only on the ROCK MASS STRENGTH of the sandstone itself, but also on the properties of the rocks exposed beneath it. Along the Arnhemland Escarpment, in northern Australia, the Kombolgie Sandstone stands in vertical cliffs where less resistant schists and granites are exposed beneath it, but forms irregular slopes of lower declivity where it occupies the full height of the escarpment. Armouring of weaker rocks by debris from sandstone outcrops also retards undercutting. In the south-west of the USA the presence or absence of thick mantles of talus seems to determine whether slopes are in equilibrium with the properties of the sandstone, or whether they are controlled by foot-slope processes and undercutting.

Plate 110 Cliffs with cavernous weathering in the Nowra Sandstone, south-east Australia

Extensive masses of sandstone may be incorporated in major failures in weaker, underlying rocks (see TOREVA BLOCK). Major rotational failures in clays on the southern part of the Msak Mallat Escarpment, in central Libya, have incorporated slabs of the Nubian Sandstone in a 3-km wide belt of mounds over 100 m high. Removal of confining stresses resulting from the incision of the Cataract Canyon of the Colorado River has allowed evaporites to slowly deform down the dip, causing fracturing and collapse of the overlying sandstones. The result is a series of spectacular graben-like depressions, which are 150 to 200 m wide and are 25 to 75 m deep. Gliding of sandstone blocks away from cliff lines is not limited to areas of highly deformable substrate. South of Sydney, Australia, the movement of large blocks of Nowra Sandstone away from cliff lines is the result of the very slow creep of underlying sandy siltstone (Plate 110).

Many outcrops of sandstone have been shaped into domes and rounded slopes, known as slickrock. In some cases curved sheeting in the sandstone roughly parallels the rounded topography. Most rounded slopes, however, appear to be the result of the granular breakdown of the sandstone and of the peeling of thin, weathered layers from the surface. This is so both in arkosic rocks, such as the Navajo Sandstone of the Colorado Plateau, and in quartz arenites, such as those which are cut into complex arrays of rounded towers in the Bungle Bungle Range of north-west Australia. Granular breakdown and slab failure also are the primary processes in the development of arches, which are quite common in some sandstones. The domes and slickrock slopes that are characteristic of many coarse conglomerates can likewise be attributed to granular breakdown and slab failure.

Weathering of sandstones varies with the mineralogy of its constituents. Where there is considerable matrix, hydration of clay is important. In arkosic sandstone, the primary process is the weathering of feldspars. Even highly quartzose sandstones are subject to extensive, though slow, CHEMICAL WEATHERING, with the order of decay generally being first the siliceous, intergranular cement, then the quartz grains, and finally the quartz overgrowths. The presence of sodium chloride, either as sea spray or as aerosols in rain, has a strong accelerating effect on the dissolution rates of silica. These various processes are dependent on the ability of water to penetrate the sandstone. The opening of joints and generation of microfractures by initial release of confining pressures create numerous pathways for seepage. Even the alteration from fresh to slightly weathered sandstone can result in a reduction by a factor of two or three in mean strength, and of about six in deformability, as porosity and moisture content increase. The advance of weathering depends very much on the permeability of the sandstone, for, where the original pore spaces are filled by overgrowths or siliceous cement, active weathering is essentially limited to a thin surface layer. Nevertheless, prolonged weathering can produce deep solutional features even in quartzites.

The removal of the more soluble constituents in sandstone, and the precipitation of siliceous and iron-rich materials, may result in case hardening of the surface layer. Some case-hardened surfaces develop distinctive polygonal patterns of cracks, known as 'elephant skin'. The cracking seems to be the result of fatigue in the case-hardened layers and of changing surface stresses, similar to the crazing of glazed pottery surfaces. Fire is an important agent in the breakdown of sandstone surfaces in well-vegetated areas. Near Sydney, such surfaces are likely to be exposed to fire once every ten to twenty years. Fire moving rapidly over sandstone produces minor spalling and granular disintegration, while intense fires may cause spalling to depths of 2 cm. Sand blasting of sandstone surfaces during high winds has been overrated, especially as a cause of CAVERNOUS WEATHERING in humid areas, but is very important in arid lands. Shattering by crystal pressures developed in fractures due to freezing is particularly important under cold climates, especially in quartzites, which, though very resistant to abrasion, are brittle rocks that can withstand only slight strain deformation before they rupture explosively.

Initial classifications of sandstone landforms at the regional scale were based on the presumed controlling effect of climate. Climate is undoubtedly the dominant control in the amazing wind eroded YARDANG landscape of the hyper-arid Boukou region of the Sahara. Pseudokarst assemblages, like the impressive array of deep dolines and caves etched into the quartzites of the Roraima region of Venezuela, are certainly best developed in the humid tropics. But solutional features occur extensively on sandstones outside the tropics. Furthermore, some of the most common types of sandstone terrain, such as rounded slickrock relief, and the angular towers and great cliffs of ruiniform assemblages, occur under various climates. The variable properties of the rock, rather than climate, may thus be of primary importance in the shaping of sandstone landforms.

Further reading

Howard, A.D., Kochel, R.C. and Holt, H.E. (eds) (1988) *Sapping Features of the Colorado Plateau, A Comparative Planetary Geology Field Guide*, Washington: NASA.

McNally, G.H. and McQueen, L.B. (2000) The engineering properties of sandstones and what they mean, in G.H. McNally and B.J. Franklin (eds) *Sandstone City, Sydney's Dimension Stone and Other Sandstone Geomaterials*, 178–196, Sydney: Geological Society of Australia.

Mainguet, M. (1972) *Le modelé des grès: Problèmes généraux*, Paris: Institut Geographique National.

Oberlander, T.M. (1977) Origin of segmented cliffs in massive sandstones of southeastern Utah, in D.O. Doehring (ed.) *Geomorphology of Arid Regions*, 79–114, Boston: Allen and Unwin.

Wray, R.A.L. (1997) A global review of solutional weathering forms on quartz sandstone, *Earth Science Review* 42, 137–160.

Young, R.W. and Young, A.R.M. (1992) *Sandstone Landforms*, Berlin: Springer.

R.W. YOUNG

SAPROLITE

Saprolite is weathered rock in which the fabric of the original rock is preserved. The word derives from the Greek sapros (σαπρος or σαπροσ) meaning rotten and was coined in 1895 by Becker. The term is generally applied to chemically altered rocks where all or part of the primary minerals are changed to new-formed minerals. Most often saprolite is referred to granitic rocks, but the term applies to the weathered component of any rock type.

The body of *in situ* weathered rock referred to as saprolite may comprise a number of zones (horizons, layers) depending on the relative rates of weathering, erosion and the composition and hydromorphic characteristics of the regolith. Figure 141 summarizes data on saprolite and various terms used by various authors for various parts of weathering profiles and their saprolite component.

Starting from the bottom of a weathering profile *saprock* is the part of the profile closest to the weathering front (Figure 141). Saprock is rock that has begun to weather, but only about 20–30 per cent of the primary minerals are chemically altered. This is hard to estimate and a preferable definition is that the material requires a sharp hammer blow to break it. The zone of saprock may contain boulders, beds or other masses of unweathered bedrock. At the weathering front, which may be very sharp or gradual, rocks change from fresh to partly weathered, alteration of feldspars to clay minerals is seen and ferromagnesian minerals release Fe^{2+} that is oxidized to red-yellow coloured Fe^{3+} oxihydroxides or oxides.

The form of the WEATHERING FRONT may be relatively flat or very irregular, the latter being more common. Its shape is primarily dependent on the nature of the rocks being weathered. In massive igneous bodies the form is generally more regular – but with isolated sub-spherical boulders or corestones of fresh rock in the saprock – than that in dipping sedimentary rocks or metamorphic rocks. The number and orientation of joints, cleavage and bedding planes that form the initial conduits for the weathering solutions mainly control this. Plates 111–115 are examples of different types of saprolite and weathering fronts.

Moving up profile the saprock gradually gives way to saprolite where the majority of labile (more easily chemically weathered) minerals are altered and replaced by new minerals formed by the chemical weathering. The most common minerals in saprolite are clays, iron oxihydroxides and oxides and any minerals largely resistant to weathering (e.g. quartz, magnetite or pre-existing clay minerals). The clay minerals generally gradually change from smectite ($Si/Al = < 1$) lower in the profile, where drainage is generally poorer than higher where the dominant clay is kaolinite ($Si/Al = 1$). Kaolinite forms, as drainage of the weathering solutions is more efficient.

Above the saprolite is collapsed saprolite or the 'mobile zone'. In this zone (Figure 141) the

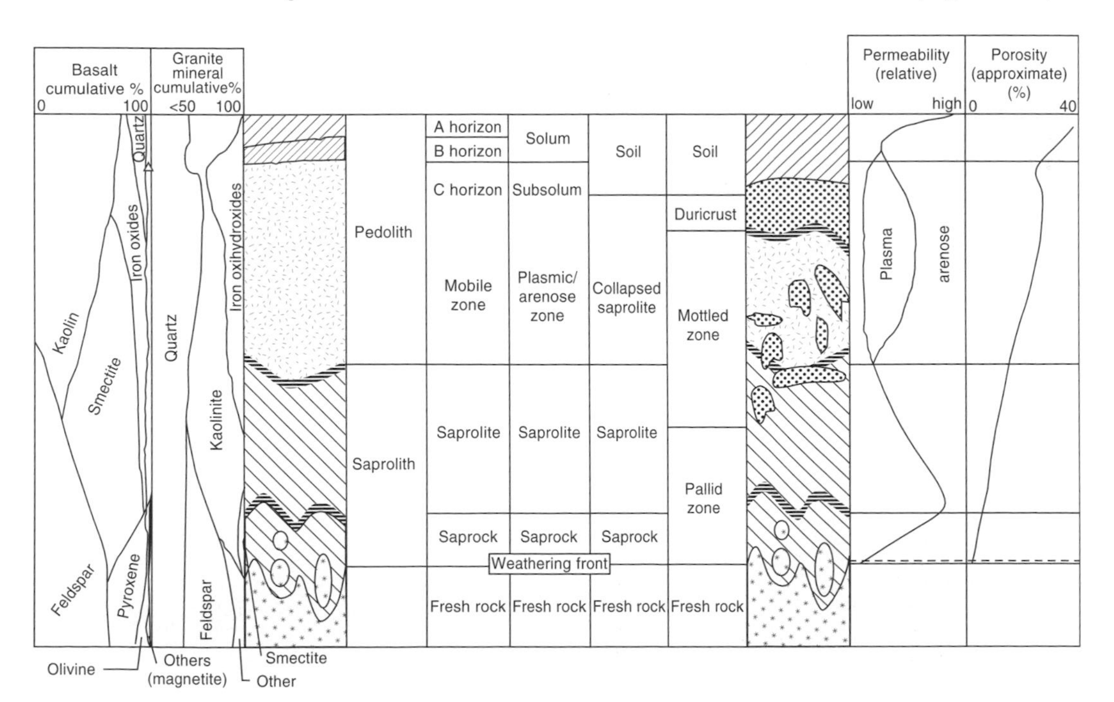

Figure 141 Some of the various terminologies used to describe weathering profiles including saprolite. On the left are two sketch examples of how mineralogy varies through the saprolite and on the right is an example of how the hydraulic properties of the weathering profile and saprolite vary with depth. (modified from Taylor and Eggleton 2001)

Plate 111 A small erosional remnant of a 'typical' weathering profile developed in felsic gneiss on the Yilgarn near Kalgoorlie in Western Australia. It is about 3 m high and shows ferruginous cap about 0.25 m developed over a 0.5 m thick mottled saprolite that grades downward into 3 m of gneissic saprolite. The saprolite is composed of relict quartz grains, and kaolinite with minor haematite. The composition of the mottled saprolite is much the same with increased quantities of haematite. The ferruginous crust is mainly haematite with quartz and minor goethite (photo Ian Roach)

Plate 113 A very deep weathering profile formed in ultra-mafic rocks at Marlborough in central Queensland, Australia. About 40 m of the profile is exposed, but no fresh rock occurs in this pit. The total profile depth is up to about 100 m. The bulk of the material in the photo is saprolite composed of clay minerals, mainly nontronite and talc with Fe^{3+} oxides, goethite and haematite and secondary silica as chrysoprase and chalcedony. The upper part of the profile is transported debris that in places has filled karstic channels in the upper saprolite. The hill is capped by siliceous duricrust derived from silica mobilization during the weathering of the ultra-mafic rocks

Plate 112 A deep weathering profile on the Yilgarn Craton of Western Australia showing similar features to those described in Plate 111. The profile here is some 30 m thick and at the base of the pit an irregular weathering front between fresh and weathered felsic granites can be observed. The original rock structures (joints) are clearly preserved in the saprolite in this profile

Plate 114 A 10 m section through a 20 m weathering profile formed in Ordovician intermediate volcanic rocks at Northparkes in New South Wales, Australia. This profile is unusual in that the pallid or bleached saprolite occurs above the mottled zone in the right of the photograph. It comprises mainly kaolinite in the bleached zone with minor secondary quartz and calcite and gypsum. The lower mottled zone has similar mineralogy with the addition of haematitic mottles. These mottles have been dated as Carboniferous (Pillans *et al.* 1999). The saprolite has been partly eroded and covered in part by palaeochannel sediments (left of photo) that are themselves weathered significantly and now form sedimentary saprolite with large haematitic and goethitic mottles. These mottles date from the Miocene. The whole profile is overlain by up to 2 m of Quaternary alluvium with a red-brown earth formed in it

Plate 115 A 7-m thick section through Proterozoic quartzite unconformably overlain by saprolite formed from Cretaceous labile sandstones and Quaternary transported cover at Darwin, Australia. The Proterozoic rocks are fresh. The Cretaceous has been completely transformed to a quartz/kaolinite saprolite and the whole sequence overlain by ferruginous lag gravel, sand and clays. It is interesting that the upper surface of the saprolite is karstic and that the overlying lags, etc. have filled the karstic channels

saprolite has been chemically eroded (weathered) to such a high degree that the original rock fabric is no longer self-supporting and it has collapsed. This process of collapse may also be enhanced by processes of bioturbation (e.g. tree roots, termites) and/or pedoturbation (e.g. wetting and drying, and illuviation). It is also from this point in the profile that the REGOLITH may move down slope (Plate 116) under the influence of gravity. Also within this zone, quartz sand and clays may begin to separate into clots of clay surrounded by sand (Taylor and Eggleton 2001: 161) by processes of pedoturbation.

Plate 116 Overturned Proterozoic metasediments moving down a slope of about 1.5° at the Mary River, Northern Territory, Australia

Plate 117 Close-up photograph of mottles on a wave cut platform at Darwin, Australia. The hammer provides scale. These mottles are up to 15 cm across and are composed of haematite, providing the red colour, cementing a saprolitic matrix of kaolinite and quartz. The intervening bleached saprolite is identical, except that it does not contain haematite

Secondary overprints may affect the appearance of the weathering profile and the saprolite. The most widely observed modification is the formation of iron oxihydroxide aggregations called mottles. Mottles (Plate 117) may extend throughout the profile, but are mostly observed in the upper collapsed saprolite. Another common feature in profiles developed mainly on felsic rocks is a bleaching or the removal of Fe-oxihydroxides from the lower parts of the saprolite and saprock. This zone is referred to by many as a pallid zone, particularly in what are described by some geomorphologists as a 'lateritic profile'.

Figure 141 (p. 909) shows idealized examples of how some physical and mineralogical properties of saprolite change through a profile. One point of interest here is the commonly observed addition of quartz to the upper parts of saprolite developed from basalt, which of course contains no quartz when fresh. This indicates the addition of 'foreign' material to saprolite profiles may occur either by overtopping of the saprolite or by other deposits from aeolian accession. This is a very common feature of most saprolite profiles but is most readily observable over weathered mafic rocks.

References

Becker, G.F. (1895) Gold fields of the southern Appalachians, *Annual Report of the United States Geological Survey, Part III. Mineral Resources of the United States, Metallic Products* 16, 251–331.

Pillans, B., Tonui, E. and Idnurm, M. (1999) Palaeomagnetic dating of weathered regolith at Northparkes Mine, NSW, in G. Taylor and C.F. Pain (eds) *Regolith '98, New Approaches to an old Continent, Proceedings*, 237–242, Perth: CRC LEME.

Taylor, G. and Eggleton, R.A. (2001) *Regolith Geology and Geomorphology*, Chichester: Wiley.

GRAHAM TAYLOR

SASTRUGI

Sharp irregular ridges, mounds or dunes. They form on ice sheets, ice caps, sea ice and tundra (typical of Antarctica and Greenland), and are composed of ice and compacted snow. Originating from the Russian word *zastrugi*, they are formed by aeolian erosion and the deposition of drifting snow (Gow 1965), and typically are 1–2 m long and 10–15 cm in height (though exceptional cases can reach 1.5 m in height and hundreds of metres long). Sastrugi align longitudinally with the predominant wind direction, making it possible to infer the prevailing wind direction at the time of sastrugi development from their configuration. They often form after a blizzard on the hard ice surface, becoming larger and harder as the blizzards blow across them throughout the winter months. Sastrugi are usually found in the lee of obstacles but are also known to exist in open conditions.

Reference

Gow, A.J. (1965) On the accumulation and seasonal stratification of snow at the South Pole, *Journal of Glaciology* 5, 467–477.

Further reading

Warren, S.G. and Brandt, R.E. (1998) Effect of surface roughness on bidirectional reflectance of Antarctic snow, *Journal of Geophysical Research – E: Planets* 103 (E11), 25,779–25,788.

STEVE WARD

SCABLAND

A scabland is an erosional landscape formed by a catastrophic flood and is generally applied to the effects of Jökulhlaups. It was first introduced by Bretz (1923) to describe the erosion and stripping of the Columbia Plateau Basalts by floods from Glacial Lake Missoula in eastern Washington, USA. Bretz adopted a term that had been used by local farmers to describe the 'scabby' terrain: 'The terms "scabland" and "scabrock" are used in the Pacific Northwest to describe areas where denudation has removed or prevented the accumulation of a mantle of soil, and the underlying

Table 41 Bedforms in the Channeled Scabland

	Scoured in rock	Scoured in sediment	Depositional
Macroforms (scale controlled by channel width)	Pool and riffle sequence, Quadrilateral residual forms in channel Anastomosis	Large-scale streamlined residual forms	Longitudinal bars (a) Pendant bars (b) Alternate bars (c) Expansion bars Eddy bars
Mesoforms (scale controlled by channel depth)	Longitudinal grooves Pot-holes Inner channels Cataracts	Scour marks	Large-scale transverse ripples (giant current ripples)
Microforms	Scallop pits	Not preserved	Small-scale ripple stratification (restricted to slack water facies)

Source: From Baker (1978a)

rock is exposed or covered largely with its own coarse, angular debris' (Bretz 1923: 617).

Channeled Scabland

The formal physiographical region known as the 'Channeled Scabland' is located in the northern portion of the Columbia Plateau in eastern Washington, USA and comprises an area of approximately 40,000 km^2. It is a spectacular channel complex eroded deeply into loess and basalt bedrock. The large flood discharges spilled over pre-flood divides into adjacent valleys and produced the effect of channels dividing and rejoining to form anastomosing (see ANABRANCHING AND ANASTOMOSING RIVER) complexes. These divide crossings are several hundred feet above valley floors.

A typical scabland complex includes erosional and depositional forms. Baker (1978a) has adopted a hierarchical classification of bedforms for the Channeled Scabland (see Table 41). The erosional landforms include grooves, pot-holes, rock basins, inner channels and cataracts. Bretz *et al.* (1956) ascribed several scabland features to differences between various basalt flows. Cataracts, such as Dry Falls, formed as one group of basalt flows was stripped from underlying resistant flows. Where they were exposed by the floods, the columnar jointed basalts exerted a strong joint control and the plucking action by floodwater yielded boulders >30 m diameter. The most spectacular of the depositional forms are the streamlined channel deposits, some superimposed by giant current ripples 0.5 to 7 m high and 18 to 130 m in chord length. They are composed predominantly of gravel and boulders. Slackwater deposits accumulated in low velocity areas including re-entrants to the major valleys and in pre-flood tributary valleys.

The scablands were formed by discrete outbursts from a range of sources. These included an enormous subglacial reservoir that extended over much of central British Columbia (conservative estimates of water volume are 10^5 km^3) (Shaw *et al.* 1999) and Glacial Lake Missoula. This lake impounded 2,184 km^3 of water during its maximum extent (O'Connor and Baker 1992). The last major period of scabland flooding is placed approximately between 18,000 and 13,000 years BP. Facies analysis of sedimentary sequences suggest that there may have been as many as forty floods (Waitt 1985) each separated by decades or centuries. Shaw *et al.* (1999) propose that there were fewer floods and that many of the variations in the sedimentary sequences can be ascribed to pulses within a flood caused by the input of multiple sources of floodwaters during these long duration flows (up to 1,000 days).

High water marks along the scabland channels have been used to reconstruct the maximum flood stages and water surface gradients. These include eroded channel margins, depositional features, ice rafted ERRATICs and divide crossings. Discharges as large as 21.3×10^6 m^3 sec^{-1} were conveyed through the channel scabland (Baker 1978b). Some constricted channels reached velocities as high as 30 m sec^{-1}. These high velocities were

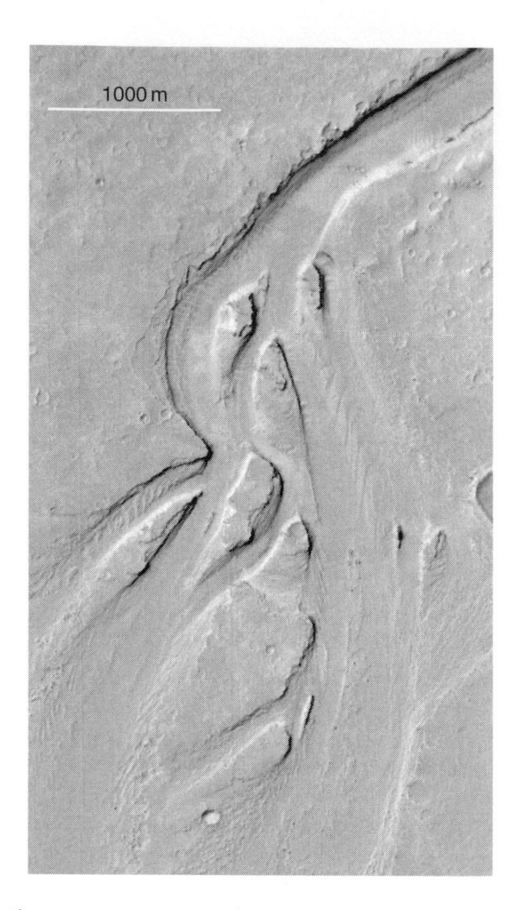

Plate 118 Portion of MOC image M2101914 which is centred near 7.89°N, 153.95°E, pixel resolution is 4.4 m. (Malin *et al.* 2001). This image shows anastomosing channel pattern and multiple streamlined forms in a flood channel emanating from a fissure. For detailed description see Burr *et al.* (2002)

possible because of the combination of great flow depth (60 to 120 m) and very steep water surface gradients 2 to 12 m/km.

On Mars, data from orbiting satellites have detected stripped zones on the floors of outflow channels in the Chryse Basin. These anastomosing complexes are 100 km wide and flow over 2,000 km across the planet's surface. They are usually initiated from collapsed zones. Other examples show multiple and asynchronous flows emanating from geothermal fissures in recent Martian history (see Plate 118) (Burr *et al.* 2002). By analogy to the scablands on Earth, it is generally accepted that the Martian outflow channels were also formed by catastrophic floods. Martian outflow channels include a distinctive assemblage of scabland landforms: regional and local anastomosing patterns, expanding and contracting reaches associated with flow constrictions, streamlined hills, inner channels with recessional headcuts, pendant forms (bars or erosional residuals) on the down current sides of flow obstacles, longitudinal grooves, irregular 'etched' zones on channel floors and scour marks around obstacles (Baker 1982).

References

Baker, V. R. (1978a) Large-scale erosional and depositional features of the Channeled Scabland, in V.R. Baker and D. Nummedal (eds) *The Channeled Scabland*, 81–115, Washington, DC: NASA.

—— (1978b) Paleohydraulics and hydrodynamics of Scabland floods, in V.R. Baker and D. Nummedal (eds) *The Channeled Scabland*, 59–79, Washington, DC: NASA.

——(1982) *The Channels of Mars*, Austin: University of Arizona Press.

Bretz, J.H. (1923) The channeled scablands of the Columbia Plateau, *Journal of Geology* 31, 617–649.

Bretz, J.H., Smith, H.T.U. and Neff, G.E. (1956) Channeled Scabland of Washington; new data and interpretations, *Geological Society of America Bulletin* 5, 957–1,049.

Burr, D.M., Grier, J.A., McEwen, A.S. and Keszthelyi, P. (2002) Repeated aqueous flooding from the Cereberus Fossae: evidence for very recently extant, deep groundwater on Mars, *Icarus* 159, 53–73.

Malin, M.C., Edgett, K.S., Carr, M.H., Danielson, G.E., Davies, M.E., Hartmann, W.K., *et al.* (2001) *M21–01914*, Malin Space Science Systems Mars Orbiter Camera Image Gallery (http://www.msss.com/moc_gallery/). (http://photojournal.jpl.nasa.gov/).

O'Connor, J.E. and Baker, V.R. (1992) Magnitudes and implications of peak discharges from Glacial Lake Missoula, *Geological Society of America Bulletin* 104, 267–279.

Shaw, J., Munro-Stasiuk, M., Sawyer, B., Beaney, C., Lesemann, J., Musacchio, A., Rains, B. and Young, R.R. (1999) The channeled scabland: back to Bretz? *Geology* 27, 605–608.

Waitt, R.B.J. (1985) Case for periodic, colossal jokulhlaups from Pleistocene glacial Lake Missoula, *Geological Society of America Bulletin* 96, 1,271–1,286.

MARY C. BOURKE

SCANNING ELECTRON MICROSCOPY

The Scanning Electron Microscope (SEM), sometimes used in association with Energy Dispersive Spectrometry (EDS), has been used for studying the surface textures (and chemistry, with EDS) of sediments (especially quartz grains) since 1962. The use of the SEM has had many implications for geomorphology, including the determination of the origin of depositional landforms, the provenance of sediments, the energy of environments and processes of diagenesis and weathering and their development through time. Examples of the use of the SEM include the separation of till from glacilacustrine and glacifluvial grains within the glacial sedimentary environment, studies of the origin of fine silt and clay particles in the geological column, examination of aeolian and other environmental fracture–abrasion mechanisms in the field and the laboratory, and analysis of grain modification under different weathering regimes. A full discussion of grain textures associated with aeolian, fluvial, mass wasting, glacial, tectonic, impact, weathering and diagenetic processes is provided by Mahaney (2002).

Reference

Mahaney, W.C. (2002) *Atlas of Sand Grain Surface Textures and Applications*, New York: Oxford University Press.

A.S. GOUDIE

SCHMIDT HAMMER

A concrete test hammer originally designed by E. Schmidt in 1948 for carrying out *in situ* tests on the hardness of concrete. The instrument measures the distance of rebound of a controlled impact on a rock surface. Because elastic recovery (the distance of repulsion of an elastic mass upon impact) depends on the hardness of the surface, and hardness is related to mechanical strength, the distance of rebound (R) gives a relative measure of surface hardness or strength.

In the Schmidt Type N hammer (which weighs 2.3 kg) the energy of impact (0.224 mkg) is obtained by releasing a spring-controlled plunger. The R value is shown by a pointer on a scale on the side of the instrument (range 10–100). This value represents the rebound distance as a percentage of the forward movement.

The Schmidt Hammer Type N is light and portable and allows *in situ* tests to be made in the field. By enabling quantitative comparison of the hardness of materials it provides a useful tool for geomorphologists. Among its uses have been the description of *nari* (calcrete) profiles, case hardening on tropical karst surfaces, and various types of aggregate resources (Day and Goudie 1977). It has also been much used to assess degree of weathering and the ages of geomorphic features upon which weathering phenomena occur (Ballantyne *et al.* 1989; McCarroll 1991).

Schmidt hammer rebound values have been found to correlate well with other measures of rock strength, including Young's Modulus and Uniaxial Strength (Katz *et al.* 2000)

References

Ballantyne, C.K., Black, N.M. and Finlay, D.P. (1989) Enhanced boulder weathering under late-lying snowpatches, *Earth Surface Processes and Landforms* 14, 745–750.

Day, M.J. and Goudie, A.S. (1977) Field assessment of rock hardness using the Schmidt Test Hammer, *BGRG Technical Bulletin* 18, 19–29.

Katz, O., Reches, Z. and Roegiers, J.-C. (2000) Evaluation of mechanical rock properties using a Schmidt Hammer, *International Journal of Rock Mechanics and Mining Sciences* 27, 723–328.

McCarroll, D. (1991) The Schmidt Hammer, weathering and rock surface roughness, *Earth Surface Processes and Landform* 17, 477–480.

A.S. GOUDIE

SCREE

The terms scree and talus are synonymous, the former being preferred in England and the latter (equivalent to the French word for slope) used predominantly in North America and elsewhere. Both terms describe accumulations of loose, coarse, usually angular rock debris at the foot of steep rock slopes. The terms are used to describe both the landform and the material of which it is composed. The debris accumulations forming screes must be of sufficient thickness to develop a characteristic morphology independent of the underlying slope. Simple debris veneers only a few particles thick are termed 'debris mantled slopes' (Church *et al.* 1979).

Scree slopes occur in a wide range of environments but most significantly in environments where physical weathering processes dominate. The production and/or accumulation of debris must be greater than its subsequent weathering or removal. The coarseness of most scree deposits makes them resistant to subsequent erosion: they are often stable long-lasting elements of the landscape and may be preserved as fossil forms.

The basic characteristics of scree slopes depend primarily on the morphology (and thereby geology) of the flanking cliffs and the geomorphic processes involved in their development. Although the dominant process is usually assumed to be ROCKFALL, scree modification and accumulation may be the result of several different processes acting singly or in combination and several distinctive types of scree may be recognized. The plan morphology of screes depends on the form of the cliffs supplying the debris and the morphology of the surface on which the debris accumulates. Relatively simple cliffs, straight in plan view, tend to produce sheet (straight) rockfall talus slopes lacking significant lateral variation in their characteristics. As cliffs become more dissected, deposition is focused below couloirs (gullies) leading to the development of cones. As well as funnelling rockfalls, couloirs channel other rockwall processes (surface stream flow, snow avalanches (see AVALANCHE, SNOW), etc.) down the cliff, sometimes resulting in significant modification of the scree below. Therefore the cones developed below dissected cliffs are rarely simple single-process forms. In alpine environments debris cones can be significantly modified by snow avalanche activity. DEBRIS FLOW generation may also occur during heavy rainstorms when drainage from the cliff zone is focused by gullies onto finer grained materials at the head of the scree.

Rockfall screes result from the accumulation of discrete rockfall events over long periods of time. The basic characteristics are fall sorting (a logarithmic increase in mean grain size downslope) and a straight slope, often with a well-developed basal concavity. There is continuing debate about whether these straight slopes reflect the angle of repose (see RESPOSE, ANGLE OF) of coarse cohesionless material at about 35° (Carson 1977). Measured profiles of the upper part of many rockfall screes range between 32° and 37° but locally reach 40°. The degree of basal concavity varies with the length of slope and the characteristics of the basal zone.

Fall sorting on rockfall scree slopes, though not universal, primarily results from two mechanisms. Larger boulders have greater momentum and therefore tend to travel further downslope. Second, the frictional resistance (roughness) of the surface over which the boulder slides, rolls or bounces is defined by the relationship between the size of the moving boulder and the irregularities of the surface (boulders and voids) over which it is passing. Big boulders are only effectively trapped in large 'holes' or when they impact other large boulders. The degree of sorting depends on the slope length, cliff height and the size and shape of dominant particles. Locally random effects or differences in particle shapes may result in the absence or anomalous patterns of sorting. On scree slopes modified by snow avalanches loose surface material is swept from the upper slopes to the end of the avalanche track at or beyond the base of the scree, in extreme cases forming AVALANCHE BOULDER TONGUEs. These screes show little size sorting on upper slopes but a rapid increase in mean grain size towards the base despite the presence of an unstable scattering of loose rock debris on the surface of larger clasts. Avalanche modified screes have strong basal concavities. Many scree cones have gullies formed by fluvial activity at their head and may have significant debris flow forms (levees, channels and terminal lobes) extending across the scree slope. In extreme cases these have been termed 'alluvial talus'. Scree-like forms produced by the breakup of single large rockfalls generally lack fall sorting and have more complex long profiles. In alpine areas large multiprocess scree cones may display complex surface characteristics associated with the interaction of these processes (Plate 119).

Most scree slopes, despite their coarse debris veneer, have considerable quantities of interstitial fine material at depth or exposed on the higher parts of the slope. They are not exclusively formed of coarse cohesionless material as many early models assumed. The upper parts of the scree slopes undergo 'talus' creep which is the aggregate movement caused by rockfall and other impacts, FREEZE–THAW CYCLE or frost heave activity, percolating flows, animals, etc. Loose material on very steep sections may undergo dry avalanching but such failures are usually small. Little if any talus creep seems to occur on coarse basal screes.

Plate 119 Multiprocess screes, Bow Lake, Alberta, Canada. Complex sheet and scree cones showing the varying influence of alluvial, debris flow and snow avalanche activity on the detailed morphology of these landforms predominantly created by rockfall. The basal area of these cones show evidence of permafrost creep and incipient (or arrested) rock glacier forms

Screes are most frequently studied in periglacial environments. At some of these sites a permanent snow (firn) patch may develop at the base of the slope. Debris landing on this icy surface slides to the base accumulating as a ridge (PROTALUS RAMPART or nivation ridge). In alpine areas, thick talus accumulations may develop PERMAFROST. Subsequent deformation and creep of this rock/ice mixture leads to ROCK GLACIER development.

References

Carson, M.A. (1977) Angles of repose, angles of shearing resistance and angles of talus slopes, *Earth Surface Processes* 2, 363–380.

Church, M., Stock, R.F. and Ryder, J.M. (1979) Contemporary sedimentary environments on Baffin Island, N.W.T., Canada: debris slope accumulations, *Arctic and Alpine Research* 11, 371–402.

Further reading

Luckman, B.H. (1988) Debris accumulation patterns on talus slopes in Surprise Valley, Alberta, Canada, *Géographie physique et Quaternaire* 42, 247–278.

Rapp, A. and Fairbridge, R.W. (1968) Talus fan or cone: scree and cliff debris, in R.W. Fairbridge (ed.) *The Encyclopaedia of Geomorphology*, 1,107–1,109, New York: Reinhold.

SEE ALSO: frost and frost weathering; geomorphological hazard; grèze litée; hillslope, form; hillslope, process; mass movement

BRIAN LUCKMAN

SEA LEVEL

Sea level is the divide between the marine and terrestrial realms; above it, the world is dominated by air, erosion and creatures that contend with gravity; below lies a submerged world dominated by sedimentation and neutrally buoyant animals. Relative to the landmasses, the position of sea level has fluctuated throughout the geologic past, owing to changes in both the quantity of ocean water and the geometry of the ocean basins. Although the total quantity of water in the hydrosphere probably has changed little since Archean times, the fractions held in land reservoirs – glaciers, lakes, ground water and, in particular, continental ice sheets – have fluctuated significantly. For example, if the present Antarctic ice sheet were to melt, sea level would rise by about 55 m; at the height of the last ice age, which is only one of many such events to have punctuated Earth history, sea level was about 120–130 m below its present position.

The proportions of land and sea are basically determined by the fact that continents, being composed of rock lighter than oceanic crust, stand about 4.5 kilometres above the ocean floor (in contrast, hot and oceanless Venus has no features resembling great ocean basins). However, owing largely to slow variations in the rates of seafloor spreading and plate tectonics, the average depth of the ocean basins has varied throughout geological time and shorelines have periodically advanced and retreated across the continental shelves. Figure 142 illustrates examples of observed sea level change on different time scales, from about a year to 10^8 years. On the longer timescale, sea level changed globally with amplitudes up to several hundred metres, largely owing to plate-tectonic changes in ocean basin geometry (Figure 142a). On timescales of tens to hundreds of thousands of years, periodic exchanges of mass between the ice sheets and oceans caused sea-level changes of tens to over a hundred metres in amplitude (Figure 142b).

Global changes of sea level caused by changes of the volumes of seawater or the ocean basins are referred to as *eustatic*. Superimposed on these global signals are more regional and local changes. At decadal, annual and shorter intervals, meteorology – and tide-driven changes become important and vary from place to place (Figure 142c). Over longer times, the relative positions of land and sea are affected by uplift or subsidence of the coastal zone. Observations vary substantially from site to site, even over relatively short distances such as in Scandinavia, where sea level at Ångerman has fallen nearly 200 m in the past 9,000 years while at Andøya the level 9,000 years ago was near its present position. In southern England, levels have risen slowly over the past 7,000 years, but along the Australian margin they have fallen by a few metres during the same interval (Figure 143).

Many factors that contribute to changes in sea level are linked. When ice sheets melt, the resulting sea-level change is spatially variable because the Earth's surface deforms under the changing ice and water load, and because the gravitational potential of the Earth–ocean–ice system also changes. The combined deformation–gravitational effects are referred to as the *glacio-hydro-isostatic* contributions to sea level, and it is they that cause the spatial variability illustrated in Figure 143.

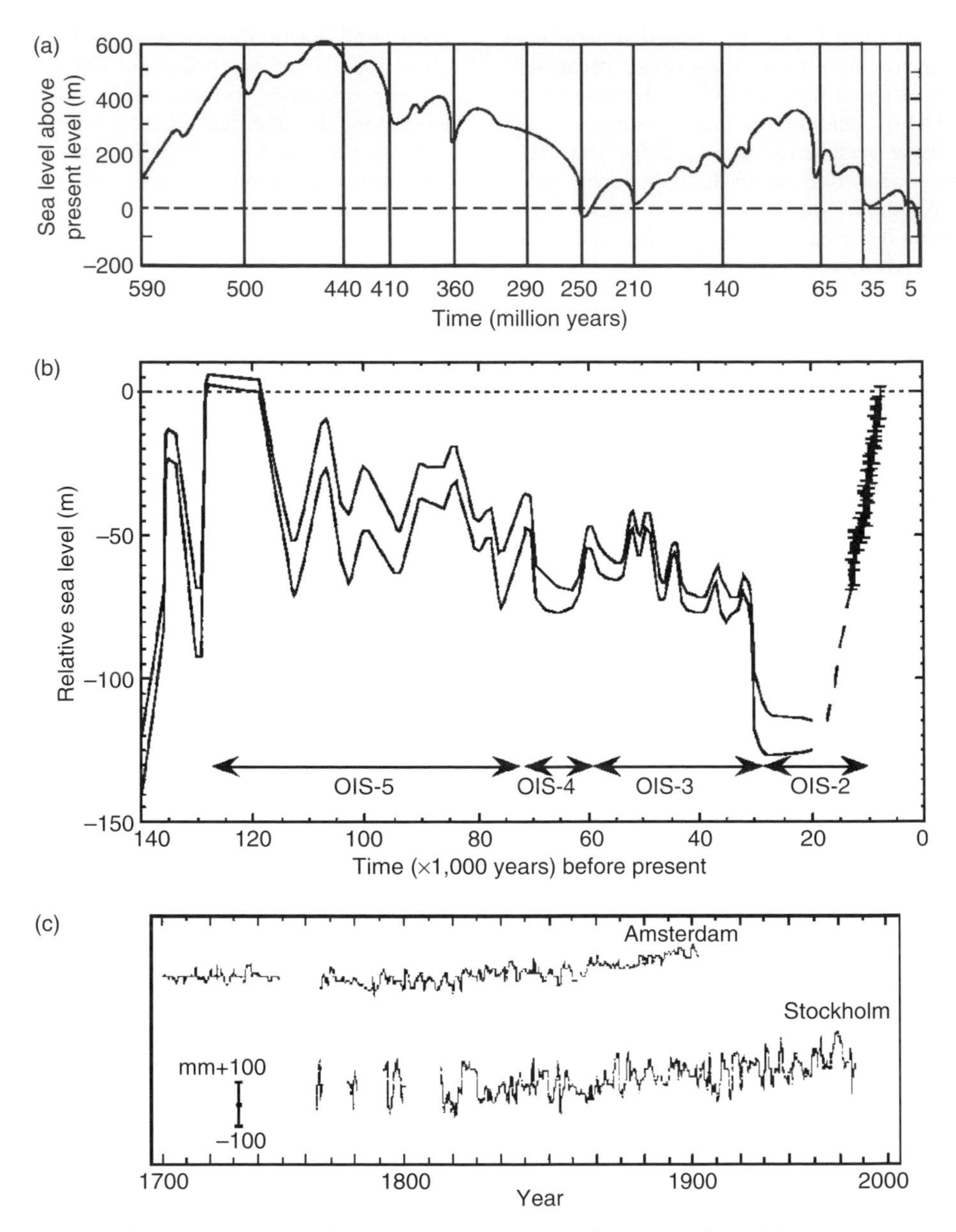

Figure 142 Sea-level variation at three timescales. (a) ~10^8 years, inferred from seismic sequence stratigraphy (Hallam 1992; Haq *et al.* 1988). The higher frequency changes reflect both global and local effects; the large, slow changes reflect continental breakup and changes of ocean ridge systems. (b) ~10^5 years of relative sea level at Huon Peninsula (HP), Papua New Guinea, driven by changes in northern continental ice sheets. Bars show marine oxygen isotope stages (OIS) discussed in text. (c) 10^0–10^2 years, measured by tide gauges from Amsterdam and Stockholm (a secular fall of ~4 mm/year has been removed from the Stockholm record): these changes are primarily of climatic origin and the apparent small rise starting AD ~1880 may reflect global warming (from Lambeck and Chappell 2001)

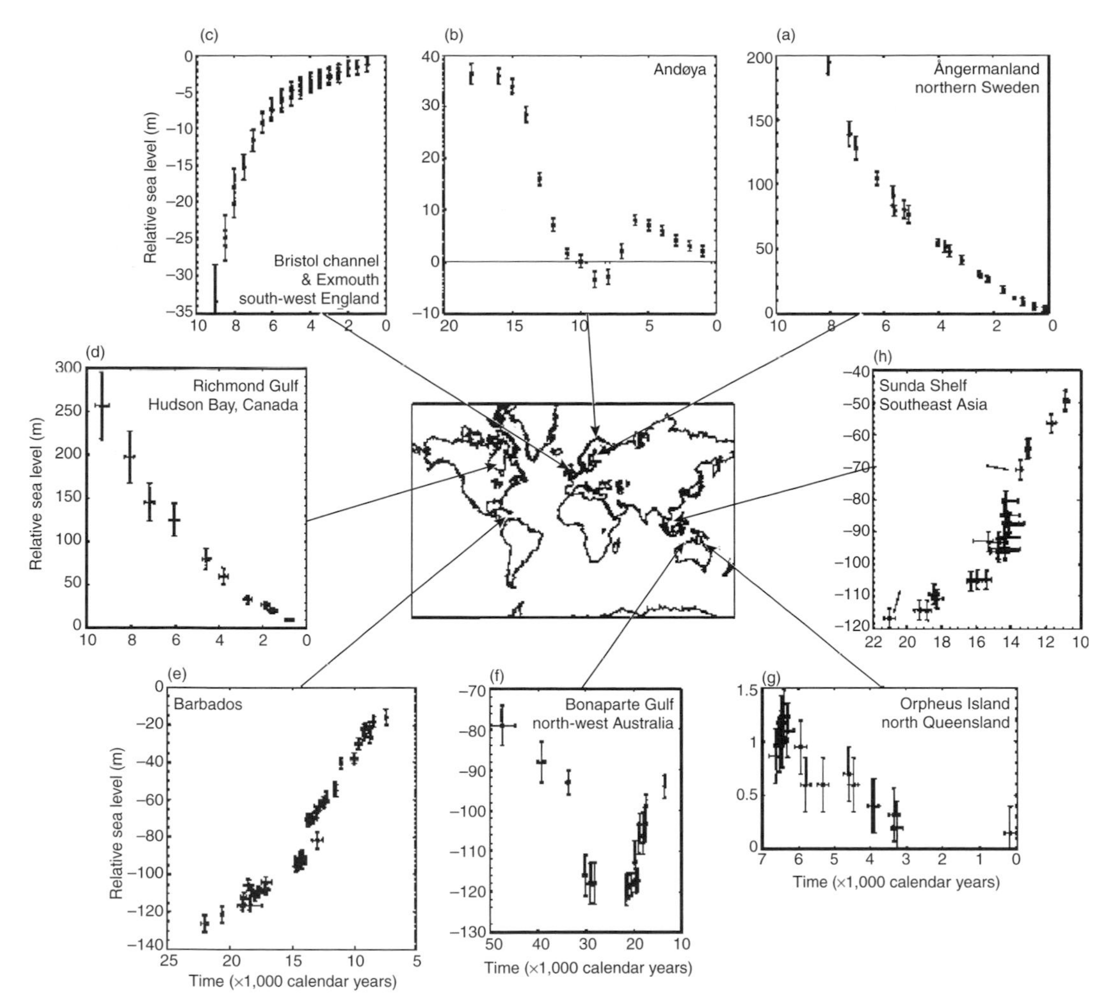

Figure 143 Observed variability of sea-level change in last 20,000 years from tectonically stable areas or sites where the tectonic rate is known and has been removed. (a) Ångerman, Gulf of Bothnia, Sweden. (b) Andøya, Nordland, Norway. (c) South of England. (d) Hudson Bay, Canada. (e) Barbados. (f) Bonaparte Gulf, north-west Australia. (g) Orpheus Island, north Queensland, Australia. (h) Sunda Shelf, Southeast Asia. (Note: scales differ from graph to graph). (Data sources given in Lambeck and Chappell 2001).

Isostatic warping and subsidence also occurs in sedimentary basins, in response to the accumulation over millions of years of sediments many kilometres deep. Furthermore, *tectonic* movements drive mountain uplift, enhancing landscape denudation and leading to rapid accumulation in sedimentary basins, and through coastal uplift or subsidence also contribute to changes of relative sea level.

Knowledge of the complex history of relative changes of sea level has applications in fields as diverse as understanding climate changes, analysing the structure of petroleum-bearing sedimentary basins, and determining deep-earth properties such as the viscosity of the Earth's mantle. Once comprehensive sea-level models are developed, it becomes possible to test hypotheses about the migrations of flora and fauna across shallow seas that are now covered by the ocean. Finally, to understand future sea-level rise under atmospheric greenhouse conditions, the background 'natural' signal must be known. The success of the outcomes of the various sea-level studies depends very much on the ability to separate the different contributions – eustatic, isostatic and tectonic – in the observational record.

Observational evidence

Evidence for historical sea-level changes comes from tidal marks and gauges, usually in ports and harbours, as well as buildings and other structures in littoral towns such as Venice. For the geologic past, the evidence occurs mostly in the form of sediments and biohermal reefs that were formed in coastal and nearshore situations. Upper Quaternary sea-level changes are usually pieced together from raised or submerged shoreline features, including shallow-water coral reefs (Figure 144). Further back in time, relative sea-level changes can be deduced from sedimentary basins, in which eustatic variations are registered in cyclic sediments (cyclothems) (Figure 145), and by alternating transgressive and regressive sediment tracts. Using the methods of *sequence stratigraphy* based on seismic and drillhole or outcrop data, the locus of coastal-zone sediments can be traced throughout a given basin sequence, allowing the course of sea-level changes relative to the basin to be identified (Hallam 1992; Haq *et al.* 1988).

A relative sea-level curve for a given area can be established from age–height data for a series of ancient shorelines, or other *indicator deposits* such as shallow marine sediments for which the depth of deposition relative to sea level is determinable. Terrestrial deposits within a sequence, such as peats and floodplain sediments, may usefully indicate levels not reached by the sea. Various methods are used to establish the ages. For deposits less than ~40,000 years old, *radiocarbon* dating is used widely, although *uranium-series* dating is preferable in the case of coral formations and has a greater time range, extending to ~0.5 Myr. Thermoluminescence (*TL*) and optically stimulated luminescence (*OSL*) methods are increasingly used for dating Upper Quaternary shoreline deposits, and amino acid racemization has proved to be useful despite its low precision. Sea-level indicator deposits also are correlated to marine oxygen isotope records, described later, that have a chronology based on slow variations in the Earth's orbit, which affect the seasonal receipts of solar radiation and acted as an ice-age pacemaker. *Orbital chronology* rests on astronomical observations and has been extrapolated several million years into the past. Finally, the dating of older deposits and sedimentary basin sequences generally rests on *geomagnetic reversal*

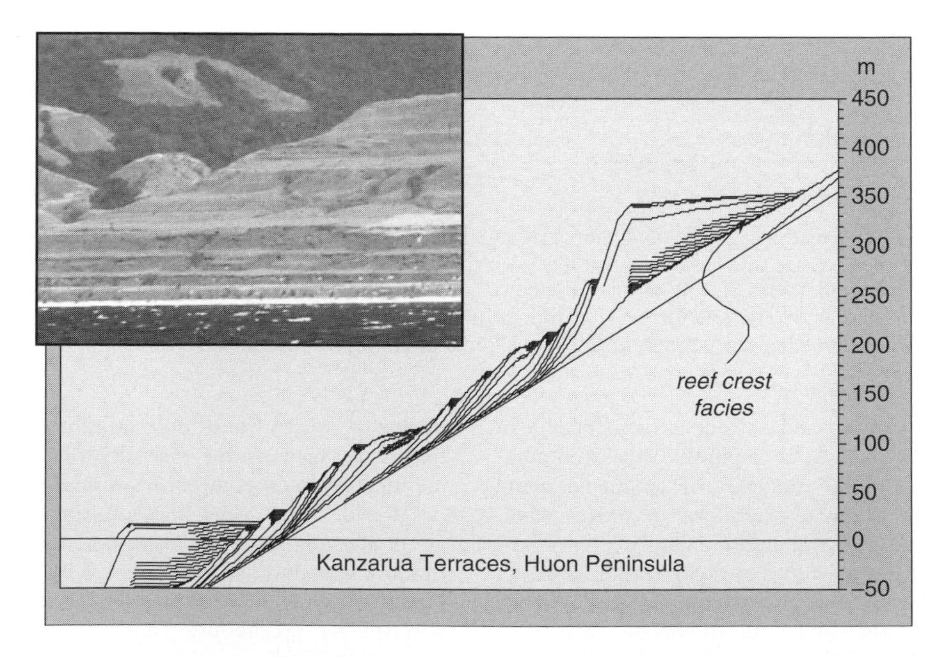

Figure 144 Typical expression of Late Quaternary sea-level changes superimposed on tectonically rising terrain: coral terraces Huon Peninsula, Papua New Guinea. The top terrace at right of the inset photo is 350 m above sea level and formed at the beginning of the Last Interglacial, ~127,000 years ago. Each of the smaller reef structures within the downstepping stratigraphic sequence was formed during a sea-level oscillation within the last glacial cycle. Subtracting the effect of uplift yields the sea-level curve shown at Figure 142b

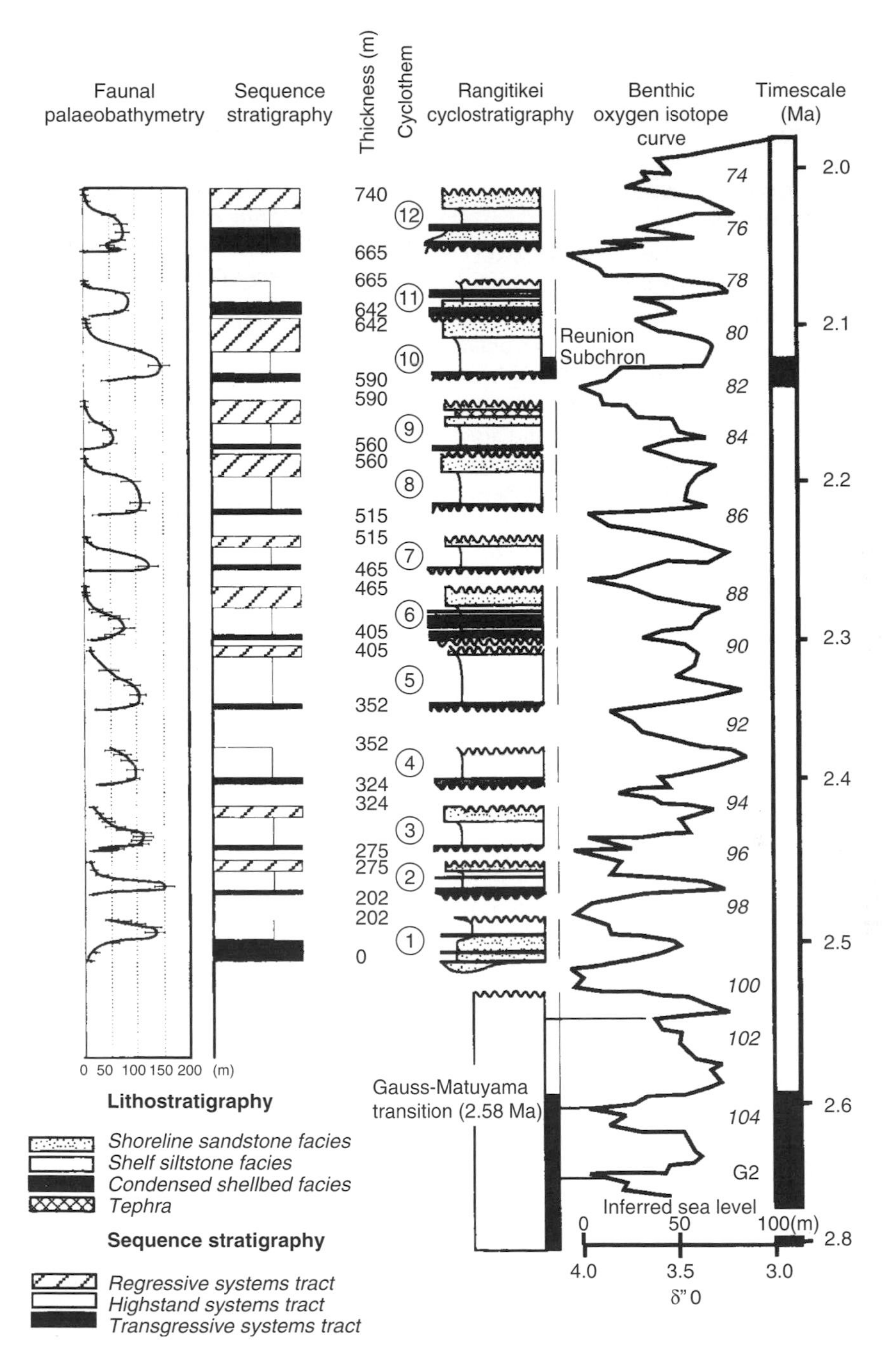

Figure 145 Late Pliocene sea-level changes inferred from shallow marine cyclothems in New Zealand. Centre columns show 12 sedimentary cycles; left column shows cyclic variations of water depth at the site of sedimentation, inferred from fossil marine invertebrates. Correlations to global marine oxygen isotope cycles (Shackleton *et al.* 1995) is shown at right, tied to magnetic reversal chronology (from Pillans *et al.* 1998)

chronology or on microfossil-based correlations with standard stratigraphic sequences tied to this chronology, which in turn is secured by potassium-argon dating methods.

Separation of eustatic, isostatic and tectonic components

RAW SEPARATION OF UNIFORM UPLIFT OR SUBSIDENCE

A first step towards separation of the tectonic, isostatic and eustatic contributions is subtraction of any obvious vertical crustal movements from a relative sea-level curve. More often than not, this is done by assuming that the sea level for some reference deposit in the record is known and also that the rate of uplift (or subsidence) was constant throughout the duration of the record. With these assumptions, the local sea level S represented by a deposit of age t that accumulated at depth d below sea level and now is height H above present sea level, is given by

$$S = H + d - Ut \quad \text{with} \quad U = (S_r - H_r + d_r)/t_r \qquad (1)$$

where H_r, d_r and t_r are height, depositional depth and age of the reference deposit, and S_r is the sea level at its time of formation. For Upper Quaternary studies, the local height of the Last Interglacial shoreline is widely used to determine uplift rate U, as evidence reviewed below suggests that sea level then was little different from that of today. However, for every study, each variable in (1) has an error term arising from dating errors and uncertainties in field relationships. Moreover, in assuming a constant rate of uplift, this approach neglects reversing vertical movements that arise from the global glacio-hydro-isostatic response to advancing and retreating icesheets. Figure 142b shows the 'uplift-free' sea-level curve derived by this method from the coral terraces illustrated at Figure 144.

Similar principles are used to derive sea-level changes from sedimentary basin sequences, although here the vertical movement is downwards. In terms of sedimentary facies, individual cyclothems often are very similar from bottom to top of a thick cyclothemic sequence (Figure 145), implying fairly uniform subsidence of the basin.

GLACIO-HYDRO-ISOSTASY

When ice sheets melt, to a first approximation the sea level rises by an amount $\Delta\zeta_e(t)$ related to the land-based ice volume V_i (using the notation of K. Lambeck: see Lambeck and Chappell 2001),

$$\Delta\zeta_e(t) = -(\rho_i/\rho_o)\int(1/A_o(t)\ dV_i/dt)\,dt \qquad (2)$$

where $A_o(t)$ is the ocean surface area (which changes as sea level rises or falls) and ρ_i, ρ_o are the average densities of ice and ocean water, respectively. $\Delta\zeta_e(t)$ is the ice-volume equivalent sea-level change (or *equivalent sea-level change*), which equals eustatic change if no other factors contribute to changes in ocean volume. The relative sea-level change $\Delta\zeta_{rsl}(\varphi, t)$ at position and time t, ignoring tectonic displacements, is

$$\Delta\zeta_{rsl}(\varphi, t) = \Delta\zeta_e(t) + \Delta\zeta_i(\varphi, t) + \Delta\zeta_w(\varphi, t) \qquad (3)$$

where $\Delta\zeta_i$ and $\Delta\zeta_w$ are the glacio- and hydro-isostatic contributions. Both are functions of position and time. The water depth or terrain height, expressed relative to coeval sea level, is

$$h(\varphi, t) = h(\varphi, t_0) - \Delta\zeta_{rsl}(\varphi, t) \qquad (4)$$

where $h(t_0)$ is the present-day (t_0) bathymetry or topography at φ. Both isostatic terms in (3) are functions of Earth rheology as well as of fluctuations in the ice sheets over time.

In formerly glaciated areas, the glacio-isostatic term $\Delta\zeta_i(\varphi, t)_i$ dominates during and after deglaciation, and leads to uplift at a rate that can exceed the global eustatic rise, so that sea level locally falls (the Ångerman result: Figure 143). Rebound is smaller near the ice margin and although it may dominate initially, the global sea-level rise becomes important later. When all melting has ceased, the residual rebound leads to falling local sea level (the Andøya result: Figure 143). During ice sheet growth, mantle material beneath the loaded area is displaced outward and broad bulges develop around the perimeter, which subside when the ice melts, leading to slowly rising local sea level after melting has ceased. Much further from the ice, the water load becomes the dominant cause of planetary deformation (the hydro-isostatic contribution $\Delta\zeta_w(\varphi, t)$), producing subsidence of the seafloor and adjacent margins. The effect is most pronounced at continental margins far from the ice sheets, such as the Australian coast (Figure 143), and once melting has ceased, sea levels continue to fall at a slow but perceptible rate. The amplitude of this postglacial 'highstand' effect can vary by several metres from site to site.

Isostatic corrections and a global eustatic curve can be derived from local sea-level curves, using a rheologically appropriate Earth model to predict surface deformation in response to changing ice and water loads, with the ocean surface remaining a gravitational equipotential surface at all times. The sea-level signal at sites far from the former ice margins approximates the equivalent sea-level function to about 10–15 per cent and the isostatic contribution is mainly from water loading, which is insensitive to the details of the ice sheets, provided that the total ice volumes are correct to within about 10 to 20 per cent. Hence, through an iterative procedure, it becomes possible to estimate changes in ice volumes V_i from observed sea-level changes $\Delta\zeta_{rsl\ (obs)}$, using (2), and (5), below:

$$\Delta\zeta_{rsl} = \Delta\zeta_{rsl\ (obs)} - (\Delta\zeta_i + \Delta\zeta_w) \quad (5)$$

The use of local sea-level curves from widely separated places allows Earth rheology parameters and models of ice distribution to be evaluated. Recent models include deformation of the basins over time, movement of grounded ice across the shelves, modification of sea level by the time-dependent gravitational attraction between the solid Earth, ocean, and ice, and the effect of glacially induced changes in Earth rotation on sea level.

Sea level through the last glacial cycle

Sea-level data for the last glacial cycle are more plentiful than for earlier periods and, at Huon Peninsula, Papua New Guinea, provide a near-complete relative sea-level curve (Figure 142b), which has been used for reconstructing ice-equivalent sea level for the past 140,000 years (Lambeck and Chappell 2001). Results indicate that ice melting has varied since the Last Glacial Maximum (LGM), with two periods of rapid sea-level rise from ~16,000 to 12,500 and from 11,500 to 8,000 years ago, separated by the Younger Dryas (YD) cold episode when sea level seems to have risen less rapidly. By 7,000 years ago, the northern ice sheets except for Greenland had gone and ocean volume approached its present level, but Antarctic ice melting may have since contributed a few metres of equivalent sea level.

The Last Interglacial, when sea level was similar to the present, ended about 118,000 years ago with rapidly falling sea level, associated with the growth of northern continental ice. Cyclic sea-level changes from 118,000 to ~60,000 years reflect the effect of the 20,000-year orbital precession cycle on the ice sheets, although other fluctuations also appear. These occur repeatedly about every 6,000 years from ~60,000 to 30,000 years, with amplitudes of 10–15 m, and each rise apparently coincides with a major episode of ice-rafted sediment deposition recorded in the North Atlantic, suggesting that the rise was caused by a large, rapid discharge of continental ice.

Oxygen isotopes and long sea-level records

Long, continuous records of oxygen isotopes in calcareous foraminifera preserved in deep-sea sediments have become standard records of Quaternary sea-level and temperature changes. The oxygen isotope ratio in foraminifera, conventionally expressed as $\delta^{18}O$ (the ‰ difference of $^{18}O/^{16}O$ in a sample from an international standard), depends on the temperature and isotope ratio of the seawater in which they lived. Furthermore, the seawater isotopic ratio is related to the size of polar ice sheets, because ^{18}O is preferentially removed from atmospheric water vapour as it makes its way poleward, causing the icecaps to be depleted in ^{18}O relative to seawater by ~25–55 ‰, varying with atmospheric temperature. Thus, foraminiferal isotopes $\Delta\delta^{18}O_f$ respond to ice volume changes ΔV_i according to

$$\Delta\delta^{18}O_f \approx \delta^{18}O_i\ \Delta\ V_i/V_w + C_T\ \Delta T + \varepsilon \quad (6)$$

where ΔV_i is relative to present day V_i, V_w is the present volume of seawater (mean $\delta^{18}O$ of modern seawater = 0 $‰_{SMOW}$), C_T is the coefficient of temperature-dependent isotope fractionation for calcite (-0.23 ‰ $°C^{-1}$), ΔT is temperature change, and ε represents any local change of seawater $\delta^{18}O$ not related to ΔV_i. (Equation 6 is approximate because the mean isotopic composition of the ice sheets $\delta^{18}O_i$ is assumed constant, but the uncertainty here probably is smaller than the effects of ice volume and temperature.) Finally, the first term on the RHS of (6) can be expressed in terms of equivalent sea level $\Delta\zeta_e$: comparison of isotopes and sea levels for the last glacial cycle indicates that $\Delta\delta^{18}O_w/\Delta\zeta_e \sim -0.009$ ‰ m^{-1}.

The numerical value of $\delta^{18}O_f$ becomes increasingly positive under a fall of both sea level and temperature, which typically happens under a major ice-sheet advance. Hence, marine oxygen isotopes provide composite records of sea level and temperature: for example, Figure 145 illustrates typically close correspondence between sea level and isotopic cycles. In a longer time frame, Tertiary isotope records reveal both progressive cooling and the onset of ice-driven sea-level cycles

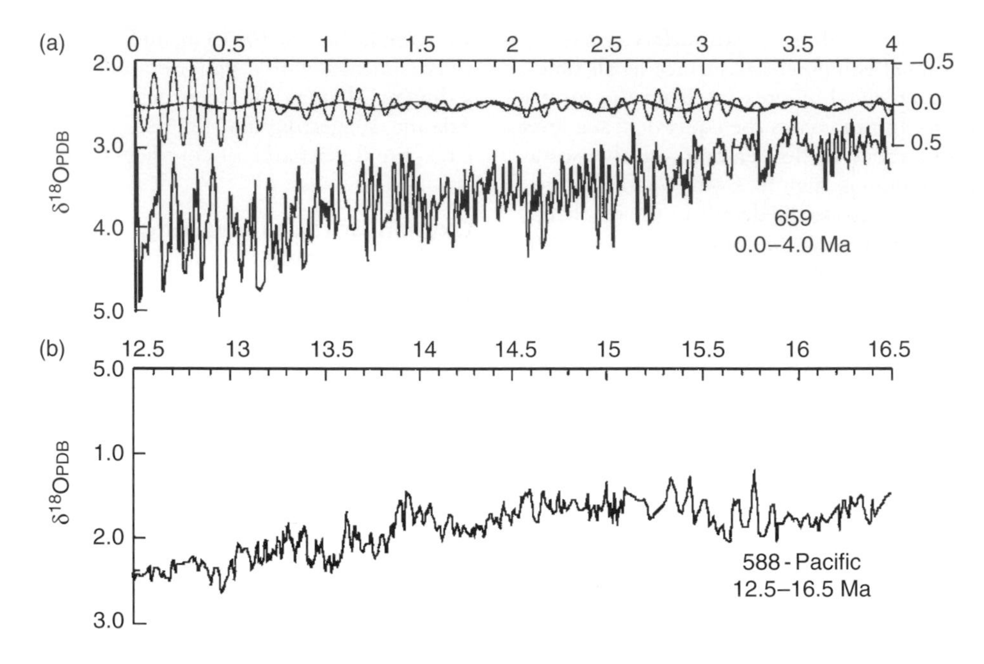

Figure 146 Contrasting records of marine oxygen isotopes from 0–4 Myr (Plio-Pleistocene) and 12.5–16.5 Myr (Miocene). The increase of amplitude in $\delta^{18}O$ cycles around 2.6 Myr represents the onset of large fluctuations of ice sheets and sea level, which become even more pronounced around 1 Myr, when they adopt a characteristic period of ~100,000 years. The upper curve in (a) shows the changing amplitude of the 100,000 year cycle (after Zachos *et al.* 2001)

in the Pliocene (Figure 146). As the history of ocean temperature becomes increasingly well determined, through trace-element analysis of microfossils and other techniques, the sea-level contribution in such records is now being isolated (see Zachos *et al.* 2001).

Shoreline reconstructions

Once a global eustatic curve $\zeta_e(t)$ is established, the course of shoreline changes through time can be predicted. Provided that the present-day shallow water bathymetry is known with high resolution, water depths for any region at any time within the range of the eustatic curve follow from (4), and the palaeo-shorelines at time t correspond to the contours $h(\varphi, t) = 0$. Thus, it becomes possible to examine the migrations of shorelines through time for intervals for which sufficient observational data exist to constrain the isostatic variables. Predictions of shorelines since the time of the LGM have been published for both global and regional reconstructions, which can provide useful insights into the interpretation of prehistoric sites. As Lambeck (1996) has shown, for example, the interpretation of post-Palaeolithic archaeology of the Aegean is intimately linked to reconstructions of shoreline changes, which were a powerful factor in trade and changing human activities in the region.

Geomorphic consequences of sea-level changes

Rising sea level at the end of the Pleistocene Period led to widespread geomorphologic changes in coastal regions. Under the influence of ice-age low sea levels, today's coastal valleys tended to become incised well below present sea level, only to turn into traps for floodplain aggradation when sea level rose, as ice sheets retreated. The alternation of incision and aggradation doubtless was repeated in each glacial cycle, throughout the Quaternary, leading to development of broad floodplains on thick sediment valley-fills, which in tectonically stable regions often extend hundreds of kilometres inland. Near-coastal limestone regions under the influence of low sea levels developed vadose karst systems below present sea level, including stream passages and speleothem formations. At coastlines, a range of landforms were the result of the postglacial sea-level rise, from drowned valleys through sand-barrier basins to estuarine and deltaic plains, the particular form depending on sediment supply, tidal range and wave energy. In tropical seas, coral reefs re-established after being reduced by subaerial erosion during glacial-age lowstands.

In tectonically stable terrain, the geomorphic expression of any given Quaternary sea-level cycle tends to overprint the record of earlier cycles. In contrast, tectonic uplift results in flights of coastal terraces built at times of high relative sea level (e.g. Figure 144) that often pass inland into flights of river terraces. However, even though a terraced river valley may grade to present sea-level, not all its terraces necessarily relate to sea level highstands. Where continental shelves are broad, a river carrying sufficient fluvial sediment, when sea level was low and the coast far seawards of its present position, may have aggraded to a level higher than the recent floodplain, which thus becomes inset below a Pleistocene lowstand terrace. Given the tendency for all such features to become degraded and overprinted, the geomorphic legacy of past sea levels – while very sharp for recent events – becomes increasingly blurred with time. In contrast, the sedimentary record remains relatively intact.

References

Hallam, A.J. (1992) *Phanerozoic Sea-level Changes*, New York: Columbia University Press.

Haq, B., Hardenbol, J., Vail, P.R., Hardenbol, J. and Baum, G.R. (1988) Sea-level change: an integrated approach, *Society of Economic Paleontologist and Mineralogists Special Publication* 42, 71–108.

Lambeck, K. (1996) Sea-level change and shore-line evolution in Aegean Greece since Upper Palaeolithic time, *Antiquity* 70, 588–611.

Lambeck, K. and Chappell, J. (2001) Sea level change through the last glacial cycle, *Science* 292, 679–686.

Pillans, B., Chappell, J. and Naish, T. (1998) A review of the Milankovitch beat: template for Plio-Pleistocene sea-level changes and sequence stratigraphy, *Sedimentary Geology* 122, 5–21.

Shackleton, N.J., An, Z., Dodonov, A.E., Gavin, J., Kukla, G.J., Ranov, V.A. and Zhou, L.P. (1995) Accumulation rates of loess in Tadjikistan and China: relationship with global ice volume cycles, *Quaternary Proceedings* 4, 1–6.

Zachos, J., Pagani, M., Sloan, L., Thomas, E. and Billups, K. (2001) Trends, rhythms, and aberrations in Global Climate 65 Ma to Present, *Science* 292, 686–693.

JOHN CHAPPELL

SEAFLOOR SPREADING

Seafloor spreading is the process by which new oceanic crust is generated at mid-ocean ridges, the long linear belts of elevated seafloor that lie at the centres of most ocean basins (see SUBMARINE LANDSLIDE GEOMORPHOLOGY). The term was coined by Dietz (1961) for the idea (jointly proposed with Hess 1962) that new crust is formed by magmatic intrusion along the crests of mid-ocean ridges, and that conveyor belts of crust thus formed move symmetrically away from the ridges, driven by convection currents in the underlying mantle. Their suggestion provided the first viable mechanism by which continental drift might take place. Acceptance of the concept came only after the demonstration by Vine and Matthews (1963) that linear magnetic anomalies, which are symmetrical on either side of a mid-ocean ridge, record reversals of the Earth's magnetic field as newly formed crust cools and moves laterally away from the axis. The seafloor spreading hypothesis was the direct catalyst for the development of the concept of PLATE TECTONICS, the grand unifying theory of Earth sciences. This posits that the Earth is capped by a number of rigid plates, containing both continental and oceanic crust, that deform only at plate boundaries.

The large-scale motion of the Earth's plates is now believed to be driven primarily by the pull of dense slabs of oceanic lithosphere as they are subducted into the mantle at convergent margins (see WILSON CYCLE). Seafloor spreading at divergent oceanic plate boundaries may therefore be regarded as an essentially passive process: plates are pulled rather than pushed apart. Separation of the oceanic plates induces upwelling of the convecting, plastically deforming mantle asthenosphere. As it rises up and decompresses the mantle peridotite starts to melt. This generates a basaltic liquid which separates from its host and rises upward to feed a magma chamber beneath the ridge axis, thence solidifying to form ocean crust.

The rate at which seafloor spreading occurs varies from place to place along the 60,000 km long global mid-ocean ridge system, from less than 1 cm yr^{-1} at the Gakkel Ridge (in the Arctic Ocean) and parts of the Southwest Indian Ridge, to 16 cm yr^{-1} at the East Pacific Rise (west of Peru). The oldest surviving ocean floor lies in the north-west Pacific Ocean and is of Jurassic age (approximately 180 Ma).

Seismic refraction experiments indicate that the ocean crust is generally 6–7 km thick and has a layered internal structure. A seismic discontinuity (the Mohorovicic discontinuity or 'Moho') separates it from rocks with typical mantle velocities below. The seismic layering has been correlated directly with the lithological layering observed in ophiolites, which are regarded as on-land fragments of ocean crust and shallow mantle. This has led to the conventional view of ocean crustal structure as a simple, uniform 'layer-cake' internal structure composed (from top to bottom) of: basaltic lavas, usually with lobate, pillowed morphology indicative of submarine eruption; a parallel 'sheeted' swarm of dolerite dykes that fed the lava flows; and coarse-grained gabbros and ultramafic (olivine-rich) plutonic rocks that crystallized slowly in the magma chamber.

In early models for the generation of ocean crust these magma reservoirs were envisaged as huge bodies the height of the entire lower crust and tens of kilometres wide beneath spreading axes. However, seismic reflection experiments at the fast-spreading East Pacific Rise in the late 1980s showed that no such magma body exists: purely molten material is instead restricted to a thin lens (probably of the order of 100 m thick) in the middle crust, overlying a much larger region in the lower crust that appears to be a partially molten mush composed of crystals with small amounts of interstitial melt. At slower spreading rifts such as the Mid-Atlantic Ridge even this thin magma lens is normally absent, implying that the lower crust there probably freezes solid in between melt delivery events from the mantle, and that magma supply in these environments may be relatively reduced (e.g. Sinton and Detrick 1992).

Ridges may be offset by several hundred kilometres by transform faults, across which lateral motion occurs as seafloor spreads in opposite directions on either side. Smaller scale offsets or discontinuities of the spreading axis between transform faults define the boundaries between individual spreading segments, which may be regarded as elongate volcanoes more or less aligned along the ridge crest (Macdonald *et al.* 1988).

The morphology of the region around the ridge crest is very much dependent upon spreading rate. Whereas fast-spreading ridges are characterized by relatively smooth seafloor with an elevated ridge crest, slower spreading ridges (less than ~5 $cm\,yr^{-1}$) are instead marked by a very rough seafloor and an axial valley that may be more than 2 km deeper than the surrounding walls. Mantle peridotite, now altered by the action of seawater to serpentinite, is commonly recovered in these axial valleys. The rough seafloor results from extensional faulting, which helps to accommodate separation of the plates when magma supply to the ridge is low. Some of the fault planes have very low dip angles and can accommodate tens of kilometres (in excess of a million years' worth) of displacement on a single structure. These structures, termed 'detachment faults', provide a mechanism by which mantle and lower crustal rocks may be exhumed onto the seafloor. Mantle rocks may also be exposed at some slow-spreading ridges because magma supply to the ridge axis was so low that a continuous layer-cake magmatic crustal layer was never produced in the first place (Cannat 1993). In places, therefore, the seismically defined crustal layer may be composed partly or completely of serpentinite, which has a velocity much lower than fresh peridotite but similar to that of basalt or gabbro. Ocean crustal structure, particularly at slow-spreading ridges, is now understood to be far more heterogeneous than originally suggested on the basis of the seismic refraction experiments. Fast-spread ocean crust may have a more regular 'layer-cake' architecture.

Total crustal production by seafloor spreading at mid-ocean ridges is estimated at 18 $km^3\,yr^{-1}$, generating a thermal flux equivalent to ~50 megawatts per kilometre of spreading ridge worldwide. This heat is extracted from the newly formed crust primarily by hydrothermal convection: seawater descends through cracks in the crust, heats up and is eventually vented back into the water column as 'black smoker' fluid. This fluid is hot (up to 400 °C) and rich in metals that have been stripped from the basaltic crust. Volumes are such that the equivalent of the entire world ocean is believed to circulate through the crust every seven million years or so. Hydrothermal circulation therefore plays an essential role in regulating the composition of seawater on geological timescales.

Fluid circulation and cooling persists away from the ridge crests. It causes the lithosphere – the mechanically rigid plate – to increase progressively in thickness as it moves away from the ridge. As the uppermost mantle cools below ~1,000 °C it ceases to be able to flow in the convecting asthenosphere and moves with the overlying crust, in effect being attached or welded to the base of the ocean crust. Mature ocean lithosphere is up to 100 km thick, all but the uppermost few kilometres being rigid mantle.

On a broad scale the ridge crest and surrounding seafloor is elevated relative to the abyssal plains because of the lower density of hot, partially molten asthenospheric mantle flowing upward beneath the ridge crest. Moving the lithosphere away from the region of upwelling and thickening it causes the seafloor to subside (see ISOSTASY). The rate of subsidence is proportional to the square root of the age of the crust and is independent of spreading rate, explaining why the width of the elevated region around the East Pacific Rise is far broader than that around the Mid-Atlantic Ridge.

References

Cannat, M. (1993) Emplacement of mantle rocks in the seafloor at mid-oceanic ridges, *Journal of Geophysical Research* 98, 4,163–4,172.
Dietz, R.S. (1961) Continent and ocean basin evolution by the spreading of the sea floor, *Nature* 190, 854–857.
Hess, H.H. (1962) History of the ocean basins, in A.E.J. Engel, H.L. James and B.F. Leonard (eds) *Petrologic Studies: A Volume to Honor A.F. Buddington*, Denver: Geological Society of America.
Macdonald, K.C., Fox, P.J., Perram, L.J., Eisen, M.F., Haymon, R.M., Miller, S.P., *et al.* (1988) A new view of the mid-ocean ridge from the behaviour of ridge-axis discontinuities, *Nature* 335, 217–225.
Sinton, J.M. and Detrick, R.S. (1992) Mid-ocean ridge magma chambers, *Journal of Geophysical Research* 97, 197–216.
Vine, F.J. and Matthews, D.H. (1963) Magnetic anomalies over ocean ridges, *Nature* 99, 947–949.

Further reading

Kearey, P. and Vine, F.J. (1996) *Global Tectonics*, 2nd Edition, Oxford: Blackwell.
Nicolas, A. (1995) *The Mid-Oceanic Ridges: Mountains Below Sea Level*, Berlin: Springer-Verlag.
Open University Course Team (1998) *The Ocean Basins: Their Structure and Evolution*, 2nd Edition, Oxford: Butterworth-Heinemann.
Oreskes, N. and Le Grand, H. (eds) (2001) *Plate Tectonics: An Insider's History of the Modern Theory of the Earth*, Boulder, CO: Westview Press.

CHRIS MACLEOD

SEDIMENT BUDGET

A sediment budget is a quantitative accounting of the rates of production, transport and discharge of detritus in a geomorphic system such as a DRAINAGE BASIN, coastal BEACH, offshore zone, hillslope, river channel, GLACIER, or any landscape unit around which boundaries can be drawn. Sediment budgets apply the principle of conservation of mass to geomorphic systems, an approach that became popular in the 1970s. Sediment budgets provide a tool for research geomorphologists to judge the relative importance of sediment sources, storage sites and transfer processes, including how they change over time. Sediment budgets are also useful tools for resource management when it becomes important to distinguish human impacts on geomorphic systems from those that would have occurred without human interference (Reid and Dunne 1996). Studies have been reported that use sediment budgets to document the effect of agriculture, forestry, road construction, urbanization, DAMs, wildfires and mining. Sediment budgets have been used to construct a more complete picture of the distribution of sediment sources within a drainage basin (e.g. Marston and Dolan 1999) and the heavy metals that can be associated with the sediment (e.g. Marcus *et al.* 1993).

Four basic steps must be followed to construct a sediment budget (Lehre 1982): (1) define the boundaries of the geomorphic system; (2) identify the processes and sites of EROSION, transport and storage (deposition) in the geomorphic system, including the linkages between them; (3) quantify the contribution of each over space and time; and (4) set up an accounting sheet that balances the sediment production, sediment yield (see SEDIMENT LOAD AND YIELD), and storage. The first step depends on the ability to recognize the boundaries of the geomorphic system to be studied. Most sediment budgets have been prepared for drainage basins and sandy beaches, both of which have readily recognized boundaries. However, sediment budgets have also been attempted for KARST and glacier systems, where subsurface passages transport significant amounts of sediment that is difficult to trace and measure. Sediment budgets generated by wind in the form of sandstorms and DUST STORMs have been rare, although this component has been used to explain the discrepancies between inputs, change of storage and outputs – a dangerous practice unless all transfer and storage components have been measured with great accuracy (Hill *et al.* 1998). The second step requires training, experience and expertise with the full range of field, lab and office techniques in geomorphology (e.g. aerial photography and historical maps) to discern evidence of the range of possible processes and storage sites. A flow chart may help to visualize the processes, storage sites, outputs of sediment and linkages between them (Figure 147). The third step requires measurement of these processes using an appropriate choice of techniques. Direct measurement of processes is beset by problems, although controlled field experiments can help. In some cases, ^{137}Cs measurements have been used in combination with more traditional techniques to compile tracer-based sediment budgets (e.g. Walling *et al.* 2002). Geomorphologists might also use predictive equations, examine soil profiles, undertake photogrammetric measurements from aerial photographs, or conduct lab measurements of soil and sediment characteristics. Changes in sediment storage can be derived from geophysical surveys, morphometric modelling or field surveys using anchor chains.

One of the classic sediment budget studies in understanding the importance of sediment storage was undertaken by Trimble (1983). He compiled a sediment budget for the 360 km^2 Coon Creek watershed in Wisconsin for two periods: 1853–1938 when soil erosion was severe because of poor land management, and 1938–1975 following the implementation of SOIL CONSERVATION practices. Sediment yield at the mouth of the WATERSHED was essentially identical for the two periods, from which one might conclude that soil conservation practices failed to have the desired effect. However, when one examines the sediment budget data, it becomes apparent that upland erosion had been reduced by 26 per cent. Sediment yield figures remained high even after soil conservation measures had been implemented because sediment that had been stored in tributary valleys and the upper main valley during the early period had been mobilized during the latter period. This removal from storage offset the reduction in hillslope erosion, with the net effect that overall sediment yields remained essentially unchanged. One of the main lessons of erosion studies has been learned from Trimble's study: sediment yield values alone are not a good indicator of spatial and temporal changes in erosion within a watershed.

Reneau and Dietrich (1991) used a sediment budget approach in the southern Oregon Coast Range to demonstrate that a steady state condition exists between sediment production from hillslopes and sediment yield from the watershed. The authors suggest that erosion rates were spatially uniform and perhaps even diagnostic of a landscape in geomorphic equilibrium. In earlier work elsewhere in the Oregon Coast Range, Dietrich and Dunne (1978) found that the rate that bedrock in the Rock Creek watershed was converted to soil, with the associated increase in volume by a factor of 1.17, was equivalent to the rate of DENUDATION calculated for the watershed over the same time interval. Thus, they conclude that reduction in watershed relief by denudation was balanced by the increase in volume of hillside materials; relief is not changing over time, without accounting for tectonics.

Graf (1994) compiled a fluvial sediment budget for the northern Rio Grande drainage system in New Mexico for the express purpose of tracking deposits likely to contain plutonium. Early development of nuclear weapons at Los Alamos National Laboratory had led to discharge of highly concentrated plutonium wastes to arroyos that had carried the plutonium adsorbed onto sediment into one of the major waterways of western North America. Graf coupled extensive and detailed field mapping with simulation modelling of plutonium movement to demonstrate the importance of large FLOODs in delivering pulses of plutonium downriver. Fortunately, the concentrations of plutonium in stored sediments are generally low, but do vary considerably. With large floods, concentrations will decline over time to approximate levels from atmospheric fallout; without large floods, the portion of plutonium that is associated with bedload will persist. Graf was also able to demonstrate the significance of plutonium storage behind Cochiti Reservoir along the Rio Grande, a site that is now closed for fishing and swimming.

Madej (1987) showed how the residence time of stored sediment could be calculated as part of fluvial sediment budget studies if one knows the sediment storage, expressed as cubic metres per metre length of valley sediment, and the sediment transport rate, expressed in units of cubic metres per year. When the storage is divided by the transport rate, the residence time is derived, in units of years per cubic metre. This strategy was used by Madej to calculate the residence time of sediment in various reaches of Redwood Creek in north-west California. Comparing the residence time to the rate of sediment delivery allows one to determine whether the system is at/near geomorphic equilibrium or likely to experience progressive change over time. If a reach receives

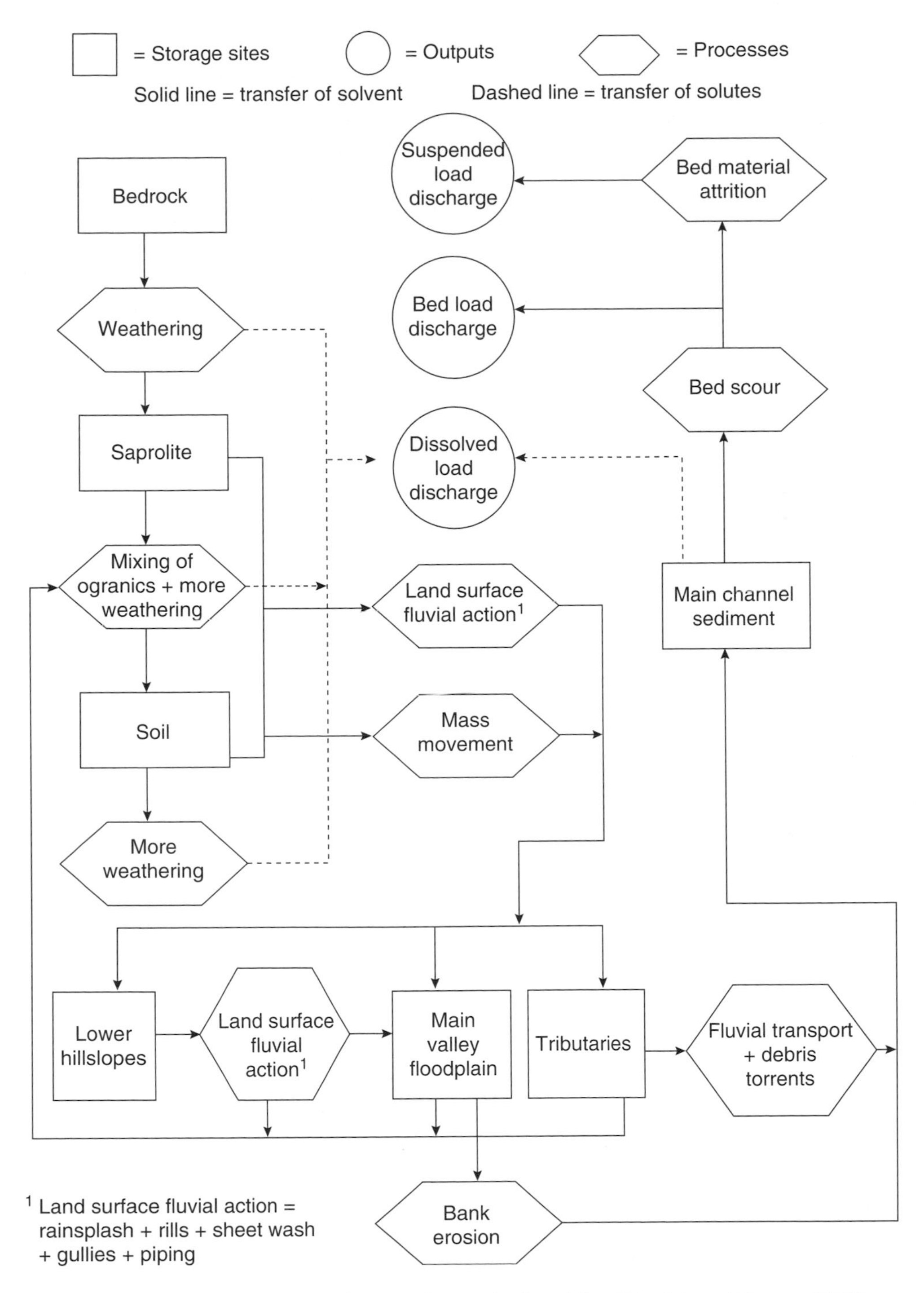

Figure 147 Flow chart for a drainage basin sediment budget (after Dietrich and Dunne 1978)

a volume of sediment from storms with a recurrence interval of fifty years and that sediment is likely to reside in the stream for a hundred years, the stream will experience progressive AGGRADATION. The recovery of Redwood Creek from catastrophic floods and associated MASS MOVEMENT was found to vary depending on the residence time, which in turn varies with landscape position in the valley floor. Indeed, sediment budgets have been used to compare the importance of frequent, low magnitude events with infrequent, catastrophic events in the transport of sediment (e.g. Springer *et al.* 2001).

Sediment budgets have proved especially useful in examining changes in beaches over time. Consider a sediment budget for a sandy beach between two rocky headlands. Sediment sources could include eroding cliffs (see CLIFF, COASTAL), onshore transport, marine erosion of beach material, supply from dunes (see DUNE, COASTAL), subsurface erosion, fluvial input, BEACH NOURISHMENT and LONGSHORE (LITTORAL) DRIFT. Sediment storage could occur on the beach and adjacent inland dunes, as well as in offshore banks, SPITS and bars (see BAR, COASTAL). Sand could be lost through offshore transport, longshore drift, aeolian erosion (see AEOLIAN PROCESSES), and dredging. The effects of various shoreline protection measures can be effectively measured with this approach (Cooper *et al.* 2001). One noteworthy study demonstrated that the loss of sand along beaches of southern California could be attributed to the construction of sediment detention basins in the mountains surrounding the Los Angeles Basin. When it was discovered that valuable recreational beaches were being deprived of river-delivered sand and diminishing in size, artificial nourishment was required at great expense (Cooke 1984).

Sediment budgets are utilized over a wide range of spatial scales, from small plot studies (Duijsings 1987) to continental-scale assessments. New techniques are being developed to construct more accurate sediment budgets. The sediment budget approach is being applied to an ever-expanding variety of geomorphic environments and for discerning the effects of human activities in these settings. Because sediment budgets deal with the sources, storage, throughflow and outputs of sediment in a geomorphic system, they have been characterized as a fundamental method in understanding cascading process systems.

References

Cooke, R.U. (1984) *Geomorphological Hazards in Los Angeles*, London: Allen and Unwin.

Cooper, N.J., Hooke, J.M. and Bray, M.J. (2001) Predicting coastal evolution using a sediment budget approach: a case study from southern England, *Ocean and Coastal Management* 44, 711–728.

Dietrich, W.E. and Dunne, T. (1978) Sediment budget for a small catchment in mountainous terrain, *Zeitschrift für Geomorphologie, Supplementband* 29, 191–206.

Duijsings, J.J.H.M. (1987) *Streambank Contribution to the Sediment Budget of a Forest Stream*, Publication No. 40, Amsterdam: Fysisch-Geografisch en Bodemkundig Laboratorium van de Universiteit van Amsterdam.

Graf, W.L. (1994) *Plutonium and the Rio Grande: Environmental Change and Contamination in the Nuclear Age*, New York: Oxford University Press.

Hill, B.R., Decarlo, E.H., Fuller, C.C. and Wong, M.F. (1998) Using sediment fingerprints to assess sediment-budget errors, North Halawa Valley, Oahu, Hawaii, *Earth Surface Processes and Landforms* 23, 493–508.

Lehre, A.K. (1982) Sediment budget of a small coast range drainage basin in north-central California, in F.J. Swanson, R.J. Janda, T. Dunne and D.N. Swanson (eds) *Sediment Budgets and Routing in Forested Drainage Basins*, General Technical Report PNW-141, 67–77, Portland, OR: USDA Forest Service.

Madej, M.A. (1987) Residence times of channel-stored sediment in Redwood Creek, northwestern California, in R.L. Beschta, T. Blinn, G.E. Grant, G.G. Ice and F.J. Swanson (eds) *Erosion and Sedimentation in the Pacific Rim*, Publication No. 165, Wallingford: International Association of Hydrological Sciences.

Marcus, W.A., Neilsen, C.C. and Cornwell, J. (1993) Sediment budget analysis of heavy metal inputs to a Chesapeake Bay estuary, *Environmental Geology and Water Sciences* 22, 1–9.

Marston, R.A. and Dolan, L.S. (1999) Effectiveness of sediment control structures relative to spatial patterns of upland soil loss in an arid watershed, Wyoming, *Geomorphology* 31, 313–323.

Reid, L.M. and Dunne, T. (1996) *Rapid Evaluation of Sediment Budgets*, Reiskirchen: Catena Verlag.

Reneau, S.L. and Dietrich, W.E. (1991) Erosion rates in the southern Oregon Coast Range: evidence for an equilibrium between hillslope erosion and sediment yield, *Earth Surface Processes and Landforms* 16, 307–322.

Springer, G.S., Dowdy, H.S. and Eaton, L.S. (2001) Sediment budgets for two mountainous basins affected by a catastrophic storm: Blue Ridge Mountains, Virginia, *Geomorphology* 37, 135–148.

Trimble, S.W. (1983) A sediment budget for Coon Creek basin in the Driftless Area, Wisconsin, *American Journal of Science* 283, 454–474.

Walling, D.E., Russell, M.A., Hodgkinson, R.A. and Zhang, Y. (2002) Establishing sediment budgets for two small lowland agricultural catchments in the UK, *Catena* 47, 323–353.

RICHARD A. MARSTON AND MARCUS E. PEARSON

SEDIMENT CELL

A *sediment cell* is a section of the coastal zone where the sediment inputs, throughput and outputs may be considered part of a closed system. Given the vast range of coastal environments and the nature and sources of sediments, a wide range of sediment cells exist, both in form, and temporal and spatial scale. A more general definition is, therefore, a coastal environment where the input, accumulation and output of sediments are part of an interrelated flow of sediments, some of which may be derived from, and/or exported to, adjacent sediment cells.

Komar (1998) discusses sediment cells using the concept of coastal SEDIMENT BUDGETs, based on the example of the Southern California littoral cells. Each cell has a sediment source (river, cliff erosion, etc.), longshore transport, and finally sinks which include submarine canyons, loss to dunes and transport to downdrift cells. Each cell therefore contains a cycle of sediment input, transport and sedimentation, the latter either as accumulation within the cell, or loss onshore, offshore or longshore. He presents a more generalized budget of littoral sediments which includes sediment credit, debit and a balance, the latter resulting in positive beach deposition or negative beach erosion.

Once established cells can undergo *temporal variation* in response to changes in the boundary conditions. This can include a change in the sediment budget, as has occurred on many coasts following the postglacial marine transgression, as shelf sediment supplies have become exhausted. It could also be human induced through construction of dams or shoreline structures which interrupt or stop sediment supply. It can also be produced by changes in the driving processes, such as wave climate, wave refraction and climatic conditions, all of which can change both the magnitude and direction of sediment transport within the cell.

Cells can also reach *equilibrium*, when there is essentially zero sediment transport and zero change to the sediment budget over a given time span. This can occur on swash aligned beaches where the wave crest is in equilibrium with the shoreline alignment. On a drift or current aligned shore it would infer zero longshore sediment transport, which is possible but unlikely. On a graded shore the sediment texture is arranged so that it is in equilibrium with the prevailing processes so that no further regrading is required. Swash aligned beaches may take the form of a log spiral (see LOG SPIRAL BEACH) or zeta shape, in locations where waves are refracted around a headland and the shoreline aligns to the spiral of the refracted wave crest. So long as there is no littoral drift they may be considered an equilibrium or closed cell. However on many coasts such systems do release sediment downdrift through pulse boundaries.

Davies (1980) approaches sediment cells from the concept of a sediment store, which has input, throughput and outputs, as well as internal biological supply and loss from attrition. The nature of the store or cell is dependent on the nature, scale and rate of the boundary components and internal dynamics.

Carter (1988) provides the most advanced treatment of what he calls a 'coastal cell', within which is a recognizable compartmentalization of the sediment budget. He goes on to say that this is easy to identify where the cell is restricted to a bay, estuary, river mouth or pocket beach, but many coastal cells have leakages longshore to adjacent systems, onshore to dunes and estuaries or offshore to the inner continental shelf. The cell boundaries may therefore be free or fixed (Table 42). *Fixed* boundaries include impermeable morphological structures such as headlands, shoals, inlets, river mouths. *Free* boundaries are more transparent and therefore less recognizable, as they may result from a change in wave field and direction of transport, rather than a distinctive morphological feature. Both boundary types may either cause the cells to 'divide', 'meet' or 'pulse'. The *divide* boundary occurs at the updrift limit of shoreline erosion, while the *meet* boundary occurs at the downdrift limit of sediment deposition. The *pulse* boundary permits sediment exchange between cells.

Sediment cells can also vary considerably in size and relation to neigbouring cells. Cells may be either independent (e.g. small pocket beach) with fixed boundaries and no leakages, or be nested or cascaded within a series of interconnected (leaking) subcells (e.g. deltaic systems or series of interconnected beaches).

Sediment cells can be identified as a *morphological* unit, particularly when they have fixed boundaries, e.g. an embayed or pocket beach or estuary. They can also be identified based on *sediment texture*, either through the presence of one sediment type, e.g. a coastal dunefield, or

Table 42 Sediment cell definitions and budget

Sediment cell – a closed and balanced sediment budget	
Boundaries	
fixed morphological structure (e.g. headland, inlet, river mouth, foreland)	
free dynamic divide induced by processes, e.g. change in direction of littoral drift	
Boundary types	
divide–	updrift, limit of erosion or sources of sediment
meet–	downdrift, limit of sediment deposition, sink
pulse–	leakage and exchange across boundary
Sediment inputs	
External	
	Terrestrial – river supply, cliff erosion
	Biological – carbonate detritus
	Updrift longshore transport – from neighbouring cell
	Headland bypassing via dunes or subaqueous sand pulses
	Onshore transport from inner shelf, esp. during sea level transgression
Internal	
	Biogenic production
	Chemical production, e.g. ooids and cements
	Beach nourishment
Sediment outputs	
Onshore	barrier – dunes
	– overwashing
	estuary infilling (flood tide deltas)
Longshore	longshore sediment transport
Offshore	inner shelf sand bodies
	submarine canyons
Internal	attrition, solution, cementation (e.g. beachrock)
	beach mining
Sediment balance	
Inputs–outputs = cell erosion, deposition or stable	

Sources: After Komar (1998), Davies (1980) and Carter (1988)

a gradation in the sediment texture resulting from selected longshore transport of size factions. Finally they can be defined by the rates and scale of sediment *transport* within the system, with the boundaries having little or no transport.

References

Carter, R.W.G. (1988) *Coastal Environments*, London: Academic Press.

Davies, J.L. (1980) *Geographical Variation in Coastal Development*, 2nd edition, London: Longman.

Komar, P.D. (1998) *Beach Processes and Sedimentation*, 2nd edition, Upper Saddle River, NJ: Prentice Hall.

SEE ALSO: log spiral beach; longshore (littoral) drift; sediment budget

ANDREW D. SHORT

SEDIMENT DELIVERY RATIO

The rate of sediment yield at a specified point in a channel network, expressed as a fraction of the rate of erosion in the contributing catchment, is termed the sediment delivery ratio. Sediment delivery ratios are widely used to adjust SOIL EROSION estimates to account for deposition of sediment as it is transported from its point of origin to, and through, the stream network.

Sediment delivery ratios have been widely used to estimate stream sediment loads from erosion rates predicted by the Universal Soil Loss Equation (USLE) and its successor, the Revised USLE (RUSLE). These empirical equations are designed to predict gross rates of erosion at the soil surface. Because the USLE and RUSLE were developed from studies of small test plots, they define 'erosion' as the movement of soil particles from one location to another – but, importantly, not necessarily from their point of origin to a stream channel. A fraction of the sediment mobilized by surface erosion will be intercepted (for example, in densely vegetated zones or low-gradient footslopes) before it reaches the channel network. Of the sediment that reaches the channel network, a further fraction will be deposited on the floodplain or stored in the channel. The proportion that is delivered to a sampling point in the channel network – rather than intercepted on the soil surface, deposited on the floodplain, or stored in the channel – is the sediment delivery ratio.

Sediment delivery ratios are commonly estimated from the measured sediment yield (from sediment gauging methods or accumulation in a sediment trap) at a given point in the channel network. This is then divided by the estimated rate of erosion in the surrounding catchment (derived from the USLE/RUSLE or, in some cases, direct field measurements). Thus sediment delivery ratios will not only reflect sediment interception, storage and deposition, but will also reflect any errors made in estimating sediment yields or rates of surface erosion; both are subject to significant uncertainties (Meade 1988; Trimble and Crosson 2000).

Sediment delivery ratios reported in the literature range from over 100 per cent to less than 1 per cent. This variability arises from differences in geomorphic characteristics between catchments, as well as from variations in erosion rates and sediment yields through time at any individual site. Sediment delivery is often highly episodic, and measurements of sediment yield – even when averaged over decades – can be significantly higher or lower than long-term rates of sediment supply to the channel network (e.g. Clapp *et al.* 2000; Kirchner *et al.* 2001).

Nonetheless, systematic relationships have been observed between sediment delivery ratios and catchment morphology and processes. Sediment delivery ratios tend to be higher in catchments where channel slopes and valley sideslopes are steep, and where relief and drainage density are high. Conversely, sediment delivery ratios tend to be lower where sediment sources are far from channels, or are separated from them by sediment-trapping zones (typically characterized by low gradients and dense vegetation). Sediment delivery ratios also tend to be lower where sheet and rill erosion predominate, and higher where gully erosion predominates, because gullies tend to be more directly connected to the channel network.

Sediment delivery ratios also generally decrease as drainage area increases, ranging from roughly 30–100 per cent in 0.1 km^2 catchments to roughly 2–20 per cent in 1,000 km^2 catchments (e.g. Novotny and Olem 1994). This is consistent with the fact that as one moves downstream, channel and valley gradients typically become gentler and floodplains and footslopes typically become wider. All these trends provide greater opportunities for sediment storage, both on hillslopes and in the fluvial system. As one might expect, the lowest sediment delivery ratios are typically observed where rivers emerge from steep mountain fronts and flow out across broad depositional basins.

Sediment yield predictions are often generated by combining USLE or RUSLE estimates of surface erosion with sediment delivery ratios plucked from the literature. This approach is problematic, because sediment delivery ratios vary widely and are not always consistently defined. The denominator of the sediment delivery ratio is sometimes the gross rate of sediment mobilization on the soil surface (as in the USLE/RUSLE); in this case the ratio reflects sediment interception en route to the channel as well as net deposition and storage in the channel network. Alternatively, the denominator is sometimes the rate of sediment supply to the channel network (excluding sediment interception during overland transport, but including sediment production from channel incision or bank erosion); in this case the sediment delivery ratio reflects only the transmission efficiency of the fluvial system. Because sediment delivery ratios may be conceptually defined or operationally measured differently from one study to the next, they should be interpreted with caution.

References

Clapp, E.M., Bierman, P.R., Schick, A.P., Lekach, J., Enzel, Y., and Caffee, M. (2000) Sediment yield exceeds sediment production in arid region drainage basins, *Geology* 28, 995–998.

Kirchner, J.W., Finkel, R.C., Riebe, C.S., Granger, D.E., Clayton, J.L., King, J.G. and Megahan, W.F. (2001) Mountain erosion over 10-year, 10,000-year, and 10,000,000-year timescales, *Geology* 29, 591–594.

Meade, R.H. (1988) Movement and storage of sediment in river systems, in A. Lerman and M. Meybeck (eds) *Physical and Chemical Weathering in Geochemical Cycles*, 165–179, Dordrecht: Kluwer.

Novotny, V. and Olem, H. (1994) *Water Quality: Prevention, Identification, and Management of*

Diffuse Pollution, New York: Van Nostrand Reinhold.
Trimble, S.W. and Crosson, P. (2000) U.S. soil erosion rates – Myth and reality, *Science* 289, 248–250.

JAMES W. KIRCHNER

SEDIMENT LOAD AND YIELD

Normally, sediment discharge is reported in units of mass per unit time, kilograms per second, megagrams per day or megagrams per year. A megagram is equivalent to a tonne, so frequently loads are reported in tons per day or tons per year. Sediment concentration is reported in milligrams per litre or grams per cubic metre of water. The size of particulate material transported by rivers ranges between fine clay and colloidal particles of less than 0.5 micrometres in diameter to large boulders moved during flood events. A distinction is commonly made between bedload and suspended load. This is a distinction based on mode of transport. Bedload consists of the coarser sediment particles that roll, hop or slide along the bed, with more or less continuous contact with the bed, while the suspended load consists of sediment particles that are supported in the flow by the upward components of turbulent currents. The total sediment load is the sum of the bedload and the suspended load. Engineers prefer the distinction between bed-material load and wash load where sediment particles that are finer than those contained in the bed are called wash load and sediment particles that are the same size range as those found in the bed are called bed-material load. The implication is that bed-material load is primarily a function of transporting capacity, boundary shear stress or force available to initiate movement, whereas wash load is commonly a function of variable sources of supply. In either case, there is overlap between the two categories. Church (1983) recommends 1 mm as the appropriate boundary value between bedload and wash load; Walling (1988) recommends 0.2 mm as the cutoff between bedload and suspended load.

Fluvial sediment transported as bedload consists almost entirely of inorganic material, except in steep, forested watersheds where large organic debris is a highly significant component. The inorganic bedload normally resembles the local bedrock in terms of mineral composition except in glacially disturbed watersheds. The finer sediment transported in suspension commonly incorporates a proportion of organic material. The mineralogy of these sediments may bear little resemblance to the local bedrock because of selective detachment and transport processes, chemical weathering processes which disintegrate the rock and secondary redistributed glacigenic sediments. Also the size distribution of the suspended sediment will be considerably enriched in clay-sized particles and organic matter when compared with the source material. Enrichment in fine material has important implications for transport of contaminants which tend to be adsorbed onto finer particles as these finer particles have greater specific surface areas and cation exchange capacities.

Sediment transport capacity load or non-capacity load is a classification of the load of sediment being carried by a stream. If a stream has excess or unsatisfied capacity it is said to be carrying a non-capacity load; if it is carrying as much material as available stream energy permits, it is carrying a capacity load. The enormous capacity for the transport of fine sediment is exemplified by rivers carrying hyperconcentrated flows. Almost all streams carry a non-capacity load of fine sediments. For particles larger than some critical diameter, streams carry a capacity load. A commonly accepted critical particle diameter is around the silt/sand boundary (or about 0.063 mm).

Sediment transport formulae are still evolving, especially bedload formulae. Bedload transport commences when the rate of water flow reaches a magnitude adequate to exert a critical tractive force. Du Boys introduced a formula for movement of bedload of the form

$$Q_b = C_b \cdot \tau(\tau - \tau_c)$$

where Q_b = bedload transport per unit width per unit time, C_b = a special sediment parameter, τ = shear stress, and τ_c = critical shear stress for start of material movement.

Suspended load transport can be considered as an advanced stage of bedload transport by which particles in saltation are caught by the upward component of the turbulent velocity and are kept in suspension. The fundamental equation of sediment suspension by fluid turbulence is

$$c\omega = -\beta \cdot \varepsilon \cdot dc/d\gamma$$

where c = concentration of suspended load in dry weight per unit volume, ω = settling velocity of the particles, β = sediment transfer coefficient, ε = exchange coefficient for suspended load, and γ = vertical distance from the stream bed.

There are problems of application because ε varies through the vertical, and the expression for the variation in suspended sediment concentration with depth is fairly complex. Nevertheless, suspended sediment load can be predicted more accurately than the bedload.

It is also possible to combine bedload and suspended load transport formulae in one expression for total load. The fundamental reference is Einstein (1950) but there are numerous papers still appearing in *Water Resources Research* debating improvements in these formulae.

Milliman and Syvitski (1992) admit that no one knows exactly how much sediment is discharged to the ocean but they suggest that it is probably of the order of 20 billion tonnes per year. This is 50 per cent higher than the estimate by Milliman and Meade (1983) and is accounted for by the fact that the contribution of small mountainous rivers was overlooked in the earlier estimate. On the other hand, it seems probable that, because of the proliferation of dam construction during the second half of the twentieth century, that a smaller total is in order; they also note that an unknown amount is stored in subaerial parts of sinking deltas. The suggestion is that prior to widespread farming and deforestation, the export of sediment was about 10 billion tonnes per year.

Sediment yield is defined as the total mass of particulate material reaching the outlet of a drainage basin per year. Specific sediment yield is the sediment yield per unit area and, making suitable adjustments for sediment specific gravity and density of packing, this figure can be converted into a depth of sediment removed from the whole surface of the basin per year. A unit of measurement known as the Bubnoff (Fischer 1969), which is equivalent to 1 mm of denudation per 1,000 years has been found to be useful for regional comparisons, though it should always be borne in mind that variable thicknesses of sediment are removed from different parts of each basin and the Bubnoff is only an average statistic. Documented values range from 0.09 to over 13,000 Bubnoffs (0.26 to 36,000 $\mathrm{t\,km^{-2}}$). The two extreme cases are from Finland and Taiwan respectively. The total sediment yield should include both bedload and suspended load, but only rarely are measurements of bedload available. In such a case, 10 per cent is normally added to the suspended load. In steep mountainous basins, this is clearly an underestimate and in lowland rivers, this is an overestimate.

Where the sediment load of a river is deposited in a lake or reservoir it may be possible to estimate the total yield quite accurately. The sediment yield from a drainage basin is commonly only a small proportion of the gross erosion in the basin. Much of the eroded material is deposited before it reaches the outlet and the ratio of sediment yield to gross erosion is termed the sediment delivery ratio. The magnitude of the sediment yield is a function of climate, topography, sediment availability, lithology, vegetation cover and land use.

The sediment budget is defined as a method of accounting of sources, sinks and redistribution pathways of sediments in a unit region over unit time (Swanson *et al.* 1982). The method provides a bridge between studies of fluvial sediment transport and of the associated deposits in storage in the basin. The transport of sediment out of the basin is then seen as a residual term after deposition and storage of the sediment within the basin. In general, the storage term gets larger with increasing size of basin. Sediment on its way from source to outlet of a basin gets side tracked in a number of ways, not only into lake, floodplain and channel storage, but also into other hillslope locations. It is in this context that the concept of virtual velocity of sediment becomes important (Church 2002). Virtual velocity can be defined as the velocity of sediment through a reservoir and is simply the inverse of residence time per metre of reservoir length. Residence time per metre equals the mapped volume of sediment per metre divided by the bedload discharge rate.

It is estimated for the United States that the amount of sediment delivered to the oceans is only about 10 per cent of the total amount eroded off the uplands and 90 per cent is stored between the erosion sites and the sea. They provide illustrative data on the importance of short-term scale, decade to century scale and longer timescales of storage. The short-term pattern on the lower Mississippi River shows sediment being deposited and stored on the river bed at lower flows and being resuspended and flushed out to sea at higher flows. The decade to century scale effect is illustrated by the movement of hydraulic mining debris through the Sacramento River system in California. The slug of sediment (or wave according to Gilbert 1917) created by gold mining during 1855–1885 took almost a century to reach the sea and caused major changes in river bed levels from temporary storage during that period. A classic study by Trimble (1983) examines the long-term effects of sediment storage resulting from forest clearance and accelerated erosion in the uplands of Coon Creek basin in Wisconsin in the 1850s. During 1853–1975, 80 million tons of sediment were eroded off the uplands and delivered to storage sites whereas only 5 million tons were transported out of the basin. In British Columbia, Church and Slaymaker (1989) estimate that sediment stored in paraglacial fans, valley fill and floodplains will require several tens of thousands of years to be transported to the sea.

References

Church, M. (1983) Concepts of sediment transfer and transfer on the Queen Charlotte Islands, *Fish/Forestry Interaction Program Working Paper* 2/83, Victoria, BC: Ministry of Forests and Ministry of Environment, Lands and Parks.

——(2002) 'Fluvial sediment transfer in cold regions', in K. Hewitt, M.L. Byrne, M. English and G. Young (eds) *Landscapes of Transition*, 93–117, Dordrecht: Kluwer.

Church, M. and Slaymaker, O. (1989) Disequilibrium of Holocene sediment yield in glaciated British Columbia, *Nature* 337, 452–454.

Einstein, H.A. (1950) The bed load function for sediment transportation in open channel flow, *US Department of Agriculture Paper* No. 1,028.

Fischer, A.G. (1969) Geological time-distance rates: the Bubnoff unit, *Geological Society of America Bulletin* 80, 549–552.

Gilbert, G.K. (1917) Hydraulic mining debris in the Sierra Nevada, *US Geological Survey Professional Paper* 105, Washington, DC: US Geological Survey.

Milliman, J.D. and Meade, R.H. (1983) World wide delivery of river sediment to the oceans, *Journal of Geology* 91, 1–21.

Milliman, J.D. and Syvitski, J.P.M. (1992) Geomorphic/tectonic control of sediment discharge to the ocean: the importance of small mountainous rivers, *Journal of Geology* 100, 525–544.

Swanson, F.J., Janda, R.J., Dunne, T. and Swanston, D.N. (1982) Sediment budgets and routing in forested drainage basins, *US Forest Service General Technical Report*, PNW-141, Portland, OR: US Department of Agriculture.

Trimble, S.W. (1983) A sediment budget for Coon Creek basin in the Driftless Area, Wisconsin, 1853–1977, *American Journal of Science* 283, 454–474.

Walling, D.E. (1988) Erosion and sediment yield research: some recent perspectives, *Journal of Hydrology* 100, 113–141.

Further reading

Dietrich, W.E. and Dunne, T. (1978) Sediment budget for a small catchment in mountainous terrain, *Zeitschrift für Geomorphologie Supplementband* 29, 191–206.

Hallet, B., Hunter, L. and Bogen, J. (1996) Rates of erosion and sediment evacuation by glaciers: a review of field data and their implications, *Global and Planetary Change* 12, 213–235.

Hinderer, M. (2001) Late Quaternary denudation of the Alps, valley and lake fillings and modern river loads, *Geodinamica Acta* 14, 231–263.

Hovius, N., Stark, C.P. and Allen, P.A. (1997) Sediment flux from a mountain belt derived by landslide mapping, *Geology* 25, 231–234.

SEE ALSO: sediment rating curve; sediment routing

OLAV SLAYMAKER

SEDIMENT RATING CURVE

A sediment rating curve represents the relation between suspended (see SUSPENDED LOAD) sediment concentration (or discharge) and water discharge at a stream measurement station. It is used for estimating suspended sediment discharge averaged over a period of flow record. The relation may concern 'instantaneous' values or may relate average values over (e.g. daily) time intervals.

In theory, suspended sediment concentration should increase with water discharge because the associated increase in turbulence increases the capacity of the river to carry suspended sediment. In practice, the concentration of suspended sediment, particularly the silt-clay fractions, tends to be more influenced by the sediment supply to the channel from hillslope and riparian erosion processes, which can be 'patchy' in space and time. During runoff events, the relation can vary due to sediment exhaustion and also to phase lags between the sediment and water peaks passing the measurement station. The relation can also vary due to seasonal controls on the supplies of sediment and water, to changes in basin land use, or following extreme hydrological events.

Recognizing that there is no unique relation between suspended sediment concentration and water discharge, a sediment rating curve thus aims to model the conditional mean sediment concentration (as a function of water discharge). This is estimated by sampling a series of concurrent measurements of water discharge, Q, and discharge-weighted sediment concentration, C. Ideally, these samples should be collected over a wide range of water discharge, at rising and falling stages, over all seasons, and over many years so that all the factors inducing variance in the relation are represented in an unbiased fashion.

The traditional approach to fitting a rating curve has been to plot C against Q on log-log graphs. These plots accommodate the large ranges of Q and C observed in rivers, the data-scatter tends to be homoscedastic (i.e. independent of discharge), and the underlying relation often shows a simple power form $C = aQ^b$ (a and b are empirical coefficients) which is linear on a log-log plot and easily modelled with linear regression methods. Note that by using log data, the regression procedure models the geometric conditional mean, rather than the desired arithmetic conditional mean, thus a correction factor is required when transforming the rating curve back from log values (Ferguson 1986).

The power law approach should be applied with caution (e.g. Walling and Webb 1988), since large errors can arise because the least-squares curve, while appearing to fit the overall dataset well, may fit poorly at high discharges, and it is these discharges that usually transport the bulk of the sediment load. In such cases, other curve-fitting techniques such as non-linear regression, Locally Weighted Scatterplot Smoothing (LOWESS), or simply segmented curves based on subsets of the data perform better.

References

Ferguson, R.I. (1986) River loads underestimated by rating curves, *Water Resources Research* 22, 74–76.

Walling, D.E. and Webb, B.W. (1988) The reliability of rating curve estimates of suspended sediment yield: some further comments, *International Association of Hydrological Sciences Publication* 174, 337–350.

D. MURRAY HICKS

SEDIMENT ROUTING

The process through which sediment is transported downstream following a specific path or route. The sediment is fluvial sediment, which includes both bedload and suspended load. The path or route, of the sediment may be the course of the natural channel, an artificial canal or a restored channel. Sediment routing is closely related to sediment transport. Where sediment transport may be concerned with the details of sediment movement, sediment routing defines the path of that sediment on a channel reach or watershed scale.

Sediment routing models are used in measuring the SEDIMENT BUDGET for a watershed. The routing model quantifies sediment sources and sinks in the watershed, and how much and how fast that sediment is transferred downstream by river reaches. One of the first steps in measuring the sediment budget in a watershed is the creation of a map of the paths travelled by the sediment.

Sediment routing is affected by landscape changes. Where roads are built in forests, there is an immediate effect on the route travelled by the fluvial sediments. Where roads cut across channels, the path of the water and sediment often changes to follow the road instead of continuing down the natural channel. The result may be either deposition on the road surface or increased road surface erosion. Either way, the path of the channel has been changed, altering the sediment routing processes in the watershed.

Artificial canals are built to route sediment through a city without causing damage. Trapezoidal concrete channels route both water and sediment through residential areas without causing flooding or other damage during high flow events. These types of canals are most common where towns are situated next to mountains, for example, Los Angeles, California and Albuquerque, New Mexico. When a large flow event occurs, the canals are filled with material flowing off the mountain side.

Sediment routing is an important consideration when planning a river restoration project, and restorations are often undertaken when the transport of sediment through the channel has been determined to be too slow or too fast, resulting in either erosion or sedimentation of the channel bed. Natural channels are altered to change the sediment route through the addition of sediment sinks and bedforms. Often, both the path and slope of the channel may be altered in an attempt to change the rates of sediment and water transport. In the case of the Florida Everglades, a meandering channel was straightened. The route that the sediment travelled went from a winding path to a straight line. Subsequent effects to the ecosystem were disastrous, and current work is attempting to re-create the original route.

Further reading

Benda, L. and Dunne, T. (1997) Stochastic forcing of sediment routing and storage in channel networks, *Water Resources Research* 33(N12), 2,865–2,880.

Madej, M.A. (1993) Development, implementation, and evaluation of watershed rehabilitation in Redwood National Park, in *Watershed and Stream Restoration Workshop: Shared Responsibilities for*

Shared Watershed Resources, Symposium Proceedings of the American Fisheries Society, 75–82, Portland, OR.

SEE ALSO: sediment budget

JOANNA C. CURRAN

SEDIMENT WAVE

A sediment wave is a transient zone of sediment accumulation in a river channel that is created by sediment input and does not originate solely from variations in channel topography. Similar terms are 'sediment slug' or 'pulse'. Sediment waves have a minimum spatial scale measured in channel widths and a minimum volumetric scale corresponding to major bars (see BAR, RIVER); they exist over a number of hydrographic events (Nicholas *et al.* 1995; Lisle *et al.* 2001). A sediment wave is not necessarily a body of bed material moving *en masse* downstream, but evolves as a disturbance in the interactions between flow, channel topography and the transport of bed material that can be contributed from any part of the basin, including the input immediately responsible for forming the wave.

Sediment waves evolve by various degrees of dispersion and translation, depending on the form and sedimentology of the channel. Dispersion is the spreading of the wave as its apex lowers and remains stationary and its trailing edge remains stationary or is accreted upstream. Translation is the downstream advancement of the wave, including its apex and trailing edge. A bed material wave that evolves only by dispersion is still regarded as a wave. Dispersion dominates the evolution of bed material waves in quasi-uniform, gravel-bed channels (see GRAVEL-BED RIVER), where the Froude number at significant sediment-transporting flows is generally high (<1); translation of waves can be important in sand-bed channels where the Froude number is low ($\ll 1$) (Lisle *et al.* 2001). Well-documented examples of these contrasting behaviours are provided by Sutherland *et al.* (2003) and Meade (1985). Wave material that is finer than ambient bed material can promote wave translation, but sand waves in steep, gravel-bed channels tend to evolve primarily by dispersion (Lisle *et al.* 2001).

The foregoing applies to quasi-uniform channels, but other features and processes in natural channels may promote wave translation. DEBRIS FLOWS, for example, translate sediment downstream during single events. The downstream spread of a sediment wave through a series of sedimentation zones (Church 1983), where deposition is locally enhanced by valley-scale topography, could manifest wave translation. Advancing zones of increased transport at the leading edge of a wave may activate sediment stored in unstable reaches and propagate a more pronounced wave downstream (Wathen and Hoey 1998).

The relative dominance of dispersion and translation is important for river resources. Dispersion of sediment inputs attenuates but prolongs sediment impacts downstream, whereas translation propagates pronounced sequences of impact and recovery.

References

Church, M. (1983) Pattern of instability in a wandering gravel bed channel, in J.D. Collinson and J. Lewin (ed.) *Modern and Ancient Fluvial Systems*, 169–180, Oxford: Blackwell Scientific.

Lisle, T.E., Cui, Y., Parker, G., Pizzuto, J.E. and Dodd, A.M. (2001) The dominance of dispersion in the evolution of bed material waves in gravel bed rivers, *Earth Surface Processes and Landforms* 26(13), 1,409–1,420.

Meade, R.H. (1985) Wavelike movement of bedload sediment, East Fork River, Wyoming, *Environmental Geology Water Science* 7(4), 215–225.

Nicholas, A.P., Ashworth, P.J., Kirkby, M.J., Macklin, M.G. and Murray, T. (1995) Sediment slugs: large-scale fluctuations in fluvial sediment transport rates and storage volumes, *Progress in Physical Geography* 19(4), 500–519.

Sutherland, D.G., Hansler, M.E., Hilton, S. and Lisle, T.E. (2003) Evolution of a landslide-induced sediment wave in the Navarro River, California, *Geological Society of America Bulletin*, 114, 1036–1048.

Wathen, S.J. and Hoey, T.B. (1998) Morphologic controls on the downstream passage of a sediment wave in a gravel-bed stream, *Earth Surface Processes and Landforms, 23*, 715–730.

SEE ALSO: sediment routing

THOMAS E. LISLE

SEDIMENTATION

The term sedimentation refers to the settling of solids from suspension in a fluid. In geomorphology the fluid is typically water (fluvial, lacustrine or marine sedimentation) or air (aeolian sedimentation). The fundamental process of sedimentation is described by Stokes' law which describes the settling of spherical particles from a still fluid.

This is expressed as $V = (2gr^2)(d_1 - d_2)/9\eta$ where: V is the particle fall velocity (cm s^{-1}), g is acceleration due to gravity (cm sec^{-2}), r is the radius of the particle (cm), d_1 is the density of the particle (g cm^{-3}), d_2 is the density of the fluid (g cm^{-3}) and η is the viscosity of the fluid (dyne sec cm^{-2}). In natural systems this simple relation is complicated by the non-spherical nature of most sedimentary particles, and by the fact that particles are typically deposited from a moving fluid column. In a moving fluid the vertical component of turbulent eddies transfers momentum to the particle that may exceed the velocity of gravitational settling so that the particle remains suspended. As flow velocities decline successively finer particles settle out of the column so that theoretically the typical sedimentary structure associated with sediments deposited from a waning flow is a normally graded bed which fines upwards. In practice, the nature of the source sediment, temporal and spatial variability in flow, and post depositional modification, significantly complicate the nature and interpretation of deposits. For a good review of controls on sedimentation and the interpretation of deposits see Leeder (1992).

As a consequence of the difficulties of developing physical models of sediment transport and deposition, geomorphologists have developed empirically based generalizations to describe sedimentation in particular systems. Examples include the Hjulström curve (Hjulström 1935) which predicts entrainment and deposition threshold velocities for sediments of varying size in fluvial systems. More work has focused on defining conditions for the initial entrainment of sediment than subsequent sedimentation. (See for example Shields (1936) diagram or work by Bagnold (1941) on aeolian entrainment.) Threshold velocities for deposition are lower than those for entrainment due to the role of inertia and bed packing in limiting initial motion. For example bedload sediments in Turkey Brook, England show depositional thresholds that are only 35 per cent of entrainment thresholds (Reid and Frostick 1994).

Geomorphological studies of sedimentation can be divided into those which aim to analyse and model contemporary sedimentary processes, and those which draw on this understanding to interpret past environments and processes from sedimentary evidence. The former, in the tradition of the work by Shields and Bagnold, constitute a central part of modern process geomorphology. The insights of the latter approach, often classified as Quaternary science or historical geomorphology, are equally necessary for the proper understanding of contemporary landscapes.

Within geomorphology a broader definition of sedimentation is generally accepted which includes the deposition of sediments from beneath glaciers and ice sheets (glacial sedimentation) and deposition by mass movement processes on slopes. Sedimentation in glacial environments is complex, ranging from purely glacial deposition of sediment, where there is a close link between style of deposition and the dynamics of the glacial ice, through waterlain tills deposited below floating ice, to extensive fluvial and aeolian sedimentation of glacially derived material. A good review of glacial sedimentation is given by Hambrey (1994).

In the broader sense referred to above, sedimentation is taken to be synonymous with deposition of sediment and is consequently central to the study of a wide range of depositional landforms and environments. Sedimentation is the process which defines the end point of the sequence of sediment EROSION, sediment transport and sediment deposition. As such, accumulation of sediment in sedimentation zones provides an integration of upstream/upglacier/upwind geomorphic action. Where system stability allows that the locus of sedimentation remains constant over time the resulting depositional sequence provides a valuable stratigraphic archive of changing rates of sediment flux. These records are of particular importance in determining process rates over geomorphologically significant timescales, extending beyond the few years of most monitored datasets. Lake sediments are a good example of a stable locus of sedimentation and numerous studies of lake sedimentation have inferred long-term catchment sediment yields from lake sediment volumes and chronological control on stratigraphy (e.g. Desloges and Gilbert 1998). In appropriate contexts lake sediments can be used not only to measure sediment flux, but also to identify changes in style of sedimentation which indicate change in the balance of sediment delivery processes. For example, Menounos (2000) used distinct coarse layers in the sediments of a high alpine lake in Colorado to reconstruct the frequency of debris flow activity in the catchment.

Church and Jones (1982) used the term sedimentation zones to describe river reaches where sediment characteristically accumulates, which are separated by reaches where sediment transport dominates. Sedimentation zones are large-scale stores of sediment within the landscape system. The concept can be to some extent generalized. Sedimentation is not uniform across landscape systems but is typically concentrated in zones where the landscape morphology is such that sedimentation is favoured. In the strict definition of sedimentation these areas are those where lower fluid velocities or physical retention of sediment promote sedimentation, for example lakes, or areas of preferential aeolian sedimentation around major topographic obstacles. A good example of the latter are the sand dune systems of Great Sand Dune National Monument in Colorado where sands deflated from alluvial surfaces accumulate against a downwind mountain front (Plate 120). Taking the broader definition of sedimentation, breaks of slope, which promote the accumulation of slope deposits, may be identified as areas of preferential sedimentation. In alpine environments for example cirque basins can be identified as morphological units characterized by local sedimentation.

One common usage in the literature is accelerated sedimentation referring to increased rates of sedimentation usually as a consequence of human action. Accelerated sedimentation is the necessary consequence of enhanced erosion upstream due to human modification of land use, or of increased deposition associated with human modification of the fluid flow. The most common impacts include mining and intensification of upland agriculture. The classic study of mining impacts was carried out by Gilbert (1917) on the impacts of hydraulic gold mining in the Sierra Nevada. Between 1853 and 1884 over 1 billion m^3 of sediment was produced by mining leading to major fluvial aggradation of upland valleys. This sediment continued to be reworked through the Sacramento valley throughout the twentieth century (James 1999). Xu (1998) records a 25 fold increase in sedimentation rates in the Lower Yellow River in China spanning the past 2,200 years associated with a 40 fold increase in population upstream and consequent intensification of agriculture and engineering works on the river. Similarly, O'Hara *et al.* (1993) suggested that significant increases in lake sedimentation rates in a Mexican highland lake were due to over-intensification of pre-Hispanic agricultural practices.

Plate 120 Alluvial sediments in the foreground provide the sediment source for aeolian sedimentation of the dune systems against the mountain front, Great Sand Dune National Monument, Colorado

Phases of accelerated sedimentation are not confined to fluvial systems. Severe desertification in north–central China is occurring as a result of human modification of land use leading to accelerated aeolian sedimentation due to encroachment of dune systems (Fullen and Mitchell 1994).

In general the most dramatic increases in sedimentation rates across a range of different environments, are typically seen within the past 200 years, associated with mechanization and increased rate of land use change in response to rapid population growth.

A related term often used in this context, particularly with reference to infill of river channels and reservoirs, is siltation. The strict definition of siltation is the sedimentation of silt-sized particles but the term is also used more generally to refer to the infill of channels and basins with fine-grained sediment. In areas of rapid erosion reservoir siltation can significantly reduce reservoir capacity and represents a significant economic cost (Palmieri *et al.* 2001) necessitating erosion control measures in the catchment. Fine-grained sedimentation causes particular problems in upper reaches of salmon streams (Hartman *et al.* 1996) as the sedimentation clogs the gravel interstices and inhibits spawning of the returning fish. In areas of the Pacific North West of North America where commercial logging coexists with economically important salmon runs the fine-grained sedimentation associated with logging-related slope failures is a major source of land use conflict.

Long-term changes in sedimentation rate also occur naturally without direct human impact. One of the main drivers is climate change, either directly through impact on weathering and erosion rates, or indirectly through climate driven vegetation change (e.g. Evans 1997; Xu 1998). One major shift in sedimentation rates characteristic of formerly glaciated areas is a peak in sedimentation associated with deglaciation. PARAGLACIAL sedimentation, conditioned by the former presence of glacial ice, commonly occurs at rates of at least an order of magnitude above subsequent equilibrium rates of sedimentation (Hinderer 2001).

Increases in sedimentation rate at a point have two fundamental causes. One is changes in the nature of erosion so that the sediment load is increased and rates of sedimentation increase without necessary changes in the style or location of sedimentation. The second main cause is associated with anthropogenic changes to the nature of the landscape system which change the balance of sedimentation and sediment transport for a particular location. An understanding of sedimentation and sedimentary processes are important to the geomorphologist in many ways. There are many practical applications in controlling and mitigating anthropogenically induced sedimentation, and the sediments are valuable archives of past rates and styles of sedimentation. Fundamentally, however, sedimentation is necessarily linked to erosion (the source of the sediment) and it is the balance of erosion and sedimentation across space and over long timescales which determines the geomorphology of contemporary landscapes.

References

Bagnold, R.A. (1941) *The Physics of Blown Sand and Desert Dunes*, London: Methuen.

Church, M. and Jones, D. (1982) Channel bars in gravel-bed rivers, in R.D. Hey, J.C. Bathurst and C.R. Thorne (eds) *Gravel-bed Rivers*, Chichester: Wiley.

Desloges, J.R. and Gilbert, R. (1998) Sedimentation in Chilko Lake: a record of the geomorphic environment of the eastern Coast Mountains of British Columbia, Canada, *Geomorphology* 25, 75–91.

Evans, M. (1997) Temporal and spatial representativeness of alpine sediment yields: Cascade mountains, British Columbia, *Earth Surface Processes and Landforms* 22, 287–295.

Fullen, M.A. and Mitchell, D.J. (1994) Desertification and reclamation in North–Central China, *Ambio* 23(2), 131–135.

Gilbert, G.K. (1917) Hydraulic mining debris in the Sierra Nevada, *US Geological Survey Professional Paper* 105, Washington, DC: US Geological Survey.

Hambrey, M. (1994) *Glacial Environments*, Vancouver: UBC Press.

Hartman, G.F., Scrivener, J.C. and Miles, M.J. (1996) Impacts of logging in Carnation Creek, a high energy coastal stream in British Columbia, and their implication for restoring fish habitat, *Canadian Journal of Fisheries and Aquatic Sciences* 53, 237–251.

Hinderer, M. (2001) Late Quaternary denudation of the Alps, valley and lake fillings and modern river loads, *Geodinamica Acta* 14, 231–263.

Hjulström, F. (1935) Studies of the morphological activity of rivers as illustrated by the river Fyris, *Bulletin of the Geological Institute, University of Uppsala* 25, 221–257.

James, A. (1999) Time and the persistence of alluvium: river engineering, fluvial geomorphology and mining sediment in California, *Geomorphology* 31, 265–290.

Leeder, M.R. (1992) *Sedimentology: Process and Product*, London: Chapman and Hall.

Menounos, B. (2000) A Holocene debris-flow chronology for an alpine catchment, Colorado Front Range, in O. Slaymaker (ed.) *Geomorphology, Human Activity and Global Environmental Change*, 117–149, Chichester: Wiley.

O'Hara, S.L., Street-Perrott, F.A. and Burt, T.P. (1993) Accelerated soil erosion around a Mexican highland lake caused by pre-Hispanic agriculture, *Nature* 362, 48–51.

Palmieri, A., Shah, F. and Dinar, A. (2001) Economics of reservoir sedimentation and sustainable management of dams, *Journal of Environmental Management* 61, 149–163.

Reid, I. and Frostick, L.E. (1994) Fluvial sediment transport and deposition, in K. Pye (ed.) *Sediment Transport and Depositional Processes*, 89–144, Oxford: Blackwell.

Shields, A. (1936) Anwendung der Aehnlichkeitsmechanik und der Turbulenzforschung auf die Geschiebebewegung, *Mitteilung der Preussischen Versuchsanstalt für Wasserban und Schiffban*, Heft 26, Berlin.

Xu, J. (1998) Naturally and anthropogenically accelerated sedimentation in the lower Yellow River, China over the past 13000 years, *Geografiska Annaler* 80A(1), 67–78.

SEE ALSO: alluvial fan; bar, river; dune, aeolian; erosion; glacial deposition; paraglacial

MARTIN G. EVANS

SEISMOTECTONIC GEOMORPHOLOGY

Seismotectonic geomorphology is the study of landforms produced by earthquakes. It combines the results of seismotectonic, geomorphic and palaeoseismological research. Palaeoseismology deals with the age, frequency and size of prehistoric earthquakes (Wallace 1981). The palaeoseismic record includes strong (M >6.5) and very strong (M >7.8) earthquakes, since geological effects of moderate or weak earthquakes are rarely preserved in the near-surface zone. Seismic activity is associated with active faulting. Faults are considered active when they show potential or probability of future displacements in the present tectonic setting, or may have displacement within a future period of concern to humans, i.e. they have ruptured during the Holocene (active faults) or the Quaternary (potentially active faults). Active faults are usually segmented, each segment showing a different history of movement. Large earthquakes of a characteristic size repeatedly rupture the same part or segment of a fault.

Earthquakes producing recognizable surface deformation are called morphogenic earthquakes, whereas deposits or landforms formed during an earthquake are described as coseismic, as opposed to delayed-response features that follow the seismic event. Geomorphic features occurring both on-fault and off-fault form either primary (resulting from coseismic slip) or secondary (produced by earthquake shaking, like rockfalls or deformed tree rings) evidence of seismicity (McCalpin 1996). Sediments deformed during seismic shaking are called seismites.

Evidence of present and past earthquakes include deformation of the ground surface along seismogenic faults (fault scarps, fissures, sag ponds, offset stream valleys, shutter ridges, folded terraces, deformed alluvial fans, river reversals, fractured cave structures, displaced beach ridges, coral platfroms, delta plains or wave-cut notches), large-scale features of sudden uplift or subsidence above plate-boundary faults (warped river terraces, elevated shorelines, drowned tidal marshes, emerged or subsided coral reefs), as well as stratigraphic or geomorphic effects of ground shaking or tsunamis far from the seismogenic fault (i.e. landslides, slumps, rockfalls, liquefaction features like mud volcanoes or sand-blow deposits).

The primary geomorphic evidence of seismic activity associated with *normal faulting* are fault scarps (see FAULT AND FAULT SCARP). These scarps range in size from mountain fronts up to 1 km high, cut on bedrock, to centimetre-scale scarplets in unconsolidated sediments. Simple (single-event) fault scarps are formed almost instantaneously, and attain heights from a decimetre to a few metres per event. In normal or reverse faulting such scarps face in the direction of slip, whereas during strike-slip faulting they face in different directions. At the base of recent fault scarps, closed basins or sag ponds may develop. Some scarplets, called earthquake rents, reverse scarplets or cicatrices, parallel the base of the scarp but face uphill. Horsts and grabens, as well as unpaired normal faults creating half-grabens (fault-angle depressions) are also very common. Coseismic displacement on normal faults is characterized by greater subsidence of the hanging wall as compared to the size of uplift of the footwall. Scarp degradation is affected by both lithologic and (micro)climatic factors. In semi-arid climate, the free face in unconsolidated sediments becomes completely destroyed in a timespan ranging from one day (1983 Borah Peak earthquake, Ms = 7.3) to one to two thousand years (Crone *et al.* 1987; Wallace 1977). Fractured bedrock scarps will gradually degrade to an angle of repose that can be maintained for one million years or so. Scarps formed by more than one earthquake are called compound, composite or multiple-event scarps. During each earthquake, a colluvial wedge is shed from the scarp, being subsequently covered in interseismic periods by soils that can be dated by different techniques. The fault-induced incision into the upthrown block after faulting can create tectonic terraces that diverge downstream and abruptly terminate against the fault scarp.

Thrust earthquakes occur frequently in fold-and-thrust belts (e.g. Algeria, 1980), at convergent continental plate boundaries (Iran, 1978; Armenia, 1988) or in regions near transpressive bends of strike-slip faults. During thrusting, hanging wall uplift usually exceeds footwall subsidence, and the area affected by coseismic deformation depends on the size of displacement, fault geometry and the rigidity of the deformed crust. The size of displacement diminishes towards the thrust tip, hence, the hanging wall is usually folded adjacent to that tip. Many M < 7 earthquakes may not be accompanied by any geomorphic expression. Typical landforms produced by thrusting are thrust fault scarps which are usually more sinuous and irregular as compared to other fault types, and are composed of short, disconnected segments, or produce a continuous but zigzag-like trace on the scale of metres. These scarps show varied morphology due to mixed low-angle faulting and folding. Seven to eight types of thrust fault scarps have been distinguished. Steeply dipping reverse faults in bedrock form simple thrust scarps, whereas those in unconsolidated sediments result in hanging wall collapse scarps. Low-angle thrust faults produce pressure ridges of shape depending on surface material rheology and the magnitude of slip. In more cohesive materials, pressure ridges have smoother fronts and may display backthrusts or represent low-angle pressure ridges, but as thrust displacement decreases below 1 m, all pressure ridge types merge into a single type of a small moletrack. An increasing oblique component of slip produces *en echelon* pressure ridges or oblique tension gashes in pressure ridge front. Surface thrusting deforming flat terrains is usually expressed as a wide gentle warp of fluvial/marine terrace surfaces. Thrust faulting in bedrock produces an overhanging scarp, but in unconsolidated deposits such overhangs collapse very quickly, creating a free face and a steep debris slope. Numerous degraded reverse fault scarps are asymmetric, with the steepest part of the scarp lying downslope of the scarp midpoint, whereas the normal fault scarps typically show symmetric cross profiles. High escarpments formed by repeated thrusting are usually obscured by a long chain of landslides. Thrust faults are difficult to recognize where cutting high relief and irregular topography, whereas broadly distributed thrusting, surface warping and folding cannot be detected unless planar geomorphic features are deformed.

Active folding is a coseismic process related to faulting on a blind thrust or other dip-slip fault at depth. It can also induce seismicity due to flexural slip during folding. Geomorphic manifestations of surface folds are deformed fluvial channels and terraces. Hanging wall ramp folds can generate distinct facets at the top of scarps, resembling those of normal fault scarps. These facets, however, may be independent of the palaeoseismic history of the fault. Some propagation folds formed at thrust tips may also produce multifaceted scarp profiles.

Major seismogenic *strike-slip faults* are associated either with plate boundaries or are located within intraplate settings, at the boundaries of continental microplates. Landforms made by (palaeo)earthquakes along active strike-slip faults include: linear valleys (they can be created by simple deflection of streams along the fault trace, even without brecciation), offset, beheaded or deflected valleys and streams, offset ridges, sags and sag ponds (related to downwarping between two strands of the fault zone), shutter ridges (formed where a fault displaces ridge crests on one side of the fault against gullies on the other side; usually occurring where a fault breaks through the pre-existing pressure ridges), pressure ridges (small warped areas formed by compression between multiple traces in a fault zone), topographic benches (elevated, flat, gently warped or tilted areas, usually formed due to the displacement between several fault segments or splays in the fault zone), fault scarps of minor to moderate height, and small-scale horsts, grabens and pull-apart basins. The fault trace is frequently composed of a wide zone of alternating tension gashes (extensional) and moletracks (compressional) that strike obliquely to the general fault strike. Landforms typically used to estimate palaeoseismic offset are: fluvial terraces, stream channels and alluvial fans. A minimum slip rate, e.g. of the San Andreas fault, California, has been estimated based on offset alluvial fans at 10–35 mm/yr (Keller *et al.* 1982). The maximum single-event stream offset of 9.5 m was produced by the 1857 Ft. Tejon, San Andreas fault, earthquake of M = 8 (McCalpin 1996).

References

Crone, A.J., Machette, M.N., Bonilla, M.G., Lienkaemper, J.J., Pierce, K.L., Scott, W.E. and Bucknam, R.C. (1987) Surface faulting accompanying the Borah Peak earthquake and segmentation of the Lost River Fault, Central Idaho, *Bulletin Seismological Society of America* 77, 739–770.

Keller, E.A., Bonkowski, M.S., Korsch, R.J. and Shlemon, R.J. (1982) Tectonic geomorphology of the San Andreas fault zone in the southern Indio Hills, Coachella Valley, California, *Geological Society of America Bulletin* 93, 46–56.

McCalpin, J.P. (ed.) (1996) *Paleoseismology*, San Diego: Academic Press.

Wallace, R.E. (1977) Profiles and ages of young fault scarps, north-central Nevada, *Geological Society of America Bulletin* 88, 1,267–1,281.

——(1981) Active faults, paleoseismology, and earthquake hazards in the western United States, in D.W. Simpson and P.G. Richards (eds) *Earthquake Prediction – An International Review*, 209–216, Washington, DC: American Geophysical Union.

Further reading

Burbank, D.W. and Anderson, R.S. (2001) *Tectonic Geomorphology*, Malden: Blackwell.

Keller, E.A. and Pinter, N. (1996) *Active Tectonics*, Upper Saddle River, NJ: Prentice Hall.

Schumm, S.A., Dumont, J.F. and Holbrook, J.M. (2000) *Active Tectonics and Alluvial Rivers*, Cambridge: Cambridge University Press.

Stewart, I.S. and Hancock, P.L. (1994) Neotectonics, in P.L. Hancock (ed.) *Continental Deformation*, 370–409, London: Pergamon Press.

Wallace, R.E. (ed.) (1990) *The San Andreas Fault System, California*, Washington, DC: US Geological Survey Professional Paper 1,515.

SEE ALSO: crustal deformation; fault and fault scarp

WITOLD ZUCHIEWICZ

SELF-ORGANIZED CRITICALITY

Self-organized criticality is an approach to understanding non-linear systems initiated by Per Bak and colleagues, as explained in his book *How Nature Works* (Bak 1997). Self-organized criticality is one of a whole host of linked new approaches, including CHAOS THEORY, complexity theory (see COMPLEXITY IN GEOMORPHOLOGY) and FRACTALS, which aim to provide better explanations of the complex behaviour of non-linear natural systems.

Self-organized criticality is used to explain the behaviour of many complex natural systems which seem to evolve into a poised or critical state, far from equilibrium. Per Bak uses the helpful analogy of a self-organized system being like a sandpile created by dropping grains of sand onto a flat surface. During the initial stages of development of the sandpile, predicting the behaviour of the pile is relatively simple and depends upon the physical properties of the individual grains. As the pile grows, however, avalanches start to occur, whose behaviour is complex and cannot be predicted by the characteristics of the individual grains. At this point the system becomes a complex, self-critical one – whose behaviour can only be understood by considering the properties of the whole pile (a holistic approach) rather than from a reductionist description of the behaviour of individual grains. There are, however, many concepts of self-organization used in science which have subtly different meanings and which are based on different interpretations of natural systems (as elucidated for geomorphology by Phillips 1999). Geomorphologists have used a range of these concepts.

In recent years geomorphologists have had a great interest in ideas such as self-organized criticality which may help explain many of the complex landscapes we see around us. Why and how, for example, do regular patterns such as river networks, rills, stone polygons, beach cusps and dune systems develop? Reductionist approaches to these questions have hoped that studying the basic physics of processes operating at the microscale could provide a general answer. However, such approaches have not often proved able to successfully link process and pattern across different scales. Could such patterns instead be seen to be examples of self-organized criticality, where regular patterns have emerged out of the complex behaviour of smaller scale processes? Many geomorphologists have used cellular models to investigate such systems, in which simple rules are applied to describe the interactions of neighbouring cells. As the models are run, patterns at a larger scale emerge from these simple rules.

Several examples illustrate the use of cellular models to investigate self-organized criticality in geomorphic systems. Werner and Fink (1993) investigated the formation of beach cusps, finding that they could be simulated from a cellular model based on the interaction of water flow, sediment transport and morphological change. Similarly, and at a larger scale, Rodriguez-Iturbe and Rinaldo (1997) use models for flow and sediment transport to develop fluvial networks. The resultant networks have fractal and multifractal properties and are seen by Rodriguez-Iturbe and Rinaldo to be the product of self-organizing processes. In a recent study, de Boer (2001) has built a cellular model to simulate the long-term evolution of a fluvial landscape. Using simple rules to model sediment transport between adjacent cells, a record of total sediment yield is created which has complex magnitude and frequency properties which cannot be predicted from the simple, local rules. Thus the sediment dynamics of the modelled landscape are an emergent property of the entire system resulting from the local interaction of individual cells. For geomorphologists, the insight that complex behaviour may be the result of internal interactions rather than external forcings is highly important, and complicates our search for the geomorphic imprint of external forcings such as climate change.

However, despite these promising model simulations it is as yet unclear whether self-organized criticality really exists in natural geomorphic systems. The behaviour of modelled systems cannot simply be applied to natural systems, for which we do not always have enough data and in which external forcings may also play a role. Werner (1999) suggests that hierarchical models might be better suited to modelling complex natural landform patterns which self-organize in temporal hierarchies.

References

Bak, P. (1997) *How Nature Works: The Science of Self-organized Criticality*, Oxford: Oxford University Press.

de Boer, D.H. (2001) Self-organisation in fluvial landscapes: sediment dynamics as an emergent property, *Computers and Geosciences* 27, 995–1,003.

Phillips, J.D. (1999) Divergence, convergence and self-organization in landscapes, *Annals, Association of American Geographers* 89, 466–488.

Rodriguez-Iturbe, I. and Rinaldo, A. (1997) *Fractal River Basins: Chance and Self-organization*, Cambridge: Cambridge University Press.
Werner, B.T. (1999) Complexity in natural landform patterns, *Science* 284, 102–104.
Werner, B.T. and Fink, T.M. (1993) Beach cusps as self-organised patterns, *Science* 260, 968–971.

Further reading

Favis-Mortlock, D. (1998) A self-organizing dynamic systems approach to the simulation of rill initiation and development of hillslopes, *Computers and Geosciences* 24, 353–372.
Hallet, B. (1990) Self-organisation in freezing soils: from microscopic ice lenses to patterned ground, *Canadian Journal of Physics*, 68, 842–852.
Kauffman, S. (1995) *At Home in the Universe: The Search for Laws of Self-organisation and Complexity*, Oxford: Oxford University Press.
Werner, B.T. and Hallet, B. (1993) Numerical simulation of self-organized stone stripes, *Nature* 361, 142–145.

HEATHER A. VILES

SENSITIVE CLAY

The basic idea of sensitivity is that the structure of the clay soil system has an effect on the properties, so that once the structure is destroyed (by remoulding) a new set of properties is observed. This is probably true in just about all clays although in heavily overconsolidated (see OVERCONSOLIDATED CLAY) systems the effect will be negligible; but in the QUICKCLAYs structure is all-important and when it is destroyed most of the strength properties disappear. The high sensitivities are of interest in applied geomorphology because of associated ground failures and landslides.

The sensitivity of a clay is the ratio of the undisturbed strength to the remoulded strength. It appears to have been first defined by Karl Terzaghi in 1944. The next development is due to Skempton and Northey (1952) and it is this paper which launched the scientific study of sensitivity. There had been Scandinavian experiences with sensitive clays for many years but the Skempton and Northey paper marks a critical beginning. They divided the sensitivity range:

About 1,	insensitive clays
1–2,	low sensitivity
2–4,	medium
4–8,	sensitive
>8,	extra-sensitive
>16,	quickclays

Much of the discussion about sensitivity in clays concerns the mechanism by which it arises. There has been considerable discussion on this topic and many factors influencing sensitivity have been proposed (see Mitchell and Houston 1969); the factors have been nicely ordered by Quigley (1979).

Factors affecting sensitivity

Factors producing high undisturbed strength and high sensitivity:

1. Depositional flocculation, including low electro-kinetic potential, high sediment concentration, divalent cation adsorption.
2. Slow increase in sediment load.
3. Cementation bonds, including carbonates and amorphous sesquioxides

Factors producing low remoulded strength and high sensitivity:

1. High water content (greater than the liquid limit), little consolidation.
2. Low specific surface of soil grains, high silt content or high rock flour content in the clay fraction. High primary mineral, low clay mineral content.
3. High electro-kinetic (zeta) potential, low salinity via leaching, organic dispersants, inorganic dispersants, high monovalent cation adsorption relative to divalent cations.
4. Low amorphous content.
5. Low smectite (EXPANSIVE SOIL clay) content.

The list is fairly comprehensive, but it includes major and minor factors. It appears that there are probably two basic populations of sensitive clays: those with relatively low sensitivities, which contain clay minerals to a significant degree; and those with high sensitivities which tend to lack clay minerals. These high sensitivity materials are dominated by primary minerals with particle sizes in the clay or very fine silt ranges.

References

Mitchell, J.K. and Houston, W.N. (1969) Causes of clay sensitivity, *Journal of Soil Mechanics, American Society of Civil Engineers* 86(SM3), 19–52.
Quigley, R.M. (1979) Geology, mineralogy and geochemistry of soft soils and their relationship to geotechnical problems, *Proceedings of the 32nd Canadian Geotechnical Conference, Quebec City; State of the Art*.
Skempton, A. and Northey, R.D. (1952) The sensitivity of clays, *Géotechnique* 3, 30–53.

Further reading

Brand, E.W. and Brenner, R.P. (eds) (1981) *Soft Clay Engineering*, Amsterdam: Elsevier.

SEE ALSO: quickclay

IAN SMALLEY

SERPULID REEF

Serpulid reefs are built by marine polychaetes which secrete hard, calcareous tubes. The reefs built by *Ficopomatus enigmaticus* (Fauvel) develop reefs which are greater than 3 m thick and up to 20 m in diameter (Fornós *et al.* 1997). Individual worm tubes are typically 100 μm thick, 4–5 mm in diameter and up to 100 mm long, with tubes being comprised of calcium carbonate interspersed with a mucopolysaccharide matrix (Rouse and Pleijel 2001). Serpulidae have a near worldwide distribution; reefs have been found in the fossil record from the Cretaceous to the Recent and they are important in palaeoenvironmental reconstruction. The global distribution of the species has increased during historical times, due to international shipping, with several species colonizing non-native habitats and industrial structures such as docks.

The reefs typically develop in brackish and marine conditions, in intertidal and shallow subtidal zones, on hard rock substrata or shell fragments. Occasionally they extend onto, and in rare circumstances they colonize, soft substrates such as mud, sand or wood. Serpulid reefs tend to develop in three phases: (1) individuals coat the surface of hard substrata; (2) sinuous, random growth and extension of colonies; and (3) development of reef structures which are primarily controlled by water turbulence and dominant current direction. Morphological forms vary from fringing reefs along rocky shorelines, to subtidal microatolls and dense patch reefs (Fornós *et al.* 1997). Where serpulid reefs develop on soft substrata they often serve an important functional role, by providing the only hard substrate for a range of species.

References

Fornós, J.J., Forteza, V. and Martínez-Taberner, A. (1997) Modern polychaete reefs in Western Mediterranean lagoons: *Ficopomatus enigmaticus* (Fauvel) in the Albufera of Menorca, Balearic Islands, *Palaeogeography, Palaeoclimatology, Palaeoecology* 128, 175–186.

Rouse, G.W. and Plejil, F. (2001) *Polychaetes*, Oxford: Oxford University Press.

LARISSA NAYLOR

SHEAR AND SHEAR SURFACE

Shearing of material occurs when it is subjected to a stress or confining pressure, acting in a particular direction, that exceeds the strength of the material. The material fractures along a plane of least resistance, which may be curved, and the mass on one side of this shear surface is displaced so that movement takes place in opposite directions on either side.

In geomorphology, shear failure of materials is most commonly encountered in MASS MOVEMENTs. In a hillslope, material near the ground surface (soil or other sediments, *in situ* weathering products, bedrock, etc.) is subjected to shear stress as it is acted on by a downslope component of gravitational force. This is resisted by the shear strength of the material, which acts in the opposite direction. If the stress exceeds the strength, a shear surface will form (within one material or between different materials depending on relative strengths across all possible planes of weakness) and the mass above the surface will move downslope. (See HILLSLOPE, PROCESS; LANDSLIDE; SLOPE STABILITY.) Fault planes within the Earth's crust can also be regarded as shear surfaces: the rocks on one side of a fault move past the rocks on the other side in response to differential stresses.

Another condition that can cause shearing is when a material is confined at depth by stresses acting in perpendicular (x, y, z) directions, and compression by the stress in one direction exceeds the strength of the material. A shear surface forms as the material is forced to the side in opposite directions either side of a plane of least resistance.

The nature of a shear surface and any displacement along it will be determined by the properties of the material(s), the magnitudes and directions of the stresses, the topographic or geophysical context, and the scale at which the shear surface is considered. A smooth plane of movement at a large scale may comprise an irregular shear surface at a smaller scale. There may be considerable frictional resistance to movement, especially if the irregularities resemble interlocking teeth. At this extreme, the shear stress will only cause movement if it is sufficient either to force the two sides far enough apart to enable movement, or to cause shearing of the asperities thus further smoothing the overall surface.

At the other extreme, very small stresses within fine-grained materials may cause individual plate-like clay mineral particles to line up with the prevailing stress direction. Over time these 'microshears' can extend, join up and form a shear surface, which is smooth at the microscopic scale (see SLICKENSIDE).

Some materials will not form a discrete shear surface, instead developing shear zones where complex displacements occur throughout measurable thicknesses, especially in materials that display plastic behaviour under stress such as certain clay formations, or highly heterogeneous sediments.

Further reading

Selby, M.J. (1993) *Hillslope Materials and Processes, 2nd Edition*, Oxford: Oxford University Press.

SEE ALSO: riedel shear

ALAN P. DYKES

SHEET EROSION, SHEET FLOW, SHEET WASH

In a general sense, the term sheet flow is used to refer to any flow of water of more or less homogeneous depth and velocity over a surface without any clear channel development. The term is used to describe flow over alluvial floodplains, coastal plains, beach surfaces as well as flow over hillslopes without noticeable channel incision. Here, we will only refer to the use of the term sheet flow in this latter sense. Sheet flow over hillslopes therefore contrasts with flow in clearly defined channels (RILLS and gullies (see GULLY)).

Sheet erosion is often defined as the removal of a thin layer of surface soil by water erosion processes without the development of noticeable channels (Plate 121). The absence of channels (see CHANNEL, ALLUVIAL) does not imply that sheet erosion is completely uniform. The presence of roughness elements such as rock fragments, clods or vegetation tussocks will still lead to variations in flow depth and velocity, even under steady flow conditions. In order to stress the absence of channels the term interrill flow (i.e. flow on areas between or without rills) is sometimes preferred. Local variations in flow depth and velocity do not necessarily lead to channel development: as Smith and Bretherton (1972) pointed out, no channels will be formed by OVERLAND FLOW when the influx of sediment into a concentrated flowline by processes such as splash and creep equals or exceeds the local transporting capacity of the flow.

Sheet erosion mainly occurs under conditions where the soil surface is insufficiently protected by vegetation cover against drop impact and crusting (see CRUSTING OF SOIL): it is therefore frequently encountered on freshly ploughed arable land, on overgrazed rangeland and on natural hillslopes in arid and semi-arid environments.

If a soil surface is only affected by sheet erosion this will result in a gradual lowering of the whole soil surface. In many landscapes, sheet erosion often occurs simultaneously with rill and gully erosion, with gully erosion dominating in HILLSLOPE HOLLOWS (concavities), while sheet and rill erosion mainly affect convex and rectilinear slope segments. When hillslope erosion mainly occurs in channels, the much more rapid lowering of the channel beds will in principle result in a dissection of the soil surface. This need not always be the case: on arable land, rills formed due to erosion are frequently obliterated by soil tillage. The repeated occurrence of sheet and rill erosion will in this case also lead to a gradual removal of a 'sheet' of topsoil over the whole field surface. In natural conditions rill beds often stabilize after a certain time period due to bed ARMOURING: rills then become conveyers of sediment eroded on interrill surfaces rather than important sediment

Plate 121 Intense sheet erosion on arable land in central Belgium, caused by a very intense rainstorm. Height of pole: *c.*1.5 m

sources. In the long term, this situation may also result in a gradual removal of the topsoil over the whole surface. The rest of this section concentrates on sheet erosion *sensu stricto*, i.e. water erosion that occurs without the formation of noticeable channels.

In the case of sheet erosion soil detachment mainly occurs by raindrop impact. The detached sediments may then be tranported by various processes: raindrop splash, raindrop-induced flow transport and flow transport. The hydraulic properties of the sheet (or interrill) flow are an important control on these processes and are therefore discussed first.

Sheet flow on hillslopes is generally characterized by flow depths between 0 and 20 mm and velocities $<0.5\,\mathrm{m\,s^{-1}}$. In general, experimental studies in the field and laboratory show that classic approaches for the prediction of flow depths and velocities which are based on the estimation and prediction of the Darcy–Weisbach friction factor or Manning's n are only moderately succesful (Abrahams *et al.* 1994). The key reason for this is that the interaction between roughness elements and flow hydraulics is fundamentally different in sheet flow. On natural surfaces where individual roughness elements are more or less randomly distributed, flow resistance tends to increase with increasing discharge until roughness elements are fully submerged. A further increase of discharge leads to a rapid decrease of flow resistance (Lawrence 1997). On tilled surfaces roughness elements are not randomly distributed and the correct prediction of flow characteristics sometimes requires separate calculations for each flowpath (Takken and Govers 2000). As the latter requires very detailed information on surface topography, simplified approaches are often used.

Raindrop detachment is basically controlled by the balance between rainfall erosivity and the soil's resistance to splash detachment. Various rainfall characteristics have been proposed as indicators of the rainfall's EROSIVITY. Rainfall kinetic energy is undoubtedly most frequently used to assess rainfall erosivity. However, there are indications that parameters that give relatively more weight to drop diameter, such as the product of rainfall momentum and drop diameter, are somewhat better predictors of splash detachment (Salles *et al.* 2000). Splash detachment is also strongly controlled by the presence and the thickness of a water layer on the surface: most studies report an exponential decline of splash detachment with water depth, with splash detachment becoming negligible when the depth of the water film exceeds *c.* one drop diameter (Torri *et al.* 1987). The resistance of a soil to splash detachment is determined by the cohesion of the soil and the weight of the soil grains. Minimum resistance values are observed for fine sandy soil materials as these materials consist of very fine grains yet they are almost totally cohesionless (Poesen 1985).

If a significant slope gradient is present, the redistribution of the detached material will result in a net downslope transport of soil. The rate of downslope transport increases approximately linearly with slope gradient. However, as splash distances are of the order of several decimetres, splash transport is generally negligible compared to (raindrop-induced) flow transport, except in the case of short, very steep slopes (e.g. unprotected terrace risers).

Although sheet flow has a limited capacity to detach (cohesive) sediment, it is in most cases the main transporting agent in sheet erosion. Two modes of sediment transport can be distinguished: (1) flow transport, whereby the flow itself is transporting the sediment and (2) raindrop-induced sediment transport whereby sediment is brought into the flow by raindrops impacting soil surface and the suspended sediment is consequently transported downslope. Raindrop-induced sediment transport flow is most important for relatively coarse (sand-sized) material and for low-energy flows with relatively low flow depths (Kinell 2001).

Sheet erosion is slope-dependent, primarily because the transporting capacity of the sheet flow increases with slope gradient. The increase of sheet erosion intensity with slope gradient is more or less linear, all other factors being constant. In practice other factors such as grain size distribution of the surface sediments, crusting and vegetation characteristics are strongly slope dependent, so that the relationship between slope and sheet erosion rates may take a completely different form. An interesting debate exists with respect to the effect of slope length on sheet erosion rates: recent experimental evidence suggests that the rapid increase of erosion rates per unit surface area which is often observed when plots of limited length are compared cannot be extrapolated to greater slope lengths. The increase of erosion rates with increasing slope length for short slopes is due to the fact that in the first metres below the divide the sediment load in the sheet flow is supply-limited and therefore sediment transport increases rapidly with slope length. Further downslope, sediment transport becomes limited by the transporting capacity of the flow. This leads to a much slower increase of sediment transport with slope length and consequently also to a much slower increase or even a decrease of the erosion rate per unit area (Rejman and Usowicz 2002).

Considering the primary role of raindrop impact in detaching and transporting sediment in sheet flow, it is no surprise that sheet erosion is strongly dependent on rainfall and rainfall energy: this is, amongst others, reflected in the fact that rainfall erosivity in the Revised Universal Soil Loss Equation which is designed to describe both sheet and rill erosion equals the product of rainfall intensity and rainfall kinetic energy during a 30-minute period.

In general sheet erosion decreases strongly with increasing cover of the soil surface by vegetation or other non-erodible elements such as rock fragments. Most studies, such as the one by Hussein and Laflen (1982), report an exponential decline, whereby a soil cover of *c.* 30 per cent reduces erosion to less than 50 per cent of the value for a bare surface. This strong, non-linear response is due to the fact that the presence of cover has an impact on various aspects of the sheet erosion process (increased infiltration, protection of the surface cover against splash, reduced flow velocities, etc.). The adequate management of soil cover is therefore the most important management strategy in order to reduce sheet erosion.

References

Abrahams, A.D., Parsons, A.J. and Wainwright, J. (1994) Resistance to overland flow on semi-arid grassland and shrubland hillslopes, Walnut Gulch, Southern Arizona, *Journal of Hydrology* 156, 343–363.

Hussein, M.H. and Laflen, J.M. (1982) Effects of crop canopy and residue on rill and interrill soil erosion, *Transactions of the ASAE* 25, 1,310–1,315.

Kinnell, P.I.A. (2001) Particle travel distances and bed and sediment compositions associated with rain-impacted flows, *Earth Surface Processes and Landforms* 26, 749–758.

Lawrence, D.S.L. (1997) Macroscale surface roughness and frictional resistance in overland flow, *Earth Surface Processes and Landforms* 22, 365–382.

Poesen, J. (1985) An improved splash transport model, *Zeitschrift für Geomorphologie* 29, 193–211.

Rejman, J. and Usowicz, B. (2002) Evaluation of soil-loss contribution areas on loess soils in southeast Poland, *Earth Surface Processes and Landforms* 27, 1,415–1,424.

Salles, C., Poesen, J. and Govers, G. (2000) Statistical and physical analysis of soil detachment by raindrop impact: rain erosivity indices and threshold energy, *Water Resources Research* 36, 2,721–2,729.

Smith, T.R. and Bretherton, F.P. (1972) Stability and the conservation of mass in drainage basin evolution, *Water Resources Research* 8, 1,506–1,529.

Takken, I. and Govers, G. (2000) Hydraulics of interrill overland flow on rough, bare soil surfaces, *Earth Surface Processes and Landforms* 25, 1,387–1,402.

Torri, D. Sfalanga M. and Del Sette, M. (1987) Splash detachment: runoff depth and soil cohesion, *Catena* 14, 149–155.

GERARD GOVERS

SHEETING

Some rocky massifs are divided by flat-lying or gently arcuate partings that at many sites are more inclined than the land surface, and plunge steeply (up to 70°). These fractures are known as sheeting or EXFOLIATION. Even though the term sheeting suggests thin layers, some are 10 m or more thick. Sheeting has been observed to depths of 100 m or more and is well developed in granitoids but also in other quite different rocks (dacite, rhyolite, sandstone, conglomerate and limestone). There are two interpretations of sheeting that fall into two categories: exogenetic and endogenetic. The exogenetic explanations are INSOLATION WEATHERING, CHEMICAL WEATHERING and offloading or pressure release, all of them imply dilation by rock volume increase.

The endogenetic explanations are all concerned with tectonic stresses. Some authors consider the sheet structure formed during the emplacement and later cooling of the granite mass and propose the same origin for the shape of the associated dome. But sheeting is well developed in sedimentary and volcanic rocks that were never emplaced and even in granites the magnetic foliation contemporaneous with the emplacement of magmatic rock is clearly discordant with the sheeting.

It has also been suggested that sheeting is associated with big thrust-shearing planes which would affect the granite as well as the other rock types where sheeting is habitually observed. This explanation is the best one, because it serves for all kinds of rocks affected by sheeting.

Further reading

Gilbert, G.K. (1904) Domes and dome structure of the High Sierra, *Geological Society of America Bulletin* 5 (15), 29–36.
Vidal-Romani, J.R. and Twidale, C.R. (1998) *Formas y paisajes graníticos*, Servicio de Publicaciones de la Universidad de Coruña, Serie Monografías 55.
Vidal-Romani, J.R. and Twidale, C.R. (1999) Sheet fractures, other stress forms and some engineering implications, *Geomorphology* 31(1–4), 13–27.

SEE ALSO: pressure release

JUAN RAMON VIDAL-ROMANI

SHIELD

The continental nuclei where Archean and Proterozoic rocks, typically burittle, rigid, granitic, gneissic and associated rocks, outcrop. They form extensive, flat, relatively stable areas (cratons) which have been relatively undisturbed since Precambrian time, except for gentle warping. The major shields are Canadian, Fennoscandian, Angaran (north-east Siberia), African, Brazilian, West Australian and East Antarctic. The shields are bordered by stable platforms, which are continental areas floored by shield extensions and covered by a relatively thin layer of sedimentary rocks.

Because they are ancient areas, they have often experienced bevelling of their rocks by erosion and are characterized by extensive erosional plains. Some shield areas have been fashioned in part by glacial erosion, as in Canada and on the Baltic Shield, but ancient saprolites may also be extensively developed (Lidmar-Bergström 1995). Because of the presence of limited relief and resistant rocks, shields tend to be areas with low rates of mechanical and chemical denudation (Millot *et al.* 2002).

References

Lidmar-Bergström, K. (1995) Relief and saprolites through time on the Baltic Shield, *Geomorphology* 12, 45–61.
Millot, R., Gaillardet, J., Dupré, B. and Allegré, C.J. (2002) The global control of silicate weathering rates and the coupling with physical erosion: new insights from rivers of the Candian Shield, *Earth and Planetary Science Letters* 96, 83–98.

A.S. GOUDIE

SHINGLE COAST

The term 'shingle' has been used for at least 400 years in Britain and some Commonwealth countries, to describe sediments composed of mainly rounded pebbles, larger in diameter than sand (>2 mm) but smaller than boulders (<200 mm). Elsewhere terms such as gravel, stone, levées de galets, playas de cantos, schotterwälle and steinstrand are used. A generalized world distribution of shingle coasts is given in Figure 148. In many locations shingle is mixed with sand, silt, clay or organic debris, resulting in a 'mixed' sediment beach (e.g. Kirk 1980), but all shingle and boulder beaches can be regarded as different types of 'coarse clastic' beach (Carter and Orford 1993).

In general shingle coasts have received less scientific attention than sandy and muddy shorelines. In part, this reflects the fact that, at a world scale, they are much less common. However, in recent decades there has been an increasing awareness of the geomorphologic, ecological and engineering significance of shingle coasts in the contexts of sea-level change, flood defence and habitat conservation. Such coasts are now recognized as an internationally important, but disappearing resource (Packham *et al.* 2001).

Shingle coasts form in WAVE-dominated locations where suitably sized material is available. At a global scale they dominate high latitudes and those areas of temperate shores which were affected by Quaternary glaciation. They are locally important in some other temperate and low latitude areas where high relief landscapes of suitable geology occur near the coast, near the estuaries (see ESTUARY) of high-energy rivers, or where coral (see CORL REEF) is present. Elsewhere they are of limited importance. Isla and Bujalesky (1993) describe shingle coast locations in Argentina, McKay and Terich (1992) and Forbes *et al.* (1995) in North America and Shulmeister and Kirk (1993) in New Zealand.

At a regional scale, lithographic composition determines shingle availability and durability. Hard materials such as flint, chert, granite, quartzite and some metamorphic materials survive much longer at this clast size than sandstones, limestones or shells. Around Great Britain some 19,000 km of shoreline have an important shingle component, with almost 3,500 km of these coasts being pure shingle (Sneddon and Randall 1993/1994). Many of the

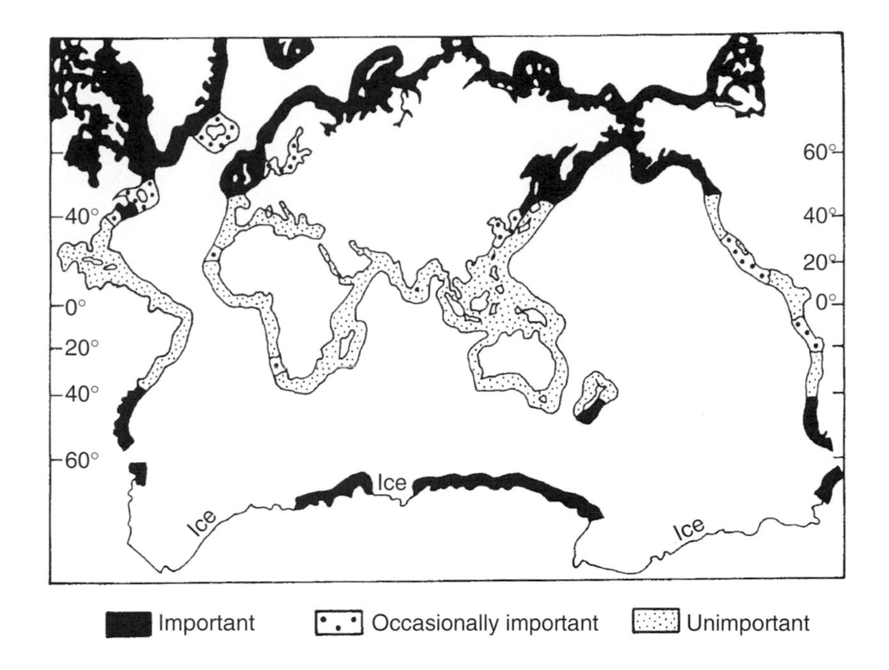

Figure 148 A generalized world distribution of shingle coasts (after Pye 2001)

shingle-barrier (see BARRIER AND BARRIER ISLAND) systems occurring on present-day coastlines were initiated during the Holocene (see HOLOCENE GEOMORPHOLOGY) marine transgression and are currently sustaining considerable morphological change as a result of increasing SEA LEVELS causing landward and longshore reworking of a finite sediment volume.

Shingle coasts can comprise several different landform types (Figure 149), which vary according to their history, mobility and oceanicity and therefore offer different habitats to vegetation, wildlife and humans (Pye 2001; Sneddon and Randall 1993/1994).

Fringing, or pocket BEACHes, are narrow strips of shingle coast in contact with the land along the top of the beach. These are usually subject to regular marine inundation. They frequently occur at the foot of sedimentary cliffs, such as chalk in southern Britain, but may also occur in front of coastal dunes or saltmarsh cliffs.

Embayment beach-ridge plains, or apposition beaches, are comprised of a series of relict storm beach-ridges and an active front ridge system which together partly or totally infill a previous embayment. Such systems may be hundreds of metres or even kilometres wide and can be transitional to CUSPATE FORELANDs or nesses.

Shingle SPITs are strips of shingle, which grow out from the coast where there is an abrupt change in the direction of the coastline. They commonly occur, therefore, along coasts which have an irregular plan. Spits often display recurved hooks along their length and at their distal ends, where the shingle is, or has been, subject to wave action from two or more directions. Indeed, in many cases, it is possible to trace the development of a spit's growth via recurved hooks, seen as lateral projections from the lee of the spit, which locate the position of the past distal points (Randall 1973; Plate 122). Paired spits are found at the entrance of several harbours on the south coast of England, including Pagham and Langstone. These may have originated as bars or TOMBOLOs, which have breached, but in other cases, independent growth of two spits may be due to bi-directional longshore drift.

On eroding coasts, shingle spits are transgressive and frequently overlay back-barrier marsh or lagoon deposits as at Shingle Street, Suffolk, and in some instances may be dissected to form barrier islands. Transgressive ridges, often composed mainly of shell-shingle, are well developed on the marsh coast of Essex. Similar features are also found in the Gulf of Mexico, where they are

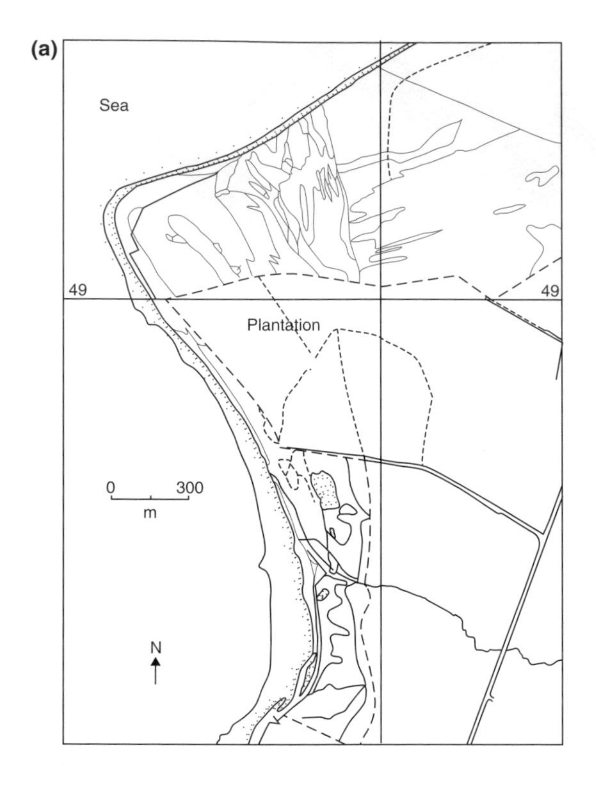

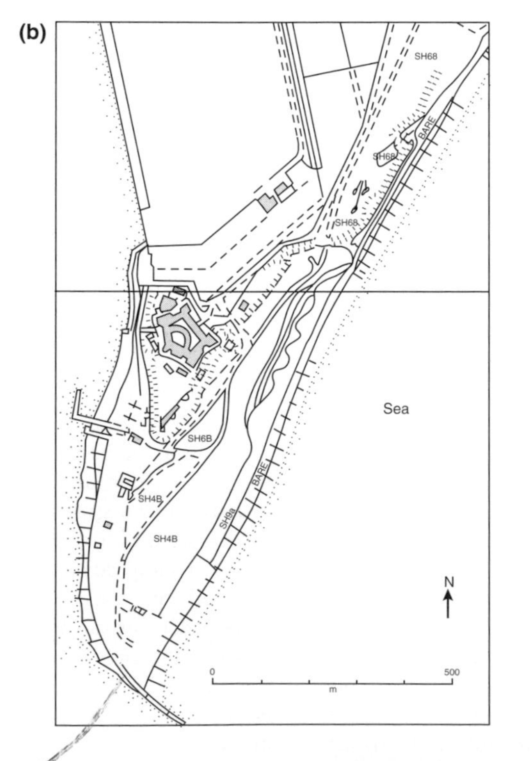

Figure 149 (a) A fringing beach at Llandulas in north Wales; (b) a shingle spit at Landguard Point, England; (c) a shingle bar at Culbin, Scotland; (d) an offshore barrier island, Scolt Head, England

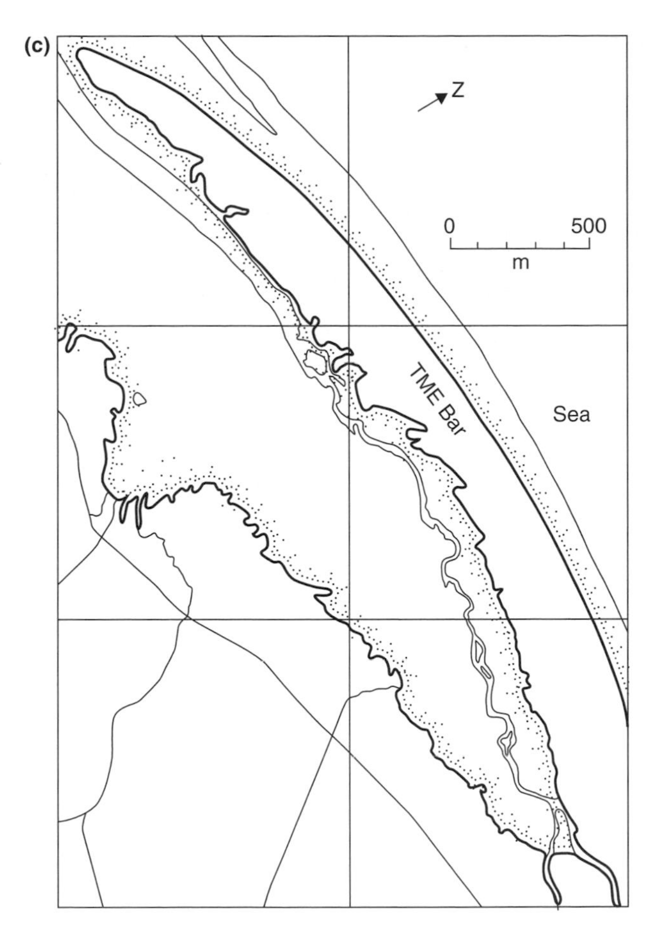

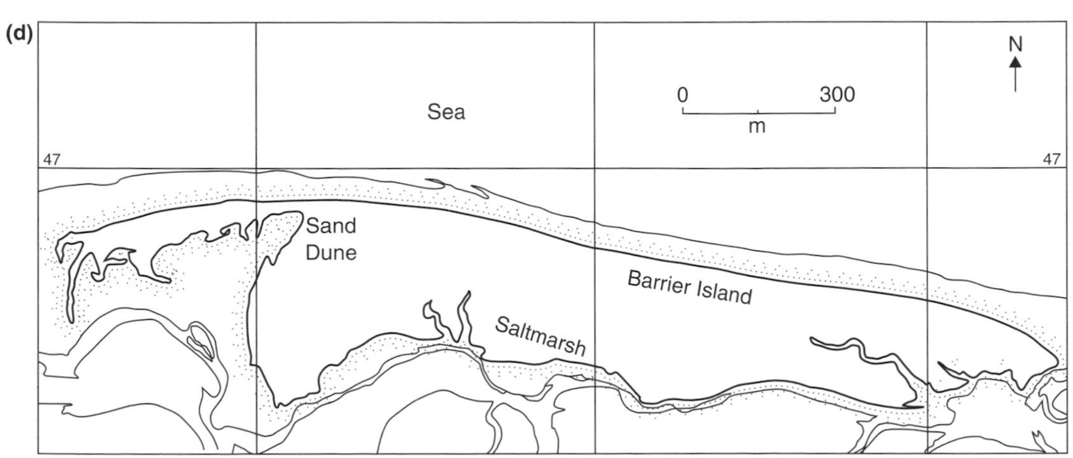

Figure 149 (Continued)

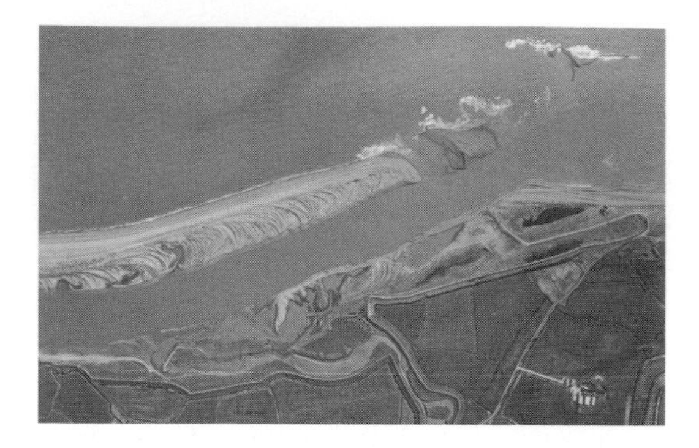

Plate 122 The shingle coast of Suffolk, UK. On the distal point of Orford Ness and on the mainland opposite there are recurved hooks enclosing lagoons. At the mouth of the estuary of the River Ore longshore drift of shingle sediments can be observed (photograph courtesy of Cambridge University Collection of Air Photographs)

known as CHENIER RIDGEs, and in Auckland Bay, New Zealand. Tombolo barriers, or bars (see BAR, COASTAL), are geomorphologically similar to spits, representing the extreme case where a spit has grown across an estuary or coastal indentation. This results in the formation of a lagoon behind the bar, which clearly affects the hydrology and ecology of the leeward slope. Chesil Beach in Dorset and Slapton Ley in Devon are prime British examples.

Rivers, which provide a source of shingle-sized sediment (see GRAVEL-BED RIVERs), may have prograded strandplains or deltas of shingle at their mouth. In Scotland, the Kingston Shingles are found at the mouth of the Spey (Sneddon and Randall 1993/1994) and South Island, New Zealand has particularly good examples, such as at the mouth of the Waitaki River.

At points of littoral drift convergence the formation of a second set of apposition ridges deposited at a different angle, will lead to the formation of a ness or cuspate foreland, a triangular mass of shingle such as Dungeness, Kent, in England, Rhunahaorine Point, Argyll in Scotland or Cape Canaveral in Florida. The Island of Rügen in Baltic Germany is effectively a cuspate foreland cut off from the mainland. Such features often support a terrestrial geomorphic system inland of the coastal ridges.

The final type of shingle formation is the offshore barrier island, formed where a large mass of shingle has been deposited offshore and which may act as the 'skeleton' for a coastal sand-dune (see DUNE, COASTAL) system. Culbin Bar, Morayshire and Scolt Head Island, Norfolk, are prime British examples.

Most shingle coasts have a steep upper beach slope and a relatively steep overall nearshore profile. Partial wave energy reflection results in the formation of edge-waves and rhythmic longshore features such as BEACH CUSPs. Shingle features are frequently of considerable ecological importance in terms of habitat diversity and play a vital geomorphologic role in determining the stability of adjoining 'soft' sediments of mudflats and saltmarshes. Unless the shingle coastal features are mobile, a partial vegetation cover is the norm. The middle and lower beach are usually kept bare by wave action, but upper beaches are vegetated. The rate and extent of plant colonization is dependent upon the degree of disturbance and shingle mobility, the presence or otherwise of a fine sediment matrix within the spaces between larger sediments and the hydrological regime of the shingle.

All shingle coasts contain a mixture of different sized sediments. Some are well sorted and consist entirely of pebbles, while others are poorly sorted and may also contain sand and/or boulders. Because there is frequently considerable temporal and spatial variation in shingle and mixed shingle/sand beaches (Kirk 1980; Schulmeister and Kirk 1993), accurate determination of average textural qualities is difficult.

Most coarse sediment coasts become coarser up-beach, because backwash and gravity can move larger clasts. Hence many locations have shingle only on the upper beach. Williams and Caldwell (1988) also comment on clast shape with discoid pebbles sorted preferentially on the upper parts of the beach with spheres and rods occurring nearer the sea. Sediment grading alongshore also occurs due to selective transport of finer sediments in the downdrift direction as at Chesil Beach. However, other sites show much more complex patterns as a result of bi-directional currents of varying magnitudes.

Shingle coast micro-relief dynamics depend upon spring to neap tidal patterns and wind and wave conditions. The upper 50–80 cm of sediment is frequently remobilized forming berms and cusps that change from one tidal cycle to another. More major changes occur seasonally as a result of spring to neap tidal fluctuations and especially at those times when storm-wave energy is higher.

The internal sedimentary architecture of shingle landforms reflects the process regime and net evolutionary trends of the structure (Randall 1973). Ridge external structures vary dependent upon whether they are vertically accreting but laterally stable, laterally migrating or developing on a seaward prograding plain. The depressions between ridge crests may be partly filled by washover and storm-tossed deposits, so that there is often a marked difference in average particle size and shape between ridge fulls (crests) and lows (Randall and Fuller 2001). Sediment grading may also occur as a result of longshore drift with selective transport of finer sediments downdrift. However, on many coasts sediment grading has been found to be complex in relation to seasonal variation in the longshore current regime (Pye 2001).

Sea-level rise has the tendency to move shingle landforms inland (Carter and Orford 1993; Forbes *et al.* 1995), but if sea-level rise is particularly rapid, shingle structures may be drowned *in situ* by overstepping. Normally, however, under moderate storm-wave activity, shingle is pushed to the top of the front-beach ridge, while in major storms the ridge is overtopped or breached, creating shingle aprons in the backbarrier area. As this pattern is repeated, so the ridge migrates landwards by rollover. Many of the major shingle formations present today formed in this way during the Holocene marine transgression, initiating at a time of lower sea-stand and reaching their present location by around 4,000 BP. Most current shingle features are relict or dependent upon erosional sediments rather than glacial debris. Hence, there is currently a shortage of sediment at the updrift end of many transport cells and increasing risk of OVERWASHING and breaching (Orford *et al.* 2001).

Traditionally in developed countries, shingle coasts have been heavily managed to retard erosion, drift and sediment cycling and, more recently, for habitat conservation. Management methods may include beach reprofiling or protection (with gabions or tetrapods) or the construction of GROYNEs and offshore breakwaters. Groynes have been used since the nineteenth century but frequently they have a negative effect downdrift by reducing longshore sediment availability. More recently beach nourishment has been seen as more cost effective and environmentally acceptable (Bradbury and Kidd 1998), but this, too, may change the natural geomorphologic character of the coast and may not be cost effective in the long term. Elsewhere gravel extraction, building developments, military activity and MANAGED RETREAT have markedly changed the landscape and landforms of shingle coasts.

At a world scale, large shingle structures are uncommon and under increasing pressure from development, mining and 'coastal squeeze' as a result of rising sea levels and coastal erosion. Most shingle structures were formed earlier in the Holocene Period and currently shingle supply is limited at the updrift end of coastal SEDIMENT CELLs. This results in the increased likelihood of breaching during STORM SURGEs. Naturally, shingle coasts are dynamic and tend to migrate landward but people prefer a static coast. For economic reasons some areas of shingle coasts have to be protected, but in less developed areas space should be left for natural dynamic coastal change. Wherever possible shingle structures should be left entirely alone, since in the majority of circumstances, geomorphologic change promotes environmental diversity (Randall and Doody 1995).

References

Bradbury, A.P. and Kidd, R. (1998) *Hurst Spit Stabilisation Scheme – Design and Construction of Beach Recharge*, Proceedings of the 33rd MAFF Conference of River and Coastal Engineers, Keele University, July 1998, 1.1.1–1.1.13.

Carter, R.W.G. and Orford, J.D. (1993) The morphodynamics of coarse clastic beaches and barriers: a short and long term perspective, *Journal of Coastal Research* 15, 158–179.

Forbes, D.L., Orford, J.D., Carter, J.W.G., Shaw, J. and Jennings, S.C. (1995) Morphodynamic evolution self-organisation and instability of coarse-clastic barriers on paraglacial coasts, *Marine Geology* 126, 63–85.

Isla, F.I. and Bujalesky, G.G. (1993) Saltation on gravel beaches, Tierra del Fuego, Argentine, *Marine Geology* 115, 263–270.

Kirk, R.M. (1980) Mixed sand and gravel beaches: morphology, processes and sediments, *Progress in Physical Geography* 4, 189–210.

McKay, P. and Terich, R.A. (1992) Gravel barrier morphology, Olympic National Park, Washington State USA, *Journal of Coastal Research* 8, 813–829.

Orford, J.D., Jennings, S.C. and Forbes, D.L. (2001) Origin, development, reworking and breakdown of gravel-dominated coastal barriers in Atlantic Canada: future scenarios for the British Coast, in J.R. Packham *et al.* (eds) *Ecology and Geomorphology of Coastal Shingle*, 23–55, Otley: Westbury Publishing.

Packham, J. R., Randall, R.E., Barnes, R.S.K. and Neal, A. (eds) (2001) *Ecology and Geomorphology of Coastal Shingle*, Otley: Westbury Publishing.

Pye, K. (2001) The Nature and Geomorphology of Coastal Shingle, in J.R. Packham *et al.* (eds) *Ecology and Geomorphology of Coastal Shingle*, Otley: Westbury Publishing.

Randall, R.E. (1973) Shingle Street, Suffolk: an analysis of a geomorphic cycle, *Bulletin of the Geological Society of Norfolk* 24, 15–35.

Randall, R.E. and Doody, J.P. (1995) Habitat inventories and the European Habitats Directive: the example of shingle beaches, in M.J. Healy and J.P. Doody (eds) *Directions in European Coastal Management*, 19–36 Cardigan: Samara Publishing.

Randall, R.E. and Fuller, R.M. (2001) The Orford Shingles, Suffolk, UK: evolving solutions in coastline management, in J.R. Packham *et al.* (eds) *Ecology and Geomorphology of Coastal Shingle*, 242–260, Otley: Westbury Publishing.

Shulmeister, J. and Kirk, R.M. (1993) Evolution of a mixed sand and gravel barrier system in Canterbury, New Zealand, during the Holocene sea-level rise and still stand, *Sedimentary Geology* 87, 215–235.

Sneddon, P. and Randall, R.E. (1993/1994) *Shingle Survey of Great Britain: Final Report Appendix 1, Wales; Appendix 2, Scotland; Appendix 3, England*, Peterborough: Joint Nature Conservation Committee.

Williams, A.T. and Caldwell, N.E. (1988) Particle size and shape in pebble beach sedimentation, *Marine Geology* 82, 199–215.

SEE ALSO: coastal geomorphology; raised beach

ROLAND E. RANDALL

SHORE PLATFORM

Shore platforms are rock surfaces created by the erosion and retreat of coastal cliffs (Figure 150) (see CLIFF, COASTAL). Although geological and other factors are responsible for enormous variations in their morphology within fairly small areas, the distinction has often been made between subhorizontal, supra-, inter-, or subtidal platforms that terminate abruptly seawards in a low tide cliff, and gently sloping, largely intertidal, platforms, with gradients between about 1° and 5°, that continue below the low tidal level without a major break in slope (Plate 123). Subhorizontal platforms have generally been associated with Australasia, although they are common in much of the tropical and subtropical world, whereas sloping platforms have been described most frequently from the stormy waters of the northern Atlantic.

The inherent complexity of shore platforms and other rocky coastal systems has made it difficult to determine how they are formed or how they develop through time. The physical resistance of

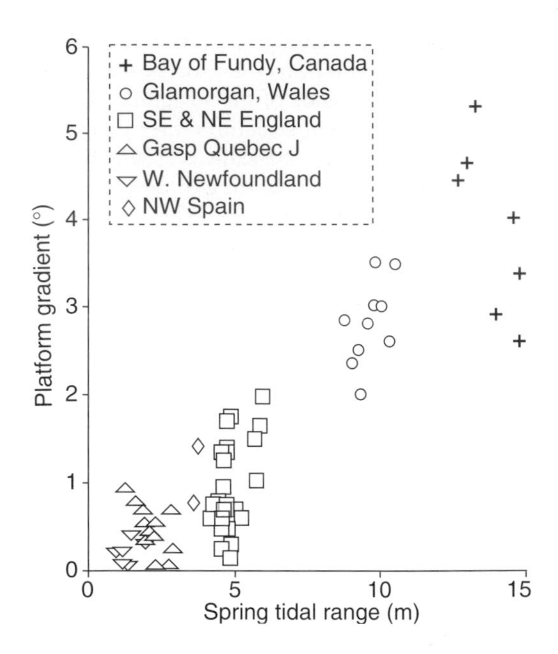

Figure 150 Shore platform in Liassic limestones and shales at Monknash, south Wales

Plate 123 Gently sloping and quasi-horizontal shore platforms

rocks depends upon their chemical composition, angle of dip, strike, bed thickness, joint (see JOINTING) pattern and density, degree of WEATHERING and a myriad of other factors. A wide range of mechanisms also operate on shore platforms, including WAVES, tides, bio-erosional and bio-constructional organisms, frost, chemical and salt weathering, WETTING AND DRYING WEATHERING and MASS MOVEMENTS. The relative and absolute importance of these processes have varied through time, with changes in relative SEA LEVEL and climate, and rock coasts often retain vestiges of environmental conditions that were quite different from today.

Much of the debate over the past one hundred years has been concerned with the relative roles of marine and subaerial processes in platform development, and, more recently, on the relationship between platform morphology and tidal range. A number of mechanisms have been proposed for platform formation:

- Platforms are cut in weathered or unweathered rock by waves. This produces sloping platforms (wave-cut platforms) in macrotidal regions and horizontal platforms, at the 'level of greatest wear' in areas with a small tidal range.
- Old Hat platforms develop in very sheltered areas, at the level of permanent seawater saturation. Above this level, weak waves wash away the fine, weathered debris, exposing the top of the resistant, unaltered rock below.
- Platforms can be formed by differential wave erosion of cliffs consisting of weak, weathered rocks above the saturation level, and more resistant, unweathered rock below. Evidence is lacking, however, for the existence of a permanent intertidal level of saturation in coastal rocks.
- Horizontal platforms, often with ridges or ramparts at their seaward ends, develop by WATER-LAYER WEATHERING and other weathering processes levelling and lowering uneven, rough, sloping or subhorizontal platforms that were originally cut by waves.
- Horizontal and sloping platforms are the result of alternate wetting and drying, which is responsible for cliff erosion and platform downwearing.
- Some workers have proposed that frost and possibly sea ice produces shore platforms in cool climatic regions.

It is increasingly evident that shore platforms are the product of mechanical wave erosion, weathering and bioerosional activity, although their relative importance depends upon climate, geology, wave and tidal conditions, and stage of development. Air compression in rock crevices and other mechanical wave erosional processes are most effective on steep uneven platform surfaces. As platforms become wider, smoother and more gently sloping, mechanical wave erosion must become less effective because of wave attenuation and the lack of rock scarps or upstanding beds of steeply dipping rock, and weathering processes must therefore become, at least relatively, more important. MICRO-EROSION METER (MEM) data from a variety of environments suggest that shore platforms are being lowered by weathering at rates that are frequently between about 0.5 to 1 mm yr^{-1}. It is difficult to accept that these high rates can be sustained indefinitely, however, and there is some MEM data to support the contention that they must decrease through time as platforms are reduced in elevation and therefore experience progressively longer periods of inundation, shorter periods of exposure and less frequent cycles of wetting and drying.

There is a moderately strong global relationship between mean regional platform gradient and tidal range (Figure 151). For wave-cut shore platforms this can be attributed to the way that tides control the expenditure of wave energy within the intertidal zone. The strong correlation between

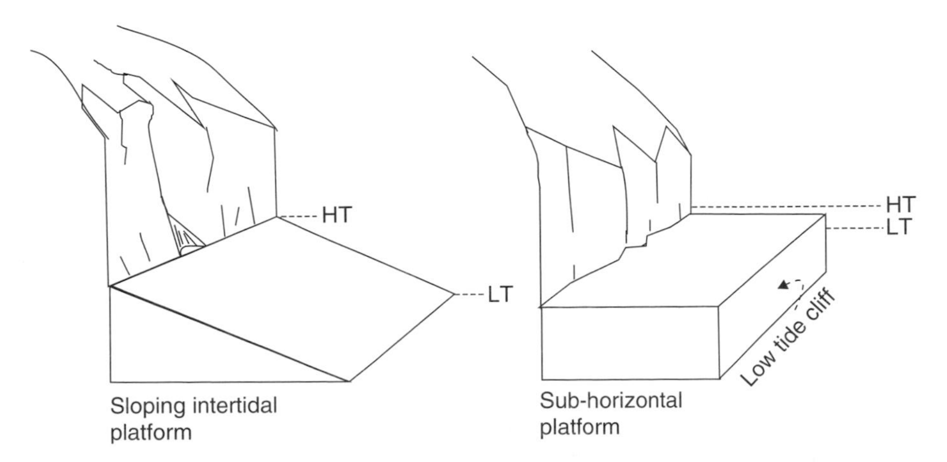

Figure 151 Relationship between mean regional shore platform gradient and spring tidal range

wetting and drying frequency distributions and tidal range also provides a possible explanation for the gradient–tidal range relationship in areas where weathering rather than wave action is dominant. Nevertheless, there is a basic problem with all theories that attribute platform formation entirely to subaerial or intertidal weathering, while relegating the role of waves to removal of fine-grained debris. This is concerned with the apparent lack of a mechanism to place a limit on maximum platform width, if wave strength and attenuation are not important factors.

Rock coast workers need to determine whether, or to what degree, shore platforms and related elements of rock coasts have been inherited from interglacial stages (see ICE AGES) when sea level was similar to today's. In hard, resistant rocks, the platforms often seem to be too wide to have developed in the few thousand years since the sea reached its present level, and they are frequently backed by ancient composite cliffs with glacial, periglacial or other terrestrial deposits, RAISED BEACHes and erosional ledges at their base. Although these coasts often lack datable materials, a variety of techniques has been used to show that at least in some areas, shore platforms, caves and other features have been inherited from one or more interglacial stage. It is particularly difficult to assess the possible role of INHERITANCE in areas of fairly weak rock, because platform width is less anomalous with regard to present rates of erosion than in more resistant rock areas, and because faster rates of erosion could account for the general lack of till covers, ancient beach deposits and structural remnants. Lacking evidence to the contrary, most workers have concluded that shore platforms in fairly weak rocks are entirely postglacial features. There is abundant evidence, however, of coastal inheritance from the last and older interglacial stages on the fairly weak rock coast of Galicia in northwestern Spain. Thick fluvio-nival and geliflucted slope deposits covered this coast during the latter part of the last glacial stage, and the ancient coast was then exhumed and inherited as sea level rose to its present position during the Holocene. The Galician evidence has important implications for the possible role of inheritance in the development of shore platforms in other areas (Trenhaile *et al.* 1999).

Modelling provides one of the few ways to study the long-term evolution of slowly changing rock coasts. The earliest MODELS were qualitative and structured within evolutionary cycles of erosion. More recent models are mathematical, but although field evidence suggests that most mechanical wave erosion occurs through water hammer, air compression in joints and ABRASION in shallow water, processes that are closely associated with the water surface, models have generally been concerned with submarine erosion in tideless seas. Nevertheless, a model has been developed that considers rates of wave attenuation and the long-term distribution of wave energy within the intertidal zone. This model has been used to study the long-term evolution of shore platforms with Quaternary changes in sea level on stable and tectonically mobile coasts. The model indicated that whether an ancient shore platform is subsequently inherited, modified or completely replaced by a contemporary platform depends upon the complex interaction of a multitude of factors that determine the erosive efficacy of the waves. It suggested that intertidal and subtidal surfaces trend towards a state of static equilibrium under oscillating sea level conditions, as attenuated waves become increasingly unable to continue eroding the rock, although they can be in a temporary, though possibly long-lasting, state of DYNAMIC EQUILIBRIUM. Most modelled surfaces were, at least in part, inherited from one, or in many cases more, interglacial stages when sea level was similar to today's (Trenhaile 2001). In future, platform models must also consider the effects of downwearing by weathering as well as backwearing by waves within the intertidal zone. In turn, reliable modelling is dependent on the acquisition of quantitative field data. Although the number of sites and the length of the records are quite limited, the micro-erosion meter has provided useful information on platform downwearing by weathering and corrasion. We are still unable, however, to measure the effect of joint block and large rock fragment detachment by wave quarrying and other mechanisms.

References

Trenhaile, A.S. (2001) Modelling the Quaternary evolution of shore platforms and erosional continental shelves, *Earth Surface Processes and Landforms* 26, 1,103–1,128.

Trenhaile, A.S., Pérez Alberti, A., Martínez Cortizas, A., Costa Casais, M. and Blanco Chao, R. (1999) Rock coast inheritance: an example from Galicia, northwestern Spain, *Earth Surface Processes and Landforms* 24, 605–621.

Further reading

Stephenson, W.J. (2000) Shore platforms: remain a neglected coastal feature', *Progress in Physical Geography* 24, 311–327.
Sunamura, T. (1992) *The Geomorphology of Rocky Coasts*, Chichester: Wiley.
Trenhaile, A.S. (1987) *The Geomorphology of Rock Coasts*, Oxford: Oxford University Press.

ALAN TRENHAILE

SICHELWANNE

Crescent-shaped grooves formed by action of a glacier. The term is derived from Germany, meaning 'sickle tubs', but is part of a collective set of features known as plastically sculptured forms, or p-forms. Sichelwannen (plural) occur in crystalline rocks set in glacial environments, and range in size from 1–10 metres in length, 5–6 metres wide, and from millimetres up to several metres in depth. They are a transverse type of p-form, commonly displaying striations, and with the horns of the crescent shape pointing down-glacier. The origin of sichelwannen is unclear, and several methods of formation have been suggested in the literature. The most plausible manner of formation is erosion by high pressure subglacial meltwater. Similar forms to sichelwannen have been produced by water in less resistant rock, where up-slope topographic obstructions force subglacial channels to bifurcate, producing the characteristic horseshoe pattern on the stoss-side of the obstacle. Erosion by glacier ice has also been proposed, supported by the presence of striations on sichelwannen. However, advocates of a fluvial origin believe the distinctively patchy nature of the striations prohibits formation by ice.

Further reading

Benn, D.I. and Evans, D.J.A. (1998) *Glaciers and Glaciation*. London: Arnold.
Shaw, J. (1994) Hairpin erosional marks, horseshoe vortices and subglacial erosion, *Sedimentary Geology* 91, 269–283.

STEVE WARD

SILCRETE

A terrestrial geochemical sediment arising from low temperature near-surface physico-chemical processes operating within the zone of WEATHERING in which silica has accumulated in, and/or replaced, a pre-existing soil, sediment, rock or weathered material. Silcretes are defined as containing >85 per cent silica by weight, with some pure examples consisting of >95 per cent silica (Summerfield 1983). They can be distinguished from other silica-cemented sedimentary rocks such as orthoquartzites as they often exhibit a porphyroclastic, as opposed to even-grained, texture when viewed in microscopic thin-section (Watson and Nash 1997). Silcretes commonly consist of hard, silica-cemented quartz sand or brittle quartzitic material with a conchoidal fracture. The cement can consist of a range of silica minerals, of which opal, chalcedony, cryptocrystalline silica and quartz are the most widely documented. The presence of other minerals may affect the silcrete colour, with grey, brown and green varieties reported.

Although not as common as many other DURICRUSTs, silcretes are widely distributed in low latitude and other environments, particularly those areas which did not undergo Late Tertiary and Quaternary glaciation. They are found on every continent except Antarctica but are most widespread in inland and southern Australia, and in the Kalahari and Cape coastal regions of southern Africa. Other locations with significant areas of silcrete include Britain (where they are termed 'sarsens') Europe, the Sahara, Tanzania, USA, Uruguay and Brazil. Given their hardness and chemical stability, they are extremely resistant to erosion and often act as CAPROCKs and influence INVERTED RELIEF development. Silcretes most commonly occur as distinct horizons, but may also form a coating on rock outcrops or lenses within other duricrusts such as CALCRETE. Well-developed silcrete horizons are between 0.5 and 3 m thick, although thicknesses of >10 m have been recorded. A variety of terms have been used to describe profiles including massive, columnar, nodular, glaebular and mammilated, reflecting the numerous surface morphologies and modes of origin of many silcretes.

Silcrete development can take place via a variety of pedogenic and non-pedogenic processes, but all require a silica source, a means of transferring this silica to the site of formation, and a mechanism to trigger precipitation. The most significant source is CHEMICAL WEATHERING of silicate-rich rocks, particularly those containing clay minerals. Silica released in this way may then be available for transport in solution. Highly alkaline conditions, such as those found in arid zone lakes, can also lead to extensive dissolution of silicate minerals. Quartz is only weakly soluble in neutral pH terrestrial surface waters (around 10 ppm at 25 °C), but solubility increases dramatically above pH 9.0. Other silica sources include replacement of quartz during carbonate precipitation, dissolution of volcanic and other dust, and biological inputs from silica-rich plants and micro-organisms. Silica from these sources may be transferred in solution to the point of precipitation via lateral or vertical movements of ground water, pore water and surface water, with a range of local and far-travelled silica potentially contributing to silcrete formation. Silica precipitation may also be initiated by a variety of factors, of which the most important are evaporation, cooling, organic processes, absorption by solids, reactions with cations and changes in pH (particularly a shift to below pH 9.0 in alkaline environments).

Silcrete formation by pedogenic processes involves the accumulation of silica from downward percolating soil water during a succession of cycles of leaching and precipitation. Such silcretes commonly consist of a nodular base overlain by a columnar section containing illuviation structures, capped by a brecciated component. Silica mineralogy often varies throughout the profile, with more ordered forms of silica cement in the uppermost sections and less ordered forms towards the base (Milnes and Thiry 1992). Pedogenic silcretes may develop over large areas, are often semi-continuous and relate directly to palaeosurfaces. Non-pedogenic models embrace formation in a variety of settings, including zones of water table fluctuation or groundwater outflow, locations marginal to drainage lines, as well as lakes and seasonal pools. Silica precipitation in these environments is usually controlled by evaporation, pH shifts or organic processes, with the resulting silcretes forming localized sheets or lenses. Non-pedogenic silcretes are usually simple in terms of their macro- and micromorphology, although a range of silica cements may be present. Significantly, they are almost always devoid of the organized profile associated with pedogenic silcretes. Non-pedogenic silcretes may also form at considerably greater depths than pedogenic types and therefore do not normally represent a palaeosurface.

Perhaps the greatest controversy surrounding silcrete is the degree to which environmental controls determine formation, a critical issue if silcretes are to be used as palaeoenvironmental indicators (Nash *et al.* 1994). Silcrete has been suggested to form under climates ranging from semi-arid to monsoonal on the basis of geochemical, mineralogical and micromorphological evidence, and by comparison of the geographic and stratigraphic distribution of silcrete with contemporary and past climate. The situation is made more complex by the fact that most silcretes are relict. Contemporary silcrete formation has only been documented from one location, a hypersaline lake in the Kalahari (Shaw *et al.* 1990). Unfortunately, silicification at this site is driven by organic silica fixation so the resultant silcretes are not an ideal modern analogue. The only safe conclusion that can be made is that the presence of silcrete should not be considered diagnostic of specific environmental conditions. It is essential to establish the mode of origin of any silcrete and view formation within the context of other climate proxies before using it as a palaeoenvironmental indicator. Pedogenic silcretes, such as those in southern South Africa, may have taken hundreds of thousands of years to form and, as a result, have integrated climatic effects over considerable time periods. In contrast, some non-pedogenic groundwater silcretes, such as those in the Paris Basin, developed in a few tens of thousands of years under steady groundwater outflow.

References

Milnes, A.R. and Thiry, M. (1992) Silcretes, in I.P. Martini and W. Chesworth (eds) *Weathering, Soils and Palaeosols*, 349–377, Amsterdam: Elsevier.

Nash, D.J., Thomas, D.S.G. and Shaw, P.A. (1994) Siliceous duricrusts as palaeoclimatic indicators: evidence from the Kalahari Desert of Botswana, *Palaeogeography, Palaeoclimatology, Palaeoecology* 112, 279–295.

Shaw, P.A., Cooke, H.J. and Perry, C.C. (1990) Microbialitic silcretes in highly alkaline environments: some observations from Sua Pan, Botswana, *South African Journal of Geology* 93, 803–808.

Summerfield, M.A. (1983) Silcrete, in A.S. Goudie and K. Pye (eds) *Chemical Sediments and Geomorphology*, 59–91, London: Academic Press.

Watson, A. and Nash, D.J. (1997) Desert crusts and varnishes, in D.S.G. Thomas (ed.) *Arid Zone Geomorphology*, 69–107, Chichester: Wiley.

DAVID J. NASH

SINGING SAND

Two types of sand that emit audible sounds with a coherent wave pattern, on being sheared by wind or other mechanical means, have been reported throughout the world over the past century. They are known as squeaking or singing (found on beaches) in booming (found in sand dunes) sands. The exact mechanism by which these coherent sounds are produced from these sand materials is still unknown.

Many acoustical sands are well-sorted materials with sizes in the range of 100–500 microns, roughly rounded particles with a high content of quartz particles. However, booming sand materials are also found in Kauai, Hawaii with a high calcareous content.

Various material science techniques have been used to show that there is a thin rind-like layer on the particles of these acoustical sand materials. Direct Scanning Electron Microscopic examination of particles that have been sliced by grinding show this layer in both pure silica-gel singing materials and the calcareous Kauai sands. In the former case, the rind layer is composed of amorphous silica and in the latter, an alumino-silicate clay-like material. The rind width is about 5 microns.

Fourier Transform Infra Red (FTIR) analysis has shown that acoustical sand exhibit a broad absorption band in the range of 2,800 cm^{-1}. This band is due to clusters of water in the rind-like layer.

Etching of acoustical sands with hydrofluoric acid (HF) removes the surface layer and causes the sand to become silent. In the case of the Kauai sands, sonication is sufficient to remove this layer and this technique also silences the booming property.

Sand materials that do not sing can be made to do so by grinding the grains in a mill, periodically removing the fines and renewing the water, leaving a well-sorted and highly polished material which 'sings' and exhibits the characteristic FTIR 3,400 cm^{-1} broad band.

Finally, excess heating of these materials also removes this singing or booming sound.

Interest in finding the underlying mechanism of the production of this coherent wave phenomenon in granular materials is due to its possible use in a saser device for sonic hammering.

Further reading

Goldsack, D.E., Leach, M.F. and Kilkenny, C. (1997) Natural and artificial 'Singing Sands', *Nature* 386, 29.

DOUGLAS GOLDSACK AND MARCEL LEACH

SINUOSITY

Few natural rivers are straight for more than a few channel widths. Rivers that are not completely straight are sinuous, even if they are not clearly MEANDERING in the sense of having more or less regular oscillations in direction. Schumm (1963) introduced a quantitative definition of sinuosity as the distance along a river between two points A and B, divided by the valley distance between A and B. It is therefore a dimensionless ratio with a minimum value of 1.0 and a realistic maximum of around 3 to 4, after which neck cutoffs occur. It can be calculated for a single bend (when A and B are successive crossovers), a series of bends, or a longer reach.

The sinuosity of a modern channel, or a well-preserved palaeochannel, is readily determined from a map or aerial photograph. Some ambiguities of operational definition must be kept in mind when comparing values quoted by different authors. Is valley distance defined as a straight line, or does it allow for large-scale valley bends? If the river is braided (see BRAIDED RIVER), is the sinuosity computed for the centreline, the biggest channel, or the sum of all the channels as suggested by Richards (1982)?

Sinuosity can alternatively be assessed using variograms and FRACTAL concepts (Nikora 1991; Lancaster and Bras 2002), which can reveal any scale dependence as one moves from single simple bends to compound loops and multiple loops. The sinuosity of a fragmentary palaeochannel can be estimated from the variance of channel direction at places where this can be reconstructed (Ferguson 1977).

Evidently sinuosity varies spatially according to the straight, meandering, or other channel pattern of different reaches. Sinuosity can also fluctuate over years or decades as bends of an actively meandering channel grow and are cut off, and it may change progressively in the event of climate change, flow regulation, or other disturbance of the system.

Sinuosity has significance for fluvial processes and channel regime because it can be written not as a ratio of distances but of slopes: the valley slope divided by the channel slope. Since valley slope is largely inherited, an increase in the meandering tendency of a river leads not only to greater sinuosity but also reduced channel slope. This has consequences for velocity, shear stress, and BEDLOAD transport. Sinuosity is therefore regarded by many geomorphologists and other fluvial scientists as one of several channel properties that can adjust if the bedload supply to a reach is not the same as the transport capacity.

References

Ferguson, R.I. (1977) Meander sinuosity and direction variance, *Geological Society of America Bulletin* 88, 212–214.
Lancaster, S.T. and Bras, R.L. (2002) A simple model of river meandering and its comparison to natural channels, *Hydrological Processes* 16, 1–26.
Nikora, V.I. (1991) Fractal structures of river plan forms, *Water Resources Research* 27, 1,327–1,333.
Richards, K. (1982) *Rivers: Form and Process in Alluvial Channels*, London: Methuen.
Schumm, S.A. (1963) Sinuosity of alluvial rivers on the Great Plains, *Geological Society of America Bulletin* 74, 1,089–1,100.

ROB FERGUSON

SKERRY

A term that describes the low rocky islets common to many mid and high latitude coastlines. Many skerries may be covered at high tide and are subject to wave processes with the result that they may carry the signature of marine quarrying and abrasion. However, in spite of a marine influence, many skerries have also been shaped by past glacial erosion and, within the constraints of the geological structure of the host rock, may have inherited a moulded bedform or even roche moutonée shape. For example, many skerries in Finland, Sweden and Norway, particularly in sites sheltered from severe wave activity, still retain striations etched onto glacially moulded bedforms. This trait can also be seen in the rocky islets of formerly glaciated lakes such as in Loch Lomond in Scotland and in Lake Nasijarvi in Finland. On the coast, fields or chains of skerries frequently occur offshore of areally scoured surfaces that have been partly submerged by Holocene sea-level rise. Good examples occur in the Outer Hebrides of Scotland and in arctic Canada, and along the STRANDFLAT coasts of Norway, Sweden and Finland in the Baltic and in western Iceland.

Further reading

Bird, E.C.F. and Schwartz, M.L. (eds) (1985) *The World's Coastline*, New York: Van Nostrand Reinhold.

JIM HANSOM

SLAKING

The disintegration of a loosely consolidated material following the introduction of water or exposure to the atmosphere (Plate 124). Clays and shales (mudrocks) are especially prone to this form of failure, especially in the presence of saline waters. Materials with high Exchangeable Sodium Percentages (ESP), including some colluvia, may be susceptible, and slaking is an important process on many badland surfaces, including DONGAS. Various tests are available for determining the durability of slaking-prone materials (Czerewko and Cripps 2001) (see WETTING AND DRYING WEATHERING).

Reference

Czerewko, M.A. and Cripps, J.C. (2001) Assessing the durability of mudrocks using the modified jar slake index test, *Quarterly Journal of Engineering Geology and Hydrology* 34, 153–163.

A.S. GOUDIE

Plate 124 Materials with a high Exchangeable Sodium Percentage, such as this colluvial deposit from central Swaziland, are prone to severe erosion as a result of their propensity to slaking following rain events

SLICKENSIDE

A polished, striated rock surface on a fault or bedding plane caused by the frictional movement between one rock mass sliding over another. Slickensides are a common feature on fault planes, and though sometimes can be featureless, they commonly display prominent parallel ribbing or striation. These striations may be exhibited on mineral coatings, such as quartz and calcite, as well as on the rock itself, and can provide an indication of the direction of fault movement from their orientation (they form parallel to the direction of fault displacement). However, striations can often be erased by subsequent fault movement, and thus should only be considered as a record of the most recent fault movement. Additionally, from analysis of a suite of slickenside striations, an estimation of the magnitude of the *in situ* stress field can be established. The origin of slickenside striations is uncertain. Some may be grooves formed by outcropped resistant rock on the opposite fault block, or mineral lineations that grow with their long axis parallel to the direction of fault movement. Slickensides with striations often contain small steps oriented in one direction similarly to the striations.

The term slickenside also refers to natural crack surfaces along planes of weakness in soils, resulting from the movement of one mass of soil against another. This is commonly by the swelling and contraction of soils with high clay levels able to swell (e.g. montmorillonite). Slickenside also refers to the polished surface produced by the passage of a mudflow.

Further reading

Doblas, M. (1998) Slickenside kinematic indicators, *Tectonophysics* 295, 187–197.
Tjia, H.D. (1964) Slickensides and fault movements, *Geological Society of America Bulletin* 75, 638–686.

SEE ALSO: fault and fault scarp

STEVE WARD

SLOPE, EVOLUTION

Landscapes change over time in response to the internal redistribution of sediment, usually with some net export of material to rivers or the ocean. The way in which landscapes and slopes evolve depends on their initial form, the slope processes (see HILLSLOPE, PROCESS) operating and the boundary conditions which determine where and how much sediment is removed. This discussion will mainly focus on two-dimensional slope profiles, but some of the aspects which can only be addressed in 3-D will also be discussed below.

Slope evolution can be described in conceptual development sequences, and much of the history of early twentieth-century geomorphology was concerned with championing alternative conceptual models (Chorley *et al.* 1973), under the banners of W.M. Davis, Walther Penck and others (see CYCLE OF EROSION). More recent approaches have focused increasingly on the application of MODELS in geomorphology, and it is instructive to compare the various development sequences in these terms, to understand the conditions under which each is most appropriate.

Although G.K. Gilbert was not primarily a theoretician, his work in the Henry Mountains (1877) repeatedly interpreted slope profiles in the context of the interaction between form and process, introducing the term DYNAMIC EQUILIBRIUM to describe this balance. He correctly attributed the convexity of divides to SOIL CREEP and similar diffusive processes; and the concavity of the lower slopes to SOIL EROSION processes.

W.M. Davis (1909) spent much of his life canvassing the concept of the Geographical Cycle, which described what he perceived as the 'normal' sequence of erosional landforms, strongly based on his experience of humid temperate conditions. The cycle assumed the rapid uplift of a low relief landscape, and its erosion during a period of tectonically stable conditions. Generally the landscape is taken as soil-mantled, with three life stages. In *youth*, rivers incise deeply into the landscape, and hillsides gradually encroach upon the original surface, parts of which may survive as an *erosion surface*. Slopes become *mature* when the original surface has been consumed, and the highest points begin to undergo appreciable erosion. During maturity, slopes decline in gradient everywhere, forming a connected or *graded* system conveying material to the lowest point. Eventually maturity gives way to *old age* as the surface is reduced to a low relief surface, or PENEPLAIN, which may still retain a few remnant hills, or *monadnocks*. This sequence of ever-reducing relief could be interrupted by relative falls in sea level, which might *rejuvenate* the landscape, perhaps creating flights of partial erosion surfaces which preserved the morphology of previous interrupted cycles. Davis, and his disciples like Johnson and Wooldridge, used the methodology of the geographical cycle to reconstruct former sea levels and the history of the landscape from the inferred remains of high level erosion surfaces and sequences of river and coastal terraces. The two key assumptions of Davis's approach were first that uplift was rapid and followed by long periods of tectonic stability, and second that hillsides were essentially soil-mantled.

Walther Penck (1953 [1924]), working in the much more tectonically active area of the Andes, considered that slope forms responded primarily to the rate of uplift, which he assumed to be a continuous, if episodic, process. In perhaps his central conceptual model, he considered that convex slopes were produced where the rate of uplift was accelerating, and concave slopes where uplift rates were falling. The normal situation, of steady uniform uplift, was associated with slopes of uniform gradient, and these would retreat parallel to themselves while the uplift continued. This approach differs from Davis's in two important respects: first in the assumption of active tectonics, and second in coupling evolution of the slope to conditions at the slope base. Both these assumptions are most relevant where slopes have little or no regolith cover, and all detached material is immediately removed.

Lester King (1953), working in the semi-arid climate of South Africa, proposed a morphologically intermediate conceptual model, in which the steep and rocky upper slopes retreat parallel to themselves, and undergo replacement by a PEDIMENT, which is a lower gradient surface with some regolith cover. Pediments gradually coalesce and eventually consume any residual mountain masses. The combination of rocky and regolith-covered slope elements is significant in this model, while the tectonic assumptions return to the stability advocated by Davis.

In comparing these conceptual models, one of the important distinctions concerns the presence or absence of regolith on the slopes. Where there is a deep regolith, slope processes are able to act at their full capacity, and removal is said to be transport-limited or flux limited. Where the regolith is thin or absent and slopes are steep, material is removed as fast as it is detached by weathering or entrainment, and removal is said to be weathering-limited or supply limited (see WEATHERING-LIMITED AND TRANSPORT-LIMITED). In this case the actual transport rate is well below the *transporting capacity* of the sediment transport processes. Slope evolution is radically different according to which of these two regimes is active, and there is the possibility of an intermediate regime, in which both detachment rates and transport capacities are important. The nature of the regime can be seen from the *travel distance* of material during a transport event. Where travel distance is short compared to the slope length, for example under soil creep, rainsplash, bedload transport or soil erosion, removal is transport-limited; whereas where travel distance is long, for example under rock fall, debris flows, washload transport or movement of solutes, removal is essentially weathering-limited.

Several conceptual models have also been proposed to describe the conditions and forms associated with equilibrium or quasi-equilibrium forms for slope profiles. The essence of these approaches is that the profile form is considered to be independent of the initial surface form which is eroded. J.T. Hack (1960) has been associated with a particular form of DYNAMIC EQUILIBRIUM, in which the landform remains essentially fixed in form as it erodes. This form may strictly occur only where balanced by equal and opposite tectonic uplift, but Hack argued that, during the stage of Davisian maturity, much of the Appalachian landscape eroded with little change in form except close to base level, so that dynamic equilibrium provided a working approximation to observed conditions. M.J. Kirkby (1971) proposed the concept of *characteristic forms*, in which slope profiles retain their form, with more and more subdued relief as they decline in elevation, corresponding to the transition from Davisian maturity to old age. This is seen as a quasi-equilibrium in which the profile form is characteristic of the ensemble of processes acting on it.

By comparing these qualitative concepts with simple slope models, the conditions for all these simplified forms can be compared in a quantitative way. For a slope profile, evolution is controlled by four sets of constraints. First, mass must be conserved; second, evolution takes place from an initial profile form; third, evolution is subject to boundary conditions, which define, for example, conditions at the top (divide) and bottom (stream) of a hillslope; and fourth, evolution occurs through sediment transport by one or, usually, more slope processes. The transporting capacity for each process varies in some way with topography, usually with distance from the divide and gradient; and removal is also subject to transport or weathering-limited conditions, which can be defined by a fuller specification of the slope processes.

The Mass Balance equation for a slope profile states that:

Input − Output = net increase in Storage

or in the simplest case of a simple slope profile from divide to stream:

$$\frac{\partial z}{\partial t} + \frac{\partial S}{\partial x} = 0 \quad (1)$$

where S is sediment transport per unit width, z is elevation above an arbitrary datum, x is horizontal distance measured from the divide, and t is elapsed time.

The cases of weathering and transport-limited removal can be distinguished using a sedimentation equation:

$$\frac{dS}{dx} = D - \frac{S}{h} \quad (2)$$

where D is the rate of sediment detachment, and h is the travel distance (the mean distance travelled by sediment in an event). The second term of the right-hand side is the rate of deposition. Where sediment detachment balances sedimentation, the flow is said to be at its travel capacity, $C = D \cdot h$.

Where the travel distance is small (in relation to the slope length), then removal is essentially transport-limited and $S = C$, so that equation (1) can be used to define slope evolution, with C replacing S. Where the travel distance is long ($h \gg x_0$), removal is weathering-limited and slope evolution is approximated by equation (2) with the second term on the right-hand side negligible. Between these extremes, it is necessary to retain both equations (1) and (2) to determine how slopes evolve.

The simplest boundary conditions are of a summit fixed in horizontal position (at $x = 0$), and a stream fixed at some point $x = x_0$.

For several processes, capacity sediment transport rates can be written in the form:

$$C \propto x^m \Lambda^n \quad (3)$$

For example simple but empirically acceptable formulations are shown in Table 43, although EQUIFINALITY allows some range of possible exponent values. Mass movements may also be included in a similar scheme, but with two threshold gradients: a

Table 43 Exponents for some sediment transport processes in equation (3)

Process	Travel distance	m	n
Soil creep	Small	0	1
Rainsplash	(<1 m)	0	1
Solifluction	(<1 m)	0	1
Tillage erosion	(<1 m)	0	1
Rillwash	~10 m	2	1–2
Solution	≫1 km	1	0

lower threshold below which no movement occurs, and an upper threshold above which material will never stop. The lower threshold Λ_T corresponds to the maximum stable slope gradient under saturated conditions, and the upper threshold Λ_0 to the angle of repose for debris. A simple formulation then takes the form:

$$D \propto \Lambda(\Lambda - \Lambda_T);\ h \propto 1/(\Lambda_0 - \Lambda);$$
$$C \propto \Lambda(\Lambda - \Lambda_T)/(\Lambda_0 - \Lambda) \quad (4)$$

with the capacity, C, defined only in the range $\Lambda_T < \Lambda < \Lambda_0$, and travel distances generally of the order of the total slope length. By treating mass movement as a continuous process, the feedbacks produced by the size of an individual slide event are ignored, so that this formulation works best for small and shallow slides, and for rockfalls from cliffs.

Without entering into a full analysis of these equations, some results can be quoted here without proof. First, the conditions for downcutting at a steady uniform rate, T corresponding to a strict application of Hack's dynamic equilibrium, is that:

$$S = Tx \quad (5)$$

This is necessarily true because, in the steady state, the slope processes must carry away all the material eroded upslope.

For any transport-limited removal process (i.e. $S = C$), equation (3) gives:

$$C = Tx \sim x^m \Lambda^n, \quad \text{or} \quad \Lambda \sim Tx^{(1-m)/n} \quad (6)$$

Thus, for steady-state downcutting, hillslopes are convex (gradient increasing downslope) when the distance exponent $m < 1$, and concave if $m > 1$. This means that slopes become convex for soil creep, rainsplash, solifluction and tillage erosion, and become concave for rillwash. More realistically, a combination of processes is acting, for example rainsplash and rillwash. Adding these terms the combined transporting capacity may be written as:

$$C = k\Lambda\left[1+\left(\frac{x}{u}\right)^2\right] \tag{7}$$

where u is the distance beyond which the rillwash term (the second term on the right-hand side) becomes greater than the rainsplash term (first term on RHS). Thus for $x<u$, rainsplash is the dominant process, and for $x>u$, rillwash is dominant.

Solving for constant downcutting as before,

$$\Lambda = \frac{Tx}{\left[1+\left(\frac{x}{u}\right)^2\right]} \tag{8}$$

In this case, gradient increases for $0<x<u$, and gradient decreases for $x>u$. In other words the constant downcutting form is convex where rainsplash is the dominant process and concave where rillwash is the dominant process (Figure 152).

This relationship is not, however, universal, and if downcutting decreases downslope, the concavity expands slightly into the rainsplash-dominated zone. A simple representation of Davisian decline can be made by assuming, instead of constant downcutting, a rate of downcutting which is directly proportional to height above the basal point. With this assumption, the whole hillslope must eventually erode to a flat uniform base-level plain, and equation (3) is replaced by:

$$-\frac{\partial z}{\partial t} = \frac{\partial C}{\partial x} = \alpha z \tag{9}$$

for an appropriate constant α. Although there is not always a simple analytical solution to this equation, the difference between this Davisian hillslope and the constant downcutting form is quite clear from Figure 153. In each case, the divide convexity begins the same, but, for the declining form, the rate of increase in gradient is less than for the constant downcutting form, and the concavity, where rillwash is present, is broader. It can be shown that the shape of both the declining and constant downcutting forms depends primarily on the combination of processes operating, and the effect of the initial form is progressively obliterated over time. This means that physical remains of former eroded land surfaces (erosion surfaces and terraces) only survive for a limited period of time before they disappear from the landscape, usually surviving longest in flat areas and along divides, where denudation is initially least.

Where travel distances are long, then removal of material is approximately *weathering-limited*. In this case slope development follows equation (2) above, with the final term negligible (because h is large), giving, when combined with equation(1):

$$-\frac{\partial z}{\partial t} = \frac{dS}{dx} = D \tag{10}$$

For the case of mass movements, using equation (3) and writing Λ as $-\frac{\partial z}{\partial x}$:

$$\frac{\partial z}{\partial t} = (\Lambda - \Lambda_T)\frac{\partial z}{\partial x} \tag{11}$$

The solution to this equation shows lateral retreat of steep slopes at a horizontal rate of $(\Lambda - \Lambda_T)$. This evolution of the landscape essentially describes a parallel retreat of the landscape

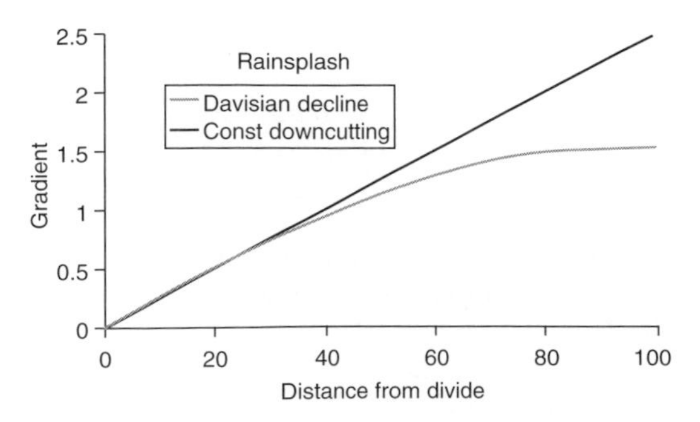

Figure 152 Slope evolution where rainsplash is dominant

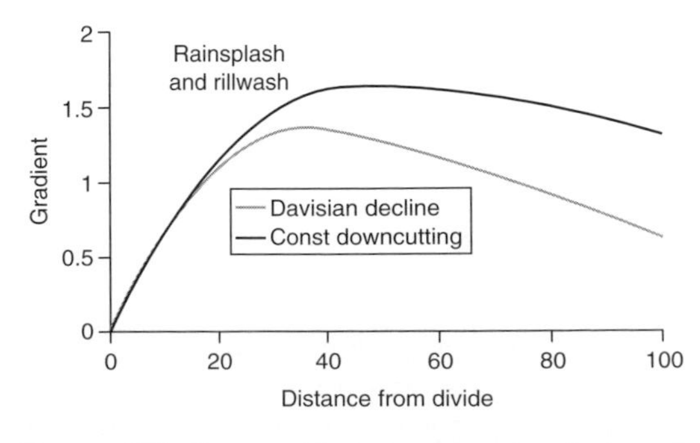

Figure 153 Slope evolution with rainsplash and rillwash

in a way which is close to the conceptual models of Penck and King. Although there is no space to fully develop this argument here, there is a strong association between *transport-limited* removal and Davisian decline of slopes, and between *weathering-limited* removal and lateral retreat of landforms. Because this distinction is closely linked to the presence or absence of a regolith deep enough to allow transport processes to operate at their capacity rates, there is also a general association between these two sets of conditions and both climate and tectonics. There is a link to climate because, in dry climates, there is little water flowing through the soil, consequently little leaching and slow bedrock weathering. Most weathering products are therefore removed by erosional processes before fine-grained and deep soils can develop. On the contrary, humid climates allow more rapid weathering, converting parent materials to fine-textured soils before it is eroded. Steep slopes increase erosion processes, but have much less effect on weathering rates, so that soil is thinner and stonier than on gentle slopes. Active tectonic uplift also plays an important role, by creating and maintaining steep slopes.

From this discussion it may be seen that the various qualitative conceptual models have been significantly shaped by the experiences of their authors. Davis, working in the humid-temperate areas of north-east USA, focused on transport-limited processes, and slope decline towards eventual peneplains. Penck and King, working in more tectonically active and more arid areas, saw the lateral retreat of steep slopes, generally with shallow regolith. It is clear, however, that, although many details are still poorly understood, a single set of principles can be widely applied, and takes in the early conceptual models as special cases.

Figure 154 provides cartoons of 'typical' slope evolution from a plateau with a steeply incised stream, but without further tectonic uplift as material is removed by the basal stream. It can be seen that elements of both transport and weathering-limited removal occur in both settings while slopes are steep, but that weathering-limited elements survive much longer under semi-arid conditions.

Although most of the features of a hillside may be described in a slope profile, the whole scale of the landscape is a problem which can only be addressed in three dimensions, defining the typical length of a single slope, and the DRAINAGE DENSITY of the landscape, which are related by the relationship:

$$\text{Mean slope length} = 1/(2\ \text{DD}) \qquad (12)$$

Following the ideas developed by Smith and Bretherton (1972), it is argued that, where sediment transport processes increase more than linearly with catchment area, any small irregularity in the landscape will tend to grow with positive feedback until it develops into a valley. The threshold at which this occurs determines the drainage density of the landscape. A 3-D *landscape model* is able to demonstrate this behaviour. Near the divides, any small irregularities in the landscape become smoother over time, whereas downslope, some hollows grow into valleys. The form of the sediment transport equations, such as equations (3) and (4) above, determine the threshold of this valley instability. For example, if the combination of rainsplash and wash is put in the form (note that this is a different form from equation (7) above):

$$S = C = k\Lambda\left[1+\Lambda\left(\frac{x}{u}\right)^2\right] \qquad (13)$$

then it can be shown that the critical distance for hollow enlargement is:

$$x = u / \sqrt{\Lambda} \qquad (14)$$

Table 44 Conditions associated with transport-limited and weathering-limited removal

	Transport-limited	*Weathering-limited*
Regolith	Deep enough to allow transport processes to operate	Shallow and generally stony
Climate	Humid temperate	Semi-arid
Gradient	Gentle	Steep
Tectonics	Inactive	Active
Dominant erosion processes	Creep, rainsplash, rillwash	Mass movements
Ratio of weathering to erosion	High	Low

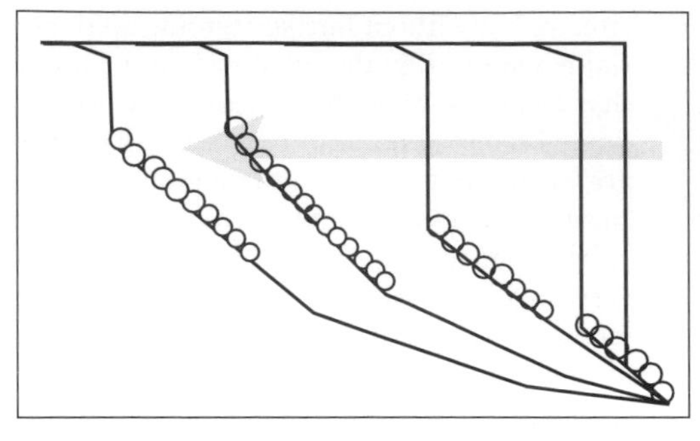

Semi-arid slope evolution
Upper plateau influenced by rainsplash etc. Steep escarpment (often defined by lithology) and boulder slope influenced by mass movements. This is the only section undergoing parallel retreat. Lower concavity modified by rillwash and stream incision.

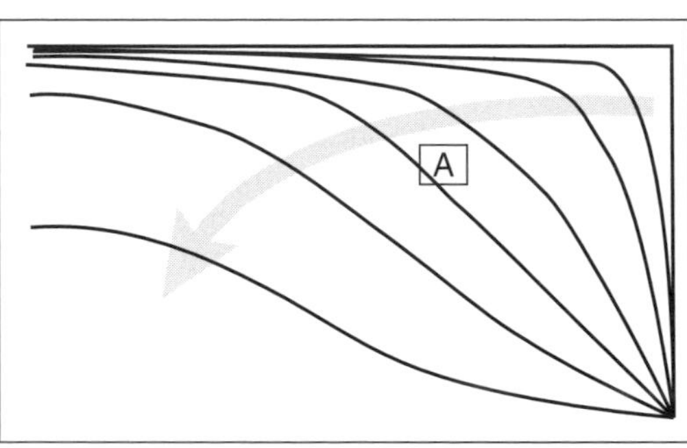

Humid-temperate slope evolution
Upper plateau influenced by creep or solifluction. Until A, lower slope dominated by mass movements, at reducing rate as slope towards a landslide-stable gradient. After A, slope dominated by creep and rillwash, declining towards base-level peneplain.

Figure 154 Slope evolution from a plateau with a steeply incised stream under semi-arid and humid-temperate conditions

This expression suggests that drainage density should be greater in steeper areas, and this forecast is supported by empirical evidence, particularly Dietrich and co-worker's data from California (Dietrich and Dunne 1993). Thus the spacing of streams, and so the whole scale of the landscape, is also set by the balance between slope and stream processes, and valleys occur where the processes driven by water flow begin to predominate over processes driven mainly by gradient, such as creep or rainsplash. This view of the landscape indicates not only that drainage density varies over the landscape in relation to steepness, but that it evolves through time as relief changes through erosional lowering or tectonic uplift.

References

Chorley, R.J., Beckinsale, R.P. and Dunn, A.J. (1973) *The History of the Study of Landforms or the Development of Geomorphology. Volume 2. The Life and Work of William Morris Davis*, London: Methuen.

Davis, W.M. (1909) *Geographical Essays*, Boston: Ginn; (1954), New York: Dover.

Dietrich, W.E. and Dunne, T. (1993) The Channel Head, in K. Beven and M.J. Kirkby (eds) *Channel Network Hydrology*, 175–219, Chichester: Wiley.

Gilbert, G.K. (1877) *Report on the Geology of the Henry Mountains*, Washington, DC: US Geological and Geographical Survey.

Hack, J.T. (1960) Interpretation of erosional topography in humid temperate regions, *American Journal of Science* 258A, 80–97.

King, L.C. (1953) Canons of landscape evolution, *Geological Society of America Bulletin* 64, 721–752.

Kirkby, M.J. (1971) Hillslope process-response models based on the continuity equation, *Institute of British Geographers Special Publication* 3.

Penck, W. (1953) *Morphological Analysis of Landforms*, trans. by H. Czech and K.C. Boswell, London: Macmillan.

Smith, T.R. and Bretherton, F.P. (1972) Stability and the conservation of mass in drainage basin evolution, *Water Resources Research* 8, 1,506–1,529.

MIKE KIRKBY

SLOPE STABILITY

Slope stability and its corollary slope instability, are defined as the propensity for a slope to undergo morphologically disruptive processes, especially landsliding (Plate 125) (see LANDSLIDE). Slow distributed forms of MASS MOVEMENT such as soil creep are generally not considered sufficiently disruptive to be included in this definition. From a hazard and engineering perspective, assessments of slope stability are focused on periods ranging from days to decades. However, slope stability may also be treated as a component of landform evolution and therefore its significance can only be judged by taking into account much longer periods of time.

In every slope, there are stresses which tend to promote downslope movement of material (shear stress) and opposing stresses which tend to resist movement (shear strength). In order to assess the degree of stability, these stresses can be calculated for a failure surface within the slope and compared to provide a FACTOR OF SAFETY (defined as the ratio of shear strength:shear stress). In a static slope, shear strength exceeds shear stress and the factor of safety is greater than 1.0 whereas for slopes on the point of movement shear strength is just balanced by shear stress and the factor of safety is 1.0.

While engineering codes of practice may specify a particular factor of safety for completed earthworks, it is not the most meaningful representation of slope stability. Two slopes that have the same factor of safety but large absolute differences in excess strength (i.e. strength minus shear stress) can be used to illustrate the limitations of the factor of safety. For example, the strength to stress ratios, in unspecified stress units for a slope (A) of 400/200 and for a slope (B) of 200/100 both yield a factor of safety of 2.0. However, slope (A) has an excess strength of 200 units while slope (B) has an excess strength of only 100 units. As excess strength is the quantity that must be entirely reduced (by reduction in strength or increase in shear stress) in order to produce failure, it represents the 'margin of stability' or inherent resistance to failure. Instability, however, is determined not only by the margin of stability of the existing slope but also by the magnitude of (external) destabilizing forces which may affect the slope to reduce that margin.

Plate 125 Actively unstable slopes, subject to deep-seated earthflows, Poverty Bay, New Zealand. Photo: Ministry of Works, New Zealand

Slopes can therefore be viewed as existing at various points along a stability spectrum ranging from high margins of stability with low probabilities of failure at one end to actively failing slopes, with no margin of stability, at the other. It is useful to define three theoretical stability states along this spectrum based on the ability of dynamic external forces to produce failure. First is the 'stable state', defined as slopes with a margin of stability which is sufficiently high to withstand the action of all dynamic destabilizing forces likely to be imposed under the current environmental/geomorphic regime. Second is the 'marginally stable state', represented by static slopes, not currently undergoing failure, but susceptible to failure at any time that dynamic external forces exceed a certain threshold. Third is the 'actively unstable state', represented by slopes with a margin of stability close to zero and which undergo continuous or intermittent movement. The margin of stability possessed by a slope is a measure of its landslide susceptibility and, together with the frequency and magnitude of dynamic destabilizing factors, provides a measure of probability of failure. In turn, the probability of failure together with its magnitude provides a measure of landslide hazard.

The concept of three stability states offers a useful framework for understanding the causes and development of instability. In this context four groups of destabilizing factors can be identified on the basis of function (Figure 155).

1 *Preconditions* (predisposing factors) are static, inherent factors which not only influence the margin of stability but more importantly in this context act as catalysts to allow other dynamic destabilizing factors to operate more

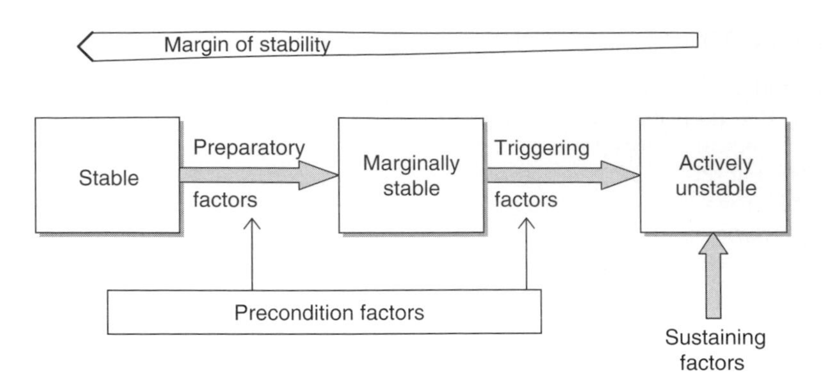

Figure 155 Stability states and destabilizing factors

effectively. For example, slope materials that lose strength more readily than others in the presence of water predispose the slope to failure during a rainstorm; or a particular orientation of rock structure may enhance the destabilizing effects of undercutting.

2 *Preparatory factors* are dynamic factors that by definition decrease the margin of stability in a slope over time without actually initiating movement. Hence, facilitated by preconditions, they are responsible for shifting a slope from a 'stable' to a 'marginally stable' state. Some factors, such as reduction in strength by weathering, climate change and tectonic uplift, operate over long periods of geomorphic time whereas others may be effective in shorter time periods e.g. slope oversteepening by erosional activity, deforestation, or slope disturbance by human activity.
3 *Triggering factors* are those factors which initiate movement, i.e. shift the slope from a 'marginally stable' to an 'actively unstable' state. The most common triggering factors are intense rainstorms, seismic shaking and slope undercutting.
4 *Sustaining* factors are those that dictate the behaviour of an 'actively unstable' slope e.g. duration, rate, and form, of movement.

As most slopes are stable for most of the time, the onset of slope instability from a geomorphic perspective represents an effective hillslope response to a destabilizing change in the boundary conditions, enabling a rapid adjustment and eventual return (or tendency toward) more persistent landscape forms. Thus, given sufficient time, the process of landsliding tends to stabilize a slope by reducing slope angle, height or weight, or by removing susceptible material.

The concept of slope instability may usefully be broadened to include any significant adjustment in processes or forms that tend to change the equilibrium conditions on a slope. For example, a slope where mature soils are being depleted by the onset of gullying can be considered to be undergoing a phase of slope instability. Whether that adjustment is viewed as a perturbation of a system in dynamic equilibrium or a change to another equilibrium state depends on the timescale being considered.

Further reading

Crozier, M.J. (1989) *Landslides: Causes, Consequences, and Environment*, London: Routledge.
Selby, M.J. (1993) *Hillslope Materials and Processes*, Oxford: Oxford University Press.

MICHAEL J. CROZIER

SLOPEWASH

The term 'slopewash' is not well defined in geomorphological literature. It has been used interchangeably with terms such as 'surface wash', 'rainwash', 'unconcentrated wash' and 'rillwash', now largely replaced by more precise terms such as 'overland flow', 'sheet wash', 'rainflow', 'rain-impacted flow', and 'rilling' or 'rill erosion'. The process was first described by McGee (1897), and defined by Bryan (1922: 29) as 'the water from rain, after it has fallen on the surface of the ground and before it has concentrated into definite streams'. Surface wash and rainwash were extensively invoked in early geomorphological literature as processes responsible for concave hillslope profiles, for hillslope profiles with a marked concave 'break', and for the formation of rock-cut pediments in southern Africa (King 1949). However, few field studies (e.g. Schumm 1956) attempted to measure the process, and it was not until Emmett's (1970) study of the hydraulics of overland flow on hillslopes that the complexity of the process was recognized. Subsequent studies on agricultural soils have identified the precise interacting processes involved, which include rainflow, sheet wash and rill erosion. The processes combined in slopewash are frequent on disturbed agricultural soils but their significance on natural hillslopes is less clear. The extensive thin sheets of water described by McGee are rare, even in dry regions where conditions are most favourable. In all areas, overland flow occurs most frequently on bare, relatively impermeable rock surfaces. Depending on the rock, such surfaces will usually yield some fine-grained material which can be transported by shallow flows, but the dominant transport in most cases is probably solution. Overland flow generated where rainfall exceeds surface infiltration capacity (Hortonian overland flow) occurs quite frequently on regolith-covered hillslopes, but is usually localized, reflecting patchy vegetation cover, microtopography, variations in rainfall intensity and marked variations in surface infiltration capacity. This produces a complex, varied interaction of processes along the hillslope, rather than the orderly transition from rainsplash to sheet wash to rill erosion envisaged in earlier literature. Instead, on most hillslopes patches of erosion, erosional lag deposits and sedimentation are interspersed in complex patterns. Often these are random and irregular, but on dryland hillslopes, the patterns are often regular, caused by the shrink–swell characteristics of smectite clays, the moisture requirements of shrubby plants, or selective transportation of material by pulsatory flows. These features may also occur where flow is not Hortonian, but generated as saturation excess, as return flow or as seepage. Such flows are more predictable and usually more uniform across the hillslope and therefore may more frequently cause profile concavity.

References

Bryan, K. (1922) Erosion and sedimentation in the Papago Country, Arizona, *United States Geological Survey Bulletin*, 730B, 19–20.

Emmett, W.W. (1970) The hydraulics of overland flow on hillslopes, *United States Geological Survey Professional Paper* 730B.

King, L.C. (1949) The pediment landform: some current problems, *Geological Magazine* 86, 245–250.

McGee, W.J. (1897) Sheetflood erosion, *Geological Society of America Bulletin* 8, 87–112.

Schumm, S.A. (1956) The role of creep and rainwash on the retreat of badland slopes, *American Journal of Science* 254, 693–706.

RORKE BRYAN

SLUSHFLOW

Slushflows, also called slush avalanches, are water-saturated snow masses flowing principally along a first-order stream channel (Larocque *et al.* 2001). Their formation is associated with an increase in the water content of a snowpack through rainfall, or through rapid snowmelt or through a combination of both. At a critical point instability occurs and snow mass is released. They are widespread in artic, subarctic and alpine environments but may, unlike snow avalanches (see AVALANCHE, SNOW), be initiated on relatively low angle slopes (Nyberg 1989). They are capable of transporting a high debris load over long distances, and also of causing substantial abrasion and erosion. The material they deposit can form such features as boulder tongues. They can also cause exceptional rates of glacial ablation (Smart *et al.* 2000) and pose a hazard to engineering structures.

References

Larocque, S.J., Hétu, B. and Filion, L. (2001) Geomorphic and dendroecological impacts of slushflows in Central Gaspé Peninsula (Québec, Canada), *Geografiska Annaler* 83A, 191–201.

Nyberg, R. (1989) Observations of slushflows and their geomorphological effects in the Swedish Mountain area, *Geografiska Annaler* 71A, 185–198.

Smart, C.C., Owens, I.F., Lawson, W. and Morris, A.L. (2000) Exceptional ablation arising from rainfall-induced slushflows: Brewster Glacier, New Zealand, *Hydrological Processes* 14, 1,045–1,052.

A.S. GOUDIE

SOIL CONSERVATION

Society can try to minimize SOIL EROSION by soil conservation practices, which may be 'active' or 'passive'. Active soil conservation includes positive actions to decrease erosion rates, such as terracing or contour farming. 'Passive' conservation is avoiding actions, such as ploughing on steep slopes or overgrazing erodible soil. Passive conservation is just as valid and often much cheaper than active conservation. We also need to distinguish between soil conservation and sediment control. Soil conservation is taking action to prevent erosion and keep soil *in situ*. Sediment control deals with soil which has already been eroded and transported, keeping it within fields or removing it from water courses.

Conservation strategies can be complex and varied. However, Morgan (1995) proposed that soil is conserved by decreasing EROSIVITY, decreasing ERODIBILITY, improving vegetation protection, or any combination of these. Ancient features, such as the terraced fields of South-East Asia, show that soil conservation has long been employed. However, soil conservation technologies were considerably improved by the US Soil Conservation Service.

The US Midwest suffered severe soil erosion in the 1930s or 'dirty thirties'. By this time, many of the organic soils had been cultivated for over eighty years, so soil organic contents were falling, thus increasing soil erodibility. Many erodible soils were brought into cultivation due to increased grain prices during the First World War. Then, in the 1930s, there was a severe drought. This combination of circumstances led to severe erosion, especially wind erosion, in the Great Plains States, which became labelled 'The Dust Bowl'. The social upheaval associated with these events was graphically portrayed in the novel *The Grapes of Wrath* (Steinbeck 1939), which tells the tragic struggle of a family from Oklahoma.

In response to these problems, an agricultural engineer, Hugh Hammond Bennett, led a campaign to promote soil conservation. Bennett was almost evangelical in his campaigning among farmers and politicians. During one presentation to the US Congress, a dust storm blew into Washington, DC. Bennett declared, pointing out the window, 'there, gentlemen, goes the State of Oklahoma!'. Congress then allocated funding for the foundation of the US Soil Conservation Service in 1934.

Various soil conservation techniques have been proposed and conservation projects can adopt one or a combination of techniques. The choice depends on many factors, including environmental (climate, topography and soil type) conditions and social, economic and political circumstances.

Windbreaks can be very effective in decreasing wind velocity and thus erosivity. They are usually aligned perpendicular to the direction of the most frequent erosive winds and are effective for 10–20 times their height downwind. Dense windbreaks greatly decrease wind velocity, but velocities soon increase in their lee. Better effects are achieved with more permeable windbreaks; velocity is not decreased so much, but their effectiveness in the lee is greater. For maximum benefit, windbreaks with a porosity of about 50 per cent and a mix of heights are recommended.

Hedgerows are effective windbreaks and their large-scale removal from the British countryside since the 1940s has contributed to increased wind erosion of arable soils. Hedgerows also protect against water erosion, by dividing slopes into shorter sections. Their removal allowed erosive runoff to operate over effectively longer slopes, thus increasing erosion risk. The restoration of all hedgerows is not feasible, as modern farm machinery cannot operate efficiently on fields as small as those which dominated the British countryside before the Second World War. However, it is important to identify 'key hedgerows', that protect areas exposed to predominant wind directions and convex slope sections susceptible to water erosion. Their retention or establishment should separate large catchment fields from lower convex–concave slopes. The absence of hedgerows integrates fields into geomorphological systems that are vulnerable to water erosion.

Slope management is a key component of soil conservation strategies. Simply leaving steeper erodible soils with a vegetation cover is a cheap, but effective, form of passive conservation. 'Set-aside', which involves taking areas out of crop production and leaving them with a permanent vegetative cover, was originally a means of decreasing grain surpluses. Carefully targeted on steeper erodible slopes, it could be an effective means of achieving both agricultural and environmental objectives.

Terracing is the most spectacular form of soil conservation and involves dividing slopes into a series of steps, cultivating the flatter sections and protecting the 'riser' with vegetation or masonry walls. In South-East Asia, terraces are extensively used to grow rice. To retain water, small earth walls or 'bunds' are built on the lower sides of terraces. However, terracing poses several problems. First, the risers take up about 10 per cent of land, though this is usually compensated by increased crop yield associated with increased soil water storage. Second, construction and maintenance are costly in terms of human resources; most of the world's areas of extensive terraces were constructed over many generations. Third, it is often difficult to operate machines efficiently on terraces.

Controlled colluviation is particularly applicable in semi-arid countries with a distinct rainy season. Lines of stones are laid out along the contour. When the seasonal rains arrive, soil is eroded and sediment deposited on the upslope side of the stones. Over a few rainy seasons, a fine silty moisture-retentive colluvial soil accumulates and, in semi-arid climates where soils tend to become saline, seasonal flushing with water can desalinize the COLLUVIUM, making it relatively fertile and suitable for crops. The technique is simple, cheap and thus affordable in poorer countries. Check dams operate in a similar way. Walls are constructed across gullies, sediment collects behind them and is periodically removed. Additionally, obstacles (such as straw bales and willow fences) are often placed in gullies to impede runoff and thus encourage sedimentation.

Contour farming involves orientating agricultural operations along the contour, rather than up-and-down slope and is particularly applicable on gentle uniform slopes used for mechanized agriculture (e.g. the Prairies and the Steppes). Complex slopes tend to limit the applicability of contour cultivation in northern Europe. However, slopes must not be too steep (>15°), as water can accumulate in furrows and eventually breach the ridges between them, causing even higher soil erosion rates than the more common up-and-down slope cultivation. Moreover, farm machinery cannot operate safely or efficiently along the contours on steep slopes.

In strip cropping systems, alternate strips of land are arranged perpendicular to the relevant erosive agent, wind or water. The crops themselves protect the soil, braking the velocity of the erosive agent and trapping sediments. Temporary grassland (leys) can form part of a strip cropping system and also increase soil organic matter, lowering soil erodibility. Usually, strips are 15–45 m wide, becoming wider as erosion risk increases.

Rotation is a well-established agronomic technique, by which different crops are grown in an established sequence. A temporary grass ley is usually an integral component of rotational systems, allowing natural recovery. Twentieth-century development and mass production of chemical fertilizers allowed continuous arable production, without temporary leys. Fertilizers provided ample supplies of macronutrients needed for crop production but, on many soils, allowed soil organic contents to decrease, so that erodibility increased. The increased incidence of erosion on arable soils in much of North America and Western Europe has been attributed to long-continued arable cultivation.

Addition of organic matter can decrease soil erodibility. The most common is farmyard manure (FYM) and there are many commercial organic manures. 'Green manures' are crops such as clover and mustard which grow quickly, producing much biomass, but rapidly decompose to increase the soil organic matter. In developing countries, human waste is used. This is usually transported and applied at night and is known as 'night soil'.

Mulching is the use of vegetative or other material on the soil surface, to simulate the protective effects of vegetation. Often, residues from the previous crop are applied, such as straw on wheat fields. Many studies have shown mulching is very effective, especially in the tropics and subtropics. However, in temperate environments, mulches can insulate the soil and prevent it warming in spring.

Hydromulching is particularly applicable to engineered slopes, such as road cuttings and construction sites. Mulch consists of a mixture of materials, such as fibre, straw, paper and shredded wood and is sprayed with seed, protecting the soil surface while seeds germinate and establish a protective vegetation cover. Hydromulching is expensive and used only in high value projects.

Geotextiles are cloths used to protect soil surfaces. Usually, they consist of biodegradable material, such as jute, have a coarse mesh and last for a short time, usually less than two years. This is long enough for seed mixtures to establish protective plant communities. There are also non-biodegradable geotextiles, which are used for permanent stabilization of, for example, channel banks.

Compaction affects the hydrological and thus erosional behaviour of soil, decreasing the size and interconnectivity of soil pores and impeding infiltration, so that runoff and erosion are increased. Increasing use of heavy farm machinery is exacerbating compaction problems. Compaction by animals also poses problems, as they produce fairly small compact hoof imprints. On wet soils, this weakens soil structure, and gives soil a 'puddled' appearance, a process known as poaching.

There are many methods of minimizing soil compaction. Trafficking wet soil should be avoided, though a shallow tillage tool mounted behind the tractor wheel can break up compacted soil. Tramlines, along which all passes for agricultural operations are made, limit compaction to narrow zones. Even within tramlines, it is possible to diminish compaction by using larger tyres or low-ground pressure vehicles (LGPV) to spread the load. Ploughing in very wet conditions and/or with blunt plough shares can compact subsoils. This 'plough pan' must then be disrupted by 'subsoiling', using a deep blade with a 'shoe-like' structure at the end, mounted behind a powerful tractor.

Several crop production methods (known variously as crop residue management systems, no-tillage, minimum cultivation, conservation tillage or direct drilling) minimize compaction, runoff and loss of organic matter by avoiding ploughing and preserving a cover of crop residues. The various terms partly reflect the diversity of systems in use. Residues from the previous crop are left on the soil surface, to simulate the protective effects of vegetation. Further crops are then sown into the residue with minimum disturbance and eventually provide a protective vegetative cover. These systems allow crop production on steeper slopes without increasing erosion risk. Increased soil organic matter also improves moisture retention and encourages earthworms, which increase infiltration rates. There are disadvantages, as lack of tillage can allow weed infestation, particularly by perennial grasses. As compaction can increase without tillage, especially on weakly structured silty soils with little organic matter, a crucial factor for success is soil faunal activity, especially earthworm activity. A rich earthworm population can effectively till the soil on behalf of the farmer. This is why conservation tillage systems became popular on organic-rich soils in both North and South America. There is a gradation from full 'no-till' to traditional ploughing. The selected method depends on the various advantages and disadvantages. A minimum cultivation system, with occasional full ploughing, can be an acceptable balance.

Chemical soil conditioners can decrease soil erodibility, binding particles together, improving aggregation and increasing infiltration rates. Conditioners, such as 'Krilium', 'Flotal' and 'Glotal', were developed in the 1950s and 1960s, and were followed by many ionic and non-ionic conditioners. Field and laboratory experiments showed they were effective, but high cost restricted their use to high value crops or specialist engineering applications (e.g. stabilizing road cuttings, engineered slopes, oil-well heads, temporary helipads and airfields).

Successful soil conservation is not simply an engineering problem, but is a complex amalgam of agronomic, social, economic and political considerations. It is essentially a team effort, requiring the participation of national, provincial and municipal government, farmers, scientists, extension workers and agricultural advisers. Also, it is essential that an effective dialogue develops between soil conservationists and local people, as their involvement, support and participatory agreement are crucial.

References

Morgan, R.P.C. (1995) *Soil Erosion and Conservation*, 2nd edition, London: Longman.

Steinbeck, J. (1939) *The Grapes of Wrath*, New York: Milestone Editions.

MICHAEL A. FULLEN AND JOHN A. CATT

SOIL CREEP

Slow MASS MOVEMENT processes are generally grouped together under the term soil creep (Kirkby 1967). A distinction is made between continuous creep, which is driven directly by downslope shear stresses, and seasonal creep which is the downslope component of movements which are either randomly directed, or primarily perpendicular to the soil surface.

Continuous creep may extend to depths of 5 m or more, and is driven by gravity shear stresses against the frictional and cohesive strength of the soil. The relationship between stress and strain for soils usually shows three zones. At low stress there is no movement; at high stress there is a more or less linear deformation, at a rate proportional to stress above an apparent failure threshold; at stresses below this threshold there is slow and non-linear increase in strain which bridges between these two behaviours (Figure 156). Continuous creep occurs in this non-linear transitional zone. Clearly continuous creep can occur over only a narrow range of stress–strain conditions, and is, in most cases, associated with the larger mass movements which occur when stresses appreciably exceed the threshold. In

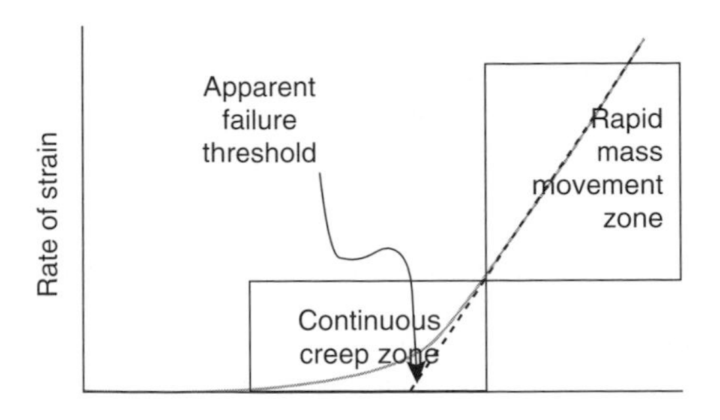

Figure 156 Stress–strain relationship in continuous creep

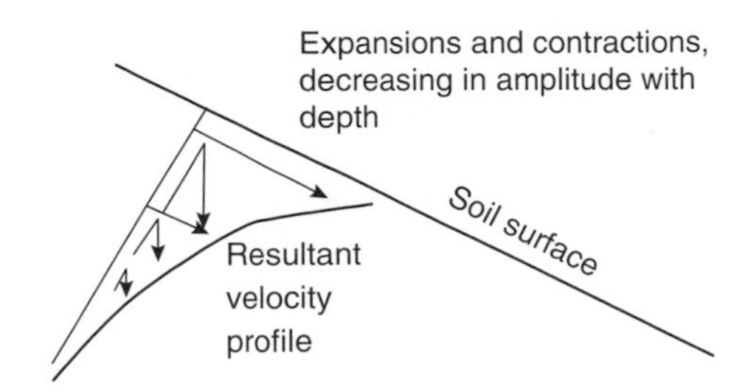

Figure 157 Zigzag seasonal creep movement

many soil materials, the creep itself changes the stress–strain relationship as the soil is reworked by movement.

Seasonal soil creep is driven by a number of processes, some of them biogenic and others driven by the climate. Biogenic movements are more or less random in direction, and include wedging by plant roots, movement of soil by burrowing animals such as gophers, earthworms and termites. Random movements lead to a net migration of material from zones of high soil concentration to zones of lower concentration, and may be considered as a form of diffusion.

The net direction of this random migration is towards the free upper surface of the soil, and it is eventually balanced by resettlement under gravity. The two main climate drivers are WETTING AND DRYING WEATHERING cycles and FREEZE–THAW CYCLEs. These cycles lead to an upward expansion or heave during wetting or freezing, followed by a downward settlement during drying or thawing. On a slope these expansions are at right angles to the slope surface, while settlement tends to occur more nearly along a vertical (Figure 157). Thus both random movements and heaves driven by climate produce a net movement which consists of zigzags which move material slowly downslope. To a first approximation the rate of this movement should be proportional to the slope gradient, but this theoretical inference has never been completely validated, because of the large observed variability in creep rates over small areas. All these seasonal soil creep processes are commonly referred to as *diffusive processes*, and have in common a linear dependence of transport rate on slope gradient, and little or no dependence on distance from divide or catchment area because they do not depend on the flow of water. This balance between outward diffusion and settlement under gravity not only moves material downslope but is also one of the main processes responsible for the observed increase in pore space towards the surface of uncultivated soils.

It can be seen that the rate of seasonal soil creep is limited by the depth and amplitude of these random and heave movements, which are always small. Thus there can never be a transition from seasonal soil creep to larger and more rapid mass movement, and it rarely extends to depths where the movement reworks material along potential mass movement failure surfaces. In different areas, the dominant processes reported to drive seasonal soil creep can be almost any of those listed above, according to the faunal and climatic activity in the soil. Seasonal soil creep rarely penetrates to depths of more than 30 cm, and surface translational movements reported generally lie in the range of 2–5 cm per year. If movement is proportional to gradient, then a widely quoted diffusivity for soil creep is $10^{-3}\,m^2\,yr^{-1}$. This means that on a 10 per cent slope, the actual transport will be $10\% \times 10^{-3}\,m^2\,yr^{-1}$, or 1 $cm^2\,yr^{-1}$. These values are low, of the same order of magnitude as those quoted for rainsplash (see RAINDROP IMPACT, SPLASH AND WASH), whereas the highest values reported for SOLIFLUCTION are 10–100 times greater. Both these processes are commonly included with soil creep in the category of diffusive processes, at least to a first approximation. Another important anthropogenic diffusive process is termed *tillage erosion*, in which ploughing and other forms of cultivation produce considerable net downslope movement of material, even if successive passes of the plough are along the contour, and alternately turn the soil up- and down-slope. Rates of tillage erosion are currently 10–1,000 times greater than for soil creep or rainsplash, and have increased greatly with the more widespread use of heavy farm machinery.

G.K. Gilbert (1909) recognized the importance of diffusive processes in creating convex hilltops, which are found on almost all soil-covered slopes. The convex area is usually relatively smooth, and may give way to an area where rills or larger channels begin, or may lead directly into an area where slopes become more uniform in gradient. The width of this convexity varies widely, primarily with climate and lithology, and generally lies in the range 1–1,000 m. Typically narrow convexities are associated with high DRAINAGE DENSITY and broad convexities with low drainage density. The relationship between diffusive processes and hillslope convexity is based on the assumption that transport by diffusive processes increases with gradient, and little or not at all with distance from the divide (or catchment area). If the area around the divide is eroding in the long term, then the volume of material which must be transported to remove the erosion products necessarily also increases continuously away from the divide. The diffusive processes can, by definition, only transport this increase in material through an increase in gradient. If the REGOLITH is deep enough not to constrain movement of material, it necessarily follows that gradient must also increase continuously from the divide, i.e. that the slope profile is convex.

This argument makes very few assumptions, so that the conclusion – that divides are convex – has great generality, and requires only that diffusive processes are dominant close to the divide. It is also fair to invert the argument, since this is a necessary and sufficient condition, and state that divide convexity demonstrates the dominance of diffusive processes close to the divide, and indeed roughly outlines the zone in which they are dominant. In the simplest realistic case, denudation of the area around the divide may be assumed to be constant, at rate T. At distance x from a ridge crest, an amount $T \cdot x$ must be exported per unit length of ridge crest to carry away the eroded material. If the rate of transport by creep processes $= K \cdot s$, where s is slope gradient and K is the rate constant, then to balance the erosion against the transport rate, $T \cdot x = K \cdot s$ or the slope must evolve until the gradient $s = T/K \cdot x$. In other words the gradient must eventually increase linearly with distance from the ridge crest.

It is useful to consider what happens if the assumptions of this argument are not met. First, on a long valley side, there may be deposition at the base of the slope, so the assumption of uniform erosion breaks down, and diffusive processes may be associated with concavity at the slope base. Second, if the regolith is thin over a resistant rock layer, then the K value may drop where the soil is thin, and rise again below the resistant layer, so that there is exaggerated convexity over the rock band, and a short concavity below, as the K value rises to its previous value.

At high gradients, there is some departure from linearity for creep and other diffusive processes. As the gradient approaches a threshold of stability (say 35° or 70 per cent), then material will roll downslope with minimal disturbance, and the rate of sediment transport increases rapidly with increasing gradient. Instead of a linear dependence on gradient, s, sediment transport can be estimated as:

$$\frac{K \cdot s}{(1 - s/s_*)^m}$$

where s_* is the threshold gradient, and the exponent $m = 1 - 2$. Applying the same argument for the slope profile, non-linear diffusion generates a summit convexity which, if there is sufficient relief, straightens out downwards towards a uniform slope at gradient s_*.

In the summit area where diffusive processes are dominant, SLOPE EVOLUTION creates a stable regime in which any minor irregularities in the hillslope surface tend to be obliterated rather than enlarged. Thus the divide area is generally smooth at a scale larger than that of microtopography, and any initial irregularities are progressively removed by the diffusive processes. This broad-scale smooth convex form may sometimes be observed on slopes where there is no soil, particularly on limestones in semi-arid and tropical climates. It may generally be inferred that the convexity was developed under a continuous soil cover, which has since been removed by erosion or washed into joints enlarged by solution.

Seasonal soil creep normally extends to depths of 30–50 cm, and the transport processes act at their full capacity provided that the regolith exceeds this depth. To maintain the rate of soil creep, WEATHERING processes must replace the soil lost by denudation, which is likely to be at about $100\,\mu m\,yr^{-1}$ for typical rates and slope forms, increasing with convexity and with the diffusive rate K. Weathering in this context refers to the physical breakdown of rock or *saprolite* (regolith which has been chemically altered *in situ*, retaining the original parent material structure), usually by biogenic or cryogenic processes. Dating of the cosmogenic (see COSMOGENIC DATING) isotope Beryllium-10, from the base of soils in California where bioturbation was by rodents (pocket gophers), Heimsath *et al.* (2001) showed a rate of weathering which decayed exponentially with soil depth, of approximately $280 \exp(-z/30)$ $\mu m\,yr^{-1}$ at the base of a soil of depth z (cm). To maintain equilibrium between this rate of biogenic weathering and a denudation rate of $100\,\mu m\,yr^{-1}$, the soil should therefore be approximately 31 cm deep. This equilibrium analysis correctly forecasts, for example, that, in comparing the gentle convexities typical of temperate landscapes with the much sharper and narrower convexities of semi-arid badlands, generally thinner regolith is developed on the semi-arid divides.

In summary, soil creep is defined as any slow mass movement. Continuous creep generally occurs in soils which are also subject to rapid mass movements, at shear stresses just below their failure threshold. Seasonal soil creep occurs in a wide range of soils, due to biogenic activity, freeze–thaw and wetting–drying, and the rate of sediment transport is usually assumed to increase linearly with slope gradient. Together with other diffusive processes, it is responsible for developing bulk density profiles and summit convexities. In equilibrium with weathering processes, seasonal creep also determines the depth of the regolith on convexities.

References

Gilbert, G.K. (1909) The convexity of hilltops, *Journal of Geology* 17, 344–350.

Heimsath, A.M., Dietrich, W.E., Nishizumi, K. and Finkel, R.C. (2001) Stochastic processes of soil production and transport: erosion rates, topographic variation and cosmogenic nuclides in the Oregon Coast range, *Earth Surface Processes and Landforms* 26, 531–552.

Kirkby, M.J. (1967) Measurement and theory of soil creep, *Journal of Geology* 75, 359–378.

MIKE KIRKBY

SOIL EROSION

Water and wind can erode, transport and eventually redeposit soils. The initial impact of raindrops can break soil aggregates into primary particles, by the translation of kinetic energy from

Plate 126 A simulated raindrop hitting a moist sandy surface. The kinetic energy of the impacting raindrop causes the formation of an impact crater, around which the water rebounds as a corona. The process occurs very quickly – this photograph was shot at 1:2,000 of a second

the drops to the soil aggregates, a process known as SLAKING (Plate 126). Due to the influence of gravity, more soil particles are splashed downslope than upslope, and detached particles are splashed further downslope. The cumulative effect is a net downslope transfer of soil particles, known as splash erosion.

When rainfall intensity exceeds soil infiltration capacity, runoff occurs. Initially, a sheet of water of fairly uniform thickness flowing over the surface causes quite uniform sheet erosion (see SHEET EROSION, SHEET FLOW, SHEET WASH). However, this state is unstable, as flowing water concentrates in surface depressions and incises into the soil. Where these channels are shallow, the process is RILL erosion. However, if rill erosion continues gullies develop.

The distinction between rill and GULLY erosion is problematic. Early guidelines from the US Soil Conservation Service stated a gully is too wide for a prairie dog to jump across! Later definitions stated a gully would need to be mechanically infilled for agricultural activities to proceed. Others argue that rills are incised into the topsoil (the A horizon), whereas gullies incise into the subsoil or parent material (the B or C horizons).

Eroded soil is eventually redeposited. When this occurs on slopes, the redeposited material is termed COLLUVIUM. It accumulates on concave sections of slopes or against obstructions, such as walls and hedges, as decreasing flow velocities decrease the transport capacity of the runoff water. Sediment can also enter water courses, where it is reworked to form ALLUVIUM. Sediments derived from farmland soils are often rich in agrochemicals, such as phosphate, nitrate and pesticides and can pollute water and damage aquatic ecosystems.

Wind erosion (AEOLIAN PROCESSES) can transport fine sediment considerable distances. For instance, at the Mauna Loa Observatory in Hawaii, the onset of the Chinese spring planting season results in increased dust fallout, even though Hawaii is 5,000 km east of China (Parrington *et al.* 1983). Finer particles, especially silt, are usually transported in suspension above the land surface; sand is usually transported by SALTATION (particles bouncing along the surface).

The severity of soil erosion varies markedly. Langbein and Schumm (1958) proposed a model relating water erosion to rainfall. In very dry climates, there is usually little water erosion, but a slight increase in rainfall often causes a large increase in erosion rates. The occasional rainstorms tend to be very intense convectional storms, which have considerable energy to cause erosion. Also, there is little protection from vegetation, which is sparse in semi-arid environments. In more humid climates, greater vegetation cover protects the soil surface from erosion.

The Langbein and Schumm model has been critically assessed in many parts of the world. Certainly, in the semi-arid tropics, erosion rates are very high. It is estimated that some 6,000 million tonnes of soil per year are washed off the croplands of India and that 1,600 million tonnes per year are transported by the River Ganges to the Bay of Bengal. The Mediterranean environment also has high rates because of the semi-arid climate and long human occupancy of the landscape, which has promoted deforestation (Plate 127).

Plate 127 The effect of a stone protecting soft silty sediments from splash erosion in the Tabernas Badlands of south-east Spain

There are areas where erosion rates should be low, according to the Langbein and Schumm model, but actually are high. This largely reflects human activities, particularly vegetation removal. For instance, destruction of tropical rainforests has greatly increased erosion rates. Rainforest soils contain little organic matter, unlike those of more temperate environments, and removal of vegetation can lead to further losses. Organic matter stabilizes soil structure, so its loss increases erosion risk.

Temperate continental interiors (Prairies of North America, Steppes of Central Asia, Pampas of Argentina) should have a natural vegetation cover of grassland. Under these conditions, the soils are thick, black organic Chernozems. Over the past 100–130 years many of these grasslands have been converted to arable use, particularly for cereal production. The soil organic content has decreased and the soils have become more ERODIBLE. Some 4,000 million tonnes of soil per year have been eroded from the continental USA since the 1930s. This would fill a freight train long enough to encircle the equator twenty-four times!

There is increasing evidence of soil erosion in northern and western Europe. In the Langbein and Schumm model, these areas should experience little erosion. However, the damage to soil structure imposed by increasingly mechanized and intensive agriculture, particularly since the 1940s, is believed to have increased erosion rates. Erosion directly related to cultivation is often termed tillage erosion.

Soil erosion can be described as 'a quiet crisis, one that is not widely perceived. Unlike earthquakes, volcanic eruptions or other natural disasters, this human-made disaster is unfolding gradually' (Brown 1984). Brown estimated that the maximum rate of soil formation is 2 tonnes of soil per hectare of land per year, but that the average soil erosion rate is about 20 $t\,ha^{-1}\,yr^{-1}$. The calculated global excess of soil erosion over formation is 25,730 million tonnes, roughly equivalent to the amount of topsoil in Australia's wheatlands. These estimates are very tentative, but they indicate the scale of the problem.

It is difficult to define what is 'acceptable' as a rate of soil erosion. 'Natural' or 'geologic' erosion is a natural process by which uplands are denuded over geological time. This is different from 'accelerated' erosion, where human activities are increasing erosion rates much above their background levels. The commonest activity is vegetation removal, exposing soils to erosion.

To define 'acceptable' soil erosion, many soil scientists suggest that the soil system should be in a state of dynamic equilibrium; that is, the rate of loss by erosion should not exceed the rate of soil formation. We know very little about rates of soil formation, except that it is a slow process. Soils form by the WEATHERING of material at the soil–parent material interface, or by deposition of sediment on the surface. The parent material is usually rock, but often includes soft sediments. It normally takes 1,000 years to weather 10 cm of material at the B/C horizon interface. By scraping the bedrock and increasing soil aeration and microbial activity, tillage can accelerate the process up to a maximum of about 10 cm in 100 years (i.e. $1\,\mathrm{mm\,yr^{-1}}$). If erosion does not exceed this value, the soil system is in a state of dynamic equilibrium. This value has been labelled a 'tolerable' or 'T value' and is used in planning SOIL CONSERVATION strategies.

The $2\,\mathrm{t\,ha^{-1}\,yr^{-1}}$ 'T value' may be too high, as it assumes an unlimited supply of weatherable material and ignores the complexity of soil-forming processes. Soils mature through time, usually incorporating organic matter into the topsoil. Erosion preferentially removes topsoil, often with serious implications for soil fertility. The topsoil is the 'seat' of most biological activity and contains most of the soil fauna, organic matter and nutrients, both natural and applied. Therefore, erosion involves more than the loss of physical components of the soil system. Often, the most fertile material is lost and any new soil formed at the base of the profile is much less fertile. The topsoil may be completely stripped away, deposited downslope and then less fertile subsoil or parent material deposited on top. This 'soil profile inversion' diminishes soil fertility. All criticisms suggest the $2\,\mathrm{t\,ha^{-1}\,yr^{-1}}$ 'T' value is too high. However, soil erosion rates can exceed it by orders of magnitude and, in extreme cases, completely remove the soil. In the long term, even low erosion rates can be damaging.

Numerous techniques exist for measuring soil losses. For water erosion a simple technique is to establish runoff plots. These are bounded on three sides (usually by wood, metal or plastic) and runoff and eroded sediment are collected at the downslope end. This approach was established in Germany by Ewald Wollny in the 1880s. Runoff and erosion rates are measured in precisely defined conditions of soil type, slope and vegetation cover. However, it has been criticized, mainly because plot boundaries interfere with erosion processes, for example impeding rill development. Runoff plots were nevertheless adopted by the US Soil Conservation Service, following its establishment in 1934. Standardized 0.01 acre plots were constructed in a range of agricultural environments, principally east of the Rockies and led to the development of the 'UNIVERSAL SOIL LOSS EQUATION' (USLE).

Since the USLE was introduced, many soil erosion equations and models have been developed, including the revised USLE (RUSLE) and the Water Erosion Predictive Equation (WEPP). The 'Erosion Productivity Impact Calculator' (EPIC) attempts to predict the long-term effects of erosion on soil properties and crop productivity, simulating erosion for up to hundreds of years. EUROSEM is an attempt to predict erosion rates in European conditions and LISEM is a model to predict erosion on LOESS soils (Plate 128).

Models are also used extensively in wind erosion research. The most notable example is the Bagnold Equation, first published in 1937 by Major R.A. Bagnold. During military service in

Plate 128 High sediment concentrations in the Yellow River of China. It is estimated that the river transports some 1,800 million tonnes of sediment per year to the Yellow Sea. Much of this sediment is entrained when the river flows through the Loess Plateau, an area of erodible soils derived from wind-blown silt

the British army in Libya, Bagnold studied the dynamics of sand dunes. In the equation, sediment transport is related to wind velocity:

$$Q = 1.5 \times 10^{-9} (V - Vt)^3$$

where Q = total sediment load ($t\,m^{-1}h^{-1}$), V = velocity at measuring height ($m\,s^{-1}$), and Vt = fluid threshold velocity for sand movement ($m\,s^{-1}$).

Vt is the wind velocity necessary to initiate effective sand transport. Once that velocity is exceeded, sediment transport increases as the cubic power of wind velocity, so slight increases in wind velocity above Vt cause marked increases in wind erosivity. The Bagnold Equation has formed the basis of other predictive equations, especially in North America. Morgan (1995) comprehensively reviewed erosion models.

There are many techniques to monitor erosion processes in the field and laboratory (De Ploey and Gabriels 1980). A simple technique is to insert erosion pins vertically into the soil, allowing accurate measurements of changing surface levels to show how much erosion (surface lowering) or accumulation (surface raising) has occurred. Simple traps, often referred to as Gerlach 'troughs', can be used in the field to collect sediment. The dynamics of sediment movement can be studied by 'tagging' soil particles with a tracer and using tracer movement to indicate soil movement. Tracers have included painted stones and soil particles, fluorescent dyes, radioactive isotopes and magnetic materials. The fallout of isotopes from nuclear explosions or accidents, such as Chernobyl, has been used to quantify erosion rates. The most widely used isotopes are Cs^{137} and Cs^{134}. These are positively charged and, as they fall to Earth, become attached to negatively charged clays and organic particles. As a site is eroded, it becomes depleted in the isotope so, comparing the radioisotope content of eroded soils with non-eroded soils, such as in a woodland, indicates approximate erosion rates. Another method is to investigate the distribution of fallout isotopes in colluvial deposits. Loughran (1989) summarized field methods to quantify erosion rates. Geomorphologists have also simulated soil erosion processes in the laboratory. However, there are difficulties with this approach, such as how realistically laboratory studies simulate field conditions.

Soil erosion is the product of many complex and interacting factors. For instance, in a laboratory study to assess erosion risk on fifty-five soil samples from the US Cornbelt, Wischmeier and Mannering (1969) found twenty-two soil and surface properties were necessary to explain 95 per cent of soil loss variance. However, Morgan (1995) offered a useful qualitative simplification, stating that soil erosion results from the dynamic interaction of:

1. The energy of the water or wind in causing erosion (EROSIVITY),
2. The inherent resistance of soil to detachment and transport (ERODIBILITY),
3. The protection factor of vegetation.

Many studies have attempted to relate rainfall erosivity to erosion rates. Rainfall erosivity is a function of its intensity and duration and the mass, diameter and velocity of raindrops. The energy of water flowing over the soil surface also affects erosion and is related to its velocity, volume, turbulence and shear stress (Plate 129). Slope angle, length and shape profoundly affect the erosivity of running water. Generally, as slope angle increases, so does erosion risk, because runoff velocity and energy increase. Usually, erodible soils on slopes >10° are particularly susceptible to rill erosion. Rills are hydraulically efficient systems for transporting soil and promote high erosion rates. Increased slope angle also increases the efficacy of splash erosion processes. On longer slopes, runoff has more time to accelerate and thus achieve greater erosivity. Slope

Plate 129 Tu Lin (the 'Soil Forest') in Yunnan Province, China. Highly erosive summer monsoonal rains have incised into soft Tertiary sediments. Human agency has also played a role in promoting erosion. For instance, in the top-centre note the cultivation of melons right at the edge of the gullied area

shape is also important, as runoff tends to accelerate rapidly over convex slope segments, achieving higher velocities and thus greater erosivity, but decelerates over concave slope sections, often leading to sediment deposition. Thus, the typical convex–concave morphology of slopes in humid temperate environments makes them particularly vulnerable to water erosion.

Soil erodibility is influenced by many factors, principally texture and soil organic content. The most erodible materials are silts and sands; which are not cohesive and can be transported by the flow rates characteristic of rills. Soils with low organic matter contents are highly erodible, particularly those with <2 per cent organic matter by weight. As organic matter increases above 2 per cent, soil erodibility decreases to a minimum at about 10 per cent organic matter, a typical value for a deciduous forest topsoil in a humid temperate environment, such as the British Isles. Soils with >20 per cent organic matter tend to be more erodible, as there is less clay to form aggregates with organic matter, and organic material is very light and easily transported. Organic particles have densities typically about $0.8\,g\,cm^{-3}$ compared with $>2.5\,g\,cm^{-3}$ for most mineral particles. Low density is the main reason why very organic soils, such as the peaty soils in the Fens of East Anglia, are susceptible to wind erosion.

Vegetation protects the soil from erosive forces by reducing, braking and filtering runoff. Many studies agree that >30 per cent plant cover protects the soil from erosion. Vegetation surfaces dissipate raindrop energy and prevent slaking. This is particularly true for short vegetation. A fall of about 8–9 m is required to achieve terminal velocity, so raindrops falling from tree canopies can regain their erosivity.

References

Brown, L.R. (1984) The global loss of topsoil, *Journal of Soil and Water Conservation* 39(3), 162–165.

De Ploey, J. and Gabriels, D. (1980) Measuring soil loss and experimental studies, in M.J. Kirkby and R.P.C. Morgan (eds) *Soil Erosion*, 63–108, Chichester: Wiley.

Langbein, W.B. and Schumm, S.A. (1958) Yield of sediment in relation to mean annual precipitation, *Transactions of the American Geophysical Union* 39, 257–266.

Loughran, R.J. (1989) The measurement of soil erosion, *Progress in Physical Geography* 13(2), 216–233.

Morgan, R.P.C. (1995) *Soil Erosion and Conservation*, 2nd edition, London: Longman.

Parrington, J.R., Zoler, W.H. and Aras, N.K. (1983) Asian dust: seasonal transport to the Hawaiian Islands, *Science* 8 April, 195–197.

Wischmeier, W.H. and Mannering, J.V. (1969) Relation of soil properties to its erodibility, *Soil Science Society of America Proceedings* 33, 131–137.

MICHAEL A. FULLEN AND JOHN A. CATT

SOIL GEOMORPHOLOGY

Soil geomorphology is the scientific study of the processes of landscape evolution and the influence that these processes have on soil formation and distribution on the landscape. Soil geomorphology provides a unique framework for an integrated Earth surface assessment. This discipline couples knowledge from soils, surficial deposits, stratigraphy and sedimentation, and parent material with the process-oriented approach of geomorphology in a three-dimensional framework. The soil geomorphology discipline in the United States has provided the primary foundation for interpreting the relation of soils and palaeosols to the landscape.

Soil geomorphology, an interdisciplinary construct, merges two scientific fields: pedology, the study of soils, and geomorphology, the study of landforms and surficial processes. Robert V. Ruhe was one of the first to both quantify landscape form and process in space and time and to integrate these concepts with soil science (e.g. Ruhe 1956, 1960, 1969). Olson (1989, 1997) describes the event chronology leading to the emergence of soil geomorphology as an independent discipline.

Soil geomorphology provides the framework for understanding the geomorphic history of a landscape. A soil-geomorphic approach requires three components: (1) knowledge of the surficial stratigraphy and parent material, (2) geomorphic surfaces defined in time and space and (3) correlation of soil patterns and properties to landscape features. These three components are assessed independently and their results subsequently integrated to produce a soil-geomorphic interpretation or soil-geomorphic model of the landscape and the processes leading to its evolution. A soil-geomorphic model can represent the landscape at defined scales from the hillslope to the continental.

Soil landscape models

A common ground is required to study and understand landscape evolution and the processes that shape the land surface and its soils. The foundation for our current understanding of soils and landscapes lies in the historic concepts of William Morris Davis and Walther Penck. The 'CYCLE OF EROSION' a time-dependent landscape evolution model (Davis 1899), describes landscape progression by downwearing through stages from youth through maturity to old age. In developing this model, Davis was heavily influenced by early theories of evolutionary biology and the concepts of contemporaries including John Wesley Powell and G.K. Gilbert. Slope reduction occurred by uniform downwearing in which geologic differences became insignificant with time as landscapes advanced through the cycle. Process was not a part of this model, a serious flaw. In contrast to the Davisian model, Penck (1924) emphasized backwearing and parallel slope retreat. Penck's model meshes well with our understanding of soil distribution on the landscape and provides a better foundation for soil geomorphology.

Although the Davis and Penck concepts form the basis for general landscape evolution models, they did not emphasize hillslope development in relation to soils. Milne (1936a, b) was one of the first to introduce the CATENA concept to illustrate soil patterns on a hillslope. Here, soil properties depend on topography and are repeated relative to each other from the hillslope summit to the adjacent valley floor. Milne's catena model recognized that the processes of erosion and deposition on the various hillslope positions directly affect the distribution of soil properties. Two variations were recognized: (1) all soils of a catena are formed in a single parent material and (2) soils of a catena are formed in two or more materials. Soils of a catena differ in case (1) because of 'drainage conditions, differential transport and deposition of eroded material and leaching, and translocation and redeposition of mobile chemical constituents' (Milne 1936a). Milne's statement implies that pedogenic processes together with hydrologic properties define the soil landscape and are not restricted to a point on the landscape. In the second case (2), a geologic factor for multiple-parent materials is included. Milne's catena model has evolved into a more limiting model known as a toposequence. Numerous studies have continued to follow Milne's catena model in evaluating soil landscape relations today. The importance of the catena concept to soil science was recognized quickly in the USA. In the USA Soil Survey today, the toposequence is in practice a hydrosequence, i.e. a series of soil profile colours used as indicators of water-table elevation changes along a hillslope.

Wood (1942) and King (1953) created 'the fully developed hillslope' model. Ruhe (1956, 1960, 1975) formulated a soil-geomorphic hillslope model that integrated soil properties with hillslope models and modified the Wood and King models by proposing hillslope elements: summit, shoulder, backslope, footslope and toeslope. These elements are now widely used to describe hillslope positions. Conacher and Dalrymple (1977) extended the two-dimensional hillslope approach to a drainage basin. This nine-unit model, based on form and geomorphic and pedogenic processes, attempted to integrate the components of a hillslope by considering material and water movement.

Landscape morphology

To fully evaluate landscapes and their relations to soils, an understanding of landscape morphology is important. Minimum parameters are gradient, aspect, and vertical and horizontal curvature. Ruhe (1975) described slope curvature using three components: slope gradient, slope length and slope width. A matrix of nine basic forms was used to represent changes in curvature. Huggett (1975) added surface flow lines to the basic slope shapes and Pennock and others (1987) combined hillslope elements and curvature to identify seven hillslope positions.

Soil landscapes and water movement

Water movement is one of the most important mechanisms in landscape evolution and soil development. Water movement is governed by a complex set of interrelated factors both internal (e.g. soil properties) and external (e.g. climate) to the soil-landscape system. Some of the geomorphic models above included water movement, with an emphasis on overland flow or vertical infiltration at a given hillslope position rather than flow through the landscape. This early emphasis reflected the influence of Horton's studies on OVERLAND FLOW in the 1930s. However, in addition to surface runoff and infiltration, throughflow and groundwater recharge are equally important paths for water flow on the landscape. In recent years, the importance of subsurface water movement and transport through the soil landscape has received more attention.

Subsurface-flow landscape models including flownet analysis have become important particularly in wetlands studies.

Soils and geomorphic surfaces

Soils and geomorphic surfaces are closely aligned. A geomorphic surface is that part of the landscape specifically defined in space and time with definite geographic boundaries (Ruhe 1956, 1969; Daniels *et al.* 1971). When a soil-geomorphic study is undertaken, the geomorphic surfaces are delineated based on geomorphic principles. Separately, soils in the study area are mapped. The results are compared and the linkages established. Soil boundaries do not necessarily coincide with geomorphic surface delineations and one or more soils may occur on the same geomorphic surface (Ruhe 1975). The critical point to the interpretations is that the pattern of soils and surfaces should be predictably repeated throughout a drainage basin and represents the landscape as it is at a given time. If three different soils occur on a geomorphic surface on one side slope, this soil complex should recur on similarly situated side slopes throughout the watershed. Geomorphic surfaces represent a time sequence but environment and other factors vary independently of one another through time. The latter phenomena affect the types of soils found on any given geomorphic surface. However, the soils will have a common degree of soil development. This relationship provides a valuable tool for understanding the chronological framework of landscape evolution in a soil geomorphic investigation.

Palaeosols and soil geomorphology

Palaeosols are soils formed on landscapes of the past. Ruhe (1975) defined three types: buried, exhumed and relict soils. Buried palaeosols are soils later covered by younger sediment or rock. Exhumed palaeosols are those buried but later re-exposed on the land surface, and relict palaeosols are those formed on a pre-existing landscape but never buried. As researchers began to demonstrate the close interdependence of soils and landscapes (e.g. Ruhe *et al.* 1967; Ruhe 1969), palaeosols also became indicators for understanding past landscapes. Often, buried soils are known as stratigraphic markers and are especially useful as keys to past environments (e.g. Follmer 1982).

Quantifying soil landscape models

Landscape models have shown that landscapes are predictable and have a large non-random variability component. This non-random component is useful in predicting soils on the landscape if the processes that govern landscape development are quantifiably described. Many new and exciting approaches seeking to capture and quantify soil landscape relations encompass geostatistical techniques, digital terrain models, fuzzy logic, neural networks and expert systems and produce visualizations and interactive, virtual immersive modules previously unavailable (e.g. <http://grunwald.ifas.ufl.edu> and <www.soils.wisc.edu/virtual_museum>).

References

Conacher, A.J. and Dalrymple, J.B. (1977) The nine unit landscape model: an approach to pedogeomorphic research, *Geoderma* 18, 1–154.

Daniels, R.B., Gamble, E.E. and Cady, J.G. (1971) The relation between geomorphology and soil morphology and genesis, *Advances in Agronomy* 23, 51–88.

Davis, W.M. (1899) The Geographical Cycle, *Geographical Journal* 14, 481–504.

Follmer, L.R. (1982) The geomorphology of the Sangamon surface: its spatial and temporal attributes, in C. Thorn (ed.) *Space and Time in Geomorphology*, 117–146, Boston: George Allen and Unwin.

Huggett, R.J. (1975) Soil landscape systems: a model of soil genesis, *Geoderma* 13, 1–22.

King, L.C. (1953) Canons of landscape evolution, *Geological Society of America Bulletin* 64, 721–752.

Milne, G. (1936a) A provisional soil map of East Africa, *Amani Memoirs* No. 28. Eastern African Agricultural Research Station, Tanganyika Territory.

——(1936b) Normal erosion as a factor in soil profile development, *Nature* 138, 548.

Olson, C.G. (1989) Soil geomorphic research and the importance of paleosol stratigraphy to Quaternary investigations, Midwestern USA, *Catena Supplement* 16, 129–142.

——(1997) Systematic soil-geomorphic investigations – contributions of R.V. Ruhe to pedologic interpretation, *Advances in GeoEcology* 29, 415–438.

Penck, W. (1924) *Die Morphologische Analyse* (*Morphological analysis of landforms*), trans. by K.C. Boswell and H. Czech (1953), New York: St Martin's Press.

Pennock, D.J., Zebarth, B.J. and E. deJong (1987) Landform classification and soil distribution in hummocky terrain, Saskatchewan, Canada, *Geoderma* 40, 297–315.

Ruhe, R.V. (1956) Geomorphic surfaces and the nature of soils, *Soil Science* 82, 441–445.

——(1960) Elements of the soil landscape, *Transactions of the 7th International Congress of Soil Science* 4, 165–170.

——(1969) *Quaternary Landscapes in Iowa*, Ames: Iowa State University Press.

Ruhe, R.V., (1975) *Geomorphology, Geomorphic Processes and Surficial Geology*, Boston: Houghton Mifflin.
Ruhe, R.V., Daniels, R.B. and Cady, J.G. (1967) Landscape evolution and soil formation in southwestern Iowa, *US Dept. of Agriculture Technical Bulletin* 1,349.
Wood, A. (1942) The development of hillside slopes, *Geological Association Proceedings* 53, 128–138.

Further reading

Birkeland, P.W. (1999) *Soils and Geomorphology*, 3rd edition, New York: Oxford University Press.
Carson, M.A. and Kirkby, M.J. (1972) *Hillslope Form and Process*, New York: Cambridge University Press.
Gerrard, A.J. (1981) *Soils and Landforms, An Integration of Geomorphology and Pedology*, Boston: George Allen and Unwin.
Gile, L.H., Hawley, J.W. and Grossman, R.B. (1981) Soils and geomorphology in the Basin and Range area of southern New Mexico – guidebook to the Desert Project, *New Mexico Bur. Mines and Mineral Resources Mem.* 39.
Hall, G.F. and Olson, C.G. (1991) Predicting variability of soils from landscape models, in M.J. Mausbach and L.P. Wilding (eds) *Spatial Variabilities of Soils and Landforms*, 9–24, Special Publication No. 28, Madison: Soil Science Society of America.
Kirkby, M.J. (ed.) (1978) *Hillslope Hydrology*, New York: Wiley.
Richardson, J.L. and Vepraskas, M.J. (eds) (2001) *Wetland Soils. Genesis, Hydrology, Landscapes, and Classification*, New York: Lewis Publishers.
Ruhe, R.V. (1965) Quaternary paleopedology, in H.E. Wright, Jr and D.G. Frey (eds) *The Quaternary of the United States*, 755–764, Princeton: Princeton University Press.
Ruhe, R.V. and Walker, P.H. (1968) Hillslope models and soil formation I. Open systems, *Transactions of the 9th International Congress of Soil Science* 4, 551–560.
Walker, P.H. and Ruhe, R.V. (1968) Hillslope models and soil formation. II. Closed systems, *Transactions of the 9th International Congress of Soil Science* 4, 561–568.
Wright, H.E. Jr, Coffin, B.A. and Aaseng, N.E. (eds) (1992) *The Patterned Peatlands of Minnesota*, Minneapolis: University of Minnesota Press.
Yaalon, D.H. (ed.) (1971) *Paleopedology: Origin, Nature, and Dating of Paleosols*, Jerusalem: Israel University Press.

SEE ALSO: hillslope, form; hillslope, process; Horton's Laws; hydrological geomorphology; palaeosol

CAROLYN G. OLSON

SOLIFLUCTION

Slow downslope movement of soil mass usually associated with FREEZE–THAW CYCLEs and FROST HEAVE, first termed to refer to the movement of saturated unfrozen soil in the Falkland Islands (Andersson 1906). MASS MOVEMENTs of soils in periglacial regions involve localized rapid soil failures, which result in active-layer detachment slides or flows, and more widely operating slow pre-failure movements (Ballantyne and Harris 1994: 114). The latter is referred to as solifluction as a collective term. In a broad sense, solifluction consists of frost creep (SOIL CREEP associated with freezing) resulting from nearly vertical settlement of soils heaved normal to the slope and gelifluction representing downslope displacement of ice-rich soils during thawing (Washburn 1979: 201). The two components often operate together, displacing soils downslope generally at rates of 0.5–10 cm a^{-1}. Where the ground is underlain by PERMAFROST, solifluction occurs within the ACTIVE LAYER and is distinguished from permafrost creep that originates from plastic deformation of frozen debris. The long-term operation of solifluction results in low-relief, gentle slopes having a number of lobes and sheets.

Processes

Frost creep is subdivided into needle-ice creep, diurnal frost creep and annual frost creep in terms of the time span of operation and the vertical extent of movement. Annual frost creep is often accompanied by gelifluction. In cold permafrost regions, a large part of annual frost creep and/or gelifluction may occur near the base of the active layer, producing a plug-like velocity profile.

Diurnal freeze–thaw cycles lead to either needle-ice creep or diurnal frost creep (Figure 158). The former occurs when nocturnal cooling to just below 0 °C is followed by daytime thawing. Only the uppermost few grains are heaved by ice needles and resettled on thawing by toppling or rotational movement. The resulting downslope movement is superficial but relatively rapid, accelerating approximately with the second power of the slope gradient.

More intensive nocturnal freezing can produce ice lenses at a depth of a few centimetres, which heave the overlying soil layer normal to the slope. On thawing, the heaved soil resettles vertically or, in reality, with some upslope component due to cohesion. Such a freeze–thaw alternation induces diurnal frost creep, the movement of which is proportional to the slope gradient and shallower than 10 cm. The velocity profile of diurnal frost creep is typically concave downslope, in response to the increasing number of freeze–thaw cycles

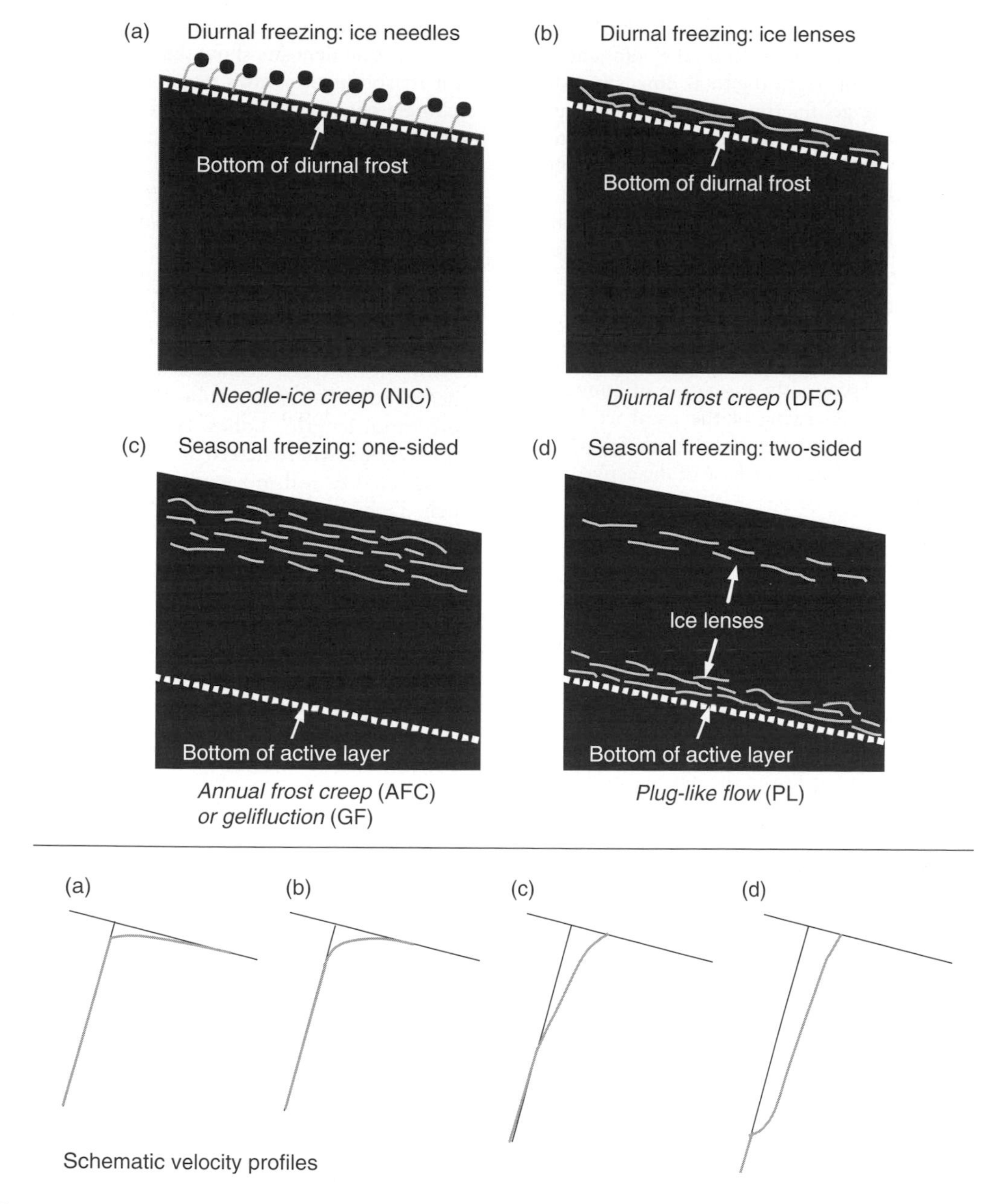

Figure 158 Types of solifluction
Source: Matsuoka (2001)

towards the ground surface. Where needle-ice creep dominates, a gap in velocity occurs between the uppermost grains and the underlying soil (Matsuoka 2001).

Annual freeze–thaw cycles cause frost heave of soils either by one-sided freezing or by two-sided freezing. In regions subjected to seasonal frost or underlain by warm permafrost, one-sided freezing from the ground surface in winter often produces ice lenses within the uppermost few decimetres of soil. During seasonal thawing, the resettlement of the heaved soil results in annual frost creep and/or gelifluction. Reflecting the depth of the ice lenses, the base of movement typically lies at 30–50 cm depth. The relative contribution of gelifluction is estimated by subtracting the potential frost creep, which is the product of the amount of frost heave and the tangent of the slope gradient, from the total downslope displacement. The gelifluction component increases with the silt plus clay content, because excess pore water during thawing reduces frictional strength in the soil (Harris 1996).

Where the permafrost temperature is low, winter freeze-back of the active layer may progress both downward from the top and upward from the base. Such two-sided freezing may produce ice lenses both near the surface and near the permafrost table. The high moisture availability at the latter location favours the formation of numerous thick ice lenses. Thawing of the basal ice lenses induces plug-like flow that shows a velocity profile convex downslope near the base of the active layer (Mackay 1981).

Rates

Field measurements in a wide range of cold regions indicate that rates of solifluction, expressed by either the surface velocity or the volumetric velocity, significantly vary with climate (Matsuoka 2001). In the polar, cold permafrost regions, where the mean annual air temperature (MAAT) is $-6\,°C$ or lower, the paucity of diurnal frost creep restrains the surface velocity to below $5\,cm\,a^{-1}$, while plug-like flow allows movement of the soil mass deeper than 50 cm. In the alpine, shallow seasonal frost regions, the predominance of diurnal frost creep (including needle-ice creep) raises the surface velocity up to $100\,cm\,a^{-1}$, whereas the soil movement is confined largely within the uppermost decimetre; as a result, the volumetric velocity is very low despite the high surface velocity. The volumetric velocity reaches a maximum in regions with warm permafrost or deep seasonal frost (MAAT between -6 and $0\,°C$), which are located in both subpolar and alpine settings, because diurnal and annual processes combine to dislocate a 30–50-cm thick soil mass with moderate surface velocities.

The slope gradient is another significant control on solifluction rates. Increasing rates with inclination have been reported from a number of polar slopes. In lower latitude alpine areas, however, other factors like soil frost susceptibility, moisture distribution and freeze–thaw frequency obscure the overall dependence of solifluction rates on inclination (Harris 1981: 123–125). Gelifluction can occur on gradients as low as 1°.

Radiocarbon ages of organic materials buried by solifluction deposits show that long-term variation in solifluction rates generally corresponds to climate change during the Holocene, although the responsible climatic factors are not unequivocal (Matthews *et al.* 1993).

Landforms

Solifluction produces characteristic surface features involving lobes, terraces and sheets. Where vegetation is sparse or absent, sorted stripes (a kind of PATTERNED GROUND) often develop on the tread of these features. The most widespread features are lobes 2–50 m in both width and length. Lobes typically occur in alpine regions where heterogeneous surface conditions localize soil movement, whereas sheets, which lack lateral margins, dominate polar slopes that experience more uniform movement (Plate 130). The downslope edge of these features terminates in a steep riser 0.2–2 m in height, where vegetation, coarse debris or downslope declination acts as a brake on soil movement. As a result, the riser is commonly turf- or stone-banked (Benedict 1970). Terraces may develop under the interaction between wind, vegetation and soil movement (Ballantyne and Harris 1994: 261–267). Isolated boulders on slopes subject to solifluction often move as PLOUGHING BLOCKS AND BOULDERS.

The height of lobes reflects the maximum depth of soil movement. The predominance of shallow diurnal frost creep results in small lobes about 0.2 m in height. Such lobes occur mainly

Plate 130 Turf-banked solifluction lobes on a limestone slope, Swiss National Park (2,400 m ASL)

on alpine near-crest locations where thin REGOLITH can only respond to diurnal freeze–thaw action (Matsuoka 2001) or on tropical high mountain slopes where the lack of seasonal variation in temperature highlights the effect of diurnal freeze–thaw action (Bertran *et al.* 1995). Larger and higher lobes develop where solifluction originates mainly from annual freeze–thaw action. The length of lobes depends partly on the time of process operation. Well-developed lobes require several thousand years of activity (Matthews *et al.* 1993).

Numerical simulations suggest that long-term predominance of solifluction, accompanied by subsurface debris production and comminution due mainly to frost weathering, eventually leads to gentle convex-upward slopes, often referred to as CRYOPLANATION surfaces, near mountain crests with possible occurrence of summit TORS (Anderson 2002).

References

Anderson, R. (2002) Modeling the tor-dotted crests, bedrock edges, and parabolic profiles of high alpine surfaces of the Wind River Range, Wyoming, *Geomorphology* 46, 35–58.

Andersson, J.G. (1906) Solifluction, a component of subaerial denudation, *Journal of Geology* 14, 91–112.

Ballantyne, C.K. and Harris, C. (1994) *The Periglaciation of Great Britain*, Cambridge: Cambridge University Press.

Benedict, J.B. (1970) Downslope soil movement in a Colorado alpine region: rates, processes, and climatic significance, *Arctic and Alpine Research* 2, 165–226.

Bertran, P., Francou, B. and Texier, J.P. (1995) Stratified slope deposits: the stone-banked sheets and lobes model, in O. Slaymaker (ed.) *Steepland Geomorphology*, 147–169, Chichester: Wiley.

Harris, C. (1981) *Periglacial Mass-Wasting: A Review of Research*, BGRG Research Monograph 4, Norwich: Geo Abstracts.

——(1996) Physical modelling of periglacial solifluction: review and future strategy, *Permafrost and Periglacial Processes* 7, 349–360.

Mackay, J.R. (1981) Active layer slope movement in a continuous permafrost environment, Garry Island, Northwest Territories, Canada, *Canadian Journal of Earth Sciences* 18, 1,666–1,680.

Matsuoka, N. (2001) Solifluction rates, processes and landforms: a global review, *Earth-Science Reviews* 35, 107–134.

Matthews, J.A., Ballantyne, C.K., Harris, C. and McCarroll, D. (1993) Solifluction and climatic variation in the Holocene: discussion and synthesis, in B. Frenzel (ed.) *Solifluction and Climatic Variation in the Holocene*, 339–361, Stuttgart: Gustav Fisher Verlag.

Washburn, A.L. (1979) *Geocryology: A Survey of Periglacial Processes and Environments*, London: Edward Arnold.

SEE ALSO: freeze–thaw cycle; frost heave; periglacial geomorphology; mass movement

NORIKAZU MATSUOKA

SOLUBILITY

Minerals, and the chemical elements which constitute them, vary in the degree to which they dissolve in water. During the dissolution process, the constituents of the chemical compounds in minerals split up, or dissociate, into water. It follows that the solubility of such compounds can be assessed by measuring the concentrations of the constituent elements in the water after a period of time. Minerals generally display increased solubility with higher temperatures and with increasing acidity, notable exceptions to the latter being silica (as quartz) which increases in solubility above pH 10 and aluminium (as gibbsite or kaolinite) which is least soluble at around pH 6 but increases both as pH rises and falls from this value. The solubility of gypsum (as measured by Ca in solution) does not vary with pH. For gypsum, the rate of water flow is a governing factor.

Maximum concentrations are reached asymptotically over time but nevertheless some practical measures of solubility can be found. The Nernst equation describes the dissolution trend over time:

$$C = C_{max}\,(1 - e^{-kt})$$

where C is concentration at time t, C_{max} = maximum concentration and k is a rate constant for each solute.

Highly soluble compounds have a high value of k and exhibit steep rise in concentration over time. The process of dissolution involves the redistribution of energy – and for dissolution to occur there must be a decrease in free energy – and if a solute and a solvent are composed of similar molecules in terms of structure and electrical properties, then high solubility is favoured (see further Davidson 1978: 83–85).

Water flow rate is also important in determining the solubility of minerals in the field and there are also many other factors which limit the transfer of *in vitro* observations of solubility to the field (Casey *et al.* 1995). For water flow, the rate of dissolution of the more rapidly dissolving minerals becomes transport-limited as the products of dissolution readily accumulate and further dissolution is then only facilitated by the removal of weathering products and the arrival of further chemically unsaturated water. For slowly dissolving minerals, the rate of dissolution tends to be slow as compared to the flow of water and so the overall solutional losses tend to become rate-limited. Thus water flow rates, and, for bare surfaces, duration of rainfall and thus time of wetting, become important controls on solution rates.

Additionally, laboratory measurements of solubility are usually made under controlled conditions, whereas field conditions tend to be variable and possess different combinations of factors which are in practice difficult to separate, like freezing, thawing, salt and biological weathering as well as dissolution (Trudgill and Viles 1998).

References

Casey, W.H., Banfield, J.F., Westrich, H.R. and McLaughlin, L. (1995) What do dissolution experiments tell us about natural weathering? *Chemical Geology* 105, 1–15.

Davidson, D. (1978) *Science for Physical Geographers*, London: Arnold.

Trudgill, S.T. and Viles, H.A. (1998) Field and laboratory approaches to limestone weathering, *Quarterly Journal of Engineering Geology* 31, 333–341.

STEVE TRUDGILL

SOLUTE LOAD AND RATING CURVE

All natural waters contain organic and inorganic material in solution; these are called solutes. Spatial and temporal variation in solute content has been studied as a means of investigating CHEMICAL WEATHERING processes, rates of chemical denudation, evaluating nutrient cycling and elucidating the variable pathways followed by water through the drainage basin. The major solutes of interest are calcium, magnesium, sodium, potassium, chloride, bicarbonates, sulphate, nitrate and silica.

According to the classification of Gibbs (1970), the major natural mechanisms controlling world surface water chemistry are: (a) atmospheric precipitation, both composition and amount; (b) rock weathering; and (c) evaporation and fractional crystallization. When he plotted total dissolved solids (t.d.s.) in milligrams per litre against the weight ratio of sodium to sodium plus calcium for the major rivers of the world, he showed three domains: (1) high t.d.s. and high Na to Na + Ca ratio, where the evaporation/crystallization processes are dominant; (2) average t.d.s. and low Na to Na + Ca ratio; and (3) low t.d.s. and high Na to Na + Ca ratio. From this classification, climate and hydrology are the dominant controls of solute concentration at high and low t.d.s. values and lithology becomes dominant at average t.d.s. values. Meybeck (1988) has provided a more detailed breakdown of river water solutes according to climate, hydrology and relief, on the assumption that the residence time of water in a basin will depend strongly on relief.

Meybeck (1987) estimated the global average chemical denudation rate for crystalline igneous and metamorphic rocks and sandstones and shales at 18–19 tonnes/km^2/year and volcanic rocks at 1.5 times higher. Denudation rate for carbonate rocks is 100 and for evaporites is 423 tonnes/km^2/year. This is in the context of a world average of 42.

The effect of relief on chemical denudation rates varies with geology as carbonate rocks dissolve easily regardless of local topography. No effect of hillslope steepness or extent of recent glaciation has been detected on chemical weathering fluxes in small granitoid watersheds. This implies that physical erosion rates are not critical in influencing chemical weathering of silicates and, by further implication, geology and climate are probably more important than relief on a global basis.

There are further factors that are important. Of these, the anthropogenic and the biotic factors are the most crucial. Vegetation increases chemical weathering by supplying carbon dioxide and organic acids to the soil and it increases water contact time with minerals in the soil by retaining moisture and by locally accelerating water recycling via evapotranspiration-enhanced rainfall. Vegetation also affects weathering by stabilizing soil against erosion and thereby increasing weathering rates in regions of high physical erosion (Berner and Berner 1996). Nutrients, organic carbon and dissolved trace elements will be most affected by anthropogenic and biotic controls, and will, at times, vary quite independently from the topographic, geologic, climatic and hydrologic factors.

The most geomorphically relevant ways of analysing solute data depend on the scale of interest. At individual site and slope scale, diffusive and convective equations have been used to model variations in solute flux (e.g. Carson and Kirkby 1972; Berner 1978), but it is probable that the buried tablet technique (Trudgill 1977) will provide the most reliable relative weathering rate information. At river basin scale, input–output budgets and solute budgets, combined with variable runoff source analysis, are the most popular (e.g. Zeman and Slaymaker 1978; Laudon and Slaymaker 1997). At global scale, Meybeck (1982, 1988) has made major contributions through the use of solute budgets.

Precipitation inputs to the land surface contain solutes in dry and wet fallout. The magnitude and composition of this solute content varies with distance from the ocean. As water moves through the vegetation canopy, the soil and the rock of the drainage basin, the solute concentration and composition will change. Solute concentrations within the soil are influenced by precipitation, interactions with the soil matrix, release of solutes through chemical weathering and biotic uptake and release of nutrients.

The solute content of stream flow will therefore reflect the characteristics of the upstream basin, including its geology, topography, vegetation cover and the variable pathways and residence time associated with water movement through the basin. Concentrations will vary with time in response to hydrologic conditions and will frequently exhibit a dilution effect during storm runoff events.

The precise hydrologic pathway has important implications for residence time and hence on solute enrichment. Identification of the pathway gives an understanding of the magnitude and timing of solute fluxes from different hydrologic reservoirs in the landscape and is therefore essential for the understanding of variations of the stream water chemistry. Hydrograph separation of rain-driven storm flow into its storm and pre-storm components can be carried out by collecting stream water samples for stable isotope analysis before, during and after a storm runoff event. Four assumptions are made: (1) rain isotopic content can be characterized by a single isotopic value; (2) pre-storm component, ground water and vadose water can be characterized by the base flow with a single isotopic value; (3) isotopic content of precipitation will be significantly different from pre-storm runoff values; and (4) contribution of stored surface water to the stream is negligible. Under certain hydrologic conditions, alternative hydrologic tracers (such as silica and electrical conductivity) can be used. The specific advantage of electrical conductivity is that it can be continuously monitored and stored in dataloggers.

Measurements of the solute input into a drainage basin and the output in streamflow provide a means of establishing a solute budget for the basin. The net solute yield (output–input) reflects the production of solutes within the basin. This production may be related to chemical weathering, the uptake of carbon dioxide by weathering reactions and the mineralization of organic material. On a global basis, approximately 50 per cent of the solutes found in river water are the products of chemical weathering, but this value is highly variable depending on lithology.

Rating curves are used to describe relations between solute transport and water discharge. If the rating curve is stable, the water discharge can then be used to predict solute concentration and load. The characteristics of these plots, including slope, degree of scatter and intercept are frequently used to characterize the solute response of a drainage basin.

References

Berner, E.K. and Berner, R.A. (1996) *Global Environments*, Upper Saddle River, NJ: Prentice Hall.

Berner, R.A. (1978) Rate control of mineral dissolution under Earth surface conditions, *American Journal of Science* 278, 1,235–1,252.

Carson, M.A. and Kirkby, M.J. (1972) *Hill Slope Form and Process*, Cambridge: Cambridge University Press.

Gibbs, R.J. (1970) Mechanisms controlling world water chemistry, *Science* 170, 1,088–1,090.

Laudon, H. and Slaymaker, O. (1997) Hydrograph separation using stable isotopes, silica and electrical conductivity: an alpine example, *Journal of Hydrology* 201, 82–101.

Meybeck, M. (1982) Carbon, nitrogen and phosphorus transport by world rivers, *American Journal of Science* 282, 401–450.

——(1987) Global chemical weathering of surficial rocks estimated from river dissolved loads, *American Journal of Science* 287, 401–428.

——(1988) How to establish and use world budgets of riverine materials, in A. Lerman and M. Meybeck (eds) *Physical and Chemical Weathering in Geochemical Cycles*, Dordrecht: Kluwer.

Trudgill, S.T. (1977) Problems in the estimation of short-term variations in limestone erosion processes, *Earth Surface Processes and Landforms* 2, 251–256.

Zeman, L.J. and Slaymaker, O. (1978) Mass balance model for calculation of ionic input loads in atmospheric fallout and discharge from a mountainous basin, *Hydrological Sciences Bulletin* 23, 103–117.

Further reading

Caine, N. (1992) Spatial patterns of geochemical denudation in a Colorado alpine environment, in J.C. Dixon

and A.D. Abrahams (eds) *Periglacial Geomorphology*, 63–88, Chichester: Wiley.
De Boer, D.H. and Campbell, I.A. (1990) Runoff chemistry as an indicator of runoff sources and routing in semi-arid badland drainage basins, *Journal of Hydrology* 121, 379–394.
Paces, T. (1986) Rates of weathering and erosion derived from mass balance in small drainage basins, in S.M. Coleman and D.P. Dethier (eds) *Rates of Chemical Weathering of Rocks and Minerals*, 531–550, Orlando: Academic Press.
Velbel, M.A. (1986) The mathematical basis for determining rate of geochemical and geomorphic processes in small forested watersheds by mass balance, in S.M. Coleman and D.P. Dethier (eds) *Rates of Chemical Weathering of Rocks and Minerals*, 439–451, Orlando: Academic Press.

SEE ALSO: chemical denudation; denudation

OLAV SLAYMAKER

SPALLING

Spalling is the peeling off of platy fragments from the surface of rock. The resulting 'spalls' vary from a few centimetres to several metres in scale but their thickness is usually from 1–5 cm. The under surface of spalled fragments may be very irregular. Spalling can be transitional with EXFOLIATION or SHEETING though the latter usually occurs on a much larger scale. The term 'flaking' is also sometimes used.

Spalling can be attributed to several processes. Fundamentally, differential stresses in the outer layer of rock cause separation. The source of this differential stress may be from the growth of salt or ice crystals. Chemical change may be a source of differential stress as secondary minerals are precipitated. These occupy a greater volume and therefore may exert an outwards force on the rock surface. Expansion and contraction due to thermal change (e.g. forest fires or insolation) may also produce sufficient differential stress (Gray 1965). Change of internal stress equilibrium (e.g. due to erosion) may cause separation of plates from an intact rock mass. Spalling may be accompanied by surface alteration or by CASE HARDENING.

Reference

Gray, W.M. (1965) Surface spalling by thermal stresses in rocks, *Proceedings of the Rock Mechanics Symposium, Toronto*, Department of Mines and Technical Surveys, Ottawa, 85–106.

Further reading

Ollier, C. (1984) *Weathering*, London: Longman.
Yatsu, E. (1988) *The Nature of Weathering: An Introduction*, Tokyo: Sozosha.

DAWN T. NICHOLSON

SPELEOTHEM

'Speleothem' (Greek: *spele* – cave, *them* – make) – a general term for minerals precipitated in CAVES. More than 250 different minerals are known (Hill and Forti 1997). Calcite deposited in limestone caves is overwhelmingly predominant, accumulating in both air-filled (vadose) and water-filled (phreatic) conditions. 'Travertine' and 'sinter' are alternative terms, or 'tufa' (see TUFA AND TRAVERTINE) when precipitated on organic frameworks. Aragonite is second in abundance. Gypsum is third, found in gypsum caves and also in limestone caves with gypsum interbeds or where H_2SO_4 can react with the rock. Hydrated carbonates and sulphates (e.g. hydromagnesite, epsomite) are quite common, usually as pastes or powders in small amounts. Other minerals are more localized, associated with particular source conditions in bedrocks or clastic fillings; they include native sulphur, many oxides and hydroxides, halides, nitrates, phosphates, silicates, vanadates and a few organic minerals. There is perennial ice in many cold caves.

$CaCO_3$ (calcite and aragonite) speleothems

Calcite occurs mostly as coarse crystals with c axes oriented to growth ('length fast') or in microcrystalline form with c axes oriented across the direction of growth ('length slow'; see Railsback 2000). Many vadose speleothems display alternations of coarser and finer crystals, often with temporary cessations due to drying or dissolution, zones with dust, mud or organic grains: existing crystals may grow through these or new crystals form upon them. Subaqueous deposits are more homogeneous. Pure calcite is translucent, or opaque white due to fluid inclusions. Colour banding (yellow, brown, red-brown) is common, due chiefly to incorporation of fulvic acid chromophores from soil waters (van Beynen *et al.* 2001). Metals such as iron (red), copper (blue) and nickel (bright green) also provide colour, but rarely in visible concentrations (Cabrol and Mangin 2000).

Aragonite generally occurs as needle-like clusters or massive stalagmitic aggregates. Its deposition instead of calcite is attributed to enrichment of Mg^{2+} ions in the feedwater, due to presence of dolomite ($CaMg{\cdot}2CO_3$) or to evaporative effects. Aragonite speleothems are more common in warmer climates, e.g. wet–dry seasonal alternations of calcite and aragonite are reported in Botswana.

Vadose speleothem shapes are created by gravity, or by growth and capillary forces. Principal gravity types are dripstones (stalactites, stalagmites), and flowstone sheets on floors and/or walls. A 'column' is a stalactite–stalagmite pair grown together.

The fundamental form is the 'straw' stalactite, a monolayer crystal sheath enclosing a feedwater canal and growing downwards only. Leakage from the canal may overplate the sheath, creating tapered (carrot-like) stalactites up to one metre in diameter and several in length. Accelerated deposition on protruberances can add a myriad of subsidiary forms such as crenulations, corbels, drapes and lesser stalactites. 'Curtains' grow downwards where feedwater trickles down a sloping ceiling.

The most simple stalagmite is a 'candlestick' adding all new growth at the top under a nearly constant drip. Varying drip or greater fall height causes terraced or corbelled thickening; at the extreme the form is like a pile of soup plates. More common are conical or tapered forms, broadening into domes with flowstone sheets around them. Some stalagmites are >30 m high and domes may be 50 m or more in diameter.

Flowstones are deposited from film flow and accrete roughly parallel to the host surface. They may extend tens to hundreds of metres downstream of their sources and accumulate to thicknesses of several metres. 'Gours' or 'rimstones' are dams building upwards from irregularities in stream channels or on flowstone surfaces. The greatest impound water to depths of several metres. Rims are often strikingly crenulated.

'Helictites' or 'excentrics' grow where crystal or capillary forces predominate, skewing c-axes to create narrow, curvilinear tubes extending out, up and down from rock or parent stalactites, etc. Most are short, <10–20 cm in length. Dense clustering can form tangled masses like the Medusa's hair. 'Anemolites' grow upwind into prevailing drafts. Clusters of needles fanning outwards ('frostwork') are the principal aragonite excentrics.

'Cave pearls' are spheroidal accretions about a nucleus such as grit agitated by water dripping into a pool. 'Popcorn' ('cave coral') describes semi-spherical accretions on flowstone or other surfaces, often in dense, multilayered clusters.

Subaqueous calcite may precipitate from thermal or meteoric waters. The principal forms are spar linings, e.g. most of the 150+ km of passages in Jewel Cave, South Dakota. Deposition extends from the water table to a limiting depth determined by pressure and Ca^{2+} saturation state. Aggregate thicknesses of one metre or more are known. More complex crystal structures and rounded microcrystalline 'clouds' form in static pools. Water surfaces are marked by shelfstone around the edges and floating rafts of calcite accreted to dust particles.

Distribution and abundance

Speleothems can occur as isolated individuals, in clusters, aligned along fractures, or broadcast. Density can increase until all surfaces are covered. Vadose speleothems grow most readily at shallow depths beneath soils rich in CO_2 in tropical and temperate conditions that permit year-round deposition; the largest individuals and greatest densities are found in these settings.

Many caves have several levels. Speleothems are often fewer and smaller in the lower levels, which are usually younger. In any setting speleothem deposition is largely prohibited where there are impermeable beds, e.g. shales, above a cave.

Growth rates, age and environmental studies

Under optimal conditions (large excess of Ca^{2+} ions, high drip rate and evaporation) straws may extend several centimetres in one year. Normal accretion rates in other speleothems probably range between ~1.0 mm/10^3 yr in cold climate flowstones to >1.0 m/10^3 yr in warm cave entrances. Some speleothems grow at constant rates while others vary by factors of ten or more. Many contain hiatuses caused by drought, cold or change of groundwater routing.

Many speleothems can be dated accurately (+/−1 per cent error) by the $^{230}Th/^{234}U$ method if they are less than ~550 kyr in age. Variations of $^{18}O/^{16}O$ isotope ratios during growth may indicate palaeotemperature changes and $^{13}C/^{12}C$ ratios suggest changes of vegetation amount or type. Where present, annual or event banding revealed by u/v fluorescence and other techniques now permits very high resolution reconstructions of past conditions above caves (Hill and Forti 1997: 271–284).

Gypsum speleothems

Gypsum is deposited in three principal modes: (1) as evaporitic growths within bedrocks or cave sediments, which they rupture – 'evapoturbation'. (2) As scattered encrustations or excentric extrusions on rock, sediments or calcite speleothems. Most frequent are 'flowers', extruded, twisting fibrous bundles up to 50 cm in length. Needles grow from sediments and 'hair' from roofs. Larger, bifurcating stalactites are known in a few caves. (3) As regularly bedded floor or wall encrustations in evaporating pools: thicknesses of several metres occur in Carlsbad Caverns, New Mexico, where much is reprecipitated from alteration crusts formed by H_2SO_4 reacting with the limestone walls.

References

Cabrol, P. and Mangin, A. (2000) *Fleurs de pierre*, Lausanne: Delachaux et Niestle.
Hill, C. and Forti, P. (eds) (1997) *Cave Minerals of the World*, 2nd edition, Huntsville, AL: National Speleological Society of America.
Railsback, L.B. (2000) *An Atlas of Speleothem Microfabrics*, http://www.gly.uga.edu/railsback/speleoatlas/SAindex1.html
Van Beynen, P.E. and Bourbonnière, R., Ford, D. and Schwarcz, H. (2001) Causes of color and fluorescence in speleothems, *Chemical Geology* 175(3–4), 319–341.

DEREK C. FORD

SPHEROIDAL WEATHERING

Spheroidal weathering is the phenomenon whereby joint blocks within the regolith become rounded as a result of the separation of concentric layers of block surfaces. Spheroidal weathering is common in basalt and granite but is also found in dolerites, andesite and some sandstones (Heald *et al.* 1979).

There are two main schools of thought concerning the cause of spheroidal weathering. The first envisages that the separation of shells occurs due to residual stress from cooling and contraction. However, this would not explain the presence of corestones in sedimentary rocks such as sandstones. Ollier (1971) also makes the case that spheroidal weathering is a constant volume process, i.e. there is no accompanying expansion or contraction as is the case with EXFOLIATION and SPALLING.

The second school of thought envisages chemical activity, primarily the process of HYDROLYSIS, leading to migration of mineral elements and their concentration in separate layers. Thus distinct bands of accumulation and depletion become established (Augustithus and Ottemann 1966). Activity is most prevalent at corners and edges and so the tendency is for angular joint-bounded blocks to become rounded, or spheroidal in shape.

Ollier (1984) argues that use of the term spheroidal weathering should be restricted to situations where the weathering front attacks the block from all sides, i.e. the block must be beneath the ground surface. Clearly boulders originally rounded by spheroidal weathering may subsequently become exposed at the surface due to erosion. At this point, the style and rate of weathering are likely to change significantly. If the REGOLITH is subsequently completely removed by erosive agents, landforms known as boulder fields and boulder tors may remain on the surface.

The shape and size of corestones and boulders are determined by the spacing of joints in the intact mass. Spheroidal weathering is likely to be more effective in more closely jointed rocks with wide apertures (gaps between blocks). In this way, the opportunity for permeation of the rock with chemical solutions is optimized.

References

Augustithus, S.S. and Ottemann, J. (1966) On diffusion rings and spheroidal weathering, *Chemical Geology* 1, 201–209.
Heald, M.T., Hollingsworth, T.J. and Smith, R.M. (1979) Alteration of sandstones as revealed by spheroidal weathering, *Journal of Sedimentary Petrology* 49, 901–909.
Ollier, C. (1971) Causes of spheroidal weathering, *Earth-Science Reviews* 7, 127–141.
——(1984) *Weathering*, London: Longman.

Further reading

Chapman, R.W. and Greenfield, M.A. (1949) Spheroidal weathering of igneous rocks, *American Journal of Science* 247, 407–429.
Selby, M.J. (1993) *Hillslope Materials and Processes*, New York: Open University Press.
Yatsu, E. (1988) *The Nature of Weathering: An Introduction*, Tokyo: Sozosha.

SEE ALSO: weathering

DAWN T. NICHOLSON

SPIT

An easily recognized deposition coastal landform, which belies the potential complexity of the formation processes (Gilbert 1890; Davis 1896; Zenkovich 1967; Carter 1988). Spit structure and formation are best analysed through a plan-view perspective, as the coastal configuration in which the feature is developed is crucial to the spit's formation. Spits are found on an irregular coastline where sediment availability and wave power allow a constructional smoothing of the coastline by maintaining open coast longshore beach direction (in the form of a spit) into coastal re-entrants/bays.

Spits are essentially narrow depositional embankment-type features that show a dominance of longshore sediment deposition (growth) over cross-shore sediment movement. A spit's elongation relative to width is an indication of both the coastal sediment availability and net longshore-directed wave-generated transport potential. Such sediment can be from a broad size range, though sand-dominated spits are the most common. Gravel-dominated spits are more likely in mid-upper latitudes where gravel is a major component of coastal sediment availability. As spits are essentially a product of breaking wave activity, mud-dominated spits are unlikely to be observed. A spit's presence generates a back-spit energy lee with low-energy currents (tidal and small wave) and fine sediment stores (tidal banks and marshes).

A spit's plan-view shape is linked to the plan view of breaking-wave crests (and hence longshore transport vectors) determined by near-shore bathymetry. Spits tend to develop where wave refraction cannot accommodate to sudden changes of coastal trend and rapid reduction in breaker approach angle reduces the longshore drift rate to zero at this point. This allows beach deposition to overshoot the directional shift in coastline. The spit builds from this depositional nucleus, its orientation a function of wave refraction accommodating to the changing near-shore bathymetry induced by the presence of the spit.

Spits show a sequence of planform changes that are related to variation in both sediment supply and longshore transport potential, and are best developed when near-shore wave approach is angled along the spit. Spits can occur within a re-entrant when wave crests approach parallel to the re-entrant mouth. The regularity of the spit's plan-view form is a function of wave direction and refraction consistency. A spit is connected to the coast and its proximal sediment source by the neck, while a spit extension occurs at the spit's distal end or terminus. A spit per se, is usually only the subaerial (superstructure) expression of a larger submarine feature (spit platform), the distal position of which is the spit ramp. Ramp deposition controls spit growth and usually has a high fine-sediment proportion, even in gravel-dominated spits, related to wave-generated currents. As most of the sediment for the spit platform is supplied by longshore transport, it mimics sediment availability to the superstructure, though tending towards finer sediment. The spit platform requires an increasing sediment volume as the spit progressively builds into deeper water and as the volume of the superstructure generally remains the same, spit elongation rates will decline over time if the longshore sediment supply rate does not increase. Thus sediment supply rate is a major control on spit development.

There is rapid wave shoaling and landward curvature of the breaking wave crest at the spit terminus given its steep bathymetric gradients. This leads to curvature of the distal structure against the general trend of the spit. Curvature, correlating with decreasing breaker height and sediment fining, is enhanced at times of diminished longshore sediment supply. When supply is reinstated, then the spit can extend in line with its original plan and the recurve is isolated. High volume, but episodic, sediment supply can lead to drift-aligned spits where the spit plan outline is essentially rectilinear despite overlapping recurves (Carter and Orford 1991). This scenario is often associated with the initial formation of spits in a disjointed coastline where sediment supply is formed from isolated finite sediment sources (e.g. a drumlin coastline: Orford *et al.* 1996). Once the spit's sediment source is exhausted then continuing wave power starts to rework the existing spit sediment (cannibalization) leading to a thinning of the spit neck, the increased potential of the superstructure to rollover (as bigger waves break closer to the shore) and the landward movement of the spit into swash-alignment. Cannibalization leads to extension or stabilization of the distal position which then tends to act as a down drift hinge position for control of spit form's swash alignment. Spits generally retreat under a rising relative sea level through overwashing and hence rollover. Retreat evidence is provided by truncated back-spit recurves and back-barrier organics emerging on the seaward face.

Tidal currents can influence the spit terminus, especially when spit growth squeezes coastal inlet width. As inlet tidal hydraulic efficiency increases, any ebb-tidal asymmetry leads to protection of the spit terminus by sand shoals related to delta formation. These in turn influence wave refraction at the terminus and allow swash bars to drive onshore and build up the terminus beach. This can be a source for distal aeolian dunes, even given sediment depletion elsewhere on the spit. Changes to delta deposition due to back-barrier reclamation can affect terminus growth: a decrease in sediment supply can lead to a withering of the spit terminus, thinning of the spit and spit beheading by overwash. The control by ebb deltas on wave refraction patterns can lead to apparent opposing spit growth across tidal inlets despite a dominant single direction of sediment supply. Spits that seal off re-entrants form barriers (see BARRIER AND BARRIER ISLAND).

References

Carter R.W.G. (1988) *Coastal Environments*, New York: Academic Press.

Carter, R.W.G. and Orford, J.D. (1991) The sedimentary organization and behaviour of drift-aligned gravel barriers, *Coastal Sediments '91*, American Society of Civil Engineers 1, 934–948.

Davis, W.M. (1896) The outline of Cape Cod, *Proceedings of the American Academy of Arts and Science* 31, 303–332.

Gilbert, G.K. (1890) Lake Bonneville, *US Geological Survey, Monograph* 1, 47–48, Washington.

Orford, J.D., Carter, R.W.G. and Jennings, S.C. (1996) Control domains and morphological phases in gravel-dominated coastal barriers, *Journal of Coastal Research* 12, 589–605.

Zenkovich, V.P. (1967) *Processes of Coastal Development*, 409–447, Edinburgh: Oliver and Boyd.

JULIAN ORFORD

SPRING, SPRINGHEAD

Springs are points where ground water, recharged at higher elevations, emerges at the surface. Depending on the nature of the recharge and of the storage/transmission characteristics of the aquifer through which the water has flowed, they may be permanent (perennial), seasonal or intermittent. In karst areas reversing springs called estavelles are found, particularly in association with poljes. Another common feature of karst areas is the presence of permanent 'underflow' springs and higher, intermittent, 'overflow' springs. Spring flow may show little variation over time or may respond rapidly to recharge, varying over several orders of magnitude. Some springs where the outflow is controlled by a siphoning reservoir system exhibit regular ebbing and flowing with a typical period of minutes to hours. Geysers are periodic hydrothermal springs in which a pressurized body of water is warmed to boiling point and explosive spontaneous boiling occurs as pressure is released.

Springs are found at many elevations from high in mountains to beneath sea level, the vrulja of the Mediterranean being an example of the latter. Spring discharges range over seven orders of magnitude, from seeps to large springs with average flows exceeding $20\,m^3\,s^{-1}$ and instantaneous flows of several hundred $m^3\,s^{-1}$. Most of the largest springs are karstic and only those from fractured volcanic rocks rival their output. The largest is thought to be the Tobio Spring in Papua New Guinea with a mean annual discharge of $85–115\,m^3\,s^{-1}$.

Springs which discharge only percolation water are called exsurgences while those that discharge a mixture of percolation water and water from sinking streams are called resurgences. The term 'rising' is commonly used by speleologists as a synonym for a spring. Those springs where water rises from depth are pressure springs, sometimes termed 'Vauclusian' springs after the Fontaine de Vaucluse in France which has been explored to a depth of 315 m. Where the internal hydraulic head in an aquifer greatly exceeds that required to drive the flow of water, springs exhibit a marked upwelling and are commonly termed 'artesian'.

Karst springs are the output points from a dendritic network of conduits, some of which may be large enough for human exploration (caves). They therefore tend to be both larger and more variable in quantity and quality than springs that emerge from coarse granular or fractured media. The latter may result from the convergence of flow lines in a depression or from the concentration of flow along open fractures such as faults, joints or bedding planes.

Where a spring with a moderate to large discharge has emerged at the same point for a long period a marked springhead or steephead may form. Valleys that begin abruptly at the springhead are termed pocket valleys or reculées. Most are short but some are many tens of metres in height. They may form by headward recession, as water from the spring undermines the rock above it, or by cavern collapse.

Further reading

Ford, D.C. and Williams, P.W. (1989) *Karst Geomorphology and Hydrology*, London: Unwin Hyman.

Jennings, J.N. (1985) *Karst Geomorphology*, Oxford: Blackwell.

LaMoreaux, P.E. and Tanner, J.T. (2001) *Springs and Bottled Waters of the World*, Berlin: Springer-Verlag.

JOHN GUNN

STACK

Stacks are isolated pillars of rock that form when part of a retreating coast is separated from the mainland, usually along joints (see JOINTING) or faults (see FAULT AND FAULT SCARP). Stacks also develop because of folding, tropical KARST submergence, solution pipes, induration and variable rock types. They form in fairly strong rocks with well-defined planes of weakness and are uncommon in weak or thinly bedded rocks with dense joint systems. Stacks can develop from the collapse of arch (see ARCH, NATURAL) roofs, but many form directly from erosion of the cliff face.

Further reading

Trenhaile, A.S. (1987) *The Geomorphology of Rock Coasts*, Oxford: Oxford University Press.

ALAN TRENHAILE

STEP-POOL SYSTEM

Step pools are characteristic bedforms that dominate the channel morphology of steep mountain streams (Chin 1989). Steps are generally composed of cobbles and boulders; they are separated by finer materials forming the pools. Steps and pools alternate to produce a repetitive sequence of bedforms, with a longitudinal profile resembling a staircase. The step-pool morphology similarly develops in bedrock channels and in vegetated basins where channels incorporate woody debris to produce log steps. Step pools are part of a continuum of coarse-grained bedforms that includes POOLS AND RIFFLES (Montgomery and Buffington 1997). Despite external influences, step pools commonly occur with sufficient regularity to produce a rhythmic streambed. They represent a type of meandering in the vertical dimension (Chin 2002).

Step pools are functionally important because they provide hydraulic resistance. Steps induce water to plunge into pools below, promoting tumbling flow where much of the flow's kinetic energy is dissipated by roller eddies. By causing a vertical drop in the water surface elevation as water flows from step to pool, steps also decrease potential energy that otherwise would be available for conversion to a longitudinal component of kinetic energy used for erosion and sediment transport. Steps provide the ability to counteract steep slopes, thereby preventing excessive erosion and channel degradation. The role of step pools is especially important in confined mountain streams where lateral adjustments and energy dissipation by meandering and braiding are prohibited.

Step pools form an integral part of the hydraulic geometry of mountain streams. Consistent relations exist between step length, step height and channel gradient. Step length increases and height decreases with a decrease in slope. Such relations are found regardless of substrate type and the presence of woody debris, suggesting that step characteristics are controlled, at least in part, by flow energy expenditure (Wohl *et al.* 1997). Step pools represent a means of adjusting boundary roughness. They evolve toward a condition of maximum resistance, which is apparently achieved when the ratio of mean step height to mean step length to channel slope is between 1 and 2 (Abrahams *et al.* 1995).

Step pools are mobilized by high-magnitude, low frequency floods on the order of fifty years or more (Grant *et al.* 1990). The specific generating mechanism is incompletely understood. Laboratory experiments suggest that steps may originate as antidunes under high flows (Whittaker and Jaeggi 1982). However, although limited field data support the antidune model, the theory cannot explain step-pool formation in all cases. For example, step pools develop in some channels where flows are unlikely to completely submerge clasts and form antidunes. Alternative explanations focus on flow instabilities and the random movement of large particles.

The step-pool bed configuration controls hydraulics and sediment transport in distinct ways.

For example, velocity and flow resistance fluctuate between step and pool and along with increasing discharge (Lee and Ferguson 2002). Sediment transport is episodic, characterized by alternating transport steps and intervals of non-movement (Schmidt and Ergenzinger 1992). Thus, prediction of sediment transport in step-pool streams using standard equations is problematic. New data from instrumented watersheds (e.g. Lenzi 2001) have the potential to yield considerable insights for bedload transport and associated step-pool processes.

References

Abrahams, A.D., Li, G. and Atkinson, J.F. (1995) Step-pool streams: adjustment to maximum flow resistance, *Water Resources Research* 31, 2,593–2,602.

Chin, A. (1989) Step pools in stream channels, *Progress in Physical Geography* 13, 391–407.

——(2002) The periodic nature of step-pool mountain streams, *American Journal of Science* 302, 144–167.

Grant, G.E., Swanson, F.J. and Wolman, M.G. (1990) Pattern and origin of stepped-bed morphology in high-gradient streams, western Cascades, Oregon, *Geological Society of America Bulletin* 102, 340–352.

Lee, A.J. and Ferguson, R.I. (2002) Velocity and flow resistance in step-pool streams, *Geomorphology* 46, 59–71.

Lenzi, M.A. (2001) Step-pool evolution in the Rio Cordon, northeastern Italy, *Earth Surface Processes and Landforms* 26, 991–1,008.

Montgomery, D.R. and Buffington, J.M. (1997) Channel-reach morphology in mountain drainage basins, *Geological Society of America Bulletin* 109, 596–611.

Schmidt, K.H. and Ergenzinger, P. (1992) Bedload entrainment, travel lengths, steplengths, rest periods – studied with passive (iron, magnetic) and active (radio) tracer techniques, *Earth Surface Processes and Landforms* 17, 147–165.

Whittaker, J.G. and Jaeggi, M.N.R. (1982) Origin of step-pool systems in mountain streams, *Journal of the Hydraulic Division*, ASCE 108, 758–773.

Wohl, E., Madsen, S. and MacDonald, L. (1997) Characteristic of log and clast bed-steps in step-pool streams of northwestern Montana, USA, *Geomorphology* 20, 1–10.

Further reading

Wohl, E. (2000) *Mountain Rivers*, Washington, DC: American Geophysical Union.

SEE ALSO: gravel-bed river

ANNE CHIN

STERIC EFFECT

An effect in which the molecular dimensions of the material controls the rate or path of a physical or chemical reaction. A steric effect on a rate process may lead to a rate increase (steric acceleration) or a decrease (steric retardation). The most significant steric effect in geomorphology is SEA LEVEL change. In sea water, steric effects are driven by changes in temperature and differences in salinity (i.e. density). As heat is able to exchange freely with the atmosphere, temperature is the dominant steric parameter, particularly over longer timescales (thousands of years) where salinity remains fairly constant in oceans. The change in seawater level as a result of steric effects is referred to as the steric height. This is defined as the height of the sea as the integral of the specific volume from a specified pressure level to the ocean surface. However, steric heights are hard to calculate and trends difficult to attain, as changes occur over varied spatial and temporal scales.

Steric changes occur on seasonal and intra-annual time periods, mostly as a result of steric effects in the upper 500 m of the oceans where heat exchange is much more rapid. It is estimated that an increase in temperature of 1 °C throughout the uppermost 500 m will result in a sea-level rise of approximately 100 mm. In comparison, colder deeper waters show slower heat exchange, and it has been argued that they should be ignored in steric height calculations. However, a parcel of water at 4 °C at 2,000 m depth has a thermal expansion coefficient (which increases with temperature and pressure) 60 per cent as large as that of a parcel at the ocean surface at 20 °C (Roemmich 1990). A warming of the entire seawater column of 1 °C would raise the sea level by about 0.2 m, whereas a warming of 10 °C would lead to an increase in sea level of about 8 m (as the coefficient of thermal expansion increases with temperature) (Knutti and Stocker 2000).

Steric height variations have been shown to influence global sea-level fluctuations on short-term timescales in particular (seasonal, intra-annual and decadal periods on the order of 10 cm), yet steric effects are not great enough to account for sea-level variations over greater timescales (for instance, the 150 m rise in sea level since the last glacial period) (Roemmich 1990). However, thermal contraction of sea water could account for a sea-level rise between the climatically warm Cretaceous period (144–66 Ma) and the onset of the ice-dominated system in the Cenozoic.

Steric effects hold great contemporary importance with regards to global warming, and several modelling studies have investigated past changes and the likely response of the oceans to future changes (e.g. Knutti and Stocker 2000). A rise in sea level of between 10–50 cm is projected due to steric changes over the next century.

References

Knutti, R. and Stocker, T.F. (2000) Influence of the thermohaline circulation on projected sea level rise, *Journal of Climate* 13, 1,997–2,001.

Roemmich, D. (1990) Sea level and the thermal variability of the ocean, in National Research Council *Sea Level Change (Studies in Geophysics)*, Washington, DC: National Academy Press.

STEVE WARD

STONE-LINE

Stone-lines are synonymous with 'stone-layers', *nappes de gravats* (French) and *Steinlagen* (German) and are common stratigraphic features within many tropical soils and weathering profiles. They form striking, mainly undulating and approximately downslope-oriented discontinuities in soils, consisting of stringers of resistant, largely unweathered coarse clasts at different depths below the groundsurface. In thickness they may range from several centimetres up to a metre or even more. These layers of coarse material separate the overlying loamy to sandy topsoil, or hillwash, from the strongly chemically weathered subsoil or SAPROLITE. In most cases the substrate is not well sorted and may be angular to subangular or possibly well rounded in shape, comparable to gravels and pebbles. Although usually dominated by quartz, pisoliths, iron nodules and larger fragments of lateritic crusts, as well as human artefacts can occur.

A three-stepped stratigraphic subdivision of the stone-line complex was developed for Central Africa, distinguishing the cover (hillwash) or α-layer, the stone-line or β-layer, and the weathered rock (saprolite) or γ-layer (Stoops 1967). The α-layer is typically a few centimetres to some metres thick. It consists of loose but structured material, of sandy to clayey texture, practically devoid of elements coarser than 4 mm (with the exception of some loose iron concretions, mainly at the surface). It shows no stratification. The hillwash follows the general slope of the hill, albeit with a locally more sinuous path with introversions. The sinuosity amplitude varies between 2 and 4 m, and the height-differences seldom reach 50 cm. At the transition from the hillwash (α-layer) to the stone-line (β-layer), slightly coarser material up to 1 cm in diameter occurs. In striking contrast to the covering hillwash, the stone-line often shows a vertical zonation dominated by weathered vein-quartz and rolled pebbles with iron coatings, which are reminiscent of alluvial gravel. Prehistoric implements have been identified in some locations (β_1-layer). Along some sections only, a gradual transition to the lower, so-called β_2-layer takes place. This part of the stone-line consists essentially of fragments of *in situ* weathered quartz veins and chert bands showing a subangular to angular shape. Surface coatings of clasts are rare. Thickness of this layer can reach several centimetres to two metres. Below the stone-line a profound weathering zone characterized by kaolinitic clay with a subsequent transition to bedrock (mottled and pallid zone) is recognizable. A further subdivison of the γ-layer relating to soil colour and micromorphological properties, in connection with successive alteration of the soil profile (e.g. γ_1-, γ_2-, γ_3-layer) may be present.

Due to a huge variety of stone-line phenomena in the tropics an extensive literature on these stratigraphic features exists, indicating that several theories and geomorphic processes may be considered in explaining their morphogenetic origin. Early in stone-line research in the late 1940s it was thought that stones, originally dispersed over the whole depth of the soil profile, were able to sink through the matrix of fines, and would finally concentrate on top of the underlying bedrock. However, this conception proved wrong as the bearing capacity of a soil generally remains high enough to support stones. Today, it is generally accepted that stone-line formation is closely linked to a catenary context, especially to the domain of hillslope processes and slope morphology. It is equally related to long-term weathering and morphodynamics of landscapes. Nevertheless the interpretation of different stone-line features, in particular whether they are the result of an *in situ*, autochthonous formation by down-weathering, or whether their origin lies in a combination of laterally active morphodynamic processes due to former palaeoenvironmental modifications of the landscape (allochthonous formation) is still controversial.

Stone-line formation

The following explanations of stone-line formation are most likely:

1 Stone-lines are residual surface accumulations (palaeopavements) which were later covered by finer sediments. This process results from selective erosion by episodic sheet wash (see SHEET EROSION, SHEET FLOW, SHEET WASH), soil creep and the formation of slope pediments with retreating scarps, causing the hillslope-oriented accumulation of colluvial- or hillwash-like fine material. This landscape instability is effected by a climate change to drier, arid conditions (Rohdenburg 1969).
2 Stone-lines occuring near river plains can be understood as parts of palaeochannels (e.g. former anastomosing branches), that were formed by redistribution and concentration of gravel by surface water flows and related colluvial activity.
3 Stone-lines may be considered as the result of bioturbation by termites, ants and worms. Selective zoogenic uptake of fine material from bottom to top in a soil leads to a concentration of coarser material in greater depth.
4 It is most likely that the formation of many stone-lines has to be considered within the context of drier climatic conditions during the Last Glacial Maximum (LGM). However, there is also evidence that stone-lines can be interpreted as stratigraphic markers of a Younger Dryas event with cold and dry climatic conditions at the onset of the Holocene (Runge 2001).

Economic significance of stone-lines

The coarse material is frequently quarried for road construction as paving gravel and for mineral exploration (e.g. cassiterite, columbite, gold, monazite, zircon, rutile, ilmenite, diamonds). Due to their greater specific weight and greater mechanical and chemical weathering resistance these minerals are concentrated in stone-lines (placer deposits). Such sites are often the first to be exploited as no heavy equipment is required. The mineral content of stone-lines is often used as a pathfinder towards major hardrock orebodies (Thorp 1987).

References

Rohdenburg, H. (1969) Hangpedimentation und Klimawechsel als wichtigste Faktoren der Flächen und Stufenbildung in den wechselfeuchten Tropen, *Giess. Geogr. Schriften* 20, 57–152.
Runge, J. (2001) On the age of stone-lines and hillwash sediments in the eastern Congo basin – palaeoenvironmental implications, *Palaeoecology of Africa and the Surrounding Islands* 27, 19–36.
Stoops, G. (1967) Le profil d'altération aus Bas-Congo (Kinshasa). Sa description et sa genèse, *Pédologie* 17, 60–105.
Thorp, M.B. (1987) The economic significance of stone-lines, *Geo-Eco-Trop*, 11, 225–227.

Further reading

Alexandre, J. and Malaisse, F. (eds) (1987) Journée d'étude sur 'Stone-lines', Bruxelles, *Géo-Eco-Trop* 11, 1–239.
Thomas, M.F. (1994) *Geomorphology in the Tropics*, Chichester: Wiley.

SEE ALSO: saprolite; sheet erosion, sheet flow, sheet wash

JÜRGEN RUNGE

STONE PAVEMENT

Sometimes called 'desert pavement', a stone pavement is an armoured surface composed of a thin mosaic of rock fragments that is set in or on a matrix of finer material. In Australia the rock fragments may consist of SILCRETE fragments, locally termed *gibbers*. Pavements are important because they are a major control on surface stability. They also provide a record of the activities working on desert surfaces. In addition, if they are disrupted, accelerated erosion of the underlying finer material may occur. Pavements also act as a store for material in transit by the wind and may fundamentally affect infiltration, runoff and sediment erosion rates (Poesen *et al.* 1994). They tend to occur in areas with limited vegetation cover, and the presence of vegetation, as in wet years, can encourage increased levels of bioturbation and surface disturbance (Haff 2001). Stone pavements are not restricted to deserts, however, and occur, *inter alia*, in tundra regions.

Stone pavements may display soil horizons or they may be produced without appreciable soil development, especially in the case of immature examples developed on relatively unstable ground surfaces by superficial processes, such as deflation and runoff. In those examples with soil horizonation, there is often a vesicular A horizon of mainly silt-clay-size particles (McFadden *et al.* 1998).

There has been a great deal of discussion about the processes that lead to pavement formation.

A classic model is that of deflational sorting, whereby a lag of coarse material is left at the surface after finer materials have been removed by wind action. Another possible mode of horizontal removal of fines needs to be considered, however, namely water sorting (Cooke 1970). Upward migration processes may also play a role, for the concentration of coarse particles at the surface and at depth, and the relative scarcity of coarse particles in the upper part of the underlying soil profile, suggests that the coarse particles may have migrated upwards through the soil to the surface. This could be achieved by a range of processes, including freezing and thawing, wetting and drying, changes in salt phases and bioturbation. In addition, pavement characteristics may be much modified by the addition of aeolian materials, especially dust, from above (McFadden *et al.* 1987; Wells *et al.* 1985). Pavement characteristics also change with age, with pavement development being greater on older surfaces (Amit and Gerson 1986). In particular, the nature of older surfaces will be characterized by a greater degree of particle weathering caused by processes like salt weathering. They may also display greater degrees of rock varnish cover.

In recent years there has been great interest in the way that pavements recover from disturbance brought about, for example, by the passage of wheeled transport. In the Western Desert of Egypt, vehicle tracks from the two world wars of the twentieth century can still be clearly seen on pavement surfaces. However, in other localities pavements have been seen to heal relatively rapidly following deliberate local disruption of their surfaces. Haff and Werner (1996), working in California, found that gaps healed in around 5 years and that displacement of surface stones by small animals was a major component of the healing process. Similarly, Wainwright *et al.* (1999), using rainfall simulation experiments at Walnut Gulch in Arizona, found that raindrop erosion processes resulted in rapid surface recovery.

References

Amit, R. and Gerson, R. (1986) The evolution of Holocene (reg) gravelly soils in deserts. An example from the Dead Sea region, *Catena* 13, 59–79.

Cooke, R.U. (1970) Stone pavements in deserts, *Annals of the Association of American Geographers* 60, 560–577.

Haff, P.K. (2001) Desert pavement: an environmental canary? *Journal of Geology* 110, 661–668.

Haff, P.K. and Werner, B.T. (1996) Dynamical processes on desert pavements and the healing of surficial disturbances, *Quaternary Research* 15, 38–46

McFadden, L.D., McDonald, E.V., Wells S.G., Anderson, K., Quade, J. and Forman, S.C. (1998) The vascular layer and carbonate collars of desert soils and pavements: formation, age, and relation to climate, *Geomorphology* 24, 101–145.

McFadden, L.D., Wells, S.G. and Jercinovich, M.J. (1987) Influence of aeolian and pedogenic processes on the origin and activities and evolution of desert pavements, *Geology* 15, 504–508.

Poesen, J.W., Torri, D. Burt, K. (1994) Effect of rich fragments on soil erosion by water at different spatial scales: a review, *Catena* 23, 141–166.

Wainwright, J., Parsons, A.J. and Abrahams, A.D. (1999) Field and complete experiments on the formation of desert pavements, *Earth Surface Processes and Landforms* 24, 1,025–1,037.

Wells, S.G., Dohrenwend, J.C., McFadden, L.D., Turrin, B.D. and Mehrer, K.D. (1985) Late cenozoic landscape evolution on lava flow surfaces of the Cima volcanic field, Mojave Desert, California, *Geological Society of America Bulletin* 96, 1,518–1,529.

A.S. GOUDIE

STORM SURGE

Storm surge is a response of the ocean to changing atmospheric pressure and strong winds caused by cyclonic weather systems, that can result in higher water surface elevations than are predicted by normal astronomical tides, lasting between an hour and four days but typically in the order of 6–18 hours. Storm surge results from the combined action of extreme wind shear stress on the ocean which moves and holds water against windward coasts (wind set-up), and the inverse-barometer effect of changing atmospheric pressure that increases the mean water surface level as pressure drops (pressure set-up). Pressure set-up increases the average water surface by 1 cm per 1 hPa drop in atmospheric pressure, but in cases of extreme storm surge, wind set-up is much more significant. Wind speeds, the track and the relative position of the storm centre to the shore, the slope of the continental shelf and the configuration of the shoreline (particularly the extent of embayment) are all influential in determining the size of the storm surge. The largest storm surges result from hurricanes (otherwise known as tropical cyclones or typhoons) and raised water levels of up to 8 m have been reported. Significant storms at higher latitudes tend to produce surges in the order of 1–3 m, although on shallow continental shelves higher surges are possible.

In generally low-lying coastal areas, increases in mean water level by storm surge can result in large areas of land inundation, often coinciding with floods and other storm-related effects. The most significant regularly affected locations for storm surge damage are the Bay of Bengal, the south-east coast of the USA and the east coast of China. Surges in the shallow, funnel shaped Bay of Bengal have resulted in more than 100,000 deaths on four occasions since 1897 (Bao and Healy 2002), with the worst event occurring in 1970 resulting in the loss of approximately 300,000 lives. In the USA, although surges result in inundation of significant areas of land and high economic loss, deaths have been few due to good warning systems, and disaster response infrastructure. Considerable effort has gone into understanding and modelling storm surge (Bode and Hardy 1997).

The effects of storm surge can be exacerbated if the surge coincides with periods of high astronomical tide. Additionally, on open coasts storm surge is normally associated with increased wave set-up and the establishment of long-period wave motions in the surf zone, all of which increase water level on the beach, allowing storm waves to penetrate much further inland which can lead to significant coastal erosion.

References

Bao, C. and Healy, T. (2002) Typhoon storm surge and some effects on muddy coasts, in T. Healy, Y. Wang and J.-A. Healy (eds) *Muddy Coasts of the World: Processes, Deposits and Function*, 263–278, Amsterdam: Elsevier Science.

Bode, L. and Hardy, T.A. (1997) Progress and recent developments in storm surge modeling, *Journal of Hydraulic Engineering* 123(4), 315–331.

SEE ALSO: continental shelf; overwashing; wave

KEVIN PARNELL

STRANDFLAT

The word 'strandflat' is a name used for the low country and shallow sea along the western Norwegian coast, and also along coasts in Arctic and Antarctic areas that have been covered by ice sheets during the Quaternary ice age. Apart from long stretches of the west coast of Norway where the strandflat is an almost continuous feature, the strandflat has also been recognized in areas as far apart as the South Shetland Isles, Alaska and western Scotland. The low areas of strandflat often appear as broad glacially moulded coastal rock platforms sometimes as much as 80 km in width and backed by high cliffs. However, these shore platforms generally exhibit considerable local relief thus making it difficult to assign a precise altitude to any individual area of platform.

The strandflat was first described by Reusch (1894) while its possible origins were first considered in detail by Nansen (1922). The various processes of strandflat formation are well summarized by Larsen and Holtedahl (1985) and include marine abrasion, subaerial weathering, glacial erosion, frost shattering and cold climate shore erosion. Larsen and Holtedahl proposed that the strandflat was primarily the result of sea-ice erosion and frost shattering during the Quaternary and that most surfaces had been later modified by marine erosional and glacial erosional processes. They also noted that the Norwegian strandflat surfaces exhibit glacio-isostatic tilting and are therefore likely to have been produced during periods of Quaternary glaciation rather than during temperate interglacial periods.

It is difficult to determine precise ages for the formation of the various strandflat surfaces around the world although recent developments in cosmogenic isotope dating techniques represent one possible way forward of dating individual rock surfaces. Consideration of the Quaternary marine oxygen isotope record and of the glacio-isostatic changes that have affected land areas buried by Quaternary ice sheets implies that the position of relative sea level in any area is unlikely to have remained stationary for any significant length of time, certainly no less than *c.*10,000 years. Accordingly, this would seem to point to the conclusion that individual strandflat surfaces having been produced by cold climate shore processes, must have been repeatedly overwhelmed by ice sheets and subject to marine processes during numerous intervals of cold climate throughout the Quaternary.

References

Larsen, E. and Holtedahl, H. (1985) The Norwegian strandflat: a reconsideration of its age and origin, *Norsk Geologiske Tidsskrift* 65, 247–254.

Nansen, F. (1922) The strandflat and isostasy, *Videnskapelkapets Skrifter 1. Math.-Naturw. Kl. (Kristiana)*, 11.

Reusch, H. (1894) Strandfladen et nyt traek I Norges geografi, *Norges Geologiske Underssokelse* 14, 1–14.

ALASTAIR G. DAWSON

STREAM ORDERING

Stream ordering is a technique for characterizing the constituent parts of a drainage network. Ordering can start from the outlet and move in the upstream direction or it can start from each source and move downstream. The most successful have been those ordering systems which move in a downstream direction. The upstream moving systems require a series of subjective decisions about which upstream extension is the master stream. Horton (1932, 1945) introduced the following ordering system:

(a) Channels that originate at a source, and have no tributaries are defined to be first-order streams;
(b) When two streams of order x join, a stream of order x + 1 is created;
(c) When two streams of different order join, the channel segment immediately downstream of the junction takes the higher order of the two combining streams;
(d) When the highest order stream segment (n) has been defined, then the upstream extensions of that segment are deemed to have the same order (n) all the way to the source. Similarly, stream segments of order (n−1) are extended back to their source and so on.

This hybrid system (first downstream and then upstream) incorporates the subjectivity mentioned and therefore Strahler (1952, 1957) revised Horton's scheme by removing step (d) above. This so-called Strahler system (or Horton-Strahler ordering system) is the most commonly used in hydro-geomorphology.

A third-ordering system is the link magnitude system proposed by Shreve (1966). Source streams or links have magnitude 1. At a bifurcation, the downstream link takes the magnitude of the sum of the magnitudes of the two incoming links. The magnitude of each link is therefore equivalent to the number of sources in the network draining into that link.

The theoretical basis for Shreve's ordering system is his view of the river basin as a random topological structure. Terminology used includes: a node is the one outlet furthest downstream; n sources are the points furthest upstream and there are n − 1 junctions. Edges of the network are links; exterior links emanate from sources; interior links emanate from junctions. A network with n sources has 2n − 1 links, n of which are exterior and n − 1 are interior links.

The most important measuring device that Horton identified was that involving stream ordering. The idea originated with a German hydraulic engineer Gravelius (1914). Chorley (1995) noted that there were two important corollaries of Horton's stream ordering: (a) it placed the emphasis on analysis based on the identification of individual drainage basins. The latter thus emerged as rational, clearly defined topographic units whose geomorphic status was expressed by their order and which prompted geometrical comparisons from one location to another; and (b) the procedure generated a nested hierarchy of drainage basin forms, each of which could be viewed as an open physical system in terms of inputs of precipitation and outputs of discharge and sediment load. A third corollary could be added to Chorley's list. The application of increasingly refined statistical analysis to problems of watershed geomorphology was facilitated by ordering the channels: issues of sample size and representativeness became central considerations in geomorphology.

Perhaps most importantly, the concept of drainage density, which was to become the most important geometric indicator in the work of the Columbia school of quantitative geomorphology, was encouraged by the ordering of streams. Melton (1957) explored the relation between drainage density and stream frequency, as well as the ratio of the two as a measure of the completeness with which a channel system fills a basin outline and as a possible evolutionary index of drainage basins.

Horton's ordering also generated the laws of drainage composition. These were exponential relations between stream order and (a) number of streams of a given order; (b) average length of streams of each order; (c) total stream lengths of each order; (d) basin areas of each order; and (e) average stream slopes of each order. Each of Horton's laws generated a ratio (e.g. the bifurcation ratio from the first law) and these ratios were shown to lie within quite narrow ranges except where differential geological controls were important. Horton's laws provided a topologically and geometrically logical set of procedures for the analysis of fluvially dissected terrain. Much geomorphic work has subsequently been based on systems of stream ordering.

References

Chorley, R.J. (1995) Classics in physical geography revisited: Horton, R.E., 1945, *Progress in Physical Geography* 19, 533–554.

Gravelius, H. (1914) *Flusskunde*, Berlin: Goschensche Verlagshandlung.
Horton, R.E. (1932) Drainage basin characteristics, *American Geophysical Union Transactions* 13, 350–361.
——(1945) Erosional development of streams and their drainage basins: hydrophysical approach to quantitative morphology, *Geological Society of America Bulletin* 56, 275–370.
Melton, M.A. (1957) An analysis of the relations among elements of climate, surface properties and geomorphology, *Office of Naval Research Project NR 389–042, Technical Report 11*, New York: Columbia University Press.
Shreve, R.L. (1966) Statistical law of stream numbers, *Journal of Geology* 74, 17–37.
Strahler, A.N. (1952) Hypsometric (area-altitude) analysis of erosional topography, *Geological Society of America Bulletin* 63, 1,117–1,142.
——(1957) Quantitative analysis of watershed geomorphology, *American Geophysical Union Transactions* 38, 912–920.

SEE ALSO: allometry; drainage density; dynamic geomorphology

OLAV SLAYMAKER

STREAM POWER

Power is the rate of doing work (force × distance) and is expressed in Watts which are Joules per second ($J s^{-1}$). Stream power is the rate at which a stream can do work, especially in the transport of its sediment load, and is usually measured over a specific length of channel. It expresses the rate of energy expenditure in flowing water and, as such, could provide a basic integrating theme within the physical environment (Gregory 1987). In hydraulics and fluvial processes, attempts to analyse the processes involved have used a number of variables whereas expressing energy expenditure as stream power is a more fundamental approach. The potential energy that water possesses at a particular location is proportional to its height above some datum which can be sea level or a lake level; this potential energy is converted into kinetic energy as the water flows downhill under the influence of gravity.

Three important aspects of stream power are: how it is expressed, what controls it, and how has it been utilized and applied. Stream power (ω) was first expressed (Bagnold 1960) as the product of fluid density (ρ), discharge (Q), acceleration due to gravity (g) and slope (s) in the form:

$$\omega = \rho Qgs$$

This expression for power can of course be applied to any fluid, and Bagnold used a similar approach in relation to wind movement over the Earth's surface. Bagnold's (1960) definition has subsequently been applied to the rate of sediment transport (Bagnold 1977) expressed as the amount of energy expended per unit area of the bed. Such unit stream power could be obtained per unit channel width (w) or bed area as:

$$\omega = \frac{\rho Qgs}{w}$$

which, because Q = wdv, is simplified to $\omega = \rho gdvs$ thus including depth (d) and velocity (v), and this is often referred to as specific power. Stream power is measured in Joules per second ($J s^{-1}$) (Watts) and unit stream power is expressed per square metre ($J m^{-2} s^{-1}$ or $W m^{-2}$). Unit stream power is expressed per unit length of channel or per unit channel area, and when comparing results from several areas it is important to differentiate between the results achieved by several methods. Values of unit stream power ω range from less than $1\,W m^{-2}$ in flow between rills to $>12{,}000\,W m^{-2}$ in riverine flood flows in India, up to $18{,}582\,W m^{-2}$ for large flash floods, and up to $3 \times 10^5\,W m^{-2}$ for the Missoula flood (Baker and Costa 1987) in the Quaternary, the largest known discharge of water on Earth.

Controls upon stream power can be deduced from its component elements, of which g is constant and ρ, Q and s are variables. Along a single channel, slope may tend to decrease downstream, whereas discharge will increase, and there can be significant variations in water quality and sediment transport which affect the value of ρ. Along a single river unit, stream power tends to peak in the middle parts of some basins; it is lower upstream, where discharges are relatively small, and lower downstream, where river gradients have their lowest values. Stream power can be calculated for the discharge in the channel at any one specific time or it can be calculated for the estimated channel capacity flow or for some flood flow value; the pattern of stream power distribution down valley may vary somewhat in each case.

Applications of stream power have now been made to many aspects of analysis of river systems. Most usefully in relation to *sediment transport*, stream power has been used instead of stream discharge, velocity or bed shear stress to relate to sediment motion and transport, especially that of bedload (e.g. Allen 1977). This approach is arguably more geomorphological than using hydrological parameters. It has also been used more generally as a means of considering the efficiency of sediment transport; by comparing the power needed to transport the sediment along a particular reach with the power actually available, *critical power* was defined as the power just sufficient to transport the sediment through the reach (Bull 1979, 1991). Interest in the significance of large flood events has led to the estimation of *flood power*, including that related to palaeofloods (Baker and Costa 1987), and a threshold for catastrophic modification of the channel or fluvial landscape has been suggested as a unit stream power of $300\,\mathrm{W\,m^{-2}}$ (Magilligan 1992).

Variations in unit stream power along a river channel have been used to explain patterns of the pool riffle sequence; to determine bedform type for specific sediment size; to relate to the channel HYDRAULIC GEOMETRY; and to explain river channel patterns. Such patterns have been classified according to amount and size of bedload and stream power (Schumm 1981) and channel SINUOSITY has been related to stream power (Schumm 1977). Three major types of floodplain (Nanson and Croke 1992) have been differentiated according to stream power values, including High energy ($\omega > 300\,\mathrm{Wm^{-2}}$), Medium energy ($10 < \omega < 300\,\mathrm{Wm^{-2}}$), and Low energy ($\omega < 10\,\mathrm{Wm^{-2}}$), and stream power variations have also been related to the pattern of the river long profile. These are all ways in which stream power can be related to aspects of *channel morphology*, showing how knowledge of spatial variations of stream power can be the basis for useful applications. In the case of British rivers, Ferguson (1981) demonstrated a thousand fold range in the values of specific power with a clear distinction between values of 100 and $1{,}000\,\mathrm{Wm^{-2}}$ in the high runoff, steep slope areas of the west, contrasting with the values between 1 and $10\,\mathrm{Wm^{-2}}$ in the low slope, low runoff areas of the south and east. Relations with channel morphology and their spatial variations can be employed to provide explanations for the pattern of long profiles or channel patterns, for example, by considering downstream variations in power to be with minimum unit stream power expenditure, or between equalizing power expenditure and minimizing power expenditure.

Variations in power can also be useful in *river management* problems, as employed by analysing channel adjustments downstream from river channelization works (Brookes 1987); in this case the relationship between bankfull discharge per unit width and water slope was subdivided according to lines of equal specific stream power, showing that eroded sites had specific powers in the range $25\text{–}500\,\mathrm{Wm^{-2}}$ whereas the remaining sites without change had specific powers between 1 and $35\,\mathrm{Wm^{-2}}$. Such examples can be the basis of general guidelines for river managers, with stream power per unit bed area as a criterion for stability in stream restoration projects; a simple classification for guiding river restoration has been developed (Brookes and Sear 1996) by using the ratio between available stream power and erodibility of the substrate. In Denmark, straightened channels tend to recover naturally above a threshold stream power of $35\,\mathrm{Wm^{-2}}$ and it is only channels with very high energies that regain some or all of their original sinuosity, so that thresholds could easily be developed for other river environments. Types of river channel adjustment have therefore been related to thresholds of stream power (Brookes 1990), indicating how temporal variations of stream power can be used to understand *variations over time* that have occurred. In a study of the arroyo systems of the northern part of the Henry Mountains in south central Utah (Graf 1983) it was shown that, whereas stream power decreased in the downstream direction during a deposition period which occurred before 1896 when channels were small and meandering, after 1896 total stream power increased in the downstream direction because channels were in the floors of arroyos that confined discharges and resulted in channel erosion and throughput of sediment. In 1980, however, the rate of downstream change in total power was intermediate between the depositional conditions of the 1890s and the erosional conditions of 1909, with deposition occurring in the smallest and largest channels but not in the mid-basin areas. Stream power can also be used as a unifying theme for analyses of urban fluvial geomorphology (Rhoads 1994) and although values are not always easy to calculate it remains a very important variable in fluvial geomorphology.

References

Allen, J.R.L. (1977) Changeable rivers: some aspects of their mechanics and sedimentation. in K.J. Gregory (ed.) *River Channel Changes*, 15–45, Chichester: Wiley.

Bagnold, R.A. (1960) Sediment discharge and stream power: a preliminary announcement, *US Geological Survey Circular* 421.
——(1977) Bedload transport by natural rivers, *Water Resources Research* 13, 303–312.
Baker, V.R. and Costa, J.E. (1987) Flood power, in L. Mayer and D. Nash (eds) *Catastrophic Flooding*, 1–21, Boston: Allen and Unwin.
Brookes, A. (1987) River channel adjustments downstream from channelization works in England and Wales, *Earth Surface Processes and Landforms* 12, 337–351.
——(1990) Restoration and enhancement of engineered river channels: some European experiences, *Regulated Rivers: Research and Management* 5, 45–56.
Brookes, A. and Sear, D.A. (1996) Geomorphological principles for restoring channels, in A. Brookes and F.D. Shields (eds) *River Channel Restoration. Guiding Principles for Sustainable Projects*, 75–101, Chichester: Wiley.
Bull, W.B. (1979) Threshold of critical power in streams, *Geological Society of America Bulletin* 90, 453–464.
——(1991) *Geomorphic Responses to Climatic Change*, Oxford: Oxford University Press.
Ferguson, R.I. (1981) Channel forms and channel changes, in J. Lewin (ed.) *British Rivers*, 90–125, London: George Allen and Unwin.
Graf, W.L. (1983) Downstream changes in stream power in the Henry Mountains, Utah, *Annals of the Association of American Geographers* 73, 373–387.
Gregory, K.J. (1987) The power of nature-energetics in physical geography, in K.J. Gregory (ed.) *Energetics of Physical Environment. Energetic Approaches to Physical Geography*, 1–31, Chichester: Wiley.
Magilligan, F.J. (1992) Thresholds and the spatial variability of stream power during extreme floods, *Geomorphology* 5, 373–390.
Nanson, G.C. and Croke, J.C. (1992) A genetic classification of floodplains, *Geomorphology* 4, 459–486.
Rhoads, B.W. (1994) Stream power: a unifying theme for urban fluvial geomorphology, in E.E. Herricks (ed.) *Urban Runoff and Receiving Systems: An Interdisciplinary Analysis of Impact, Monitoring and Management*, Proceedings of the Engineering Foundation Conference, Mt Crested Butte, Colorado, 4–9 August, 1991, 84, 91–101.
Schumm, S.A. (1977) *The Fluvial System*, New York: Wiley.
——(1981) Evolution and response of the fluvial system, sedimentologic implications, *Society of Economic Palaeontologists and Mineralogists Special Publication* 31, 19–29.

KENNETH GREGORY

STREAM RESTORATION

Stream restoration is the changing of physical, chemical and biological characteristics of a lotic system to match those of a former natural stream condition, or one that has not been disturbed by humans. One common definition of stream restoration is the 'reestablishment of the structure and function of a stream ecosystem' (National Research Council 1992: 17). Ecological restoration, in general, assists the recovery of an ecosystem that has been degraded, damaged or destroyed and restores its historical trajectory (SER 2002).

Other terms in use include stream rehabilitation, reclamation, reconstruction, mitigation, and 'creation' of new functions and values. Reconstructed channels include a new ecological structure so that the desired flora and fauna can return. Rehabilitated or reclaimed streams are partially restored in that only selected functions and values are returned, primarily to serve a human purpose (e.g. flood control, water supply, land stabilization). Most stream restoration projects are actually partial rehabilitations. It is exceedingly difficult to restore all functions and values of the original stream.

Restoring streams that have been routed through pipes is known as 'daylighting'. Many urban (see URBAN GEOMORPHOLOGY) streams have been paved or lined with large rocks, to carry more water faster and without erosion. Efforts to restore these streams are complex, involve many people, lengthy planning times and are very costly (Plate 131).

Plate 131 Concrete-lined channels and restrictive boundary conditions complicate urban stream restorations

Planning: restore to what conditions?

Stream restoration efforts usually cannot achieve pristine or prehistoric conditions because of the massive cultural changes that would be required. Many streams have been dammed, leveed or channelized to convey floodwaters. Full 'restoration' of these streams would require returning processes like flooding, meander migration, channel avulsion, formation and destruction of LARGE WOODY DEBRIS jams, and backwater sedimentation.

Where management approaches alone cannot achieve the desired ecological functions and values, channel reconstructions using large equipment may be needed. With continued monitoring and adaptive management, reconstructed channels can return some lost ecological functions and values.

Streams that function within natural ranges of flow, sediment movement, temperature, channel migration, and other variables, are said to be in dynamic equilibrium. Restoration projects attempt to restore and maintain DYNAMIC EQUILIBRIUM and ecological integrity. Streambank erosion may be part of this dynamic equilibrium if it is balanced and overall dimensions of the channel remain the same. Successful stream restorations integrate geomorphic processes into their designs. Measurement of channel-forming flow conditions is critical to restoring a stable form, pattern and profile (Rosgen 1996).

Restoration design

Design approaches may be based on stream classification and regional curves for hydraulic geometry (Riley 1998; Rosgen 1996), regime and tractive force equations, and reference reaches (Newbury and Gaboury 1993; Rosgen 1996), but the most rigorous approaches feature more sophisticated hydraulic engineering. Risk of failure can be minimized by incorporating the appropriate levels of management (removing disturbance factors) and engineering controls (e.g. weirs, riprap). Planning and design are best accomplished with an interdisciplinary approach (USDA *et al.* 1998), involving knowledgeable fluvial scientists, ecologists and land users.

'Natural channel design' uses reference reaches in dynamic equilibrium with desired ecological functions and values. Stream classification systems have been developed to assist this process and to promote accurate stream morphological descriptions (Rosgen 1996). When used with experience and scientific design approaches, natural channel designs can result in successful restorations. When used as a 'carbon copy' or cookie-cutter approach, however, the results can be less than successful.

Most stream restoration projects involve modifying an existing stream's location, alignment, meander pattern (see MEANDERING) cross-section dimension, longitudinal profile, or aquatic or terrestrial habitat. Streambank erosion, sediment transport, flooding and sediment deposition are processes that support ecological functions. Having these physical processes become self-sustaining is what separates stream restoration from stream stabilization, stream reconstruction or stream rehabilitation projects.

Soil bioengineering

Soil bioengineering can be an important stream restoration component (Figure 159). Live plants, plant materials and man-made materials are used as systems that interface with Earth materials to create stable streams and banks, and achieve the desired ecological functions and values. Practical application of soil bioengineering techniques requires knowledge of overall performance criteria, design flow and sediment conditions, and habitat suitability.

References

National Research Council (1992) *Restoration of Aquatic Ecosystems: Science, Technology, and Public Policy*, Washington, DC: National Academy Press.

Newbury, Robert W. and Marc N. Gaboury (1993) *Stream Analysis and Fish Habitat Design, A Field Manual*, Gibsons, British Columbia Newbury Hydraulics Ltd, Canada.

Riley, Ann L. (1998) *Restoring Streams in Cities: A Guide for Planners, Policymakers, and Citizens*, Washington, DC: Island Press.

Rosgen, Dave (1996) *Applied River Morphology*, Wildland Hydrology, Minneapolis, MN: Printed Media Companies.

SER (2002) *The SER Primer on Ecological Restoration*, Society for Ecological Restoration Science and Policy Working Group, www.ser.org/

USDA et. al. (10/1998) *Stream Corridor Restoration: Principles, Processes, and Practices*, the Federal Interagency Stream Restoration Working Group, www.usda.gov/stream_restoration

Further reading

Brookes, A. and Shields, F.D. Jr (eds) (1996) *River Channel Restoration: Guiding Principles for Sustainable Projects*, Chichester: Wiley.

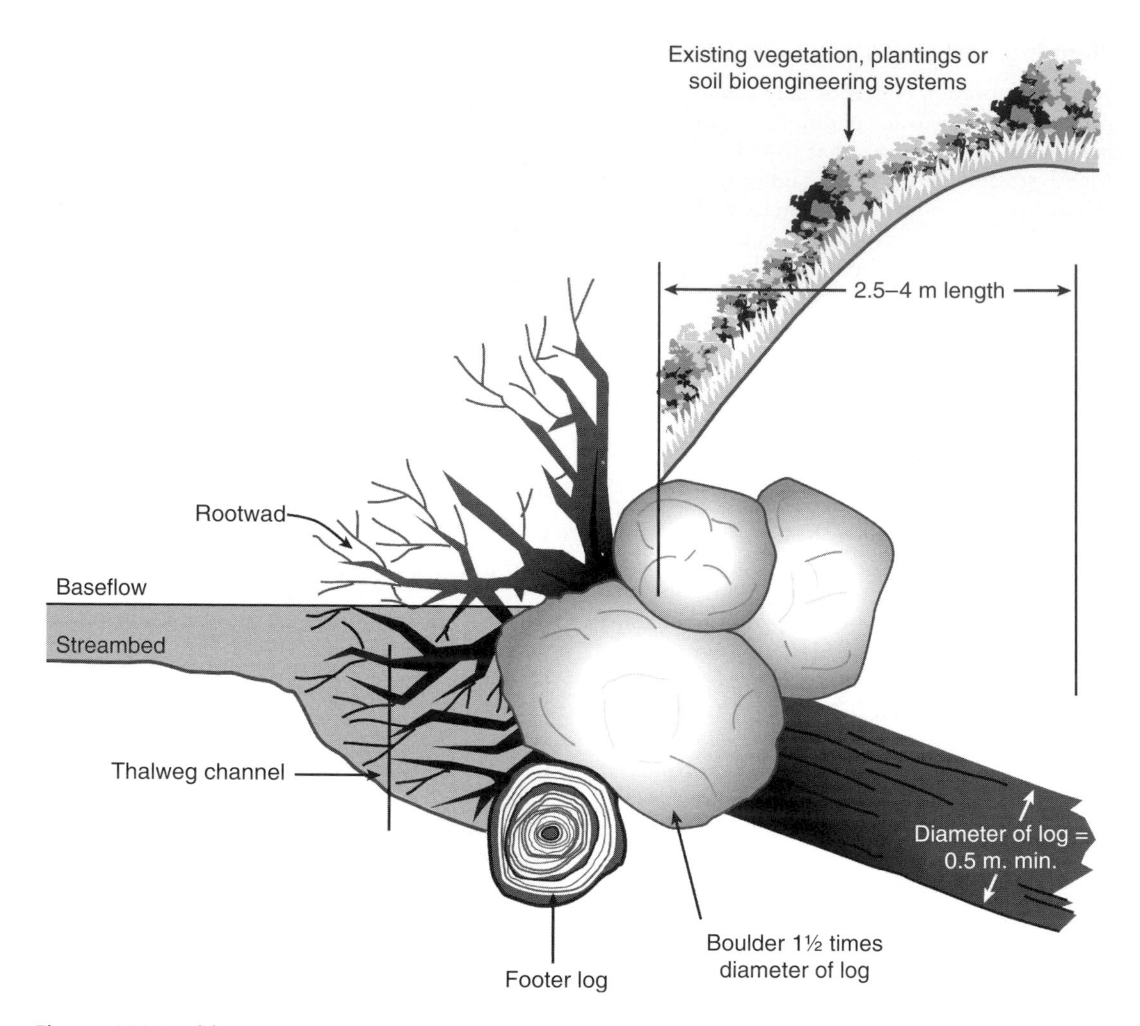

Figure 159 Soil bioengineering example: rootwad

Dunne, T. and Leopold, L.B. (1978) *Water in Environmental Planning*, New York: W. H. Freeman.

Lane, E.W. (1955) The importance of fluvial geomorphology in hydraulic engineering, *American Society of Civil Engineering, Proceedings* 81, paper 745, 1–17.

Leopold, L.B. (1994) *A View of the River*, Cambridge: Harvard University Press.

Raine, A.W. and Gardiner, J.N. (1995) *Rivercare: Guidelines for Ecologically Sustainable Management of Rivers and Riparian Vegetation*, LWRRDC Occasional Paper Series No. 03/95, Canberra: Land and Water Resources Research and Development Corporation.

Ward, D. and Holmes, N.; Paul José (eds) (1995) *The New Rivers and Wildlife Handbook*, National Rivers Authority, Royal Society for the Protection of Birds, and the Royal Society for Nature Conservation: The Lodge, Sandy, Bedfordshire, UK.

SEE ALSO: bankfull discharge; biogeomorphology; fluvial geomorphology; mining impacts on rivers; riparian geomorphology; river continuum; river restoration; sediment budget; step-pool system; stream ordering; stream power

JERRY M. BERNARD

STRIATION

Striations are shallow scratches or grooves cut by brittle impact into rock surfaces, boulders or pebbles. Striations may be up to a metre or more in length. They occur widely in areas of former glacial erosion where rock fragments, sand and silt grains transported in the basal ice have impacted surfaces as the ice moved forward jerkily by basal sliding. Some striations have nail head or wedge shapes, with the broad section being at the down-ice end. Striations occur frequently in glacierized areas, especially on fine-grained physically hard rocks such as quartzites and massive limestones. Glacial polish often occurs with striae. Crossing striations may reflect local ice flow variations, changes in ice flows from different glacial centres during one glaciation, or multiple stages of glaciation. Striations may also form where sea-ice, lake-ice, snow banks, screes, large rock masses and debris creep, flow, avalanche, slide, fall or shear over or onto rock surfaces. ICEBERGs may striate the upper surfaces of clasts in current-winnowed boulder lag horizons of glacimarine deposits. Detritus carried by high velocity water flow in jökulhlaups can also form striations in channels eroded in rock.

Further reading

Benn, D.I. and Evans, D.J.A. (1998) *Glaciers and Glaciation*, London: Arnold.

Bennett, M.R. and Glasser, N.F. (1996) *Glacial Geology: Ice Sheets and Landforms*, Chichester: Wiley.

ERIC A. COLHOUN

STROMATOLITE (STROMATOLITH)

A term first used by Kalkowsky (1908) to describe some sedimentary structures in the Bunter of North Germany. A more modern definition (Walter 1976: 1) is that they are 'organosedimentary structures produced by sediment trapping, binding and/or precipitation as a result of the growth and metabolic activity of micro-organisms, principally cyanophytes'. They can develop in marine, marsh and lacustrine environments and, though they form today where conditions permit, they reached the acme of their development in the Proterozoic (Hofman 1973). The largest known forms are mounds several hundreds of metres across and several tens of metres high. Gross morphologies vary in the extreme and range from stratiform crustose forms, through nodular and bulbous mounds and spherical oncoids, to long slender columns, erect to inclined, and with various styles of branching. Classic examples are known from coastal regions like those at Shark Bay in Western Australia and from pluvial lake shorelines in areas like the Altiplano of Bolivia where they form massive calcareous encrustrations and bioherms (Rouchy *et al.* 1996).

References

Hofman, H.J. (1973) Stromatolites: characteristics and utility, *Earth-science Reviews* 9, 339–373.

Kalkowsky, E. (1908) Oolith und Stromatolith im norddeutschen Buntsandstein, *Zeitschrift Deutsche Geologische Gesselschaft* 60, 68–125.

Rouchy, J.M., Servant, M., Fournier, M. and Causse, C. (1996) Extensive carbonate algal bioherms in upper Pleistocene saline lakes of the central Altiplano of Bolivia, *Sedimentology* 43, 973–993.

Walter, M.R. (1976) Stromatolites, *Developments in Sedimentology* 20.

A.S. GOUDIE

STRUCTURAL LANDFORM

Structural landforms are those which in their appearance reflect, and are adjusted to, geological structure of underlying bedrock. This effect is achieved through direct or indirect control which structural elements exert on the course and intensity of exogenic processes shaping the landforms. 'Structure' here is usually understood *sensu lato*, i.e. it encompasses such diverse phenomena as facies differentiation, lithological contrasts, fracture patterns, faults and folds, tectonic disposition of strata, geometries of intrusive and extrusive bodies, etc. In general, structural landforms develop through differential weathering and erosion which exploit the structures and emphasize unequal resistance of adjacent rock complexes, whereas relief features created by direct action of endogenic forces fall into the category of tectonic landform (see TECTONIC GEOMORPHOLOGY).

Following the above definition, structural landforms may be subdivided into several categories. There are landform assemblages specific for certain rock types, for example granite (see GRANITE GEOMORPHOLOGY) or carbonate rocks (see KARST). In many cases, their development and appearance are controlled by jointing patterns, in the way that joints of regional extent (master joints) and zones of dense fracturing are exploited towards topographic depressions, whereas more massive compartments are left standing as residual hills, uplands or big boulders. Thus, joint-aligned valleys, rows of sinkholes, basins at joint intersections, and many TORs and domed INSELBERGs may be regarded as structural landforms. A number of small-scale features, such as KARREN on carbonate rock outcrops or deep clefts, may develop along joints and are therefore also structural. In igneous rocks in particular, many structural features originating at the consolidation stage subsequently become avenues for weathering and become decisive for the shape of minor and medium-scale landforms (Twidale and Vidal-Romani 1994).

Another group includes landforms reflecting the variable dip of sedimentary strata and, at the same time, the unequal resistance of consecutive layers against exogenic agents. Undeformed, horizontal layers give rise to plains, if at low altitude, or plateaux, if elevated and bounded by marginal escarpments. A plain or a plateau surface is usually underlain by a resistant rock layer, such as quartz sandstone or massive limestone. Dissection of a plateau may expose underlying strata of variable resistance, the stronger of which will support structural benches on valley sides, such as in the Grand Canyon of the Colorado River. Tilting of strata induces differential denudation, in the course of which rock complexes of lower strength are eroded into valleys or rolling plains, whereas more resistant rocks give rise to parallel ridges, escarpments or mid-slope ledges. If the dip is less than 10°, highly asymmetric ridges called CUESTAs develop. With the dip in the range 10°–30°, less asymmetric monoclinal (or homoclinal) ridges form, whereas in the case of even steeper tilt a symmetric ridge named a HOGBACK will originate. In areas built of dipping sedimentary rocks, drainage patterns usually show much adjustment to structure too. Dendritic patterns are typical for negligible dip, whereas with increasing differential erosion they will evolve into trellis patterns. The spatial arrangement of ridges depends on the regional pattern of deformation. If a simple tilt is involved, ridge axes will generally follow straight lines, perpendicular to the dip. If a central part of a former sedimentary basin is downwarped, concentric ridges facing outwards will develop (the Paris Basin is an example), whereas inward facing ridges will typify domed structures with breached central parts.

In mountain areas built of folded sedimentary rocks typical structural landforms are anticlinal ridges and synclinal valleys or, in the case of INVERTED RELIEF, anticlinal valleys and synclinal ridges. Asymmetric homoclinal ridges and hogbacks occur frequently too, but the degree of structural deformation in mountains is usually so high that the spatial extent of these landforms is limited.

A further category includes landforms built of igneous rocks, the present-day appearance of which may reflect the way of magma emplacement. Large granite intrusions in post-orogenic settings may assume the form of large-radius domes (laccoliths) which, after unroofing, are reflected in the topography as upland terrains, sloping in all directions. Dartmoor upland in south-west Britain is one example. Rising igneous domes induce updoming of overlying strata which then may be differentially eroded towards triangular faces called FLAT IRONs (Ollier and Pain 1981). Smaller linear intrusions, i.e. dykes and sills, are typically composed of material more resistant than the host rock, hence differential denudation leaves them appearing as laterally extensive, vertical, jagged ridges (for dykes) or topographic steps (for sills). Denudation of a former volcano may expose deeper parts of the vent filled with solidified, resistant magma, which then becomes a steep-sided conical hill, called a neck. Necks often display impressive columnar jointing of rock, the most famous example being perhaps Devil's Tower in Wyoming, USA.

Many landforms built of extrusive rocks may also be regarded as structural, including rhyolitic domes and plugs, sloping flanks of shield volcanoes, or plateaux underlain by horizontal or gently inclined lava flows (traps). Subsequent tilting and erosion of a multiple lava flow area may produce an assemblage of landforms similar to those developed on tilted sedimentary rocks.

Structural landforms are of various sizes, from features of regional extent to local manifestations of small-scale structures. Mega-scale examples are extensive plains developed upon flat-lying strata or dome mountains. Medium-scale landforms include plateaux, cuesta ridges, domed hills and intermontane basins. Even smaller are tors and mid-slope benches, whereas joint-aligned pools in a bedrock river bed would indicate the most localized structural control.

References

Ollier, C.D. and Pain, C.F. (1981) Active gneiss domes in Papua New Guinea: new tectonic landforms, *Zeitschrift für Geomorphologie* 25, 133–145.

Twidale, C.R. and Vidal-Romani, J.R. (1994) On the multistage development of etch forms, *Geomorphology* 11, 157–186.

Further reading

Gerrard, J. (1986) *Rocks and Landforms*, London: Unwin Hyman.
Peulvast, J.-P. and Vanney, J.-R. (2001) *Géomorphologie structurale – Terre, corps planétaires solides, tome 1, 2*, Paris: Gordon and Breach
Yatsu, E. (1966) *Rock Control in Geomorphology*, Tokyo: Sozosha.

PIOTR MIGOŃ

STURZSTROM

A sturzstrom is a high volume of mostly dry rock material caused by the collapse of a slope or cliff created by large falls and slides moving at high velocities and for long distances, even on a gentle slope (Hsü 1975). Sturzstroms can reach velocities of over $50\,\mathrm{m\,s^{-1}}$ and can travel over distances of kilometres. The accumulation volume may exceed 1 million m^3, covering a total surface of over $0.1\,\mathrm{km}^2$. In relation to its velocity and dimensions, this kind of landslide can be extremely costly in terms of human lives and damage.

Alternative terms for sturzstrom are rock avalanche, rockfall avalanche or rock-slide avalanche (Angeli *et al.* 1996). Examples of historic events are the Elm sturzstrom of 1881 in Switzerland (Heim 1932), the Valpola rock avalanche of 1987 in the Italian Alps and the Frank landslide of 1903 in Canada. In Europe the highest concentration of these phenomena is found in the Northern and Southern Calcareous Alps (Abele 1974).

A sturzstrom can develop (1) by the fall or slide of a rock body which during movement progressively looses its cohesion by turning into dry debris and, thus, continues its advancement as a debris avalanche, (2) by the sudden mobilization of a debris deposit by a debris avalanche or DEBRIS FLOW, either because of the fall of an overhanging rock mass or by seismic shocks.

Although several investigations have been carried out, no universally accepted explanation has yet been proposed. The mechanical analysis of sturzstroms includes two stages: the initial failure and subsequent streaming. Explanations include turbulent grain flow with dispersive stresses arising from momentum transfer between colliding grains (Cruden and Varnes 1996), fluidization of particles caused by incorporating air, the existence of a cushion of trapped air, acoustic fluidization or rock melting by frictional heat (Erismann and Abele 2001).

In alpine regions the ice retreat of the last alpine glaciation caused significant stress relief on mountain slopes. This unloading is one of the causes of Holocene sturzstroms. However, there is evidence that the process not only took place shortly after the ice retreat (as found between 12,000 and 10,000 BP) but that the loss of shear strength needed much more time to cause the process. The Eibsee sturzstrom (Zugspitze, German Alps, volume: $400 \times 10^6\,\mathrm{m}^3$) has been dated at 3,700 years BP which may fit this hypothesis. The exact triggering causes which may enable realistic predictions of sturzstroms requires further attention.

Sturzstroms are highly destructive. Once an event has occurred it is important to discover whether the mass has dammed the valley. If a lake has formed, maximum effort must be given to take control of the breaching dam, because the resulting flow may cause a second disaster.

References

Abele, G. (1974) *Bergstürze in den Alpen*, Wissenschaftl, Alpenvereinshefte, 25, München.
Angeli, M.G., Gasparetto, P., Menotti, R.M., Pasuto, A., Silvano, S. and Soldati, M. (1996) Rock avalanche, in R. Dikau, D. Brunsden, L. Schrott and M.-L. Ibsen (eds) *Landslide Recognition*, 190–201, Chichester: Wiley.
Cruden, D.M. and Varnes, D.J. (1996) Landslide types and processes, in A.K. Turner and R.L. Schuster (eds) *Landslides. Investigation and Mitigation*, 36–75, Washington: National Academy Press.
Erismann, T.H. and Abele, G. (2001) *Dynamics of Rockslides and Rockfalls*, Heidelberg: Springer.
Heim, A. (1932) *Bergsturz und Menschenleben*, Zürich: Fretz und Wasmuth (English trans. N. Skermer, (1989) *Landslides and Human Lives.*, Vancouver, Canada: BiTech Publishers.
Hsü, K.J. (1975) Catastrophic debris streams (sturzstroms) generated by rockfall, *Geological Society of America Bulletin* 86, 129–140.

RICHARD DIKAU

SUBAERIAL

Refers to all conditions or processes occurring in the open air or on the land surface (e.g. subaerial weathering) as opposed to those occurring in submarine (underwater) or subterranean (underground) environments. The term also applies to materials and features that are created and/or located on the Earth's surface (e.g. subaerial aeolian dunes, subaerial volcano, etc.), and sometimes is inclusive of fluvial forms (of rivers). The notion that all things in the landscape are created by subaerial processes and conditions is termed subaerialism.

STEVE WARD

SUBCUTANEOUS FLOW

The lateral transfer of water in the subcutaneous (or epikarstic) zone. The subcutaneous zone is a highly weathered region in well-developed KARST environments lying in the upper part of the percolation zone, between the soil and the relatively unweathered and permeably saturated phreatic zone below. When water stored in the subcutaneous zone is full (e.g following precipitation) a potentiometric surface (epikarstic water table) is produced. Any further input is subsequently transferred laterally, with movements of water occurring along preferred pathways (neighbouring joints and shafts) down the hydraulic gradient in the epikarstic water table (Williams 1983). Corrosion is intensified where flow routes converge above the more efficient paths, resulting in differential surface lowering, accentuating over time and by irregular distribution of solution. Uniform percolation through the subcutaneous zone will result in uniform surface lowering. Rates of subcutaneous flow vary considerably, commonly between 2–10 weeks, but are hard to gauge (especially in old, well-developed karsts). The term is also used to refer to the process of piping in soils.

Reference

Williams, P.W. (1983) The role of the subcutaneous zone in karst hydrology, *Journal of Hydrology* 61, 45–67.

Further reading

Gunn, J. (1981) Hydrological processes in karst depressions, *Zeitschrift für Geomorphologie* 25(3), 313–331.

SEE ALSO: epikarst

STEVE WARD

SUBGLACIAL GEOMORPHOLOGY

Subglacial geomorphology has the most profound and indelible imprint on any glaciated landscape. Subglacial geomorphic processes are among the most complex, yet least understood set of glacial processes. Our restricted knowledge stems from the inaccessible nature of what occurs beneath an ice mass and the limited extent of modern analogues. No other aspect of glacial geomorphology impacts on the daily lives of millions of people to such a degree as does subglacial geomorphology, for example, in terms of foundations, roads, railroads, waste disposal sites, agriculture, aquifers and construction materials.

The subglacial environment is that glacial subsystem directly beneath an ice mass that includes cavities and channels that are not influenced by subaerial processes. Subglacial geomorphology considers all aspects of topographic change beneath ice masses as a result of erosional and depositional processes (Figure 160). The subglacial environment is a boundary interface where complex sets of processes interact altering the morphological, thermal and rheological states of the interface between the ice mass and its bed. The key to understanding subglacial geomorphology lies in the mechanics of this interface. This boundary zone migrates across the landscape with every ice advance and retreat. All terrains covered by glaciers have been affected and altered by the passage of this interface. This interface is a function of the prevailing basal ice and bed conditions; a complex relationship between basal ice dynamics, sediments and bedrock, subglacial hydraulics and the glacier bed ambient temperature. Changes in basal ice and/or bed conditions may be widespread or local, and develop rapidly or slowly. These fluctuating conditions may be of enormous magnitude or simply be minor variations at the interface, the former being detectable whilst the latter may leave little or no imprint on subglacial geomorphology.

Subglacial geomorphological processes are constrained by the temperature at the ice–bed interface. Basal thermal regimes can be either temperate or polar but in all likelihood most ice masses are polythermal. In temperate, wet-based glaciers, typical of almost all ice masses today and most ice masses during the Pleistocene and earlier, basal temperatures are found at −1 to −3 °C. In polar, cold, dry-based glaciers, typical of a few isolated central areas of East Antarctica today and possibly the central areas of the vast Pleistocene ice sheets, the basal parts of an ice mass are frozen to the bed with temperatures of −13 to −18 °C (Van der Veen 1999). Thermal conditions at the ice–bed interface are more

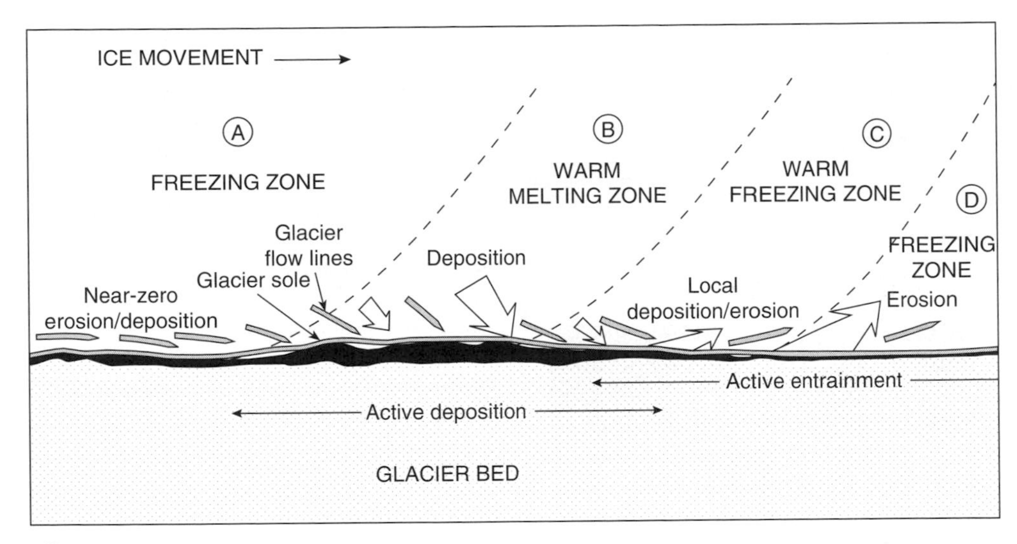

Figure 160 Models of subglacial thermal regimes, their spatial relationships and processes of subglacial erosion, transportation and deposition

complex than implied by these two thermal states, in fact basal ice temperatures vary temporally and spatially producing polythermal conditions (Menzies and Shilts 2002).

In considering ice–bed interface temperature fluctuations, the following parameters converge to establish the specific thermal conditions: (a) rate of snow accumulation and snow temperature; (b) geothermal heat flux; (c) mean annual surface air temperature; (d) ice surface velocity; (e) basal ice velocity; (f) subglacial meltwater flux and temperature; and (g) the imprinted 'memory' of previous thermal conditions (Figure 161a). The relevance of these two extreme thermal states is that under temperate conditions, meltwater occurs at the ice–bed interface, basal ice containing debris can be released, and meltwater processes and/or saturated debris moving as a deforming layer can be accomplished thereby permitting various landforms to develop. Under polar conditions, with the ice mass frozen to its bed, no meltwater is present and only ice movement by plastic deformation occurs thereby geomorphological processes are constrained but not altogether suspended.

It is apparent that polythermal bed conditions prevail in the long term under any ice mass, the switch from wet to dry and back again repeatedly results in localized erosion and deposition of much glacial sediment. Although long distance transport does occur beneath and within ice masses the dominant form of transport is short distance of <10 km resulting in subglacial sediments being typically locally derived, transported and deposited. Figure 160 illustrates the probable relationship between areas of dominantly erosive glacial action and depositional processes under ice sheet conditions. Beneath valley glaciers, due to much shorter distances and thinner ice cover (lower basal ice pressures), a similar, but less complete, set of conditions may prevail (Benn and Evans 1998).

Subglacial erosional landforms

Subglacial erosion processes are pervasive at the active ice–bed interface. In some terrains under certain subglacial conditions, erosive processes become dominant. Almost any description of the impact on the land surface of past glaciers focuses on the grandeur and size of FJORDS and the sculpted bedrock features of once glaciated terrains. However, our understanding of how these distinctive features were fashioned remains limited. Erosional forms exist at an immense range of

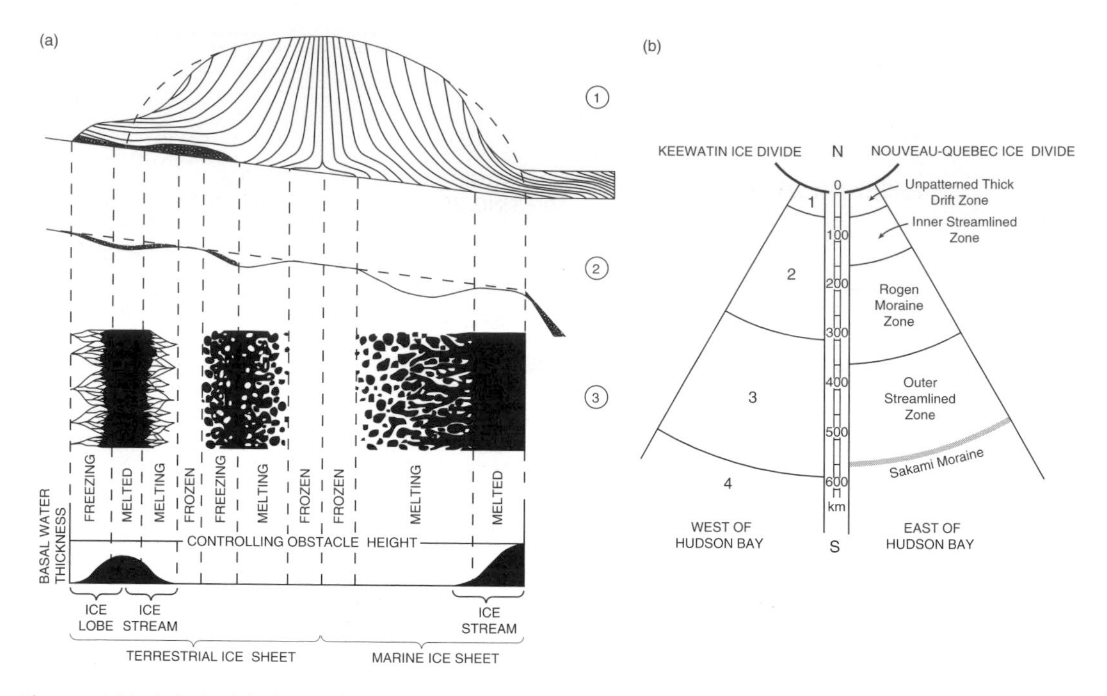

Figure 161 Subglacial thermal zonation in relation to subglacial processes beneath a steady-state ice sheet with terrestrial and marine margins. 1, internal ice flow trajectories; 2, subglacial topography created by subglacial erosion; 3, areal distribution of subglacial meltwater (reprinted with modification after Hughes 1995)

scales from surface microscopic features on particles to the megascale landforms. Regional erosional features can be subdivided into regional and local forms. However, some forms, such as ROCHES MOUTONÉES, occur at all scales. Regional erosional features are either areal (spatially pervasive) or linear (spatially discrete) landscape types. In the former case, scoured terrain develops where limited debris existed at the ice–bed interface and dominantly polar bed conditions prevailed. In contrast, linear erosion processes are differential in their impact upon terrain being confined within specific areas. Linear erosional forms are indicative of subglacial bed states in which rapid but spatially restricted basal ice movement and/or meltwater channelling has occurred.

Terrains of regional areal erosion typically exhibit low relief amplitude, limited sediment deposition and have a moulded and scoured appearance. Geological structure has often been partially exhumed in these terrains and irregular depressions and small roches moutonnées occur. These terrains, known as 'knock and lochan' in north-west Scotland, are typical of this form of GLACIAL EROSION. Similar landscapes exist in shield terrains in Canada (Plate 132a) and Fennoscandia, along the edges of the Greenland and Antarctic Ice Sheets, in Patagonia, and South Island, New Zealand. The form of this regional erosional landscape appears related to relatively slow-moving ice masses under polar-bed conditions with limited debris present. A range of wear processes operate across bedrock surfaces where occasional protuberances lead to ice pressure melting and meltwater discharges often under high pressure. Within this landscape, at a lower scale, crag and tails, and roches moutonnées are prevalent. Typically, P-forms and associated forms related to rapid subglacial meltwater flow are common.

Beneath specific zones within ice sheets, regional linear erosion appears to occur probably as a consequence of ice streaming and fast-moving, but spatially restricted, basal ice. Under these conditions, major linear forms of glacial erosion develop, such as incised bedrock troughs, fjords and tunnel valleys. At meso and microscale fluted bedrock, striae and grooves are witness to this style of selective erosion.

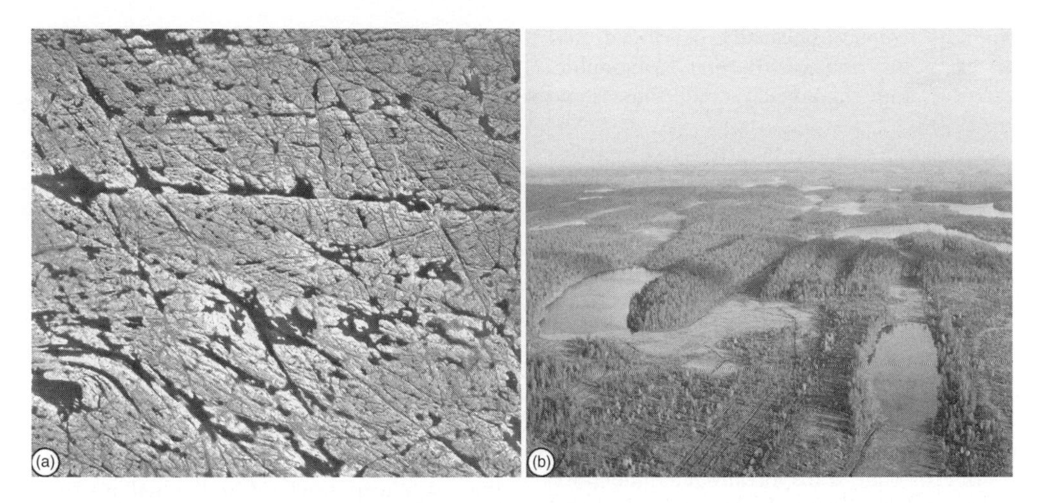

Plate 132 (a) Areally scoured glaciated landscape of the Canadian Shield near La Troie, Quebec (photo courtesy Government of Canada); (b) subglacial depositional landscape of drumlins of the Kuusamo drumlin field, Finland (photo courtesy of R. Aario)

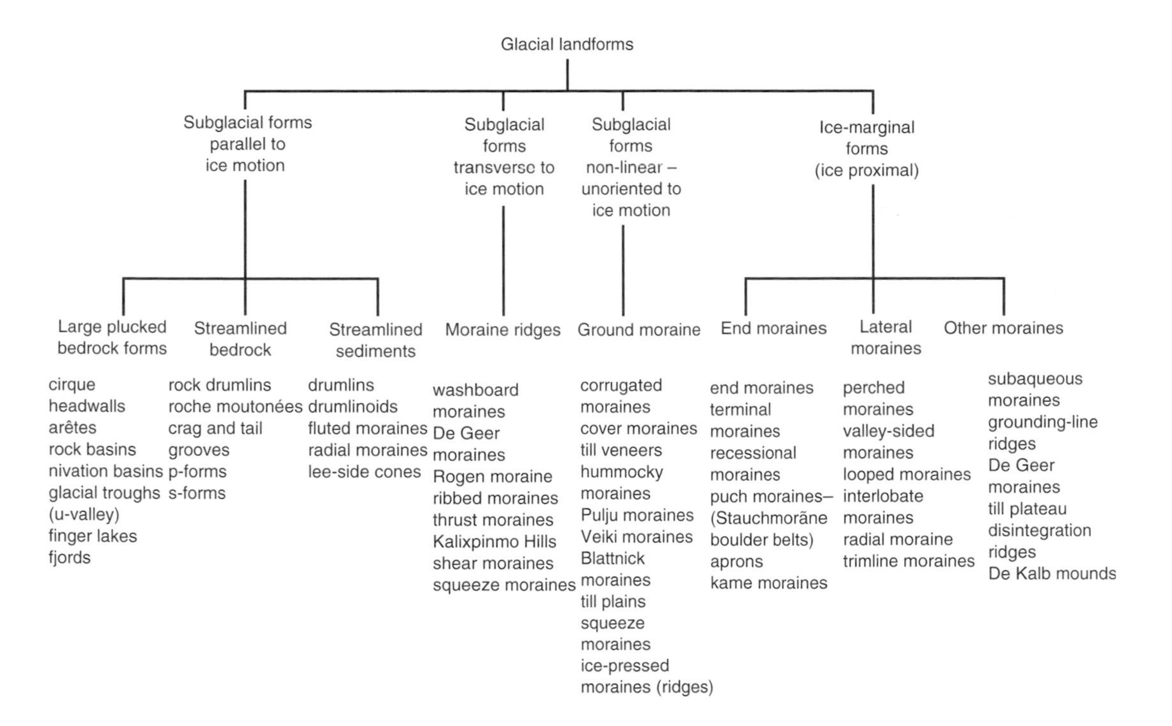

Figure 162 Subglacial landform types

Subglacial depositional landforms

Subglacial landforms can be subdivided into those developed under (1) active ice flow (advance and active retreat phases); and (2) passive or 'dying' ice flow (retreat phase). It is apparent that at least one group of landforms, previously considered as separate entities, may be related to each other as variant bedforms in a continuum of landforms at the subglacial interface. Thus drumlins, fluted moraine and Rogen moraine may be a spectrum of forms subglacially developed that evolve as a function of thermal, topgraphic, glaciological and rheological conditions (Plate 132b).

Suites of landforms can be attributed to active ice and to indirect or passive ice action (Figure 162). Most of the forms are constructional under active and/or indirect passive ice action while others may be partially erosional following ice retreat and the influence of subaerial processes. The location and type of subglacial landforms reflect the complex patterns of processes beneath the ice in differing locations under varying glacidynamic conditions (Figure 161b).

Subglacial geomorphology is a reflection of a myriad of erosional, transportational and depositional processes that produce a complex and varied glaciated landscape within which the influence of underlying topography may be muted due to the effects of deposition or sharpened due to erosional processes. The uniqueness, for example, of fjord landscapes or the gently rolling plains of the Prairies in North America stand as remarkable testaments to the impact on the Earth's surface of subglacial geomorphology.

References

Benn, D.I. and Evans, D.J.A. (1998) *Glaciers and Glaciation*, London: Arnold.

Hughes, T.J. (1995) Ice sheet modelling and the reconstruction of former ice sheets from glacial geo(morpho)logical field data, in J. Menzies (ed.) *Modern Glacial Environments*, 77–100, Oxford: Butterworth-Heinemann.

Menzies, J. and Shilts, W.W. (2002) Subglacial environments, in J. Menzies (ed.) *Modern and Past Glacial Environments*, 183–278, Oxford: Butterworth-Heinemann.

Van der Veen, C.J. (1999) *Fundamentals of Glacier Dynamics*, Rotterdam: A.A. Balkema.

JOHN MENZIES

SUBMARINE LANDSLIDE GEOMORPHOLOGY

The term submarine landslide is often used as a generic descriptor for an erosive and/or depositional feature, of sediment or rock, preserved on the seafloor, or below it (if ancient) after a MASS MOVEMENT. Less generic definitions of seafloor failures such as DEBRIS FLOWS, avalanches, spreads, and the like suggest a rheology of the material as it fails. This is often hard to discern when only a scar is left as evidence of erosion, and as evidence of deposition only the rare sediment or debris pile that can be discerned by remote sensing techniques. An excellent review of submarine landslides (Plate 133) can be found in Hampton *et al.* (1996).

Detection

Most submarine landslides are old and may take place unnoticed if contemporary. Therefore, evidence of an event is obtained by remote sensing techniques such as seismic reflection, side-scan sonar and multibeam bathymetry. Data are then analysed using a seismic interpretation package that can keep track of individual horizons, or assembled in a Geographic Information System (GIS) database.

Remote sensing techniques used in the ocean employ the reflection of acoustic waves to measure the depth and physical properties of material on the surface and in the subsurface. In general, higher frequency acoustic energy yields higher resolution returns, but the energy will not propagate as deep. A 12 kHz (kilohertz) signal is used to measure water depth, and has very little sub-seafloor penetration, whereas a 3.5 kHz signal can penetrate over 100 m in soft sediment, and a high power, low frequency airgun source can penetrate many kilometres below the seafloor. Once a potential feature is identified, groundtruthing and geotechnical analysis of the material can be done using a variety of coring devices.

Types

Submarine mass movements have many characteristics similar to subaerial landslides, and much of the terminology is the same. Submarine failures tend to be much larger, and often better preserved over the timescale of tens of thousands of years due to slow diffusion processes in deepwater environments. Despite their size and preservation, submarine failures are hard to detect, and there is seldom a witness to describe the dynamics of the slide as it occurs. Furthermore, there is a tendency for submarine sediment failures to disintegrate, leaving a deposit that is difficult to discern on the surface. We are left to make inferences based on the resulting morphology of the erosive and depositional features, and physical properties of the surrounding material.

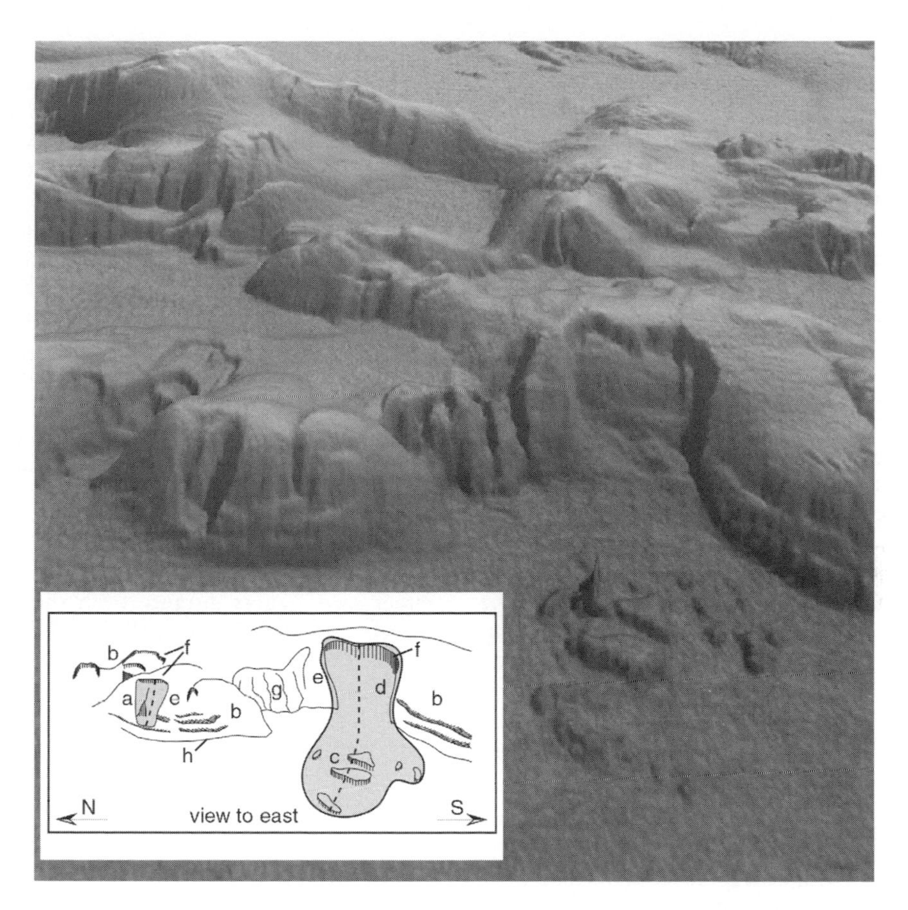

Plate 133 Submarine landslides imaged using multibeam bathymetric data from the base of the convergent continental margin, offshore Oregon, USA (modified from McAdoo *et al.* 2000). Approximately 3× vertical exaggeration, view is looking east with north to the right. The field of view at the base of the slope is approximately 30 km. The letters in the inset line drawing are referred to below. Disintegrative slides (a) have characteristic scar morphology, but lack bathymetric evidence of failed material at the base of the slide. Blocky slides have cohesive blocks of material (c) that can be related to a particular slide scar (d). The unfailed slope adjacent to the failure scar (e) is used as a proxy for the pre-failure slope conditions. Regions of the slope where material has clearly been eroded (g) are difficult to classify as a landslide as they lack characteristic evidence of discrete slide events such as a distinct headscarp (f). A series of curious terraces (h) rim the very base of the slope, and often coincide with the lowermost terminus of the slides, resulting in a 'Hanging Slide' (a). Notice the very steep seafloor with little visible evidence of erosion between slides on the westernmost ridge (closest) and on the other background ridges (b)

Failed material can either remain cohesive (a 'blocky' slide) or disintegrate into a flow. *Mass movements* (translational and rotational slides and slumps) move downslope under the force due to gravity and other body forces. The physics of *mass flows* (TURBIDITY CURRENTs, avalanches, debris flows, etc.) is governed by tractive stresses associated with a fluid.

A displaced mass of sediment or rock originates from and moves along a *rupture surface* (failure or slide surface), and portions of that failed mass may remain in contact with the rupture surface. If the rupture surface is curved (concave upward), the displaced mass rotates, and the failure is often termed a *rotational* slide or slump. If the failure plane is flat, say a bedding plane, then the slide is described as *translational*. As a block of failed material moves downslope it may disintegrate and become either a debris flow (large clasts supported by a fine-grained matrix) or *turbidity current* (turbulent flow of fluid supported sediment; Morgenstern 1967). LIQUEFACTION occurs when repacking of sediment grains causes a buildup of fluid pressure to the point where individual grains lose contact with each other, reducing the material's shear strength to nil. This often occurs during earthquakes, and mass flows can occur on very low inclines.

Mechanics

Failure occurs when the downslope-oriented shear stress exceeds the shear strength of the slope material. As stresses for any given environment are likely to be similar (with the possible exception of earthquakes), there is a tendency to focus on the strength side of the equation, which can range from unconsolidated sediment to well-lithified igneous rock.

The strength can be defined by the effective cohesive strength (C_0), the stress acting normal to the failure surface (σ_n), the PORE-WATER PRESSURE (p_f) and the angle of internal friction (ϕ)

$$\tau_f = C_0 + (\sigma_n - p_f) \tan \phi$$

Many submarine landslides occur on slopes well below the angle of repose (see REPOSE, ANGLE OF), therefore require a mechanism to reduce the strength of the sediment at the locus of failure. For a material with a given C_0 and ϕ, the reduction in strength can come from either an increase in p_f or a reduction in σ_n. Sediment strength is not affected by water depth, as hydrostatic conditions prevail to the depths of most slides. Therefore, rock or sediment on the seafloor will not fail on slopes less than ϕ without a transient of some sort.

There are several processes that increase p_f which in turn reduces the EFFECTIVE STRESS ($\sigma_n - p_f$), hence reducing the strength of the material. Gas hydrates (a methane-ice compound found in the sub-seafloor of many continental margins) dissociate with a fall in sea level (reduction of pressure) or an increase in temperature, and release bubble-phase methane which can get trapped, leading to an increase in p_f. Large failures that occurred during sea-level lowstand and may be related to hydrate dissociation include the Storegga slide offshore Norway, where 5,000 km^3 of material failed 8,000 years ago (Bouriak *et al.* 2000), and numerous slides in the Beaufort Sea (Kayen and Lee 1991). Earthquake generated seismic waves propagating through sediment can also increase pore-water pressures as the waves pass rapidly enough to prevent pressure dissipation (Lee and Edwards 1986). Accelerations from earthquakes can cyclically reduce and increase σ_n, therefore the resulting increase in p_f (which is frequency dependent) may be as important as a 'critical acceleration' required for failure. In shallow water, storm waves may also generate increased pore pressures due to cyclic loading, and failure can be initiated in zones of increased shear stress halfway between the crests and troughs of the waves (Seed and Rahman 1978). These failures will occur in shallow water (continental shelf and upper slope), and should not be deep-seated.

Post-failure evolution

Depending on the stresses present, sediment properties, and slope morphology, a failed mass can either stop a short distance from the source, or travel very long distances. One method of determining the likelihood of a sediment failure to disintegrate or remain as cohesive packages under cyclic loading is to consider the pre-failure water content and steady-state effective stress (see Hampton *et al.* 1996). For material with a high water content and effective stress, pore pressure is likely to increase during a transient loading event, and the effective stress will decrease, resulting in a tendency towards disintegrative failures ('contractive' sediment, as the fabric tends to collapse). In contrast, sediment that already has a high effective stress and lower water content, cohesive failures are more likely to occur as the sediment dilates, reducing the effective stress.

The post-failure dynamics are a complex combination of viscous and plastic elements. Norem *et al.* (1990) proposed an approach to modelling submarine landslides with a combination of these plastic and viscous terms:

$$\tau = \tau_c + \sigma_n \, [1 - (p_f/\rho_s H)] \tan \phi + m \rho \, (dv/dy)^r$$

where τ is the total shear strength mobilized during the flow, τ_c is the yield strength (similar to C_0 in the static case), $p_f/\rho_s H$ is the ratio of pore pressure to sediment density (ρ_s) and failure thickness (H), with an additional viscous term where m is the Bingham viscosity, ρ is the fluid density and a vertical velocity gradient term $(dv/dy)^r$ where r ranges between 1 for macroviscous and 2 for inertial flows. This formulation may be able to predict the final shape of a failure deposit, but does not explain the incredible mobility of slides such as the Nuuanu slide off Hawaii, where individual blocks travelled hundreds of kilometres. A hydroplaning model, where the failed material traps a layer of water above the rupture surface, helps explain the mobility of some failures (Mohrig *et al.* 1998)

Location

Submarine landslides are ubiquitous throughout the oceans, occurring on ACTIVE MARGINs and PASSIVE MARGINs, continental shelves and slopes, in the heads of FJORDs and near active deltas, on the seaward side of barrier reefs (see BARRIER AND BARRIER ISLAND), and on the flanks of mid-ocean ridges, seamounts, GUYOTs, or volcanic islands. Triggers of offshore landslides include earthquakes, large storms, SEA LEVEL change, rapid sedimentation, gas hydrate dissociation, or channel erosion in SUBMARINE VALLEYs and CANYONs among other things. Listed below are environments particularly susceptible to failure.

RIVER DELTAS

As regions of rapid sedimentation and erosion in canyons and channels, RIVER DELTAs offer zones of steep slopes and high fluid pressure that facilitate failure. When deposited on the seafloor, fine-grained sediment often has high water content yet low permeabilities, which can lead to fluid overpressure and underconsolidation. When deposited, organic-rich sediment decays producing bubble phase hydrocarbons, which subsequently aids overpressure.

The steady accumulation of fluid overpressures, and the increasing slopes are not sufficient to generate a noticeable failure – a transient such as a storm or earthquake is a necessary trigger. In 1969, large swells from Hurricane Camille in the Gulf of Mexico cyclically loaded the seafloor sediment yielding a build-up of fluid pressure that caused a failure in 100 m of water. In 1980, a magnitude 7.0 earthquake caused liquefaction of overpressured sediment on a 0.25-degree slope in the Klamath River delta, offshore northern California.

FJORDS

The offshore portion of deltas at the heads of FJORDs is a region highly susceptible to landslides. The high, often organic-rich, sediment loads of streams that drain glaciers can create elevated fluid pressures in the delta due to rapid sedimentation and decomposition that creates gas. The combination of fine-grained rock flour and coarse sediment, fluid overpressuring, and the steep nature of fjords make for a highly susceptible environment for failure during earthquakes, which are common in isostatically rebounding or tectonically uplifting regions where glaciers are likely.

The very nature of a fjord increases the landslide hazard. Fjords are an attractive place for human settlement as they offer good deepwater ports, flat land, and fresh water supply in mountainous regions close to the sea. Despite the fact that the Port of Valdez, Alaska (located at the head of a fjord) was destroyed following a submarine landslide that retrogressed onto land following an earthquake, the Trans-Alaska pipeline (which is responsible for bringing all of Alaska's North Slope crude oil to market) ends there. Were a similar event to have happened twenty years later, an unprecedented ecological disaster may well have occurred. To add to the hazard, the funnel-shaped geometry of fjords helps to focus tsunami waves. In an extreme example, a subaerial landslide at the head of a fjord in Lituya Bay, Alaska in 1958 triggered a 525 m high wave witnessed only by a few lucky boaters.

CANYONS

Submarine canyons are the primary transport avenues for sediment moving from land to the deep water. As such, these dynamic environments tend to have significant landslides. Canyon heads sequester the sediment that moves in littoral currents on the shelf. Following the 1989 Loma Prieta earthquake in central California, a failure occurred in the head of Monterey Canyon. These canyon-head failures likely fluidize (see FLUIDIZATION) into erosive turbidity currents that eroded the canyon floor, and are deposited on flat seafloor as turbidites. Successive incision events can eventually lead to large, slope-clearing landslides on the canyon walls (Densmore *et al.* 1997).

Regularly spaced canyons that occur on continental slopes but do not breach the shelf break, and are isolated from downslope erosive flows, are termed 'headless' (Orange and Breen 1992; see also GROUND WATER). Elevated seepage forces generated by the topography focusing groundwater flow to the seafloor can assist in failure, especially during transient events such as earthquakes (McAdoo *et al.* 1997). These failures are most likely small, but are a key factor in shaping the seafloor landscape.

CONTINENTAL SLOPE

Submarine landslides occur on continental slopes regardless of tectonic activity, latitude or sedimentation histories, however the temporal frequency of occurrence may vary substantially. The continental slopes are the regions in the ocean where the steepest sedimented slopes tend to exist, therefore it is a likely locus for landslide activity. The slope gradients, however, rarely approach the *angle of repose*, except in *canyons*, where *overconsolidated* (well-*lithified)* material is exposed at the seafloor, therefore either an increase in pore pressure or a significant transient is required for failure.

Many consider earthquakes the de facto trigger for slides on continental margins, even on passive margins where seismic events are less common. Curiously, one of the better concrete examples of an earthquake-triggered landslide occurred after the magnitude 7.2 Grand Banks earthquake in 1929 offshore from Canada's passive North Atlantic margin, and there have been many large earthquakes in regions with little evidence of substantial landsliding.

Landslides on open continental slopes may be associated with changes in sea level. Gas hydrate dissociation occurs during times of falling sea level, and may provide enough free gas to reduce the strength (increasing p_f) of the overburden, causing failure. Sedimentation rates are likely to be higher on the continental slope when sea level is low. Rivers bypass the sediment sinks in estuaries (see ESTUARY)and on the CONTINENTAL SHELF, and deposit material directly on the continental slope and in deep water by way of canyons. Rapid sedimentation not only increases pore pressures within the sediment, but provides a source of material to fail. Layers of finer grained mechanically weaker material deposited during highstands may provide a sliding surface following low stand sedimentation events. Continental slopes may well enter an equilibrium (see EQUILIBRIUM SLOPE) condition not long after a lowstand has provided material on the slope for which to fail.

VOLCANIC ISLANDS

Some of the largest and potentially most devastating submarine landslides occur on the flanks of volcanic islands and the mid-ocean ridges. Steep slopes are made of often highly fractured, hydrothermally altered and thermally stressed material. Earthquakes, rapid surface changes and rapidly expanding gases come with movement of subsurface magma. A block from prehistoric Nuuanu slide off Oahu, Hawaii is so large (30 km long, 17 km wide and 1.8 km thick) and far from its source (almost 100 km) it was initially identified as a seamount. A slide such as this may have been responsible for a tsunami that deposited material hundreds of metres high on the Hawaiian island of Lanai. Another slide on the Canary Islands in the Atlantic Ocean covered an area of 2,600 km^2 and 150 km^3 of material. It has been speculated that the tsunami from this slide was responsible for depositing a car-sized boulder some 20 m above sea level on Eleuthera Island, Bahamas on the other side of the ocean.

Hazard

Most submarine landslides are small, and are not likely to be noticed. The *hazard* associated with offshore slides is of primary concern to structures such as oil drilling rigs and production facilities, seafloor pipelines and cables. In 1969, Hurricane Camille in the Gulf of Mexico caused a drilling platform offshore from Louisiana to be buried in mud and moved 30 m downslope. Some slides begin in the offshore and can retrogress headward towards land. Following the 1964 Alaska 'Good Friday' earthquake, the Port of Valdez waterfront slumped into the ocean, killing thirty people.

TSUNAMI

Another significant *hazard* that may result from submarine landslides is TSUNAMI. There are two features that are unique to submarine landslide-generated tsunami. First, if the landslide was triggered by an earthquake, the tsunami generated tends to be larger than one might expect for the earthquake alone. Second, whereas earthquake-generated tsunami tend to affect the very large areas, and even cause significant damage throughout the ocean basin, landslide-generated tsunami tend to have more localized affects. On 17 July 1998 in Papua New Guinea, a tsunami with waves up to 15 m high followed a magnitude 7.0 earthquake, killing over 3,000 people. This was an unusually large tsunami given the size of the earthquake, and the region of devastation was limited to a small (~40 km) stretch of coast. Detailed multibeam and seismic surveys of the offshore revealed a large *amphitheatre* where a landslide likely triggered by the earthquake was initiated. This landslide probably contributed significantly to the amplitude of the tsunami.

The magnitude 7.2 earthquake offshore from Canada in 1929, triggered a large submarine landslide. This landslide quickly transformed into turbidity currents that travelled over 700 km into the abyssal plains, severing trans-Atlantic cables. Based on the timing of the cable breaks, the turbidity current velocity was calculated to be 55 km hr^{-1}. The seafloor displacement from the landslide caused a tsunami that hit Newfoundland's Burin Peninsula, killing twenty-eight people. Canada's worst earthquake-related disaster occurred on a passive margin because of a tsunami set off by a submarine landslide.

References

Bouriak, S., Vaneste, M. and Saoutkine, A. (2000) Inferred gas hydrates and clay diapers near the Storegga slide on the southern edge of the Voring Plateau, offshore Norway, *Marine Geology* 163(1–4), 125–148.

Densmore, Alexander L., Anderson, R.S., McAdoo, B.G. and Ellis, M.A. (1997) Hillslope evolution by bedrock landslides, *Science* 275(5,298), 369–372.

Hampton, M., Lee, H. and Locat, J. (1996) Submarine landslides, *Reviews of Geophysics* 34, 33–59.

Kayen, R. and Lee, H. (1991) Pleistocene slope instability of gas hydrate-laden sediment on the Beaufort Sea margin, *Marine Geotechnology* 10, 125–141.

Lee, H. and Edwards, B. (1986) Regional method to assess offshore slope stability, *Journal of Geotechnical Engineering, ASCE* 112, 489–509.

McAdoo, B., Orange, D., Screaton, E., Lee, H. and Kayen, R. (1997) Slope basins, headless canyons, and submarine palaeoseismology of the Cascadia accretionary complex, *Basin Research* 9, 313–324.

McAdoo, B., Pratson, L. and Orange, D. (2000) Submarine landslide geomorphology, US continental slope, *Marine Geology* 169, 103–136.

Mohrig, D., Whipple, K., Hondzo, M., Ellis, C. and Parker, G. (1998) Hydroplaning of subaqueous debris flows, *Geological Society of America Bulletin* 110(3), 387–394.

Moore, J. and Moore, G. (1984) Deposit from a giant wave on the island of Lanai, *Science* 226, 1,312–1,315.

Morgenstern, N.R. (1967) Submarine slumping and the initiation of turbidity currents, in A.F. Richards (ed.) *Marine Geotechnique*, 189–210, Urbana: University of Illinois Press.

Norem, H., Locat, J. and Schieldrop, B. (1990) An approach to the physics and the modeling of submarine landslides, *Marine Geotechnology* 9, 93–111.

Orange, D. and Breen, N. (1992) The effects of fluid escape on accretionary wedges II: seepage force, slope failure, headless submarine canyons and vents, *Journal of Geophysical Research* 97, 9,277–9,295.

Seed, H. and Rahman, M. (1978) Wave-induced pore pressure in relation to ocean floor stability of cohesionless soils, *Marine Geotechnology* 3, 123–150.

BRIAN G. McADOO

SUBMARINE VALLEY

Submarine valleys (canyons), sometimes tightly linked to fluvial systems, represent one of the main components that contribute to modelling the sea-bottom. They are one of the main ways for sediments to move from coastal zones to the deep oceanic basins, strongly cutting long bands of submarine scarp. Submarine valleys, localized on the main continental margins, show different features and evolutionary patterns according to whether they have a subaerial or submarine origin (De Pippo *et al.* 1999).

The characters and forms of some valleys confirm the hypothesis of their prevalently erosive origin as continental valleys. This hypothesis, even if it is acceptable for important palaeovalleys now located in shallow waters and filled with recent sediments, can hardly explain, except for particular cases, the characters and the evolution of active canyons. In fact, the bottom of most submarine valleys is a thousand metres deep, below the lowest depth (−200 m) reached by fluvial erosion during the maximum lowstand of sea level in the last glacial.

It is possible to distinguish the origin of submarine valleys on the base of certain main factors, such as eustatism and tectonics, that control their genesis and evolution. In particular the presence of valleys with a subaerial origin is more frequent in the less steep and deep areas that have experienced important sea-level changes, with the partial emersion of the continental shelf. Deep valleys, instead, are cut on steep continental margins where a big sedimentary supply activates big turbiditic flows that cut the trace of the primordial canyon.

In elevated depth areas, where the contrast between the sea waters and the waters filled with solutes and suspended sediments ceases, the mixture of water and sediment begins to slide offshore, also favoured by the shelf slope. The high sediment supply allows the erosion of the sediments previously deposited in the channels during the periods of insufficient solid supply by moving the sediments in the delta, from which turbiditic flows can originate following tectonic and/or seismic events. These flows are able to strongly erode the substratum, producing a deep incision on the shelf, where a valley could be already present. In shallow waters, instead, where the continental shelf temporarily emerges as a consequence of eustatic sea-level changes, the river directly affects the substratum, originating a submarine valley.

The tectonic characteristics of the area in which a canyon is present play a fundamental role in its genesis and evolution through time. At the same time the ampleness and the inclination of the shelf and continental slope, with the sedimentary mechanisms that regulate the transfer of the sediments from the shelf towards deeper basins, influence the evolutionary pattern of submarine valleys. In fact in basins fed by important fluvial systems, the sedimentary circulation is much more active than in areas where the only supply source is represented by gravitational processes. These processes are influenced, as well, by the eustatic sea-level changes, becoming more frequent in the lowstand periods than in the highstand ones.

The different morphologies observed in the canyons are tightly linked to different genesis and sedimentary processes. The cross section is one of the main morphological factors to be considered in the analysis of a canyon: the shape gives decisive information to define canyon evolution. A submarine valley with a V-shaped cross section generally points to dissection because of very speedy flows, while a U-shaped section points to the presence of slow and sporadic flows or to a relatively recent formation of the valley. The prevalence of depositional over erosional processes is another factor that contributes to canyon evolution by favouring the formation of a U-shaped, or at least, of a flared cross section. This causes the filling of the submarine valley bottom, with the consequent evolution of the canyon from a V towards a U-shape section. On the other hand, it is more frequent that the canyon presents steep walls and therefore a V-shaped section when erosion prevails over deposition. Slow phenomena, such as subsidence, can contribute to the evolution of the submarine valley cross section. In fact when canyons lie on slightly subsident basins they show a flared shape, while in more tectonically active areas, valleys frequently show more V-shaped sections.

The erosive processes in a canyon are strongly influenced by lithology. Canyons cut on very resistant bottom rocks, will tend in time to fill rather than become deeper. That happens in submarine valleys located along forearc systems, where the valley bottom shows a high resistance to erosion with a consequent widening of the valley section.

Another important aspect to be taken into account for the analysis of a canyon is its longitudinal profile. It can generally be pseudo-rectilinear to meandering with a high sinuosity index. In the areas that are not excessively steep that have been affected by tectonic events, but which are now stable, rectilinear canyons prevalently develop. The meandering ones, instead, generally develop in tectonically active areas and with more elevated gradients. The meandering of a canyon is function of its own form and longitudinal development. In fact the bending ray of a canyon and the meander wavelength increase both with width and inclination of the longitudinal profile. The presence of deep meandering channels points out the existence of transport processes characterized by frequent, continuous and slow turbiditic flows tightly connected to the inclination of the valley. In particular the local increment of canyon slope gradient is compensated by the increment of the profile sinuosity to maintain constant the solid supply.

It is possible to establish a relationship between the increment of slope gradient and the increment of the canyon sinuosity by the analysis of morphological parameters. Therefore the rectilinear or meandering profile of a submarine valley is mainly correlated both to the slope gradient value and to transport processes, and also to the nature of the material that flows inside.

The very rectilinear profile of a canyon can indicate that it has been cut on an important tectonic feature. Nevertheless valleys with a good rectilinear layout can change after tectonic events, because of continental shelf gradient variations, or also after eustatic sea-level changes.

In particular slope gradient variation owing to tectonic, tilting and subsidence processes, can cause the valley to evolve from a rectilinear towards a meandering profile. The evolution of a single submarine valley can also be strongly influenced both by the convergence and the feeding of the numerous canyons existing on the continental slope, causing consequent valley deepening and widening.

Reference

De Pippo, T., Hardi, M. and Pennetta, M. (1999) Main observations on genesis and morphological evolution of submarine valleys, *Zeitschrift für Geomorphologie* 43, 91–111.

SEE ALSO: canyon; rejuvenation; sea level

TOMMASO DE PIPPO

SUBMERGED FOREST

Submerged forests are former land surfaces on which *in situ* rooted tree stumps and associated organic deposits are found in the intertidal zone and on the continental shelves. They have been described by Geikie (1885), Reid (1913), Godwin (1943), Heyworth (1978) and Tooley (1979) in north-west Europe and by Krishnan (1982) and Mascarenhas (1997) in India.

Submerged forests owe their existence to a variety of causes: sea-level rise, coastal erosion, SUBSIDENCE and land uplift. The submerged forest at Rossall Beach on the Lancashire coast, United Kingdom, is the basal organic deposit of a kettle hole, the ramparts of which have been removed by erosion. Whereas at Hartlepool in north-east England (Tooley 1979) the submerged forest, which contains a Neolithic age human skeleton, is associated with estuarine clays. At Formby on the Lancashire coast the submerged forest formed in palaeo-dune slacks (Tooley 1979) and associated with them are human and animal footprints of Neolithic to Bronze Age (Huddart *et al.* 1999).

References

Geikie, A. (1885) *Textbook of Geology*, London: Macmillan.
Godwin, H. (1943) Coastal peat beds of the British Isles and North Sea, *Journal of Ecology* 31, 199–237.
Heyworth, A. (1978) Submerged forests, in J. Fletcher (ed.) *Dendrochronology in Northern Europe*, British Archaeological Reports S-51, 279–288.
Huddart, D., Roberts, G. and Gonzalez, S. (1999) Holocene human and animal footprints, Formby Point, NW England, *Quaternary International* 55, 29–41.
Krishnan, M.S. (1982) *Geology of India and Burma*, New Delhi: CBS.
Mascarenhas, A. (1997) Significance of peat on the western continental shelf of India, *Journal of the Geological Society of India* 49, 145–152.
Reid, C. (1913) *Submerged Forests*, Cambridge: Cambridge University Press.
Tooley, M.J (1979) Sea-level changes during the Flandrian stage, *Proceedings of the 1978 International Symposium on Coastal Evolution in the Quaternary*, Sao Paulo, Brazil, 502–533.

MICHAEL TOOLEY

SUBSIDENCE

Ground subsidence occurs only in specific environments, where any of three distinctive processes can occur. Compaction of porous and deformable clay, peat or silt causes surface subsidence as the soils restructure with declining pore space, normally accompanied by abstraction or expulsion of interstitial ground water. Collapse or deformation of the ground into natural caves occurs mainly in limestone, gypsum and basalt, and comparable dissolution occurs on salt. Large-scale processes include crustal sag, delta compaction, earthquake movements and volcanic deflation. Mining subsidence and the collapse of old mines are entirely artificial processes that are not a part of geomorphology.

Clays are deformable because their water is loosely bonded to the chemically complex clay mineral particles. They may therefore compact (and cause subsidence) when the water is squeezed out under load or in response to water abstraction. Clay subsidence may be regional or localized, but this ranks as the most widespread and most destructive subsidence process. Entire cities have subsided, with Venice (Plate 134), Mexico City, Tokyo, Shanghai and Bangkok among the better known examples.

The natural process of clay compaction causes water to be squeezed out during slow consolidation. Artificial removal of the water, by abstraction pumping, induces more rapid compaction. The subsidence hazard lies in alluvial sequences with alternating beds of poorly consolidated sand and clay beneath large cities. Convenient water supplies are pumped from the incompressible

Plate 134 Boardwalks across the Piazza San Marco in Venice, which has now subsided below the level of most winter high tides

sand aquifers, and the overall decline in pore-water pressures causes the adjacent clays to compact. The amount of subsidence is proportional to the groundwater head decline (Poland 1984), but greater subsidence occurs on younger clays that are less consolidated by self-weight, and also on those with higher contents of unstable smectite. This clay mineral forms primarily by weathering of volcanic rocks in wet tropical environments, and is a factor behind the major subsidence of Mexico City.

The role of pressure decline within over-pumped aquifers is now recognized worldwide, and is clearly seen in the subsidence record of Venice. Long-term subsidence of the entire delta region combines with rising sea levels to create a continuing problem at Venice, but subsidence was greatly accelerated when groundwater abstraction caused clay compaction in response to head decline (Figure 163). Pumping controls have allowed head recovery, but 90 per cent of the induced compaction is irreversible, and the unabated natural subsidence continues to cause increasingly frequent flooding of the city.

Groundwater withdrawals can be matched by extraction of petroleum, natural gas and steam as causes of major ground subsidence, where the loss of hydraulic support may also affect rocks other than clay. Oilfield subsidence includes the classic case at Wilmington, USA, which caused the Long Beach harbour area of Los Angeles to subside by nearly 9 m.

Shrinkage due to water loss causes subsidence in any clay at shallow depths. Climatic changes

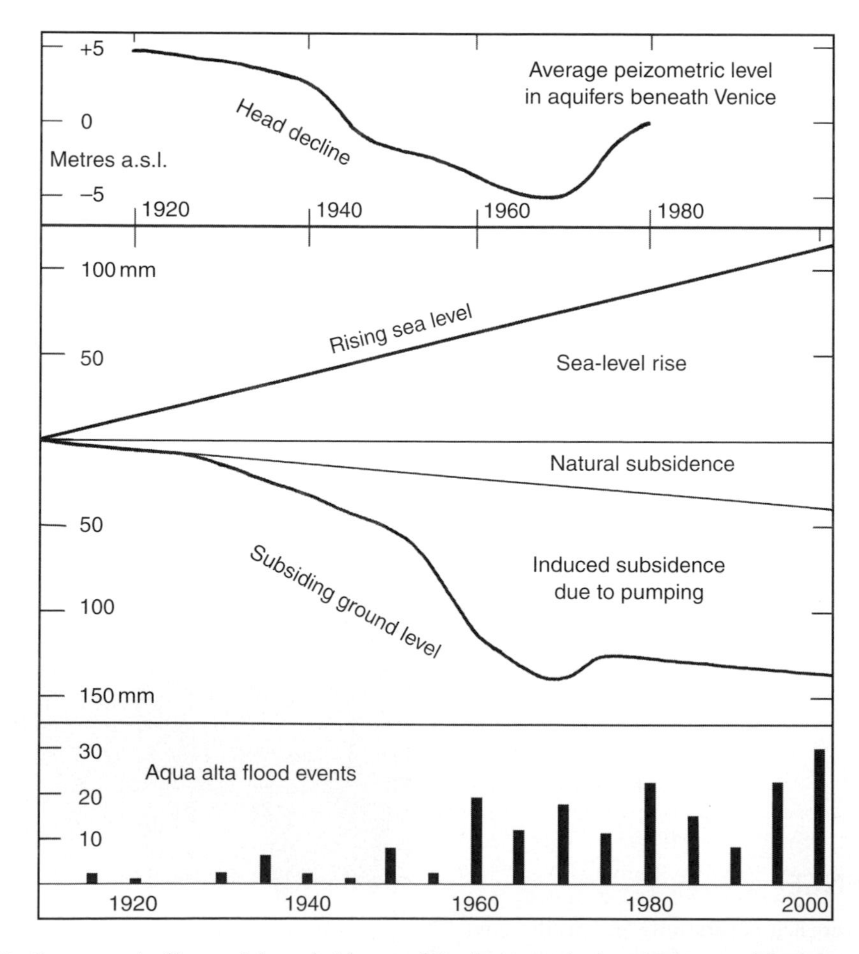

Figure 163 Causes and effects of the subsidence of Venice over the last 100 years. The bar graph shows, for each 5-year period, the numbers of flood events when high tides reach more than 600 mm above the level that initiates flooding in the lowest part of Piazza San Marco

affect large areas. Britain's dry summers since 1976 have produced a wave of subsidence damage to older houses on foundations so shallow that clay beneath them suffered first-time shrinkage in the new regime of drier climates. Subsidence is greater when trees extract excessive amounts of soil water during times of drought.

Subsidence on organic peat soils is induced by loading and/or drainage, and is similar to clay except that the compressibility of peat is far greater. Drainage causes immediate subsidence of peat, by about half the head decline, as recorded by the Holme Post in the English Fens (Hutchinson 1980). Through multiple phases of land drainage, the Post has emerged from the subsiding peat because it is founded on the stable clay beneath. After the first major movements, subsidence was about a quarter of head decline in each subsequent phase of drainage. Following the initial drainage and compaction, subsidence on peat continues due to wastage. This is loss to the atmosphere of the peat left above the water table and thereby exposed to microbial oxidation, causing annual surface lowering of 5–100 mm. In the drained English Fens, rivers have to be maintained between high banks across the subsidence bowls, land drainage water has to be pumped up into the elevated rivers, and buildings on piles progressively rise from the peat.

After clay, limestone KARST provides the world's most widespread subsidence environment, wherever it is at outcrop or beneath a soil cover. Caves constitute a subsidence hazard by threatening total ground collapse over small areas. The roof of a cave develops a natural arch in compression, which is stable when it is thinned down to about a quarter of its span (by surface lowering and roof stoping). Cave roofs need to be about double this thickness to take loads imposed by engineering, and variable degrees of fracturing and fissuring make cave roof stability difficult to assess. Also, locations of unseen caves are generally unpredictable, so cave collapse is a random and potentially destructive subsidence hazard. Most of the world's natural caves are in limestone, while gypsum and basalt are the other significantly cavernous rocks. However, collapses of rock over caves are very rare events; almost none of the world's limestone gorges originated as a collapsed cave.

A more widespread hazard is the development of subsidence dolines in soil covers that are washed into caves or fissures in underlying limestone. There are various types of DOLINES (alternatively known as sinkholes), each distinguished by its processes of erosion and collapse. Suffosion and dropout dolines (known collectively as subsidence dolines) form entirely within the soil profile (Plate 135); infiltrating rainfall washes soil down into pre-existing rockhead fissures at rates which can be significant to engineered structures. Slow slumping of non-cohesive sandy soils produces suffosion dolines that may damage structures but are not life-threatening. In a cohesive clay soil, cavitation initiates over a rockhead fissure, and grows slowly beneath an arched soil roof. It propagates upward until the surface collapses instantly and without warning; such a 'dropout' can be a major subsidence hazard in soil-covered karst, and individual failures can be up to 100 m across.

Subsidence sinkholes are the most frequent subsidence geohazard in karst terrains, especially the rapidly formed dropouts. Sites of new failures are impossible to predict, except that buried limestone boundaries with allogenic water input are recognizable subsidence zones, notably where soil cover is 1–15 m thick (though ground collapses have occurred where the cavernous limestone lies beneath 100 m of soil cover). Dropouts are caused by water flow, and therefore occur most frequently during rainstorms, and/or where drainage paths have been modified or water tables have declined due to over-abstraction. Most doline collapses are induced by engineering activity, and are therefore avoidable (Newton 1987).

Plate 135 A dropout doline at Ripon, Yorkshire. Only the soil profile is exposed after weak cover rocks and soil failed into cavernous gypsum beneath

Rock salt is rapidly soluble in natural waters. Total removal of salt beds can develop within a few years, orders of magnitude faster than mineral losses in limestone, and ground subsidence occurs at significant rates where salt is being removed by circulating ground water. In lowland sites, including the Cheshire saltfield of England, dissolution at the rockhead leaves insoluble material in unstable residual breccias beneath the drift cover. Rates of dissolution and subsidence are a function of groundwater flow patterns, and overall natural subsidence is generally $>0.1\,\mathrm{mm\,yr^{-1}}$; during the Pleistocene, this created subsidence hollows that are now occupied by the mere lakes. All salt subsidence is vastly accelerated by any brine-pumping operations that draw in new supplies of chemically aggressive fresh water to replace the abstracted brine. Traditional wild brining targets the 'brine streams' that are zones of enhanced groundwater flow just above rockhead, and thereby causes the linear subsidences above them to deepen by $100\,\mathrm{mm\,yr^{-1}}$ or more, ultimately creating new ribbon lakes known as 'flashes'.

It is significant that all the widespread subsidence processes on clay, limestone, gypsum and salt are induced or accelerated by human activities. Subsidence over active and abandoned mines is entirely due to humans. The implication therefore stands that subsidence is a process and a geohazard that can largely be controlled. The same applies to other, less common types of subsidence.

Collapsing soils are sediments prone to internal structural collapse when water is added to them. Weak clay bonds between the grains of loosely packed, silt sediments are broken by the introduction of water; the soil then densifies by repacking under self-load or imposed load, in the process known as hydrocompaction. Potentially collapsible soils are wind-deposited LOESS and some alluvial silts that were rapidly deposited and then desiccated in large basinal fans. Hydrocompaction can promote ground subsidence of up to 5 m over wide areas in semi-arid regions where the soils have not been previously wetted.

Clays and silts are generally weak when wet and saturated, but become solid when frozen in PERMAFROST. Subsidence is then inevitable where ice-rich soils are thawed by stripping vegetation or placing warm buildings directly on the ground, though sands and gravels are largely thaw-stable.

The only subsidence that is totally natural, and therefore not controllable, is that originating with deep-seated processes. True tectonic subsidence occurs over boundary zones of subduction and also where plates are necked and thinned in zones of tension; the London area is subsiding by more than $2\,\mathrm{mm\,yr^{-1}}$ due to thinning and sinking of the North Sea Basin. Such rates of subsidence may be critical at coastal sites, especially when combined with the current rise in world sea levels (at 1–2 $\mathrm{mm\,yr^{-1}}$), largely due to global warming that has been continuous for about the last 500 years.

Major deltaic basins are sites of very slow subsidence due to a combination of factors. Within the sediment piles, soft clays are consolidated into mudstones with reduced porosities, and this compaction causes subsidence. At the same time crustal sag of the overburdened basin floor provides a second component of subsidence. Venice lies towards the margin zone of the Po Valley deltaic basin, where long-term mean subsidence is $0.4\,\mathrm{mm\,yr^{-1}}$ due to both compaction and sag; this is the natural, uncontrollable component of Venice's continued subsidence (Figure 163).

Accelerated tectonic subsidence is accompanied by earthquakes when crustal deformation is transferred to fault displacement. Large areas in Alaska subsided by up to 2.5 m during the 1964 earthquake, causing permanent drowning of coastal forests. Subsidence is more widespread as a secondary effect of earthquakes, due to liquefaction of unconsolidated sand soils. During the period of earthquake vibration, sands may temporarily behave as a liquid, so that structures subside into them. The vibrations also cause densification of loose sands by improved grain packing, and this causes permanent subsidence of the ground surface.

Entire volcanoes, and their immediate surrounds, can deflate due to migration or retreat of magma from beneath them. Parts of the Naples region subsided by 10 m due to 1,200 years of deflation of the Campi Flegri volcanic centre. This is one case where subsidence is welcome, as the reverse uplift is due to volcanic inflation that normally precedes an even more destructive eruption.

References

Hutchinson, J.N. (1980) The record of peat wastage in the East Anglia fenlands at Holme Post, 1846–1978 AD, *Journal of Ecology* 68, 229–249.

Newton, J.G. (1987) Development of sinkholes resulting from man's activities in the eastern United States, *US Geological Survey Circular* 968, 1–54.

Poland, J.F. (ed.) (1984) *Guidebook to Studies of Land Subsidence due to Groundwater Withdrawal*, Unesco Studies and Reports in Hydrology 40.

Further reading

Carbognin, L. and Gatto, P. (1986) An overview of the subsidence of Venice, *International Association Hydrological Sciences Publication* 151, 321–328.
Cooper, A.H. and Waltham, A.C. (1999) Subsidence caused by gypsum dissolution at Ripon, North Yorkshire, *Quarterly Journal of Engineering Geology* 32, 305–310.
Culshaw, M.G. and Waltham, A.C. (1987) Natural and artificial cavities as ground engineering hazards, *Quarterly Journal of Engineering Geology* 20, 139–150.
Holzer, T.L. (1991) Nontectonic subsidence, *Geological Society America Centennial Special Volume* 3, 219–232.
Prokopovich, N.P. (1986) Origin and treatment of hydrocompaction on the San Joaquin Valley, USA, *International Association Hydrological Sciences Publication* 151, 537–546.
Stephens, J.C., Allen, L.H. and Chen, E. (1984) Organic soil subsidence, *Geological Society America Reviews Engineering Geology* 6, 107–122.
Waltham, A.C. (1989) *Ground Subsidence*, Glasgow: Blackie.

TONY WALTHAM

SUFFOSION

An erosional process, occurring in areas where well developed KARST is overlain by unconsolidated superficial materials (usually loess or till). Suffosion is associated with PIPES AND PIPING, and is also known as ravelling, describing both catastrophic and gradual collapse of the superficial material into the bedrock cavity. The sediments slump down into widened joints and cavities in the bedrock surface, producing an irregular land surface, often exhibiting dimpling and multiple suffusion dolines (also termed shakeholes). The unconsolidated sediments can be susceptible to collapse compression upon saturation, with infiltrating water beneath the regolith creating subsoil KARREN and widened joints connected with deeper cavities. Fine materials are often eroded internally by a combination of solution and downwashing (mechanical suffosion), suggesting that these processes are related to karst underwatering. Layers of more compacted material may arch over the collapsing soil, and these are often attributed to lowering of the water table. Suffosion dolines can form within seconds and are the most widespread problem encountered in karst terrains in the construction industry, save pollution of aquifers. The depressions thereby formed typically range from 1 m diameter and 0.5 m deep, up to 100–200 m diameter and 10–50 m deep (though these are less frequent) (Ford and Williams 1989: 525).

Reference

Ford, D. and Williams, P. (1989) *Karst Geomorphology and Hydrology*, London: Unwin Hyman.

SEE ALSO: doline

STEVE WARD

SULA

Rocky zones along stream courses, especially in the tropics, where channels divide into anastomosing channels associated with cataracts (Zonneveld 1972). They have been explained in terms of rivers crossing an irregular weathering front, having limited amounts of abrasive material to enable incision, the induration of outcrops with ferromanganese compounds, and possible inheritance of braided conditions from Quaternary dry periods. The view that tropical rivers lack abrasive tools because of the speed and thoroughness of weathering reducing the amount and size of sediment load has been advanced by many climatic geomorphologists (e.g. Büdel 1982), but reliable data on this are sparse.

References

Büdel, J. (1982) *Climatic Geomorphology*, Princeton: Princeton University Press.
Zonneveld, J.I.S. (1972) Sulas and sula complexes, *Göttinger Geographische Abhandlungen* 60, 93–101.

A.S. GOUDIE

SULPHATION

A reaction between sulphur dioxide from the atmosphere and building materials (including stone) to form gypsum (calcium sulphate). It is often regarded as a feature of polluted urban atmospheres. Sulphur dioxide becomes oxidized (sometimes in the presence of catalysts such as soot or metal-rich particles) on moist surfaces to form sulphuric acid. The sulphuric acid then reacts with the stone in the following way:

$$CaCO_3 + H_2SO_4 + 2H_2O \rightarrow CaSO_4{\cdot}2H_2O + CO_2 + H_2O$$

According to Cooke and Gibbs (1993), where little moisture is present two alternative series of reactions may occur. Hydrated calcium sulphate may form first from the reaction of calcium carbonate and sulphur dioxide, as represented in the following reaction:

$$CaCO_3 + SO_2 + 2H_2O \rightarrow CaSO_4{\cdot}2H_2O + CO_2$$

Subsequently, this hydrated calcium sulphate may become oxidized to form gypsum in the presence of catalysts. Alternatively, sulphur dioxide may react with bicarbonate solutions formed by reaction with rainfall containing dissolved carbon dioxide as follows:

$$Ca(HCO_3)_2 + SO_2 \rightarrow CaSO_3 + 2CO_2 + 2H$$

The deposition of sulphates on material surfaces, often in the form of a normally dark gypsum crust, can be seen as both a consequence and a cause of weathering. This is especially true of limestones, where such a crust may be harder than the material on which it has developed. Other properties may also be different and may cause an acceleration in the speed of stone degradation. Amoroso and Fassina (1983: 264) identify three mechanisms that may account for this. First, a variation in volume occurs because gypsum has a greater volume than the quantity of calcite it replaces. One volume of calcium carbonate forms over two volumes of hydrated calcium sulphate. This causes expansive stresses in pores and cracks. Second, calcite and gypsum have different thermal expansion characteristics. The linear coefficient of the thermal expansion of gypsum is about five times that of calcite. This difference may be further increased when blackened crusts develop because they tend to absorb a larger amount of radiation than white surfaces. Third, the development of a crust will reduce the permeability of the material, which will in turn increase the water retention beneath.

Sulphation can lead to blister development and lamination, possibly through a combination of the factors just described.

Gypsum crusts in urban environments can also develop on non-carbonate rocks such as sandstone, either as a result of sulphation of calcium derived from an external source (e.g. the mortar surrounding the stone), or because of the accumulative role of lichens, algae, bacteria, etc., or because of chemical reactions with certain minerals within the rock (McKinley *et al.* 2001).

References

Amoroso, C.G. and Fassina, V. (1983) *Stone Decay and Conservation*, Materials Science Monographs 11, Amsterdam: Elsevier.

Cooke, R.U. and Gibbs, G.B. (1993) *Crumbling Heritage? Studies of Stone Weathering in Polluted Atmospheres*, London: National Power and Powergen.

McKinley, J.M., Curran, J.M. and Turkington, A.V. (2001) Gypsum formation in non-calcareous building sandstone: a case study of Scrabo Sandstone, *Earth Surface Processes and Landforms* 26, 869–875.

A.S. GOUDIE

SUPRAGLACIAL

Supraglacial refers to the surface zone of glaciers and ice sheets where there is distinctive drainage, sediment sources, transport and deposition and a set of landforms characteristic of the supraglacial association and landsystem (Paul 1983). Many of the landforms produced are ephemeral and do not survive deglaciation but some do, particularly where associated with stagnant ice.

Supraglacial debris enters glacial transport from a number of sources such as mass movements from nearby mountains, tephras and from subglacial sediment carried to the surface. Minor inputs include meteorites, pollutants from anthropogenic sources and sea spray salts. Usually in high mountain glaciers where the surface ice is adjacent to valley slopes, or where isolated peaks (nunataks) protrude through the ice sheet, mass movements from the slopes are the dominant surface sediment supplier. These processes include rockfall, slides, snow and ice avalanching, debris flows, slushflows and streamflows. The supplied supraglacial debris is incorporated into the glacier by snow and ice burial in the accumulation area and by falling into crevasses and MOULINS (cylindrical vertical shafts). This is then transported by high-level, passive debris transport and suffers little modification, retaining its primary, parent material characteristics. These are angular to very angular particle shape, with common elongate to slabby particles, and coarse-grain size, with little fines.

Surface snowmelt allows downward water percolation through the snowpack and if the melting exceeds refreezing, water accumulates as slush swamps. Usually water drains down a gradient forming rills and eventually a dendritic drainage network. If discharge is high enough surface channels form, being well developed on the ablation area ice because of its low permeability. These channels are smooth-sided, offering minimal water flow resistance, hence high velocities and may follow structural weaknesses, like foliation. On warm ice the surface channels are usually short and water is diverted into the glacier via crevasses and moulins. On cold glaciers supraglacial channels commonly flow towards and alongside the margins. Supraglacial lakes can form during the early ablation season but in temperate glaciers they drain as the englacial drainage opens. In cold ice these lakes can persist and where there are large amounts of supraglacial debris, differential melting causes widespread formation of water-filled hollows. Supraglacial lakes, or ice-cauldrons, also form in depressions created by geothermal melting and subsidence.

Supraglacial debris is spatially variable and it has an important control on ablation dynamics. It acts as insulation and retards surface melting. Differential ablation causes dirt cone development (Drewry 1972) and glacier tables can form by boulders protecting a pedestal of ice from melting. Supraglacial lateral MORAINES are ice-cored, debris accumulations at valley glacier margins, often formed by scree cone coalescence. Ablation reduction by thick debris cover means that these landforms stand above the adjacent relatively clean ice. Medial moraines are the supraglacial expression of a vertical medial debris septum which can extend to the glacier base but the surface debris is usually more concentrated and laterally extensive than that within the glacier because glacier ablation redistributes the debris. They have been classified into several types by Eyles and Rogerson (1978), including ice stream interaction medial moraines and ablation-dominated medial moraines. The former are created by lateral moraine confluence at the junction of two glaciers where the medial moraine marks the suture line between the glaciers. The second type form where ridges of englacial debris are revealed downglacier by ablation. They can form in two ways. A rockwall on a nunatak can supply debris to a small glacier area forming a linear debris plume downglacier from that point. This debris descends in the accumulation zone in the flow direction as it becomes buried. In the ablation zone it is revealed as a supraglacial medial moraine. Alternatively, folding of debris-rich, ice stratification may occur, particularly where ice flows into a restricted channel (Hambrey *et al.* 1999). The axes of these folds are usually parallel to the ice flow direction. When the debris-rich ice reaches the ablation zone, surface melting reveals the anticlinal crests of the longitudinal, debris-rich folds. If the folding is relatively open then a series of small medial moraines may define the axis of a fold.

Some glaciers have much supraglacial debris mantling the lower part of their ablation zone, particularly where mass wasting delivers large debris volumes in both the accumulation and ablation areas, as in high relief mountains, or where topography allows the movement of basal debris to the surface along shear planes. Uneven reworking and debris deposition in multiple events during ablation is responsible for a distinctive landform assemblage which can lead to topographic inversion on the glacier surface. The result is a complex sediment assemblage of faulted and folded fluvial, lacustrine and mass movement sediment (Benn and Evans 1998; Huddart 1999) which is finally deposited on the substrate. This can include supraglacial ESKERs which can be let down from supraglacial and englacial channels, ice-walled lake plains (Huddart 1983) and several KAME types. Supraglacial debris can also contribute to the GLACIMARINE environment as part of morainal banks and from iceberg rafting.

References

Benn, D.I. and Evans, D.J.A. (1998) *Glaciers and Glaciation*, London: Arnold.

Drewry, D.J. (1972) A quantitative assessment of dirt-cone dynamics, *Journal of Glaciology* 11, 431–446.

Eyles, N. and Rogerson, R.J. (1978) A framework for the investigation of medial moraine formation: Austerdalsbreen, Norway and Berendon Glacier, British Columbia, *Journal of Glaciology* 20, 99–113.

Hambrey, M.J., Bennett, M.R., Dowdeswell, J.A., Glasser, N.F. and Huddart, D. (1999) Debris entrainment and transfer in polythermal valley glaciers, *Journal of Glaciology* 45, 69–86.

Huddart, D. (1983) Flow tills and ice-walled lacustrine sediments, the Petteril Valley, Cumbria, England, in E.B. Evenson, C. Schlüchter and J. Rabassa (eds) *Tills and Related Deposits*, 81–94, Rotterdam: Balkema.

——(1999) Supraglacial trough fills, southern Scotland: origins and implications for deglacial processes, *Glacial Geology and Geomorphology*, 1–16, http://boris.qub.ac.uk/ggg/papers/full/1999/rp041999/rp04.html

Paul, M.A. (1983) The supraglacial landsystem, in N. Eyles (ed.) *Glacial Geology*, 71–90, Oxford: Pergamon Press.

SEE ALSO: esker; glacial deposition; ice stagnation topography; kame; moraine; moulin

DAVID HUDDART

SURGING GLACIER

Surging glaciers are often instantly recognizable from their 'looped medial moraines' or teardrop-shaped loops of debris on the glacier surface. These loops record cyclic changes in velocity between a trunk GLACIER and its tributaries. Additionally, folding, thrust faulting and severe crevassing of ice at the snout is induced by compressive flow. A glacier is said to surge when it exhibits major fluctuations in velocity over timescales that range from a few years to several centuries, swinging between phases of rapid and slow flow. The phase of rapid motion is called the 'surge' or 'active' phase during which ice is transferred from the upper part of the glacier (the reservoir area) to the snout. This results in a dramatic advance of the snout and a concomitant thinning of the reservoir area. The period of slow flow between surges is termed the 'quiescent phase' and is characterized by the build-up of ice in the reservoir area and the mass stagnation of the snout. This gradually increases the surface gradient of the glacier until the next surge is initiated. Glacier flow velocities in the surge phase can be ten times faster than those of the quiescent phase. The surge and quiescent phases are usually of constant length for each individual glacier, resulting in a periodic cycle of surges. However, cycle length varies greatly between glaciers and glaciated regions. Glacier surge velocities also vary considerably, ranging from 50 m per day on the Variegated Glacier in Alaska to maximum speeds of only 16 m per day on Svalbard. The length of the quiescent phase also varies, maximum periods of time being 50–500 years for Svalbard glaciers compared to 20–40 years for most other regions.

Geographically, surging glaciers tend to cluster in Alaska, Yukon and British Columbia in North America, Svalbard and Iceland in the North Atlantic and the Pamirs in western Asia. Further examples have, however, been identified in the Canadian high arctic, Greenland, the Caucasus, Tien Shan and Karakoram mountains in Asia and in the Andes. Several glacier characteristics in these regions have been correlated with surging behaviour, such as glacier length and overall gradient. Bedrock types also appear to be linked to surging.

Glacier surges are not triggered by climatic fluctuations but instead result from changes in the internal dynamics of the glacier system. Variations in basal sliding appear to drive the changes in glacier flow velocity in a surging glacier, and this is driven by reorganizations of the subglacial drainage system. Based upon the few intense studies undertaken on surging glaciers, rapid sliding appears to be initiated by rising water pressures at the glacier bed. This is a function of the trapping of subglacial meltwater by conduit closure due to ice creep. Although this mechanism of fast flow initiation applies well to temperate glaciers it does not explain fully surges by subpolar glaciers. A study on Trapridge Glacier in the Yukon, which is frozen to its bed at the snout, suggests that deformable sediment beneath the warmer ice further up glacier plays a significant role in surging (Clarke *et al.* 1984). A wave-like bulge develops on the glacier surface at the boundary between cold and warm-based ice. This bulge moves down the glacier mostly by subsole deformation. This is initiated by changes in the subglacial water flow through the deformable substrate.

A linkage between surging behaviour and regional climate has been suggested by Budd (1975), who indicates that the concentration of surging glaciers in specific regions, and the variability of surge velocities and phases between regions, must reflect some climatic control. Specifically, in a continuously fast-flowing glacier the annual mass balance can be discharged by normal flow velocities. In contrast, in a surging glacier the mass throughput is too great to be discharged by slow flow alone but too small to sustain fast flow over long periods. Consequently, surging glaciers build up mass slowly until fast flow is triggered. This fast flow drains and exhausts the supply from the reservoir, thereby reinitiating slow flow. This means that the velocity of a surging glacier is constantly out of equilibrium with climate.

References

Budd, W.F. (1975) A first simple model of periodically self-surging glaciers, *Journal of Glaciology* 14, 3–21.

Clarke, G.K.C., Collins, S.G. and Thompson, D.E. (1984) Flow, thermal structure and subglacial conditions of a surge-type glacier, *Canadian Journal of Earth Sciences* 21, 232–240.

Further reading

Benn, D.I. and Evans, D.J.A. (1998) *Glaciers and Glaciation*, 169–175, London: Arnold.

Clarke, G.K.C., Schmok, J.P., Ommaney, C.S.L. and Collins, S.G. (1986) Characteristics of surge-type glaciers. *Journal of Geophysical Research* 91, 7,165–7,180.

Dowdeswell, J.A., Hamilton, G.S. and Hagen, J.O. (1991) The duration of the active phase of surge-type glaciers: contrasts between Svalbard and other regions, *Journal of Glaciology* 37, 388–400.
Fowler, A.C. (1987) A theory of glacier surges, *Journal of Geophysical Research* 92, 9,111–9,120.
Jiskoot, H., Murray, T. and Boyle, P. (2000) Controls on the distribution of surge-type glaciers in Svalbard, *Journal of Glaciology* 46, 412–422.
Kamb, B., Raymond, C.F., Harrison, W.D., Engelhardt, H., Echelmeyer, K.A., Humphrey, N., Brugman, M.M. and Pfeffer, T. (1985) Glacier surge mechanism: 1982–1983 surge of Variegated Glacier, Alaska, *Science* 227, 469–479.
Lawson, W., Sharp, M. and Hambrey, M.J. (1994) The structural geology of a surge-type glacier, *Journal of Structural Geology* 16, 1,447–1,462.
Meier, M.F. and Post, A.S. (1969) What are glacier surges? *Canadian Journal of Earth Sciences* 6, 807–819.
Murray, T. and Porter, P.R. (2001) Basal conditions beneath a soft-bedded polythermal surge-type glacier: Bakaninbreen, Svalbard, *Quaternary International* 86, 103–116.
Raymond, C.F. (1987) How do glaciers surge? A review, *Journal of Geophysical Research* 92, 9,121–9,134.
Raymond, C.F. and Harrison, W.D. (1988) Evolution of Variegated Glacier, Alaska, USA, prior to its surge, *Journal of Glaciology* 34, 154–169.
Sharp, M.J. (1988) Surging glaciers: behaviour and mechanisms, *Progress in Physical Geography* 12, 349–370.

DAVID J.A. EVANS

SUSPENDED LOAD

The suspended load of a river comprises mineral and organic matter that is dispersed through the flow by turbulence. Typically, the mineral load is dominant and consists of grains ranging in size from clay ($<4\,\mu$m) up to sand grade (<2 mm). Often, suspended particles exist as 'flocs' or clumps of finer grade particles linked by weak chemical forces. The suspended load is quantified in terms of its concentration (mass of sediment per unit volume of water), discharge (sediment mass flux per unit time – also referred to as the 'load'), and particle-size distribution (proportions of the load in given size fractions).

The clay-silt fractions (often termed 'washload') are largely sourced from erosion processes outside the river channel; being more easily suspended, they are well mixed through the flow and travel long distances in suspension. Sand requires more intense turbulence to remain suspended, thus it tends to be concentrated near the stream bed and tends to be sourced from the stream-bed material. In cobble/boulder-bed rivers, the suspended load is also augmented by fine particles generated by abrasion of cobbles/boulders as they roll, hop and slide along the bed as part of the BEDLOAD. Downstream through a drainage basin, the suspended load increasingly dominates over the bedload.

Since the fine washload is easily suspended, its concentration generally depends only on the relative rates of supply of sediment and water to the channel (i.e. rather than being limited by the physical capacity of the stream to suspend it). Indeed, when the concentration of mud-rich washload becomes sufficiently large, the fluid properties change from those of a water flow to those of a hyperconcentrated or debris flow. While turbulence intensity does limit the suspended sand concentration in sand-bed streams, in gravel or rock-bed channels the primary limitation on suspended sand concentration may be sand availability. Thus in most rivers, the suspended load tends to be 'under-supplied' with respect to the river's potential capacity to transport it. Because of this, it cannot be determined by physics-based formula but must be measured.

Accurate suspended load measurement at a river section requires measurements of both sediment concentration and water velocity. One approach is to collect point samples of water and to make point velocity measurements at intervals over the flow depth, then plot profiles of concentration and velocity. Special 'point-samplers' are used for this purpose. These cable-suspended samplers, comprising a brass bomb and an internal sample bottle, have a solenoid-operated valve to control the water-sample capture and they are designed so that they sample isokinetically (i.e. they accept a sample at the ambient water velocity). A second, simpler approach is to use a 'depth-integrating' sampler. With this, the inlet nozzle is kept open while the sampler is traversed from the water surface to the bed and back again, so performing a mechanical integration of the concentration and velocity profile. Details about samplers and methods developed in North America can be found in Edwards and Glysson (1999).

A single suspended sediment measurement is time consuming, and where a continuous record is required it is usually obtained by collecting 'index' samples at one location and establishing a relationship between the index sample concentration and the cross-section mean concentration. Index samples may be collected manually, but in remote locations or in 'flashy' small basins they are more often collected by an automatic pumping-sampler or by measuring a surrogate property such as water turbidity. Turbidity sensors have been widely used to date (e.g. Gippel 1995). They may be either transmissivity (attenuance) or back-scattering (nephelometric) types. Since they do not sense sediment concentration directly, a further calibration relation must be established between the turbidity sensor and suspended sediment concentration. The turbidity signal depends both on sediment concentration and particle characteristics, notably particle size and shape. Light scattering is greatest from clay particles and least from sand grains, thus turbidity sensors are more sensitive to the washload. Suspended sediment concentrations at-a-site tend to increase with water discharge, and so a relatively simple way to estimate the suspended discharge over a period of time is with a SEDIMENT RATING CURVE that empirically predicts sediment concentration at a given water discharge.

Globally, average annual river suspended sediment discharges vary greatly. The primary factors causing this variation include basin area, topography, precipitation, lithology, tectonics, vegetation cover, land use, floodplain sequestration and sediment entrapment in reservoirs. On a unit area basis, the largest sediment discharges occur in tectonically active, high-rainfall, steeplands around the western Pacific basin (e.g. Taiwan and New Zealand, where sediment discharges can exceed $20{,}000\,t\,km^{-2}\,yr^{-1}$). According to Milliman and Syvitski (1992), the largest sediment discharges from single river basins to the oceans are from the Amazon River ($1.2 \times 10^9\,t\,yr^{-1}$, due to its vast area) and China's Huanghe River ($1.1 \times 10^9\,t\,yr^{-1}$, due to its large areas of landuse-affected erodible loess terrane). Milliman and Meade (1983) estimated the total worldwide delivery of suspended sediment to the oceans at $13.5 \times 10^9\,t\,yr^{-1}$. Milliman and Syvitski (1992) revised this figure and considered that it might have been at least $20 \times 10^9\,t\,yr^{-1}$ before the proliferation of dams during the twentieth century, which have intercepted a significant fraction of the sediment discharge. They considered that global suspended sediment discharges have more than doubled as a result of widespread deforestation and farming over recent millennia.

References

Edwards, T.K. and Glysson, G.D. (1999) Field methods for measurement of fluvial sediment, *US Geological Survey Techniques of Water-resources Investigations* Book 3, Chapter C2.

Gippel, C.J. (1995) Potential of turbidity monitoring for measuring the transport of suspended solids in streams, *Hydrological Processes* 9, 83–97.

Milliman, J.D. and Meade, R.H. (1983) World-wide delivery of river sediment to the oceans, *Journal of Geology* 91, 1–21.

Milliman, J.D. and Syvitski, J.P.M. (1992) Geomorphic/tectonic control of sediment discharge to the ocean: the importance of small mountainous rivers, *Journal of Geology* 100, 525–544.

Further reading

Hicks, D.M. and Gomez, B. (2003) Sediment transport, in G.M. Kondolf and H. Piegay (eds) *Tools in Fluvial Geomorphology*, 425–461, San Francisco: Wiley.

D. MURRAY HICKS

SYNGENETIC KARST

This refers to karst that has developed concurrently with the diagenesis and consolidation of the host karst rock. It is most well known in calcareous dune limestones, rocks which are largely of biogenic origin, being comminuted shell fragments consolidated into calcareous AEOLIANITE, although there may be significant admixtures of quartz and other non-calcareous minerals. Such rocks are usually found within 40° north and south of the Equator and karst development in them is especially well known in the coastal regions of South and West Australia (White 2000), but is also found in other regions such as the eastern and southern Mediterranean and on Caribbean islands.

The dunes developed during glacio-eustatic oscillations during the Quaternary that exposed large amounts of shell material in coastal regions. Dunes were blown inland and accumulated in a series of ridges roughly parallel to the coast. The first stage in diagenesis was fixing by vegetation, the growth of which promoted soil development and production of biogenic carbon dioxide in the root zone. This encouraged dissolution of the dominantly aragonitic sands by infiltrating rain water. Saturation with respect to calcium carbonate is achieved close to the surface, under conditions that are open system with respect to carbon dioxide. Further percolation of this saturated solution down into the sands away from the root zone results in degassing of carbon dioxide and evaporation, with the result that the water becomes supersaturated and calcite is precipitated in interstitial pores in the dune, thus cementing it. Cementation produces a surface case-hardened crust over the dune that advances progressively downwards from the surface. Vertical solution pipes with case-hardened rims often perforate the dune crust. Cementation also occurs in the water saturated (phreatic) zone, and eventually the entire dune becomes calcreted.

During this process of dune cementation streams flow from inland areas towards the coast, but are blocked in their progress by dune ridge barriers that lie parallel to the coast. The streams pond on the upstream side and their waters permeate the porous sands to emerge on the seawards side. As the dune sands become indurated, stream flow is restricted to defined routes which eventually develop into blind valleys leading into caves many of which have flat water-table controlled roofs. The cave ceilings are also case hardened. Some cave ceilings collapse and give rise to large chambers that become profusely decorated with speleothems. Further collapse develops collapse dolines.

Reference

White, S. (2000) Syngenetic karst in coastal dune limestone: a review, in A.B. Klimchouk, D.C. Ford, A.N. Palmer and W. Dreybrodt (eds) *Speleogenesis: Evolution of Karst Aquifers*, 234–237, Huntsville, AL: National Speleological Society.

PAUL W. WILLIAMS

SYSTEMS IN GEOMORPHOLOGY

Students of the Earth's surface have been referring to 'systems' for nearly three centuries: but when Buffon (1749) did so, he was concerned with the theoretical structures (which we should now term cosmogonies) of the likes of Whiston, Burnet and Woodward; whereas Playfair (1802: 102) in his famous discussion of the 'system of vallies' was dealing with concrete entities. These two aspects were brought together in A.N. Strahler's enormously influential paper on the dynamic basis of geomorphology (1952). Here Strahler considers (pp. 934–935) that the 'fullest development' of geomorphology will occur 'only when the forms and processes are related in terms of dynamic systems and the transformation of mass and energy are considered as functions of time'. He goes on (p. 935): 'Many of the geomorphic processes operate in clearly defined systems that can be isolated for analysis.'

Strahler's view derived from early papers by the biologist, Ludwig von Bertalanffy, notably that published in *Science* in 1950. The move towards the use of systems as a theoretical device, unifying all sciences and extending to fields such as economics, was christened General Systems Theory (GST) by von Bertalanffy (1950). Both the theoretical and the practical aspects of GST were adopted by several of Strahler's students, most notably by the British geomorphologist R.J. Chorley (1960; Chorley and Kennedy 1971) and the American S.A. Schumm (1977). The two versions of 'system' can be seen side by side in the Twenty-third Binghamton Symposium (Phillips and Renwick 1992).

So what *are* systems? and what has been their dual role in geomorphology since 1952? According to Hall and Fagan, 'A system is a set of objects together with relationships between the objects and between their attributes' (1956: 18). They acknowledge that this is an imprecise definition, but say 'This difficulty arises from the concept we are trying to define; it simply is not amenable to complete and sharp description' (*idem*). Further, they acknowledge the existence of abstract or conceptual systems (pp. 19–20) as well as natural systems (pp. 23–24), which latter they consider as either open (exchanging 'materials, energies or information') with their surroundings; or closed (with no such imports and exports). Chorley and Kennedy (1971: 2) modified this distinction by relabelling Hall and Fagan's 'closed' systems as 'isolated' and identifying a new category of 'closed' system which exchanged energy (or, presumably, information) but not mass with its environment. In this sense, there must be very few true isolated systems studied by geomorphologists, but there may be several categories of effectively closed systems, including the planet Earth itself. Chorley and Kennedy went on (pp. 4–10) to identify four classes of system of relevance to physical geography: morphological, cascading, process-response and control. They discuss each in turn, devoting most attention to the process-response examples together with related considerations, including input and output, equilibrium, thresholds (see THRESHOLD, GEOMORPHIC), trends and simulation models. Finally, Chorley and Kennedy consider control systems, a topic taken up and elaborated by Bennett and Chorley (1978) in a book ranging far beyond geomorphology.

So how have these ideas – of both the conceptual and the natural systems – actually been employed by geomorphologists? Chorley (1960) followed Strahler's (1952) lead in emphasizing the distinction between historically focused geomorphology (explicitly, the ideas of the American colossus, W.M. Davis and, in particular, his concept of the geographical cycle, see CYCLE OF EROSION), which he termed 'closed system' thinking: and the 'open system' approach deriving from GST and the study of process. Chorley lists (1960: B8 and 9) seven advantages which will accrue from the adoption of the 'open system' view. They are:

1. Emphasis will be placed on the 'tendency to adjustment' between form and process.
2. Emphasis will be directed towards 'the essentially multivariate character' of geomorphic processes.
3. It will allow consideration to be given both to progressive changes and to stasis or abrupt changes (i.e. both concepts such as DYNAMIC EQUILIBRIUM and thresholds).
4. It will foster 'a less rigid view regarding the aims and methods of geomorphology' than (he contends) is held by those concerned with studying the historical development of landscapes and landforms.
5. Emphasis will be placed upon the whole landscape, rather than on 'the often minute elements ... having supposed evolutionary significance' (by this he clearly has in mind the 1950s' obsession with terrace and PENEPLAIN remnants).
6. It will encourage 'rigorous' work in areas where there are few or no traces of erosional history.
7. There will be major implications for geography as a whole, especially by directing attention 'to the increasingly hierarchical differentiation which often takes place with time'.

More than half a century on, it is not evident that these advantages have been realized. One very important drawback to the use of 'systems thinking' was identified by the historical geographer, Langton (1972). If we accept Hall and Fagan's definition, then it is evident that we can identify or create as many systems in thought or reality as we wish. This becomes, then, fundamentally an exercise in classification. It should then be clear that we need to be able to rate the value of different classifications and this, Langton contends, can only be done if we can identify the optimum outcome from the system we define. Whilst this may be feasible, say, in evaluating the water yield from a DRAINAGE BASIN in terms of irrigation, or as a site for a hydro-power operation, it is not at all clear that a slope, a PINGO or even the drainage basin can be legitimately thought to possess a 'goal'. This problem is, it seems, a fundamental one. And very closely linked to this inherent difficulty is that of identifying unique and non-overlapping systems in the real world. This is absolutely crucial for process-response systems if their operation is to be fully understood. Consider a drainage basin. Can its physical watershed be truly and exactly fixed? Can we be certain that the watersheds for mass or energy inputs and outputs coincide with the topographic limits? (Consider the question of subsurface water and solute movement.) If our designation of discrete basins may seem accurate, can we be as content with the identification of 'hillslopes' or even 'channels'? If we do not have this certainty – so different from the identification of (say) a household 'hot water system' – then it must surely cast doubt on the validity of our conclusions about the systems' operations. As a result, by the end of the twentieth century, the use of the term 'system' in geomorphology seemed really loosely, if at all, related to the specific ideas of von Bertalanffy, Strahler and Chorley. An example would be the text by Allen (1997) which opens with a chapter entitled 'Fundamentals of the Earth surface system' (pp. 1–50) with very little in the way to indicate direct descent from – say – Chorley and Kennedy, in terms of vocabulary or illustrations.

Before 1952, it was possible to think of 'systems' at all levels, both conceptual and practical. What might be termed the 'systems boom' of the 1950s–1970s in geomorphology argued for a situation in which a particular theoretical vision would inform the whole choice of research topics as well as the vocabulary in which results would be couched. In general the link between philosophical concept and practice was never really made. As a result, we have a residual terminology which really owes more to the vernacular than to either Strahler or Chorley. And we also see far more use of systems as a research tool in precisely those applied situations where a desired outcome can be specified, thus confirming Langton's assessment.

References

Allen, P.A. (1997) *Earth Surface Processes*, Oxford: Blackwell.

Bennett, R.J. and Chorley, R.J. (1978) *Environmental Systems: Philosophy, Analysis and Control*, London: Methuen.
Buffon, J.M.L. (1749) *Histoire naturelle, générale et particulière, avec la description du Cabinet du Roi*, Paris: L'Imprimerie Royale.
Chorley, R.J. (1960) Geomorphology and General Systems Theory, *US Geological Survey, Professional Paper* 500-B.
Chorley, R.J. and Kennedy, B.A. (1971) *Physical Geography: A Systems Approach*, London: Prentice Hall.
Hall, A.D. and Fagan, R.E. (1956) Definition of system, *General Systems Yearbook* 1, 18–28.
Langton, J. (1972) Potentialities and problems of adopting a systems approach to the study of change in human geography, *Progress in Geography* 4, 125–179.
Phillips, J.D. and Renwick, W.H. (eds) (1992) *Geomorphic Systems*, Amsterdam: Elsevier.
Playfair, J. (1802) *Illustrations of the Huttonian Theory of the Earth*, London: Cadell and Davies. Reprinted in facsimile, G.W. White (ed.) (1964), New York: Dover.
Schumm, S.A. (1977) *The Fluvial System*, Chichester: Wiley.
Strahler, A.N. (1952) Dynamic basis of geomorphology, *Geological Society of America Bulletin* 63, 923–938.
von Bertalanffy, L. (1950) The theory of open systems in physics and biology, *Science* 3, 23–29.

SEE ALSO: Cycle of Erosion; drainage basin; flow regulation systems; integrated coastal management; land system; peneplain; step-pool system; threshold, geomorphic

BARBARA A. KENNEDY

T

TAFONI

Tafoni (singular tafone) are cavernous weathering forms which typically are several cubic metres in volume and have arch-shaped entrances, concave inner walls, overhanging margins (visors) and fairly smooth gently sloping, debris-covered floors (Mellor *et al.* 1997). They occur in many parts of the world (see Goudie and Viles 1997: table 6.1), including polar regions, but may be especially well developed in coastal and dryland situations. They occur in a wide range of rock types, but especially in medium and coarse-grained granites, sandstones and limestones. Indeed it is only rocks with relatively closely spaced discontinuities (bedding planes, foliation, joints) such as shales and slates, that seem to be relatively unaffected by these cavernous weathering forms.

The cavernous hollows of tafoni are believed to result largely from flaking and granular disintegration caused by a range of possible weathering processes that include hydration, salt crystallization, and chemical attack by saline solutions. Some workers have found clear evidence of SALT WEATHERING being involved, while others have not. The role of case hardening in their foundation is also the subject of debate, but can help to explain the formation of the visor. It is also possible that tafoni develop through a positive feedback effect, in that once a hollow is initiated it creates an environment in which weathering is favoured (Smith and McAlister 1986). For a cavity to grow there needs to be a mechanism to remove flakes and spalls. Wind may play a part, as may organisms such as pack rats. Although some early workers thought that the actual excavation of a cavity might be achieved by wind abrasion, many tafoni occur in

Plate 136 A large tafoni developed in granite near Calvi in Corsica

Plate 137 A remarkably developed tafoni in volcanic rocks in the Atacama Desert near Arica in northern Chile

environments where sand blasting does not occur or they may have an aspect (i.e. the leeward side of a boulder) or a height up a cliff face that precludes such a mechanism. It is clear that in whatever way they form, tafoni grow significantly over tens of thousands of years (Norwick and Dexter 2002).

References

Goudie, A.S. and Viles, H.A. (1997) *Salt Weathering Hazards*, Chichester: Wiley.

Mellor, A., Short, J. and Kirkby, S.J. (1997) Tafoni in the El Chorro area, Andalucia, southern Spain, *Earth Surface Processes and Landforms* 22, 817–833.

Norwick, S.A. and Dexter, L.R. (2002) Rates of development of tafoni in the Moenkopi and Kaibab formations in Meteor Crater and on the Colorado Plateau, northeastern Arizona, *Earth Surface Processes and Landforms* 27, 11–26.

Smith, B.J. and McAlister, J.J. (1986) Observations on the occurrence and origins of salt weathering phenomena near Lake Magadi, Southern Kenya, *Zeitschift für Geomorphologie NF* 30, 445–460.

A.S. GOUDIE

TALSAND

Large-scale infillings of ice-marginal valleys occurring in the Pleistocene lowlands of northern Germany and Poland. However, the term is also employed to represent any flat sandy region of glacifluvial and/or periglacial origin, set below the general level of Pleistocene uplands (Geest Plateau) (Schwan 1987). This is referred to as a talsand plain. Talsand is thus a geomorphological term that is largely restricted to north German Quaternary geology, translated as valley sand.

Talsand is composed of glacifluvial or periglacial sediments, though often an overlying bed of aeolian sands is present. The majority of talsand was deposited during the last glacial period (the Weichselian glacial *c*.70–10 ka years ago – the equivalent of the Devensian glacial in the UK, and the Wisconsin glacial in the USA), though formation may have initiated in the preceding late Saalian period. The overlying aeolian deposit typically originates from the late Weichselian and often continues into the Holocene. The transformation from fluvial to aeolian sediments is predominantly due to increased wind intensity and the lowering of ground water due to permafrost degradation (Schwan 1987).

Reference

Schwan, J. (1987) Sedimentological characteristics of a fluvial to aeolian succession in Weichselian Talsand in the Emsland (F.R.G.), *Sedimentary Geology* 52, 273–298.

STEVE WARD

TALUS

A term of French origin, which can have different meanings. For Anglo-Saxon geomorphologists it is used as a synonym for scree. It is an accumulation of weathered rock fragments under a cliff. The mechanisms which control this accumulation are, however, complex. The weathering of the underlying rock face is most often linked to frost (see FROST AND FROST WEATHERING) or earthquake action. The transit of rock debris from the cliff to the deposit can be due to one or several processes. Consequently, the morpho-sedimentological characteristics of this deposit depend on the type of these processes and the lithology.

In rare cases the accumulation is exclusively linked to rockfall. This rockfall talus presents a steep slope, a longitudinal sorting, and does not show any stratification in cross section. In most cases transit is due to a combination of different processes like avalanches or DEBRIS FLOWs. The morpho-sedimentological characteristics of the deposit will depend on the process or the dominant processes.

Further reading

Bertran, P. (ed.) (2003) *Dépôts de pente continentaux: dynamique et faciès*, BRGM.

Jomelli, V. and Francou, B. (2000) Comparing characteristics of rockfall talus and snow avalanche landforms in an alpine environment using a new methodological approach, *Geomorphology* 35, 181–192.

VINCENT JOMELLI

TALUVIUM

Taluvium is a slope deposit composed of rock fragments in a fine matrix, thereby bridging the gap between TALUS (composed of rock fragments) and colluvium (fine material only). Given time, talus will eventually change into taluvium, and subsequently into COLLUVIUM. The development of taluvium from talus has been related to weathering of the rock, although formation by the incorporation of finer airfall material in the talus lattice, such as by loess (Pierson 1982) and volcanic tephra, has also been suggested.

Taluvium deposits are stratigraphically, texturally and hydrologically heterogeneous, as talus formation and emplacement of the fine matrix is typically an episodic process. The accumulation of a fine matrix increases the total surface area of particle contact, increasing internal friction and encouraging aggregation. However, further accumulation of fines decreases the void ratio, impeding drainage and resulting in increasing pore-water pressures. Thus, the slope stability of taluvium (typically 25–28°) is predominantly influenced by its hydrological properties, with saturation resulting in a reduction in its stability angle to approximately half its dry value.

Reference

Pierson, T.C. (1982) Classification and hydrological characteristics of scree slope deposits in the northern Craigieburn Range, New Zealand, *Journal of Hydrology (New Zealand)* 21(1), 34–60.

STEVE WARD

TECTONIC ACTIVITY INDICES

Morphometric techniques used as reconnaissance tools to identify areas experiencing rapid tectonic deformation. Among the indices that have been found most useful are the following (Keller and Pinter 2002).

The hypsometric integral

The hypsometric curve describes the distribution of elevations across an area of land. It is created by plotting the proportion of total basin height (h/H = relative height) against the proportion of total basin area (a/A = relative area). The total height (H) is the relief within the basin (the maximum elevation minus the minimum elevation). The total surface area of the basin (A) is the sum of the areas between each pair of adjacent contour lines. The area (a) is the surface area within the basin above a given line of elevation (h). The value of relative area (a/A) always varies from 1.0 at the lowest point in the basin (where $h/H = 0.0$) to 0.0 at the highest point in the basin (where $h/H = 1.0$).

A simple way to characterize the shape of the hypsometric curve for a given drainage basin is to calculate its *hypsometric integral* (H_i). The integral is defined as the area under the hypsometric curve. One way to calculate the integral for a given curve is as follows [8,9]:

$$H_i = \frac{\text{Mean elevation} - \text{minimum elevation}}{\text{Maximum elevation} - \text{minimum elevation}}$$

A high hypsometric integral indicates a youthful topography. Digital Elevation Models (DEMs) make calculations easy (Gardner *et al.* 1990).

Drainage basin asymmetry

The Asymmetry Factor (AF) has been developed to detect tectonic tilting transverse to flow:

$$AF = 100\ (A_r/A_t)$$

where A_r is the area of the basin to the right (facing downstream of the trunk stream), and A_t is the total area of the drainage basin. AF values are sensitive to tilting perpendicular to the trend of the trunk stream and the greater the divergence of the AF values from 50 the greater is the degree of tilt.

Stream Length–Gradient Index (SL Index)

This is represented as:

$$SL = (\Delta H/\Delta L)L$$

where ΔH is the change in elevation of a stream reach and ΔL is the length of the reach. L is the total channel length from the midpoint of the reach upstream to the highest point on the channel. The index is used to identify recent tectonic activity by identifying anomalously high index values on a particular lithology.

Mountain-front sinuosity

This is an index that reflects the balance between erosional forces that tend to cut embayments into a mountain front and the tectonic forces that tend to produce a straight front coincident with an active range-bounding fault (see FAULT AND FAULT SCARP). It is expressed as:

$$S_{mf} = L_{mf}/L_s$$

where S_{mf} is the mountain-front sinuosity, L_{mf} is the length of the mountain front along the foot of the mountain at the pronounced break in slope, and L_s is the straight-line length of the mountain front. Given that mountain fronts associated with active tectonics and uplift are straight, they have low values of S_{mf}.

Ratio of valley floor width to valley height

This may be expressed as:

$$V_f = 2V_{fw}/[E_{ld} - E_{sc}) + (E_{rd} - E_{sc})]$$

Where V_f is the valley floor width-to-height ratio, V_{fw} is the width of the valley floor, E_{ld} and E_{rd} are elevations of the left and right valley divides respectively, and E_{sc} is the elevation of the valley floor. High values of V_f are associated with low uplift rates that have enabled streams to cut broad valley floors. Low values of V_f are associated with uplift.

The use of indices such as these, particularly in combination, enable the production of a relative tectonic activity class designation for an area.

References

Gardner, T.W., Sasowsky, K.C. and Day, R.L. (1990) Automated extraction of geomorphometric properties from digital elevation data, *Zeitschrift für Geomorphologie* 80, 57–68.
Keller, E.A. and Pinter, N. (2002) *Active Tectonics. Earthquakes, Uplift and Landscape*, 2nd edition, Upper Saddle River, NJ: Prentice Hall.

A.S. GOUDIE

TECTONIC GEOMORPHOLOGY

As Burbank and Anderson (2001: 1) remark, 'The unrelenting competition between tectonic processes that tend to build topography and surface processes that tend to tear them down represents the core of tectonic geomorphology.' Immersed in small-scale process geomorphology, geomorphologists have been remarkably slow to explore both the significance of the conceptual advances provided by the PLATE TECTONICS model and the range of new techniques and data sources relevant to the quantification of long-term denudation rates (Summerfield 2000: 3). Nonetheless, over recent years the enormous growth in topographic data (see, for example, DIGITAL ELEVATION MODEL) and in geochronologic data (see, for example, COSMOGENIC DATING; FISSION TRACK ANALYSIS) have provided new opportunities to address long-standing questions of landscape evolution at the regional and continental scales (Morisawa and Hack 1985).

Plainly there are many Earth features that owe their form in large measure to tectonic activity (see ACTIVE AND CAPABLE FAULT; FAULT AND FAULT SCARP; FOLD; MANTLE PLUME; PULL-APART AND PIGGY-BACK BASIN; RING COMPLEX OR STRUCTURE; TECTONIC ACTIVITY INDICES). Equally there are miscellaneous types of tectonic activity (see CYMATOGENY; DIASTROPHISM; EPEIROGENY; ISOSTASY; SEAFLOOR SPREADING; WILSON CYCLE). At the large scale some major landscape features are associated intimately with various types of plate boundary (see ACTIVE MARGIN; ESCARPMENT; MOUNTAIN GEOMORPHOLOGY; PASSIVE MARGIN; SEISMOTECTONIC GEOMORPHOLOGY; VOLCANO, etc.)

The scope of modern tectonic geomorphology can be appreciated by the context of the recent text by Burbank and Anderson (2001). They start with a consideration of *geomorphic markers* (features or surfaces that provide a reference frame against which to gauge differential or absolute tectonic deformation). Next they consider dating methods, which are vital to determine rates at which faults move or surfaces deform. Then they pass on to stress, faults and folds, before analysing geodetic methods for analysing short-term deformation (e.g. GPS). This is followed by a consideration of the evidence for, and consequences of, palaeoseismology. Then they look at the balance between erosion and uplift, and deformation at various timescales from the Holocene to the Late Cenozoic. They conclude with a discussion of numerical modelling of landscape evolution.

References

Burbank, D.W. and Anderson, R.S. (2001) *Tectonic Geomorphology*, Oxford: Blackwell Science.
Morisawa, M. and Hack, J.T. (eds) (1985) *Tectonic Geomorphology*, Boston: Unwin Hyman.
Summerfield, M.A. (ed.) (2000) *Geomorphology and Global Tectonics*, Chichester: Wiley.

A.S. GOUDIE

TERMITES AND TERMITARIA

Termites, insects of which there are several thousand species, are members of the Isoptera order, and about four-fifths of the known species belong to the Termitidae family (Harris 1961). They vary in size according to their species, from the large African *Macrotermes*, with a length of around 20 mm and a wingspan of 90 mm, down to the Middle Eastern *Microcerotermes* which is only about 6 mm long with a wingspan of 12 mm.

They occur in great numbers – 2.3 million ha^{-1} in Senegal and 9.1 million ha^{-1} in the Ivory Coast

(UNESCO, UNEP, FAO 1979). The vast majority of termite species are found in the tropics, though their distribution is wider than this; they extend to 45–48°N and to 45°S.

Termite mounds and hills are the most striking manifestations of termite activity, and have a large range of sizes and morphologies (Goudie 1988). The heights of termite constructions vary considerably according to species. There are records in the literature of mounds attaining heights in excess of 9 m, though most are less than this. Among the species that create the tallest mounds are *Bellicositermes bellicosus*.

In general the densities of mounds vary considerably according to both environmental conditions (e.g. soil type) and termite species. So, for example, the density of the very large mounds produced by *Macrotermes bellicosus*, *Macrotermes subhyalinus*, *Macrotermes falciger*, *Bellicositermes bellicosus* and *Nasutitermes triodiae* tends to be less (often around 2–10 ha^{-1}) than those for the smaller types of mound (often around 200–1,000 ha^{-1}).

Undoubtedly soil characteristics are an important control of mound formation. Mounds tend to be rare on sands (where there is insufficient binding material), on deeply cracking self-mulching clays (which are unstable), or on shallow soils (where there is a shortage of building material) (Lee and Wood 1971). Soil drainage may also be important, as are human activities. Although in most cases it is obvious that particular mounds have been produced by particular species of termites there have been some arguments about the origin of some mounds found in the tropics and elsewhere (see MIMA MOUND). For example, Cox and Gakaha (1983) have argued that certain mounds in Kenya are created by the mole-rat, *Tachyoryctes splendens*, whereas Darlington (1985) argues that the same mounds are produced by a type of termite – *Odontotermes*.

Plate 138 A large termite mound developed by *Macrotermes* in the Mopane woodland of northern Botswana

Whereas termites live in hidden subsoil chambers or in the conspicuous mounds just discussed, they have a significant effect on soils, partly because of their mechanical activities and partly because of their feeding habits. They can, for example, cause soils to become rich in calcium, play a major role in nutrient cycling, remove organic litter from soil surfaces and mobilize particular soil fractions (e.g. clay). Whether they contribute to laterite formation or lead to its degradation, however, has been the subject of debate (Runge and Lammers 2001).

Termites can contribute to accelerated rates of soil denudation. Lee and Wood (1971) identify three main ways in which termites can do this:

1 by removing the plant cover;
2 by digesting or removing organic matter which would otherwise be incorporated into the soil, and thus making the soil more susceptible to erosion;
3 by bringing to the surface fine-grained materials for subsequent wash and creep action.

The huge numbers of termites and their large total biomass in favoured localities ensures that these three mechanisms are important. The live weight biomass of termites can be comparable to the live weight biomass of large mammalian herbivores in tropical areas. However, in addition to the potential for erosion and sediment yield caused by mound formation and abandonment it is important to remember the other major consequence of termite-caused soil translocation. This is the construction of covered runways or 'sheetings'. These are constructed of soil particles cemented together with salivary secretions (Bagine 1984).

Through their effects on denudation termites may have more widespread effects on fluvial systems. Drummond (1888: 158) postulated that while

Egypt was the gift of the Nile, that river's sediments resulted from 'the labours of the humble termites in the forest slopes about Victoria Nyanza'.

References

Bagine, R.K.N. (1984) Soil translocation by termites of the germs Odontotermes (Holmgren) (Isoptera: Macrotermininae) in an arid area of northern Kenya, *Oecologia* 64, 263–266.
Cox, G.W. and Gakaha, C.G. (1983) Mima mounds in the Kenya Highlands: significance of the Dalquest-Scheffer hypothesis, *Oecologica* 57, 170–174.
Darlington, J.P.E.C. (1985) The underground passages and storage pits used in foraging by a nest of termite Macrotermes michaelseni in Kajiado, Kenya, *Journal of Zoology*, London 198, 237–247.
Drummond, H. (1888) *Tropical Africa*, London: Hodder and Stoughton.
Goudie, A.S. (1988) The geomorphological role of termites and earthworms in the tropics, in H.A. Viles (ed.) *Biogeomorphology*, 166–197, Oxford: Blackwell.
Harris, W.V. (1961) *Termites: Their Recognition and Control*, London: Longman.
Lee, K.E. and Wood, T.G. (1971) *Termites and Soils*, London and New York: Academic Press.
Runge, J. and Lammers, K. (2001) Bioturbation by termites and Late Quaternary landscape evolution in the Mbomou Plateau of the Central African Republic, *Palaeoecology of Africa* 27, 153–169.
UNESCO, UNEP, FAO (1979) *Tropical Grazing Land Ecosystems*, Paris: UNESCO.

A.S. GOUDIE

TERRACE, RIVER

A river terrace is the planar surface that remains after the river, which formed it, incised its former valley floor. River terraces are abandoned river channels and floodplains. Their presence in river valleys throughout the world provides a record of changes in the flow regimes of rivers and the sediment supplied to them over time.

The flat surface of a terrace, called the *tread*, represents the highest elevation of the valley floor before incision occurred. Terrace treads, also called benches or platforms, are composed of alluvium, bedrock, or bedrock covered with a thin deposit of alluvium. They dip downvalley, recording the gradient of the channel that formed them, unless tectonic or isostatic uplift has subsequently altered their slope. The steep slope between treads or between the tread and the active floodplain is called the *riser*. Flights of terraces represent multiple, discrete episodes of downcutting, punctuated by periods of stability or AGGRADATION. Terraces may be continuous along a valley, or discontinuous if portions of the same terrace have become separated by tributary entrenchment or other geomorphic processes. More recent deposits, including those of mass movements, alluvial fans, volcanic ash, or wind-blown fines, may bury terrace treads. In a tectonically active area, faulting can alter the height relationships between terraces.

Genetically, terraces are considered to be either depositional (fill) or erosional (cut) landforms. Depositional terraces form as a result of the aggradation and later entrenchment of alluvium. They are abandoned floodplains, and stratigraphically show vertical and lateral processes of sediment accretion. Erosional terraces are surfaces formed by the erosional removal of bedrock or alluvial fill from the former valley floor. The term river terrace is used inclusively to describe the landform without specifying the materials (bedrock or fill) or the genetic processes (erosional or depositional) responsible for a specific feature. When the ages or relative ages of a flight of terraces are known, terrace surfaces are usually identified numerically, with the lowest number (1) used for the oldest surface. By recording changes in the flow regime of rivers, terraces are important indicators of tectonic, climatic, and even anthropogenic environmental history.

Terrace-forming processes

Different processes and circumstances, acting alone or in combination, cause rivers to incise, stranding their former valley floors above the active channel. Incision may begin gradually or catastrophically. A river cuts down through its own deposits when greater flow energy, lower BASE LEVEL, and/or less sediment load increase its erosional capacity. Discharge and stream energy can be increased by changes in climate and by processes, including base-level lowering, that steepen the channel gradient. Climatically driven incision occurs when climate becomes wetter, when ice melts (warmer climate), or when upstream climate–vegetation–soil relationships lead to conditions of flashier rainfall runoff. The latter would occur where decreasing precipitation causes a marginally semi-arid area to become more arid such that vegetation becomes sparser, soil erosion accelerates, infiltration rates decrease and the proportion of rainfall flowing to the river as storm runoff thereby increases. Anthropogenic activities, including forest removal and road building, also increase the flashiness of runoff. Factors that cause the channel gradient to steepen increase the energy of the flow, allowing the river to entrain materials previously deposited. Increased gradients result from tectonic and isostatic uplift, headward propagation of knickpoints, faulting, or lowering the erosional base level. Many of the terraces in contemporary landscapes are Pleistocene or Holocene in age and record changes in river regimes due to shifts in climate and in geomorphic process regimes between glacial and interglacial periods. Among the multiple factors leading to Pleistocene river incision and terrace formation was the lowering of sea level during glacial periods, when more water was stored on land as ice. Glacial period sea levels dropped more than 100 m, lowering erosional base levels and steepening continental river channel gradients accordingly (see GRADE, CONCEPT OF).

Changes in sediment supply tip the balance between aggradation and degradation in a river. A river's capacity to transport sediment derives from the volume and calibre of sediment supplied to it and the energy available to move the sediment. In a period of little environmental change, the channel geometry of a river, including its gradient, represents the discharge, energy and sediments characteristic of that environment at that time. When active glaciers, for example, provide an abundant sediment supply, the resulting channel will become steep, swift, shallow and probably braided. Reducing the sediment supply leaves such a river, adapted for heavy sediment loads, with excess energy. Thus, reducing the sediment supply upstream leads to more aggressive erosion and downcutting downstream, a condition that will continue until the flow regime becomes more in balance with the available energy and sediment. The supply of sediment to a river can be diminished by changes in climate or land management practices that increase vegetative cover or decrease mass movements and wind erosion on upstream land surfaces. Dams stop sediment, too. Such changes increase the erosional energy of the river downstream, possibly to the point that it will begin to incise its own deposits. RIVER CAPTURE of a higher gradient, sediment-laden headwater river by a lower gradient piedmont river causes the piedmont river to gain additional sediment input from the high gradient headwaters, and the headwater river to adjust to a lower local base level. Steepening gradients in the capturing stream can cause flow energy to increase, and the discharge of the capturing stream may also increase due to the increased size of its drainage basin. Finally, catastrophic events, e.g. outburst flooding from a glacial lake or a landslide-dammed lake, can trigger catastrophic incision and initiate terrace formation.

Depositional terraces

A depositional (fill, aggradational) terrace is a former floodplain that has been incised by the river. It differs from an active floodplain by being too far above the river channel to convey the overbank flows of the mean annual flood. Massive alluvial deposits in large terraces represent conditions of abundant sediment supply to the river (Figure 164a). The sediments in depositional terraces were built up by the accretion of alluvial sediment, either vertically or laterally, during a period in which that surface was the active floodplain of the river. For floodplains built by vertical accretion, a cross section of the terrace would reveal horizontal stratification, with flood deposits on top of flood deposits sorted by size and fining upward. Gravels in the deposits would be rounded and probably imbricated and aligned to point downstream. If the floodplain had built up by lateral accretion, the deposits would have the stratigraphy of POINT BARS. Depositional terraces may be metres to kilometres wide, and follow a present or former course of the river for thousands of kilometres, often on both sides of the channel. Flights of depositional terraces occur where rivers have developed new floodplains between separate episodes of downcutting.

In many examples around the world, repeated episodes of valley filling followed by major evacuation of sediment have created younger (inner) depositional terraces developed on fill material which was emplaced after the valley was scoured out between higher, older terraces (Figure 164b). Depositional terraces can be distinguished from erosional terraces in that (1) the tread surface of a depositional terrace represents the uneroded surface of a valley fill and (2) the underlying rock surface (if its topography can be determined seismically or viewed in a transverse cut) may be very irregular.

Erosional terraces

The surface of an erosional river terrace was levelled by lateral fluvial erosion before the river

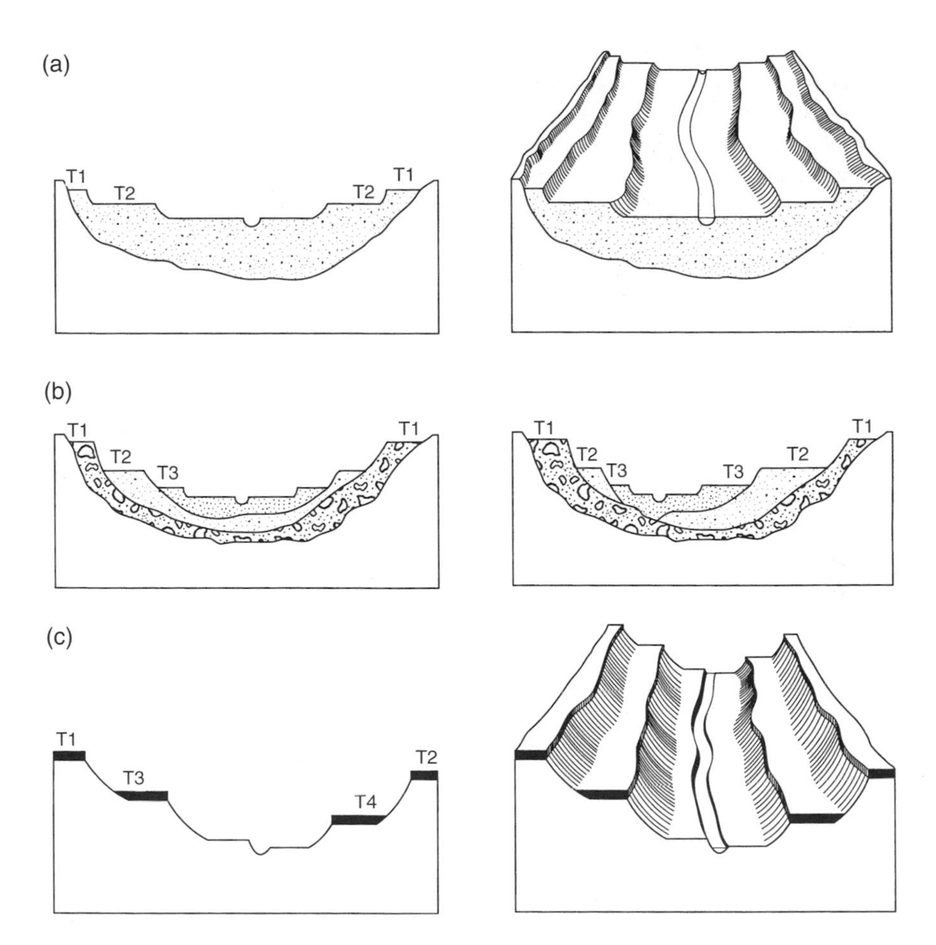

Figure 164 Valley cross section and block diagrams illustrating (a) paired depositional terraces, (b) depositional terraces formed from multiple episodes of valley fill and entrenchment, and (c) unpaired strath terraces

incised to form the terrace. Like depositional terraces, erosional terraces are remnants of an older valley floor, into which the river has cut down. Erosional terraces scoured from bedrock are called *bench, strath*, or *rock-cut* terraces (Howard *et al.* 1968: 1,117). The term 'strath' is a Scottish word meaning wide valley. Rivers carve straths by lateral corrasion, with abrasive alluvial sediments eroding the channel bed from side to side as meanders migrate downstream and/or the meander radius increases. The underlying bedrock surface, as nearly planar as channel beds and valley bottoms are today, lies parallel to the surface of the thin cap of alluvium (Figure 164c). The thickness of the alluvial cap indicates the depth of scouring in the former river regime: the alluvium must be thin enough so that the river could have been in erosional contact with the bedrock valley bottom. A thick valley fill, on the other hand, protects the valley bottom from erosion. The maximum thicknesses of alluvial lag deposits, usually gravel, on strath terraces depends on the size and energy level of the river. Mackin (1937: 828) reported 2.5 m of gravel with a cover of silt in the Shoshone Valley (Wyoming, USA); Wegmann and Pazzaglia (2002: 734) suggest 3 m as a typical maximum depth.

Not all erosional terraces are strath terraces with eroded bedrock at or just below the surface. A valley floor composed of unconsolidated fill may also become truncated by lateral erosion and subsequently stranded by incision to become a river terrace. Such terraces are called *fill*-cut (Bull 1990: 355) or *fill-strath* (Howard *et al.*

1968: 1,119). *Structural* terraces are erosional benches of resistant bedrock resulting from differential erosion rather than from changes in the river flow regime. Structural terraces abound in the Grand Canyon of the Colorado River, where contrasting erodibilities of the nearly horizontally bedded sedimentary strata cause the resulting canyon walls to appear as rock steps.

Paired and unpaired river terraces

Where terraces occur at the same elevation across a valley, they are called *paired* (Figure 164a); otherwise they are *unpaired*. Paired and unpaired terraces can be erosional or depositional; they can be formed on bedrock or on alluvial fill. Paired terraces represent conditions in which downcutting predominates over lateral cutting. A river flowing across the middle of a thick floodplain would respond to a drop in base level by incision. The result would be a lower channel between a pair of terraces. If the base level remained at the same low position for an extended period of time, the period of downcutting, accompanied by mass movements of channel banks, would be followed by a period of aggradation in which the river might form a new floodplain along the entrenched channel. A second episodic event that renewed the greater erosional capability of the river could initiate a second phase of downcutting, causing the river to abandon the newer floodplain and incise a second, inner set of paired terraces. Such pairs of depositional terraces are assumed to be of equal age.

Unpaired river terraces reflect conditions in which downcutting is slow and lateral erosion is occurring at the same time (Ritter 1986: 269). Lateral migration of the river channel can erode and eliminate some older terraces, leaving an asymmetrical record of depositional or erosional surfaces. Unpaired terraces are unlikely to be of equal age, and likely to be erosional in origin. Strath terraces are typically unpaired.

Terraces as evidence of environmental change

Terraces may record tectonic events, changes in climate and other environmental changes that alter the erosional capacity and sediment load of a river. Dating terrace surfaces enables researchers to calculate incision rates and better understand the climate or tectonic history of a region. The age of a terrace is determined by its relative position in the valley and by other evidence present in its biological, chemical and anthropogenic record. Alluvial particles on a strath terrace are synchronous with the time period of erosion, whereas alluvial particles in a depositional terrace predate the time of terrace formation by the time required for a slug of sediment to travel to that location in the river system and the time lapse between deposition and incision (Bull 1990: 360). Organic matter buried in the terrace can be dated with radiocarbon (^{14}C) dating techniques (e.g. Wegmann and Pazzaglia 2002: 734), and investigators have begun to date clasts on terrace surfaces based on the concentration of cosmogenic isotopes such as ^{10}Be and ^{26}Al (e.g. Hancock *et al.* 1999: 47) (see DATING METHODS). Terrace ages can sometimes be inferred from the environmental setting portrayed in the biological record. The presence of sub-Arctic molluscs and fossil ice-wedge casts in terraces along the River Thames indicates deposition of terrace material during cold periods, interpreted as glacial periods in the Pleistocene (Goudie 1984: 292). Terraces from the late Tertiary or Pleistocene can be distinguished from Holocene terraces, not only by their biota and their position in the landscape, but by the degree of weathering of terrace materials. In some older terraces, formerly unconsolidated terrace materials have become cemented by carbonate, silica or iron oxides (Costa and Baker 1981: 161). Human artefacts, including ruined structures and Roman coins (Judson 1963: 899), have helped date Late Holocene terrace surfaces.

In some investigations, terrace formation is attributed to a single causative factor. Born and Ritter (1970: 1,240), for example, attributed the flight of terraces in the Truckee River above Pyramid Lake (California, USA) to an anthropogenically lowered base level. Alternatively, terrace formation can represent a complex landscape response to one change (e.g. a climatic factor) or the integrated landscape response to a set of changes (e.g. climatic and tectonic). The terrace record of the River Rhine, which records the climatic and erosional history of the region, has also been affected by uplifts in the middle and subsidence in the lower portions of the valley (Fairbridge 1968: 1,131).

Evidence provided by river terraces can be complex and challenging to interpret. Erosional episodes can remove older terraces; in fact, terrace sequences are rarely found completely intact. Terraces in one drainage basin may respond more to local factors than to regional climatic or tectonic controls. As an example of this, Brakenridge (1981: 75) found terrace formation along the Pomme de Terre River in southern Missouri (USA) to not match either the sea-level history for the Gulf of Mexico or the glacial chronology of the upper Missouri–Mississippi basin. In other examples, Ritter (1982: 352) found terraces in one valley in the Alaska Range to have formed after a moraine-dammed lake overtopped a drainage divide, and Wegmann and Pazzaglia (2002: 740) did not find base-level control responsible for terrace formation in the Clearwater River (Olympic Mountains, USA), even though the river flows directly into the Pacific Ocean. They suggest that the Clearwater River operates at or near its capacity for sediment transport and responds to changes in upstream sediment production, some of which is likely to be earthquake-related, with episodes of vertical (terrace-forming) or lateral (valley-widening) incision. Their interpretation is supported by Schumm's (1975: 77) work showing COMPLEX RESPONSE of a fluvial system could lead to small terrace formation without external forcing variables. Bull (1990: 352) emphasizes a scale difference between major terraces, particularly climatically caused aggradation surfaces and large tectonically caused straths, and minor terraces, which develop in response to local factors.

Although terrace tread formation is generally thought to reflect relatively stable conditions over long periods of time, terraces are also known to have formed catastrophically or over intervals of only a few years. From photographic and historical evidence, Born and Ritter (1970: 1,240) documented the formation of at least six well-developed river terraces upstream of Pyramid Lake (California, USA) in the forty-four years since water diversion began to lower the lake level.

Tread surfaces of terraces in the contemporary landscape may not have been disturbed by erosion or deposition since the time of their abandonment by the active river channel. A flight of such terraces presents a CHRONOSEQUENCE of weathering and soils, and terraces of known age present special opportunities for studying soil development (Bull 1990: 352). In inhabited areas, opportunities for the study of terrace soils are commonly constrained by human disturbance. Younger terraces are sought as sources of sand and gravel, and terrace surfaces are often chosen for agriculture, urbanization, and the location of highways and airports. Their relative flatness in high-relief environments and their position above elevations of frequent flooding and poor drainage makes terrace surfaces attractive for human occupation.

References

Born, S.M. and Ritter, D.F. (1970) Modern terrace development near Pyramid Lake, Nevada, and its geologic implications, *Geological Society of America Bulletin* 81, 1,233–1,242.

Brakenridge, G.R. (1981) Late Quaternary floodplain development along the Pomme de Terre River, southern Missouri, *Quaternary Research* 15, 62–76.

Bull, W.B. (1990) Stream-terrace genesis: implications for soil development, *Geomorphology* 3, 351–367.

Costa, J.E. and Baker, V.R. (1981) *Surficial Geology. Building with the Earth*, New York: Wiley.

Fairbridge, R.W. (1968) Terraces, fluvial–environmental controls, in R.W. Fairbridge (ed.) *Encyclopedia of Geomorphology*, 1,124–1,138, New York: Reinhold.

Goudie, A. (1984) *The Nature of the Environment*, Oxford: Basil Blackwell.

Hancock, G.S., Anderson, R., Chadwick, O. and Finkel, R. (1999) Dating fluvial terraces with ^{10}Be and ^{26}Al profiles: application to the Wind River, Wyoming, *Geomorphology* 27, 41–60.

Howard, Arthur D., Fairbridge, R.W. and Quinn, J.H. (1968) Terraces, fluvial – Introduction, in R.W. Fairbridge (ed.) *Encyclopedia of Geomorphology*, 1,117–1,123, New York: Reinhold.

Judson, S. (1963) Erosion and deposition of Italian stream valleys during historic time, *Science* 140, 898–899.

Mackin, J.H. (1937) Erosional history of the big Horn Basin, Wyoming, *Geological Society of America Bulletin* 48, 813–893.

Ritter, D.F. (1982) Complex terrace development in the Nenana Valley near Healy, Alaska, *Geological Society of America Bulletin* 93, 346–356.

——(1986) *Process Geomorphology*, 2nd edition, Dubuque, IA: William C. Brown.

Schumm, S.A. (1975) Episodic erosion: a modification of the geomorphic cycle, in W. Melhorn and R. Flemal (eds) *Theories of Landform Development*, 69–85, London: George Allen and Unwin.

Wegmann, K. and Pazzaglia, F. (2002) Holocene strath terraces, climate change, and active tectonics: the Clearwater River basin, Olympic Peninsula, Washington State, *Geological Society of America Bulletin* 114, 731–744.

Further reading

Selby, M.J. (1985) *Earth's Changing Surface*, Oxford: Oxford University Press.

SEE ALSO: floodplain; fluvial geomorphology; sediment load and yield; valley

CAROL HARDEN

TERRACETTE

A miniature unvegetated step-like feature that forms on hillslopes. Terracettes extend across the slope in a parallel manner, though predominantly following the contours of the land. They commonly form on ground that is fairly unconsolidated, particularly pasture, possessing moderate to steep hillslope gradients. Terracettes are rarely greater than 0.5 m in height and depth, with a spacing of about 1 m, and may extend laterally for tens of metres. They may form in a variety of climatic environments. The vertical drop exhibited in a terracette is known as the riser whereas the horizontal platform is called the tread. Additionally, the base of the riser is referred to as the foot of the terracette, while the point where the tread meets the riser is termed the crown.

Terracettes are irregular and often anastomosing forms, and may feature as intermittent steps or a whole network of steps covering a hillside. On a loessic slope of 24°, it has been estimated that terracettes may form on up to 11 per cent of the ground surface, and as much as 40 per cent on a slope angle of 37° (Selby 1993). Further terms for a terracette include pseudo-terraces or false terrace. Angle of slope is crucial for terracette formation. An average slope angle of about 30° dictates the boundary for terracette formation. Below this angle the slope face will not break, but instead will exhibit a distinctive undulating surface. Surficial material (based on grass) will begin to crack at angles greater than 30°, while at angles greater than 50° cracks occur at the back of each tread and small-scale slumping occurs, thus developing the characteristic step-like feature (Selby 1993: 258).

The origin of terracettes is contentious, and several explanations of their development have been produced. However, it is likely that the following mechanisms of formation are interrelated, and that each example is the result of a mix of mechanisms, unique to the site. The dominant mechanism of terracette formations is by soil creep, and occurs when hillslope angles are greater than that of which the unconsolidated mantle material can remain stable, resulting in slippage. Terracette formation is not limited to open grassland, and may develop in forested land as soil is washed downslope and leaf litter accumulates in the wake of tree roots. Additionally, erosion may be provoked or enhanced by the trampling of land by livestock. The hooves of both cattle and sheep can remove and wear down the surficial materials, particularly upon frequently used tracks and walkways (termed cattle or sheep tracks).

Terracettes can also result from near-surface faulting, which may break the overlying land into a series of small terracettes. Additionally, they can also be ablation products, man-made aids for cultivation on hillslopes (lynchets), as products of solifluction and antiplanation, or as miniature fluvial terraces.

Reference

Selby, M.J. (1993) *Hillslope Materials and Processes*, Oxford: Oxford University Press.

Further reading

Rahm, D.A. (1962) The terracette problem, *Northwest Science* 36, 65–80.

Vincent, P.J. and Clarke, J.V. (1976) The terracette enigma – a review, *Biuletyn Peryglacjalny* 25, 65–77.

STEVE WARD

TERRAIN EVALUATION

The term *terrain evaluation* has been used to describe a wide range of geomorphological techniques and no single definitive meaning has been established. In its narrowest definition terrain evaluation is regarded as synonymous with mapping of LAND SYSTEMs, a procedure for classifying the landscape by dividing it into landform assemblages with similarities in terrain, soils, vegetation and geology (Mitchell 1973). In a slightly broader definition, Lawrance *et al.* (1993) regarded terrain evaluation for engineering projects as a method for summarizing the physical aspects of a landscape initially through classification and then including an assessment of ground conditions in terms of engineering requirements. Griffiths and Edwards (2001) use the expression *land surface evaluation* as an alternative to terrain evaluation in ENGINEERING GEOMORPHOLOGY and engineering geology because the varied usage had created confusion and led to misunderstanding. However, in practice terrain evaluation and land surface evaluation are synonymous and therefore the definition proposed by Griffiths and Edwards (2001) is the most appropriate to use. This states that it is the evaluation and interpretation of land surface and near-surface features using techniques that do not involve ground exploration by excavation (except using small hand-dug pits or hand auger holes) or geophysics. Based on this definition terrain evaluation can be regarded as integral to the development of the ground model proposed by Fookes (1997) as central to all successful civil engineering construction. The definition would also be suitable when terrain evaluation is used as a technique of APPLIED GEOMORPHOLOGY in planning and environmental studies (see Smith and Ellison 1999). In these studies, the terrain evaluation procedure would include an evaluation of soils, vegetation, land use, materials, drainage and human activity in addition to geomorphological processes and landforms.

The techniques that can be employed in terrain evaluations include: GEOMORPHOLOGICAL MAPPING, geological mapping, engineering geological mapping, remote sensing interpretation, analytical photogrammetry, land systems mapping, natural hazard and risk assessment, and the use of Geographical Information Systems (GIS). The output from a terrain evaluation is usually a suite of maps either held as hardcopy or as a suite of overlays in a GIS. The map data can be classified into three categories:

1. Factual or element maps that record the actual ground conditions (Table 45a).
2. Derivative maps that are obtained by either combining element maps or are based on an interpretation of the element maps (Table 45b).
3. Summary maps that pull together a range of derivative and element maps to identify combinations of hazards, resources or land use issues that act either as constraints to any development or indicate the potential of the land for exploitation (Table 45c).

Table 45 Terrain evaluation map categories

Terrain evaluation map category	Examples of typical maps
(a) Element maps	• morphology • topography • bedrock geology • superficial geology • lithology • vegetation • pedology • land use • geotechnical properties • location of sites of special scientific interest • exploratory holes and wells • hydrology
(b) Derivative maps	• slope steepness that uses topography to classify maps into distinct groups based on morphology and topography • depth to bedrock that utilize data from the geological maps • geomorphology • hydrogeology • depiction of various types of resources, such as sand and gravel or brick clay • foundation conditions for engineering structures • geotechnical zoning, i.e. areas of homogeneous ground conditions • hazards, such as subsidence, landslides, flooding or contaminated land • previous industrial usage
(c) Summary maps	• development potential • potential resources • planning constraints, including statutory protected land • construction constraints

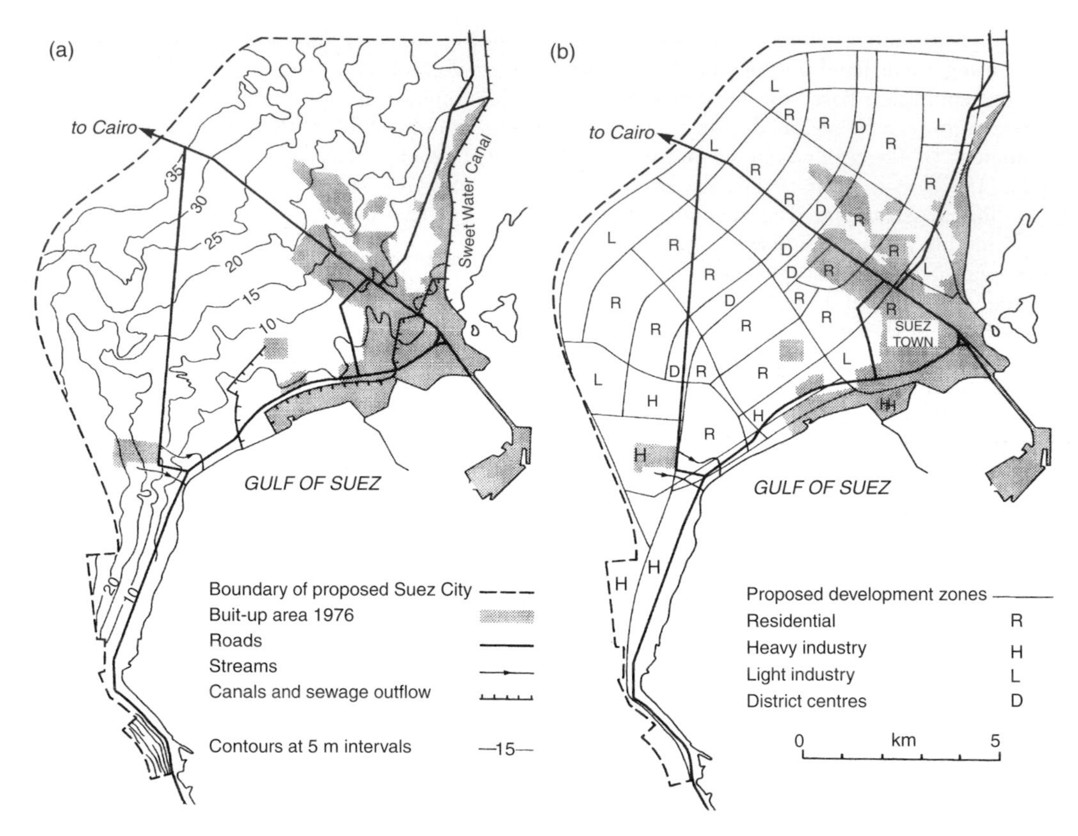

Figure 165 Maps of Suez, Egypt: (a) topography and extent of urban area in 1976; (b) proposed layout of an enlarged urban area (after Jones 2001, reprinted with permission of the Geological Society of London)

An extended legend is normally attached to each of the maps and most terrain evaluation studies would include an interpretative report that explains the basis for the development of the derivative and summary maps.

Terrain evaluation is best illustrated through a case study and Jones (2001) provides a classic example of its use for flood hazard assessment. In this study of a proposed development for Suez New City (Figure 165) geomorphological mapping was initially undertaken based on aerial photograph interpretation and field mapping. This resulted in the production of a geomorphological map, originally at a scale of 1:25,000, that established the distribution of a suite of marine, fluvial and bedrock features in addition to existing areas of urban development (Figure 166). These data were then combined with an evaluation of WADI catchment areas and channel form, based on further aerial photograph interpretation and field data analysis, to produce an interpretative map of flood hazard utilizing an ordinal scaling system (Figure 167). The final hazard map provided the base for both planning development and identifying the areas requiring flood protection.

References

Fookes, P.G. (1997) Geology for engineers: the geological model, prediction and performance, *Quarterly Journal of Engineering Geology* 30, 290–424.

Griffiths, J.S. and Edwards, R.J.G. (2001) The development of land surface evaluation for engineering practice, in J.S. Griffiths (ed.) *Land Surface Evaluation for Engineering Practice*, Geological Society Engineering Geology Special Publication No. 18, 3–9.

Jones, D.K.C. (2001) Ground conditions and hazards: Suez City Development, Egypt, in J.S. Griffiths (ed.) *Land Surface Evaluation for Engineering Practice*, Geological Society Engineering Geology Special Publication No. 18, 159–170.

Lawrance, C.J., Byard, R.J. and Beaven, P.J. (1993) *Terrain Evaluation Manual*, Transport Research

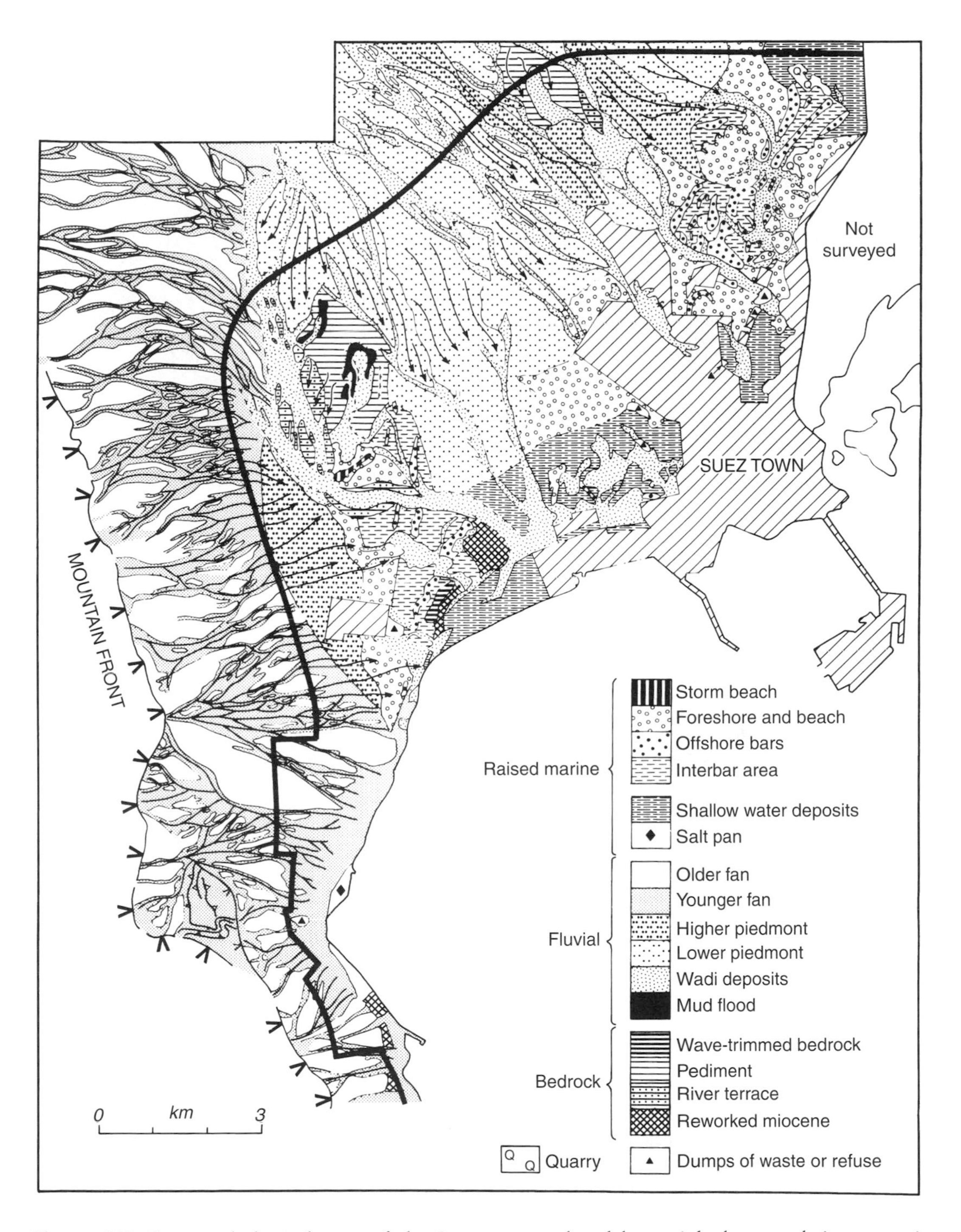

Figure 166 Geomorphological map of the Suez area produced by aerial photograph interpretation and ground mapping (after Jones 2001, reprinted with permission from the Geological Society of London)

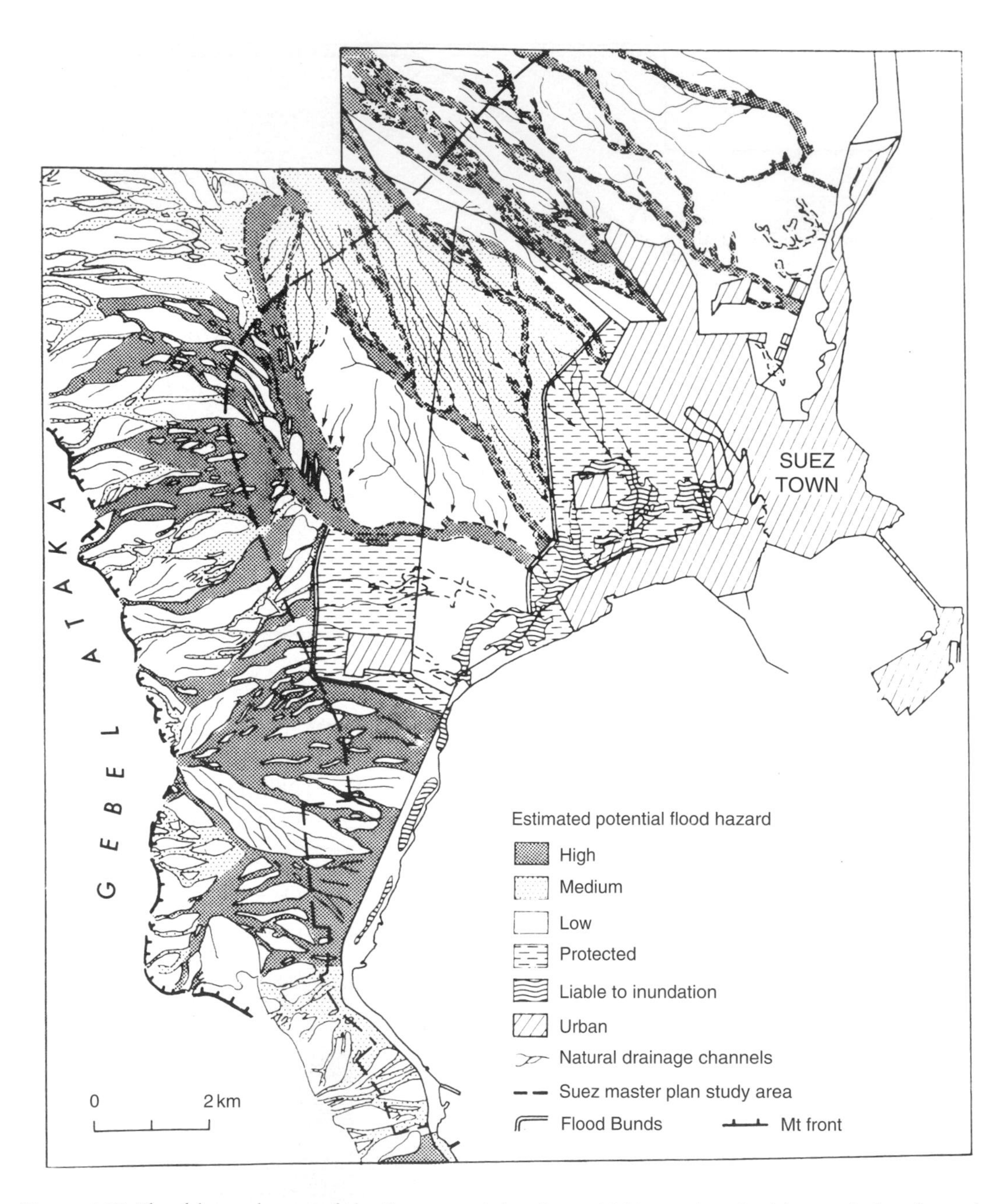

Figure 167 Flood hazard map of the Suez area (after Jones 2001, reprinted with permission from the Geological Society of London)

Laboratory, State of the Art Review 7, London: HMSO.

Mitchell, C.W. (1973) *Terrain Evaluation*, London: Longmans.

Smith, A. and Ellison, R.A. (1999) Applied geological maps for planning and development: a review of examples from England and Wales 1983 to 1996, *Quarterly Journal of Engineering Geology* 32, S1–S44.

Further reading

Cooke, R.U. and Doornkamp, J.C. (1990) *Geomorphology in Environmental Management*, 2nd edition, Oxford: Oxford University Press.

Edwards, R.J.G. (2001) Creation of functional ground models in an urban area, in J.S. Griffiths (ed.) *Land Surface Evaluation for Engineering Practice*,

Geological Society Engineering Geology Special Publication No. 18, 107–113.
Fookes, P.G., Lee, E.M. and Sweeney, M. (2001) Pipeline route selection and ground characterization, Algeria, in J.S. Griffiths (ed.) *Land Surface Evaluation for Engineering Practice*, Geological Society Engineering Geology Special Publication No. 18, 115–121.
Griffiths, J.S. (ed.) (2001) *Land Surface Evaluation for Engineering Practice*, Geological Society Engineering Geology Special Publication No. 18.
Waller, A.M and Phipps, P. (1996) Terrain systems mapping and geomorphological studies for the Channel Tunnel rail link, in C. Craig (ed.) *Advances in Site Investigation Practice*, 25–38, London: Thomas Telford.

JAMES S. GRIFFITHS

THERMOKARST

The term 'thermokarst' was created in 1932 by a Russian, Yermolayev (Czudek and Demek 1970: 103), to describe an uneven morphology (thermokarst terrain) with some similarities to karstic morphology and resulting from soil subsidence due to the melting of ice within PERMAFROST. Later, the same word was used mainly to refer to the processes of soil subsidence resulting from thaw.

A prerequisite to thermokarstic terrain is excess ice within permafrost; in other words, the volume of ice must exceed the volume of the pores that water, under natural conditions, can fill in the soil. If excess ice is present, some change, even a small one, in the surface conditions (e.g. vegetation) can induce the melting of the upper part of the permafrost; such melting recurs during several summers and causes collapse that, cumulatively, can be important. Such collapses will continue until the ACTIVE LAYER has regained a sufficient thickness to protect the permafrost from further melting. Excess ice does not occur everywhere in permafrost. It is mainly found in the low ground of valley bottoms and in coastal lowlands containing abundant silty clays – locations of thaw-sensitive permafrost. Thermokarst also refers to collapse resulting from the disappearance of glacial ice. The forms resulting from melting of ice differ according to the distribution of the ice in the soil and, thus, according to the types of ice. The principal forms of thermokarst are summarized below.

Thermokarst lakes are widespread in the coastal lowlands of Siberia, Alaska and the Mackenzie Delta area (Canada), where excess ice is abundant in the soil. Such lakes are rounded depressions and are rather shallow. They progressively enlarge and their diameter can reach 1–2 km.

In some regions, thermokarst lakes are elongated and under influence of the direction of prevailing winds. They are called 'oriented lakes'. The mechanism leading to such a shape is not quite clear, but it now seems that the longer axis of the oriented lakes is perpendicular to the direction of the prevailing summer winds.

Thermokarst lakes also develop at the expense of ice wedges (see ICE WEDGE AND RELATED STRUCTURES). If a pool appears above an ice wedge (often as a result of a topographic change induced by the growth of the ice wedges), this pool warms the underlying soil and melts the top of the permafrost, i.e. the ice wedge. Because of this melting, the pool grows above the ice wedge, forming linear and polygonal troughs separating centres of the polygons (Plate 139). The centres, which contain less ice, are brought out into relief and evolve into conical mounds called 'thermokarst mounds'. Related to melting ice wedges, beaded streams may develop; these are characterized by narrow reaches linking pools or small lakes. The pools occur at the junction of the ice wedges.

In Siberia, well-known thermokarst forms are the ALASes, thermokarstic depressions with steep sides and flat bottoms covered by grassland, where shallow lakes often exist. Such depressions are generally round or oval, 3–40 m deep and 0.1–15 km long. They occupy 40–50 per cent of the surface of

Plate 139 Thermokarst in Siberia. Vegetation was cut near the road and a part of the active layer was stripped off, causing melting of a polygonal net of ice wedges and consequent formation of thermokarst mounds

the Lena and Aldan terraces in central Yakutia (Washburn 1979: 274) and they result from the melting of exceedingly ice-rich permafrost developed in the silty cover of the terraces.

Melting PINGOs induce formation of closed depressions surrounded by ramparts; such depressions are also thermokarstic. In contemporary permafrost zones, they are called 'collapsed pingos' or 'pingo remnants'. Generally, pingos begin to melt at their top because of the cracks relating to their growth.

Melting PALSAs also give birth to depressions that are visible only with regards to the areas raised by permafrost. When permafrost has disappeared, only very short-lived, shallow depressions remain marked by vegetation different from that on both previously unfrozen peatlands and the remnants of the peaty permafrost landforms.

Lithalsas (the same forms as palsas, but without any cover of peat, and developed in mineral soil), after melting leave, like pingos, depressions surrounded by ramparts. However, unlike pingos, but in the same way as palsas, lithalsas appear as numerous and almost adjacent forms.

On slopes, the most spectacular thermokarst forms are the 'retrogressive thaw slumps'. Described from the boreal forest as well to the High Arctic, they result from a process initiated by thawing of ground ice. It begins by the slide of the active layer on the permafrost table, which acts as a lubricated slip plane for movement and controls the depth of the failure plane. This process produces semicircular hollows opening downslope and usually less than 2 m high (French 1996: 119). Further thawing of permafrost produces steep slopes as much as 8 m high.

A peculiar thermokarstic phenomenon is fluviothermal erosion, namely erosion by the water of rivers, lakes or sea, attacking the permafrost not only by mechanical erosion processes, but also by its warmth which melts the ice. Such erosion is rapid and causes undercutting of riverbanks, particularly in sandy layers. It forms thermo-erosional niches at the floodwater level.

Thermokarstic phenomena have climatic or local causes. Climate warming, by increasing the thickness of the active layer, leads to the melting of the upper permafrost and to thermokarstic phenomena. However, response to climate warming is complex: besides temperature, snowfall also varies and changes in vegetation occur. The immediate response is not obvious. Nevertheless because of the global change, thermokarstic phenomena are finally to be feared.

But more often than not, permafrost is melting because of local causes. Thermokarstic phenomena, for instance, result from destruction or change of the vegetation cover as a consequence of forest fires or human action. Vegetation plays a complex role, generally protecting soil against warmth more than against cold. It acts in winter by trapping snow between branches and impeding insulation of the soil, in summer by protecting the soil by its shade, or by decreasing air circulation, by increasing evaporation, etc.

On areas weakened by excess ice in the soil, buildings, roads, airports, pipelines, etc. pose severe problems that engineers try to solve by impeding the melting of the permafrost. Houses and pipelines are built on piles; roads and airstrips are embanked above the ground in order to lift the permafrost surface; sometimes cooling devices are put into the ground to radiate its warmth to the surface.

Surveying fossil thermokarstic forms in regions where permafrost existed in the past has aroused interest in many scientists. Such forms already mentioned are pingo and lithalsa scars, recognizable by the ramparts surrounding the depressions (Pissart 2000: 344). Traces of thermokarstic collapses have disappeared because of the general subsidence induced by the melting of the whole permafrost. Only remnants of peculiar sediments deposited in those temporary hollows reveals their former existence. However thermokarstic explanations are often invoked without such observations. Their origin remains uncertain.

References

Czudek, T. and Demek, J. (1970) Thermokarst in Siberia and its influence on the development of lowland relief, *Quaternary Research* 1, 103–120.

French, H.M. (1996) *The Periglacial Environment*, Harlow: Longman.

Pissart, A. (2000) Remnants of Lithalsas of the Hautes Fagnes, Belgium: a summary of present-day knowledge, *Permafrost and Periglacial Processes* 11(4), 327–355.

Washburn, A.L. (1979) *Geocryology. A Survey of Periglacial Processes and Environments*, London: Edward Arnold.

Further reading

Ballantyne, C.K. and Harris, C. (1994) *The Periglaciation of Great Britain*, Cambridge: Cambridge University Press.

Harris, S.A., French, H.M., Heginbottom, J.A., Johnston, G.H., Ladanyi, B., *et al.* (1988) *Glossary of Permafrost and Related Ground-Ice Terms*, Ottawa: National Research Council Canada, Technical Memorandum no. 142.

ALBERT PISSART

THRESHOLD, GEOMORPHIC

A geomorphic threshold can be defined as the critical condition at which a landform abruptly changes. The change can be the result of an external variable, that exceeds the stability of a landform at an extrinsic threshold, or the change at an intrinsic threshold can be the result of a progressive change of the landform itself.

Extrinsic thresholds have been recognized in many fields. Perhaps the best known is the threshold velocity required to set in motion sediment particles of a given size. With a continuous increase in velocity, a threshold velocity is reached at which sediment movement commences, and with a progressive decrease in velocity, a threshold velocity is encountered at which sediment movement ceases. The best known thresholds in hydraulics are described by the Froude and Reynolds numbers, which define the conditions at which flow becomes supercritical or turbulent. Particularly notable are the changes of BEDFORM characteristics at threshold values of stream power. In these examples, an external variable changes progressively thereby triggering an abrupt change within the affected system at an extrinsic threshold. That is, the threshold exists within the system, but it will not be crossed and change will not occur without the influence of an external variable. The word, threshold, describes the critical range of conditions over which these transitions occur.

Thresholds can also be exceeded when the external variables remain relatively constant, yet a progressive change of the landform itself renders it unstable, and failure occurs at an intrinsic geomorphic threshold. An example is long-term progressive weathering, that reduces the strength of slope materials until eventually there is slope adjustment and MASS MOVEMENT. Another example of an intrinsic threshold is provided by a typical sequence of morphologic changes resulting in the collapse of sandstone-capped cliffs. Beneath a vertical cliff of sandstone is a gentler slope of weak shale. Through time, the basal shale slope is eroded, which produces a vertical shale cliff beneath the sandstone cap. At some critical height, the cliff collapses and the cycle begins again. The episodic retreat of this type of escarpment is the result of the change in cliff morphology under essentially constant climatic, base level and tectonic conditions. Similarly, a meander can increase in amplitude until a cutoff occurs under constant hydrologic conditions.

Field and experimental work supports the concept of geomorphic thresholds, which have been used to explain the distribution of discontinuous gullies in semi-arid valleys. Discontinuous gullies, short gullied reaches of valley floors, can be related to the slope of the valley-floor surface. For example, the beginning of GULLY erosion in these valleys tends to be localized on steeper reaches of the valley floor, which are the result of sediment storage. For a region of uniform geology, land use and climate, a critical intrinsic threshold of valley slope exists above which the valley floor is unstable and subject to incision during floods.

Similar relationships can be established for other alluvial deposits. For example, trenching of ALLUVIAL FANS is common, and the usual explanation for fan-head trenches is renewed uplift of the mountains or climatic fluctuations. However, as the fan grows through continual deposition, the fan-head steepens until it exceeds a threshold slope, when trenching occurs. Experimental (see EXPERIMENTAL GEOMORPHOLOGY) studies of alluvial-fan growth confirm that periods of trenching alternate with deposition at the fan-head. Therefore, the fan-head trenching can occur as a result of the oversteepening of the fan-head, and it is the result of the exceeding of an intrinsic geomorphic threshold.

A similar example is provided by damaging debris-flow events along the Wasatch Mountains in Utah. In 1993, a storm triggered debris flows from some canyons, but not all. It was suggested that debris basins be constructed at the mouths of the active canyons. However, further investigation revealed that the active canyons had been flushed of sediment and were not a threat, whereas the inactive canyons were storing sediment, which at some future time would produce damaging debris flows. The storage and flushing of sediment in these canyons is similar to the storage of sediment and its incision in semi-arid valleys and on alluvial fans as an intrinsic threshold is exceeded.

The identification of an intrinsic geomorphic threshold has significant practical applications.

If, as in the study of discontinuous gullies and alluvial-fan trenches, the critical slope is identified (intrinsic threshold), then unfailed, but sensitive valley floors and fan-heads can be identified. In this way, preventive conservation can be practised, thereby preventing erosion rather than attempting to control it after incision has occurred.

The concept of intrinsic geomorphic thresholds, which involves landform change without a change in external controls, challenges the well-established geomorphic thesis that relatively abrupt landform change is the result of some climatic, base level, or land-use change. Therefore, the significance of the intrinsic geomorphic threshold concept for geomorphologists is that it makes them aware that abrupt erosional and depositional changes can be inherent in the normal development of a landscape and that a change in an external variable is not always required for a geomorphic threshold to be exceeded and for a significant geomorphic event to result.

Further reading

Begin, Z.B. and Schumm, S.A. (1984) Gradational thresholds and landform singularity, *Quaternary Research* 21, 267–274.

Coates, D.R. and Vitek, J.D. (eds) (1980) *Thresholds in Geomorphology*, London: Allen and Unwin.

Patton, P.C. and Schumm, S.A. (1975) Gully erosion, northwestern Colorado: a threshold phenomenon, *Geology* 3, 88–90.

Phillips, J.D. (2001) The relative importance of intrinsic and extrinsic factors in pedodiversity, *Annals of the Association of the American Geographers* 91, 609–621.

Schumm, S.A. (1977) *The Fluvial System*, New York: Wiley.

——(1979) Geomorphic thresholds: the concept and its applications, *Transactions of the Institute of British Geographers* 4, 485–515.

Schumm, S.A., Harvey, M.D. and Watson, C.C. (1984) *Incised Channels: Morphology, Dynamics, and Control*, Littleton, CO: Water Resources Publications.

Westcott, W.A. (1993) Geomorphic thresholds and complex response of fluvial systems: some implications for sequence stratigraphy, *Bulletin American Association of Petroleum Geologists* 77, 1,208–1,218.

STANLEY A. SCHUMM

TIDAL CREEK

A creek is an inlet in a shoreline, a channel in a marsh, or another narrow, sheltered waterway. Creeks occur extensively on MUD FLATS AND MUDDY COASTS, on MANGROVE SWAMPS and on SALTMARSH surfaces (Eisma 1998).

Tidal creeks often have a high drainage density because of the large volumes of water that they drain. Saltmarsh creek densities may be 40 km/km^2 (Pethick 1984). The morphology of the creeks is also often distinctive. Although some may bear a superficial resemblance to dendritic river channel networks, flow along them is bi-directional (French and Stoddart 1992; Pestrong 1965). They have a tendency to taper upstream and flare downstream (Fagherazzi and Furbish 2001), and their discharge is determined by the tidal prism. In areas with a large tidal range or rapid seaward progradation, creek systems may be markedly linear in form. In areas with cohesive sediments creeks have steep edges, whereas in sandier areas they tend to be shallower and wider.

References

Eisma, D. (1998) *Intertidal Deposits: River Mouths, Tidal Flats, and Coastal Lagoons*, Boca Raton, FL: CRC Press.

Fagherazzi, S. and Furbish, D.J. (2001) On the shape and widening of salt marsh creeks, *Journal of Geophysical Research* 106, 991–1,003.

French, J.R. and Stoddart, D.R. (1992) Hydrodynamics of saltmarsh creek systems: implications for marsh morphological development and material exchange, *Earth Surface Processes and Landforms* 17, 235–252.

Pestrong, R. (1965) The development of drainage patterns on tidal marshes, *Stanford University Publications in Geological Science* 10, 1–87.

Pethick, J. (1984) *An Introduction to Coastal Geomorphology*, London: Arnold.

A.S. GOUDIE

TIDAL DELTA

Tidal deltas are large sand bodies formed within, or in the vicinity of, tidal inlets. The latter may be associated with barrier island chains (see BARRIER AND BARRIER ISLAND) and the entrances to coastal lagoons (see LAGOON, COASTAL) or estuaries (see ESTUARY). Flood-tidal deltas form landward of the inlet mouth, under the influence of flood-tidal currents. Ebb-tidal deltas occur seaward of the inlet, predominantly under the influence of ebb-tidal currents and wave action.

The major morphological features of flood-tidal deltas typically include (after Hayes 1980): a seaward-dipping flood ramp, up which landward sand movement occurs through the migration of sand waves under the action of flood currents; subtidal flood channels, which extend into the inlet and which dissect the partly intertidal landward portion of the delta (the 'ebb shield'); marginal ebb-aligned spits; and spillover lobes formed by the action of ebb currents over the lower parts of the ebb shield.

Ebb-tidal deltas are usually comprised of: an ebb channel, maintained by strong tidal currents; linear bars, formed through wave-current interactions along the margins of the ebb channel; a terminal lobe formed at the distal (seaward) end of the ebb channel, where the tidal current diminishes; sandsheets (or 'swash platforms') formed by wave action adjacent to the ebb channel characterized by migrating swash bars; and marginal channels dominated by flood-tidal currents.

Studies of inlet morphometry have shown that the morphology of tidal deltas is related to tidal prism (itself a function of both tidal range and inlet geometry), the configuration of the inlet and adjacent shoreline (including the offshore bathymetry), wave climate, and the rate of littoral sediment transport. In micro-tidal areas, flood deltas are often better developed than their ebb counterparts, owing to the dominance of landward, wave-driven, sediment transport. Ebb delta morphology is generally more variable than that of flood deltas, owing to the importance of regional and local contrasts in wave climate (Boothroyd 1985), and due to the tighter coupling of delta processes with wider coastal morphodynamic behaviour. Ebb delta volume increases with tidal prism, and decreases with inlet width/depth ratio and wave energy. Under conditions of low wave energy, ebb deltas are typically more elongated and extend further seaward. These controls are interactive so that, for example, wave energy can be modified by the presence of a headland, which also influences both the tidal prism and the intensity of tidal flows. For a sample of seventeen natural inlets in North Island, New Zealand, Hicks and Hume (1996) showed that over 80 per cent of the variation in ebb delta volumes could be successfully predicted by an empirical equation incorporating spring tidal prism and the angle between the ebb channel and the adjacent shoreline. Other factors accounting for some of the observed variation in delta volume include wave energy, sediment grain size (finer sands are less likely to be retained in the vicinity of strong tidal current jets and are thus associated with smaller deltas), and the supply of sediment through littoral drift.

FitzGerald *et al.* (2002) have drawn attention to marked contrasts in the occurrence and morphology of tidal deltas along the coast of New England, associated with a large degree of variability in tidal range, wave energy, sediment supply and inlet origin and geometry. Flood-tidal deltas in meso-tidal inlets tend to have a classic horseshoe shape and a significant intertidal area. Those in micro-tidal inlets tend to be predominantly subtidal and digitate or multi-lobate in form, owing to the limited ability of weak ebb currents to rework their deposits. Ebb-tidal deltas are best developed in moderately sized mixed energy environments. In wave-dominated inlets, they are either absent or small and entirely subtidal.

Although flood-tidal deltas act as long-term sediment sinks, ebb-tidal deltas are more dynamically coupled to the morphodynamic adjustment of the adjacent coast. Important processes include partial wave sheltering by the delta sand body; wave refraction around the delta, causing trapping and storage of beach sediments; and sediment recirculation within the delta (Oertel 1977). Ebb-tidal deltas interrupt the continuity of alongcoast sediment movement, and bypass sand across their inlets in discrete pulses. Hicks *et al.* (1999) have shown that such inlet/ebb-delta processes can be an important source of interdecadal variability in beach behaviour. Years when the delta is accumulating sand are associated with erosion of beaches on the downdrift side of the inlet. Conversely, in years when the delta releases sand, the same beaches experience accretion associated with a migrating sand pulse.

Tidal deltas have historically been exploited as a sand resource (including mining to supply BEACH NOURISHMENT schemes). Questions are now being asked over the sustainability of this practice and its implications for beach stability. Furthermore, the correspondence between delta volumes and tidal prism means that their sand-trapping function is potentially sensitive to sea level rise. Inlets with extensive intertidal areas might experience a significant increase in tidal prism with a rise in sea-level, thus leading to increased sand storage. This may have adverse consequences for the stability of adjacent beaches, especially downdrift of the inlet.

References

Boothroyd, J.C. (1985) Tidal inlets and tidal deltas, in R.A. Davis (ed.) *Coastal Sedimentary Environments*, 2nd edition, 445–532, New York: Springer-Verlag.

FitzGerald, D.M., Buynevich, I.V., Davis, R.A. and Fenster, M.S. (2002) New England tidal inlets with special reference to riverine-associated inlet systems, *Geomorphology* 48, 179–208.
Hayes, M.O. (1980) General morphology and sediment patterns in tidal inlets, *Sedimentary Geology* 26, 139–156.
Hicks, D.M. and Hume, T.M. (1996) Morphology and size of ebb-tidal deltas at natural inlets on open sea and pocket-bay coasts, North Island, New Zealand, *Journal of Coastal Research* 12, 47–63.
Hicks, D.M., Hume, T.M., Swales, A. and Green, M.O. (1999) Magnitudes, spatial extent, time scales and causes of shoreline change adjacent to an ebb tidal delta, Katikati Inlet, New Zealand, *Journal of Coastal Research* 12, 220–240.
Oertel, G.F. (1977) Geomorphic cycles in ebb deltas and related patterns of shore erosion and acretion, *Journal of Sedimentary Petrology* 47, 1,121–1,131.

J.R. FRENCH

TOMBOLO

A sandbar, barrier or spit that joins an island with a mainland or another island, resulting from longshore drift or the migration of an offshore bar toward the coast. Tombolos are constructive features (though ultimately ephemeral due to wave erosion), occurring along shorelines of submergence that are protected from large waves and where islands are common. Sediment supply is predominantly derived from the islands, yet some may also come from erosion of the shoreline, fluvial materials, underwater reefs and offshore glacial deposits. Several types of tombolo exist including single, double, multiple, forked, parallel and complex tombolos, all of which are reflective of the coastal system (e.g. wave mechanisms) from which they are derived. For example, double tombolos (two ridges extending to shore), often form in areas with seasonal shifts in longshore drift. Tombolos can restrict flow between the sea and intertidal zone, forming a lagoon (see LAGOON, COASTAL), and altering the local ecology. An example of a tombolo is Chesil Beach, which extends northwestwards from the Isle of Portland to the coast of Dorset, south England.

Further reading

Schwartz, M.L. (1972) *Spits and Bars*, Benchmark Papers in Geology, Stroudsburg, PA: Dowden, Hutchinson and Ross.

SEE ALSO: bar, coastal; barrier and barrier island; coastal geomorphology

STEVE WARD

TOR

Linton (1955: 470) describes a tor as 'a solid rock outcrop as big as a house rising abruptly from the smooth and gentle slopes of a rounded summit or broadly convex ridge'. Tors are in fact large, freestanding, residual masses of rock (Plate 140). The word derives from the Old Welsh word *twr* or *twrr* meaning heap or pile. Tors are most common – and most well known – in granitic rocks (e.g. Haytor Rocks, Dartmoor, England), but also occur in coarse sandstones, schists, dacites and dolerites, among other lithologies. Although perhaps best known in south-west England (Devon and Cornwall), tors occur on all continents. In Africa, tors are often known as castle koppies (or kopjes). The rocks in which they occur vary considerably in age, but it is thought that most tors formed during the Tertiary or Pleistocene. Tors may occur in any position in the landscape, but are most common in summit and spur locations (Gerrard 1978; Ehlen 1991). They may occur as single, massive exposures (e.g. Middle Staple Tor, Dartmoor), or consist of groups of individual outcrops clustered together (e.g. Great Mis Tor, Dartmoor). The latter configuration is most common in summit positions. They may also appear to be piles of very large, loose boulders (core stones), but a rock core anchored in bedrock is usually present in the centres of such exposures.

Plate 140 Middle Staple Tor, Dartmoor, south-west England. This tor is approximately 18 m long, 8 m tall and 10 m deep

Occasionally, the individual blocks in large summit tors form a pattern such that there is an elongate, open space, called an avenue, at the crest (e.g. Hound Tor, Dartmoor). Summit tors tend to be the largest, and those along valley sides tend to be the tallest (e.g. Vixen Tor, Dartmoor). Spur tors are most often the smallest. Sizes range from about one metre in height and a few metres in length and width to tens of metres in each dimension. Fallen rock debris, called clitter on Dartmoor, is often present at the bases of tors as well as for some distance downslope. Much of the clitter has been moved downslope by periglacial processes (e.g. stone lines on the west side of Middle Staple Tor, Dartmoor).

It is generally accepted that tor shapes and locations in all lithologies are controlled by JOINTING. Each tor typically contains three major joint sets (and two to five minor ones), one horizontal or gently dipping set defining the top, and two vertical or steeply dipping sets defining the sides of the tor. Diagonal joints may also be present (e.g. Great Tor, Dartmoor), but they are not common. Tors have a variety of appearances depending upon the arrangement of the joints that form them. They may be quite massive blocks of rock, either very rounded or blocky in appearance (Plate 140). They can be lamellar in form (Haytor Rocks, Dartmoor). They can also be tall and narrow, almost pinnacles (e.g. Bowerman's Nose, Dartmoor). These variations in appearance are caused by changes in the distributions of the different types of joints in the tor. If vertical and horizontal joints occur with about equal frequency, the tor will be composed of joint blocks of approximately equal size and will appear massive and blocky. If horizontal or gently dipping joints occur in significantly greater numbers than vertical or steeply dipping joints, the tor will be lamellar in form. If horizontal or gently dipping joints are rare, the tor will be tall and narrow. Logan stones, joint-bounded, precariously balanced boulders that move when touched, can be found in association with all types of tors, but are most common in shorter tors where horizontal joints are closer together than vertical joints, producing an elongate, rectangular block.

Origin of tors

Various theories have been proposed for the origin of tors. One theory suggests that they are the product of atmospheric weathering – that wind, rain, freeze – thaw, salt crystallization and insolation round the shapes of exposed, angular rock outcrops produce rounded, bouldery tors (e.g. Palmer and Neilson 1962). This theory is generally discounted except in certain unusual and specific cases (e.g. Selby 1972).

A second theory for the origin of tors assumes they are the product of a two-stage process (the two-stage theory). In this theory, based on Linton's (1955) work on Dartmoor in south-west England, tors form in the subsurface by chemical weathering along joints, and are subsequently exposed by erosional stripping. Once exposed, they retain their rounded shape. Linton noted a buried tor in a small quarry near Two Bridges in central Dartmoor as unrefutable evidence of this theory. Others (e.g. Thomas 1974) have expanded Linton's theory to include multiple phases of weathering and denudation. The theory suggests that weathering progresses most rapidly where the distance between joints is narrow, and more slowly where it is wide. Once the weathered mantle, which is thickest where the joints were most closely spaced, is removed by erosion, a rock outcrop with relatively widely spaced joints remains. Ehlen *et al.* (1997) showed that this was in fact the case in the Granite Mountains, Wyoming, and Ehlen and Wohl (2002) provided additional evidence favouring this theory in their study of BEDROCK CHANNELS in the Colorado Front Range. The weathered mantle formed from granitic rocks is called growan, GRUS, or saprolite. This theory has received significant support from many workers over the years, such as Eden and Green (1971) in their work on Dartmoor and C.R. Twidale in his many papers primarily on granite landforms in Australia. The genetic nature of the two-stage theory, however, has made it difficult for many to accept.

The third theory (the scarp retreat theory), proposed by King (1949) and based on his work in southern Africa, is that tors are the product of lateral planation and pediment formation. In a later publication, King (1958) accepts that sub-skyline tors could result from chemical weathering and exhumation, and states that his 1949 theory refers only to skyline (i.e. summit) tors. Ollier and Tuddenham (1961) in their work on Australian INSELBERGS and Ojany's (1969) on Kenyan inselbergs, among others, have provided support for King's scarp retreat theory.

The final theory of tor formation (the periglacial theory) is that proposed by Palmer and

Radley (1961). Their studies of sandstone tors in north-east England, where there is no evidence of deep chemical weathering, suggest that the tors were isolated from free faces along joint planes, and then rounded and fretted by subsequent atmospheric denudation. Palmer and Nielson (1962) studied the Dartmoor tors and suggested that – because core stones are lacking, there is no evidence of a DEEP WEATHERING PROFILE, and atmospheric weathering can account for the rounded shapes – the Dartmoor tors have a periglacial origin. They proposed a three-stage theory of periglacial tor formation. The application of this theory to the Dartmoor tors, however, is generally discounted.

The favoured theories among those described above are the modified two-stage theory first presented by Linton and King's scarp retreat theory, and much work has been done since they first published their work to support one or the other theory, as noted above. But perhaps the most reasonable approach to the origin of tors is that tors form by different processes in different environments (i.e. the principle of equifinality) as suggested by Brunsden (1964) and Thomas (1974), among others.

References

Brunsden, D. (1964) The origin of decomposed granite on Dartmoor, in I.G Simmons (ed.) *Dartmoor Essays*, Exeter: Devonshire Association for the Advancement of Science, Literature and Art, 97–116.

Eden, M.J. and Green, C.P. (1971) Some aspects of granite weathering and tor formation on Dartmoor, England, *Geografiska Annaler* 53, 92–99.

Ehlen, J. (1991) Significant geomorphic and petrographic relations with joint spacing in the Dartmoor Granite, southwest England, *Zeitschrift für Geomorphologie* 35, 425–438.

Ehlen J. and Wohl, E. (2002) Joints and landform evolution in bedrock canyons, *Transactions, Japanese Geomorphological Union* 23, 237–255.

Ehlen, J., Gerrard, J. and Zen, E. (1997) Joint spacing and landform evolution: the Granite Mountains, WY, *Geological Society of America Abstracts with Program* 29, A36.

Gerrard, A.J.W. (1978) Tors and granite landforms of Dartmoor and eastern Bodmin Moor, *Proceedings of the Ussher Society* 4, 201–210.

King, L.C. (1949) A theory of bornhardts, *Geographical Journal* 112, 83–87.

——(1958) Correspondence on the problem of tors, *Geographical Journal* 124, 289–291.

Linton, D. (1955) The problem of tors, *Geographical Journal* 121, 470–487.

Ojany, F. (1969) The inselbergs of eastern Kenya with special reference to the Ukambani area, *Zeitschrift für Geomorphologie* 13, 196–206.

Ollier, C.D. and Tuddenham, W.G. (1961) Inselbergs of central Australia, *Zeitschrift für Geomorphologie* 5, 257–276.

Palmer, J. and Neilson, R.A. (1962) The origin of granite tors on Dartmoor, Devonshire, *Proceedings of the Yorkshire Geological Society* 33, 315–340.

Palmer, J. and Radley, J. (1961) Gritstone tors of the English Pennines, *Zeitschrift für Geomorphologie* 5, 37–52.

Selby, M.J. (1972) Antarctic tors, *Zeitschrift für Geomorphologie* 13, 73–86.

Thomas, M.F. (1974) Granite landforms: a review of some recurrent problems in interpretation, in E.H. Brown and R.S. Waters (eds) *Progress in Geomorphology*, Institute of British Geographers Special Publication No. 7, 13–35.

Further reading

Gerrard, A.J. (1988) *Rocks and Landforms*, London: Unwin Hyman.

Twidale, C.R. (1982) *Granite Landforms*, Amsterdam: Elsevier.

SEE ALSO: exhumed landform; granite geomorphology; inselberg; rock control; salt weathering; spheroidal weathering; weathering

JUDY EHLEN

TOREVA BLOCK

Large masses of relatively stratigraphically coherent rock that have slipped down a cliff or mountain side upon normal listric faults, and has rotated backwards toward the parent cliff (Reiche 1937). The blocks can measure beyond 600 m in thickness and lateral extent, and are end members for this landslide type. Some blocks lie close to the parent cliff whereas others may have slipped several hundred kilometres from their sources. Their emplacement age remains uncertain, though probably in different (humid) climates in the Pleistocene.

Reference

Reiche, P. (1937) The toreva-block, a distinctive landslide type, *Journal of Geology* 45, 538–548.

SEE ALSO: mass movement

STEVE WARD

TRACER

Tracer techniques enable geomorphologists to quantify the movement of Earth materials (whole systems, individual particles, water) and provide data that enable them to model the movement of these materials through a range of Earth systems. Applications have focused on four major research areas.

1 Studies of whole system behaviour (e.g. measurement of soil creep, mass movement and mudflows on hillslopes; measuring rates of movement and internal deformation of glaciers, measuring the surface deformation of volcanoes) (Meier 1960; Carson and Kirkby 1972; Anderson and Finlayson 1975; Goudie 1990).
2 Studies of coarse sediment transport on hillslopes, in fluvial systems and in littoral zones (e.g. determination of entrainment thresholds, transport distances, particle size and shape controls on sediment transport) (Sear *et al.* 2000).
3 Studies of fine particle transport and sediment provenance (e.g. airborne particles, hillslope erosion and soil redistribution, quantifying sediment-associated contaminant movement and determining the provenance of a range of dated fluvial, limnic, aeolian and marine deposits) (Foster and Lees 2000).
4 Tracing water movement and flow pathways (e.g. through soil profiles, groundwater and cave systems and estimating flood wave travel times in rivers) (Leibundgut 1995; Kranjc 1997).

While by no means exhaustive, this brief list of examples demonstrates the remarkable breadth of tracer applications in geomorphology.

Whole system tracer studies have made use of inert or active surface and subsurface markers (e.g. three dimensional Global Positioning Systems; steel/aluminium rods or spheres; distinctly painted stones or pebbles) whose movement can be monitored directly by satellite tracking, by field re-survey or by using remotely sensed images (e.g. aerial photographs).

Studies of coarse sediment transport may use 'passive' natural materials that are marked in some way to allow identification, recovery and measurement (e.g. painted pebbles, pebbles drilled with bar magnets or impregnated/coated with radionuclides). Alternatively, a range of natural or artificial materials may contain radio transmitters to allow rapid location even if the particle is buried.

Studies of fine particle movement may use 'exotic' materials that have characteristic signatures (e.g. fine-grained magnetic powders; fluorescent micro-spheres) that can be detected by field survey or by field sampling and laboratory analysis. Natural properties of environmental materials can also be used if their signatures (e.g. geochemical, mineralogical, mineral magnetic, stable isotope, radionuclide) characterize a distinct source of origin (e.g. geological or lithological units, soil types, topsoils/subsoils).

Tracing water movement and flow pathways has made use of fine-grained neutrally buoyant materials that move in the same way as the liquid (e.g. *Lycopodium* spores) or a wide range of (fluorescent) soluble dyes (e.g. Rhodamine) that can be added to cave or groundwater systems or to the surface of a soil profile and which may be detected at low concentrations.

Whichever method is appropriate to the research problem, a number of assumptions underpin the use of all tracers whatever the field of study. Before looking at an example, we need to ask:

What makes a good tracer?

There are a number of factors which need to be taken into account in order to ensure a successful outcome in tracer experiments:

THE TRACER DOES NOT INTERFERE WITH OR ALTER THE PROCESS BEING MEASURED

In many cases this is difficult to achieve. For example, digging soil pits and excavating or drilling holes in glaciers, installing metal pins to estimate rates of soil creep or ice deformation, or installing piezometer tubes in a mudflow, directly disturb the immediately surrounding environment. While widely used, these and other intrusive methods of measurement are subject to unknown and largely unknowable errors.

THE TRACER IS REPRESENTATIVE OF THE SYSTEM BEING MEASURED

Artificial coarse and fine tracers may differ in size, shape and density from the natural materials being investigated. All these factors are important. For coarse particle studies, for example, entrainment and settling velocities are functions of the mass, density and shape of the particle. In fine particle studies (especially in the silt and clay-sized fractions), particle interactions are important and many natural materials move as aggregates rather than as individual particles. Replacement of natural by artificial materials may poorly replicate particle interactions.

THE TRACER CAN BE RECOVERED AND IDENTIFIED

Historically, poor tracer recovery in coarse sediment studies has posed a major problem, especially if particles have become buried. Poor recovery leads to problems in interpreting results in a statistically meaningful way. Tracers marked with surface paints, for example, are often difficult to relocate even if only buried to a few centimetres depth. Placement of pebbles in high-energy environments can lead to abrasion and loss of paint cover, while brightly coloured pebbles on a beach undoubtedly attract the attentions of young children often leading to redistribution in a manner no natural geomorphological process could explain. More recent developments in tracer technologies have used magnetic or radio-tracers emplaced in locally derived material so that even buried particles have a better chance of being found and do not attract unwarranted attention. It is essential that, whatever the marking system used, the tracer will last the lifetime of the project, that the identity is resistant to removal and that each individual particle has an unequivocal identity (Sear *et al.* 2000).

THE TRACER IS TRANSPORTED AND DEPOSITED IN THE SAME WAY AS THE MATERIAL BEING STUDIED

In many situations, coarse particles or other large material, can be identified by marking the surface with paint on which a code is added for later identification. Painted surfaces, however, may change the surface roughness and porosity characteristics of the original material leading to changes in buoyancy over relatively short timescales. Alternatives include the use of exotic (non-local) materials in order to aid identification and recovery but again differences in shape, density, porosity, buoyancy and/or surface roughness may lead to errors in experimental results. While magnetically or radio-tagged natural or 'exotic' particles have improved recovery rates, tracer physical characteristics may not exactly match those of the natural materials whose behaviour they are manufactured to represent. Fine sediment studies that use natural tracer characteristics may also prove problematic since erosion and sediment transport processes are particle-size selective. In consequence, concentrations of many natural tracers (e.g. heavy metals, nutrients, radionuclides, mineral magnetic signatures) increase with a decrease in particle size and an increase in particle specific surface area and for which a correction factor is often required. Additional considerations must be given to changing environmental conditions during transport. This is especially true in aquatic systems where changes in pH, redox potential (Eh) and salinity may drive adsorption and/or desorption reactions during sediment transport (Horowitz 1991).

LONG-STERM STORAGE

Once deposited, fine-grained aeolian, marine, fluvial and limnic sediments can undergo a range of complex transformations. Interruptions in loess or floodplain deposition, for example, often leads to periods of soil development while changes in the trophic status of lakes and estuaries may again lead to changes in pH and Eh driving the release of many natural tracers into the water column. It cannot therefore be assumed that natural tracer properties remain unchanged with time (e.g. Foster *et al.* 1998)

An example of fine sediment tracing for interpreting erosion processes

Erosion processes operating in a grazed paddock in New South Wales, Australia were analysed by Wallbrink and Murray (1993) using a rainfall simulator and two radionuclides that adsorb strongly to sediment and label different parts of the soil profile in different ways. Figure 168a and b show the vertical distribution of ^{137}Cs and ^{7}Be in a typical soil profile. ^{137}Cs has a thirty-year half-life and has been detected in the environment since atmospheric testing of thermonuclear weapons in the early 1950s (Higgitt 1995). Following the international treaty banning atmospheric nuclear weapons testing in 1963, the southern hemisphere has received little ^{137}Cs fallout. ^{7}Be is continuously produced in the upper atmosphere by cosmic ray bombardment but has a short half-life (53 days) in comparison with ^{137}Cs. Figure 168c shows how the ^{7}Be and ^{137}Cs activities of sediment produced by four different erosion processes (sheet, gully floor, gully collapse, rill erosion) could produce suspended sediment with different combinations of the two signatures. Gully collapse would result in low activities of both nuclides since the gully wall does not receive ^{7}Be fallout and the depth penetration of ^{137}Cs in the soil profile is less than 10 cm. The majority of the gully wall sediments are not labelled with either radionuclide. By contrast, sheet erosion would produce sediment with high activities of both radionuclides. Gully floor sediments would be labelled with ^{7}Be (since it is produced continuously) but would have no ^{137}Cs (since in this case the gully developed after 1963). Rill erosion would produce sediment strongly labelled with ^{137}Cs, but the shallow depth penetration of ^{7}Be in the soil profile would lead to a dilution of ^{7}Be activities as it mixes with sediment derived from deeper in the soil profile. By collecting suspended sediment during rainfall simulation experiments generating surface runoff, Figure 168d shows that the model of high ^{7}Be and ^{137}Cs activity is supported by the experimental results.

References

Anderson, M.G. and Finlayson, B.L. (1975) *Instruments for Measuring Soil Creep*, BGRG Technical Bulletin No.16, Norwich: Geo Abstracts.

Carson, M.A. and Kirkby, M.J. (1972) *Hillslope Form and Process*, Cambridge: Cambridge University Press.

Foster, I.D.L. and Lees, J.A. (2000) Tracers in geomorphology: theory and applications in tracing fine particulate sediments, in I.D.L. Foster (ed.) *Tracers in Geomorphology*, 3–20, Chichester, Wiley.

Foster, I.D.L., Lees, D.E., Owens, P.N. and Walling, D.E. (1998) Mineral magnetic characterisation of sediment sources from an analysis of lake and floodplain sediments in the catchments of the Old Mill Reservoir

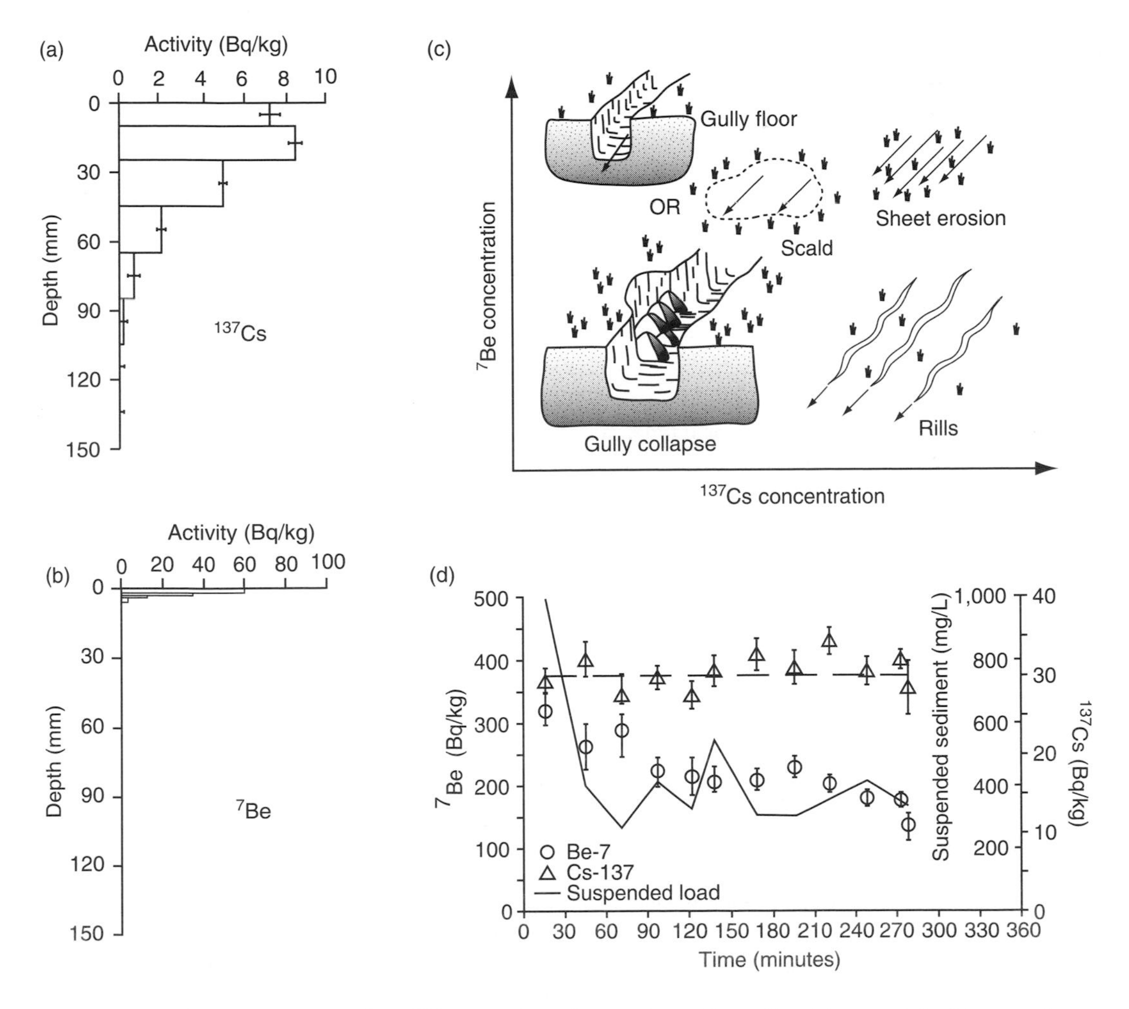

Figure 168 The distribution of (a) ^{137}Cs, and (b) ^{7}Be in the soil profile; (c) a model of how sediment would be labelled with ^{137}Cs and ^{7}Be depending on the erosion process involved; and (d) the results of a rainfall simulation experiment generating surface erosion

Source: Wallbrink and Murray (1993).

and Slapton Ley, South Devon, UK, *Earth Surface Processes and Landforms* 23, 658–703.
Goudie, A. (ed.) (1990) *Geomorphological Techniques*, London: Allen and Unwin.
Higgitt, D.L. (1995) The development and application of Cs-137 measurements in erosion investigations, in I.D.L. Foster, A.M. Gurnell and B.W. Webb (eds) *Sediment and Water Quality in River Catchments*, 287–305, Chichester: Wiley.
Horowitz, A. (1991) *A Primer on Sediment Trace Chemistry*, Chelsea, MI: Lewis.
Kranjc, A. (1997) *Tracer Hydrology 97*, Rotterdam: Balkema.
Leibundgut, Ch. (1995) *Tracer Techniques for Hydrological Systems*, IAHS Publication 229, Wallingford: International Association of Hydrological Sciences Press.
Meier, M.F. (1960) Mode of flow of Saskatchewan Glacier, *US Geological Survey Professional Paper* No 351, Washington, DC.
Sear, D.A., Lee, M.W., Oakey, R.J., Carling, P.A. and Collins, M.B. (2000) Coarse sediment tracing technology in littoral and fluvial environments, in I.D.L. Foster (ed.) *Tracers in Geomorphology*, 21–55, Chichester: Wiley.
Wallbrink, P.J. and Murray, A.S. (1993) Use of fallout radionuclides as indicators of erosion processes, *Hydrological Processes* 7, 297–304.

Further reading

Peters, N.E., Hoehn, E., Leibundgut, Ch., Tase, N. and Walling, D.E. (eds) (1993) *Tracers in Hydrology*, IAHS Publication 224, Wallingford: IAHS Press.
Rukin, N., Hitchcock, M., Streetly, M., Al Faihani, M. and Kotoub, S. (1994) The use of fluorescent dyes as tracers in a study of artificial recharge in Qatar, in E.M. Adar and Ch. Leibundgut (eds) *Application of Tracers in Arid Zone Hydrology*, IAHS Publication 232, 67–78, Wallingford: IAHS Press.
Verusob, K.L., Fine, M.J. and TenPas, J. (1993) Pedogenesis and palaeoclimate interpretation of the magnetic susceptibility of the Chinese loess-palaeosol sequences, *Geology* 21, 1,011–1,014.

IAN D.L. FOSTER

TRANSGRESSION

A transgression is the movement of mean sea level (MSL) in an upward direction, while any lowering of MSL is called a regression. A movement of MSL, as the datum for wave and tidal activity, generates the potential for change in areas above and below the datum. Changes in the position of MSL are usually relative. A transgression is not solely due to MSL moving, as it is possible that the land surface is also moving relative to MSL. Long-term MSL changes can be due to both eustatic (changes in volume of sea water in the oceans) and isostatic (vertical movement of land) causes. A transgression is specified when the net balance between these two processes resolves in a relative rise in mean sea level (RSLR). Eustatic changes generally relate to climatic changes and their control on (1) unit ocean volume contraction/expansion due to atmospheric-ocean heat exchange (the steric effect); (2) evaporation rates of sea water; and (3) variations in the hydrological cycle by which free water is either locked up as terrestrial ice, or terrestrial ice melting to release water back into the oceans.

Quaternary transgressions usually relate to warming phases by which glacier ice melts and sea surface temperatures rise, e.g. the early to mid-Holocene in which atmospheric warming contributed to a major transgressive phase occurring in the mid-latitudes (Pirazzoli 1996). Although the Late Glacial was marked by rapid positive eustatic change, it was also a period of rapid upward crustal lift in the mid-upper latitudes, due to the release of terrestrial ice pressure through ice melting. The near-exponential early rapid isostatic rise in land level meant that eustatic change was more than exceeded, in effect inducing a regression or relative fall in MSL. This was only reversed as the isostatic rate diminished and the eustatic effect maximized some time in the mid-Holocene, to induce a well-recognized transgression (e.g. the Flandrian in north-west Europe). The British Isles can be crudely characterized as showing two main zones of response in the mid-Holocene; a northern zone in which the transgression peaked above present-day MSL (*c.*5–6 ka BP) and then switched to a regressive phase; and a southern zone showing a continuing transgressive phase to the present day but associated with a decelerating RSLR rate since the mid-Holocene. Local isostatic effects tend to modulate a regional eustatic signal and complicate the general picture at any one place. Contemporary concerns with the effects of accelerating global climate change are reflected in the forecasts for rising MSL over the next century, with RSLR up to five times the current rates being forecast. This will result in a major new transgressive phase, regardless of current crustal changes, as these modern eustatic changes are linked to global climate change, while the endogenetic changes required to generate crustal isostatic responses are unaffected.

Transgressions are a common element of geological-scale change. In recent decades, litho-stratigraphies have been interpreted via sequence stratigraphy by which rising and falling sea levels have been used to link subaerial erosion sources to submarine deposition sinks. The timescale of such deposition relates to transgressions ($>10^6$ years duration) as a consequence of: (1) slow orogeny and oceanic basin volume reduction; (2) massive sediment transfers due to landscape denudation causing isostatic readjustment; and (3) probable climate change. The speed of the Holocene transgression (millennia-scale) stands in contrast to these earlier episodes, and emphasizes the idiosyncratic conditions associated with Quaternary deglaciation-induced transgressions.

A transgression leads to inundation of the coastal zone, the onshore extent of which depends on the overall coastal slope angle. The sedimentary expression of the transgression is dependent on both the non-rectilinearity of the coastal slope setting the template for deposition, and availability of free sediment. The penetration distance of the transgression is not solely that of inundation, as the transgression carries wave and tidal activity that reworks the sediment mantle beyond the initial flooding limit. Bruun (1962) has attempted to specify the predictive relationship between RSLR and shoreline retreat, through what is now controversially termed the BRUUN RULE.

The concept of the 'Erosion Front' defines the spatial variation of coastal activity experienced from the quiescent leading edge of the front that moves up estuary, through the high energy breaking wave zone, and the trailing edge that has tidal reworking of near shore and beach face (Carter *et al.* 1992). The basal extent of the transgression is identified by an erosional surface also known as a ravinement. Much interest is centred on the way in which a transgression aids sediment-reworking into distinctive coastal morphologies, e.g. the development of both sand and gravel barriers and associated back-barrier intertidal sediment stores and marshes. Some dune investigators believe that transgressive conditions were requisite for the major coastal dunes associated with the Holocene, though a counter argument exists in which dunes are also associated with regressions. Local excess deposition can generate an apparent regressive signature by outweighing transgressive tendencies, though to achieve this sediment has to be derived from elsewhere alongshore leading to overall shoreline retreat. Key issues still to be resolved relate to transgressive migration rate and coastal morphology stability. Can saltmarsh growth match fast RSLR? Can barriers rollover and maintain longshore continuity under fast RSLR (Jennings *et al.* 1998)? These are critical issues for coastal communities dependent on maintaining sustainable natural coastal morphologies as coastal defences in the face of future extreme rates of RSLR.

References

Bruun, P. (1962) Sea-level rise as a cause of shore erosion, *Journal of Waterways, Harbour Division*, American Society of Civil Engineers 88 (WW1), 117–130.

Carter, R.W.G., Orford, J.D., Jennings, S.C., Shaw, J. and Smith, J.P. (1992) Recent evolution of a paraglacial estuary under conditions of rapid sea-level rise: Chezzetcook Inlet, Nova Scotia, *Proceedings of the Geological Association* 103, 167–185.

Jennings, S.C., Orford, J.D., Canti, M., Devoy, R.J.N. and Straker, V. (1998) The role of relative sea-level rise and changing sediment supply on Holocene gravel barrier development; the example of Porlock, Somerset, UK, *Holocene* 8, 165–181.

Pirazzoli, P. (1996) *Sea Level Changes: The Last 20000 Years*, Chichester: Wiley.

JULIAN ORFORD

TREE FALL

The falling over of trees during high winds is a significant factor in the translocation of material, churning of soils (Schaetzl 1986) disruption of strata, and the development of mound and pit micro-topography (Denny and Goodlett 1956). Deep-rooting trees affect topography to a greater degree than shallow-rooting trees, and when they are blown over leave deep pits and high mounds (Veneman *et al.* 1984). In mixed hardwood forests in Ithaca, New York, USA, mounds were typically 0.48–0.60 m high and pits 0.20–0.41 m deep (Beatty and Stone 1986). Windthrow appears to be a more prevalent phenomenon on deeper soils where there is a sharp contrast between the fine soil material and the underlying stony horizons (Boyd and Webb 1981). Tree fall is also a major contributor to the development of log steps and LARGE WOODY DEBRIS in forest streams (Marston 1982).

References

Beatty, S.W. and Stone, E.L. (1986) The variety of soil microsites created by tree falls, *Canadian Journal of Forest Research* 16, 539–548.

Boyd, D.M. and Webb, T.H. (1981) The influence of the soil factor on tree stability in Balmoral Forest, Canterbury, during the gale of August 1975, *New Zealand Journal of Forestry* 26, 96–102.
Denny, C.S. and Goodlett, J.C. (1956) Microrelief resulting from fallen trees, *US Geological Survey Professional Paper* 288, 59–66.
Marston, R.A. (1982) The geomorphic significance of log steps in forest streams, *Annals of the Association of American Geographers* 72, 99–108.
Schaetzl, R.J. (1986) Complete soil profile inversion by tree uprooting, *Physical Geography* 7, 181–189.
Veneman, P.L.M., Jacke, P.V. and Bodine, S.M. (1984) Soil formation as affected by pit and mound microrelief in Massachusetts, USA, *Geoderma* 33, 89–99.

A.S. GOUDIE

TRIMLINE, GLACIAL

Reconstruction of the geometry of former ice sheets and ice caps from geomorphological evidence is based on the identification of trimlines, defined by the maximum level to which GLACIERS or ICE SHEETS have eroded or 'trimmed' bedrock or debris in a valley hillslope (Ballantyne and Harris 1994). The sharpness of this boundary depends on the effectiveness of GLACIAL EROSION, the degree of frost WEATHERING after its formation, and the downslope MASS MOVEMENT during and after DEGLACIATION. The formation of weathering boundaries and glacial trimlines are open to four possible hypotheses (Figure 169):

1 Summit blockfields (see BLOCKFIELD AND BLOCKSTREAM) are formed by *in situ* rock weathering over a longer timescale than the Holocene, representing a glacial trimline marking the maximum altitude to which glacial erosion has eroded or 'trimmed' a pre-existing cover of REGOLITH/frost-shattered debris.
2 Frost weathering on high ground may reflect more pronounced breakup of rock at high

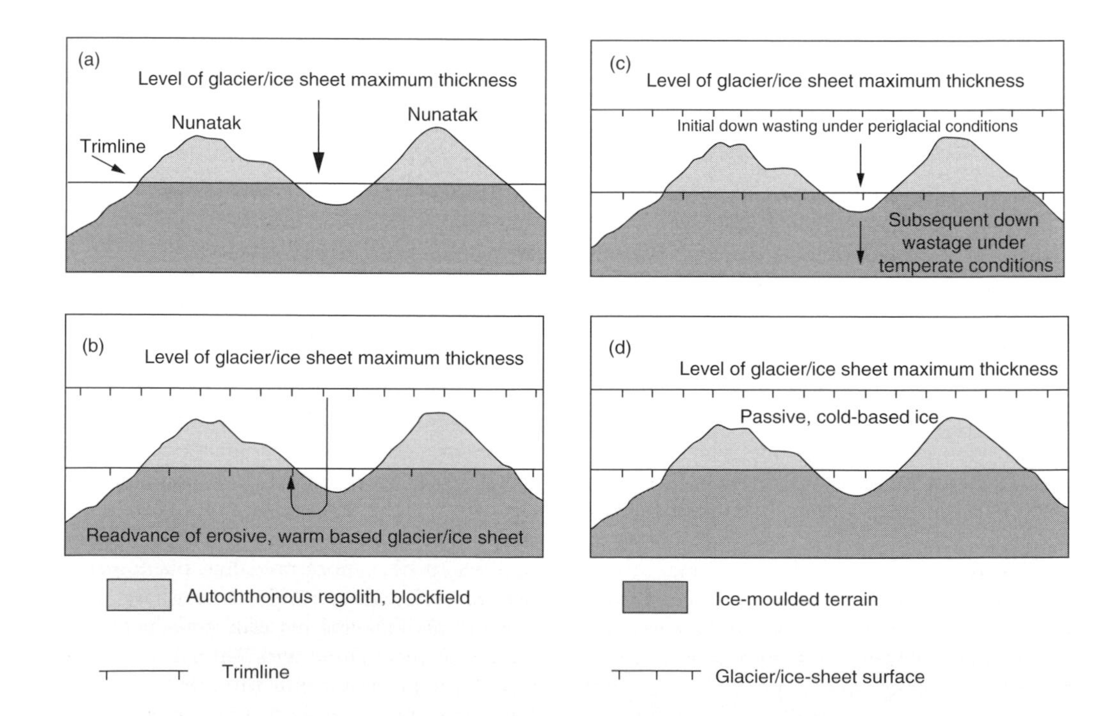

Figure 169 Four hypotheses of formation of glacial trimlines: (a) a glacial trimline representing the upper surface of an ice sheet at its maximum thickness; (b) a trimline cut by a glacial readvance during overall ice-sheet downwastage; (c) trimline formed during an initial period of ice-sheet downwastage under periglacial conditions; (d) weathering limit representing a thermal boundary between cold-based (temperature below pressure melting point) ice and a warm-based (temperature at the pressure melting point) ice within a former ice sheet (modified from Ballantyne *et al.* 1998)

altitudes under periglacial (see PERIGLACIAL GEOMORPHOLOGY) conditions, particularly during downwasting of an ice sheet and subsequent valley glaciation.

3 The initial stage of downwastage of ice sheets/ice caps may be accompanied by frost (see FROST AND FROST WEATHERING) shattering of exposed rock which ceases when the climate warms. The limit between frost-weathered and glacially abraded terrain represents the upper limit of glacier ice at the time of this thermal transition.
4 The weathering limit may represent a thermal boundary within a former ice sheet or ice cap, with *in situ* frost-weathered debris surviving under a cover of cold-based (basal temperature below the PRESSURE MELTING POINT) ice on high ground, whereas lower areas experience scouring by warm-based (basal temperature at the pressure melting point) glaciers.

Various approaches have been adopted to test these hypotheses. These involve analyses of weathering characteristics of bedrock and soils above and below the weathering limit/trimlines, and reconstruction of their altitudinal trend. Techniques employed are SCHMIDT HAMMER measurements of rock surface hardness, measurements of surface ROUGHNESS, measurements of differential relief of adjacent minerals, depth of open horizontal dilation (stress-release) joints using a graduated probe, studies of clay mineral assemblages and mineral magnetic signatures, and COSMOGENIC DATING.

References

Ballantyne, C.K. and Harris, C. (1994) *The Periglaciation of Great Britain*, Cambridge: Cambridge University Press.
Ballantyne, C.K., McCarroll, D., Nesje, A., Dahl, S.O. and Stone, J. (1998) The last ice sheet in north-west Scotland: reconstruction and implication, *Quaternary Science Reviews* 17, 1,149–1,184.

Further reading

Brook, E.J., Nesje, A., Lehman, S.J., Raisbeck, G.M. and Yiou, F. (1996) Cosmogenic nuclide exposure ages along a vertical transect in western Norway: implication for the height of the Fennoscandian ice sheet, *Geology* 24, 207–210.
Grant, D.R. (1977) Altitudinal weathering zones and glacial limits in Western Newfoundland, with particular reference to Gros Morne National Park, *Geological Survey of Canada Paper* 77–1A, 455–463.
Locke, W.W. (1995) Modelling the icecap glaciation of the northern Rocky mountains of Montana, *Geomorphology* 14, 123–130.
Nesje, A., Dahl, S.O., Anda, E. and Rye, N. (1988) Blockfields in southern Norway; significance for the Late Weichselian ice sheet, *Norsk Geologisk Tidsskrift* 68, 149–169.
Stone, J.O., Ballantyne, C.K. and Fifield, L.K. (1998) Exposure dating and validation of periglacial weathering limits, northwest Scotland, *Geology* 26, 587–590.

ATLE NESJE

TROPICAL GEOMORPHOLOGY

Tropical geomorphology has usually addressed conditions in the tropical forests and savannas, and although many areas within the tropics (23.5°N and S lat.) fall within the arid and semi-arid climates, these environments will not be included here. However, near-tropical climates extend to at least 30° lat., along the humid east coasts of Africa, Asia and the Americas and these areas have much in common with the tropics *sensu stricto* (Figure 170).

The study of geomorphology in the tropics cannot be separated from the wider history of the subject. During the early twentieth century most reports on tropical landscapes arose from geological investigations on behalf of former colonial powers: Britain, France, the Netherlands. Many observations concerned the extent and products of rock weathering (Falconer 1911; Scrivenor 1931). Reports also commented on unusual landscapes: extensive plains and great waterfalls in east and south Africa; high granite domes (or *inselberge*) in east and west Africa; the tower karst of South-East Asia. In addition, the widespread occurrence of *laterite*, long known from Buchanan's early work in India, attracted a lot of comment. Because the authors came from Europe, the exotic nature of tropical landforms was frequently emphasized.

In Davis's (1899) paper on 'The cycle of erosion' the evolution of landscapes was seen from the perspective of a humid temperate 'normality' and although Davis later addressed contrasts with semi-arid regions in western USA, the tropics were largely ignored. For fifty years the Davisian view was dominant in Britain, France and the USA and tropical landforms were seen as 'climatic accidents' (Cotton 1942) and peripheral to mainstream interests. In Germany, tropical landscapes were considered within a 'climatic geomorphology' (Thorbecke 1927; Büdel 1948), which led to specific hypotheses for landform

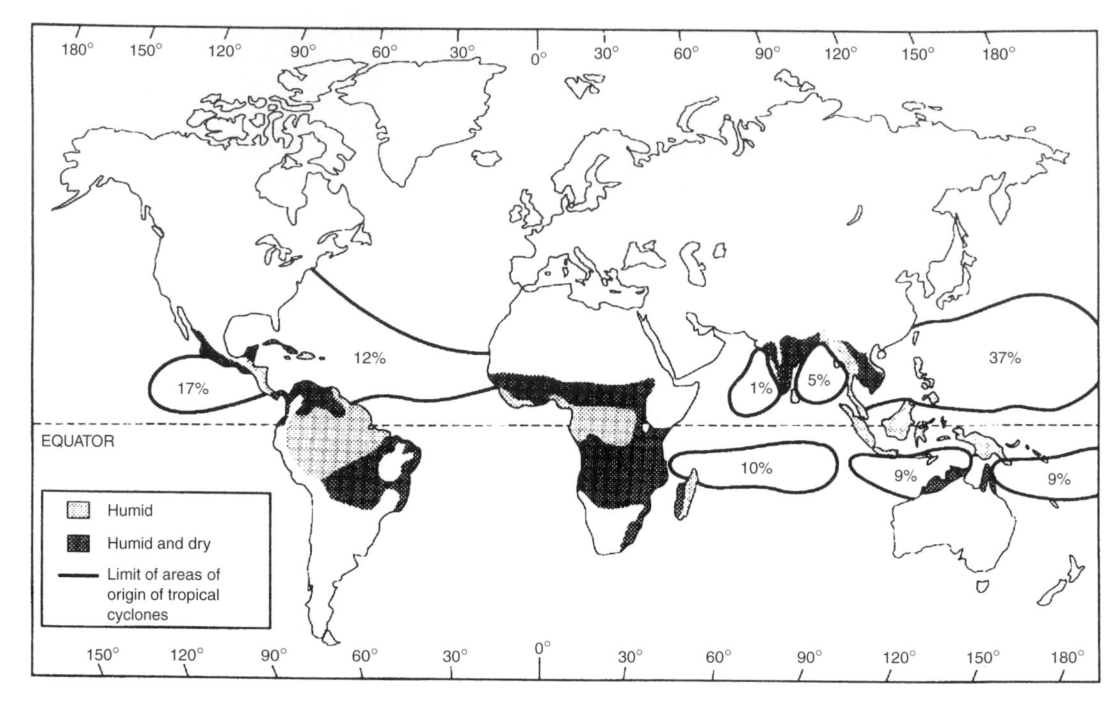

Figure 170 Tropical climates according to the Köppen system (modified by Trewartha), and occurrence of tropical cyclones shown as percentages (After WMO (1983))

development in the different climatic zones. In contrast, the parallel retreat of hillslopes, first proposed to explain some semi-arid landscapes in the USA, was extended by King (1953, 1957) as a universal model for hillslope development. King (1962) applied his ideas to all the Gondwana continents, and by implication, to most of the tropical world. This construct had enormous impact because it proposed a single hypothesis of slope development that was linked to 'continental drift' and the emerging theory of plate tectonics.

King's views contradicted the concept of a 'climatic geomorphology', but had little influence in Germany, where W. Penck (1924) had previously linked the extension of plains to tectonics, in a way that did not conflict with climatic control over geomorphic processes. In particular, the importance of a deeply weathered mantle in explanations of humid tropical relief was central to Büdel's (1957) hypothesis of *Doppelten Einebnungsflächen*, or 'double surfaces of levelling'. This paper led geomorphologists in the tropics to rediscover earlier accounts of weathering and relief development, especially in Africa (Falconer 1911; Wayland 1933; Willis 1936). These authors shared the view that plateau landscapes in the humid tropics were lowered incrementally by stripping and renewal of the SAPROLITE cover. Wayland (1933) called the resulting landscapes 'etched plains' (subsequently *etchplains*), the term later adopted by Thomas (1966, 1994) and Büdel (1982) (Figure 171). Falconer (1911) argued that much of the rocky relief in north Nigeria was due to differential weathering, and Willis (1936) argued that the prominence of granite monoliths or BORNHARDTs was due to repeated cycles of weathering and lowering of more susceptible rocks over geologic time. The isolation of bornhardts by selective weathering was also argued by Rougerie (1955) in Ivory Coast and in Brazil by Birot (1958) who stressed that *les dômes crystallines* were not inselbergs, as often described, but are found in zones of dissection. The importance of DEEP WEATHERING, followed by stripping was advocated by Ollier (1959, 1960) for Uganda, and new attempts were made to document the depth and distribution of the weathered mantle by Thomas (1966).

Linton (1955) drew on ideas about tropical weathering to explain 'the problem of tors' in

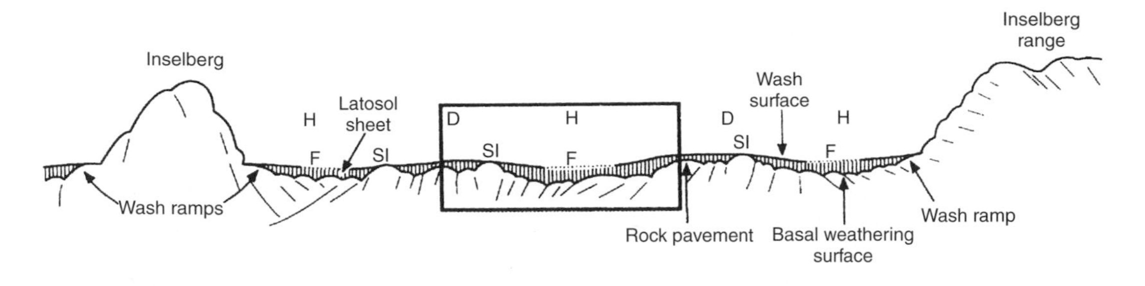

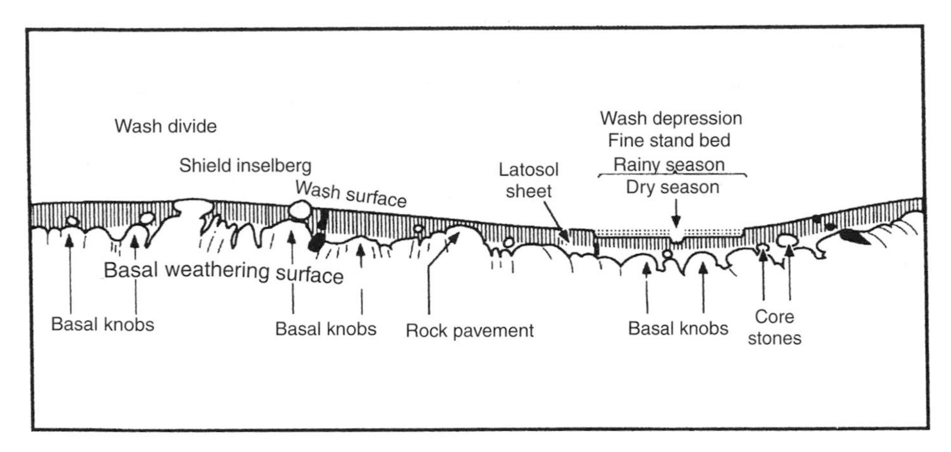

Figure 171 Characteristics of an active etchplain as described by Büdel (1982), based on studies of the Tamilnadu Plain in south India. H: wash depressions; D: wash divides; SI: shield inselbergs; F: fine sand in rainy season riverbed. Details of the wash divide and wash depression are shown in the expanded box. Lowering of the landscape is achieved by rock decay at the basal weathering surface and removal of sands and clays from the wash surface by seasonal runoff

Britain, leading to a clash with King. These ideas were applied to tor landscapes in North America, and to help explain the relief of glaciated areas (Feininger 1971). The notion of changing climates and the existence of 'relief generations' in world landscapes, was reviewed in a European context by Bakker and Levelt (1964), and became essential to a revised 'climato-genetic' geomorphology (Büdel 1968). Stoddart (1969) in an influential review of climatic geomorphology was scathing. He thought that the concept had its roots in Davisian theory, and considered that observations of landform differences between climatic zones were 'methodologically trivial' (p. 213). This view was widely accepted by those who pointed out that the same physical laws apply everywhere, and that the processes of chemical weathering and water flow are similar in all climates.

Although the Davisian paradigm prevailed for half a century, by 1950 proponents of the new 'process geomorphology' were directing attention away from geomorphic cycles in order to understand the mechanisms of landform change, rediscovering the work of G.K. Gilbert (1877, 1914). In the tropics, Branner (1896) had reported from east Brazil the roles of high rainfall and the presence of organic acids in tropical weathering; recognized the involvement of kaolinized saprolite in generating landslides under a natural forest cover, and the importance of soil fauna such as termites. By the 1950s geomorphologists in the tropics had begun to study these processes. White (1949) emphasized the importance of landslides in Oahu (Hawaii), while in West Africa studies of chemical dissolution of rocks, runoff and erosion arose from French research (Corbel 1957; Fournier 1960; Rougerie 1960). In South America, Tricart (1956, 1959) focused attention on fluvial processes, and the impacts of climate change. His 'Le modelé des régions chaudes, forêts et savanes' (Tricart 1965) was the first 'modern' study of tropical geomorphology. Tricart and Cailleux (1965, English trans. 1972) also published their 'Introduction à la géomorphologie climatique' in the same year complaining (pp. xiii) that 'the study of tropical geomorphology... has been delayed by the uniformist ideas of "normal erosion" ', which led to the systematic neglect of the tropics. This neglect was compounded by problems of access, fear of disease and, in the rainforests, a lack of inter-visibility.

By this time development involving new roads and construction projects had exposed the reality and extent of deep weathering in the humid tropics. In Hong Kong, detailed studies of weathering (Ruxton and Berry 1957) and analyses of intense tropical rainfall (Lumb 1975) showed unequivocally the importance of both to an understanding of landslide occurrence. The devastating effects of tropical storms was also reported from Rio de Janeiro (Jones 1973). On the other hand, the high permeability of many tropical soils was documented by Nye (1954, 1955) and subsequently by many others, who found that infiltration rates in *ferrallitic* soils (*oxisols, ultisols*) over gneissic rocks exceeded $200\,mm\,h^{-1}$, greater than expected rainfall intensities. Nye did not observe surface runoff in his experiments at Ibadan, Nigeria, but others have measured significant runoff on moderate slopes at intensities of 60–70 $mm\ h^{-1}$ (Morgan 1986). Outside the equatorial zone, the seasonal concentration of tropical rainfall combined with the high intensity of individual falls convinced many writers of its potential for serious soil erosion (Fournier 1960; Roose 1981). Studies of sediment yield have recorded great variation (Douglas and Spencer 1985; Thomas 1994). On undisturbed plots under rainforest, yields of $0.1–10\,t\,km^2\,yr^{-1}$ have been recorded, but small erosion plots left bare of plant cover can return sediment yields of $5{,}000\,t\ km^2\,yr^{-1}$. Small catchments in deforested highland areas have recorded from $1{,}500–3{,}000\,t\,km^2\ yr^{-1}$, but larger catchments, which have floodplains for major sediment storage, produce far less sediment per unit area and the average for Africa is just $35\,t\,km^2\,yr^{-1}$, rising to $380\,t\,km^2\,yr^{-1}$ in Asia. Rates are more influenced by catchment relief and rainfall amount than by other factors (Milliman and Syvitski 1992).

Tropical rivers are as varied as in other regions. Large rivers traversing the plateaux and plains of the southern (Gondwanaland) continents appear distinctive because of the fine calibre of their bedload (usually sand) and the interruption of their thalwegs by rock outcrops, which create rapids and waterfalls. Opinions formed on the basis of these characteristics hold that the thoroughness of tropical weathering reduces most rocks to sand and clay, or causes their total dissolution. According to writers such as Büdel (1957), Birot (1958) and Tricart (1965) this leads to a lack of abrasive tools for channel cutting; valleys are consequently open with shallow slopes, and river channels are interrupted by resistant bands of rock, which they do not erode effectively. However, wherever tropical rivers traverse hill ranges or fault scarps coarse bedload material becomes abundant and deep valleys and gorges result. Deep valleys are also cut into saprolite.

Great seasonal variations in river discharges are found in monsoon areas, and appear to lead to distinctive channel morphologies. The Auranga River, India was described by Gupta and Dutt (1989) as having a braided sandsheet with shallow (30 cm) channels in the long dry season, but is converted during the monsoon rains (1,500 mm in 4 months) to a wide meandering channel with sand and gravel point bars. By contrast rivers draining the equatorial Papua New Guinea mountains have a very low seasonal variability in flow and are subject to almost instantaneous runoff from steep slopes kept in a near saturated state by frequent rainfalls. Rainfall intensity is lower in this environment and most sediment reaches the river via frequent landslides (Pickup 1984; Pickup and Warner 1984).

Justification for a 'geomorphology of the tropics' lies in the balance of processes and their outcomes in terms of weathered materials and soil properties, erosional forms and sediments. Paradoxically, the major rock landforms, characterized by the granite domes or *bornhardts* that previously received such emphasis, owe more to petrology and structure than to specific climatic parameters. Nonetheless many illustrate the principles of selective deep weathering and stripping of regolith over very long time periods. Although it has been argued that lateritic (ferrallitic) weathering is more a product of time than of climate (Taylor *et al.* 1992), these factors are not independent. Advanced weathering occurs beneath ancient landsurface in non-tropical Australia, but is found more widely in the humid tropics, where some Neogene formations are also affected (Thomas *et al.* 1999). Weathering rates are, however, greatly influenced by water movement through the profile, so that dry climates lacking a seasonal water surplus have a much reduced potential for deep and advanced weathering. Bourgeon (see Pedro 1997) showed that deep clay-rich kaolinitic 'alterites' pass into shallow, sandy materials (arènes, grus) along an east–west transect of southern peninsular India ($P > 2{,}000\,mm\ y^{-1} - 700\ y^{-1}$), where smectite increases in proportion to kaolinite as the climate becomes drier.

Properties of the saprolite are fundamental to the understanding of tropical soils, as much for engineering as for agriculture. The permeability of tropical saprolite is increased by Fe_2O_3 adhesion to kaolin platelets, which produces larger aggregates. But these form 'cardhouse' fabrics that can collapse under loading, causing settlement of building structures. On the other hand, 2 : 1 lattice clays (smectites) found in seasonal climates, experience remarkable expansion as water content increases. This reduces permeability and causes 'heave' in many drier areas, forming GILGAI. Iron enrichment of the saprolite can arise from the flux of Fe^{2+} ions that are subsequently oxidized to Fe_2O_3. The Fe is mobilized as Fe^{2+} under conditions of low pH or by chelation involving organo-metallic complexes. This is likely to occur within poorly drained depressions in humid climates or under forest. But the fixation of Fe needs a rise in pH accompanied by drying of the deposit. The term *laterite* is often used to describe materials with high Al/Fe sesquioxide content (see Aleva 1994). It is a form of FERRICRETE that may also form in transported sediments; it becomes indurated and resistant to erosion if subject to wetting and drying, and may appear as hill cappings and valleyside benches in the landscape. Seasonality of rainfall contributes to this process, but in many cases the duricrusts have formed due to falling water tables, resulting from climate change and/or dissection. Beneath well-drained sites in the rainforest Fe does not become concentrated, but Al_2O_3 rich *bauxite* (mainly gibbsite) forms often as irregular nodules. Beneath undisturbed forests, the dominant exports are ions in solution, leading to chemical sediments offshore (Erhart 1955). In equatorial environments, where water tables are high, almost complete dissolution of silicate minerals is possible, leading to the formation of quartz-dominated 'white sands' (Thomas 1994).

In the mid-Miocene, there was a marked reduction of rainfall in the southern tropics, probably associated with the growth of the Antarctic ice sheet. Large sandsheets, such as the Kalahari Sands that penetrate the Congo Basin rainforests from the south may have originated at this time, and it has been argued that weathering systems were 'switched off' by this event, but many equatorial areas were humid throughout the Neogene. Quaternary climate change in the tropics, involved temperature reductions of 5–6 °C at the Last Glacial Maximum (LGM), and lowland areas experienced rainfall reductions of 25–50 per cent, and increased seasonality. Rainforests were converted to deciduous woodlands and/or savanna mosaics over large areas. Many small streams in the tropics left few traces of sedimentation during the LGM; other formerly meandering rivers became braided. Alluvial fans formed along many escarpments, probably due to diminished stream power and increased sediment load. Today, tropical landscapes are sensitive to environmental changes because the thick saprolites and stores of Quaternary sediment are subject to rapid erosion under intense rainfalls, if vegetation is cleared.

Accidents of history led to tropical areas lying beyond the immediate experience of geographers and geologists resident in the temperate zones of Europe and North America. This was not the case for the arid, glacial and periglacial environments, and it has led to persistent neglect of tropical geomorphology. Those who argue against the recognition of rainforests and savannas as distinctive environments for geomorphology fail to recognize the linkages between abundant rainfall, warm soil conditions, the productivity of humid tropical ecosystems, and the formative environments for soils and sediments. Ecologists and pedologists have long recognized that system outcomes depend on rates of biogeochemical cycles: of growth and decay; leaching and accumulation. Under tropical forests, rates of weathering and bioturbation can compensate for soil loss and erosion. Little known biota enter into weathering processes and organic acids can alter classic geochemistry to enhance the mobility of many cations. The frequency of intense rainfall in many areas is greater than in non-tropical climates, and the incidence of cyclones with high wind velocities is regionally important. These characteristics increase the hazards of flooding, inundation and landsliding, even in tectonically quiescent areas.

If geomorphology had developed from a tropical perspective it might have emerged with a different emphasis. But our views are also influenced by the available database, and the most serious problem in relating a geomorphology of the tropics to studies elsewhere remains the paucity of data about conditions in tropical environments.

References

Aleva, G.J.J. (ed.) (1994) *Laterites. Concepts, Geology, Morphology and Chemistry*, Wageningen: ISRIC.

Bakker, J.P. and Levelt, Th.W.M. (1964) An inquiry into the probability of a polyclimatic development of peneplains and pediments (etchplains) in Europe during the Senonian and Tertiary Period, *Publication Service Carte Géologique, Luxembourg* 14, 27–75.

Birot, P. (1958) Les dômes crystallines, *Memoires et Documents, CNRS* 6, 8–34.

Branner, J.C. (1896) Decomposition of rocks in Brazil, *Geological Society of America Bulletin* 7, 255–314.

Büdel, J. (1948) Das System der klimatischen Morphologie, *Deutscher Geographen-Tag, Munchen*, 65–100.

——(1957) Die 'doppelten Einebnungsflächen' in den feuchten Tropen, *Zeitschrift für Geomorphologie*, N.F 1, 201–288.

——(1968) Geomorphology principles, in R.W. Fairbridge (ed.) *Encyclopaedia of Geomorphology*, 416–422, New York: Reinhold.

——(1982) *Climatic Geomorphology*, trans. by L. Fischer and D. Busche, Princeton: Princeton University Press.

Corbel, J. (1957) L'érosion climatiques des granites et silicates sous climats chauds, *Revue Géomorphologie Dynamique* 8, 4–8.

Cotton, C.A. (1942) *Climatic Accidents in Landscape Making*, Wellington: Whitcomb and Tombs.

Davis, W.M. (1899) The Geographical Cycle, *Geographical Journal* 14, 481–504.

Douglas, I. and Spencer, T. (eds) (1985) *Environmental Change and Tropical Geomorphology*, London: Allen and Unwin.

Erhart, H. (1955) Biostasie et rhexistasie: esquise d'une théorie sur le rôle de la pédongenèse en tant que phénomène géologique, *Comptes Rendus Academie des Sciences française* 241, 1,218–1,220.

Falconer, J.D. (1911) *The Geology and Geography of Northern Nigeria*, London: Macmillan.

Feininger, T. (1971) Chemical weathering and glacial erosion of crystalline rocks and the origin of till, *US Geological Survey Professional Paper* 750-C, C65–C81.

Fournier, F. (1960) *Climat et Érosion: La Relation entre l'Érosion du Sol par l'Eau et les Précipitations Atmosphériques*, Paris: Presses Universitaires.

Gilbert, G.K. (1877) *Report on the Geology of the Henry Mountains*, United States Department of the Interior, Washington, DC.

——(1914) The transport of debris by running water, *United States Geological Survey Professional Paper* 86.

Gupta, A. and Dutt, A. (1989) The Auranga: description of a tropical monsoon river, *Zeitschrift für Geomorphologie, N.F.* 33, 73–92.

Jones, F.O. (1973) Landslides of Rio de Janeiro and the Serra das Araras Escarpment, Brazil, *US Geological Survey Professional Paper* 697.

King, L.C. (1953) Canons of landscape evolution, *Geological Society of America Bulletin* 64, 721–752.

King, L.C. (1957) The uniformitarian nature of hillslopes, *Transactions of the Edinburgh Geological Society* 17, 81–102.

——(1962) *The Morphology of the Earth*, Edinburgh: Oliver and Boyd.

Linton, D.L. (1955) The problem of tors, *Geographical Journal* 121, 470–487.

Lumb, P. (1975) Slope failures in Hong Kong, *Quarterly Journal of Engineering Geology* 8, 31–65.

Milliman, J.D. and Syvitski, P.M. (1992) Geomorphic/tectonic control of sediment discharge to the ocean: the importance of small mountain rivers, *Journal of Geology* 100, 525–544.

Morgan, R.P.C. (1986) *Soil Erosion and Conservation* Harlow: Longman.

Nye, P.H. (1954) Some soil-forming processes in the humid tropics. Part I: A field study of a catena in the West African forest, *Journal of Soil Science* 5, 7–27.

——(1955) Some soil-forming processes in the humid tropics. Part II: The development of the upper slope member of the catena, *Journal of Soil Science* 6, 51–62.

Ollier, C.D. (1959) A two cycle theory of tropical pedology, *Journal of Soil Science* 10, 137–148.

——(1960) The inselbergs of Uganda, *Zeitschrift für Geomorphologie N.F.* 4, 43–52.

Pedro, G. (1997) Clay minerals in weathered rock materials in soils, in H. Paquet and N. Clauer (eds) *Soils and Sediments – Mineralogy and Geochemistry*, 1–20, Berlin: Springer.

Penck, W. (1924) Die Morphologische Analyse, *Geographische Abhandlungen* 2.

Pickup, G. (1984) Geomorphology of tropical rivers I. Landforms, hydrology and sedimentation in the Fly and Lower Purari, Papua New Guinea, *Catena Supplement* 5, 1–18.

Pickup, G. and Warner, R.F. (1984) Geomorphology of tropical rivers II. Channel adjustment, sediment load and discharge in the Fly and Lower Purari, Papua New Guinea, *Catena Supplement* 5, 19–41.

Roose, E.J. (1981) Approach to the definition of rain erosivity and soil erodibility in West Africa, in M. De Broodt and D. Gabriels (eds) *Assessment of Erosion*, 143–151, Chichester: Wiley.

Rougerie, G. (1955) Un mode de dégagement probable de certains dômes granitiques, *Comptes Rendus Academie des Sciences* 246, 327–329.

——(1960) *Le Façonnement Actuel des Modelés en Côte D'Ivoire Forestière*, Memoire Institut Français Afrique Noire 58, Dakar.

Ruxton, B.P. and Berry, L. (1957) Weathering of granite and associated erosional features in Hong Kong, *Geological Society of America Bulletin* 68, 1,263–1,292.

Scrivenor, J.B. (1931) *The Geology of Malaya*, London: Macmillan.

Stoddart, D.R. (1969) Climatic geomorphology: review and assessment, *Progress in Physical Geography* 1, 160–222.

Taylor, G.R., Eggleton, R.A., Holzhauer, C.C., Maconachie, L.A., Gordon, M., Brown, M.C. and McQueen, K.G. (1992) Cool climate lateritic and bauxitic weathering, *Journal of Geology* 100, 669–677.

Thomas, M.F. (1966) Some geomorphological implications of deep weathering patterns in crystalline rocks

in Nigeria, *Transactions of the Institute of British Geographers* 40, 73–193.
——(1994) *Geomorphology in the Tropics*, Chichester: Wiley.
Thomas, M.F., Thorp, M.B. and McAlister, J. (1999) Equatorial weathering, landform development and the formation of white sands in Northwestern Kalimantan, Indonesia, *Catena* 36, 205–232.
Thorbecke, F. (ed.) (1927) *Morphologie der Klimazonen*, Breslau: Düsseldorfer Geographische Vorträge.
Tricart, J. (1956) Comparaison entre les conditions de façonnement des lits fluviaux en zone tempérée et zone intertropicale, *Comptes Rendus Academie Sciences* 245, 555–557.
——(1959) Observations sure le façonnement des rapides des rivières intertropicales. *Bulletin Section Géographique, Comité Travaux Historique et Scientifique* 68, 333–343.
——(1965) *The Landforms of the Humid Tropics, Forests and Savannas*, trans. C.J. Kiewiet de Jonge (1972), London: Longmans.
Tricart, J. and Cailleux, A. (1965) *Introduction à la Géomorphologie Climatique*, Paris: SEDES.
Wayland, E.J. (1933) Peneplains and some other erosional platforms, *Annual Report and Bulletin, Protectorate of Uganda Geological Survey, Department of Mines*, Note 1, 77–79.
White, L.S. (1949) Process of erosion on steep slopes of Oahu (Hawaii), *American Journal of Science* 247, 168–186.
Willis, B. (1936) East African Plateaus and Rift Valleys, Studies in Comparative Seismology, *Carnegie Institute, Washington, Publication*, 470.
WMO (World Meteorological Organization) (1983) Operational hydrology in the humid tropical regions, in E. Keller (ed.) *Hydrology of Humid Tropical Regions with Particular Reference to the Hydrological Effects of Agriculture and Forestry Practice*, IAHS Publication, 140, 3–26.

Further reading

Dubreuil, P.L. (1985) Review of field observations of runoff generation in the tropics, *Journal of Hydrology* 80, 237–264.
Faniran, A. and Jeje, L.K. (1983) *Humid Tropical Geomorphology*, Harlow: Longman.
Fookes, P.G. (1997) *Tropical Residual Soils*, The Geological Society, London.
Gutierrez Elorza, M. (2001) *Geomorphología Climática*, Barcelona: Omega.
Taylor, G. and Eggleton, R.A. (2001) *Regolith Geology and Geomorphology*, Chichester: Wiley.
Wirthmann, A. (2000) *Geomorphology of the Tropics*, Berlin: Springer.

MICHAEL F. THOMAS

TROTTOIR

Trottoirs (surf ledges) are narrow, subhorizontal erosional SHORE PLATFORMS with a veneer of Vermetid gastropod tubes (see VERMETID REEF AND BOILER) and encrusting coralline algae, and surfaces that often consist of tiers of VASQUES. Trottoirs occur at, or a little below, mean sea level in tropical seas and in the warmer parts of the southern and eastern Mediterreanean. They have been attributed to CORROSIONal cliff erosion in the spray zone and possibly the protection afforded to the wave-battered seaward edges of the platforms by Vermetid tubes, algae and other organic encrustations. On Curacao and other islands in the southern Caribbean Sea, trottoirs, up to 10 m width, are restricted to exposed areas where there are thick encrustations, and they are replaced by notches (see NOTCH, COASTAL) in less exposed areas. It has been suggested that trottoirs may eventually attain a state of DYNAMIC EQUILIBRIUM with the processes operating on them as their increasing width reduces the rate of cliff erosion.

Further reading

Focke, J.W. (1978) Limestone cliff morphology on Curacao (Netherlands Antilles), with special attention to the origin of notches and vermetid/coralline algal surf benches (corniches, trottoirs), *Zeitschrift für Geomorphologie* 22, 329–349.
Trenhaile, A.S. (1987) *The Geomorphology of Rock Coasts*, Oxford: Oxford University Press.

ALAN TRENHAILE

TSUNAMI

A tsunami is a Japanese word that describes a 'harbour wave' and it is used within the scientific community to describe a series of waves that travel across the ocean with exceptionally long wavelengths (up to several hundred kilometres between the wave crests in the open ocean). As the waves approach a coastline, the speed of the waves decreases as they are deformed within shallower water depths. During this process of wave deformation, the height of the wave increases significantly and as the waves strike the coastline they often cause widespread flooding across low-lying coastal areas and on many occasions cause loss of life and widespread destruction of property. Tsunamis are frequently described in the media as tidal waves. However this view is completely wrong since tsunamis have nothing to do with tides or weather. Most tsunamis in the world occur in the Pacific region. For example, during the past ten years, damaging tsunami floods have taken place in Nicaragua (1991), Flores, Indonesia (1992) and Okushiri Island, Japan (1993). The most recent destructive tsunami disaster took place in Papua New Guinea (1998) where several thousand people lost their lives.

Frequently tsunamis are described as seismic sea waves that are produced as a result of a sudden displacement of part of the seafloor. Usually a seismic disturbance associated with an offshore earthquake will cause rupturing of the ocean floor and when this happens the overlying water column is disturbed. It is during this phase of disturbed motion that tsunamis are often generated. Underwater earthquakes in tectonically active areas of the world (e.g. the Pacific rim) are the most common cause of tsunamis. For example, as recently as May 1960, a large underwater earthquake took place off the coast of Chile. The tsunami travelled across the Pacific Ocean. After twelve hours it struck the coastline of the Hawaiian island chain, while after a further twelve hours it reached the coast of Japan where it destroyed all in its path. A huge tsunami was produced as a result of a large earthquake beneath the seafloor west of Portugal on 1 November AD 1755. One observer who witnessed this tsunami was William Borlase who observed its arrival on the shore of Mounts Bay, Cornwall. He noted:

> the first and second refluxes were not so violent as the third and fourth (*tsunami waves*) at which time the sea was as rapid as that of a mill-stream descending to an undershot wheel, and the rebounds of the sea continued in their full-fury for fully two hours . . . alternatively rising and falling, each retreat and advance nearly of the space of ten minutes, till five and a half hours after it began.

Another observer of this tsunami described how, at Larmorna Cove, Cornwall:

> the sea on this occasion rushed suddenly towards the shire in vast waves, with such impetuosity, that large rounded blocks of granite from below low-water mark were swept along like pebbles, and many of them deposited far beyond high water mark. One large block, weighing probably 6 or 8 tons, was borne repeatedly to and fro several feet above the level of high water, and at length deposited about ten feet above that level in the stream, where it still lies.

Tsunamis can also be generated by underwater landslides and by landslides occurring above sea level and moving downslope into the sea (see SUBMARINE LANDSLIDE GEOMORPHOLOGY). The occurrence of such slides is quite common on a geological timescale. In 1929, an offshore earthquake led to widespread sediment slumping and the generation of turbidity currents across the seafloor off the Grand Banks of Newfoundland as well as a tsunami that locally reached up to +30 m at the coast. In the North Atlantic region a very large tsunami was generated approximately 8,000 years ago by one of the world's largest underwater landslides (the Second Storegga Slide). The tsunami caused flooding along parts of the Norwegian coastline up to levels +20 m above sea level and along the UK coastline where the highest flood levels reached up to +6 m above sea level. One of the most widely described processes of submarine slide generation is through the release of methane gases (clathrates) contained within ocean-floor sediments. It is believed that the sudden release of such gases can cause local slumping and sliding. The second process arises from low-magnitude earthquakes which in themselves have no destructive effect, but which are of sufficient intensity to induce shaking of seafloor sediments thus causing downslope slumping and sliding of sediment. Generally speaking, underwater landslides generate tsunamis with much less energy than those produced by earthquake-triggered faulting on the ocean floor. The size and energy of landslide-generated tsunamis decreases rapidly with increased distance from the source area.

Occasionally, tsunamis can be generated by the impact of meteorites onto the ocean surface. Perhaps the most famous tsunami associated with a large meteorite impact was that which took place *c.*65 million years ago. The so-called K-T impact, usually associated in the geological literature with the extinction of the dinosaurs, also created a huge tsunami. Traces of the tsunami waves can be found in areas of Mexico and Texas. Most authoritative accounts consider that, given the proportion of the Earth's surface area represented by the Pacific Ocean, a meteorite capable of striking the Pacific Ocean and generating a tsunami is in the order of 1:400,000 years.

Tsunamis can also be produced through the collapse of a volcanic crater into the sea during a major explosive volcanic eruption. Although such a phenomenon may have occurred *c.*18,000 years ago in the case of the volcanic eruption of Santorini in the Aegean Sea, such events are extremely rare. Similarly, tsunamis can be produced as a result of the collapse of the flank of a volcano into the sea. Such tsunamis may be generated by rockfalls, rockslides, rotational slips or debris flows that originate on steep slopes and move into the sea. Such an event happened during the recent eruptions on the island of Montserrat. Some authors have proposed that certain hillslopes on the flanks of some of the now dormant volcanoes in the Canary Isles may have collapsed into the sea in the past and produced large tsunamis.

Reference

Bryant, E. (2001) *Tsunami: The Underrated Hazard*, Cambridge: Cambridge University Press.

ALASTAIR G. DAWSON

TUFA AND TRAVERTINE

Tufa and travertine are terrestrial freshwater accumulations of calcium carbonate, whose formation often involves a degree of organic involvement. The names tufa and travertine can be used synonymously, but often tufa is taken to refer to a softer, more friable deposit whilst travertine refers to a harder, more resistant material frequently used as a building material. Tufa derives from tophus or tufo, used in Roman times to describe crumbly white deposits (Ford and Pedley 1996; Pentecost 1993). Travertine derives from the Latin *lapis tiburtinus*, or Tibur stone, and was originally used to describe the massive, hot spring deposits around Rome (Pentecost 1993). Some travertines are of hydrothermal origin and thus contain very limited plant material, whilst tufa and travertines precipitated by cool water typically contain the remains of micro- and macrophytes, invertebrates and bacteria. Different usages of the terms tufa and travertine are found in different countries, and there is as yet no standard international terminology although Ford and Pedley (1996) and Pentecost and Viles (1994) suggest two alternative schemes. Tufa and travertine are distinguished from CALCRETE, SPELEOTHEMS and STROMATOLITES by their environment of formation, but they share many similar characteristics and may grade into one another in some circumstances.

Tufas and travertines form in freshwater environments where thermodynamic and kinetic characteristics favour the precipitation of calcium carbonate from carbonate-rich waters. Such conditions arise where carbon dioxide is removed from the water through turbulent degassing, evaporation or biological uptake. Suitable conditions for precipitation of tufa and travertine are often found in or near KARST areas where dissolution of limestone provides high levels of dissolved carbonate, or where thermal waters rich in carbon dioxide originate in areas of recent volcanic activity. Despite difficulties in obtaining reliable dates of tufas using ^{14}C, most tufas and travertines which have formed from cool water appear to originate from the late Quaternary to early Holocene. This is certainly true in Britain and much of Europe, although in some parts of the world such tufas and travertines are still forming rapidly today. There has been much debate over the major controls on tufa and travertine formation and in particular the role of organisms in their genesis. Certainly, aquatic plants and micro-organisms can aid deposition of tufa – by providing precipitation nuclei, by removing carbon dioxide from the water and perhaps also by direct precipitation of calcium carbonate – but in some environments physico-chemical controls on precipitation outweigh any biological involvement. In many fluvial tufas within forested catchments inorganic modes of precipitation dominate in the turbulent upper reaches, whilst further down tree debris contributes to barrage architecture and macro- and microphytes, and even insect larvae can aid deposition (Drysdale 1999). Hydrothermal travertines are precipitated through largely inorganic processes, although Robert Folk (1994) has suggested that tiny nannobacteria may also play a key role.

Tufa and travertine are found on all continents except Antarctica and there are several very impressive deposits around the world, such as the barrages of the Plitvice Lakes in Croatia, the travertine complex of Antalya in south-west Turkey (Burger 1990) and the Huanglong ravine terraces in north-west Sichuan, China. Good summaries of the occurrence and nature of the major tufa and travertine deposits in Britain, Europe, China and the world are provided by Pentecost (1993), Pentecost (1995), Pentecost and Zhaohui (2001) and Ford and Pedley (1996) respectively. Many occurrences of large tufa and travertines have still not yet been properly documented in the international literature, such as the extensive deposits within the arid Naukluft area of south-central Namibia (Plate 141).

Tufas and travertines form in a range of different geomorphic settings, including fluvial, lacustrine, paludal and spring environments. Within rivers, tufas and travertines can form spectacular barrages often with waterfalls cascading over them, and with clastic tufa accumulating behind the barrage. In some fluvial environments, suites of lakes become created between barrages. Much smaller accumulations of tufa and travertine also occur in many streams, producing fluvial crusts and oncoids. In lacustrine environments, similar oncoids and crusts occur as well as larger reef-like accumulations. In marshy environments, low relief muddy tufas tend to develop, often mixed with marls and chalks. Around springs, mounds and terraces can develop, and where springs debouch on steep slopes such deposits can form huge prograding cascades. Tufas and travertines can produce great geomorphological change within a catchment, as they influence river flows through the creation of barrages and can armour slopes and change the course of springs.

Plate 141 A large cone-shaped tufa, *c.*400 m across and nearly 100 m high, developed on the edge of the Naukluft Mountains at Blasskrantz in central Namibia. It has formed where a seasonal stream cascades over a steep slope, permitting degassing to take place

Plate 142 A small tufa dam formed across a creek in the tropical Napier Range of the Kimberley district, northwestern Australia

The fabric of tufas and travertines reveals much about their mode of formation, and can also contain useful palaeoenvironmental information. Where organic influence in tufa and travertine precipitation is debated, the petrology of the deposit can provide helpful insights. Martyn Pedley has provided a useful concept of tufa and travertine as phytoherms, with different facies found in different parts of the deposits and a clear organic role played by biofilms in the formation of many zones (Pedley 1992). Some tufas and travertines possess clear laminations which can be of great use in environmental reconstruction. Tufas precipitated in association with cyanobacteria from the genus Phormidium for example, tend to display seasonal banding with alternate sparitic, light layers (representing inorganic deposition in autumn and winter) and micritic, dark layers (formed as a result of cyanobacterially mediated deposition in spring and summer).

The fact that many major cool water tufas and travertines date from the late Quaternary and early Holocene has led several authors to conclude that there is a strong climatic influence on their formation (see for example Goudie *et al.* 1993). Alternatively, or as well as such climatic control, human impacts may also be responsible for a recent decline in the accumulation of tufas and travertines in some parts of the world. Isotopic and trace element contents of tufas and travertines can be used to reconstruct palaeoenvironmental conditions within different parts of dated sequences, thus throwing light on both the environmental conditions and their influence on tufa deposition rates. Andrews *et al.* (1997) illustrate the utility of stable isotopes of oxygen and carbon in reconstructing palaeoenvironmental conditions in Holocene tufa deposits, and Matsuoka *et al.* (2001) show how trace element contents within laminated tufas can provide high resolution records of climate and catchment conditions. As Ford and Pedley (1996) conclude 'tufas have an untapped potential to provide the best land-based opportunity for accessing shorter-term Holocene environmental change'.

References

Andrews, J.E., Riding, R. and Dennis, P.F. (1997) The stable isotope record of environmental and climatic signals from modern terrestrial microbial carbonates from Europe, *Palaeogeography, Palaeoclimatology, Palaeoecology* 129, 171–189.

Burger, D. (1990) The travertine complex of Antalya/Southwest Turkey, *Zeitschrift für Geomorphologie Supplementband* 77, 25–46.

Drysdale, R.N. (1999) The sedimentological significance of hydropsychid caddis-fly larvae (order: Trichoptera) in a travertine-depositing stream: Louie Creek, Northwest Queensland, Australia, *Journal of Sedimentary Research* 69, 145–150.

Folk, R.L. (1994) Interaction between bacteria, nannobacteria and mineral precipitation in hot springs of central Italy, *Geographie physique et Quaternaire* 48, 233–246.

Ford, T.D. and Pedley, H.M. (1996) A review of tufa and travertine deposits of the world, *Earth-Science Reviews* 41, 117–175.

Goudie, A.S., Viles, H.A. and Pentecost, A. (1993) The Late Holocene tufa decline in Europe, *Holocene* 3, 181–186.

Matsuoka, J., Kano, A., Oba, T., Watanabe, T., Sakai, S. and Seto, K. (2001) Seasonal variation of stable isotopic compositions recorded in a laminated tufa, SW Japan, *Earth and Planetary Science Letters* 192(1), 31–44.

Pedley, H.M. (1992) Freshwater (phytoherm) reefs: the role of biofilms and their bearing on marine reef cementation, *Sedimentary Geology* 79, 255–274.

Pentecost, A. (1993) British travertines: a review, *Proceedings, Geologists' Association* 104, 23–39.

——(1995) Quaternary travertine deposits of Europe and Asia Minor, *Quaternary Science Reviews* 14, 1,005–1,028.

Pentecost, A. and Viles, H.A. (1994) A review and reassessment of travertine classification, *Géographie physique et Quaternaire* 48, 305–314.

Pentecost, A. and Zhaozhui, Z. (2001) A review of Chinese travertines, *Cave and Karst Science* 28, 15–28.

HEATHER A. VILES

TUNNEL EROSION

Tunnel erosion is an insidious form of land degradation that is initiated in subsoil and/or substrata and remains inconspicuous until considerable damage has occurred. This type of water erosion is found in earthworks as well as hillslopes and in the latter case refers to the hydraulic removal of subsurface material causing the formation of underground channels in the natural landscape (Boucher 1990). Tunnelling takes place primarily when the shear stress applied by flowing water enlarges an existing macropore or passageway (Bryan and Jones 1997). Corrasion of the tunnel perimeter by ensuing flow and the impact of vehicular, livestock and/or human traffic cause localized collapse of the land surface, producing sinkholes which can entrain surface runoff. The natural bridges of soil remaining between sinkholes are destroyed when the cavities merge, leaving a gully. Debris fans are most conspicuous on hillslopes, beginning from points on the land surface where the translocated sediment is debouched, and consisting of relatively coarse material which can no longer be transported by the hydraulic head. Finer particles are washed further downslope. The initial length of tunnels is uncertain but they may be a series of interconnected macropores and can extend to several hundred metres. The diameter of these features typically ranges between several centimetres and a few metres, the latter proving to be most frequent in semi-arid environments.

Tunnelling has been recorded under a wide range of climates in many different materials on landscapes subjected to diverse land-use histories, but owing to occurrence in uninhabited forest in Papua New Guinea and Eocene palaeopiping in the USA, it should be seen as a naturally occurring process. Sources of tunnelflow include rainfall, snowmelt, irrigation water and ground water. Macropores, desiccation and structural cracks, surface depressions, decayed tree roots and the burrows of insects, crabs, moles, rodents and rabbits have been shown to be points where surface water can infiltrate directly to the subsoil. The materials eroded have included soils that remained highly stable on wetting in the UK (Jones 1981), glaciolacustrine silts and permafrost in Canada as well as ignimbrite and pumice in New Zealand. Many reports of tunnelling in semi-arid climates, especially southeastern Australia, have been associated with nonsaline sodic texture-contrast soils that slake and/or disperse readily on contact with water. An important characteristic of tunnel erosion is the requirement for a layer(s) of material of relatively low permeability which act(s) as a barrier to further vertical percolation. This material can be breached by cracks, joints, tree roots and well-connected macropores. A hydraulic gradient is required to generate flow and the most common outlets occur on the hillslope proper where a hydraulic head can be seen, or comprise various types of free faces such as gully walls and stream banks.

Whilst most hydrologic research has been conducted in humid climates where ephemeral, seasonal and perennial systems are observed (e.g. Wales, UK, Jones 1981, Bryan and Jones 1997), few data are available for semi-arid environments. On grazing land in Victoria, Australia, flow in a shallow tunnel system (generally less than 1 m deep from the soil surface) responded rapidly to rainfall, and it was clear from the typically short lags between peak rainfall and peak runoff that the soil matrix had been largely bypassed. The recession was also rapid and ground water was not a component of the hydrograph (Boucher 1995). Similar characteristics were documented for shallow tunnels in an area of badlands in western Canada (Bryan and Harvey 1985) and a loess plateau in northern China. However, at least partly owing to internal collapse, the relations were more complex for deep-seated systems which were 20 m to 30 m deep in the badlands area, whilst the inlets ranged between less than 0.5 m and over 20 m in both depth and diameter in the latter catchment. The proportion of discharge passing through tunnels to the stream was estimated as up to 10 per cent and 80 per cent respectively. Therefore, tunnel erosion can be a rapid and significant form of soil and water loss, and the economic implications need investigation. Generally, reclamation of land in southeastern Australia should combine deep ripping of soil to destroy the established flowpaths, chemical amelioration of dispersive soils with gypsum in order to displace the sodium and generate an electrolyte effect, and revegetation with grasses which use up excess water.

Bryan and Jones (1997) suggested that tunnelling and piping are distinct processes which are often difficult to distinguish in practice, and that all subsurface erosion is usually combined as piping. However, from research in Australia and China it appears that the terms are still used interchangeably.

References

Boucher, S.C. (1990) *Field Tunnel Erosion: Its Characteristics and Amelioration*, Clayton, Australia: Monash University and East Melbourne: Department of Conservation and Environment.

——(1995) Management options for acidic sodic soil affected by tunnel erosion, in R. Naidu, M.E. Sumner and P. Rengasamy (eds) *Australian Sodic Soils – Distribution, Properties and Management*, 239–246, East Melbourne: CSIRO Australia.

Bryan, R.B. and Harvey, L.E. (1985) Observations on the geomorphic significance of tunnel erosion in a semi-arid ephemeral drainage system, *Geografiska Annaler* 67A, 257–272.

Bryan, R.B. and Jones, J.A.A. (1997) The significance of soil piping processes: inventory and prospect, *Geomorphology* 20, 209–218.

Jones, J.A.A. (1981) *The Nature of Soil Piping – a Review of Research*, British Geomorphological Research Group Research Monograph No. 3, Norwich: Geo Books.

Further reading

Blong, R.J. (1965) Subsurface water as a geomorphic agent with special reference to the Mangakowhiriwhiri catchment, *Auckland Student Geographer (NZ)* 1, 82–95.

Carey, S.K. and Woo, M-K. (2000) The role of soil pipes as a slope runoff mechanism, subarctic Yukon, Canada, *Journal of Hydrology* 233, 206–222.

Pickard, J. (1999) Tunnel erosion initiated by feral rabbits in gypsum, semi-arid New South Wales, Australia, *Zeitschrift für Geomorphologie N.F.* 43, 155–166.

Slaymaker, O. (1982) The occurrence of piping and gullying in the Penticton glacio-lacustrine silts, Okanagan Valley, BC, in R.B. Bryan and A. Yair (eds) *Badland Geomorphology and Piping*, 305–316, Norwich: Geo Books.

Zhu, T.X., Luk, S.H. and Cai, Q.G. (2002) Tunnel erosion and sediment production in the hilly loess region, north China, *Journal of Hydrology* 257, 78–90.

SEE ALSO: pipe and piping

STUART C. BOUCHER

TUNNEL VALLEY

Tunnel valleys are examples of large-scale, erosional glacial landforms that are formed subglacially. Morphologically, tunnel valleys are overdeepened, elongate depressions with steep, often asymmetric sides, cut into bedrock or unconsolidated glacigenic sediment. A characteristic feature is an undulating long profile, which climbs over bedrock rises and contains overdeepened areas along its floor. Long reaches of the flow path can be against the regional gradient. In dimensions, tunnel valleys can reach up to 4 km in width and over 100 km in length, and depths of erosion can be as great as 400 m below sea level. Commonly they trend oblique to the modern drainage, and many contain subglacial landforms such as drumlins, eskers or gravel dunes. In plan view, tunnel valley systems vary from individual, straight segments, to integrated, anastomosing networks of sinuous valleys. Sedimentary infills of tunnel valleys are diverse, and a wide variety of sediments associated with different depositional environments (glaciterrestrial, glacimarine, glacilacustrine and temperate) have been recorded, which reflect changing conditions during, and subsequent to, valley formation. Sediment gravity flow deposits and glacifluvial sands are particularly common.

There are three main theories of tunnel valley genesis. The first argues that they form time-transgressively, by subglacial meltwater erosion during deglaciation, at or close to the ice margin. The second theory also interprets tunnel valleys as the product of subglacial meltwater erosion but argues that the discharges involved were catastrophic channelized floods, and that tunnel valleys within anastomosing networks form synchronously. Finally, the third theory argues that tunnel valleys cut into unconsolidated sediment are due to creep of deformable sediment into a subglacial conduit from the sides and from below. This material is then removed through the conduit by subglacial meltwater.

Further reading

Ó Cofaigh, C. (1996) Tunnel valley genesis, *Progress in Physical Geography* 20(1) 1–19.

COLM Ó COFAIGH

TURBIDITY CURRENT

Turbidity currents or flows are a type of density flow in which movement on a slope occurs due to changed density between a local fluid and a surrounding fluid (Simpson 1987). In a turbidity current it is the suspended particles that cause the density of the flow to be greater than that of the surrounding fluid. The result of this is that the turbulent suspension moves down any local or regional gravity slope.

Turbidity currents in water can originate in a number of ways. One mechanism is for them to originate as sediment slides and slumps caused by scarp or slope collapse (e.g. when a slope is disturbed by earthquake shocks). Another mechanism is direct underflow of suspension-charged river water in so-called hyperpychnal plumes. These can occur during snowmelt floods in steep-sided FJORDs, in front of RIVER DELTAs, and in river tributaries whose feeder channels have extremely high loads of suspended sediments. A third mechanism is the collection of sediment by longshore drift in the nearshore heads of SUBMARINE VALLEYS and canyons (Leeder 1999).

Turbidity flows are not restricted to water. As powder snow avalanches demonstrate, for example, dense suspensions in air may flow downslope.

Submarine turbidity currents can be up to several kilometres wide and several hundreds of metres thick. They can travel as far as 4,000–5,000 km. Erosion can occur at the base of a flow, producing various scour or sole marks including mega-flutes. Deposition from a waning turbidity current produces sediment accumulations called turbidites. Associated with these may be large-scale asymmetric gravel waves and macrodunes and channel levees (Stow 1994).

References

Leeder, M. (1999) *Sedimentology and Sedimentary Basins. From Turbulence to Tectonics*, Oxford: Blackwell Sciences.
Simpson, J.E. (1987) *Gravity Currents: In the Environment and the Laboratory*, Chichester: Ellis Horwood / Wiley.
Stow, D.A.V. (1994) Deep sea processes of sediment transport and deposition, in K. Pye (ed.) *Sediment Transport and Depositional Processes*, 257–291, Oxford: Blackwell Scientific.

A.S. GOUDIE

TURF EXFOLIATION

A denudation process that is particularly prevalent in periglacial areas and leads to the destruction of a continuous vegetation cover through the removal of soil exposed along small terrace fronts. Among the processes that lead, possibly synergistically, to this phenomenon are needle-ice (pipkrake) action, desiccation, rain wash erosion, zoogeomorphic activities and aeolian deflation. It appears to be especially pronounced in high alpine regions, not least in lower latitudes (Grab 2002).

Reference

Grab, S.W. (2002) Turf exfoliation in the high Drakensberg, Southern Africa, *Geografiska Annaler* 84A, 39–50.

A.S. GOUDIE

TURLOUGH

A seasonally filled KARST depression, dependent upon water table fluctuations and tidal effects. Turloughs occur in the glacially modified carboniferous age limestones of central and western Ireland, with the term derived from the Irish *tuar loch* meaning dry lake. In times of decreasing water table (dry seasons) the level of the turlough drops, with the contents draining through the porous parts of the basin floor and via connections to swallow holes. In contrast, times of high rainfall (wetter seasons) will produce a high lake level. Theories of development (see Coxon 1986) are based upon interactions of a karstic landscape formed in the Tertiary and subsequent glaciations, with turlough form dependent upon this history. The flooding regime in turloughs varies considerably, as a result of size, depth, local water table, tidal regime and soil conditions. They can be composed of sand, clay, silt, diamicton, peat or marl, or various combinations (Coxon 1986). An example of a turlough is the Carren Turlough of the Burren Region, County Clare, Eire.

Reference

Coxon, C. (1986) *A Study of the Hydrology and Geomorphology of Turloughs*, Unpublished Ph.D. Thesis, University of Dublin, Ireland.

Further reading

Sweeting, M.M. (1972) *Karst Landforms*, London and Basingstoke: Macmillan.

STEVE WARD

U

UNDERFIT STREAM

A stream is underfit when it is much smaller than the size of its valley. Dury (1964) defines this fluvial geomorphological feature as that of a present-day stream flowing in an alluvial plain and describing free meanders far less ample than those of the enclosing VALLEY MEANDERS, which are frequently ingrown meanders of former large streams. Dury (1965) found that the ratio between the bed width of former and present-day channels (*W/w*) gives an index of underfitness, and that this ratio averages 10/1 in lowland England and near the former ice fronts in Wisconsin (USA), 5/1 and about 3/1 in the Ozarks. He also proposed an alternative index, that given by the wavelength ratios (*L/l*).

Because there is a relationship between stream size and size of meanders, it has to be assumed that some cause has operated to reduce the present stream size significantly below its former values. Davis (1913) had defined underfit streams and related their origin to the capture process taking place during the competitive development of rivers. The beheaded stream underwent shrinkage and reduced the size of the meanders which it had formerly processed. Other hypotheses proposed by subsequent researchers include: underflow through alluvium, deep percolation or cessation of meltwater discharge, overspill from glacial lakes or glacial meltwater and tidal scour. During the 1950s Dury undertook a series of field investigations of subsurface bottoms of valleys occupied by manifest underfits in several localities of the English plain in order to verify the hypothesis that, if valley bends are authentic former meanders, former large channels should have been associated with them. He demonstrated the presence of large meandering channels, up to ten times as wide as the existing channel, cut in bedrock and complete with pool and riffle sequences. Further tests on deep buried channels in the driftless area of Wisconsin, in France and elsewhere confirmed his hypothesis, that there is a sensible constancy of the wavelength ratio between valley meanders and stream meanders throughout entire regions. He rejected previous explanations on the grounds that they could only explain a fraction of the total shrinkage and they were of restricted areal application. He concluded that the required explanation of underfitness had to be a widespread one and, in order to meet this requirement, it could be no other than climatic change. In addition, Dury (1977) found that the RIVER CAPTURE hypothesis could be rejected by means of hydrologic empirical power-functional equations which relate discharge to drainage area.

Underfit streams have now been identified widely in western Europe, where perhaps at least 50 per cent of the length of second and higher order streams is underfit, and as far east as the Ukraine. They occur in all the major climatic regions of the USA, including Alaska. They are present also in the coastal drainage of Australia's Northern Territory and on the eastern coastland.

Dury (1964) distinguished several types of underfit streams (see Figure 172). A *manifestly underfit* is a stream which meanders within a more amply meandering valley, the meanders being of the ingrown type. Manifest underfits can be identified immediately from air photographs or from reliable maps. A *variant* is when a manifest underfit is enclosed by sub-parallel scarps of rock, below which weaker formations have been eroded into a broad trough. In the *underfits of the Osage type*, except for being curved around the

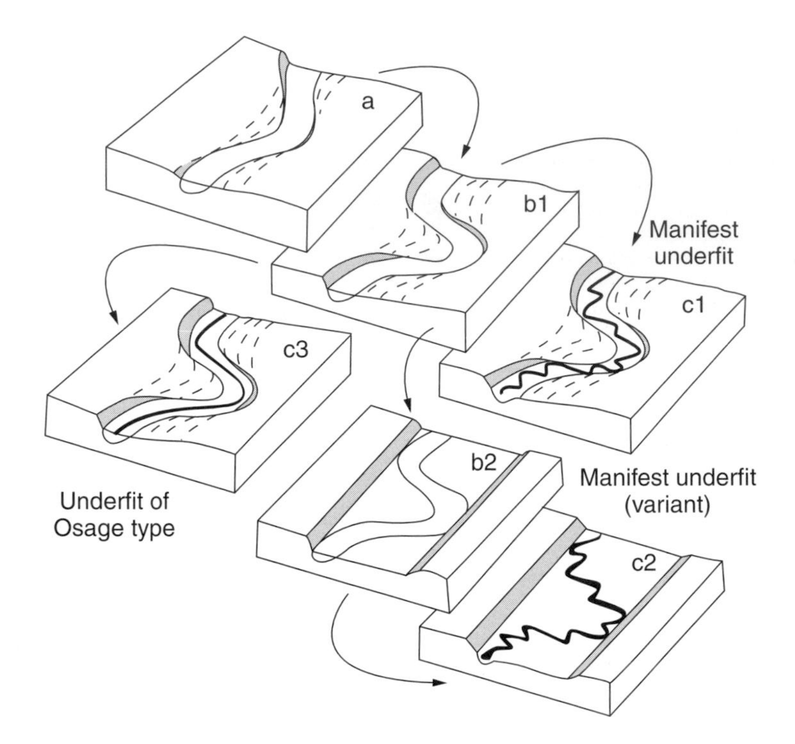

Figure 172 Partial model for the development of underfit streams

valley bends, the channel behaves as if it were straight, but the spacing of the pool and riffle sequences are more closely spaced than the valley meander wavelength would suggest. Osage-type underfits can only be suspected from their plan dimensions and have to be demonstrated from subsurface drilling of the pool and riffle sequence. Although they are named after the Osage River in Missouri, USA, they were first identified by Dury (1966) in the Colo River in New South Wales, Australia.

The development of underfit streams has two aspects: the date of origin of the rivers by which valley meanders have been cut, and the date of the shrinkage by which rivers were reduced to the underfit state. The latest date of origin for ingrown valley meanders is, in many areas, the time when incision of plateaux began. It is known that the incised meandering valleys of the Alpine Foreland and of the Hercynian massifs of Europe were initiated early in the Pleistocene. The high former discharges responsible for shaping valley meanders did not operate throughout the whole interval, but repeated episodes of shrinkage took place. Dates obtained by pollen analysis for the last major streams which are now underfit fall between about 12,000 and 10,000 BP. Evidence from river terraces also places the last main shrinkage well within last-deglacial time, when pluviosity markedly decreased. This appears to have been the time when general underfitness was finally confirmed, except in areas which were still covered by ice or by proglacial lakes.

There is a relationship between stream size and size of meanders. Meander wavelengths of valleys have been estimated to require a BANKFULL DISCHARGE of about 25 times greater than the present, and even 50 or 60 times greater when differences in channel slope, cross section and velocity are incorporated into the estimates. Stream shrinkage is then the result of a significant reduction in discharge at the recurrence interval corresponding to channel-forming flow or to the most probable annual flood. It was suggested by Dury that the channel-forming discharges required to explain the former channel patterns of streams which are now underfit could have been provided by increases in mean annual precipitation of 50 to 100 per cent. Magnitude-area-intensity analysis of precipitation shows that the means of reaching the required increase in precipitation is an increase in the frequency and power of storms.

References

Davis, W.M. (1913) Meandering valleys and underfit rivers, *Annals of the Association of American Geographers* 3, 3–28.
Dury, G.H. (1964) Principles of underfit streams, *US Geological Survey Professional Paper* 452-A.
——(1965) Theoretical implications of underfit streams, *US Geological Survey Professional Paper* 452-C.
——(1966) Incised valley meanders on the Lower Colo River, NSW, *Australian Geographer* 10, 17–25.
——(1977) Underfit streams: retrospect, perspect and prospect, in K.J. Gregory (ed.) *River Channel Changes*, 281–293, Chichester: Wiley.

SEE ALSO: bankfull discharge; river capture; valley meander

MARIA SALA

UNDRAINED LOADING

If a soil is loaded very quickly it can result in there being no time for the drainage of pore water. There may be no consequential change in volume, but pore-water pressures change and result in differential shear and normal stress at every point in the loaded material. This rapidly reduces resistance and sometimes initiates shear movement, or accelerates movement downslope. The eventual dissipation of surplus pore pressure should reinstate steady-state equilibrium in the soil. The extent of pore-water pressure change will vary from soil to soil, as it is predominantly dependent on the soil's composition and its properties. Undrained loading is therefore important in slope stability analysis, particularly for soils with low permeability and receiving rapid loading, and has been incorporated into several modelling studies (e.g. Baker *et al.* 1993). However, such conditions are hard to accommodate within models, as influential factors such as changes in PORE-WATER PRESSURE are difficult to predict.

Reference

Baker, R., Fryman, S. and Talesnick, M. (1993) Slope stability analysis for undrained loading conditions, *International Journal for Numerical and Analytical Methodology in Geomechanics* 17, 15–43.

STEVE WARD

UNEQUAL SLOPES, LAW OF

States that slopes will behave differently depending on their declivities (inclination). The law was proposed by G.K. Gilbert (1877: 140) within his paper on the geology of the Henry Mountains, USA. As rain falls on a slope, the amount of work it can do is proportional to the declivity of the slope. The steeper slope is always degraded faster, and will carry the divide towards the gentler slope. Thus, unless there are equal slope declivities, with homogenous material and identical rainfall, unequal slope activity will proceed. Eventually, a state of equilibrium will form (slope symmetry). Gilbert used this law to explain the form of BADLANDs, alongside his law of divides. This states that slopes on one side of a ridge are independent, while the law of equal declivities establishes a relationship between the slopes, the ridge crest, and the other side of the ridge. Thus, the slopes of the whole ridge behave independently over time and a landscape of unequal slopes develops.

Reference

Gilbert, G.K. (1877) *Geology of the Henry Mountains*, Washington, DC: US Geographical and Geological Survey of the Rocky Mountain Region, 140–141.

STEVE WARD

UNICLINAL SHIFTING

The gradual lateral shifting of a stream or river down-dip as a result of the slope of the underlying bedrock. When a river passes over a zone of inclined alternating hard and soft strata in a valley, it is usually easier for the channel to follow the strike of the less resistant strata, rather than to cut down into the harder rock. This may then induce a lateral shift in the channel. However, the mechanism by which uniclinal shifting occurs remains vague. Differential erosion, and rock permeability have been suggested as important factors, alongside the initial orientation of the channel. An example of uniclinal shifting is found in the Middle Thames Valley, England. Here an extensive flight of aggradation terraces exist on the northern side cascading southwards towards the London Basin syncline, yet there is almost nothing on the southern side that records the river's former route. This suggests the lateral 'uniclinal shifting' of the channel southwards, believed to have occurred in the Pleistocene (Bridgland 1985). Alternative terms to uniclinal shifting are down-dip shifting and 'down-dip migration'.

Reference

Bridgland, D.R. (1985) Uniclinal shifting: a speculative reappraisal based on terrace distribution in the London Basin, *Quaternary Newsletter* 47, 26–33.

STEVE WARD

UNIFORMITARIANISM

Uniformitarianism is a mode of thought which was conventionally defined as 'the present is the key to the past' (Geikie 1905). The names of James Hutton and Charles Lyell are permanently associated with having brought uniformitarian thinking into the main stream of geology, in opposition to the earlier catastrophist thinking, associated with names like Georges Cuvier and Abraham Werner. Prevailing schools of thought in the late eighteenth century had two main tenets: (a) the general belief that God has intervened in history, which therefore has included both natural and supernatural events and (b) the particular proposition that Earth history consists in the main of a sequence of major catastrophes, usually considered as of divine origin. Uniformitarianism as expressed by Hutton (1788) embodied two propositions that were contradictory to these catastrophist views: (a) Earth history can be explained in terms of natural forces still observable as acting today and (b) Earth history has not been a series of universal or quasi-universal catastrophes but has in the main been a long, gradual development. The most obvious example of the confrontation between catastrophists and uniformitarians at the time is provided by the contradictory views on the origin of valleys. Valleys gradually formed by rivers still eroding the valley bottoms were juxtaposed against valleys that had opened up as clefts through divinely controlled revolution. Although this dichotomy made much sense at the time (late eighteenth and early nineteenth century) (Gillispie 1960), there is considerable controversy over the use of the term uniformitarianism at present (Goodman 1967; Shea 1982; Schumm 1991). Hooykaas (1970) says that the use of uniformitarianism should be restricted to a view that states that geological forces of the past differ neither in kind nor in energy from those now in operation. The past should be reconstructed on the assumption that all geological causes of the past were of the same kind and intensity as those of the present.

Gould (1967) subdivided the confusing issues surrounding the definition of uniformitarianism into two components:

(a) a substantive uniformitarianism, which postulates uniformity of kinds and rates of processes (Hooykaas 1970) and
(b) a methodological uniformitarianism comprising a set of two procedural assumptions which are basic to historical enquiry in any empirical science: the principle of the uniformity of natural laws and the principle of simplicity.

Because (a) cannot possibly be true, and was not what Lyell had in mind (Kennedy 2000) when he popularized the term, and because (b) is standard procedure for historical science, the substantive content of the term is superfluous.

Shea (1982) identified twelve fallacies associated with the use of the term uniformitarianism: that uniformitarianism (a) is unique to geology; (b) was first conceived by James Hutton; (c) was named by Charles Lyell, who established its definitive modern meaning; (d) should be called actualism because it refers to the actual or real events and processes of Earth history; (e) holds that only currently acting processes operated during geologic time; (f) holds that the rates or intensities of processes are constant through time; (g) holds that only gradual, non catastrophic processes have occurred during Earth's history; (h) holds that conditions on Earth have changed little through geologic time; (i) holds that Earth is very old; (j) is a theory or hypothesis and can be tested; (k) applies only as far back in history as present conditions existed and only to Earth's surface or crust; (l) holds that the laws governing nature are constant through space and time. He recommends abandoning the term.

Kennedy's definition is, in the last analysis, the most satisfying

> uniformitarianism is a practical tenet held by all modern sciences concerning the way in which we should choose between competing explanations of phenomena. It rests on the principle that the choice should be the simplest explanation which is consistent both with the evidence and with the known or inferred operation of scientific laws. Uniformitarianism is therefore applicable to both historical inference and to prediction of the future outcome of the operation of natural processes (Goodman 1967).
>
> (Kennedy 2000: 502)

References

Geikie, A. (1905) *Founders of Geology*, London: Macmillan.
Gillispie, C.C. (1960) *The Edge of Objectivity: An Essay in the History of Ideas*, Princeton: Princeton University Press.
Goodman, N. (1967) Uniformity and simplicity, *Geological Society of America Special Paper* 89, 93–9.
Gould, S.J. (1967) Is uniformitarianism useful? *Journal of Geological Education* 15, 149–150.
Hooykaas, R. (1970) *Catastrophism in Geology: Its Scientific Character in Relation to Actualism and Uniformitarianism*, Amsterdam: North-Holland Publishing.
Hutton, J. (1788) Theory of the Earth, *Royal Society of Edinburgh Transactions* 1, 209–304.
Kennedy, B.A. (2000) Uniformitarianism, in D.S.G. Thomas and A. Goudie (eds) *The Dictionary of Physical Geography*, 502–504, Oxford: Blackwell.
Schumm, S.A. (1991) *To Interpret the Earth: Ten Ways to be Wrong*, Cambridge: Cambridge University Press.
Shea, J.H. (1982) Twelve fallacies of uniformitarianism, *Geology* 10, 455–460.

Further reading

Simpson, G.G. (1963) Historical science, in C.C. Albritton (ed.) *The Fabric of Geology*, 24–48, Stanford, CA: Freeman, Cooper.

SEE ALSO: actualism; catastrophism; neocatastrophism

OLAV SLAYMAKER

UNIVERSAL SOIL LOSS EQUATION

The Universal Soil Loss Equation is a method for estimating annual SOIL EROSION on the basis of soil loss from a field or hillslope. It was empirically derived from data collected over a twenty-year period from runoff plots at experimental stations established in the 1930s in the United States by the Soil Conservation Service under H.H. Bennett. The object was to measure soil erosion rates under natural rainfall on different soils, slope conditions, cropping and tillage practices, as a basis for SOIL CONSERVATION recommendations. Eventually data were available for twenty-three soils between the Rocky Mountains and the US east coast. Continuing attempts to develop a reliable equation to predict soil erosion culminated in the USLE in 1958 (Wischmeier *et al.* 1958).

The metric version of the equation is:

$$E = R \cdot K \cdot L \cdot S \cdot C \cdot P$$

where E is mean annual soil loss ($t\ ha^{-1}$), R is annual rainfall erosivity ($10^7 J ha^{-1}$), K is soil erodibility (relative to a control soil without vegetation cover), L is slope length (relative to a standard slope length of 22.6 m), S is slope gradient (relative to a standard 9 per cent slope), C is crop management (relative to a cultivated bare field), and P is a conservation practices factor (relative to a bare surface without conservation measures).

The most complex and critical factor is annual rainfall EROSIVITY, based on regression analysis of rainfall characteristics to determine those most strongly correlated with soil loss from the runoff plots. The most effective measure is a composite measure involving the total kinetic energy (E; $J\ m^2$) during a rainstorm and the maximum rainfall intensity recorded over a 30-minute period during the storm (I_{30}; $mm\ h^{-1}$). Annual rainfall erosivity is the sum of EI_{30} for all storms during a year, divided by 1,000. Calculations should be based on records spanning at least twenty-five years, but there are not many locations where such long-term records of rainfall intensity exist. Available data show the highest values from humid tropical areas like the Gold Coast of West Africa, where erosivity exceeds 1,700 (Roose 1977), and the lowest values in temperate and arid regions.

Other data were obtained by direct measurement at the original research stations and extrapolation procedures were proposed for areas without complete data. These are described in detail in *Agricultural Handbook 282* (Wischmeier and Smith 1965). This includes a nomograph for estimating soil erodibility (K) from soil texture, structure and organic content, graphical solutions for combined slope length and gradient (SL), and values for 128 crop combinations and cropping practices (C). The C factor can be subdivided for variations in cover protection through the cropping cycle. Finally, values of the conservation factor (P) are provided, based on tests at the experimental stations comparing techniques such as contour ploughing or terracing.

The USLE was designed as a conservation guide for farmers to estimate erosion hazard, to identify the most significant contributing factors and to predict the potential reduction in soil loss from introduction of conservation practices. It has been used widely in the United States, and has been effective in the area where original data are available. Subsequent modifications have been incorporated as understanding of soil erosion processes has increased, including, for example, a seasonally adjusted K factor, to reflect changes in soil structure caused by rainfall and weathering.

The modified equation was published in 1991 as the R (Revised) USLE.

The USLE has been useful for soil conservation in the central USA, and as an aid in instruction about the factors which control soil erosion. Unfortunately, its conceptual simplicity has encouraged use in areas for which it was not designed, such as steeply sloping forested lands, or in areas where inadequate rainfall records are available. It is purely empirical, with no proven physical foundation for extrapolation, and it can lead to extremely inaccurate predictions. Attempts have been made to accumulate appropriate data for wider use particularly in the tropics. Where measured data are available it can be used with some confidence, but elsewhere it is not reliable, particularly as a basis for expensive, socially disruptive or contentious measures. It has also been criticized on theoretical grounds as understanding of soil erosion processes has increased. The K factor, for example, is very simplistic, with no attention to important processes such as surface sealing, or to the effect of soil chemistry on erosion resistance. It certainly cannot be used with confidence to predict important soil erosion effects such as contaminant transport, or nutrient enrichment in lakes or streams. Much research has been directed to development of a sound physically based erosion equation, e.g. WEPP (Water Erosion Prediction Project) in the United States and EUROSEM (European Soil Erosion Model) in Europe. Although promising, these are conceptually complex, require data which are often unavailable, and are not yet sufficiently reliable for widespread use.

References

Roose, E.J. (1977) Application of the Universal Soil Loss Equation of Wischmeier and Smith in West Africa, in D.J. Greenland and R. Lal (eds) *Soil Conservation and Management in the Humid Tropics*, 177–187, London: Wiley.

Wischmeier, W.H. and Smith, D.D. (1965) Predicting rainfall-soil erosion losses from cropland east of the Rocky Mountains, *Agricultural Handbook 282*, Agricultural Research Service, United States Department of Agriculture.

Wischmeier, W.H., Smith, D.D. and Uhland, R.E. (1958) Evaluation of factors in the soil-loss equation, *Agricultural Engineering* 39, 458–462, 474.

Further reading

Hudson, N.W. (1981) *Soil Conservation*, London: Batsford.

Morgan, R.P.C. (1995) *Soil Erosion and Conservation*, London: Longman.

RORKE BRYAN

UNLOADING

'The removal by denudation of overlying material' (Yatsu 1988: 140) was suggested by Gilbert (1904) as a mechanism that produced exfoliation domes in the granites of the Sierra Nevadas, western USA. According to this theory, granites are buried under a deep cover of older rock and so are subjected to compressive stress. That compressive strength was balanced by internal expansive stress competent to cause actual expansion if the external pressure was removed by denudation. This in turn may produce sheeting (EXFOLIATION) that is broadly conformable to topography. However, considerable controversy surrounds the issue of sheeting – does the sheeting produce the topography or is it *vice versa* as the unloading model suggests.

In addition to unloading being a potential cause of joint development, the term unloading is also used to describe the process of weight removal from the crust. *Glacial unloading*, resulting from the wastage of ice caps leads to postglacial faulting and seismicity, and to isostatic compensation. Equally, *erosional* or *denudational unloading* is an important factor in determing uplift and erosion in situations like passive margins (Clift and Lorenzo 1999). *Mechanical unloading*, associated with lithospheric extension, can contribute to flexural uplift of rift flanks (Weissel and Karner 1989).

References

Clift, P.D. and Lorenzo, J.M. (1999) Flexural uploading and uplift along the Côte d'Ivoire–Ghana transform margin, Equatorial Atlantic, *Journal of Geophysical Research* B, 104, 25,257–25,274.

Gilbert, G.K. (1904) Domes and dome structure of the high Sierra, *Geological Society of America Bulletin* 15, 29–36.

Weissel, J.K. and Karner, G.D. (1989) Flexural uplift of rift flanks due to mechanical unloading of the lithosphere during extension, *Journal of Geophysical Research* B, 94, 13,919–13,950.

Yatsu, E. (1988) *The Nature of Weathering*, Tokyo: Sozosha.

A.S. GOUDIE

(URANIUM-THORIUM)/HELIUM ANALYSIS

(Uranium-thorium)/helium analysis in apatite is currently the lowest temperature thermochronometer for providing detailed information on the thermal history of the crust for low (shallow) crustal temperatures. The technique is based on the alpha decay of U and Th in apatite to yield the daughter ^4He. (U-Th)/He analysis was the first radiometric dating system developed, by Ernest Rutherford at the beginning of the twentieth century. Independent geological evidence indicated that the ages it returned in its early applications were generally too young, and so the technique was abandoned as a dating tool. It was realized in the 1980s that (U-Th)/He analysis generally returned ages that are too young because the daughter ^4He diffuses out of the apatite above *c.*80 °C, which is well below the 'formation' temperatures that analysts were attempting to date in the technique's early applications. This behaviour means, however, that the (U-Th)/He system can be used to date rock cooling below *c.*80 °C.

The analytical procedures for (U-Th)/He in apatite are relatively straightforward, involving heating the apatite grains to drive off the daughter ^4He for measurement, followed by dissolution of the grains to measure the amounts of the parent U and Th using an Inducting Coupled Plasma Mass Spectrometer (ICPMS). The standard age equation for radioactive decay of parent element(s) to daughter element is then applied. Sample preparation is also relatively straightforward via standard mineral separation procedures, except for the final stage involving microscope handpicking of apatite grains for the analyses. This careful selection of grains is necessary to avoid the analysis of apatites with inclusions of U-bearing minerals, such as zircon. These U-bearing inclusions generate ^4He in the apatite but the inclusions' resistance to dissolution may mean that their corresponding U and Th contents remain unmeasured, thereby confounding the age calculation. A correction to the calculated age must be applied to account for the loss of ^4He from the outer edges of the grain by recoil (i.e. complete expulsion of ^4He from the grain during the alpha decay). This so-called 'recoil correction' requires that grains being analysed be of standard shape and of known size, which is a further reason for careful handpicking and characterization of the crystals to be analysed.

In the same way as there is a temperature range in which fission tracks are only partially retained (the partial annealing zone in FISSION TRACK ANALYSIS), the temperature range between *c.*80 °C and 40 °C defines the ^4He partial retention zone (PRZ). All of the daughter ^4He diffuses out of the apatite grain above *c.*75 °C, but at temperatures cooler than this the He is increasingly retained. Partial retention is grain-size dependent, providing the potential in (U-Th)/He analysis for an analytical tool comparable to the track-length distribution in fission track analysis. The potential of this grain-size effect is still to be fully explored.

Geomorphological applications of (U-Th)/He analysis largely focus on the rates of denudation required to bring apatite-bearing rocks to the Earth's surface from the crustal depths corresponding to a temperature of *c.*80 °C, at which the ^4He daughter product of alpha decay of U and Th begins to be retained. This denudation may be regional continental denudation required to bring the apatite to the Earth's surface (with the only 'tectonic' component being passive denudational isostatic rebound) or the denudation may be driven by tectonic rock uplift. In the latter case, and on the assumption that cooling through the PRZ is coeval with the tectonic uplift, the (U-Th)/He age corresponds to the age of onset of tectonic uplift. The assumption that denudation is coeval with uplift is extremely important. The assumption is reasonable in settings in which agents of subaerial incision, such as fluvial and glacial processes, are efficiently connected to the 'external' reference plane for tectonic uplift (e.g. a local base level or global sea level), and uplift-driven disequilibrium in the drainage net, due to relative lowering of base level, is rapidly transmitted through the drainage net to trigger incision and denudation throughout the catchment. Not all high elevation uplifting areas are well connected to the external base level which is the reference plane for surface uplift (e.g. the Tibet plateau, the Andes Altiplano). In these cases, the low-temperature thermochronological ages of rocks now at the surface of these landscapes may bear little relationship to the onset of uplift.

A further geomorphological application of low-temperature thermochronology, especially (U-Th)/He thermochronology in apatite, relies on the fact that the thermal structure of the shallow crust (upper few kilometres) is deformed by the long-wavelength Earth surface topography. The shallow crustal isotherms mirror topography of length scales of tens of kilometres. Thus the thermal structure of the crust beneath long-lived major valleys, for example, mirrors these valleys. If (U-Th)/He ages along a transect at constant elevation across these valleys (at a mountain front, for example) mimic the topography then the topography must have existed when the apatites were passing through the PRZ. That is, if the (U-Th)/He ages are older on that part of the transect coinciding with the interfluves and younger where the transect coincides with the valleys, the (U-Th)/He ages effectively provide minimum ages for the long wavelength topography.

PAUL BISHOP

URBAN GEOMORPHOLOGY

Urban geomorphology examines the geomorphic constraints on urban development (Cooke 1984) and the suitability of different landforms for specific urban uses; the impacts of urban activities on Earth surface processes, especially during construction; the landforms created by urbanization, including land reclamation and waste disposal; and the geomorphic consequences of the extractive industries in and around urban areas (McCall *et al.* 1996).

Constraints on urban development

The original founders of towns and cities carefully chose their sites for defensive, strategic, resource exploitation, navigation or cultural reasons. Great attention was given to finding sites with adequate water supplies and protection from obvious environmental hazards. However, the growth of these settlements often led to the spread of urban development on to less suitable ground and overstretched the capacity of the local environment to support the community. Many environments have particular conditions that make conditions for modifying slopes or establishing foundations difficult (Table 46).

Application of geomorphological mapping to the classification of the suitability of land for different types of urban development is now part of the work of geological and soil surveys in many countries. Such mapping considers the steepness of slopes, their colluvial and weathered mantles, their drainage and depth to bedrock and provides guidance on the type of development suited to different parts of the slope.

Knowledge of landform evolution is particularly important, as modern earthmoving can reactivate features inherited from past conditions, such as fossil periglacial landslides in Europe and North America. Loading of peat with urban structures formed after the retreat of ice sheets can result in significant subsidence and building damage. Karstic features formed when sea levels were lower in the Quaternary, but now buried under alluvium, can pose severe problems for the foundations of high-rise buildings.

Many present-day conditions constrain urban development. Widespread soils rich in montmorillonitic clays are subject to 'shrink–swell' cracking clay phenomena which require special foundations if buildings are not to become unstable. Climate change is likely to shift the areas where these problems are severe, if summers become drier. Mobile dune systems and sources of wind-blown sand pose problems for the siting of many structures. ALLUVIAL FANs may normally be inactive with the local stream confined to a narrow channel, but they may be reactivated, flooded and covered with debris if an extreme flood descends from the adjacent mountains. Building in PERMAFROST areas has to isolate the heated structures from the frozen ground and be careful not to disturb the permafrost during the construction process.

Geomorphic impacts during urban construction

Urban construction involves removal of the natural vegetation cover and excavation of the topsoil and often much of the underlying weathered rock and bedrock layers. In new urban developments, small streams are often diverted into culverts or urban drains and minor depressions and valleys are filled in. Steep hillsides may be terraced into a series of home sites by cut-and-fill operations. Major rivers may be embanked and artificially straightened. In extreme cases, as in Palma de Mallorca, Spain and Winnipeg, Canada, large new flood channels may be built around the urban centre to divert flows away from the city. The new features replacing the original landforms are often designed to direct water away from the new developments more effectively, so producing off-site, downstream consequences.

The earthmoving operations during urban construction frequently lead to severe erosion problems and consequent channel modifications

Table 46 Geomorphological problems for urban development

Environment	Chief problems
A Climatic	
Periglacial	Permanently frozen ground and overlying active layer require special types of construction and foundations for buildings and infrastructure
Arid	Water supply problems; wind erosion; flash floods; possibility of salt weathering of building materials and foundations
Humid tropical	Rapid weathering and decomposition of building materials; deep, uneven weathering of most rocks in tectonically stable areas; frequent rain events causing rapid water erosion of exposed ground surfaces
B Topographic	
Mountainous	Risk of unstable slopes, rockfalls, debris flows and avalanches; potential for flash floods
Floodplains	Liable to periodic flooding; variable foundation conditions over former, buried river channels and alluvial deposits
Coastal plains	Storm surge and flooding risk likely to increase with rising sea levels; complex ground conditions reflecting former shorelines and old drainage channels; possible salt penetration in ground water affecting foundations
Coasts with weak rock cliffs	Liable to rapid coastal erosion, cliff undercutting and collapse; eroded debris often deposited in ports and harbours causing dredging expenditure
Islands	Particular storm-surge, rising sea level and salt water penetration risks on low-lying atolls and coastal plains
C Tectonic/lithological	
Active plate margins	Major risks associated with coastal urban developments, especially on Pacific rim, special foundation requirements on filled areas, lake sediments and other unconsolidated materials; major earthquake-triggered landslide hazards; volcanic debris and lahar risks requiring awareness of flow pathways on lower volcanic slopes likely to have urban settlements
Shrink-swell clays	Cracking clay problems likely to be accentuated by climate change
Karst	Buried karst a major problem for foundations of tall buildings and for sinkhole development; need for knowledge of buried karst plains and effects of lowered Quaternary sea levels

Sources: Based on data in Marker (1996); McCall *et al.* (1996); and Bennett and Doyle (1997)

(Table 47, Figure 173). Erosion control guidelines suggest that construction should be carried out in phases to avoid disturbing too much of the land at any one time. No unnecessary clearing should be undertaken. Immediately below any cleared area, detention ponds should be constructed to retain any sediment washed off the site and to hold back stormwater runoff so that peak discharges in streams below are not increased.

Increased sediment loads and peak storm discharges lead to channel modifications (Figure 173) with formerly meandering streams becoming braided, steeper and shallower. Sometimes these changes are controlled by modifying the channel, often with expensive structural works. However, even these are not always successful as siltation of the channel can follow, with large accumulations of weeds and silt building up if the stream receives discharges with high nutrient loadings. Further downstream, rivers may adjust in response to upstream channelization, eroding their banks, developing new gravel bars and threatening bridge abutments and riverine structures.

Landforms of extraction

Meeting the demand for construction materials changes the land surface, by creating pits and

Before construction – meandering channel

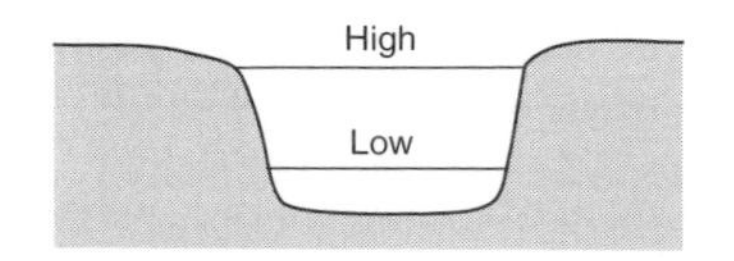

Design channel, stonework encased in concrete

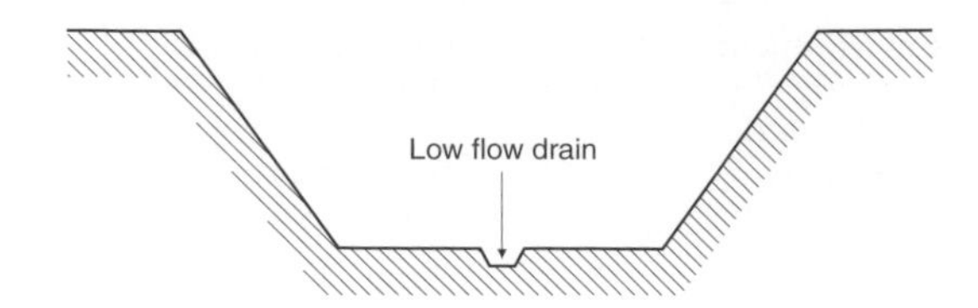

After land clearance and during construction – braided channel

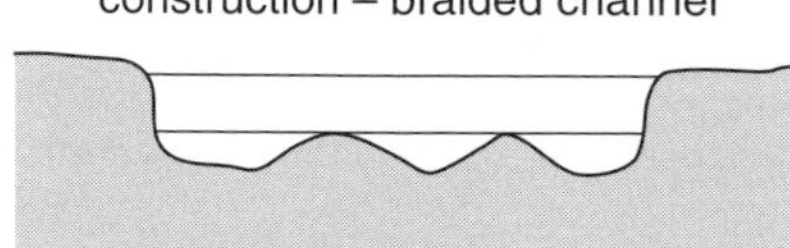

Design channel, after two years or more

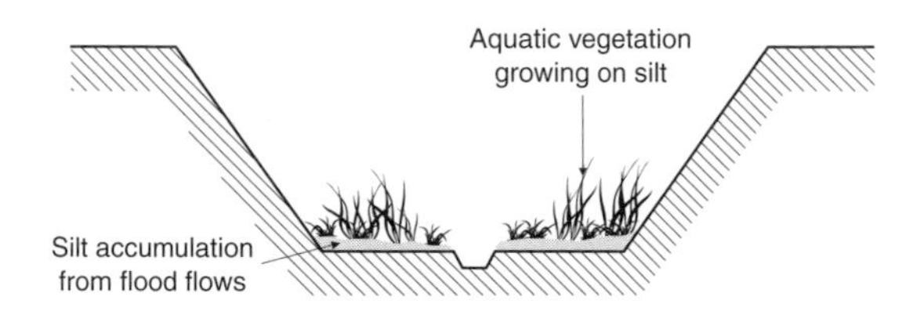

Figure 173 Channel changes to the Sungai Anak Ayer Batu at Jalan Damansara between 1960 and 1990 due to urban development

Table 47 Sequence of fluvial geomorphic response to land use change: Sungai Anak Ayer Batu, Kuala Lumpur

Land cover/land use	Channel condition
Forest	Narrow, meandering with low sediment load
Rubber plantation	Gullying during clear weeding; peak discharge increased; channel slightly widened; later stabilized; few cut-offs
Urban construction	High sediment yield; high peak discharge; metamorphosis to wider, steeper, shallower braided channel
Channelization and stable urban built-up area	Higher peak discharge; less sediment load; channel enlargement downstream; bank erosion, minor channel incision; loss of fine bed material by scour
Post channelization siltation	Where large quantities or organic debris enter concrete channels and are deposited, vegetation can become established and build up deposits that reduce channel capacity

quarries. The largest excavations take up many square kilometres of the land surface, often creating areas of brick pits that sometimes are used for waste disposal, or gravel pits that become peri-urban wetlands, often serving combined recreational and flood control purposes. Not all the filling of former opencut mines is without incident. In the past, escaping methane gas from landfills in old opencast coal pits has caused problems for the houses built upon them. In karstic terrain, the SUBSIDENCE of filled material in chalk pits or in old tin mines overlying cavernous limestone has led to severe damage to houses and urban infrastructure.

Removal of mineral resources and water from beneath the ground leads to subsidence creating new surface topographies and, often, new water bodies. Groundwater pumping has put the historic world heritage buildings of Venice at risk. Built at sea level on the lagoon, Venice has subsided some 22 cm since 1900. Most of that surface lowering occurred between 1950 and 1970. High water (*aqua alta*) has occurred more frequently since 1970. Whilst the people-induced subsidence is part of the problem, higher extreme sea levels related to global climate change may possibly be another factor. In the Los Angeles area, extraction of oil beneath Long Beach created severe subsidence that had to be halted by the injection of water into abandoned wells.

Landforms of deposition

Much modern urban development involves land reclamation and major landform modification (Gupta and Ahmad 2000). In extreme cases, huge quantities of material are moved, for example in the development of major airport sites such as Kansai, Singapore and Hong Kong. At Kansai, the fill material has caused some subsidence of the original seabed, with allowance having to be made for this in the operation and maintenance of the airport. The problems of subsidence of the second-stage runway are expected to be more severe than in the first runway, with a prediction that after fifty years subsidence will have been 18 m compared to 11 m for the first stage. Detailed analyses have been made of the way landing aircraft cause small temporary depressions in the runway that in turn affect the drag on aircraft moving along the runway.

As disposal of solid waste moves from landfill to land raise, new hills appear on the edges of floodplains, above former gravel pits and quarries and on offshore islands. In some urban areas, waste dumps are prominent features of the landscape. Although the older dumps are the result of coal, slate and china clay production, modern land raise features dominate many low relief areas. Whilst much of this waste is deposited in disused opencast mines and quarries, land raise mounds are probably the fastest growing artificial landforms in many countries today. The greatest geomorphological impact of landfill is in river valleys, sections of which are being filled, raising the height of the ground surface well above the former floodplain level. This effectively reduces the flood storage capacity of the floodplain, shifting the flood problems downstream.

Many of the old dumps are being closed or modified, from the huge dumps on the edges of cities like Istanbul and Manila to the managed disposal areas, such as Freshkills on Staten Island, New York, which has been taking nearly all the 17,000 tons of waste the city collects each day. As events at the Payatas tip in Manila have shown, some of these urban waste mounds are unstable, prone to massive slumps and landslides. The loss of life and property that ensues is a challenge to the management of the waste disposal and the application of geomorphology to the construction of land raise mounds.

Urban regeneration itself involves creating new landforms as the old buildings are demolished and construction and demolition waste is used for on-site fill or is taken short distances to sites that have to be raised to be above known flood levels. Many historic city centres have been so rebuilt that the average level of streets is now above that of the entrances to medieval buildings. These changes in landform may often be individually small, but collectively they are the result of two of the main human drivers of global environmental change: increasing urban development and the mining and quarrying which supplies minerals for industry and the construction materials required to build all the infrastructure, homes, offices and factories of cities. Urban geomorphology is thus a key element in supplying the guidance needed to achieve a better quality of urban life and working towards more sustainable use of resources.

References

Cooke, R.U. (1984) *Geomorphological Hazards in Los Angeles*, London: George Allen and Unwin.

Bennett, M.R. and Doyle, P. (1997) *Environmental Geology: Geology and the Human Environment*, Chichester: Wiley.

Gupta, A. and Ahmad, R. (2000) Geomorphology and the urban tropics: building an interface between research and usage, *Geomorphology* 31, 133–149.

McCall, G.J.H., De Mulder, E.F.J. and Marker, B.R. (eds) (1996) *Urban Geoscience*, Rotterdam: Balkema.

Marker, B.R. (1996) The role of the earth sciences in addressing urban resources and constraints, in G.J.H. McCall, E.F.J. De Mulder and B.R. Marker (eds) *Urban Geoscience*, 163–179, Rotterdam: Balkema.

Further reading

Coates, D.R. (ed.) (1976) *Urban Geomorphology*, Geological Society of America Special Paper 174.

Cooke, R.U., Brunsden, D., Doornkamp, J.C. and Jones, D.K.C. (1982) *Urban Geomorphology in Drylands*, Oxford: Oxford University Press.

Douglas, I. (1983) *The Urban Environment*, London: Arnold.
Leggett, R.F. (1973) *Cities and Geology*, New York: McGraw-Hill.

IAN DOUGLAS

URSTROMTÄLER

Broad (2–25 km wide), pronounced depressions which are aligned parallel with the margins of the Pleistocene ice sheets in Europe and North America; also called *ice marginal valleys, ice-marginal streamways*. They were first studied in North Germany (Wolsted 1950). These forms can be traced across the North European Plain from Russia to the North Sea. Marshy, flat-floored trenches usually have steep erosive external scarps, 20–40 m high and, often, systems of lateral terraces (Galon 1961). Their courses relate to particular stages of the Scandinavian ice sheet limits. Their origin is usually explained as the product of both erosion and sedimentation by the conjoined waters from proglacial *meltwater channels* (see MELTWATER AND MELTWATER CHANNEL) and rivers which drained the ice-free areas in the south. Almost certainly, large volumes of water must be involved in their production; at least some might be derived from catastrophic outbursts from ice-dammed lakes. Jahn (1975) considered that intense thermal erosion of riverbanks under permafrost conditions was also a generic factor.

References

Galon, R. (1961) *Morphology of the Notec-Warta (or Torun-Eberswalde) Ice Marginal Streamway*, Warszawa: Prace Geograficzne PAN, No. 29.
Jahn, A. (1975) *Problems of the Periglacial Zone*, Warszawa: PWN-Polish Scientific Publishers.
Woldsted, P. (1950) *Norddeutschland und angrenzende Gebiete im Eiszeitalter*, Stuttgart: Koehler Verlag.

JACEK JANIA

V

VALLEY

'A depression sloping in one direction over all its length' (Von Engeln 1942: 7), which tends to be longer than it is wide. Valleys have a range of sizes and a multiplicity of names – gully, draw, defile, ravine, gulch, hollow, run, arroyo, gorge, canyon, dell, glen, dale and vale (Huggett 2002: 193).

Valleys are normally regarded as the products of a range of fluvial processes such as corrasion, abrasion, pot-holing, cavitation, corrosion and weathering. They widen by lateral stream erosion and by weathering, mass movements and fluvial processes on their sides. They lengthen by such processes as HEADWARD EROSION or by building new land (e.g. RIVER DELTAs) at their bottom ends. In planform they develop networks (see HORTON'S LAWS) and they have a variety of DRAINAGE PATTERNs, including ALIGNED DRAINAGE. Some valleys develop in bedrock (see BEDROCK CHANNEL) while others develop in superficial materials, such as alluvium. In general big rivers have big valleys, but there are cases where small, misfit rivers occupy large valleys. This can be due to river capture, which can divert large amounts of water from that valley into another river system. Alternatively, it can be due to major climatic changes that have decreased the flows of water through the valley meander systems (Dury 1997) while some valleys may have no channels in them at all (see DRY VALLEY). Many valleys are accordant to geological structures, while others are discordant as a result of antecedence or superimposition.

There is a huge diversity to valley forms (see ARROYO; BEHEADED VALLEY; BLIND VALLEY; BOX VALLEY; BURIED VALLEY; CANYON; DAMBO; DELL; MEKGACHA; TUNNEL VALLEY; WADI). While most valleys are of subaerial type, there are also SUBMARINE VALLEYs. Some valleys have regular cross sections, while others, for structural or aspect-related microclimatic reasons, display asymmetry. Equally, normal fluvial valleys are often perceived as having a tendency towards V-shaped cross profiles (though this is far from universal), while glacially excavated valleys are often perceived as giving U-shaped cross profiles with truncated spurs.

The origin of valleys has proved troublesome. In the early nineteenth century they were sometimes regarded as the result of the Noachian deluge (see DILUVUALISM). There was also a great deal of debate as to what extent they were essentially tectonic features, related to fracturing of the Earth's crust. It was not easily recognized that they were the result of rain and rivers. These debates are well reviewed by Chorley *et al.* (1964). However, as Kennedy (1997: 67) has pointed out,

> *Any* process which creates topographic irregularities will cause the subsequent concentration of any available surface moisture and, potentially, a stream-and-valley. Moreover, since valleys are exceptionally durable features . . . we must face the fact that many networks will contain portions which owe both their ultimate origin and also their persistence to *different* processes.

References

Chorley, R.J., Dunn, A.J. and Beckinsale, R.P. (1964) *The History of the Study of Landforms or the Development of Geomorphology*, Vol. 1, London: Methuen.

Dury, G.H. (1997) The underfit meander problem. Loose ends, in D.R. Stoddart (ed.) *Process and Form in Geomorphology*, 46–59, London: Routledge.

Huggett, R.J. (2002) *Fundamentals of Geomorphology*, London: Routledge.

Kennedy, B.A. (1997) The trouble with valleys, in D.R. Stoddart (ed.) *Process and Form in Geomorphology*, 60–73, London: Routledge.

Von Engeln, O.D. (1942) *Geomorphology: Systematic and Regional*, New York: Macmillan.

A.S. GOUDIE

VALLEY MEANDER

Meanders which are usually cut in bedrock and usually have a greater wavelength than that of the contemporary river pattern. These meanders shape valleys winding rather symmetrically between hills, and they are much wider than the meanders of the river flowing in the alluvial plain or alluvial meanders. The two types of meanders tend to be geometrically similar, the only real difference being that meanders in bedrock are commonly ingrown, whereas meanders on a floodplain are not. Entrenched meanders, which have cut vertically down without enlarging themselves in the lateral and axial directions, are uncommon. They represent one end of a series which extends through the intermediate range of normally ingrown meanders to meanders on a floodplain.

Davis (1906) described bedrock meanders in relation to lateral corrasion and placed their origin during the youth stage of the cycle of erosion. Dury (1954, 1977) thoroughly studied meandering valleys in order to ascertain whether they were cut by a stream larger than the present-day one. From many examples, he showed that valley meanders were produced during periods of higher runoff and higher peak discharges, that is, before stream shrinkage which led to contemporary UNDERFIT STREAMS, and that these larger discharges in past times were associated with Pleistocene climate.

The relation of meander length to valley width in valley meanders in rock shows more scatter compared to those in alluvium, but it is apparent that the length is directly proportional to the channel width in both cases. In the rock meanders wavelength is 15 to 20 times valley width. On the other hand, study of individual bends of valley meanders suggests that differences in geologic structure and lithology lead to differences in wavelength of meanders in rock.

Because of the difficulty of visualizing how a channel could maintain a regular sinuous pattern while cutting across hard-rock strata, it has often been assumed that the meandering pattern was initiated in an overlying sedimentary cover and superimposed on the tougher rock below as the river entrenched itself into the strata. Very often these presumed overlying strata have been eroded away, and hence the hypothesis is difficult to verify. In most cases there appears to be no need for such a two-cycle hypothesis. Other meanders in bedrock suggest that the river was antecedent to uplift. That is, the river appears to have maintained its course, trenching the structure as the latter formed. In the absence of stratigraphic evidence it is impossible to distinguish between an antecedent river and one which was superimposed from an overlying cover.

References

Davis, W.M. (1906) Incised meandering valleys, *Bulletin of the Geological Society of Philadelphia* 4, 182–192.

Dury, G.H. (1954) Contribution to a general theory of meandering valleys, *American Journal of Science*, 252, 193–224.

——(1977) Underfit streams: retrospect, perspect and prospect, in K.J. Gregory (ed). *River Channel Changes*, 281–293, Chichester: Wiley.

Further reading

Leopold, L.B., Wolman, M.A. and Miller, J.P. (1964) *Fluvial Processes in Geomorphology*, San Francisco: W. H. Freeman.

SEE ALSO: underfit stream

MARIA SALA

VASQUE

Limestone and AEOLIANITE SHORE PLATFORMS in Mediterranean and tropical regions commonly consist of a series of low terraces formed by wide, flat-bottomed pools or vasques. The pools are separated from each other by narrow, winding ridges that can be: built by calcareous algae, Vermetids (see VERMETID REEF AND BOILER), or even Serpulids (see SERPULID REEF); the residual CORROSIONal pinnacles of lapiés; or a combination of the two. Plates-formes à vasques are covered at high tide and washed by breaking waves at low tide, with the return flow cascading into successively lower pools (Plate 143).

Plate 143 Rimmed pools, vasques, developed in eroded aeolianites at Treasure Beach near Durban, South Africa

Further reading

Trenhaile, A.S. (1987) *The Geomorphology of Rock Coasts*, Oxford: Oxford University Press.

ALAN TRENHAILE

VENTIFACT

A term introduced by Evans (1911) to describe wind-faceted stones, the surfaces of which are flattened such that they intersect at sharp angles. They include the brazil-nut shaped 'Dreikanter' of German workers. For their formation three conditions are required: strong, generally unidirectional winds; the presence of loose materials (sand, dust, snow, etc.) that are available for transport in suspension or saltation; and the presence of pebbles, boulders and bedrock outcrops projecting into the wind stream. However, there has been considerable debate as to the relative importance of abrasion by dust and by sand (see Breed *et al.* 1997 for a review) and to the precise mechanisms that produce flattened surfaces on three or more sides of many ventifacts.

Ventifacts have been noted on a wide range of lithologies, including basalt, granite, dolerite, aplite, andesite, chert, marble, dolomite and limestone. They occur in a range of exposed environments, including deserts, periglacial and coastal settings. They also occur on Mars (Bridges *et al.* 1999). Some ventifacts are relicts of former tundra conditions subsequent to ice recession but before vegetation establishment in the Late Pleistocene. Such ventifacts have been used to infer palaeowind directions (Schlyter 1995), with strong easterly winds being present in Denmark and southern Sweden. In Ireland, some coastal ventifacts may have formed during the Little Ice Age of the Late Holocene, when there were increased offshore winds, waves, sediment fluxes and periods of sand dune construction (Knight and Burningham 2001).

References

Breed, C.S., McCauley, J.F., Whitney, M.F., Tchakerian, V.P. and Laity, J.E. (1997) Wind erosion in drylands, in D.S.G. Thomas (ed.) *Arid Zone Geomorphology: Process, Form and Change in Drylands*, 437–464, Chichester: Wiley.

Bridges, N.T., Greeley, R., Haldemann, A.F.C., Herkenhoff, K.E., Kraft, M., Parker, T.J. and Ward, A.W. (1999) Ventifacts at the Pathfinder landing site, *Journal of Geophysical Research* 104(E), 8,595–8,615.

Evans, J.W. (1911) Dreikanter. *Geological Magazine* 8, 334–345.

Knight, J. and Burningham, H. (2001) Formation of bedrock-cut ventifacts and Late Holocene coastal zone evolution, County Donegal, Ireland, *Journal of Geology* 109, 647–660.

Schlyter, P. (1995) Ventifacts as palaeo-wind indicators in southern Scandinavia, *Permafrost and Periglacial Processes* 6, 207–219.

A.S. GOUDIE

VERMETID REEF AND BOILER

CORNICHES and other organic REEFS in the northwestern Mediterranean usually consist of calcareous algae and Serpulid (see SERPULID REEF) worms. Higher temperatures in the southern Mediterranean are favourable for large Vermetid populations, but whereas they only form veneers on TROTTOIRS in fairly easily eroded limestones and sandstones, there are purely constructional Vermetid corniches on fairly resistant substrates. Vermetids also contribute to the development of boilers or cup reefs in the Mediterranean and western Atlantic. Boilers, which are awash at low tide, are up to 12 m in height and a few tens of metres in diameter. They resemble MICROATOLLS with a central depression or micro-lagoon, up to a few metres in depth, surrounded by raised rims. Boilers in Bermuda consist entirely of coralline algae, Vermetid gastropods and the encrusting coral *Millepora*, but similar forms in the Mediterranean are merely veneers of Vermetids and algae over eroded AEOLIANITE blocks.

Further reading

Ginsburg, R.N. and Schroeder, J.H. (1973) Growth and submarine fossilization of algal cup reefs, Bermuda, *Sedimentology* 20, 575–614.
Trenhaile, A.S. (1987) *The Geomorphology of Rock Coasts*, Oxford: Oxford University Press.

ALAN TRENHAILE

VISOR, PLINTH AND GUTTER

CORROSIONal notches (see NOTCH, COASTAL) at the cliff foot sometimes have protruding visors above them and plinths below them. They have been described on AEOLIANITE SHORE PLATFORMS in southern Australia (Hills 1971), and from western Australia, Hawaii, Bermuda and northwestern India. The visor consists of a band of hardened, indurated rock, which may form when fresh rain water deposits calcium carbonate where it comes into contact with rock that is saturated with sea water. This might explain why the height of the visor declines as it is traced into sheltered areas, although it is questionable whether seawater saturation in the high and supratidal zones can be maintained during low tidal periods. The plinth is a slight prominence which is attached to the outer edge of the notch base. Hills proposed that the plinth develops at the height to which water is drawn by capillary action above the platform surface. The gutter, or moat, is a channel, occasionally found at the base of a ramp (see RAMP, COASTAL), that has been eroded by sand, pebbles and small boulders.

Further reading

Hills, E.S. (1971) A study of cliffy coastal profiles based on examples in Victoria, Australia, *Zeitschrift für Geomorphologie* 15, 137–180.

ALAN TRENHAILE

VOLCANIC KARST

Karst-like features occur in volcanic terrains and they have been classified into four types (Reffay 2001). These are shown in Table 48.

The type *pseudokarst sensu lato* consists of structural landforms in lava flows that are unrelated to denudational processes. They include lava tubes, lava speleothems, and pseudodolines and shafts generated by collapse of lava flow roofs. The type *pseudokarst sensu stricto* is created by piping processes in loose volcanic material (e.g. pyroclastic deposits). The type *orthokarst* develops as a result of dissolution of carbonatites. The type *parakarst*, develops as a result of dissolution of minerals other than carbonates.

Reference

Reffay, A. (2001) Types de karst en terrain volcanique: revue bibliographique, *Géomorphologie* 2001(2), 121–126.

A.S. GOUDIE

VOLCANO

A volcano can be defined as the site on the surface of a planet or moon through which gaseous, liquid and/or solid materials are expelled or erupted, usually through the action of internal thermal processes. Eruptions often discharge magma – molten igneous material composed of a silicate or other liquid, with variable quantities of crystalline phases and gas bubbles, though this can be transformed dramatically in violent, explosive eruptions into a stream of rock fragments and hot gases. An eruption can also result from steam explosions that blast out near-surface

Table 48 Volcanic karst classification

Pseudokarst, S.L.	Lava	Tunnels, speleothems, etc.	Structural. Related to lava emplacement
Pseudokarst; S.S.	Pyroclastics; Andosols	Pipes, holes, canyons, dry valleys	Suffosion
Orthokarst	Carbonatites	Closed depressions, megalapiés	Dissolution of carbonates
Parakarst	Basalt	Closed depressions, lapiés, speleothems, travertine and sinter	Dissolution of Ca, Mg, Na and silica. Precipitation of $CaCO_3$ and SiO_2

Source: Modified from Reffay (2001)

rocks without any accompanying fresh magma. The erupted products typically accumulate around the eruptive vent or vents, and can, if eruptions are sustained or repeated, construct mountains of very considerable volume. Olympus Mons, the highest volcano in the Solar System, rises some 24 km above the surrounding martian plains, and has a volume of around 3×10^6 km^3 (Plate 144). Volcanism, past and present, is one of the fundamental geological processes of the Solar System.

On Earth, volcanoes are broadly distinguished as either active or extinct. The term 'active' is used in different senses. Often it is used to indicate a volcano actually in eruption. However, it is also applied to all volcanoes known to have erupted in the Holocene period (last 10,000 yr). This broader definition obviously includes many volcanoes that have not erupted for centuries or even millennia, and which may actually be extinct (incapable of future eruption) but it helpfully covers many more volcanoes, which may experience long repose periods (the intervals between eruptions), and which can be considered dormant and capable of further eruption. Around 1,500 volcanoes are known or suspected to have erupted during the Holocene,

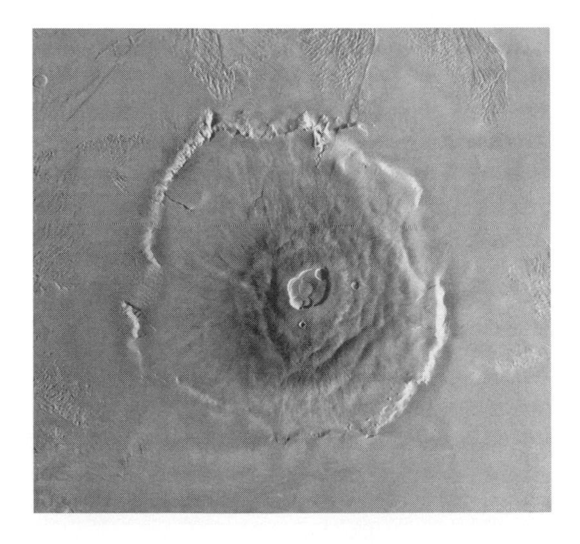

Plate 144 Olympus Mons – highest volcano in the Solar System. Note the overall shield shape, broad summit region crowned with nested calderas, and the prominent 550 km circumference, several km high basal scarp, whose origin has been the subject of much debate. Image processed by J. Swann, T. Becker and A. McEwen, and archived by NASA/NSSDC

and are therefore classed as active (Siebert and Simkin 2002). Of these, some 550 erupted in the historic period. Every year, an average of fifty to seventy volcanoes erupt, though some of these are volcanoes with long-lived eruptions spanning years or decades. On an average day, at least twenty of Earth's volcanoes will be erupting as you read this page. It is important to point out that all these figures are for volcanoes on land. Accurate figures for submarine eruptions are not available, although it is known that seafloor volcanism dwarfs the magma output of subaerial volcanism by a ratio of as much as ten-to-one.

Many dormant volcanoes discharge gases and liquids at the surface. In the case of hot springs, the flux of steam and gas is subordinate to that of liquid water. Geysers are spectacular examples of hot springs, Old Faithful in Yellowstone, Wyoming (USA) being perhaps the most famous. When gaseous emissions predominate, the term fumarole is usually applied. Emission temperatures of fumaroles therefore typically exceed the local boiling point of water. Long-lived fumarole fields are sometimes termed solfataras or soufrieres. Fumarole emissions are very often composed of both magmatic and hydrothermal gases, the latter evolving through complex chemical and physical interactions between magmatic fluids, meteoric water, seawater and rock. The emissions of some volcanoes discharge into crater lakes, which form by the condensation of the gases in the lake, as well as the capture of precipitation. The 'black smokers' associated with oceanic ridges are another important manifestation of subaqueous volatile discharge.

Some volcanoes have been in apparently continuous eruption for as long as records exist. For example, there is no evidence for any significant hiatus in the ongoing eruption of Stromboli (Italy) in over two millennia. The following volcanoes have been erupting more or less continuously for decades: Stromboli and Etna (Italy); Erta 'Ale (Ethiopia); Manam, Langila and Bagana (Papua New Guinea); Yasur (Vanuatu); Semeru and Dukono (Indonesia); Sakura-jima (Japan); Santa Maria and Pacaya (Guatemala); Arenal (Costa Rica); Sangay (Ecuador); and Erebus (Antarctica). According to the records of the Smithsonian Institution's Global Volcanism Program, the median duration of an eruption is around seven weeks. Most eruptions end within three months.

Materials erupted

Excluding some exotic but rare instances, most volcanoes erupt silicate magma of one sort or another. On eruption, these materials can be divided into lavas, which flow (with widely varying degrees of ease) across the surface in partially molten form, and pyroclastics (literally 'broken by fire'), which are expelled from volcanoes as solid fragments. Pyroclasts may be lofted in buoyant eruption columns to considerable heights in the Earth's atmosphere (exceeding 40 km altitude in exceptional cases), and can then be transported hundreds or thousands of kilometres by air currents. When they sediment to the surface they typically form ash fall deposits. These are often characterized by superposed beds of well-to-moderately sorted ash. If ash deposits harden, they are often called tuffs.

Although the term 'ash' is used widely in a loose sense, strictly it refers to pyroclasts with a diameter of 2 mm or less. The very finest ash is composed of shards, minute fragments of volcanic glass shattered apart by violent, explosive eruptions. Larger fragments (up to 64 mm across) are termed lapilli, and still larger ejecta are referred to as blocks (if solid when ejected) or bombs (partly molten when ejected). The very largest blocks and bombs tend to separate from an ascending eruption column and follow ballistic trajectories back to the surface, dropping relatively close to the vent. Fluid, coarse pyroclasts that accumulate around the vent are often termed spatter. Lapilli of mafic or intermediate composition (low or moderate silica content) that display a fragmented, bubbly texture are termed scoria. Highly bubbly pyroclasts, usually of more silica-rich (silicic) composition, are often called pumice. A common term pertaining to all pyroclastic material regardless of size is tephra.

Pyroclastic eruptions do not always produce stably convecting eruption columns in the atmosphere. They can also result in fountain collapse, in which all or part of the jet of pyroclasts leaving the vent region is unable to entrain and heat sufficient air to become buoyant. It then sinks back to the surface, where it can feed pyroclastic flows that move across the ground as a density current. Generally pyroclastic flows follow topography closely but they can gain sufficient momentum to overcome topographic barriers. The largest silicic eruptions on Earth are predominantly pyroclastic flow eruptions with volumes of a few thousand cubic kilometres (masses exceeding 10^{15} kg). Their deposits are often termed ignimbrites, and some anneal during compaction to form an excellent source of building stone. Pyroclastic flows consisting of ash-sized material are also known as ash flows, and their sediments as ash flow deposits. The finest particles can separate from a pyroclastic flow, and rise as a convecting plume to considerable heights in the atmosphere, ultimately settling out to form co-ignimbrite ash deposits.

Eruption styles

We have already referred to two very generalized eruption styles – lava (effusive) and pyroclastic (explosive) eruptions. Various schemes exist for classifying eruptions in finer detail. While they all retain some descriptive value, confusion can arise from inconsistent use of the terminology, and the fact that an individual eruption can display many different phenomena in rapid succession or even simultaneously (Plate 145). There are two basic approaches – one to describe an eruption on the basis of contemporaneous observations (i.e. the physical description of the eruptive phenomena), the other to characterize and interpret the deposits or impacts of an eruption. Clearly, the latter is the best or only available option for many eruptions in the past, and considerable efforts have gone into developing theoretical frameworks

Plate 145 Mount Etna (Italy) in eruption in August 2001. Note that several aligned vents are active simultaneously. The one disgorging dark ash clouds is at 2,550 m above sea level and produced a new cinder cone, now a tourist attraction. The vent below it is emitting a lava flow, picked out by the curtain of whitish gases rising above it

relating the depositional record to eruption physics and the atmospheric transport of ash clouds (Sparks *et al.* 1997).

The least ambiguous physical descriptors of eruptions are magnitude, intensity and duration. Intensity describes the mass eruption rate (e.g. in kg s^{-1}), and is a particularly useful parameter for explosive eruptions, as it is closely related to the height reached by the eruption column. Integrating intensity through the whole duration of an eruption yields the total erupted mass, or magnitude (in kg). This is also a useful first-order measure that enables intercomparison of sizes of different lava and/or tephra eruptions.

Eruptions do not always expel fresh magma. They can be driven by the sudden expansion of a liquid changing phase into a gas – for example, when ground water comes into contact with magma and flashes to steam – or by the instantaneous decompression of pockets of gas that have accumulated at some position within a volcano or its basement. These phreatic blasts can excavate considerable volumes of rock between the explosion source and the Earth's surface, leaving impressive holes in the ground. If new magma is also expelled in such explosions, they are termed phreatomagmatic. Such hydrovolcanic phenomena are quite common when a dormant volcano awakens – the volcano is effectively clearing its throat to make way for the passage of new juvenile magma.

A suite of more subjective terms to describe eruption style is also in common usage. These derive from particular historic eruptions (for example, 'vulcanian' refers to the 1888–1890 eruption of Vulcano; 'plinian' to Pliny the Elder's observations of the AD 79 eruption of Vesuvio, recorded by his nephew) or characteristic styles of individual volcanoes ('strombolian' refers to Stromboli volcano's propensity for modest pyrotechnic displays). Unfortunately, because one volcanologist's idea of a vulcanian eruption can be rather like another's image of a plinian, the terminology is not ideal but since it remains in widespread use it will be briefly outlined here.

Gas-rich eruptions of low viscosity magma can be sustained, generating the fire fountains characteristic of Hawaiian activity. Strombolian activity is typified by more discrete, fairly instantaneous explosions capable of propelling bombs a few hundred metres above the vent. These eruptions are formed as large bubbles of gases burst at the surface of a conduit filled with magma. Vulcanian eruptions are more violent. Here, build-up of gas pressure, often in a volcanic pipe blocked for decades or centuries since the last eruption, suddenly blows out a dense column of blocks and ashes, often composed of more old lava than new. These sometimes turn out to be throat-clearing eruptions that clean out the volcanic conduit ready for a more substantial plinian eruption. Plinian eruptions typically last for hours or days. The eruption plumes soar to heights of 20 to 40 km, and ash, gases and aerosols can circle the globe in a matter of weeks. Plinian eruptions produce well-sorted pumice and ash fall deposits. One important factor, as explosive eruptions crank up in scale, is the relationship between the duration of magma discharge and the rise time of the plume in the atmosphere. The physics of eruption columns diverge for discrete vs. sustained eruptions, with the latter capable of significantly higher ascent for a given intensity. The most intense known eruption, based on studies of its deposits, is the *c.* AD 181 outburst of Taupo in New Zealand. With an estimated intensity exceeding 10^9 kg s^{-1}, its ash cloud would have climbed to around 50 km above sea level (Carey and Sigurdsson 1989). Occasionally, as at Mount St Helens (USA) in 1980, volcanic explosions are directed more or less horizontally, and are termed lateral blasts.

Hydrovolcanic eruptions are sometimes referred to as surtseyan events. Lava dome eruptions (see LAVA LANDFORM), which often show sudden switches in behaviour between slow effusion of lava that accumulates in the dome, and explosions and dome collapses that feed pyroclastic flows, are termed peléean. Fissure eruptions have yet to earn a special name but they are quite common in some volcanic regions, especially Iceland, where magmas can rise up to the surface in vertical sheets (dykes) of considerable length. The discharge of magma quickly focuses on a number of discrete but aligned points, and the total length of the system can be up to 10 km or more.

Types of volcanoes

The landforms constructed by volcanism are very diverse, reflecting the supply rates of melt from the mantle, the storage, evolution and transport of magma in the crust, and the tectonic environment, as well as external factors such as presence of liquid water. Perhaps the simplest volcanic construct is the cinder or scoria cone. These are often monogenetic volcanoes formed as the result of a single eruptive episode. They seldom exceed 100–200 m in height and are typically composed of mafic scoria. Many much larger shield volcanoes, characterized by low angle, convex-upwards slopes and a broad summit region (Plate 144), are dotted with adventive cinder cones, which develop where branches from the central magma conduit break the surface on the flanks of the volcano. The shield profile reflects the generally low viscosity of the erupted lavas, and their ability to flow for considerable distances before solidifying. Shield volcanoes are usually crowned by nested and intersecting CALDERAS and collapse pits formed by subsidence. Calderas can also develop during large explosive eruptions as the crust founders above the emptying magma chamber.

When there is abundant ground water, hydrovolcanic eruptions can blast out wide depressions enclosed by low, circular or elliptical rims of ejecta. These features are known as maars and they may show little or no juvenile material. Tuff rings are distinguished from maars by being built on to the substrate rather than excavated into it, and typically contain more juvenile tephra. They have gentler slopes (2–10°) compared with tuff cones, which have gradients of 20–30°.

Most volcanoes are polygenetic, the result of numerous eruptive episodes. Simple cones, also called strato-volcanoes, are usually composed of interlayered lavas and tephra produced in countless eruptions over the lifetime of a volcano, which can be anywhere from a few thousand to many hundreds of thousands of years. They typically range in height between 1,000 and 3,000 m, and are often crowned by comparatively small craters (a few hundred metres in diameter). Mount Mayon (Philippines) is a good example, and justly famed for its near-perfect radial symmetry and convex-upwards slopes. Volcanoes may be subjected to major gravitational collapses during their history (see next section) but regrow by further eruption. The remodelled edifices that result are termed composite cones.

Clusters of overlapping volcanoes are sometimes called volcanic complexes, though the term is used rather loosely. Another case of more distributed volcanism is the volcanic field, formed where a single magmatic system has fed many discrete, usually monogenetic eruptions. The Michoacán–Guanajuato volcanic field in Mexico, composed of some 1,400 individual cinder cones, tuff cones and maars peppering a 200 × 250 km area, is a fine example.

Submarine volcanism has only recently received much attention, thanks to advances in deep-sea exploration. The geomorphology of ocean ridge volcanoes displays some of the characteristics of their subaerial counterparts in extensional tectonic environments but eruption in water at high pressure suppresses explosive activity and causes lava surfaces to solidify rapidly. Similar considerations apply to subglacial eruptions, common in Iceland. Here the weight of ice and presence of meltwater result in formation of tuyas in the case of central vent eruptions, or móbergs when fissure eruptions take place. Catastrophic releases of meltwater, jökulhlaups, associated with subglacial eruptions, are responsible for some of the highest discharge rates ever measured, and are capable of transporting massive quantities of sediment. In Iceland the deposits form plains known as sandur.

Erosional features

Because of their physical prominence, volcanoes are prone to all the usual agents of erosion. There can even be feedbacks between constructive and destructive processes that strongly affect the history of a volcano and its geomorphology. Measures of the degree of erosion of a volcano by wind, rain or ice, can be usefully applied to assessment of the relative age of activity. Rivers can slice rapidly through pyroclastic deposits on a volcano's flanks, while fresh tephra can even more quickly infill them. Once erosion gets the upper hand, volcanoes can exhibit well-developed planezes, triangular facets on the flanks of the cone delineated by the intersection of gully heads.

The most devastating destructive events that affect volcanoes are large-scale gravitational collapses, or volcanic landslides. The largest flank failures, sometimes called sector collapses, are a common feature of many composite cones and oceanic volcanoes, and they have generated the largest debris avalanche deposits on Earth (Moore *et al.* 1989). These are often identifiable from their characteristic hummocky topography, even when traces of the avalanche scar have been obliterated.

References

Carey, S. and Sigurdsson, H. (1989) The intensity of plinian eruptions, *Bulletin of Volcanology* 51, 28–40.

Moore, J.G., Clague, D.A., Holcomb, R.T., Lipman, P.W., Normark, W.R. and Torresan, M.E. (1989) Prodigious submarine landslides on the Hawaiian Ridge, *Journal of Geophysical Research* 94, 17,465–17,484.

Siebert, L. and Simkin, T. (2002) *Volcanoes of the World: An Illustrated Catalog of Holocene*

Volcanoes and their Eruptions, Smithsonian Institution, Global Volcanism Program Digital Information Series, GVP-3, (http://www.volcano.si.edu/gvp/world/).

Sparks, R.S.J., Bursik, M.I., Carey, S.N., Gilbert, J.S., Glaze, L., Sigurdsson, H. and Woods, A.W. (1997) *Volcanic Plumes*, New York: Wiley.

Further reading

Francis, P. and Oppenheimer, C. (2004) *Volcanoes*, Oxford: Oxford University Press.

Heiken, G. and Wohletz, K. (1985) *Volcanic Ash*, Berkeley: University of California Press.

Sigurdsson, H., Houghton, B.F., McNutt, S.R., Rymer, H. and Stix, J. (eds) (2000) *Encyclopedia of Volcanoes*, San Diego: Academic Press.

Thouret, J.-C. (1999) Volcanic Geomorphology – an overview, *Earth-Science Reviews* 47, 95–131.

SEE ALSO: caldera; lava landform

CLIVE OPPENHEIMER

W

WADI

A fluvial valley in a dryland region. Some wadis are relict landscape features that developed in former pluvials either as a result of increased overland flow or because of groundwater sapping (see DRY VALLEY). They include the *Megakcha* of the Central Kalahari, an area where there is currently little or no surface runoff. In the Sahara some of the wadis are enormous palaeodrainage systems up to 1,400 km in length (Pachur and Peters 2001). Others are more stubby features that some authors attribute to groundwater sapping (e.g. Luo *et al.* 1997).

Other desert wadis are sporadically active systems. This is, for example, the case in the Negev, where runoff and sediment yields can be high. One reason for this is the nature of rainfall events in the region. Rainfall intensities can be high (Schick 1988). In the Nahel Yael catchment, over a seventeen-year period, intensities exceeding 14 mm hr^{-1} accounted for nearly one-half of the total rain (223 mm out of 449). Of this intense rain, 37 per cent fell in intensities exceeding 2 mm min^{-1}. Extreme flooding in the wadis can follow major rainfall events, as was demonstrated by the storms that afflicted southern Israel and Jordan in 1966 (Schick 1971).

Not all desert rainfall occurs in intense storms. A major contributing factor in runoff generation is the nature of some of the desert surfaces. For example, with dry conditions and a limited vegetation cover, silty soils, associated with loess deposits, rapidly become sealed under the influence of raindrop impact, and so have diminished infiltration capacities. Even on moderate slopes, silty soils generate substantial runoff (Evenari *et al.* 1983). Another runoff generating surface type results from the presence of organic crusts. These contain Cyanobacteria which partially plug soil pore space, particularly when they swell after they are moistened by rain (Verrecchia *et al.* 1995). Yet another important type of surface for runoff generation is bare rock. Available data indicate that the threshold level of daily rainfall necessary to generate runoff in rocky areas is a mere 1–3 mm. This compares with 3–5 mm for stony colluvial soils and more than 10 mm for stoneless loess soils. Because arid areas have a greater exposure of bare rock than semi-arid their wadis may generate more runoff and sediment (Yair and Enzel 1987).

Studies of experimental catchments over several decades have indicated high rates of sediment yield. Of particular note is the magnitude of bedload transport that has been recorded in the Nahal Yatir and neighbouring areas. Reid *et al.* (1998) showed that although wadi channels are only hydrologically active for about 2 per cent of the time (*c.*7 days per year) and only have overbank flow for about 0.03 per cent of the time (3 hours per year), the bedload flux is remarkably high. Indeed, the Nahal Yatir is about 400 times more effective at transporting coarse material than its perennial counterparts in humid zones (Laronne and Reid 1993). The explanation for this (Reid and Laronne 1995) is that its bed is not armoured (see FLUVIAL ARMOUR) with coarse material. The unvegetated nature of the desert watershed provides ample supplies of sediment of all sizes and this, together with the rapid recession of the flash flood hydrographs and the extended periods of no flow, discourages the development of an armour layer. Therefore, the flux rates are not sediment-supply limited as they so often are in perennial stream channels.

References

Evenari, M., Shanan, L. and Tadmor, N.H. (1983) *The Negev: The Challenge of a Desert*, 2nd edition, Cambridge, MA: Harvard University Press.

Laronne, J.B. and Reid, I. (1993) Very high rates of bedload sediment transport by ephemeral desert rivers, *Nature* 366, 148–150.

Luo, W., Arvidson, R.E., Sultan, M., Becker, R., Crombie, M.K., Sturchio, N. and El Affy, Z. (1997) Groundwater-sapping processes, Western Desert, Egypt, *Geological Society of America Bulletin* 109, 43–62.

Pachur, H.-J. and Peters, J. (2001) The position of the Murzuq Sand Sea in the palaeodrainage system of the Eastern Sahara, *Palaeoecology of Africa* 27, 259–290.

Reid, I. and Laronne, J.B. (1995) Bedload sediment transport in an ephemeral stream and a comparison with seasonal and perennial counterparts, *Water Resources Research* 31, 773–781.

Reid, I., Laronne, J.B. and Powell, D.M. (1998) Flash-flood and bedload dynamics of desert gravel-bed streams, *Hydrological Processes* 12, 543–557.

Schick, A. (1971) A desert flood: physical characteristics; effects on man, geomorphic significance, human adaptation. A case study of the Southern Arava watershed, *Jerusalem Studies in Geography* 2, 91–155.

——(1988) Hydrological aspects of floods in extreme arid environments, in V.R. Baker, R.C. Kochel and P.C. Patton (eds) *Flood Geomorphology*, 189–203, New York: Wiley.

Verrecchia, E., Yair, A., Kidron, G.J. and Verrecchia, K. (1995) Physical properties of the psammophile cryptogamic crust and their consequences to the water regime of sandy soils, north-western Negev Desert, Israel, *Journal of Arid Environments* 29, 427–437.

Yair, A. and Enzel, Y. (1987) The relationship between annual rainfall and sediment yield in arid and semi-arid areas. The case of the Negev, *Catena Supplement* 10, 121–135.

A.S. GOUDIE

WATER-LAYER WEATHERING

Water-layer weathering refers to the accelerated geochemical weathering that occurs on SHORE PLATFORMs immediately above the platform water level. The weathering is a result of a number of interrelated processes that require an unsaturated or alternately wet and dry environment in which to operate. These processes include the combined actions of wetting and drying including thermal expansion in some rocks, the chemical action of salt spray, salt crystallization and the removal of solutions through rock capillaries. Additional processes acting in the same environment can include wave and rock abrasion, biological processes including rock boring, and frost shattering. The main role of waves however is to remove the debris provided by the weathering processes. Over time positive relief features on the platform are weathered down to the water level. Below the water level (the level of saturation) the lack of drying and free oxygen precluded the above processes.

The type and extent of individual processes will be dependent on latitude/climate, exposure to sun or orientation, lithology and rock structure, tide range and level of wave energy on the platform. Water-layer weathering will progress most rapidly on platforms exposed to regular wetting by seawater or spray, in warm temperate to tropical environments which favour drying, and in weaker and more porous sedimentary rocks which increase the depth of activity and aid removal of debris.

Further reading

Stephenson, W.J. and Kirk, R.M. (2000) Development of shore platforms on Kaikoura Peninsula, South Island, New Zealand. II: The role of subaerial weathering, *Geomorphology* 32, 43–45.

SEE ALSO: chemical weathering; shore platform; wetting and drying weathering; weathering

ANDREW D. SHORT

WATERFALL

A waterfall can be distinguished from cascades, cataracts or other sharp descents in a stream profile, by the free fall of water over a very steep rock face. The greatest single fall is 986 m at Angel Falls in Venezuela; Tugela Falls in South Africa drops 948 m; and a descent of 800 m occurs in several drops at Yosemite Falls. Although they have much smaller descents, Victoria Falls (123 m), Niagara Falls (62 m), Iguazu Falls (70 m) on the Parana River and Khone Falls on the Mekong River (22 m) are notable for their great discharges.

Many waterfalls have been initiated by tectonic uplift along continental margins, or by local tectonic disruption along a fault (see FAULT AND FAULT SCARP) or rift scarp, as at Thingvellir in Iceland. They also commonly occur where a stream profile has been steepened by glacial erosion of a valley side, as at Skykje Falls in the Hardanger Fjord of Norway, or where streams flow over sea cliffs. Some small waterfalls, like the upper Suha Falls (25 m) in Slovenia, are essentially constructional features resulting from the deposition of tufa on the bed of a stream. The great majority, however, are the result of the differential erosion of rocks of varying strength.

The example most commonly cited is Niagara Falls, probably because of the fine account written by G.K. Gilbert. There, a dolomite caprock is undercut by failure in the shale beneath it. Other notable examples of caprock waterfalls are Gullfoss and Dettifoss, in Iceland, though in these cases the fall is due to the variable resistance of basalt, and of interbasaltic sediment and breccia. But not all waterfalls are undercut. Many have vertical faces, and others are buttressed outward at the base. These types of falls are not just short-lived features, where a stream is temporarily held up by a resistant vertical barrier. Many of them have retreated considerable distances. Vertical or buttressed forms are widespread in south-east Australia, where they occur in crystalline and in sedimentary rocks; Dangar Falls and Fitzroy Falls are typical examples.

Since Gilbert's account of Niagara, undercutting of waterfalls has been widely attributed to the erosive power of water swirling back into the recessed fall face. However, in most cases this does not occur, because, especially during flood discharge, the water descends well out from the fall face. Most undercuts are quite dry, except when spray is carried in by up-valley winds. Spray may promote weathering of rock in an undercut, especially if it freezes and fractures the rock. Seepage is also important. When saturated, a rock may have only about half its normal strength. Moreover, the abrupt drop in elevation at a large waterfall may greatly increase the pore-water pressure caused by seepage within the rock at the base of the falls. For example, the large undercut at Belmore Falls, in south-east Australia, occurs not behind the falling water, but to one side of it, where the main line of seepage drops 60 m from the valley above.

Water flowing over the fall face can pluck joint-bounded blocks from the rock on the lip of the falls. During flood discharge, when velocities are great, erosion on the lip may also be caused by processes like CAVITATION. The main effect of the falling water is the excavation of a plunge pool at the base of the falls. The claim that falling water has little erosive power and that the erosion is primarily the result of the impact of transported boulders has been repeated many times, but is not true. The kinetic energy of debris-free water falling over a dam can result in serious erosion at the base. For example, water over the Kariba Dam on the Zambezi River scoured a hole 50 m deep in just four years. The maximum, or limiting, depth of erosion varies with the height of the fall and with the magnitude of the discharge, together with the strength of the rock into which the pool is cut. Debris carried over many falls, or scoured from the plunge pools, forms a protective armouring of bed at the downstream end of the pool.

The basic requirement for a waterfall to develop is that the rock incised by a stream has sufficient strength to stand in a steep face. As stress related to the weight of rock is greatest near the bottom of a steep face, a waterfall cut even in an essentially homogeneous rock, such as granite, is likely to fail at its base, and thereby retreat upstream. Even where weaker rocks occur on an undercut waterfall, they commonly are quite fresh, and rupture by brittle fracturing in response to high stress. For example, stress in the gorge walls below Niagara Falls is causing the shale beneath the dolomite to bulge outwards and to generate vertical fractures, which weaken the rock face.

The rates at which waterfalls retreat depend largely on the magnitude of their discharge and the resistance of the rock over which they tumble. It also depends on their planimetric shape, for horizontal stresses normally are such that a waterfall in a narrow slot is less stable than one flowing over a broad AMPHITHEATRE. Rates of retreat range from about 1 km per 1,000 years at Niagara and Victoria Falls, to about 0.1–2 m per 1,000 years at smaller waterfalls in south-east Australia.

References

Gilbert, G.K. (1896) Niagara Falls and their history, in National Geographic Society *The Physiography of the United States*, 203–236, New York: American Book.

Lee, C.F. (1978) Stress relief and cliff stability at a power station near Niagara Falls, *Engineering Geology* 12, 193–204.

Schwarzbach, M. (1967) Islandische Wasserfalle und eine genetische Systematik der Wasserfalle uberhaupt, *Zeitschrift für Geomorphologie* 11, 377–417.

Weissel, J.K. and Seidl, M.A. (1998) Inland propagation of erosional escarpments and river profile evolution across the southeast Australian passive continental margin, in K.J. Tinkler and E.E. Wohl (eds) *Rivers over Rock: Fluvial Processes in Bedrock Channels*, 189–206, Washington: American Geophysical Union.

Young, R.W. (1985) Waterfalls: form and process, *Zeitschrift für Geomorphologie Supplementband* 55, 81–95.

R.W. YOUNG

WATERSHED

The term watershed in British English is used for a drainage divide – watershed boundary (American English) or catchment divide (Australian English) – that defines a water parting or a line, ridge or summit of high ground separating the water flow in two different directions draining in two drainage basins or catchments. In American English and by definitions of several international organizations the term has been changed to signify the region drained by, or contributing water to, a stream, lake or other body of water and is often used synonymously with the term drainage basin or catchment (Bates and Jackson 1980). A watershed can direct water from a surrounding watershed to a point such as a watershed outlet or a sinkhole. The total area of the watershed in a horizontal projection above a discharge-measuring point (watershed outlet) is the watershed area, catchment area or basin area. The watershed can be subdivided into several subwatersheds or subcatchments that conduct water from the surrounding hillslopes from one or the other side into a channel or drain water from a tributary channel at a confluence into a larger channel of a higher order or magnitude (see STREAM ORDERING).

One source of local watershed leakage is seepage or underground flow of water from one drainage basin to an outlet in a neighbouring drainage basin during a flood. Sometimes stream flow in a karst area will flow into an underground channel or sinkhole or directly into the sea. This may also occur when a water body is involved, such as subterranean flow or seepage from a stream, swamp or a lake (drainage lake) into a neighbouring watershed, or above the surface, e.g. stream bifurcation as in deltas or braided river systems (see DRAINAGE PATTERN). In such cases it is very difficult to exactly locate the surface and subsurface watershed boundary. In mountainous regions one therefore defines a watershed that runs along the crest of the highest range as a normal watershed in contrast to an anomalous watershed that does not behave that way. The location of the watershed boundary and size is dynamic over time and space depending on the migration of the watershed boundary caused by, e.g. water body level (overtopping the divide), temperature (thermokarst), sedimentation (alluvial fan), isostatic movements or disruptions by an earthquake or MASS MOVEMENTs, that temporarily or permanently changes the flow direction (Fairbridge 1968).

A watershed can be considered as a dynamic environmental system unit that has the functional organization of interacting hydrologic and geomorphic processes, e.g. precipitation, evapotranspiration, infiltration, RUNOFF GENERATION, EROSION, transport and sedimentation. Watershed management is the administration and regulation of the aggregate resources of a drainage basin to control and regulate water quantity and quality driven by these processes, e.g. for the production of water and the control of erosion, stream flow and floods. Watershed managers use Geographical Information Systems (GIS) and a DIGITAL ELEVATION MODEL (DEM) to delineate watershed characteristics such as watershed boundary, drainage network and contributing subwatersheds for inventory and planning purposes. They also process geo-spatial information on weather and climate, topography, soils and geology, vegetation and land use as well as infrastructural data processed in a GIS to link them with environmental models to analyse and simulate the complex dynamic hydrologic and geomorphic processes for decision-making purposes (see MODELS).

References

Bates, R.L, and Jackson, J.A. (ed.) (1980) *Glossary of Geology*, Falls Church, VA: American Geological Institute.

Fairbridge, R.W. (ed.) (1968) *Encyclopedia of Geomorphology*, New York: Reinhold.

Further reading

Brooks, K.N., Ffolliott, P.F., Gregersen, H.M. and De Bano, L.F. (1997) *Hydrology and the Management of Watersheds*, Ames: Iowa State University Press.

Goodchild, M.F., Steyaert, L.T., Parks, B.O., Johnston, C., Maidment, D., Crane, M. and Glendinning, S. (ed.) (1996) *GIS and Environmental Modeling: Progress and Research Issues*, New York: Wiley.

White, I.D., Mottershead, D.N. and Harrison, S.J. (1984) *Environmental Systems: An Introductory Text*, London: Unwin Hyman.

SEE ALSO: channel, alluvial; drainage basin; hillslope, form; hillslope, process

CHRIS S. RENSCHLER

WAVE

Water waves are periodic undulations of the air-water interface (Figure 174) defined by their height (H), wavelength (L), period of oscillation (T) and speed of propagation (C). They can be *progressive* (e.g. *wind waves, tsunami*) or *standing* (e.g. lake *seiche*, ocean *tide*) and either actively *forced* (e.g. *sea*, tide) or propagate freely (e.g. *swell, tsunami*). The tide is forced by the periodic tractive forces generated by the gravitational attraction of the moon–sun system on the Earth's hydrosphere. TSUNAMI (Japanese for harbour waves) or seismic sea waves result from disturbance of the ocean by an earthquake, an explosive volcanic eruption or a mass failure of ocean sediments. Wind waves are found wherever wind blows over a significant body of water (e.g. marine or lacustrine) and constitute the most important global source of water wave energy. Wind wave size (e.g. height, length, period) depends upon the wind speed, the length of open water over which the wind blows (fetch) and the duration of time that the wind blows. When the oscillatory fluid motions under waves interact with a solid boundary an oscillatory boundary layer is formed; thus, all waves have the potential to generate stresses at the boundary and carry out geomorphological work.

Wave spectrum

By convention, complex irregular wave fields are analysed using Fourier Analysis, which provides a 2-dimensional spectrum of heights and periods. If the direction of wave propagation is included a 3-dimensional *directional spectrum* is produced (Figure 175). Water waves are classified by their frequency (1/T) or period of oscillation, the generating force and the force restoring equilibrium of the water surface (see Kinsman 1965).

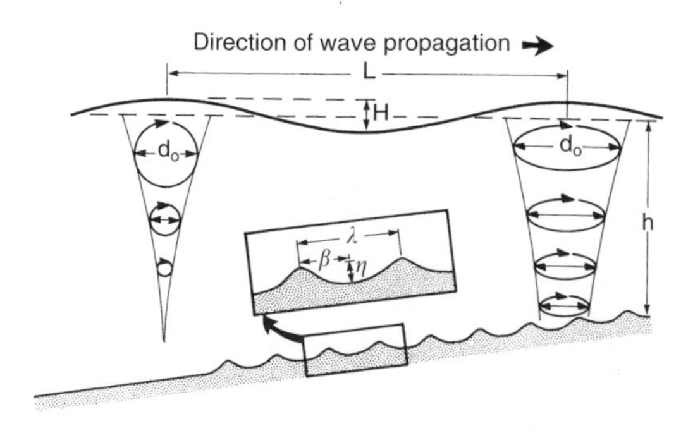

Figure 174 Schematic of wave form and water particle motion in deep and shallow water, where the wave form interacts with a rippled bed. Note: L = wave length; H = wave height; d_o = orbital diameter; h = water depth

TIDES

Tides are the characteristic semi-diurnal or diurnal rise (*flood*) and fall (*ebb*) of the Earth's mean sea level, caused by the periodic components of the gravitational tractive forces generated by the moon and sun; there are at least 360 tidal components leading to modulations of the water surface. Tidal waves are forced standing waves in the world's oceans that are resolved by Coriolis into a series of *rotary standing waves* (Kelvin waves) around the continental margins. These *amphidromic systems* have a central *amphidromic point* of no sea-level change and a *tidal range*, which increases outward to a maximum at the shoreline. Lines of equal tidal range are denoted by co-range lines, which encircle the amphidromic point. Tidal propagation is denoted by *co-tidal lines* (denoting points of equal *phase lag*), which radiate out from the amphidromic point. Such systems occur on all ocean coasts, even where tides are small. Tides are modulated at a range of timescales depending upon the changing relative positions of the Earth, moon and sun. Tides are largest (*spring tides*) when the sun, the moon and the Earth are in alignment (*conjunction* produces slightly larger tides than *opposition*); tides are smallest (*neap tides*) when the moon is in quadrature. Tidal waves are strongly influenced by interaction with the coastal boundary in terms of both their *amplitude* and *phase*, as tides attempt to penetrate into the embayments and estuaries around the coast where friction is important. In the Bay of Fundy, Canada, the tidal range may reach 18 m at springs, since the semi-diurnal forcing by the primary lunar component (M_2) is close to the *natural period of oscillation* of the basin and the shape of the basin causes a distinct topographic forcing. Tides are classified by range: (1) macrotidal >4 m; (2) meso-tidal 2–4 m; (3) micro-tidal <2 m (Davies 1973; Davis and Hayes 1984). *Surges* or meteorological tides are super-elevated tide levels (up to several metres) resulting from the combined effect on the water surface of barometric pressure differentials, wind stress on the water surface and wind wave set-up.

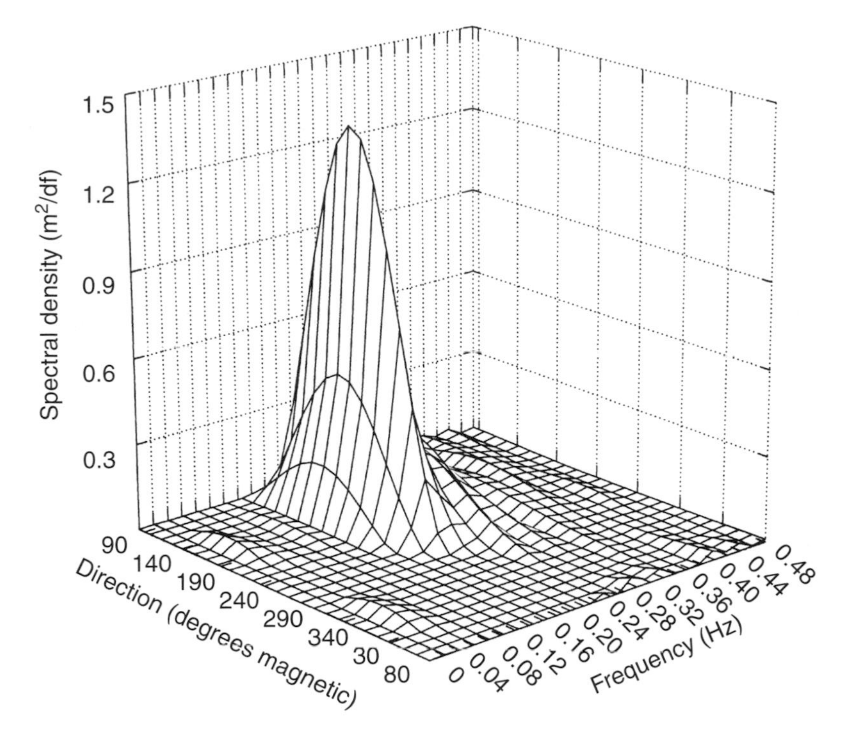

Figure 175 Directional wave spectrum recorded from southern Lake Huron, Canada in ~8 m water depth ~2 km offshore, recorded using a Falmouth Scientific 3DACM Wave Recorder. The record represents a 20-min sample of water surface elevation collected at 2 Hz. Note: a standard frequency spectrum can be obtained by simply projecting the data onto the left-hand frame. These waves have a peak period of ~5 s and a peak direction of approach between 160° and 170°

The rise and fall of mean sea level is complemented by horizontal translations of water known as *tidal currents* or *tidal streams*. Because of the long wave period, the currents are generally small, $<0.05\,\mathrm{m\,s^{-1}}$; however, where topographic forcing occurs (e.g. estuaries, straits, inlets, etc.), tidally currents can reach speeds up to $6\,\mathrm{m\,s^{-1}}$ (see CURRENTS).

TSUNAMI

In the deep ocean tsunami may be only a few centimetres or decimetres in height and may be difficult to observe, but may have periods ranging from 10–200 min, and hence extremely long wave lengths. Hence they can reach speeds of 700–900 km $\mathrm{h^{-1}}$. In shallow water, as a result of wave *shoaling*, tsunami can reach catastrophic proportions, with run up heights of 10–20 m at the shoreline. The large run-up and large current speeds produce extensive flooding, damage to shorelines and man-made structures and even loss of human life.

GRAVITY AND INFRAGRAVITY WAVES

The most important waves outside the zone of shallow-water wave breaking (*surf zone*) are gravity waves (forced by wind and restored by gravity), with periods ranging from 1–25 s. A modulation of wave height is common, and gravity waves propagate as groups of large waves separated by several smaller waves (Figure 176). This modulation forces a secondary wave, the group-bound long wave, which propagates at the group speed and has a period equal to the group period. Waves with periods of ~25 s to several minutes or longer (frequencies of 0.004–0.04 Hz) are infragravity waves (first recognized and called surf beat by Munk 1949). They are most important to water circulation and sediment transport within the surf zone (Bowen and Huntley 1984). In intermediate water depths, infragravity energy results from wave–wave interactions and consists of a mixture of forced and free wave motions (Herbers *et al.* 1995). Several mechanisms associated with wave breaking in the surf zone have been proposed for their

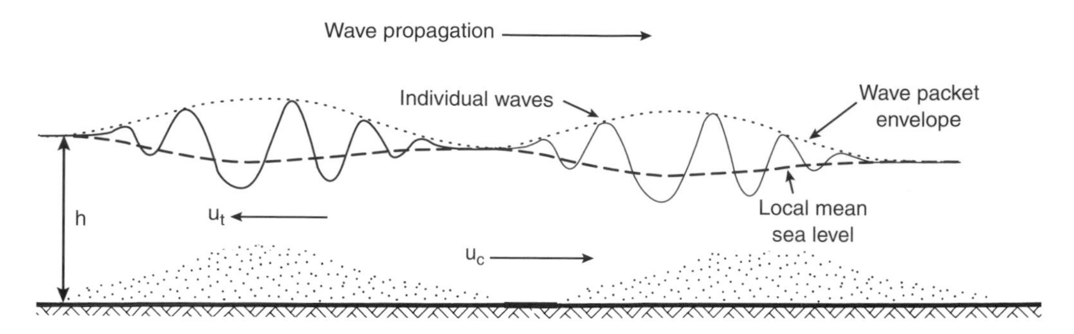

Figure 176 Wave groups and the forced group-bound long wave

generation including: (a) release of the group bound long wave (List 1992); (b) periodic shift in the position of wave breaking (Symonds *et al.* 1982); (c) persistence of groupiness into the surf zone causing a rise and fall of water at the shoreline (Watson and Peregrine 1992). Nearshore infragravity energy can take several forms including standing and progressive *edge waves* (trapped to the shoreline by refraction) and long *leaky waves* (reflected seaward without trapping; Huntley 1976). An important characteristic of infragravity waves in the surf zone is that they increase in amplitude and become more energetic as the offshore wave height increases. In contrast, gravity waves are saturated and thus decrease in importance proportionally during storms. *Shear waves* are a subset of the infragravity field (periods range from several tens of seconds to several tens of minutes) and they form within the surf zone through instabilities in the longshore current. They appear as large fluctuations in the mean current (Bowen and Holman 1989; Oltman-Shay *et al.* 1989) and can contribute significantly to both the cross-shore and alongshore transport of sediment (e.g. Aagaard and Greenwood 1995).

Gravity wave propagation and energy dissipation

Swift (1976) defined a number of *hydraulic provinces* stretching from the deep ocean to the coastline; in the innermost coastal zone (*shoreface*), both the hydrodynamics, sediment transport and morphodynamics are controlled primarily by gravity waves. Propagating gravity waves transfer energy and momentum from offshore to the shoreline. As they move into shallow water, frictional resistance results in a deformation of the orbital motions (Figure 174), a decrease in wave speed and wavelength, and an increase in wave height; this is called *shoaling*. The *specific energy density* (E, according to linear theory = $1/8\ \rho_f\ g\ H^2$, where ρ_f is fluid density, g is the gravitational constant) increases up to a point at which the wave becomes unstable and breaks. During propagation, wave energy may be redistributed laterally through convergence or divergence of the wave *orthogonals*; this is *refraction*. Further, the near-sinusoidal wave form characteristic of deep water becomes increasingly asymmetrical about the horizontal and vertical axes (*wave skewness* and *wave asymmetry*; Figure 177). These non-linearities are critical to the net transport of water and sediment by waves, as they introduce non-symmetrical motion in the oscillatory velocities. In very shallow water, waves break when $H/h \approx 0.4$–1.0. The wave may be destroyed in a *plunging* breaker, producing a large *roller vortex* of the same order as the water depth, or the wave may continue to propagate as a *spilling* breaker, with collapse of the leading face of the wave. Surf *bores* are solitary waves, propagating across the surf zone controlled only by the water depth. The surf zone is a complex hydrodynamic environment, where interactions occur between incident waves, macro and micro turbulent vortices, secondary waves and quasi-steady currents of various origins (for a review see Kobayashi 1988). Ultimately wave energy is either reflected from the beach or dissipated in the surf zone and the reversing *swash* currents on the beach face. Sediment transport in the *uprush* and *backwash* is constrained by extremely small water depths, large Froude Numbers, large pressure gradients near *bore* fronts and the *infiltration* or *exfiltration* of water from the beach face *water table* (for recent review see Butt and Russell 2000).

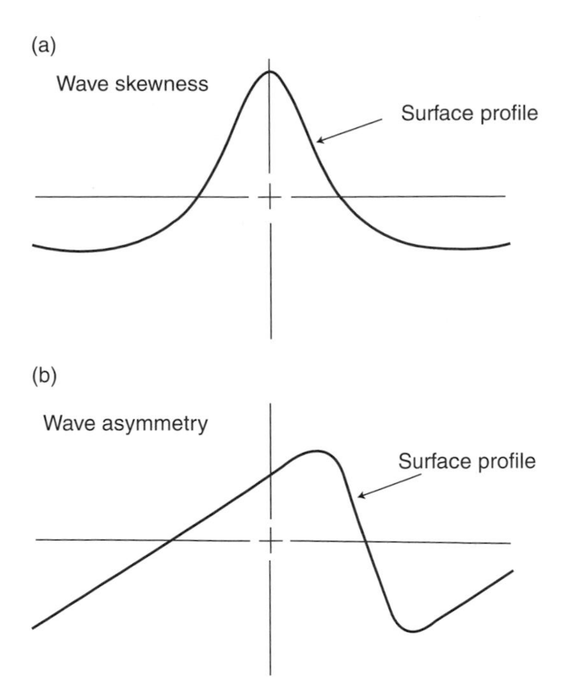

Figure 177 Asymmetrical wave forms: (a) asymmetry about the horizontal axis – wave skewness; (b) asymmetry about the vertical axis – wave asymmetry

Stresses generated by waves

Stresses generated by wave motion can be considered to take two forms: (a) *radiation stress* is the excess flux of momentum due to the presence of waves in the water and operates between the wave crest and trough. The concept was introduced by Longuet-Higgins and Stewart (1964). It explains the generation of the group-forced long wave in deep water and a number of hydrodynamic phenomena in the surf zone, such as wave *set-up* and *set-down* (i.e. the raising and lowering of the still water level at the shoreline), the *near-bed return flow* known as *undertow; longshore currents* (see CURRENTS); (b) the *boundary layer stresses* under waves are significantly larger than those of an equivalent quasi-steady current. Since the boundary layer develops and decays each half-wave cycle, it is much thinner and the resultant velocity gradients much larger. The time-varying bed shear stresses (τ_w), which have units of force per unit area ($N\ m^{-2}$ in SI units) can also be written in units of velocity using the friction or shear velocity (u_{*w} which is not a real velocity, but is used for convenience and can be related to the fluid turbulence) and the fluid mass density (ρ_f):

$$\tau_w = \rho_f u_{*w}^2$$
$$u_{*w} = (\tau_w/\rho_f)^{1/2}$$

A dimensionless form of the bed shear stress is given by the *Shields Parameter* (θ_w):

$$\theta_w = \tau_w/\{g\ (\rho_s - \rho_f)\ D\}$$

where D is the particle size on the bed. The bed shear therefore depends not only on the flow velocity but also the *bed roughness* (grain size and bedforms). When quasi-steady currents are combined with waves (usually the case), these relationships become more complex (Soulsby 1997). However, if during the wave cycle these stresses exceed that for the *initiation of motion* of particles on the bed, entrained material will begin to move and oscillate in place. However, it is rare that a perfect oscillatory motion occurs in nature and rare to find a perfectly horizontal bed where gravity has no effect; thus a net displacement of particles occurs. Usually, there is also superimposed quasi-steady current, either induced by *wave streaming* (mass transport), the *asymmetry* of the waveform, or by some other process, such as *undertow* or *longshore currents*, which will enhance the net motion.

Sediment transport by waves

A simple concept of waves 'stirring' and currents 'transporting' sediment has been used to model transport under waves (Bagnold 1966; Bowen 1980; Bailard 1981); however this is not strictly correct. Waves and currents interact in a complex, non-linear manner and there is strong feedback from the bed materials (particle size, bedforms, bed slope, etc.). This is especially true for sediment transport, where rates have been related to some power of the horizontal velocity, and exponents varying from 3 to 7 have been used (Soulsby 1997). In the presence of ripples or megaripples, there is a complex, periodic shedding of turbulent sediment-laden vortices formed by flow separation at the sharp discontinuity of the bedform crest (Sleath 1984). In general these vortices are released upwards into the flow close to the time of flow reversal, when the oscillatory currents drop to zero. However, the timing of this release is critical as the vortices can be advected either down wave or upwave depending upon the release time and the wave period.

Geomorphological response to waves

The geomorphological response to waves depends on a range of variables including: (a) the timescale; (b) the spatial scale; (c) the geological and bathymetric setting; (d) the tidal range; etc. Forms can range from small sand ripples of a few centimetres in height and perhaps 10 cm in spacing, to the upwardly concave shape of the shoreface profile out to the depth at which waves begin to shoal (called *wave base*). Coastal morphologies are often classified by the nature of the substrate: (a) *rock, cliff* or *cohesive* coasts (see Tsunamura 1992; Trenhaile 1987); (b) *sedimentary* coasts (see Davis 1985), where the grain size is a critical determinant of form (gravel, sand or mud coasts). Sedimentary coasts account for approximately 20 per cent of the world's coastline, of which sandy coasts make up 10–15 per cent, and rocky coasts make up approximately 80 per cent. The vast majority of wave studies have been restricted to sedimentary, especially sandy, coasts, where measurable change is relatively rapid compared to rocky coasts.

ROCK, CLIFF AND COHESIVE COASTS

On bedrock and cohesive coasts exposed to large *wind waves* erosion of the base of cliffs (see CLIFF, COASTAL) or bluffs may occur, producing an indentation or *wave-cut notch*. The material above the notch may become unstable and *mass wasting* processes will induce failure. The resulting debris is then entrained by the waves and transported alongshore and offshore (depending on grain size) by wave-generated currents. The factors governing basal erosion rates are: (1) the *hydraulic* and *pneumatic* action of the waves, modulated by the water level, the upper shoreface slope and any beach materials present; (2) the *resistance* of the bedrock, which is controlled by its lithology, structure and geotechnical properties (Tsunamura 1992). The hydraulic action of waves consist of: (a) *compressive* forces, which vary with the waveform (standing wave, breaking wave, broken wave) at the cliff toe; the maximum compressive pressure is exerted by a wave that breaks exactly at the cliff face; (b) *pneumatic* forces may be induced by compression of air by waves in fractures and joints or other indentations and its explosive expansive release as the wave recedes; (c) *shearing* and *tension* forces develop across the rock surface as the wave uprush and backwash generate boundary forces capable of entraining loosened debris. The total mechanical process is known as *quarrying* or *plucking*. When significant sediment load is available, waves also cause erosion through *abrasion* and *impact* forces by the mobilized sedimentary particles. No field measurements have yet been made of the forces acting against a cliff and most of the information comes from theoretical calculations or laboratory simulations. Major questions concerning rock coasts revolve around the relative importance of waves and tides and subaerial weathering, as well as the extent to which forms may have been inherited from a previous time (Trenhaile 2002). Bryant and Young (1996) cite evidence for erosion of rock cliffs up to 15 m above present sea level by *tsunami* (see also Aalto *et al.* 1999).

Typical coastal forms are: (a) *steep cliffs*; (b) *wave-cut notches*; (c) *sea caves*; (d) *rock arches* and *stacks*; and (c) *shore platforms*, with or without a seaward *rampart*. All involve wave erosion to a greater or lesser degree. Mass wasting processes, such as *rock falls* and *planar* or *rotational slides, earth* and *debris flows*, and *piping* play a major role in maintaining cliffs along both rocky and cohesive coasts. Weathering processes are also critical in making material available for wave erosion and transport. The most characteristic form is a steep cliff and shore platform, with the latter being gently sloping or subhorizontal with a sharp break at the seaward end. Here the tide plays a major role in translating the hydrodynamic zones horizontally, so controlling the location, duration and type of wave action which occurs, but it plays no active role.

SEDIMENTARY COASTS

In cohesionless and very weakly cohesive materials, the shoreface is generally upwardly concave with slopes increasing toward the shoreline. Associated with this is a general shoreward increase in grain size. The increasing slope generates gravitational forces on the sediment to balance the increasing stresses from the landward propagating waves. An equilibrium profile shape was defined empirically by Dean (1991):

$$h = A\, x^{2/3}$$

where h = water depth, A = a constant and x is the distance offshore. Using the suspended sediment transport model of Bagnold, Bowen (1980) showed that if the equilibrium profile is defined as that shape where the net sediment transport over a wave period is everywhere zero, then the exponent 2/3 also applied. Van Rijn (1998) gives an extensive review of the literature on the *profile of equilibrium* concept.

Once waves break in the nearshore, the force balance on the bed is no longer simple. The occurrence of increased turbulence (macro and micro), secondary waves at a range of frequencies, secondary quasi-steady currents, and alongshore and cross-shore pressure gradient currents lead to complex sediment flux patterns, morphological forms and highly variable grain size. On many sandy coasts the upper shoreface profile is characterized by from 1 to 13 sand ridges called bars, which may be 2 or 3-dimensional, shore-parallel or shore-normal and, in the extreme case, form classic crescentic patterns (Greenwood 2003). The generation of barred profiles, their spatial and temporal dynamics have been the focus of extensive research. In 1979, Greenwood and Davidson-Arnott classified bars by their location and the primary forcing agents. Wright and Short (1984) proposed that nearshore bars (their number, form, etc.) are simply sequential changes as wave energy increases and decreases over time from a fully dissipative beach state to a fully reflective state. Using Dean's dimensionless fall velocity parameter (Ω), based on wave height (H_b) at breaking, wave period (T) and sediment size (settling velocity, w_s):

$$\Omega = H_b /(w_s T)$$

Wright and Short (1984) linked this to six distinct beach states; a similar *model* sequence was proposed by Lippmann and Holman (1990). Recently, attention has been directed at the feedback between the shoreface topography and the hydrodynamics, leading to the concept of self-organization of morphological forms (Plant *et al.* 2001; Wijnberg and Kroon 2002).

References

Aagaard, T. and Greenwood, B. (1995) Suspended sediment transport and morphological response on a dissipative beach, *Continental Shelf Research* 15, 1,061–1,086.

Aalto, K.R., Alto, R., Garrison-Laney, C.E. and Abramson, H.F. (1999) Tsunami(?) sculpturing of the Pebble Beach wave-cut platform, Crescent City area, California, *Journal of Geology* 107, 607–622.

Bagnold, R.A. (1966) An approach to the sediment transport problem from general physics, *US Geological Survey, Professional Paper* 422-I.

Bailard, J.A. (1981) An energetics total load sediment transport model for a plane sloping beach, *Journal of Geophysical Research* 86, 10,938–10,954.

Bowen, A.J. (1980) Simple models of nearshore sedimentation: beach profiles and longshore bars, in S.B. McCann (ed.) *The Coastline of Canada*, 1–11, Geological Survey of Canada, Paper 80–10.

Bowen, A.J. and Holman, R.A. (1989) Shear instabilities of the mean longshore current, 1. Theory, *Journal of Geophysical Research* 94, 18,023–18,030.

Bowen, A.J. and Huntley, D.A. (1984) Waves, long waves and nearshore morphology, *Marine Geology* 60, 1–13.

Bryant, E.A. and Young, R.W. (1996) Bedrock-sculpturing by tsunami, south coast, New South Wales, Australia, *Journal of Geology* 104, 565–582.

Butt, T. and Russell, P. (2000) Hydrodynamics and cross-shore sediment transport in the swash zone of natural beaches: a review, *Journal of Coastal Research* 16, 255–268.

Davies, J.L. (1973) *Geographical Variation in Coastal Development*, New York: Hafner.

Davis, R.A. Jr (ed.) (1985) *Coastal Sedimentary Environments*, New York: Springer-Verlag.

Davis, R.A. Jr and Hayes, M.O. (1984) What is a wave-dominated coast, *Marine Geology* 60, 313–329.

Dean, R.G. (1991) Equilibrium beach profiles, *Journal of Coastal Research* 7, 53–84.

Greenwood, B. (2003) Wave-formed bars, in M. Schwartz (ed.) *Encyclopedia of Coastal Science*, Amsterdam: Kluwer (in press).

Greenwood, B. and Davidson-Arnott, R.G.D. (1979) Sedimentation and equilibrium in wave-formed bars: a review and case study, *Canadian Journal of Earth Sciences* 16, 312–332.

Herbers, T.H.C., Elgar, S. and Guza, R.T. (1995) Generation and propagation of infragravity waves, *Journal of Geophysical Research* 100, 24,863–24,872.

Huntley, D.H. (1976) Long period waves on a natural beach, *Journal of Geophysical Research* 81, 6,441–6,449.

Kinsman, B. (1965) *Wind Waves: Their Generation and Propagation on the Ocean Surface*, Englewood Cliffs, NJ: Prentice Hall.

Kobayashi, N. (1988) Review of wave transformation and cross-shore sediment transport processes in surf zones, *Journal of Coastal Research* 4, 435–445.

Lippmann, T.L. and Holman, R.A. (1990) The spatial and temporal variability of sand bar morphology, *Journal of Geophysical Research* 95, 11,575–11,590.

List, J.H. (1992) A model for the generation of two-dimensional surf beat, *Journal of Geophysical Research* 97, 5,263–5,635.

Longuet-Higgins, M.S. and Stewart, R.W. (1964) Radiation stress in water waves, a physical discussion with applications, *Deep Sea Research* 11, 529–563.

Munk, W.H. (1949) Surf beats, *Transactions American Geophysical Union* 30, 849–854.

Oltman-Shay, J., Howd, P.A. and Birkemeir, W.A. (1989) Shear instabilities of the mean longshore current, 2. Field observations, *Journal of Geophysical Research* 94, 18,031–18,042.

Plant, N.G., Freilich, M.H. and Holman, R.A. (2001) The role of morphologic feedback in surf zone sand bar response, *Journal of Geophysical Research* 106, 959–971.

Sleath, J.F. (1984) *Seabed Mechanics*, New York: Wiley.

Soulsby, R. (1997) *Dynamics of Marine Sands*, London: Thomas Telford.
Swift, D.J.P. (1976) Coastal Sedimentation; in D.J. Stanley and D.J.P. Swift (eds) *Marine Sediment Transport and Environmental Management*, 311–350, New York: Wiley Interscience.
Symonds, G., Huntley, D.A. and Bowen, A.J. (1982) Two-dimensional surf beat: long wave generation by a time-varying breakpoint, *Journal of Geophysical Research* 87, 492–498.
Trenhaile, A.S. (1987) *The Geomorphology of Rock Coasts*, Oxford: Oxford University Press.
——(2002) Rock coasts, with particular emphasis on shore platforms, *Geomorphology* 48, 7–22.
Tsunamura, T. (1992) *Geomorphology of Rocky Coasts*, Chichester: Wiley.
Van Rijn, L.C. (1998) *Principles of Coastal Morphology*, Amsterdam: Aqua Publications.
Watson, G. and Peregrine, D.H. (1992) Low frequency waves in the surf zone, *Proceedings of the Twenty-third International Conference on Coastal Engineering*, American Society of Civil Engineers, 818–831.
Wijnberg, K.M. and Kroon, A. (2002) Barred beaches, *Geomorphology* 48, 103–120.
Wright, L.D. and Short, A.D. (1984) Morphodynamic variability of surf zones and beaches: a synthesis, *Marine* Geology 56, 93–118.

Further reading

Komar, P.D. (1998) *Beach Processes and Sedimentation*, 2nd edition, Upper Saddle River, NJ: Prentice Hall.
Open University (1989) *Waves, Tides and Shallow-water Processes*, Oxford: Pergamon Press.
Trenhaile, A.S. (1997) *Coastal Dynamics and Landforms*, Oxford: Oxford University Press.

BRIAN GREENWOOD

WEATHERING

Weathering refers to a group of processes collectively responsible for the breakdown of materials at or near the Earth's surface. Weathering occurs because the environmental conditions under which most rock materials formed differ substantially from those which prevail near the Earth's surface. Consequently, they undergo a variety of modifications which result in more stable products under the newly imposed conditions of temperature, pressure, moisture and gaseous environment.

From a geomorphological perspective, rock weathering is extremely important. First, weathering processes prepare Earth materials for subsequent transportation by agents of erosion. Second, weathering is an essential component of soil formation at the Earth's surface. Third, weathering processes are directly responsible for land form and landscape evolution. Karst landscapes and their distinctive land form assemblages, for example, are a direct consequence of weathering processes, as are the thick regolith-mantled landscapes of the tropics and subtropics.

Weathering processes

Traditionally, weathering processes are regarded as being physical, chemical or biological in nature. In reality, the three groups of processes act synergistically making it difficult to distinguish clearly between the effects of any single one (Pope *et al.* 1995). For instance, biological processes effect both physical and chemical change within the weathering system.

PHYSICAL WEATHERING

Physical weathering collectively involves the breakdown of rock material into smaller pieces without any change in the chemistry or mineralogy of the rock. Rock material is simply disaggregated as a result of the effects of the generation of forces within the rock mass. These forces are derived either from internal expansion of rocks and minerals, or from growth of materials in voids. The principal physical weathering processes associated with internal expansion include: INSOLATION WEATHERING (thermal expansion), unloading, HYDRATION, WETTING AND DRYING WEATHERING, organic expansion and SALT WEATHERING. Growth in voids is dominated by frost action and growth of salt crystals.

Insolation weathering refers to the breakdown of rock as a result of expansion and contraction caused by frequent temperature changes. As bedrock has low thermal conductivity, the surface of rock material expands more than the interior of the rock and, as a result, stresses are set up that eventually lead to the disintegration of the rock. Its effectiveness is enhanced when rocks consist of a mixture of light and dark minerals thus facilitating marked variation in thermal conductivity with individual minerals expanding and contracting at different rates. Research from engineering and ceramics has demonstrated that relatively small temperature variations over short time periods can most effectively disintegrate rock. Thermal stress has also been shown to be intimately associated with the breakdown of rock in areas affected by fire (Ollier and Ash 1983).

Unloading, or sheeting, refers to the formation of slabs of rock parallel to the ground surface, but separated from the underlying intact rock by joints. Unloading or sheet joints are generally attributed to the reduction in compressive stress in rock masses as a result of erosion of the upper part of the rock mass which then fails in the direction of stress removal. Failure is likely to begin with the extension of an initial crack and continue with the development of sheet joints parallel to the unloading surface. While this process appears to be most effective in brittle crystalline rocks it is also widely developed in massive sandstones.

While hydration begins as fundamentally a chemical process, with the absorption of water along fractures, especially cleavage planes in minerals, its effect is primarily a physical one. Expansion of minerals as a result of the incorporation of water into the crystal lattice can result in the production of tremendous force, especially in confined spaces. The effect of mineral expansion is to disrupt adjacent mineral grains causing loss of grain boundary cohesion and ultimately bedrock disaggregation to produce friable rubble referred to as grus. Similar processes concentrated around the edges of joint blocks where water is most abundant leads to the formation of concentric shells of weathered material referred to as spheroidal weathering or EXFOLIATION. The formation of these concentric shells of weathered material is also sometimes referred to as onion skin weathering or desquamation.

Repeated wetting and drying of rock may be a significant physical weathering process. Rock breakdown by wetting and drying involves the development of internal rock stresses as a result of the progressive formation of ordered water layers which exert forces against confining walls or void boundaries. As the positively charged ends of water molecules are attracted to the negatively charged surfaces of clay minerals and colloids they form a layer of oriented water particles. With each wetting event a new layer of ordered water is added to clay particles, which remain during the drying phase. Failure appears to be most pronounced during the drying phase when negative pore-water pressure is greatest and tensile failure occurs.

Frost weathering (see FROST AND FROST WEATHERING) has long been viewed as one of the most significant physical weathering processes in cold climates. Traditionally, frost weathering has been believed to occur as a result of forces generated in association with the volume increase accompanying the phase change from liquid water to ice. However, the theoretical assumptions underlying the generation of such forces are seldom met in nature. A more realistic model, with growing empirical support, is the segregated ice model of frost weathering in which expansion is the result of the migration of unfrozen water toward growing ice lenses and only secondarily the result of volumetric change accompanying phase change. Frost weathering is thereby the result of enlargement of microcracks and relatively large pores by ice growth accompanying ice segregation (Walder and Hallet 1986).

The growth of salt crystals in voids in rock exerts forces capable of disaggregating rock material when the tensile strength of the rock is exceeded. Salt crystal growth occurs when solutions containing salts are evaporated or in some cases cooled, when water is added to salts and hydration occurs and when salts are heated. All these processes result in substantial increases in volume of salt crystals and accompanying application of force against the walls of voids, leading ultimately to rock disruption and disintegration (Goudie 1997).

Biological organisms are also effective agents of mechanical weathering. Biophysical weathering processes include root wedging, and lichen, algal and bacterial activity. It has been widely suggested that the penetration of plant roots along fractures in rocks is capable of splitting rocks apart as the root grows. Physical weathering by lichens is accomplished in two principal ways. First, by the penetration of hyphae along microcracks and grain boundaries. The penetration and growth of hyphae have been shown to exert tensile stresses which exceed the tensile strength of most rocks. Second, by the expansion of lichen thalli and hyphae due to the uptake of moisture. Such moisture uptake substantially increases the size of the thallus and hyphae thus exerting considerable pressure on the rock surface. The efficiency of this process is seen in the frequent incorporation of rock fragments into the thallus (Barker *et al.* 1997). Algae also appear to be important contributors to the breakup of rock materials. It has been found that upon wetting, the polymer sheath surrounding algal cells increases by as much as twenty times, thus exerting expansion forces sufficient to pry already weakened rock flakes from a rock surface (Hall and Otte 1990).

CHEMICAL WEATHERING

CHEMICAL WEATHERING processes are those which involve the chemical and or mineralogical transformation of rocks and minerals at/or near the Earth's surface into products that are in equilibrium with Earth surface conditions. Several principal chemical weathering processes are recognized, including solution (dissolution), HYDROLYSIS, HYDRATION, carbonation, chelation and redox reactions. Solution refers to the dissolving of minerals in the presence of water and involves the removal of atoms from mineral surfaces thus reducing the stability of minerals and enhancing their vulnerability to subsequent chemical degradation (Blum and Stillings 1995). Hydrolysis refers to the reaction of hydrogen in solution with mineral surfaces and subsequent formation of secondary clay minerals as displaced cations react with hydroxy ions in adhered water on mineral surfaces. Hydration involves the addition of water to a mineral structure to form a new mineral. Redox reactions are reduction/oxidation reactions which involve reactions with oxygen in the atmosphere. OXIDATION involves a loss of electrons while REDUCTION involves a gain of electrons: oxidation of one mineral component is achieved through reduction of another. Iron is the most commonly affected chemical species. Carbonation involves the reaction of minerals with carbon dioxide in the presence of water and is especially important in the chemical weathering of limestones and other carbonate-rich rocks. In the soil environment, where the weathering system is dominated by clay mineral–soil solution interactions, ion exchange reactions occur. Ion exchange involves the transfer of ions between solution and mineral, and generally involves the replacement of ions in the mineral interlayer though replacement within the crystal lattice can also occur. The efficiency of this process is to a large degree controlled by the strength of electrical double layer, an exposed plane of negatively charged oxygen ions and the balancing swarm of exchangeable cations (Jenny 1980).

Plants represent significant contributors to the chemical weathering of rocks and minerals, primarily due to the fact that they produce abundant quantities of organic acids. These organic acids form ring structures around a metal core with multiple bonds between the organic acid and the metal. They are responsible for the chelation of cations such as Fe and Al. In addition, during their formation they commonly release H^+ which further reacts with mineral surfaces. Most of these acids are produced in the vicinity of root tips.

Lichens, algae and bacteria produce abundant organic acids which are responsible for considerable rock and mineral weathering. These organisms are responsible for the production of two principal groups of organic compounds. These include oxalic acid and various oxalates. The former compound possesses high solubility and contributes abundant protons for mineral dissolution as well as producing ring structures for chelation. Oxalates have been shown to enhance mineral grain dissolution through proton donation (Barker *et al.* 1997).

Weathering intensity

Weathering intensity refers to the degree of decomposition of a rock or mineral; it is a measure of the amount of alteration that has occurred. Numerous indices of weathering intensity have been developed and include both descriptive measures which are based on changes in the visual appearance of regolith as it becomes more altered, as well as quantitative measures of the amount of chemical and/or mineralogical change that has occurred to the primary mineral or unaltered rock. Descriptive or qualitative measures of weathering intensity have been widely used in the fields of engineering geology and regolith geomorphology. They typically are descriptive and subjective, based on changes in visual appearance of materials associated with progressive disaggregation or loss of cohesion. These classifications typically contain a limited number of weathering intensity categories, generally consisting of categories such as fresh rock at the unweathered end and soil at the most weathered end with slightly, moderately, highly and completely weathered categories in between. The boundary between the fresh, unaltered bedrock and altered material is referred to as the WEATHERING FRONT. This boundary may be quite abrupt, but more commonly it is highly irregular.

Semi-quantitative measures of weathering intensity involve the assessment of the relative hardness of fresh and various states of altered material. The most widely used of these methods is the SCHMIDT HAMMER which provides an index of hardness based on the amount of resistance to the compressive stress applied to the rock by the hammer.

A large number of quantitative measures of the intensity of weathering are available that are based on chemical and mineralogical characteristics of the weathering rocks. In virtually all cases, these indices compare abundances of non-resistant constituents to resistant constituents. They are expressed as dimensionless ratios which generally decrease as the intensity of weathering increases: as more non-resistant materials are removed. The more commonly used ratios include silica:iron, silica:aluminium, silica:sesquioxides, silica:resistates, bases:alumina, alkalis:resitates, and alkali earths: resistates. In addition several more comprehensive chemical weathering indicies have been developed and still receive limited use including the Parker Weathering Index, and the Reiche Weathering Potential Index.

Several mineral weathering indices exist such as quartz:feldspar ratios and multiple-mineral weathering indices including those for heavy minerals such as zircon + tourmaline:amphiboles + pyroxenes which increase as weathering increases. Several methods involving the characteristics of individual mineral grains have also been developed for evaluating the intensity of weathering including surface micro-textural features of heavy minerals and the degree of etching on ferromagnesian minerals.

The intensity of weathering is influenced by numerous interacting factors which affect both the extent to which primary minerals and fresh rock have been altered as well as the rate at which alteration takes place. Intrinsic factors include the mineralogy and chemistry of the parent material, grain size and shape, porosity, and fracture abundance and openness. Extrinsic factors generally include temperature, moisture abundance and water chemistry. Traditionally it is the role of temperature and moisture that have been emphasized in controlling both rates and intensity of weathering at the global scale as portrayed in the weathering models of Peltier (1950) who used temperature and precipitation as a basis for a global model of the variability of physical and chemical weathering processes. Similarly, Strakhov (1967) portrays the arid/semi-arid and tundra regions of the world as possessing a weathering mantle dominated by no more than disaggregated bedrock with essentially no chemical or mineralogical alteration, while the hot, wet tropics/subtropics carry weathering mantles dominated by the concentration of sesquioxides. In fact, the intensity of weathering is not controlled by tropospheric climate, rather it is controlled by a combination of boundary layer and reaction site temperature and moisture. More specifically, the intensity of chemical weathering is controlled by a complex set of synergistically operating factors that control both weathering intensity and rate including: the availability and proximity of both biotic and abiotic weathering agents; mineralogy, chemistry and petrology of the parent material; structure of the parent material at multiple scales; temperature at the reaction site; hydraulics of water movement; removal of weathered materials; addition/removal of organic and inorganic fines; microtopography of both landscape surface and mineral surface; exposed surface area and the presence of any accreted surface coatings (Pope *et al.* 1995).

Deep weathering

In many parts of the world, weathering profiles reach extraordinary thicknesses. These weathering profiles are commonly referred to as deep weathering profiles. Weathering to depths in excess of 100 m is not uncommon, and cases of weathering profiles extending to a kilometre or more, not unknown. While great depths of weathering have been traditionally recognized from the tropics this does not mean that deep weathering profiles outside the tropics necessarily imply their formation under such climatic conditions. In fact deep weathering profiles can develop in a wide range of climatic settings in which weathering has been prolonged. Deep weathering profiles around the world display a marked variation in the intensity of weathering. On the ancient landscapes of Australia and Africa, deep weathering profiles are characteristically intensely weathered, with the clay mineral kaolinite dominating. These intensely weathered deep profiles simply reflect the availability of abundant moisture and efficient removal of weathered products in solution. It is important to point out that deep weathering profiles displaying little chemical and mineralogical transformation also occur extensively in Australia, Europe and North America. These profiles reflect less efficient leaching and often also reflect weakening of rock strength early in the geologic evolution of the parent material by providing exceedingly deep fracture systems which can be exploited by circulating meteoric waters.

The deep, kaolinite-dominated, weathering profiles of Australia and Africa, commonly are capped by a strongly indurated crust enriched in a variety of chemical cementing agents. Collectively, these indurated crusts are referred to as DURICRUSTS and are most frequently, but not exclusively, cemented by iron, aluminium, silica, calcium carbonate or gypsum.

References

Barker, W.W., Welch, S.A. and Banfield, J.F. (1997) Biogeochemical weathering of silicate minerals, in J.F. Banfield and K.H. Nealson (eds) *Geomicrobiology: Interactions between Microbes and Minerals*, 391–428, Reviews in Mineralogy 35, Washington, DC: Mineralogical Society of America.

Blum, A.E. and Stillings, L.L. (1995) Feldspar dissolution kinetics, in A.F. White and S.L. Brantly, (eds) *Chemical Weathering Rates of Silicate Minerals*, 291–351, Reviews in Mineralogy 31, Washington, DC: Mineralogical Society of America.

Goudie, A.S. (1997) Weathering processes, in D.S.G. Thomas (ed.) *Arid Zone Geomorphology*, 2nd edition, 25–39, Chichester: Wiley.

Hall, K. and Otte, W. (1990) A note on biological weathering of Nunataks of the Juneau Icefield, Alaska, *Permafrost and Periglacial Processes* 1, 189–196.

Jenny, H. (1980) *The Soil Resource: Origin and Behaviour*, New York: Springer Verlag.

Ollier, C.D. and Ash, J.E. (1983) Fire and rock breakdown, *Zeitschrift für Geomorphologie* 27, 363–374.

Peltier, L. (1950) The geographic cycle in periglacial regions as it is related to climatic geomorphology, *Annals of the Association of American Geographers* 40, 214–236.

Pope, G., Dorn, R.I. and Dixon, J.C. (1995) A new conceptual model for understanding geographical variations in weathering, *Annals of the Association of American Geographers* 85, 38–64.

Strakhov, N.M. (1967) *Principles of Lithogenesis*, Edinburgh: Oliver and Boyd.

Walder, J.S. and Hallet, B. (1986) The physical basis of frost weathering: toward a more fundamental and unified perspective, *Arctic and Alpine Research* 18, 27–32.

Further reading

Bland, W. and Rolls, D. (1998) *Weathering: An Introduction to the Scientific Principles*, London: Arnold.

White, A.F. and Brantly, S.L. (eds) (1995) *Chemical Weathering Rates of Silicate Minerals*, Reviews in Mineralogy 31, Washington, DC: Mineralogical Society of America.

Yatsu, E. (1988) *The Nature of Weathering: An Introduction*, Tokyo: Sozosha.

JOHN C. DIXON

WEATHERING AND CLIMATE CHANGE

Weathering plays a fundamental role in the carbon cycle and hence influences the role of CO_2 as a greenhouse gas. While volcanoes add CO_2 to the atmosphere, the major long-term process of CO_2 removal is chemical weathering of continental rocks. The prime weathering mechanism involved in the process is HYDROLYSIS. Rates of chemical weathering vary through time in response to changes in temperature and precipitation amounts. They are also affected by vegetation cover, the nature of which is also linked to temperature and vegetation conditions. Thus climate affects the global rate of weathering, but weathering has the capacity of altering the climate by regulating the rate at which CO_2 is removed from the atmosphere. Weathering may act as a negative feedback that moderates long-term climatic change (Figure 178) (Ruddiman 2000).

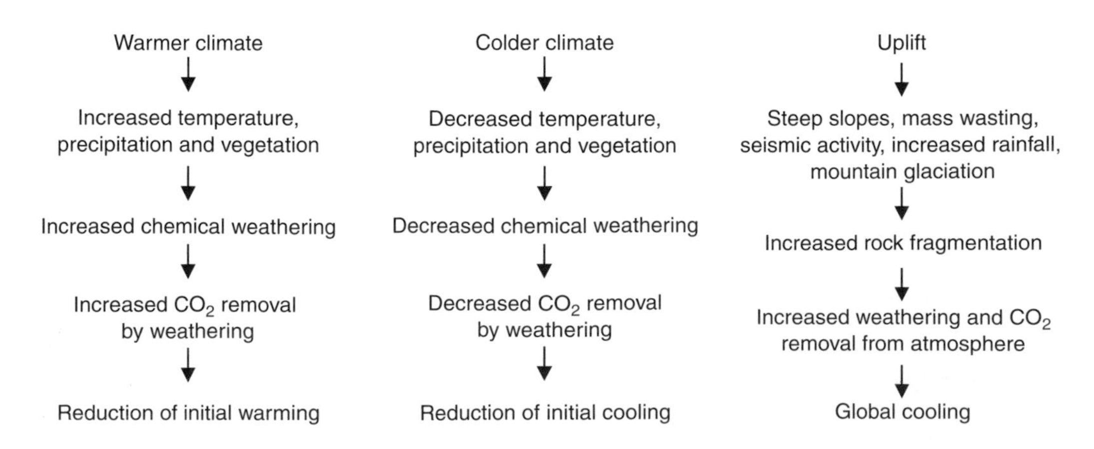

Figure 178 Negative feedbacks between weathering and climate change

However, it is also possible to see chemical weathering not only as a negative feedback that moderates climate changes, but also as an active driver of climatic change. The 'uplift weathering hypothesis' asserts that exposure of fragmented and unweathered rock is an important factor controlling the intensity of chemical weathering. It also asserts that rates of exposure of fresh rock are increased in areas of active mountain building because of seismic activity and mass wasting processes. Furthermore, uplifting areas generate more precipitation and glaciation. Mountain glaciers create pulverized bedrock.

Uplift has been active in the last few tens of millions of years (e.g. the Himalayas, Tibet, Andes, Rocky Mountains, European Alps). It is thus possible that accelerated chemical weathering has contributed to the climatic decline of the Late Cenozoic. It is also possible that differences in rates of chemical weathering between glacial and interglacial conditions in the Pleistocene occurred and contributed to the low CO_2 levels in the atmosphere during cold phases (Munhoven 2002). Karstic dissolution may also be an important component of the global carbon cycle (Gombert 2002).

The relationships between chemical weathering rates and climatic conditions are, however, complex (Kump *et al.* 2000) and the detection of variations in weathering intensity in the geologic record by the study of radiogenic isotope ratios ($^{87/86}$Sr, $^{143/144}$Nd, $^{187/186}$OS) is not without problems.

References

Gombert, P. (2002) Role of karstic dissolution in global carbon cycle, *Global and Planetary Change* 33, 177–184.

Kump, L.R., Bromtley, S.L. and Arthur, M.A. (2000) Chemical weathering, atmospheric CO_2, and climate, *Annual Reviews of Earth and Planetary Science* 28, 611–667.

Munhoven, G. (2002) Glacial–interglacial changes of continental weathering: estimates of the related CO_2 and HCO_3 flux variations and their uncertainties, *Global and Planetary Change* 33, 155–176.

Ruddiman, W.F. (2000) *Earth's Climate, Past and Future*, New York: W.H. Freeman.

A.S. GOUDIE

WEATHERING FRONT

The term pertains to subsurface weathering and is used to describe the boundary that separates solid, unweathered rock and rock that has already been weathered but remains still *in situ*. In reality, the transition between weathered and fresh rock compartments is rarely sharp. Usually there exists a transitional zone, in certain circumstances only a few centimetres thick, within which the change from unweathered mass to disintegrated or decomposed rock takes place. In some lithologies, especially in foliated metamorphic rocks and clastic sedimentary rocks, the transition is highly gradational and a well-defined weathering front may not exist. By contrast, weathering profiles in igneous rocks such as granite tend to have a very clear lower boundary.

The concept of weathering front has evolved from the notion of 'basal platform' introduced by Linton (1955) to explain the origin of TORS. Later it was shown that the boundary is hardly ever planar and the 'basal platform' is better replaced by 'basal surface of weathering' (Ruxton and Berry 1957). This implies a stable position of the rock/weathering mantle boundary and is not adequate to describe weathering profiles with abundant core stones. Therefore the present term 'weathering front' was recommended for use (Mabbutt 1961).

A weathering front is a dynamic feature, which migrates downwards and sidewards over time, as more and more rock disintegrates or decomposes. It should not be visualized as a single, continuous surface present at some depth, as each core stone is separated from the surrounding SAPROLITE by the weathering front. If the thickness of the weathering mantle is highly variable over short distances, the topography of the weathering front will accordingly be very rough.

References

Linton, D.L. (1955) The problem of tors, *Geographical Journal* 121, 470–487.

Mabbutt, J.A. (1961) Basal surface or weathering front, *Proceedings Geologists' Association* 72, 357–358.

Ruxton, B.P. and Berry, L. (1957) Weathering of granite and associated erosional features in Hong Kong, *Geological Society of America Bulletin* 68, 1,263–1,292.

SEE ALSO: deep weathering; etching, etchplain and etchplanation; saprolite

PIOTR MIGOŃ

WEATHERING-LIMITED AND TRANSPORT-LIMITED

Weathering and transport limitations have been used as an important concept in geomorphology and are basic to an understanding of hillslope development. Broadly defined, a weathering-limited system is one in which the supply of material determines the flux of mass, while in a transport-limited system, sufficient material is available at any given time for mass movement to occur.

The two terms were originally introduced to geomorphology by G.K. Gilbert in his monograph on the geology of the Henry Mountains (Gilbert 1877). In this classic work the author refers to two landscape types: one in which the overall rate of DENUDATION remains limited by the rate of rock weathering and mass movement processes are very effective in removing deposited materials as they accumulate (weathering-limited); the second is characterized by weathering processes exceeding the rate at which material can be denuded and there is a net accumulation of rock (transport-limited).

The same concept was revisited by Alfred Jahn in 1954, but remained largely unnoticed until translated from Polish into English (Jahn 1968). Jahn uses the denudational balance of slopes to classify slope processes. Although Jahn does not explicitly use the terms weathering-limited and transport-limited in his 1968 publication, he clearly separates slopes into those where the intensity of weathering is lower than that of material transport and those where the opposite is true. He further explains that the denudational balance is in equilibrium when the intensity of weathering equals that of transport.

Eighteen years after Jahn's original publication, Kirkby (Carson and Kirkby 1972) first coined the terms weathering-limited and transport-limited, leaning on Gilbert's and Jahn's early explanations and defining weathering-limited slopes as those in which the transport processes are more rapid than weathering. In contrast, transport-limited slopes are characterized by weathering rates in excess of transport rates which implies the formation of a soil cover. In other words, the thicker the soil cover, the more transport-limited a process can be classified.

Gilbert's, Jahn's and Kirkby's definitions can be expanded to other fields in geomorphology. For example, a distinction was made between weathering-limited and transport-limited watersheds to explain DEBRIS FLOW activity (Bovis and Jakob 1999). In their paper, the authors define transport-limited WATERSHEDs as those in which there is a quasi-infinite amount of material available for transport by debris flows, while weathering-limited watersheds are those in which sediment supply rates to the channel system (by ravelling, rockfall, debris slides, etc.) are low and a channel requires recharge over a given time period before a rainstorm can trigger the next debris flow. This distinction is important in predicting debris flow frequencies as well as magnitudes. In the first watershed type (transport-limited) the exceedance of a climatic threshold will likely trigger a debris flow, while in the second watershed type (weathering-limited) a debris flow will only occur at the exceedance of a climatic threshold, if sufficient material has accumulated in the channel system. A similar separation has been made by Stiny (1910) in his monograph on debris flows written in German and later translated by Jakob and Skermer into English (1997). Stiny distinguished between 'Altschuttmuren' (old rubble debris flows) and 'Jungschuttmuren' (young rubble debris flows). The first are derived from large accumulations of debris (usually morainal material) with a quasi-infinite supply of material. The latter are derived from recently accumulated debris which may be depleted in the course of the debris flow.

Recently, the concept of weathering and transport limitation has been extended to encompass the development of fluvial BEDFORM. For example, during low flow stages of sand and GRAVEL-BED RIVERs, the channel bed may be armoured and sediment from upstream will pass over the armour layer in the form of sediment supply limited (weathering-limited) bedforms such as BARCHANs and sand ribbons (Kleinhans *et al.* 2002). Fully developed RIPPLEs and dunes (see DUNE, FLUVIAL) exist when sufficient material exists for the formation of these features (transport-limited). The relevance of this distinction lies in the observation that bedform characteristics cannot be explained or predicted from local hydraulics and sediment characteristics alone. Another application is, for example, to assign controls on WEATHERING rates and solute (see SOLUTE LOAD AND RATING CURVE) fluxes on large rivers (Stallard 1985, 1995) using the same principles.

It is important to recognize that weathering-limited systems and transport-limited systems are not constant in space and time. For example, a large LANDSLIDE in a previously largely forested watershed may expose a large amount of debris which is now susceptible to transport by debris flows or other transport mechanisms. At this point the power of the stream (see STREAM POWER) network to transport the freshly accumulated and newly eroded material may have been exceeded and a previously weathering-limited watershed has now shifted to transport-limitation. Similarly, sudden climatic shifts may result in accelerated weathering rates which may not be matched by transport rates or vice versa. Finally, over time slope gradients may reach thresholds whereby the weathering-limitation may shift to transport-limitation.

The terms weathering and transport limited are somewhat misleading because they imply that the process of mechanical weathering is responsible for providing material to the system. However, other processes such as landsliding or resupply of material by fluvial processes may provide material which has not directly been supplied by weathering. Weathering is also hardly ever 'limited' compared to other geomorphic processes and is responsible for significant redistribution of material. Similarly, the term transport-limited may create confusion because it implies that it is the transport mechanism that limits the flux of mass. Rather than the transport mechanism, it is the availability of material that allows mass movement. It is therefore suggested to replace the terms weathering-limited and transport-limited by the less ambiguous terms 'supply-limited' and 'supply-unlimited'.

References

Bovis, M.J. and Jakob, M. (1999) The role of debris supply conditions in predicting debris flow activity, *Earth Surface Processes and Landforms* 24, 1,039–1,054.

Carson, M.A. and Kirkby, M.J. (1972) *Hillslope Form and Process*, Cambridge: Cambridge University Press.

Gilbert, G.K. (1877) *The Geology of the Henry Mountains*, Washington, DC: United States Geographical and Geological Survey.

Jahn, A. (1954). Denudational balance of slopes (in Polish), *Czasopismo Geograficzne* 25.

——(1968) Denudational balance of slopes, *Geographia Polonica* 13, 9–29.

Jakob, M. and Skermer, N. (translators) (1997) Stiny, J. *Die Muren*, Vancouver, BC: EBA Engineering Consultants Ltd.

Kleinhans, M.G., Wilbers, A.W.E., De Swaaf, A. and Van den Berg, J.H. (2002) Sediment supply-limited bedforms in sand-gravel bed rivers, *Journal of Sedimentary Research, Section A, Sedimentary Petrology and Processes* 72(5), 629–640.

Stallard, R.F. (1985) River chemistry, geology, geomorphology, and soils in the Amazon and Orinoco Basins, in J.I. Driver (ed.) *The Chemistry of Weathering*, 293–319, Dordrecht: Reidel.

——(1995) Tectonic, environmental, and human aspects of weathering and erosion: a global review using a steady-state perspective, *Annual Reviews of Earth and Planetary Science* 23, 11–39.

Stiny, J. (1910) *Die Muren. Versuch einer Monographie mit besonderer Berücksichtigung der Verhältnisse in den Tiroler Alpen*, Innsbruck: Verlag der Wagner'schen Universitäts Buchhandlung.

Further reading

Kirkby, M.J. (1971) Hillslope process-response models based on the continuity equation, *Transaction of the Institute of British Geographers*, Special Publication 3, 15–30.

——(1985) The basis for soil profile modelling in a geomorphic context, *Journal of Soil Science* 36, 97–121.

MATTHIAS JAKOB

WEATHERING PIT

Small closed depressions on horizontal and gently inclined rock surfaces. They are also called 'gnammas', 'Opferkessel' or 'pias'. They have been described from a range of silicate rock types, most frequently granites and sandstones. They may be similar in their broad morphology to solutional pits developed in carbonate rocks, to which the term 'kamenitza' is often applied. Goudie and Migoń (1997: Table 1) provides a list of references on weathering pits from a diverse range of morphoclimatic regions that range from polar to desert and humid tropical. The largest examples may be between 10 and 20 m long (Twidale and Corbin 1963).

There is a great deal of uncertainty concerning the processes involved in the development of pits. Chemical processes of solution are usually invoked, but other processes may include hydration, the mechanical action of frost and salt and biochemical weathering. Positive feedback mechanisms related to the ever-growing amount of water available as a pit enlarges may account for a localized high intensity of weathering (Schipull 1978).

Many weathering pits are reported to be partially infilled with debris and organic matter, yet there are also many which are empty, with bare flaky bedrock exposed on the floor. It seems that

Plate 146 A large weathering pit developed in granite in the Erongo Mountains of central Namibia. This example has a completely bare bottom and the process by which debris has been removed from it is still the subject of debate

relatively little attention has been paid to the question of how the debris gets evacuated from the pit. Three possible ways have been suggested (Smith 1941), namely solutional transport and washing out during excessive rainfalls, and deflation. Flotation is another possible mechanism, but no direct observations have been made. Moreover, the occurrence of deep closed pits devoid of any sediments (cf. Watson and Pye 1985) remains puzzling and no satisfactory explanation of their emptiness has been offered.

References

Goudie, A.S. and Migoń, P. (1997) Weathering pits in the Spitzkoppe area, Central Namib Desert, *Zeitschrift für Geomorphologie NF* 41, 417–444.

Schipull, K. (1978) Waterpockets (Opferkessel) in Sandsteinen des zentralen Colorado-Plateaus, *Zeitschrift für Geomorphologie NF* 22, 426–438.

Smith, L.L. (1941) Weathering pits in granite of the Southern Piedmont, *Journal of Geomorphology* 4, 117–127.

Twidale, C.R. and Corbin, E. (1963) Gnammas, *Revue de Géomorphologie Dynamique* 14, 1–20.

Watson, A. and Pye, K. (1985) Pseudokarstic microrelief and other weathering features on the Mswati granite (Swaziland), *Zeitschrift für Geomorphologie NF* 29, 285–300.

A.S. GOUDIE

WETTING AND DRYING WEATHERING

Fluctuations in rock moisture – wetting and drying – can cause rock weathering. In many cases these fluctuations cause weathering as a result of the rock expanding during take up of water and its inability to return to the original dimensions upon losing moisture. Moreover, high moisture contents can diminish rock strength, and through time wetting and drying cycles can reduce the bonding strength of component minerals. This in turn leads to a decrease in rock strength and possibly even failure (Pissart and Lautridou 1984). Cycles of wetting and drying can be frequent in some environments (e.g. shore platforms) and experimental simulations have shown the effectiveness of the process (Hall and Hall 1996). Micro-erosion metre studies have demonstrated the importance of surface swelling on shore platforms developed on limestones and mudstones, though it is less easy to discriminate between the importance of the growth of salt crystals and the expansion from wetting and drying (Stephenson and Kirk 2001). Indeed, wetting and drying is frequently a sine qua non for salt weathering. Wetting and drying also operates on badland surfaces developed on mudstones in semi-arid areas (Cantón *et al.* 2001), though once again it is the combined effect of moisture cycling and salt dissolution that is crucial. Nonetheless, clay-rich rocks are susceptible to SLAKING with or without the presence of salts (Gökçeuğlu *et al.* 2000), and the process can cause severe disintegration, as with the tombs in the Valley of the Kings, Egypt, which have been excavated in Esna Shales (Wüst and McLane 2000).

References

Cantón, Y., Solé-Benet, A., Queralt, I. and Pini, R. (2001) Weathering of a gypsum-calcareous mudstone under semi-arid environment at Tabernas, SE Spain: laboratory and field-based experimental approaches, *Catena* 44, 111–132.

Gökçeuğlu, C., Ulusay, R. and Sönmez, H. (2000) Factors affecting the durability of selected weak and clay-bearing rocks from Turkey, with particular emphasis on the influence of the number of drying and wetting cycles, *Engineering Geology* 57, 215–237.

Hall, K. and Hall, A. (1996) Weathering by wetting and drying: some experimental results, *Earth Surface Processes and Landforms* 21, 365–376.

Pissart, A. and Lautridou, J.P. (1984) Variations de longeur de cylinders de Pierre de Caen (calcaire bathonien) sous l'effet de séchage et d'humidification, *Zeitschrift für Geomorphologie NF* 29, 111–116.

Stephenson, W.J. and Kirk, R.M. (2001) Surface swelling of coastal bedrock on inter-tidal shore

platforms, Kaikoura peninsula, South Island, New Zealand, *Geomorphology* 41, 5–21.
Wüst, R.A.J. and McLane, J. (2000) Rock deterioration in the Royal Tomb of Seti I, Valley of the Kings, Luxor, Egypt, *Engineering Geology* 58, 163–190.

A.S. GOUDIE

WILSON CYCLE

The Wilson cycle is the hypothesis that oceans are born as rifts, grow by SEAFLOOR SPREADING, and finally close again (Wilson 1966). Six stages are identified:

1. Uplift and extension forming rift valleys (see RIFT VALLEY AND RIFTING) (like modern rift valleys in Africa).
2. Early seafloor spreading (Red Sea stage).
3. Mature ocean (Atlantic stage) with a broad ocean bounded by continental shelves and a spreading centre at the mid-ocean ridge.
4. Shrinking of the ocean, bounded by ISLAND ARCs. The Pacific is in this stage, though the Pacific is also spreading.
5. Further shrinking, compression, metamorphism and uplift of accretionary wedges to form mountains (Mediterranean stage).
6. All oceanic crust is subducted, and the continents converge on a collision-zone suture (e.g. Indus–Tsangpo suture in the Himalayas). The united plate breaks along a new rift, restarting the cycle.

Some think the cycle should be repeated by rupture along roughly the same old line of weakness, others think a new cycle can start along a new fracture. Van der Pluijm and Marshak (1997) believe that rifting may occur in warm and weak orogens. Since orogenic belts are regarded as the weaker portions of the crust, they tend to be sites of repeated rifting and collision. Rocks in a single orogen may record the effects of several phases of extension and contraction. In the eastern United States, for example, the Earth history involves two phases of rifting (Late Precambrian and Middle Mesozoic) and two phases of collision (Late Precambrian and Palaeozoic), with several subphases (van der Pluijm and Marshak 1997).

The apparent polar wander (APW) path of two continents experiencing a Wilson cycle will show parallel tracks that diverge at breakup and converge after closure to a second parallel track. Such palaeomagnetic support for the Wilson cycle has been claimed for the rifting and closure of the Atlantic (Piper 1987). Gondwanaland and Laurasia converged during Silurian times to form Pangaea. From then to the Mesozoic they had a common APW path, showing they formed a single continent. They then split up into the present continents along new lines, continuing the Wilson cycle.

The opening and closing of ocean basins up to 1,500 km wide takes about 500 Ma, but some are smaller and more short-lived. Because of the timescale, evidence for Wilson cycles comes mainly from plate tectonic interpretation of palaeomagnetic, structural and geological data. In general it has little direct effect on geomorphology, but Russo and Silver (1996) relate the formation of the Andes to the Wilson cycle.

References

Piper, J.D.A. (1987) *Palaeomagnetism and the Continental Crust*, Milton Keynes: Open University Press.
Russo, R.M. and Silver, P.G. (1996) Cordillera formation, mantle dynamics, and the Wilson Cycle, *Geology* 24, 511–514.
Van der Pluijm, B.A. and Marshak, S. (1997) *Earth Structure: An Introduction to Structural Geology and Tectonics*, New York: WCB/McGraw-Hill.
Wilson, J.T. (1966) Did the Atlantic close and then re-open? *Nature* 211, 676–681.

CLIFF OLLIER

WIND EROSION OF SOIL

Wind erosion (see AEOLATION) has for long been recognized as an important factor in the development of landforms in dryland regions, contributing to the development of such features as close depressions (PANs), streamlined hillocks (YARDANGS), deflated surfaces (STONE PAVEMENTS) and miscellaneous micro-forms (VENTIFACTS). Wind erosion may also modify the form of dunes, as, for example, by creating blowouts (see DUNE, COASTAL). The general context of wind erosion is discussed in various other entries (see AEOLIAN GEOMORPHOLOGY; AEOLIAN PROCESSES; DESERT GEOMORPHOLOGY, etc.). This entry concentrates on the erosion of soil by wind – one component of DESERTIFICATION.

Wind erosion is a natural erosion process. However, its intensity and magnitude is often increased by miscellaneous human activities such as cultivation, overgrazing, vehicular traffic, etc.

(Leys 1999). It has negative impacts on agricultural production and produces environmental pollution, including DUST STORMS. Attempts to understand and model wind erosion have been made for some decades, most notably by W.S. Chepil and colleagues from the United States Department of Agriculture (see Chepil and Woodruff 1963). Using empirical field studies and wind tunnels (see WIND TUNNELS IN GEOMORPHOLOGY), they developed a much used Wind Erosion Equation.

$$E = f(C, I, L, K, V)$$

where E is the potential loss, C is a local climatic index, I is a soil erodibility index, L is a factor relating to field shape in the prevailing wind direction, K is a ridge roughness factor for ploughed ground and V is a vegetation cover index. The equation was developed initially in the Midwest of the USA and drew attention to the factors which could be manipulated by farmers (namely, I, L, K and V).

The climatic factor (C) was a simple combination of two key climatic variables: annual wind speed and a moisture index. Plainly, dry, windy areas are likely to be most susceptible to wind erosion. The soil erodibility factor (I) is more complex and needs to be seen in terms of both individual grain size characteristics and aggregate characteristics. Fine sands and silts are likely to be most susceptible, partly because of the relatively low velocities required for their entrainment, but also because the presence of clay tends to produce wind-stable clods. The presence of large clods reduces the risk of wind erosion. The fetch distance over which the wind acts (L) is related to field size and the presence or absence of shelter belts of differing heights, spacing and permeability. The ridge roughness factor (K) is based on the experimental observation that the rougher the surface, up to about 6 cm, the lower the windspeed at the surface. Thus furrows at right angles to the wind will tend to dampen down rates of erosion. The vegetation factor (V) is absolutely fundamental, for a dense vegetation cover, especially if like grass it has short stalks and narrow leaves, does more than anything else to reduce erosion rates.

More recently predictive models of wind erosion have been developed (Shao 2000). One such model is the Wind Erosion Assessment Model (WEAM), which aims to account for the combined effect of climate, soil, vegetation and land use. The fundamental physical viewpoint of this model (Shao *et al.* 1996) is that wind erosion is a result of two opposing forces: the capacity of the wind to start and maintain erosion, and the ability of the soil to resist it. The wind's capacity to start and maintain erosion is the friction velocity u^* (the wind shear or drag on the soil), while the opposing quantity offered by the soil is the threshold velocity u^*_t (the minimum friction velocity that is required for erosion to occur). The former is determined by wind flow conditions and the surface roughness, while the latter is determined by such surface factors as soil texture, aggregate structure and moisture content.

References

Chepil, W.S. and Woodruff, N.P. (1963) The physics of wind erosion and its control, *Advances in Agronomy* 15, 211–302.

Leys, J. (1999) Wind erosion on agricultural land, in A.S. Goudie, I. Livingstone and S. Stokes (ed.) *Aeolian Environments, Sediments and Landforms*, 143–166, Chichester: Wiley.

Shao, Y. (2000) *Physics and Modelling of Wind Erosion*, Dordrecht: Kluwer.

Shao, Y., Raupach, M.R. and Leys, J.F. (1996) A model for predicting aeolian sand drift and dust entrainment on scales from paddock to region, *Australian Journal of Soil Research* 34, 309–342.

A.S. GOUDIE

WIND TUNNELS IN GEOMORPHOLOGY

Wind tunnels provide a means by which processes in AEOLIAN GEOMORPHOLOGY can be monitored with a fully controlled and regulated wind flow. This is particularly advantageous in the aeolian environment where natural fluctuations in wind velocity and direction make it difficult to perform repeatable experiments in the field.

Wind tunnels vary greatly in design characteristics. They usually consist of a fan sucking (or blowing) air through a contraction which flattens out streamlines and provides a non-fluctuating wind to a working section where experiments can be carried out. The size of the working section may vary from $0.1\,m^2$ and 1 m in length to more than $4\,m^2$ and over 10 m in length. Working velocities in wind tunnels are commonly between 1 and $20\,m\,s^{-1}$.

Geomorphological experiments in wind tunnels are normally concerned either with aeolian sand or dust transport processes (Butterfield 1998; see AEOLIAN PROCESSES) where sediment is placed in the working section, or reduced scale studies in 'clean' wind tunnels of flow over fixed models of dunes, hills or valleys (Walker and Nickling 2002). In the latter case not only must the landform be reduced in scale but also the structural characteristics of the windflow (the atmospheric boundary layer). This is often achieved through a series of upstream grids and spires which, together with a roughened working section floor, can provide scaled values of shear velocity, aerodynamic roughness and turbulence length scales which mimic full-scale values. Portable wind tunnels are commonly used to test wind erosion (SEE WIND EROSION OF SOIL) characteristics of agricultural fields. Here the floor of the tunnel working section is absent so that when placed at a test site the tunnel flow acts directly on the natural soil.

Wind flow in wind tunnels can be determined with a pitot-tube which establishes velocity through the measurement of pressure. Where highly turbulent flows are encountered then electronic hot-wire anemometers are more commonly employed. Precise measurement of shear stresses on the surface of scaled models can now be accomplished using pulse-wire anemometers (Wiggs *et al.* 1996) or particle velocimetry whilst flow visualization over scale models is achievable by seeding the flow with smoke or covering the surface of the model with oil. In experiments involving sediment entrainment and transport sand traps can be erected at the downwind end of the working section to measure the rate of sand flux.

Whilst many advances have transpired as a result of the use of wind tunnels in research on aeolian processes and wind flow over aeolian bedforms, the biggest drawback of the approach is that there is commonly insufficient field-based empirical data with which results can be verified.

References

Butterfield, G.R. (1998) Transitional behaviour of saltation: wind tunnel observations of unsteady winds, *Journal of Arid Environments* 39, 377–394.

Walker, I.J. and Nickling, W.G. (2002) Dynamics of secondary airflow and sediment transport over and in the lee of transverse dunes, *Progress in Physical Geography* 26, 47–75.

Wiggs, G.F.S., Livingstone, I. and Warren, A. (1996) The role of streamline curvature in sand dune dynamics: evidence from field and wind tunnel measurements, *Geomorphology* 17, 29–46.

GILES F.S. WIGGS

Y

YARDANG

A Turkmen word to describe wind abraded ridges of cohesive material (Hedin 1903). They have been described from a large number of arid regions (McCauley *et al.* 1977) where they have developed on a large range of materials. They have been likened in shape to the prows of upturned ships. They range in size from small centimetre-scale ridges (micro-yardangs) through to forms that are some metres in height and length (meso-yardangs), to features that may be tens of metres high and some kilometres long (mega-yardangs). The forms may go through a cycle of development and eventual obliteration (Halimov and Fezer 1989). Although they are dominantly aeolian erosion features there has been a considerable debate as to the relative importance of deflation, aeolian abrasion, fluvial incision and mass movements in moulding yardang morphology. That abrasion is important is indicated by polished, fluted and sand-blasted slopes, and the undercutting of the steep windward face and lateral slopes. It is probably the dominant process in hard bedrock yardangs whereas deflation may be important in the evolution of yardangs developed in soft sediments such as old lake beds. Fluvial erosion may provide an avenue along which wind erosion may occur but excessive fluvial erosion would tend to obliterate yardangs. Mass movements may also be significant when yardang slopes have been oversteepened by wind erosion.

One form of yardang, beloved of textbooks, but in reality not very common or significant, is the zeuge (plural zeugen). The term is derived from the German word for 'witness'. Zeugen are tabular masses of resistant rock (2–50 m high) left standing on a pedestal of softer material as a result of differential erosion by sand-laden wind.

Plate 147 Characteristic aerodynamic yardangs developed by the wind erosion of Holocene pluvial lake beds in the Dhakla Oasis, Western Desert, Egypt

Plate 148 A series of wind eroded chalk outcrops in the White Desert, Farafra, Egypt. In addition to wind erosion, note the presence of case hardening and honeycomb weathering

At the mega-scale yardangs curve round in response to the trajectories of trade winds (as, for example, around Borkou in the Central Sahara; Mainguet 1972) and may show a remarkable parallelism over hundreds of kilometres (as with the *Kaluts* of the Lut Desert in Iran).

Study of the morphometry of yardangs has revealed relationships between different parameters. Ward and Greeley (1984) found a 1:4 width to length ratio; Halimov and Fezer (1989) found that the ratios of length, width and height were 10:2:1; while Goudie *et al.* (1999) found volume, length, width, height ratios of 18.7:9.9:2.7:1.

In soft materials yardangs can form quickly. In the Sahara, Mojave and Lop Nor deserts, Holocene lake and swamp deposits have been eroded at rates of *c.*2.5 to 5 m per 1,000 years.

References

Goudie, A.S., Stokes, S., Cook, J. *et al.* (1999) Yardang landforms from Kharga Oasis, south-west Egypt, *Zeitschrift für Geomorphologie Supplementband* 116, 1–16.

Halimov, M. and Fezer, F. (1989) Eight yardang types in Central Asia, *Zeitschrift für Geomorphologie* 33, 205–217.

Hedin, S. (1903) *Central Asia and Tibet*, Scribners: New York.

McCauley, J.F., Grolier, M.J. and Breed, C.S. (1977) Yardangs of Peru and other desert regions, *US Geological Survey Interagency Report: Astrogeology* 81.

Mainguet, M. (1972) *Le Modelé des Grès*, Paris: IGN.

Ward, A.W. and Greeley, R. (1984) Evolution of yardangs at Rogers Lake, California, *Geological Society of America Bulletin* 95, 829–837.

A.S. GOUDIE

YAZOO

A tributary stream/river that flows for a long distance parallel to the main channel before joining it. This usually occurs in mature river channels where natural levees have built up on the channel banks, above the mean height of the surrounding floodplain. The tributary channel is therefore unable to penetrate the trunk channel and lakes may form. However, more often the tributary channel is bounded by bluffs beyond the floodplain, thus forcing a Yazoo-type tributary downstream parallel to the main channel. The type example is the Yazoo River, which runs south/south-west parallel to the main Mississippi River from Memphis, Tennessee (USA) for almost 400 km. The Yazoo eventually infiltrates the Mississippi channel at Vicksburg, Mississippi. The elevated bed of the Mississippi River obstructs the Yazoo tributary and forces it to flow parallel along the floodplain (35–150 km wide) and with a Yazoo channel gradient of only 11.4 cm per km. Yazoo-type channels are also evident in deep sea environments (Hesse and Rakofsky 1992).

Reference

Hesse, R. and Rakofsky, A. (1992) Deep sea channel submarine yazoo system of the Labrador Sea – a new deep-water facies model, *American Association of Petroleum Geologists Bulletin* 75(5), 680–707.

Further reading

Dorris, F.E. (1929) The Yazoo Basin in Mississippi, *Journal of Geography* 28, 72–81.

STEVE WARD

Z

ZOOGEOMORPHOLOGY

Zoogeomorphology is the study of the geomorphic effects of animals (Butler 1995), and as such can be considered as a subset of BIOGEOMORPHOLOGY. Zoogeomorphology encompasses the geomorphic effects of both free-ranging, wild animal populations as well as the geomorphic effects of domesticated animals such as cattle. It considers the geomorphic role of animals regardless of their size, ranging from small invertebrates such as ants, worms and termites (see TERMITES AND TERMITARIA) to both terrestrial and aquatic vertebrates including fish, amphibians, reptiles, birds and mammals.

Animals geomorphologically alter the Earth's surface through both direct and indirect effects. Direct effects encompass surface erosion, transportation and deposition of rock, soil and/or unconsolidated sediments by animals. Animals accomplish these effects by digging for and/or catching of food, nest building, burrowing, mound-building, wallowing, eating soil or rocks, canal excavation and dam building (accomplished by beavers; *Castor canadensis* in North America, *C. fiber* in Eurasia). Indirect effects do not specifically remove sediment from a surface, but processes such as trampling associated with overgrazing may lead to a reduced infiltration capacity in the underlying soil, in turn resulting in greater surface wash, soil creep and rainsplash detachment. Trampling and associated overgrazing by domesticated animals also contribute to widespread bank erosion, riparian degradation and unstable channel morphologies (Trimble 1994; Magilligan and McDowell 1997; Naiman and Rogers 1997; Evans 1998).

The geomorphic impacts of animals may vary spatially and temporally through the course of a year, as wild animals migrate in search of changing food sources (Baer and Butler 2000). In the Rocky Mountains of western North America, for example, grizzly bears (*Ursus arctos horribilis*) create widespread diggings during the spring on lower floodplains in search of roots and tubers. As summer progresses, they migrate upslope into areas of snowmelt where they dig for burrowing mammals such as ground squirrels as well as for roots. Late summer sees the bears on high talus slopes, where they excavate large quantities of boulders, in search of insect larvae.

The seasonality of geomorphic impacts by animals is also a function of a region's climate. Excessive heat and dryness minimize the summertime geomorphic impacts of some desert dwellers. In colder climates, frozen ground limits the seasonal ability of animals to dig into the surface. During such periods, many burrowing animals may also hibernate, effectively cutting off geomorphic impacts until the subsequent spring (Butler 1995).

Beavers have some of the most profound geomorphic impacts of any animal in North America. Beavers are large (adult mass >15 kg), semi-aquatic rodents that instinctively attempt to build dams in response to the sound of running water (Butler 1995). Prior to European settlement of North America, as many as 400 million beavers may have occupied an area of approximately fifteen million square kilometres. By the year 1900, trapping of beavers in North America had nearly extinguished the species. During the twentieth century, conservation efforts re-established the beaver throughout their native range, but at population levels of only 10 per cent of the pre-European-contact level. As beavers continue to reoccupy their former range, their geomorphic influences also grow. For example, on the Roanoke River floodplain, on the Coastal Plain of the eastern United States, beaver ponds illustrated a tenfold increase in areal coverage between the years 1984 and 1993 (Townsend and Butler 1996).

Beaver dams impound water, creating pond environments into which streams flow and deposit sediment. Individual beaver ponds may accumulate as much as 20–30 cm of sediment across the floor of a pond annually (but only during the portions of the year when the inflowing stream is not frozen). Older ponds entrap significantly more sediment than do younger ponds in topographically similar sites (Butler and Malanson 1995). Beaver dams substantially reduce stream velocity and discharge, to the point in some cases of completely eliminating surface outflow on the downstream side of a dam (Meentemeyer and Butler 1999).

Animal burrows (such as wombats in Australia) and mounds (termites) may cover as much as several cubic metres, and are visible on aerial photographs and even satellite imagery (Löffler and Margules 1980). Some of the most widespread geomorphic impacts of animals are produced on the floor of the Bering Sea by the California gray whale (*Eschrichtius robustus*). Gray whales feed almost exclusively on bottom-dwelling organisms, especially amphipod crustaceans. The whales 'slurp' deep furrows and pits into the floor of the ocean, ingesting sediment and fauna. The whales filter the food from the sediment in their baleen, and subsequently expel plumes of muddy sediment near the surface (Butler 1995). These plumes, visible from low-flying airplanes, account for an enormous amount of sediment transport in the Bering Sea. Annual whale feeding there moves a minimum of 120 million cubic metres of sediment, nearly three times the annual load of suspended sediment discharged into the Bering Sea by the Yukon River, the fourth largest sediment source in North America.

References

Baer, L.D. and Butler, D.R. (2000) Space-time modeling of grizzly bears, *Geographical Review* 90, 206–221.

Butler, D.R. (1995) *Zoogeomorphology – Animals as Geomorphic Agents*, Cambridge and New York: Cambridge University Press.

Butler, D.R. and Malanson, G.P. (1995) Sedimentation rates and patterns in beaver ponds in a mountain environment, *Geomorphology* 13, 255–269.

Evans, R. (1998) The erosional impacts of grazing animals, *Progress in Physical Geography* 22, 251–268.

Löffler, E. and Margules, C. (1980) Wombats detected from space, *Remote Sensing of Environment* 9, 47–56.

Magilligan, F.J. and McDowell, P.F. (1997) Stream channel adjustments following elimination of cattle grazing, *Journal of the American Water Resources Association* 33, 867–878.

Meentemeyer, R.K. and Butler, D.R. (1999) Hydrogeomorphic effects of beaver dams in Glacier National Park, Montana, *Physical Geography* 20, 436–446.

Naiman, R.J. and Rogers, K.H. (1997) Large animals and system-level characteristics in river corridors, *BioScience* 47, 521–529.

Townsend, P.A. and Butler, D.R. (1996) Patterns of landscape use by beaver on the lower Roanoke River floodplain, North Carolina, *Physical Geography* 17, 253–269.

Trimble, S.W. (1994) Erosional effects of cattle on streambanks in Tennessee, U.S.A., *Earth Surface Processes and Landforms* 19, 451–464.

Further reading

Butler, D.R. (1995) *Zoogeomorphology – Animals as Geomorphic Agents*, Cambridge and New York: Cambridge University Press.

SEE ALSO: biogeomorphology; termites and termitaria

DAVID R. BUTLER

Index

Page numbers in **bold** indicate references to the main entry. Please refer to the thematic list for individual entries on major topics (coastal and marine, fluvial, glacial etc.).

LIVERPOOL
JOHN MOORES UNIVERSITY
AVRIL ROBARTS LRC
TITHEBARN STREET
LIVERPOOL L2 2ER
TEL. 0151 231 4022